Geophysical Monograph Series

Geophysical Monograph Series
A. F. Spilhaus, Jr., Managing Editor

1 **Antarctica in the International Geophysical Year,** A. P. Crary, L. M. Gould, E. O. Hulburt, Hugh Odishaw, and Waldo E. Smith (editors)

2 **Geophysics and the IGY,** Hugh Odishaw and Stanley Ruttenberg (editors)

3 **Atmospheric Chemistry of Chlorine and Sulfur Compounds,** James P. Lodge, Jr. (editor)

4 **Contemporary Geodesy,** Charles A. Whitten and Kenneth H. Drummond (editors)

5 **Physics of Precipitation,** Helmut Weickmann (editor)

6 **The Crust of the Pacific Basin,** Gordon A. Macdonald and Hisashi Kuno (editors)

7 **Antarctic Research: The Matthew Fontaine Maury Memorial Symposium,** H. Wexler, M. J. Rubin, and J. E. Caskey, Jr. (editors)

8 **Terrestrial Heat Flow,** William H. K. Lee (editor)

9 **Gravity Anomalies: Unsurveyed Areas,** Hyman Orlin (editor)

10 **The Earth Beneath the Continents: A Volume in Honor of Merle A. Tuve,** John S. Steinhart and T. Jefferson Smith (editors)

11 **Isotope Techniques in the Hydrologic Cycle,** Glenn E. Stout (editor)

12 **The Crust and Upper Mantle of the Pacific Area,** Leon Knopoff, Charles L. Drake, and Pembroke J. Hart (editors)

13 **The Earth's Crust and Upper Mantle,** Pembroke J. Hart (editor)

14 **The Structure and Physical Properties of the Earth's Crust,** John G. Heacock (editor)

15 **The Use of Artificial Satellites for Geodesy,** Soren W. Henriksen, Armando Mancini, and Bernard H. Chovitz (editors)

16 **Flow and Fracture of Rocks,** H. C. Heard, I. Y. Borg, N. L. Carter, and C. B. Raleigh (editors)

17 **Man-Made Lakes: Their Problems and Environmental Effects,** William C. Ackermann, Gilbert F. White, and E. B. Worthington (editors)

18 **The Upper Atmosphere in Motion: A Selection of Papers With Annotation,** C. O. Hines and Colleagues

19 **The Geophysics of the Pacific Ocean Basin and Its Margin: A Volume in Honor of George P. Woollard,** George H. Sutton, Murli H. Manghnani, and Ralph Moberly (editors)

20 **The Earth's Crust,** John G. Heacock (editor)

21 **Quantitative Modeling of Magnetospheric Processes,** W. P. Olson (editor)

22 **Derivation, Meaning, and Use of Geomagnetic Indices,** P. N. Mayaud

23 **The Tectonic and Geologic Evolution of Southeast Asian Seas and Islands,** Dennis E. Hayes (editor)

24 **Mechanical Behavior of Crustal Rocks: The Handin Volume,** N. L. Carter, M. Friedman, J. M. Logan, and D. W. Stearns (editors)

25 **Physics of Auroral Arc Formation,** S.-I. Akasofu and J.R. Kan (editors)

Physics of Auroral Arc Formation

AGU Series

Geophysical Monograph Series
Water Monograph Series
Maurice Ewing Series
Antarctic Research Series
Geodynamics Series
Coastal and Estuarine Sciences

A woodcut by Fridtjof Nansen; Nansen depicts himself strolling on the ice under a triple curtain-like form of the aurora; the auroral arcs. (From Nansen's Nord I Takeheimen, 1911).

geophysical monograph 25

Physics of Auroral Arc Formation

S.-I. Akasofu
J.R. Kan
Editors

American Geophysical Union
Washington, D.C.
1981

Physics of Auroral Arc Formation

Based on papers presented at the Chapman Conference on "The Formation of Auroral Arcs" held at Fairbanks, Alaska, July 21-25, 1980

Library of Congress Cataloging in Publication Data
Main entry under title:

Physics of auroral arc formation.

(Geophysical monograph; 25)
Includes bibliographies.
1. Auroras--Addresses, essays, lectures. I. Akasofu, Syun-Ichi. II. Kan, J.R. III. American Geophysical Union. IV. Series.
QC971.5.P48 538'.768 81-14873
ISBN 0-87590-050-X AACR2

Printed in the United States of America.

CONTENTS

PREFACE

The American Geophysical Union has honored Sydney Chapman by designating particular topical conferences as Chapman Conferences. We are very happy that the AGU endorsed this conference on "The Formation of Auroral Arcs" as a Chapman Conference. This is particularly the case because Chapman and his wife Katherine spent a few winter months on the campus of the University of Alaska every year from about 1955 until his untimely death in 1970. I recall vividly many auroral displays we watched together during our walks on cold winter nights in 1958-70. We were often completely overwhelmed by this natural phenomenon and stood for a long time without a word. I recall that Chapman paid very much attention to the thinness of auroral arcs.

During the last decade, we have finally begun to comprehend the aurora as a result of a spectacular discharge process powered by the solar wind-magnetosphere dynamo. However, it was only the last few years during which several observations have provided us with a fairly firm indication that there is a significant potential drop along geomagnetic field lines at an altitude of 1-2 earth radii above the aurora. We had long thought that there was no potential drop along magnetic field lines in a hot and rarefied plasma. It is important to recognize that the aurora itself is an indication of breakdown of this view in astrophysics and space plasma physics which had widely held for a long time.

Meeting participants were fortunate to have this opportunity to discuss the formation of auroral arcs. The potential structure is certainly one of the most fundamental secrets of the aurora, and it is very gratifying to see that a unified physical model has finally begun to emerge on the basis of a great variety of observations and of intense theoretical studies.

This volume contains most of the invited papers and some of the papers selected from the contributed papers which were presented during the conference. It is hoped that this volume marks a milestone along the endeavor of our efforts in understanding the polar aurora.

I would like to thank Dr. J. Roederer, Director of the Geophysical Institute, University of Alaska, and University of Alaska, Fairbanks, for their support. The conference was supported also in part by the National Science Foundation, the National Aeronautics and Space Administration, Lockheed Missiles and Space Company, and the Air Force Geophysics Laboratory.

S.-I. Akasofu
Convenor

AURORAL ARCS AND AURORAL POTENTIAL STRUCTURE

S.-I. Akasofu

Geophysical Institute, University of Alaska, Fairbanks, Alaska 99701

Abstract. The aurora results from an electrical discharge process which is powered by the solar wind-magnetosphere dynamo. During the last few years, new auroral observations and the associated plasma studies have made it possible to construct physical models which can account for the formation of auroral arcs. A complete model includes the power generation on the magnetopause, the transmission of the power to the convective motion of magnetospheric plasma, the generation of field-aligned currents, the formation of the auroral potential structure, the acceleration of auroral particles and auroral light emissions. It may well be that our present models are still deficient in many ways, but it is an epoch-making advance to be able to construct such complete and quantitative models. This paper is intended to be an introduction to physics of auroral arc formation for both specialists and nonspecialists. After describing the general morphological features associated with auroral arcs, the reader will be led from the first question as to why auroral arcs appear along an annular belt surrounding the geomagnetic pole to the last question as to why the auroral potential structures develop, accelerating auroral particles. All the papers appearing in this volume are cited at appropriate points in the text.

Introduction

The aurora results from a large-scale electrical discharge process which is powered by the solar wind-magnetosphere dynamo. In order to provide a general perspective of the physics associated with auroral arc formation, it is perhaps instructive to present here an example of models which include most of the basic chain processes, from the power generation to auroral light emissions.

The electromotive force for the dynamo arises from the solar wind which flows across the normal component of the magnetic field on the magnetopause. The power generated by the solar wind-magnetosphere dynamo is transmitted to the convective ($\underline{E} \times \underline{B}$) motion of magnetospheric plasma. This situation is schematically illustrated in Figure 1. In a typical cross-section (perpendicular to the sun-earth line) of the magnetotail, the main part of the convective flow originates from the magnetopause and is directed toward the equatorial plane (which is often called the neutral sheet or the midplane). In the vicinity of the equatorial plane, this plasma flow is directed toward the sun except near the magnetopause. At a radial distance range of 4 to 10 earth radii (R_E), the plasma tends to flow around the earth. It is in this region where a dynamo process associated with the flow drives what we might call the Pedersen counterpart currents. In turn, they drive the field-aligned currents which are closed by the Pedersen current j_P in the ionosphere. The upward field-aligned currents are carried by electrons which flow down from the magnetosphere to the ionosphere. It is through such a process that the magnetospheric plasma forces the ionospheric plasma to participate in the convective motion and by which the power generated by the solar wind-magnetosphere dynamo is dissipated partly in the ionosphere by Joule heat production ($\underline{E} \cdot \underline{j}_P$).

In this way, the electric power is first transmitted to the convective motion of magnetospheric plasma. Therefore, the convective motion of magnetospheric plasma can be considered to be the so-called 'MHD-generator', which generates the field-aligned currents and the Pedersen currents.

It appears that the current density of about 10^{-6} A/m^2 is an upper limit for the upward field-aligned current which is carried by electrons, because of the atmospheric loss cone in their pitch-angle distribution and the convergence of the geomagnetic field toward the polar region. The dynamo power associated with this current density is about ~10^{18} erg/sec. When the solar wind-magnetosphere dynamo efficiency is increased, the convective motion of magnetospheric plasma will be enhanced. As a result, the efficiency of the internal MHD generator will be enhanced. Then, the intensity of the field-aligned current density must be increased beyond the limit of ~10^{-6} A/m^2. The magnetospheric plasma achieves this by develop-

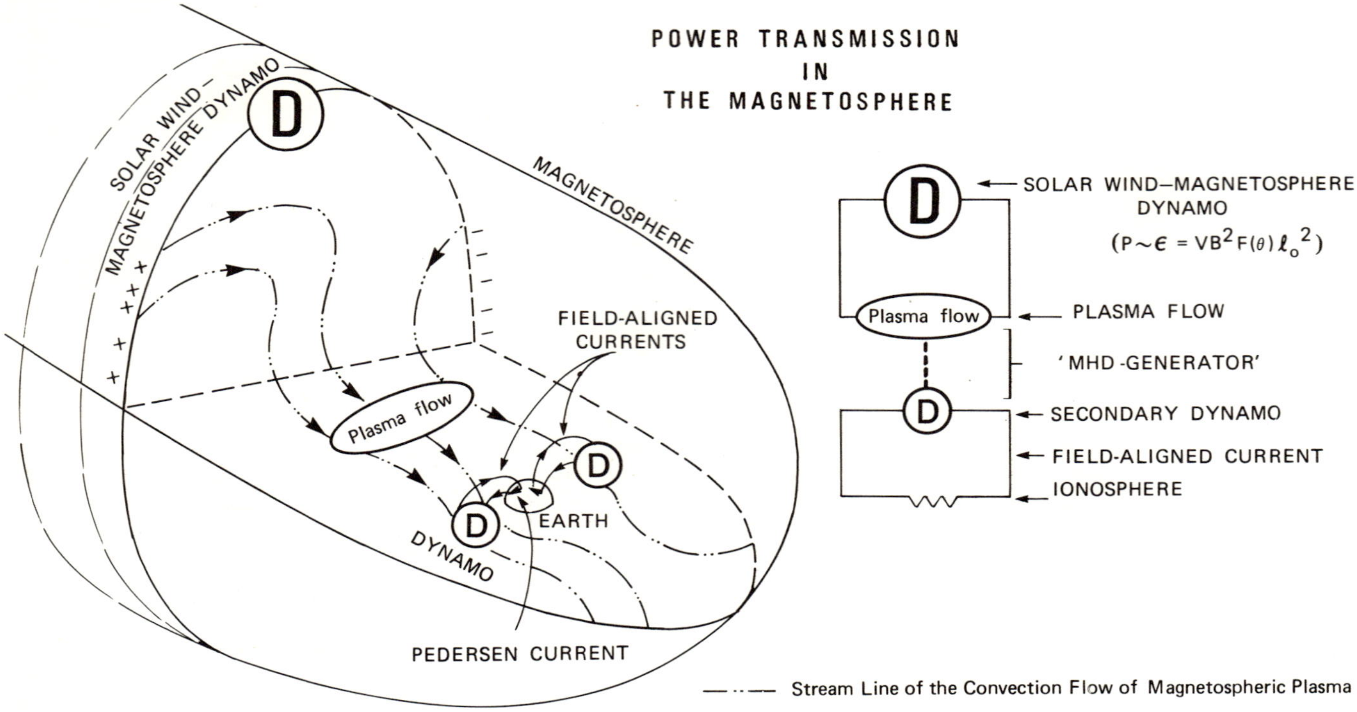

Figure 1. Schematic illustration showing how the power generated by the solar wind-magnetosphere dynamo is transmitted to the magnetosphere and the ionosphere through the convective motion of magnetospheric plasma.

ing potential structures at an altitude of 10,000 km, in which the electric field is directed upward along the magnetic fields lines. One such potential structure has a V-shaped geometry which is supported by several distinct plasma populations, such as downward streaming electrons, upward streaming ions, electrons which are trapped between the potential structure and their mirror points near the ionosphere, and electrons which are scattered above the potential structure and in the ionosphere. The last population is called the backscattered electrons. Such a potential structure satisfies most of the criteria for a double layer in which the streaming electrons and ions are accelerated downward and upward, respectively. A variety of plasma waves is generated. Some of the waves contribute in increasing the potential drop in the structure.

As the dynamo power is increased from 10^{17} erg/sec to 10^{18} erg/sec and thus the field-aligned current density reaches 10^{-6} A/m^2, the total potential drop in the structure becomes about a few kilovolts. The downward streaming electrons will thus be accelerated to a few kilovolts. Now, they are energetic enough to penetrate down to an altitude of 110 km. There they collide with upper atmospheric constituents, such as N_2, O_2, O, N, etc., causing their ionization and excitation. The 5577Å emission (the most familiar auroral emission) from excited oxygen atoms becomes intense enough to be clearly visible (~0.3 kilo-Rayleigh) for the dynamo power of ~10^{18} erg/sec. The visible auroral structure delineates the region where this interaction is taking place in the upper atmosphere. It is this visible structure which we identify as the aurora. An increase of the dynamo power beyond 10^{18} erg/sec causes a larger potential drop which in turn produces a brighter auroral form.

It should be stressed here that the above model is a "straw man model", merely intended to illustrate some of the important aspects involved in the formation of auroral arcs. Actually, many ideas have already been proposed to explain the formation of the auroral potential structure. In fact, one of the purposes of this volume is to introduce the readers to these various ideas. It is hoped that the following sections provide an introduction to many fascinating problems which are discussed in detail in this volume.

2. General Morphology of Auroral Arcs

As the frontispiece of this monograph indicates, Nansen, a great polar explorer, was accurate in illustrating that the aurora has a thin curtain-like form, often in multiples. This form is now generally called the 'auroral arc' or the 'discrete aurora'. The bottom

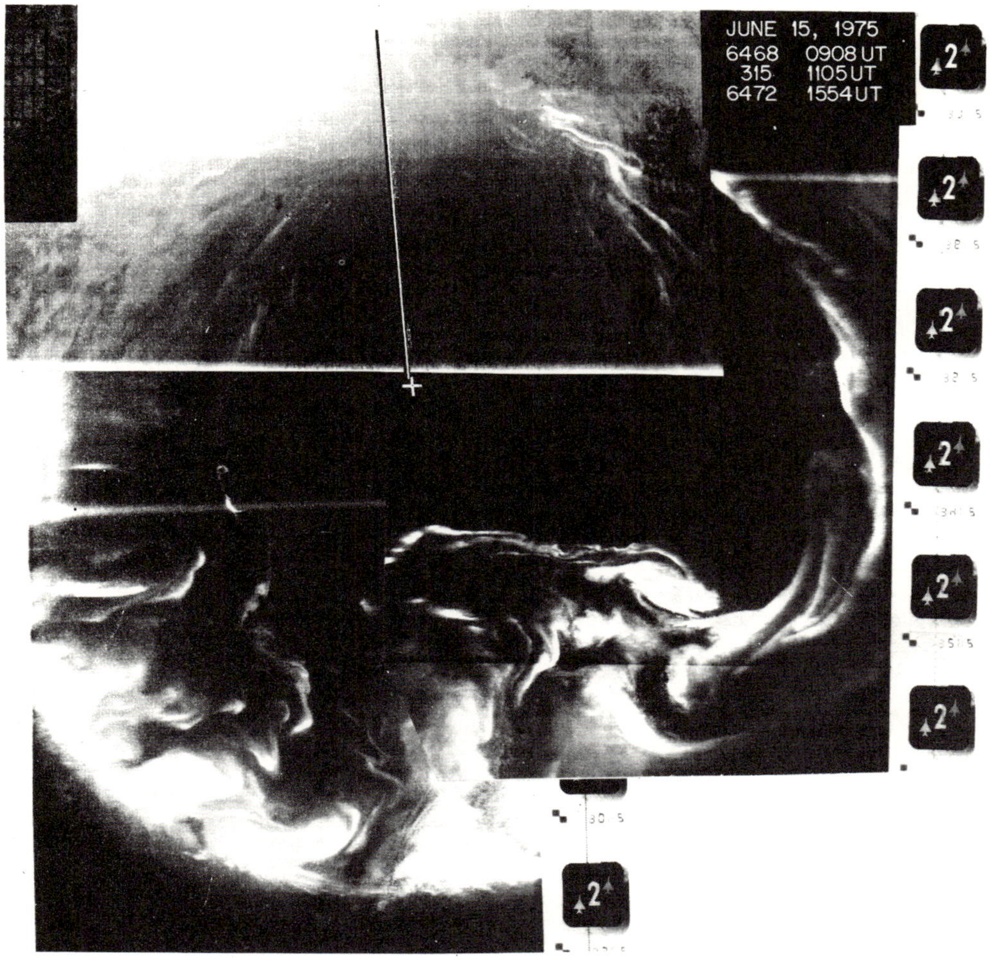

Figure 2. 'Montage' photograph of the auroral oval over the antarctic region, constructed by combining two DMSP (Defense Meteorological Satellite Program) 8531 photographs and one DMSP 10533 photograph; the cross and the line indicate the location of the South invariant pole and the magnetic noon meridian (Akasofu, 1978). Note that the morning and evening sectors are on the left and right side of the figure, respectively, as it shows the antarctic aurora.

height of the auroral curtain is typically 105 km. Its upper height is about 400 km, but varies considerably. Auroral curtains appear along an annular belt surrounding the geomagnetic pole, called the auroral oval. The radius of the oval is about 2000 km. Figure 2 is a 'montage photograph' of the auroral oval over the antarctic region, as seen by DMSP satellites. Figure 3 shows schematically some of the important features of the aurora during a fairly active condition in invariant latitude-magnetic local time (MLT) coordinates.

In Figure 3, one can see that the dayside arcs are separated from the nightside arcs by a distinct gap, although the auroral arcs form a single belt of the aurora (namely, the auroral oval). The midday part of the oval is called the cusp region. Arcs in the cusp region have very different characteristics, spectroscopically and morphologically, from those which occupy the nightside half of the oval. This topic will be discussed by J. S. Murphree et al. in Chapter 2. The auroral oval is surrounded by a belt of the diffuse aurora. In evening hours, the diffuse aurora is fairly homogeneous in brightness across the width of a few hundred kilometers. After magnetic local midnight, however, it develops dark striations within a fairly uniform luminosity oriented in the east-west direction (the so-called 'black aurora'). In mid-morning hours, each striation is widened and as a result multiple arc structures develop. In late morning hours, the equatorward half of the diffuse aurora develops the so-called 'patches' [cf. Akasofu, 1968, Chapter 2].

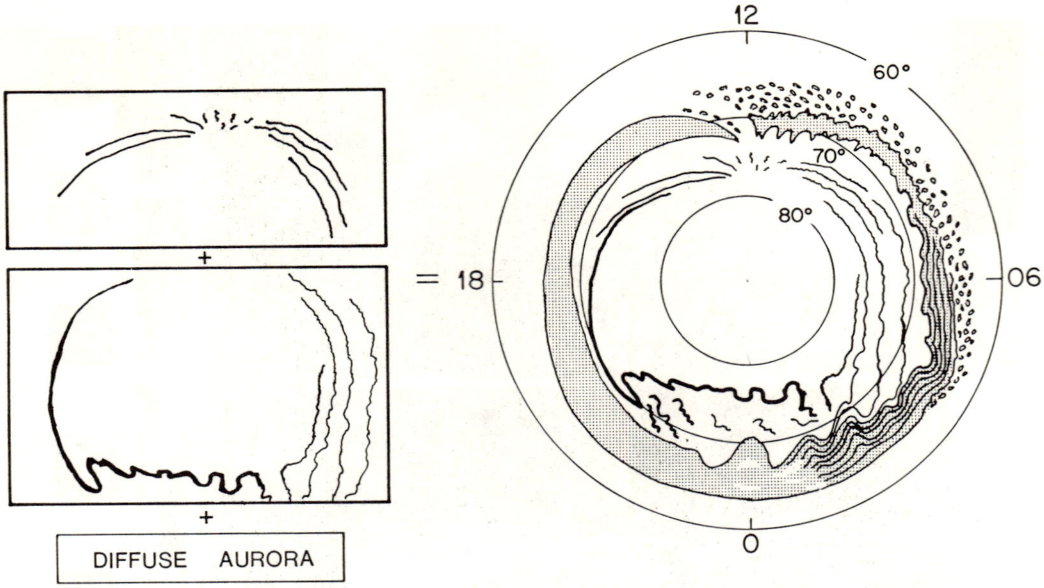

Figure 3. Schematic diagram showing the main characteristics of auroral displays, as seen from above the north geomagnetic pole. Auroral arcs are indicated by lines and the diffuse aurora is indicated by the shaded region. This is slightly modified from the original version given by Akasofu (1976).

The center of the auroral oval is not located at the geomagnetic or the invariant pole; it is displaced from the pole by several degrees toward the darkside. Thus, the relative distance between the oval and a fixed point of the earth varies as a function of local time as the earth rotates once a day. Consider a typical auroral zone station, say Fairbanks (geomagnetic latitude 65°). As Fairbanks is rotated toward the dark side from the midday sector, the distance between the oval and Fairbanks becomes progressively shorter. In evening hours, the aurora appears first near the poleward horizon and appears to shift toward the observer as evening hours progress. The auroral form near the poleward horizon in evening hours appears like an arch (the nearest part of the aurora being highest). It was this particular form which was officially classified as an "arc" by Carl Störmer [1930] who produced the first catalog of the auroral forms. This is the origin of the term "arc", although we now use it to describe the curtain-like discrete form, as we do in this Conference.

Fairbanks comes directly under the auroral oval during magnetic local midnight hours in a moderately disturbed period. When an arc appears near the zenith or, more precisely, along the direction of magnetic field lines (namely, the magnetic zenith), about 13° away from the zenith toward the southern horizon in Fairbanks, only the bottom edge of an auroral arc is seen as a line in a simple situation. The width of the curtain is less than 1 kilometer, extremely thin compared with its east-west length. In general, an auroral curtain almost always develops small-scale folds of the order of a few kilometers. As a result, the bottom of an arc is seen as a wavy band. In most cases, the waves develop into curls or eddies of scales of the order of a few kilometers, which often move rapidly along the curtain. Figure 4 shows an example of curls photographed by a TV camera. The curls are seen as vertical striations on the surface of the curtain, called the 'ray structure'. When an auroral curtain is located slightly away from magnetic zenith, both the bottom and the curtain surface will be seen. Since a typical height of the top of the auroral curtain is several hundred kilometers, the parallel rays appear to converge toward the magnetic zenith by the perspective effect, resulting in a fan-shaped spectacular display called the 'corona'. Thus, the corona is not a different form of the aurora.

An auroral curtain or arc develops also large-scale folds. Such a form is categorized as a "band" in Störmer's classification. The largest one has a scale of the order of a few hundred kilometers and is called a spiral. In an old classification, this form was called a 'drapery' or 'horse-shoe' type. There are also folds of scales of a few tenths of kilometers. These dynamical aspects of folds have been studied by using TV cameras and are reported by T. Hallinan and T. Oguti in Chapter 2.

The lights from the aurora are emitted by

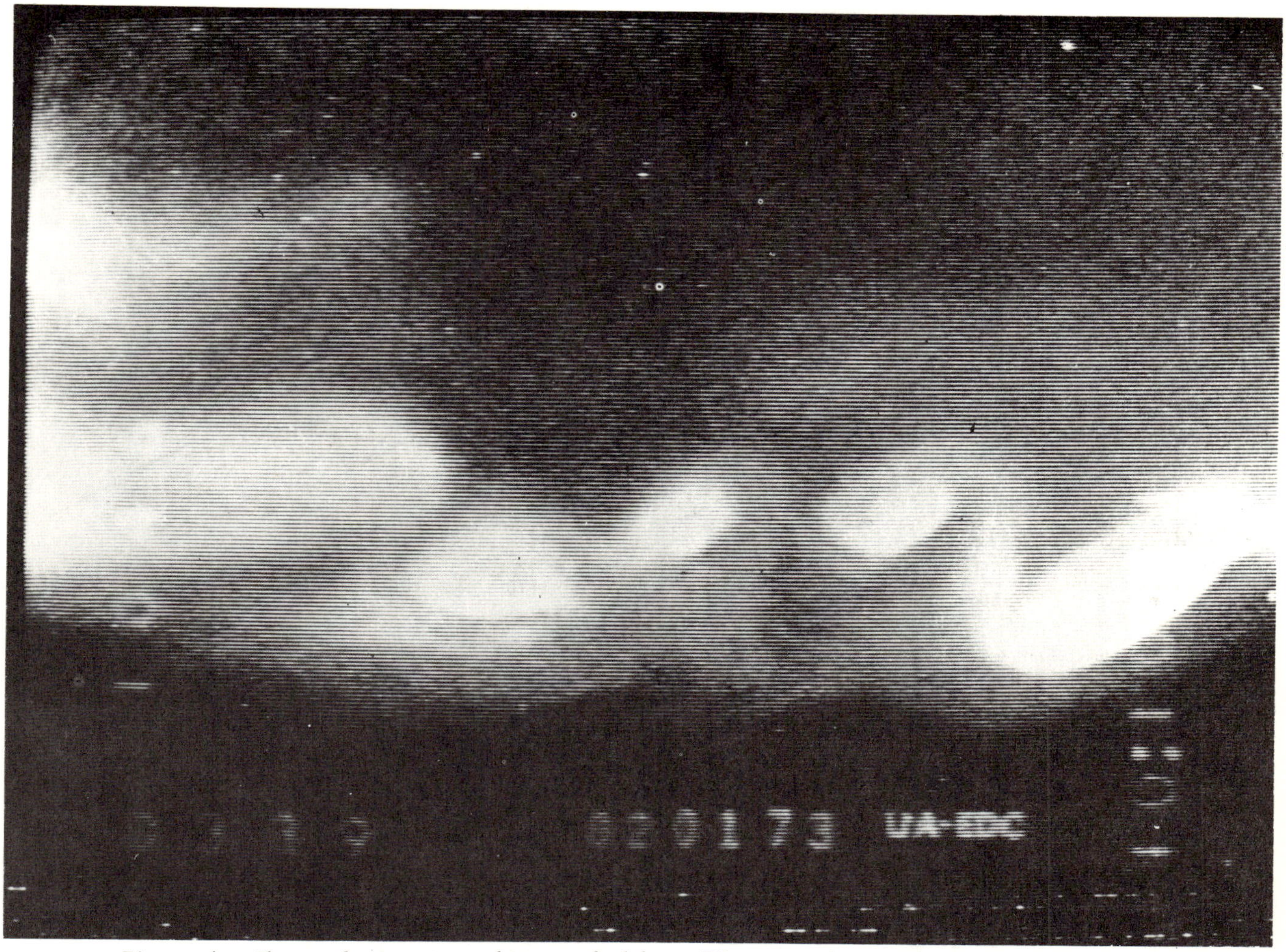

Figure 4. The curl structure photographed by a TV camera (courtesy of T. Hallinan).

upper atmospheric constituents which are produced by excitation and/or ionization by an electron sheet beam and also by complicated chains of photochemical processes which take place after the electron collisions [Chamberlain, 1961; Omholt, 1971; Vallance and Jones, 1974]. The most familiar lines and bands of the auroral arc are:

OI 5577 (the green line)
OI 6300 (the red line)
N_2^+ 3914, 4278 (the First Negative Band)

In Chapter 2, M. H. Rees discusses how the spectroscopic data can be used in obtaining vital information on the energy spectrum of auroral electrons.

3. Open Magnetosphere, the Auroral Oval and the Inverted V Structure

The magnetosphere consists of the regions of "closed" and "open" magnetic fluxes as a result of the interaction with the solar wind magnetic field. Figure 5a shows an example of the open magnetosphere. The open flux originates from both the northern and southern polar regions and constitutes the so-called 'lobe region' of the magnetotail. These field lines are connected to the solar wind magnetic field lines across the magnetopause. Figure 5b shows an ionospheric projection of the open field region for the open magnetosphere illustrated in Figure 5a. The poleward boundary of the auroral oval coincides approximately with the boundary of the foot of the open flux, and the area surrounded by the auroral oval is called the polar cap. It is very convenient that the boundary of the open flux region (namely, the polar cap) can be visible as the auroral oval, because one can monitor the amount of the open flux by observing the size of the auroral oval which varies considerably as the orientation of the solar wind magnetic field varies.

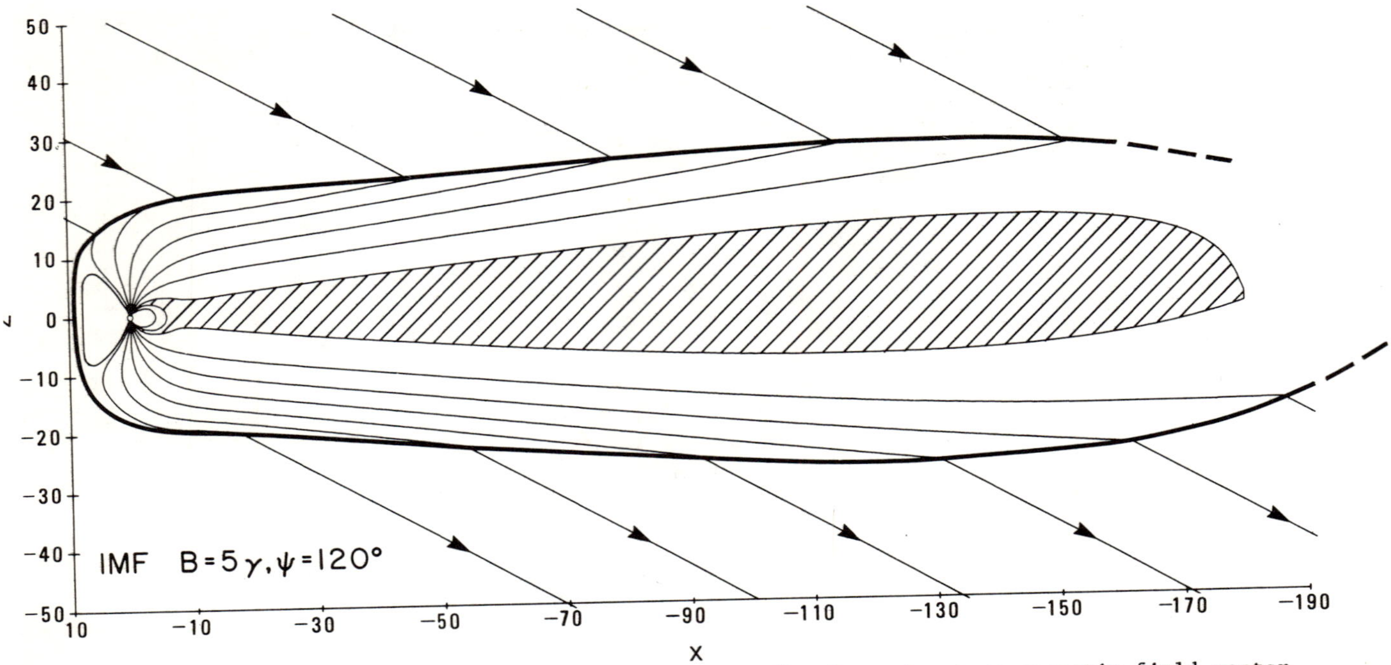

Figure 5a. An example of the open magnetosphere. The interplanetary magnetic field vector (B=5γ) is confined in the X-Z plane and is pointing 120° from the Z-axis.

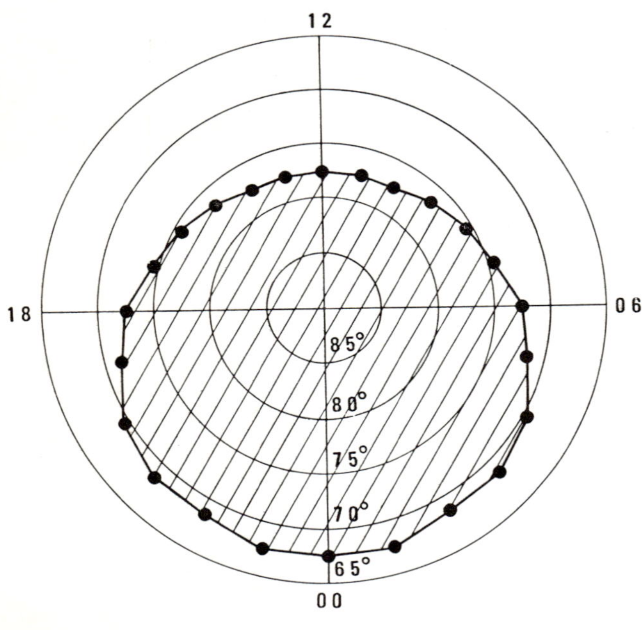

Figure 5b. The geometry of the open flux region in the polar region for the open magnetosphere model which is the same as Figure 5a, except for the IMF magnitude B = 10γ.

The auroral oval coincides with the belt into which electrons precipitate. This belt has a width of the order of a few hundred kilometers, which is much wider than individual arcs (of the order of 1 km or less). That is to say, auroral arcs are embedded in the precipitation belt. The average energy of the electrons varies in this wide precipitation belt. The typical structure of precipitation across this belt is called the inverted V structure [Frank and Ackerson, 1971]. This is because the energy of the electrons increases and then falls as one goes from the equatorward boundary to the poleward boundary of the belt, so that this structure appears like an inverted V in the energy-latitude diagram (along a path of polar orbiting satellites). Auroral arcs are imbedded in the inverted V precipitation region and they coincide with regions where the total energy flux carried by the electrons exceeds the threshold value of about ~1 erg/cm^2 sec ster.

The differential energy spectrum of the electrons in the inverted V precipitation region has the so-called 'mono-energetic peak' of a few kilovolts superposed on a power law spectrum. It is generally agreed that the 'mono-energetic electrons' are most responsible for the auroral luminosity. The bottom height of the aurora is determined by the depth of penetration of those electrons. Figure 6 shows a typical electron spectrum in an auroral arc. Details of auroral electrons will be reviewed by R. L. Arnoldy, C.-

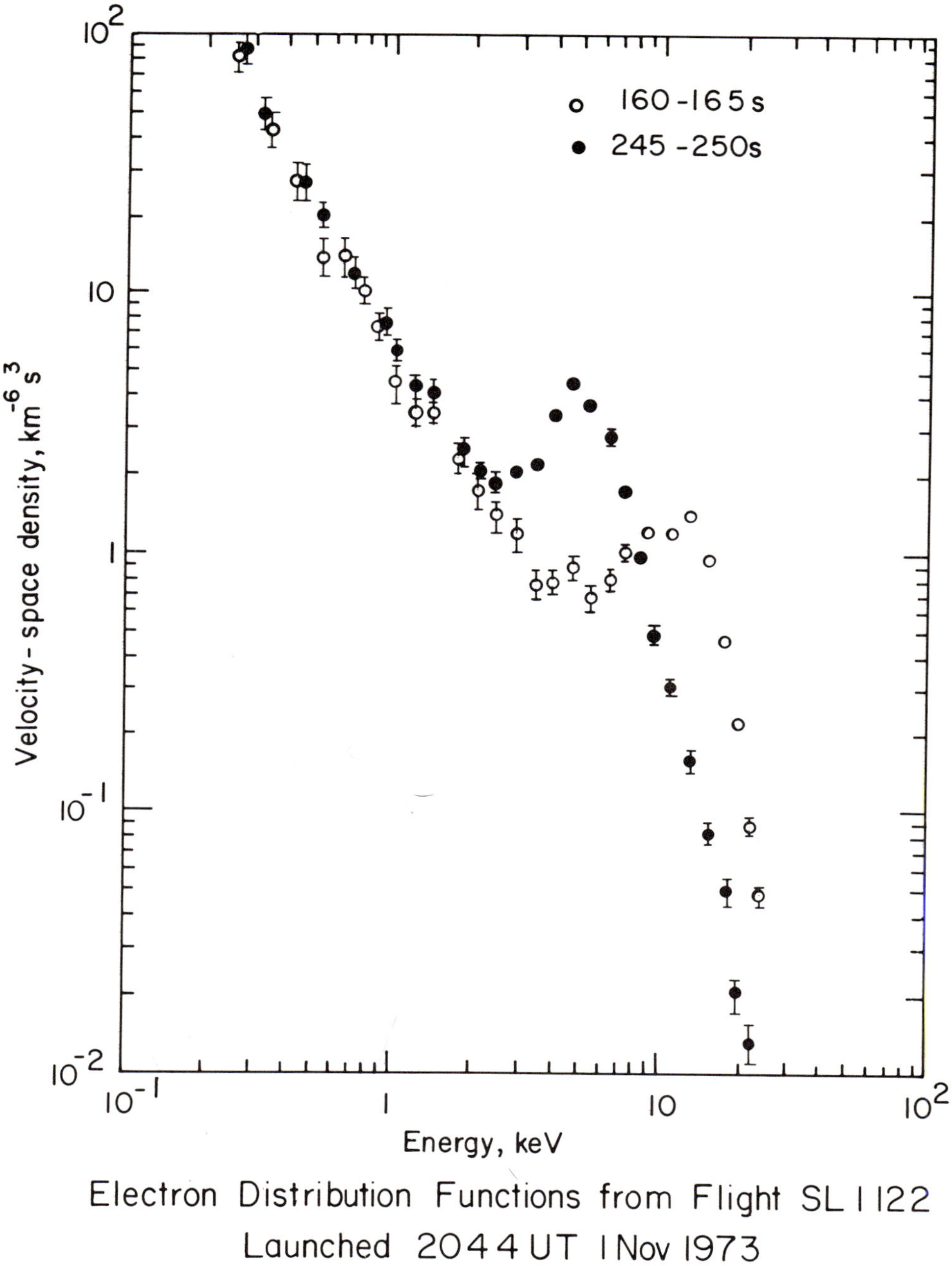

Figure 6. A typical example of energy spectrum of auroral electrons (after D.A. Bryant, 1981).

I. Meng, R. A. Hoffman and C. S. Lin, J. F. Fennell et al., and D. A. Bryant in Chapter 3. As mentioned earlier, there is another wide belt of electron precipitation (located just equatorward of the inverted V precipitation) which produces the diffuse aurora. It is generally agreed that this electron precipitation occurs as a result of wave-particle interactions which 'untrap' electrons from the radiation belt. There is no 'mono-energetic peak' in the energy spectrum of those electrons.

4. Electron Precipitation and the Field-Aligned Currents

Now, the first important question is why auroral electrons precipitate into a narrow belt

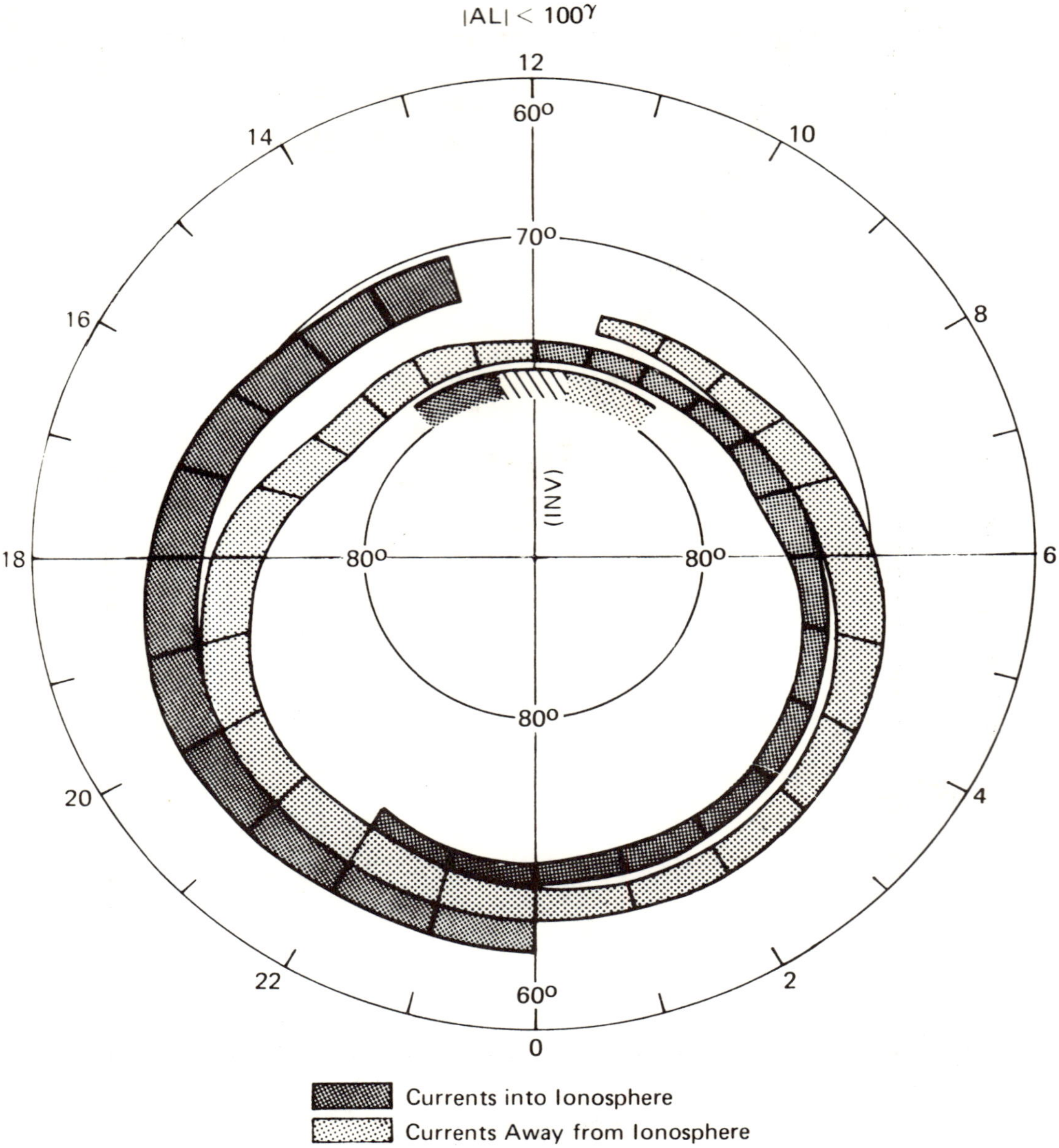

Figure 7. The distribution of the field-aligned current in invariant latitude-magnetic local time coordinates (after Iijima and Potemra, 1976).

around the polar cap. The key to this question is that field-aligned currents flow into and out from the auroral oval. A typical current intensity is of the order of 10^{-6}-10^{-5} A/m^2. The first conclusive observations of the field-aligned currents were made by Zmuda et al. [1970]. Later, the distribution of the field-aligned currents was mapped in detail by the TRIAD satellite (Figure 7). The inner and outer halves of the field-aligned current belts are called the region 1 and 2 currents, respectively. In the region 1 current region, the field-aligned current flows downward in the forenoon-morning half and upward in the afternoon-evening half. Auroral arcs are imbedded in the region where upward field-aligned currents flow, indicating the upward currents are mainly carried by downward flowing electrons. The most typical auroral arc is seen in the afternoon-evening half of the region 1 current region. Figure 8 shows a very bright auroral arc which is located at the poleward

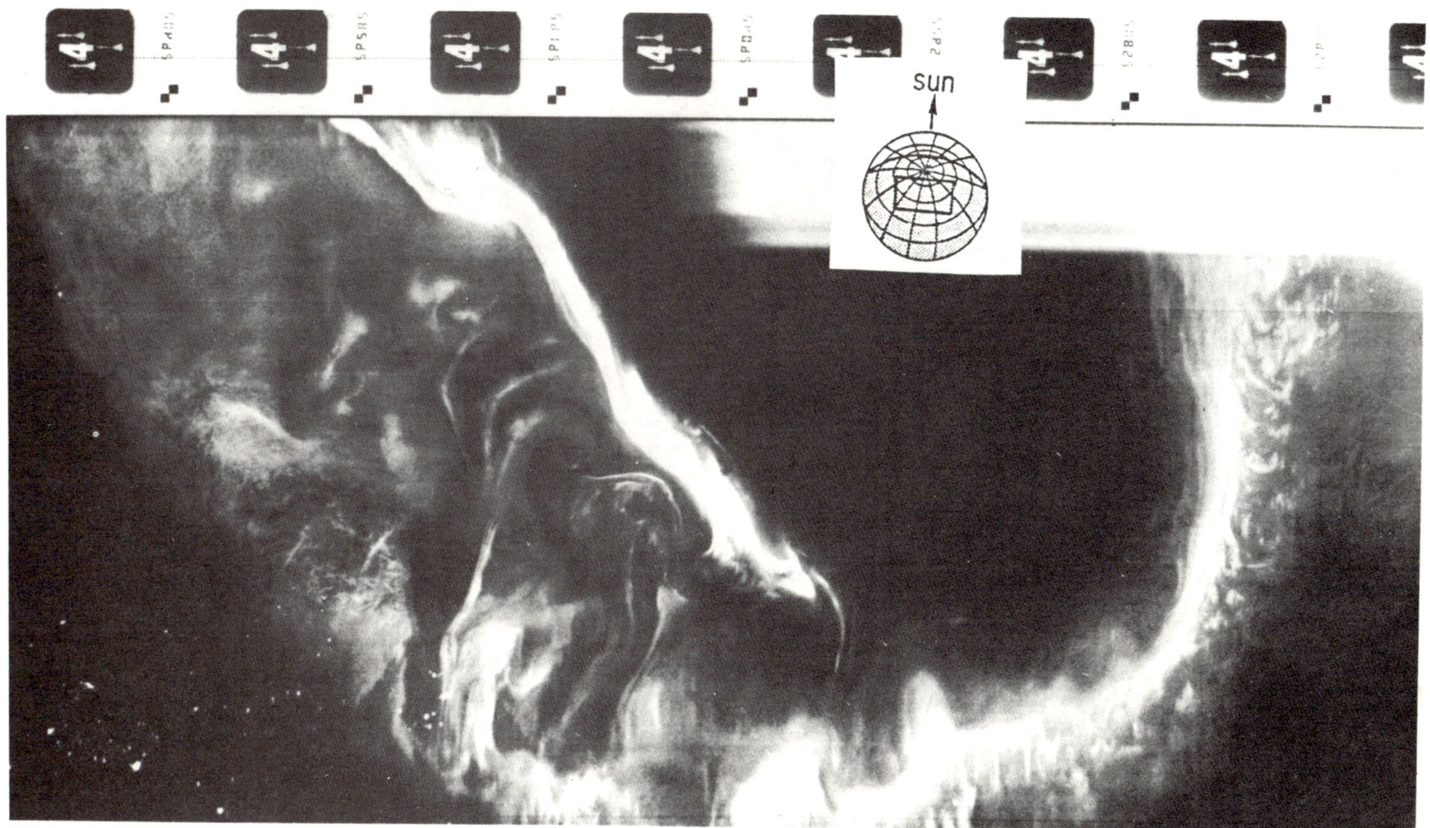

Figure 8. A very bright auroral arc in the evening sector, photographed by a DMSP satellite (2057 UT, Jan. 22, 1976). A very disturbed diffuse aurora is also seen along the equatorward boundary of the oval. An approximate location of the photograph on the globe is indicated in the insert.

boundary of the oval in the evening sector (thus in the region 1 current region).

We have already discussed briefly in Section 1 the generation of the field-aligned currents, which are closed by the Pedersen currents J_P in the ionosphere. The ionosphere has an anisotropic conductivity and as a result the Hall current also flows in the ionosphere. Both Pedersen and Hall currents constitute concentrated currents along the auroral oval, called the auroral electrojet. We can monitor roughly the intensity of the auroral electrojet by a ring of observatories along the auroral zone; the auroral electrojet index AE is derived from their magnetic data. In Chapter 4 the geometrical relationship between the field-aligned currents and the precipitating electrons is discussed by R. A. Heelis et al. and by W. J. Burke. The relationship between the ionospheric electric field, the electrojet and the field-aligned currents is discussed by R. R. Vondrak and Y. Kamide by using data from the Chatanika radar and the Alaska meridian chain of magnetic observatories. In this connection, B. A. Whalen et al. report an interesting observation of an artificially disturbed ionosphere in the auroral zone by injection of water into it by a rocket.

Since the aurora results from the discharge current generated by the solar wind-magnetosphere dynamo, one can expect a direct relationship between auroral activity and the power of the dynamo ε [Perreault and Akasofu, 1978; Akasofu, 1980]. Figure 9 shows that auroral activity depends on ε. On the basis of a study of a large number of examples, it can be shown that an auroral arc brightens considerably when ε exceeds approximately 10^{18} erg/sec or 10^{11} watts. This signals the onset of a typical auroral activity, called the auroral substorm, which is the auroral manifestation of one of the most important processes in the magnetosphere, called the magnetospheric substorm. The aurora becomes brighter as ε increases and the magnetospheric substorm intensifies. Note also that ε correlates fairly well with the AE index which is a measure of the current intensity of the ionospheric portion of the discharge circuit (the auroral electrojet), indicating also that the discharge current is driven by the solar wind-magnetosphere dynamo.

5. Field-Aligned Currents and the Auroral Potential Structures

Now, the next question is how the 'monoenergetic' auroral electrons are related to the field-aligned currents. In order for the current-carrying electrons (of energy of the order of a few hundred electron volts) of magnetospheric plasma to be able to excite and/or ionize upper atmospheric particles at an altitude of about 100 km, the electrons must be accelerated to have energies of a few kilovolts. In this Conference, we concentrate mainly on this question.

It had long been presumed that an electric field could not exist along magnetic field lines in a collisionless plasma. One of the most recent indications that there is a potential drop along the geomagnetic field line above the auroral zone came from a study of auroral particles and electric fields by Gurnett [1972]. The second indication came from the presence of the 'monoenergetic peak' and a large population of lower energy electrons. Evans [1974] was one of the first to call attention to these features of the energy spectrum and interpreted them in terms of a potential drop along the geomagnetic field lines. The electric field observations by Mozer and his group [1977] and the associated particle observations by the Aerospace and Lockheed groups have conclusively shown the presence of a potential structure at about 1~2 earth radii above the aurora (Figure 10). Their results are summarized by F. S. Mozer and R. B. Torbert et al. in Chapter 4. At about the same time, two important auroral observations were made in Alaska. The first one was an injection of barium clouds along the geomagnetic field lines by rockets [Wescott et al., 1976]. At an altitude of 3000 km, the barium ions were suddenly accelerated upward, suggesting the presence of an upward electric field there. This study is reported by E. M. Wescott in Chapter 4. The other was a detailed study of the curl structure of an auroral curtain. It suggests oppositely directed (ExB) drift motions of magnetospheric plasma across an auroral arc and thus oppositely directed electric field across the electron sheet beam [Hallinan and Davis, 1970].

It is most interesting that several different types of observations contributed almost simultaneously to enable us in envisaging the potential structure which can accelerate auroral electrons. Here, this structure is called the auroral potential structure. The potential structure also accelerates ions in the upward and transverse directions. These observations will be discussed by R. D. Sharp and D. M. Klumpar, respectively, in Chapter 3.

One should not forget a word of caution here that detailed studies of characteristics of auroral electrons by rockets do not necessarily support the idea that the entire acceleration of auroral electrons is due entirely to a potential drop along geomagnetic field lines. This important aspect of auroral electrons is also reported by rocket workers in Chapter 3. It should also be noted that there are auroral electrons of energies well above 10 keV and that their existence may require additional acceleration mechanisms. This subject is discussed by B. H. Mauk and G. K. Parks in Chapter 3.

6. Causes of the Auroral Potential Structures

Mozer and his group showed recently that there are two types of the potential geometry [1980]. One is the so-called 'S-shaped' and the other is

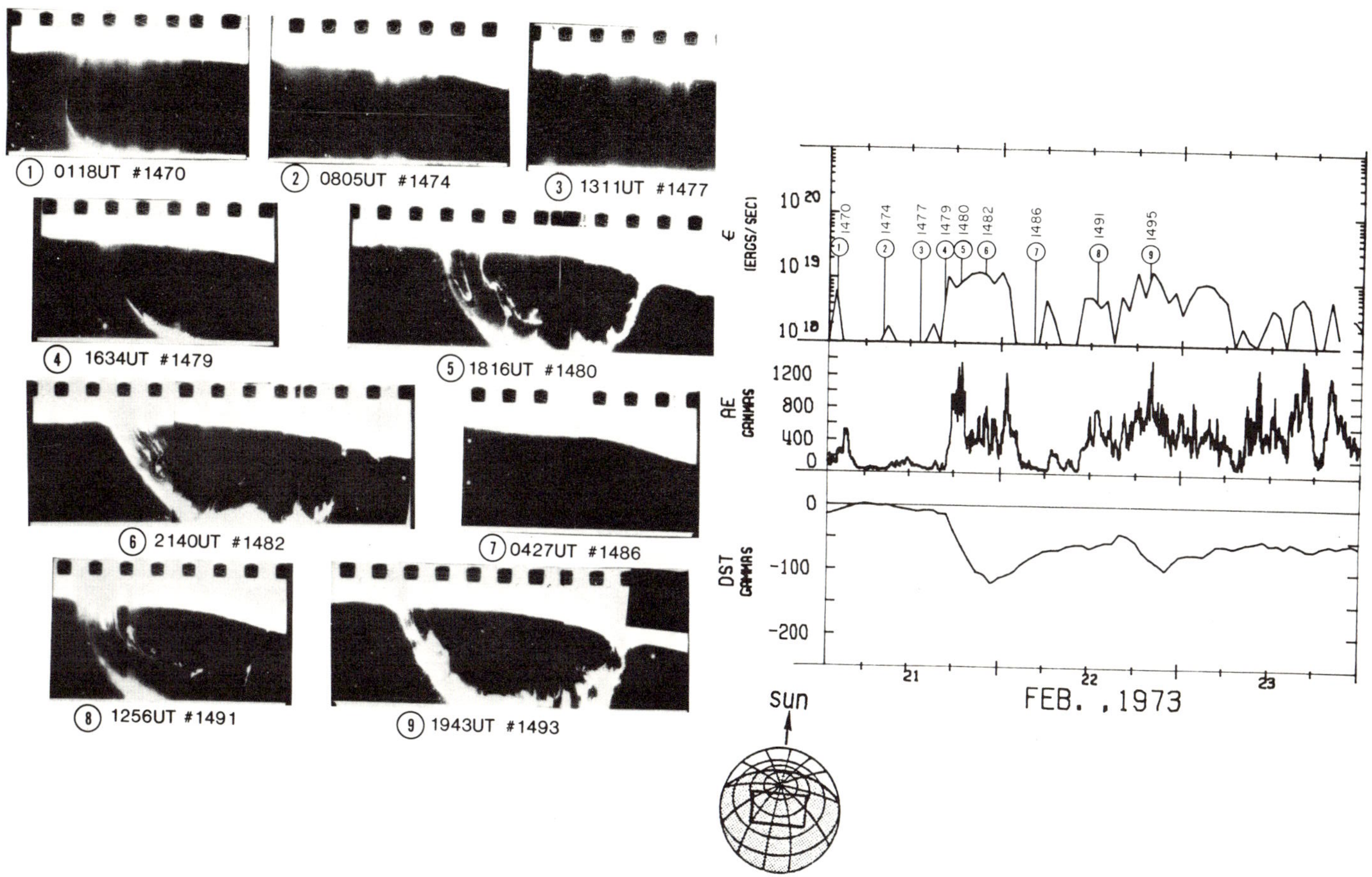

Figure 9. An example of the relationship between ε and auroral activity (February 21, 22, 23, 1973). On the left-hand side, ε and the AE, Dst indices are shown. DMSP satellite photographs during the same period are chosen. The passage times of the satellite over the northern polar region are indicated in the ε diagram. An approximate location of the photograph on the globe is indicated in the insert.

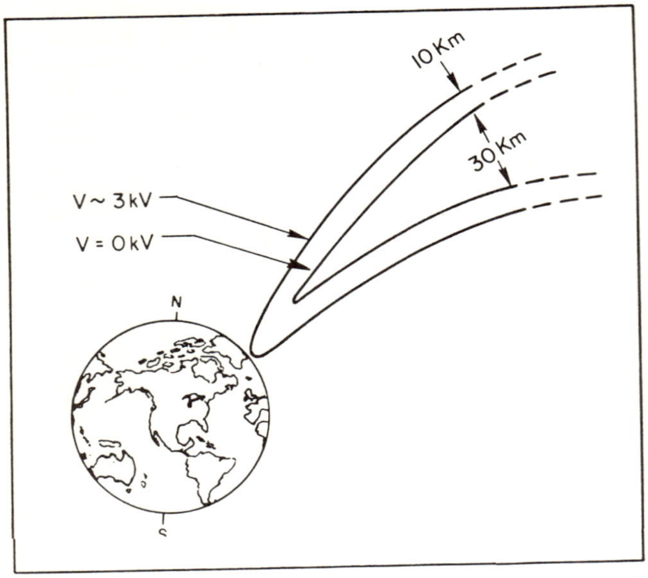

Figure 10. Approximate geometry and location of the auroral potential structure with respect to the earth (Mozer et al., 1977).

'V-shaped'. These potential structures have the north-south scale of the order of 0.01° to 0.1°, so that they are related to both the inverted V precipitation and to individual auroral arcs. Therefore, the V-shaped and S-shaped potential structures are present for both individual arcs (having the latitudinal width of about 1 km) and the inverted V precipitation (having the latitudinal width of 50 km or so). In other words, we assume that the same or similar processes are responsible for both the inverted V structure and the auroral (visible) arc formation. Most papers in this monograph treat both structures without distinguishing them. This assumption implies also that the electron current has a thin sheet-like structure (of width ~1 km) imbedded in a larger sheet structure (~100 km). However, this implication has not been proven beyond doubt. The causes of the formation of the extremely thin current sheet are briefly discussed by C. E. McIlwain in Chapter 4 and by J. R. Kan and L. C. Lee in Chapter 5.

The basic reason for the development of the potential structure on auroral field lines appears to be that magnetospheric electrons can carry only a limited amount of electric current ($\sim 10^{-6}$ A/m^2) from the magnetosphere to the ionosphere, along the converging magnetic field geometry, as a result of the presence of the atmospheric loss cone in their pitch angle distribution, as pointed out by Knight [1973], Lyons et al. [1979], and Fridman and Lemaire [1980]. As mentioned earlier, when the solar wind-magnetosphere dynamo generates the current which is more than such limiting intensities, the magnetosphere-ionosphere system has to develop a potential structure in order to force the electrons to carry a higher current density. What are the mechanisms which can establish the necessary potential structure? It is quite likely that most of the processes proposed, such as the double layer process, the electrostatic shock process, the differential pitch-angle anisotropy process, the anomalous resistivity process, and the Alfvén wave process, are operating in maintaining the parallel potential drop. It is thus of great importance to find out which one is of primary importance. So far the double layer and anomalous resistivity processes have been most extensively studied. Various models of the auroral potential structure are presented in Chapter 5 by J. R. Kan and L. C. Lee, L. P. Block, R. L. Stenzel et al., Y. T. Chiu et al., L. C. Lee and J. R. Kan, L. R. Lyons, M. K. Hudson and D. W. Potter, and W. J. Heikkila.

Plasma simulation methods have become increasingly important tools in space physics. Our subject is certainly one of the areas in which the simulation methods can be applied. It is for this reason that Chapter 6 is devoted to a general discussion of computer simulation studies of auroral plasmas by J. Dawson. His paper is followed by reports which have actually applied the simulation methods to studies of the auroral potential structure and related problems, contributed by H. Okuda et al., D. W. Swift, J. Kindel et al., J. S. Wagner et al., A. Miura and T. Sato, T. Sato and H. Okuda, and J. N. Leboeuf et al. Figure 11 shows schematically some of the interesting features associated with the potential structure, such as the kilometric radiation, VLF hiss and VLF sources.

One of the most interesting aspects of the potential structure is that a variety of plasma waves have been observed in its vicinity. These plasma waves must be related to plasma processes which operate in the potential structure. Therefore, if properly interpreted, they will provide the most valuable information on plasma processes in the potential structure. This is another important problem to be discussed during this Conference. One of the most interesting auroral wave phenomena is intense kilometric radio emissions from the region where the potential structure is expected to be present. This particular radio emission is called the auroral kilometric radiation (AKR), first observed by Gurnett [1974]. In this volume, observations of plasma waves on auroral field lines are summarized in Chapter 5 by D. A. Gurnett, M. A. Temerin, J. L. Green, and R. F. Benson. Some of the related plasma waves and other features observed in laboratory plasmas are reported by H. Böhmer and A. Lang, P. Leung et al., T. Mikkelsen et al., and R. L. Stenzel, et al. In Chapter 7 theoretical studies of auroral plasma waves are discussed by C. T. Dum, C. S. Wu, J. E. Maggs, W. Lotko and C. F.

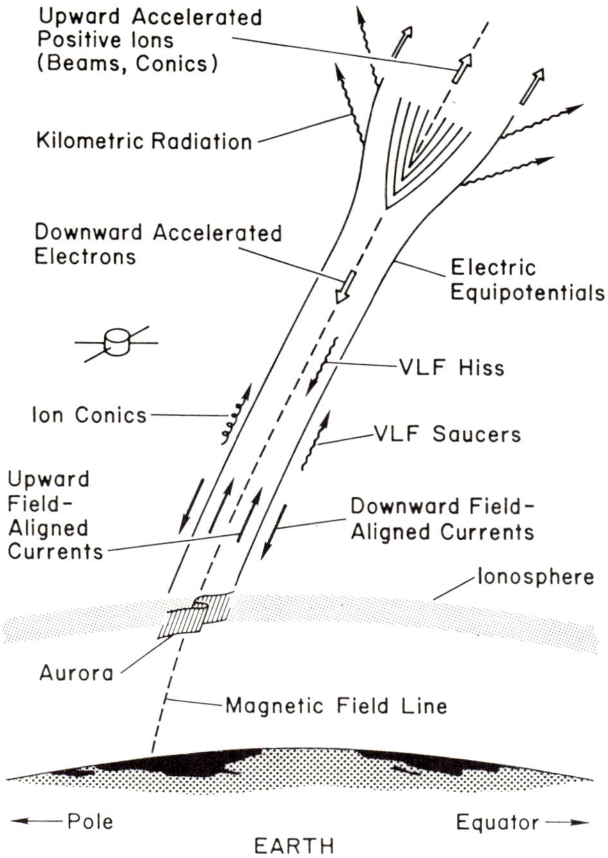

Figure 11. Schematic illustration, showing some of the interesting features associated with the auroral potential structure (P. B. Dusenbery).

Kennel, R. L. Lysak, C. K. Goertz, and P. B. Dusenbery and L. R. Lyons.

The description of auroral arcs is hardly complete without referring to their complicated motions. In spite of their variety and complexity, auroral motions can be fairly well described in the context of the auroral substorm [cf. Akasofu, 1968]. On a moderately disturbed day, auroral arcs repeatedly undergo a fairly systematic activity. The growth and decay of activity are found to be well controlled by a rise and fall of ε, as one can see in Figure 8. As mentioned earlier, typical substorm features appear when ε reaches ~10^{18} erg/sec. The active phase of a substorm continues until ε reaches a peak value (say, ~10^{19} erg/sec) and begins to decline; this phase is called the expansive phase. During the expansive phase, characteristics of the displays are different at different local times. Westward traveling surges, the poleward expanding bulge and patches are typical displays in the evening, midnight and morning sectors, respectively [cf. Akasofu, 1968]. Since both the DMSP and ISIS-II satellite take a photograph once per orbit (of period ~100 min), it is difficult to follow closely the development of the auroral substorm from their consecutive photographs. So far, a Japanese satellite, KYOKKO (Aurora), is the only satellite which can take several photographs per orbit. Their study is presented by E. Kaneda et al. in Chapter 2. The causes of much of such large-scale displays may not be directly related to the problems we intend to discuss during the Conference, although some of them are undoubtedly due to instabilities of the electron sheet current. Perhaps, the typical large-scale displays associated with the auroral substorm are primarily a manifestation of the fact that the field-aligned currents and thus the electron sheet beam shifts in systematic ways as the output ε of the solar wind-magnetosphere dynamo exceeds 10^{18} erg/sec. If this is the case, a rapid movement of auroral arcs indicates that the potential structure degenerates itself as soon as the electron sheet beam shifts away.

References

Akasofu, S.-I., Polar and Magnetospheric Substorms, D. Reidel Pub. Co., Dordrecht-Holland, 1968.

Akasofu, S.-I., Recent progress in studies of DMSP auroral photographs, Space Sci. Rev., 19, 169, 1976.

Akasofu, S.-I., Recent progress in antarctic auroral studies, Upper Atmosphere Research in Antarctica, Antarctic Research Series, Vol. 29, ed. by L. J. Lanzerotti and C. G. Park, AGU, Washington, D.C., 1978.

Akasofu, S.-I., The solar wind-magnetosphere energy coupling and magneto- spheric disturbances, Planet. Space Sci., 28, 495, 1980.

Bryant, D. A. , Rocket studies of particle structure associated with auroral arcs, (this volume), 1981.

Chamberlain, J. W., Physics of the Aurora and Airglow, Academic Press, 1961.

Evans, D. S. Precipitating electron fluxes formed by a magnetic field-aligned potential difference, J. Geophys. Res., 79, 2853, 1974.

Frank, L. A., and Ackerson, K. L., Observations of charged particle precipitation into the auroral zone, J. Geophys. Res., 76, 3612, 1971.

Fridman, M., and J. Lemaire, Relationship between auroral electron fluxes and field-aligned electric potential difference, J. Geophys. Res., 85, 664, 1980.

Gurnett, D. A., Electric field and plasma observations in the magnetosphere, Critical Problems of Magnetospheric Physics, p. 123, National Academy of Sciences, U.S.A., 1972.

Gurnett, D. A., The earth as a radio source: terrestrial kilometric radiation, J. Geophys. Res., 79, 4227, 1974.

Hallinan, T. J., and T. N. Davis, Small-scale

auroral arc distortions, Planet. Space Sci., 18, 1735, 1970.

Iijima, T., and T. A. Potemra, Field-aligned currents in the dayside cusp observed by Triad, J. Geophys. Res., 81, 5971, 1976.

Kan, J. R., and L. C. Lee, Formation of auroral arcs and inverted V precipitations: an overview (in this volume), 1981.

Lyons, L. R., D. S. Evans, and R. Lundin, An observed relation between magnetic field aligned fluxes in the vicinity of auroral forms, J. Geophys. Res., 84, 457, 1979.

Mozer, F. S., C. W. Carlson, M. K. Hudson, R. B. Torbert, B. Parady, J. Yatteau, and M. C. Kelley, Observations of raised electrostatic shocks in the polar magnetosphere, Phys. Rev. Lett., 38, 292, 1977.

Mozer, F. S., C. A. Cattell, M. K. Hudson, R. L. Lysak, M. Temerin, and R. B. Torbert, Satellite measurements and theories of low altitude auroral particle acceleration, Space Sci. Rev., (in press), 1980.

Omholt, A., The Optical Aurora, Springer-Verlag, Berlin, 1971.

Perreault, P., and S.-I. Akasofu, A study of geomagnetic storms, Geophys. J. Roy. astr. Soc., 54, 547, 1978.

Störmer, C., (Chairman), Photographic Atlas of Auroral Forms, IUGG, Oslo, 1930.

Vallance Jones, A., Aurora, D. Reidel Pub. Co., Dordrecht-Holland, 1974.

Wescott, E. M., H. C. Stenbaek-Nielsen, and T. J. Hallinan, The Skylab barium plasma injection experiment, II. Evidence for a double layer, J. Geophys. Res., 81, 4495, 1976.

Zmuda, A. J., J. C. Armstrong, and F. T. Heuring, Characteristics of transverse magnetic disturbances observed at 1100 kilometers in the auroral oval, J. Geophys. Res., 75, 4757, 1970.

Zmuda, A. J., and J. C. Armstrong, The diurnal flow pattern of field-aligned currents, J. Geophys. Res., 79, 4611, 1974.

ISIS-2 OBSERVATIONS OF AURORAL ARC SYSTEMS

J.S. Murphree, C.D. Anger, L.L. Cogger

Department of Physics, The University of Calgary, Calgary, Alberta
Canada T2N 1N4

Abstract. Using the global images from the Auroral Scanning Photometer on ISIS-2 the instantaneous discrete auroral morphology in the evening/midnight sector is studied. The most consistent feature during magnetically active times is the formation of an auroral bulge in the pre-midnight sector to which is connected an evening discrete arc system. These bulges can extend over 7° of latitude and the Magnetic Local Time of their westward leading edge has a well defined peak at 20-22 MLT. There is normally a single evening arc system which connects into the equatorward portion of the bulge. This arc system may extend into the afternoon sector and though discontinuities may exist in its morphology, the continuity of intensity and 5577Å/3914Å ratio across these discontinuities suggests a single particle source for the entire arc system. Discrete auroral arc systems, particularly these connecting to auroral bulges, occur predominantly near the poleward edge of the auroral distribution but apparently imbedded with diffuse aurora. Examples clearly show that discrete aurora can occur within the main region of diffuse auroral emission.

Introduction

Auroral observations have progressed from infrequent sightings of displays by people living in high latitude regions, to today's highly scientific studies ranging from small scale *in situ* measurements to two-dimensional optical imaging far exceeding the horizons of a ground based observer. Early studies were primarily concerned with observations of latitudinally narrow, discrete auroral arcs which display remarkable changes during auroral substorms (Akasofu, 1964). Such features have provided the starting point for an understanding of the interaction between the ionosphere and magnetosphere. Optical remote sensing instruments in satellites provide the ability to extend observations of auroral arcs over many thousands of kilometers. The resulting global images of the aurora reveal features which are, with only occasional ambiguity, separable into two classes - discrete and diffuse. Akasofu (1974) has presented some characteristics of these two types of aurora. Operationally they can be classified as follows: discrete features are normally long compared to their width and display a well defined intensity maximum. Diffuse features typically cover large regions and are > 100 km in width. Diffuse aurora in the instantaneous auroral oval is characterized by a relatively smooth and sharp equatorward boundary, by widths of 2-10° in latitude, and by extension around the entire oval.

When attempting to discuss auroral arcs using two-dimensional satellite optical images, one is confronted with the problem of relating ground based measurements of very narrow arcs (< 1 km) to the discrete features observed by satellite detectors with poorer spatial resolution. Observations by the Auroral Scanning Photometer (ASP) on the ISIS-2 satellite also prompt this question because the altitude of the satellite (1400 km) and the angular resolution (.4°) combine to produce a spatial resolution at auroral altitudes of ∿ 9 km. Averaging of data points and non-vertical viewing decrease the resolution even more. In addition, auroral arcs must be detected above a reasonably uniform background of diffuse aurora, and all observations of aurora over high albedo surfaces are subject to contamination by ground scattering. Even with these limitations a very narrow feature can be observed if it is bright enough. For example, if we take the instrumental conditions applicable to the ASP and assume vertical viewing, then in order to give a detectable signal level of 1 kR above a 1 kR background of diffuse aurora a 2 km thick arc must have an intensity of 5 kR. If data points are averaged together then the corresponding detection threshold intensity must go up as the width of the averaged region. The required intensity to detect a broader arc of course decreases. Identification of auroral arcs can be made even though their thickness is less than the spatial resolution of the ASP; however, their observed intensity will be too low. In practice the problem is minimized because auroral arcs are often multiple in character. Thus instead of a single narrow arc, several arcs, separated by a few kilometers may occur and be parallel for large distances. These discrete arc systems are what we consider to be the features that we observe.

The fact that the ASP observations are probably dominated by systems of auroral arcs rather than single arcs is offset by its ability to monitor large regions of the high latitude aurora essentially instantaneously. Figure 1 illustrates this by showing the auroral distribution in the northern hemisphere on 711215 at 0440 UT. On the left are profiles in Corrected Geomagnetic Latitude (CGL) for hourly values of Magnetic Local Time (MLT) and on the right the global images produced by the two-dimensional coordinate transformations of the same data. Because it has been established from previous work (e.g. Feldstein, 1963) that the optical aurora is magnetically ordered, we shall use exclusively the above magnetic coordinate system (see Murphree and Anger, 1980 for details of the data processing technique). The images are displayed (5577Å on top, 3914Å on bottom) as a grey scale from .5 kR (black) to 4 kR (white), with the geomagnetic pole indicated by a cross and are ordered with 24 MLT at the bottom. The white rectangular area in the late morning on the two images is due to sunlight scattering in the instrument baffle as is the long narrow feature in the post noon quadrant.

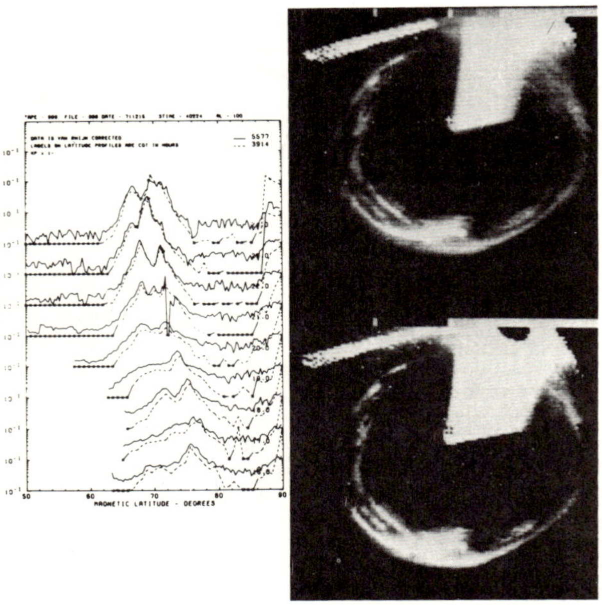

Figure 1: Latitude profiles and coordinate transformations (5577Å top; 3914Å bottom) for the 711215 pass at 0440 UT. The profiles are at hourly Magnetic Local Time (MLT) and are displaced by 1 order of magnitude each hour. Intensities are in kilorayleighs, corrected for viewing angle but not for ground scattering. The transformations are oriented with 24 MLT at the bottom. The intensity grey scale covers from .5 kR (black) to 4 kR (white). The large rectangular area in the top right and long narrow area in top left are due to sunlight scattering.

This pass is of interest because it displays several discrete auroral arc systems. In the latitudinal profiles (from 16 MLT at bottom to 24 MLT at top) they show up as intensifications in the auroral distribution. Their latitude changes with local time (higher at earlier MLT) and several systems can occur along one MLT meridian. It is clear that the information along any one meridian is insufficient to characterize the auroral oval morphology. The range of discrete auroral conditions is illustrated in Figure 2. The top row in Figure 2 (740213 at 0457 UT) illustrates what is probably the simplest discrete auroral morphology observed on ISIS-2 - a long narrow arc system throughout the evening sector (∿ 19-24 MLT) which bulges poleward at midnight. This bulge is likely the poleward expansion of a discrete arc system during the initial expansion phase of a substorm (Akasofu, 1968). Note that only one arc system at the poleward boundary of the diffuse aurora is apparently involved in this process and that the intensification of this discrete arc system occurs over many hours of local time rather than just at midnight. In contrast to this the bottom of Figure 2 (711218 at 0556 UT) shows one of the most convoluted morphologies of discrete aurora as yet observed by ISIS-2. The concept of the instantaneous auroral oval is difficult to visualize as emission can be seen at practically all latitudes and local times.

Despite the apparent randomness of discrete auroral configurations, there is a reasonably uniform overall behaviour of the discrete optical aurora, particularly during an auroral substorm (Akasofu, 1968). Such behaviour is well documented using ground based all-sky camera data so that the characterization of an auroral substorm in terms of an expansion and recovery phase is generally agreed upon (Rostoker et al., 1980). The most obvious advantage of two-dimensional optical measurements from satellites is that they permit one to remove the observations of discrete aurora from the local time frame of a ground based observer to the Universal Time (UT) frame of the magnetosphere. As a result it has become evident that the concept of the auroral substorm should be interpreted not from the viewpoint of the ground observer but from that of the magnetosphere, especially in view of the difficulty in identifying the onset time of a substorm from the ground (Rostoker et al., 1980). With 'snapshots' of the discrete auroral morphology every 110 minutes it is not possible to monitor the dynamics of a single substorm from ISIS-2. However, as we shall show, the gross characteristics of the discrete aurora in the evening/midnight sector are quite repeatable during magnetically active periods. Accordingly, our analysis will focus on those large-scale discrete auroral characteristics

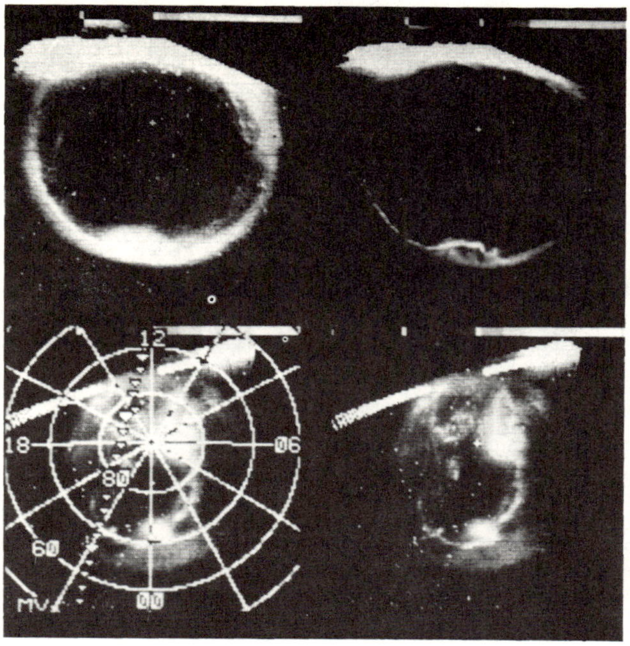

Figure 2: Transformations at 3914Å for the 740213 pass at 0457 UT (top) and 711218 pass at 0555 UT (bottom). The intensity grey scale for the top is .5-4 kR (left) and 2-10 kR (right). On the bottom it is .5-10 kR for both. The coordinate grid is the same for all the frames.

which from the satellite point of view are essential parts of optical activity during substorms.

Auroral Bulges

Perhaps the most striking discrete auroral features observed in the midnight sector are auroral bulges. These bulges are normally identified as the westward traveling surges (Anger et al., 1973; Kamide and Akasofu, 1975) which are an integral part of the substorm process (Akasofu, 1968). As such, auroral bulge characteristics and in particular the relationship with evening arc systems are important in determining the magnetospheric rather than local time effects of an auroral substorm. The expansion phase of a substorm is said to end and the recovery phase to begin when the bulge stops expanding poleward (Akasofu, 1977). Since the evolution of an auroral bulge occurs on a time scale less than the period of a satellite, little information on the dynamics over large fractions of the auroral oval are available. For instance, the ASP on the ISIS-2 satellite typically requires 2-3 minutes to image a large auroral bulge while the orbital period of the satellite is ∿ 110 minutes. Thus determination of expansion or recovery phase from the satellite pictures is not possible. Nevertheless, these pictures reveal certain well-defined features in auroral bulges which we now describe.

In Figure 3 we present the 3914Å images for 8 passes each of which illustrates an auroral bulge. Our criterion for characterizing an auroral bulge is an empirical one: there must exist a substantial north-south latitudinal shift in the poleward boundary of the discrete aurora as a function of MLT. This shift may be anywhere from 2° (Figure 3a) to 7° (Figure 3h) or even greater.

The top half of Figure 3 displays some of the variety in the shape of the leading (westward) edge of the auroral bulge. Figure 3a (740226 at 0928 UT) clearly fits the definition of a spiral given by Hallinan (1976): a large scale (in this case ∿ 400 km) counterclockwise (when seen from above) rotation of a discrete arc system. Such features can be seen in 30% of the DMSP images (Davis and Hallinan, 1976) although such a well-formed spiral in the auroral distribution is extremely unusual in the ISIS-2 data. Normally distortions in the discrete aurora are much more convoluted as the other examples in Figure 3 illustrate. It is not clear whether or not the spiral in Figure 3a should be identified as a westward traveling surge: the auroral morphology to the east of the leading edge does not display the classic break-up of typical auroral substorm theory as can be seen by comparison with the other pictures in Figure 3 or Akasofu (1976). Secondly this pass occurs during a minimum in auroral electrojet activity as defined by the index AE. This condition is not consistent with the coexistence of a substorm. Note however, that the AE index may or may not reflect the existence of an auroral bulge (see discussion of Figure 3e for example).

Figure 3b (740228 at 0655 UT) illustrates a classical westward traveling surge. As in Figure 3a a single discrete arc system exists to the west of the leading edge of the bulge. While there is only a gradual shift in the poleward boundary toward later MLT, break-up conditions behind the bulge are clearly evident with distorted discrete features appearing behind and equatorward of the bulge along with patchy looking diffuse aurora. The arc system which connects into the bulge does not do so in a continuous fashion. It apparently reverses itself twice (at 21 and 19.5 MLT) and thus the leading edge of the bulge does not appear to be a large scale spiral. Note that the connecting evening arc is located well equatorward of the poleward boundary of emissions (5 kR in this presentation) to the west of the bulge. In contrast to this well formed bulge and its obvious connection to a discrete evening arc system, Figure 3c (720111 at 0011 UT) displays a distortion in the poleward boundary with no such connection apparent. The pass again displays standard break-up characteristics behind the leading edge, but instead of a variation in latitude of the poleward boundary with MLT, it is constant at 71° CGL.

Distortions in the poleward boundary which can

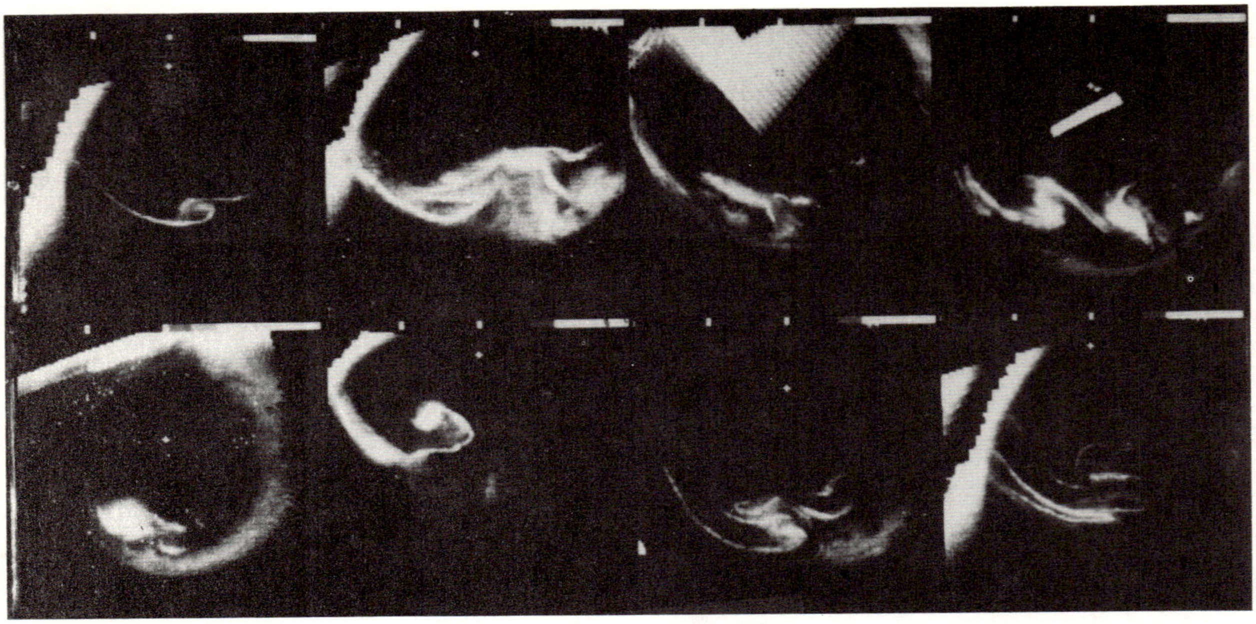

Figure 3: Midnight sector 3914Å images for 8 passes. All intensity ranges are 5-30 kR except as noted. The transforms are arranged with the MLT indicated below extending along the meridian from the geomagnetic pole (represented by a cross) to the bottom of each image. Top row (left to right): (a) 740226 at 0928 UT, 21 MLT; (b) 740228 at 0655 UT, 21 MLT; (c) 720111 at 0011 UT, 24 MLT; (d) 740121 at 1126 UT, 23 MLT; Bottom row (left to right): (e) 740123 at 0858 UT, 24 MLT; (f) 740224 at 0811 UT, 23 MLT; (g) 740129 at 1047 UT, 22 MLT; (h) 740228 at 0849 UT, 21 MLT.

be classified as auroral bulges may occur more than once in the auroral distribution. Figure 3d (740121 at 1126 UT) illustrates a sequence of two auroral bulges occurring simultanteously. The actual time between the observation of these spirals during a given spin of the satellite is only ∿ 4 seconds so that it is improbable that one has decayed and the other appeared in this time as might be consistent with the theory of Wiens and Rostoker (1975). Such a sequence of spirals have been termed "streets" by Davis and Hallinan (1976). The observation of two large scale bulges is different from an auroral bulge forming within a pre-existing bulge (Kamide et al., 1977). The interpretation of such a morphology in terms of separate substorm intensifications (McPherron, 1979) depends on whether the bulges evolved simultaneously at different local times or whether they formed one after another at the same local time.

The examples presented thus far clearly show the variety which can occur in the morphology of the auroral bulge. Aside from the distortion in the poleward boundary on all passes (i.e., the selection criterion) and the high intensities (all examples have large regions of > 30 kR 3914Å), the morphologies appear to be somewhat random. There are two ways however, in which auroral bulges can be characterized: 1) the number of evening arcs which connect into the bulge; 2) the MLT of the leading edge of the bulge.

The classical theory of substorm development (Akasofu, 1964) has the westward traveling surge propagating continuously along an arc system. In contrast to this Wiens and Rostoker (1975) suggest a stepwise formation in the northwesterly direction during the expansion phase of substorms. The results of Sergeev and Yahnin (1979) are also consistent with this, at least for the larger expansions. It should be pointed out that studies of bulges using all-sky cameras are limited in the spatial extent which they can see (3.5° latitude assuming ± 60° of zenith and 100 km altitude for the emissions). Such a distance is clearly inadequate to study the majority of bulges which have been presented thus far.

The number of evening arc connections can vary from 0 (Figure 3c and 3e) to at least 3. In addition the arc system can become distorted but yet remain continuous. This is illustrated in Figure 3f (740224 at 0811 UT). A very long evening arc system extends from before 17 MLT to ∿ 21 MLT at basically a constant geomagnetic latitude. After this MLT, however, the arc system turns poleward and actually reverses toward earlier MLT where it ends in a small spiral feature at ∿ 80° CGL and 19.5 MLT. There is obviously a large distortion in the arc system over many kilometers even though what might be considered a "bulge" occurs only at its end. Such a widespread (in latitude) distortion of the arc system is unusual but is similar to the examples shown in Figure 3b and 3h.

The vast majority of bulges (∿ 90%) connect only to one evening arc system. The example shown in Figure 3e of a very confined bulge at high latitudes occurs during a period of time when AE has returned to baseline values after a very minor intensification. The implication is that little westward electrojet is associated with this bulge and/or its latitude is too high. Examples displaying no connection (Figures 3c and 3e) might be due to dissipation of the evening arc system during the recovery phase of a substorm (consistent with AE for Figure 3e but not necessarily for 3c). On the other hand it may be that no connection to an evening arc ever existed for these cases. If a single arc system with propagating bulge is the norm, then the infrequent observations of two or more evening arcs suggest that the stepwise formation of bulges (Wiens and Rostoker, 1975) is relatively uncommon. From the two-dimensional view presented in these global images it seems probable that not only is a new bulge formed, but also the evening arc system which connects to it (see Figure 3g).

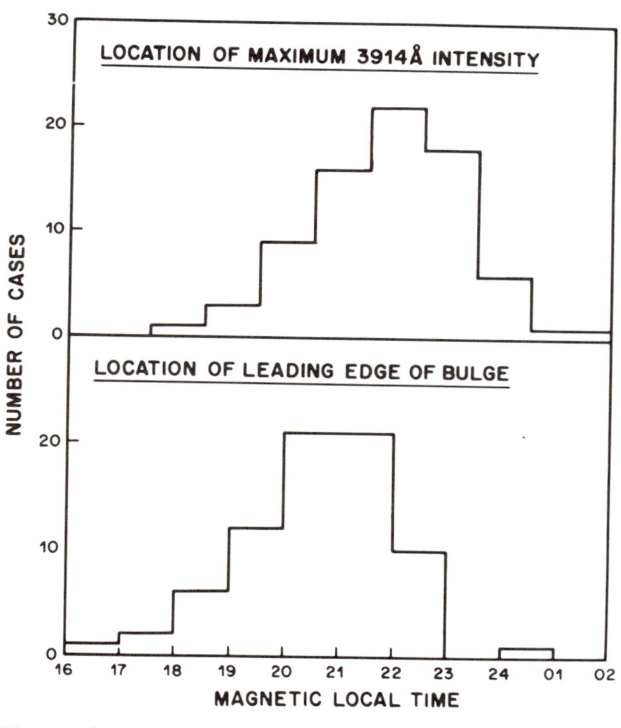

Figure 4: Distribution of bulge characteristics with MLT for all large scale auroral bulges observed during the period 1971-1975: top, location of maximum 3914Å intensity (> 5 kR) in passes when the entire 18-24 MLT sector is observed; and bottom, the location of the leading (westward) edge of the bulge.

The MLT distribution of the leading edge of large scale auroral bulges observed on ISIS-2 is shown in Figure 4 along with the distribution of peak 3914Å intensity within the midnight sector. These distributions show a prominent peak between 20 and 22 MLT. Even allowing for the expected westward propagation of surges during the life of a substorm, the peak seems surprisingly far removed from midnight and unexpectedly sharp. Such peaks in the distribution may be due to several factors such as variations in propagation speed of the surge with MLT (slower near 20 to 22 MLT) or substorm phase, variation with MLT in bulge size etc. It is interesting to note that the observations of Davis and Hallinan (1976) on much smaller scale features also peak in this time sector.

In summary, auroral bulges are striking distortions in the poleward boundary of the auroral distribution. Their latitudinal extent ranges from greater than 7° CGL down to distortions of the spiral variety of ∿ 50 km as defined by Davis and Hallinan (1976). Snapshots of bulges can be found which are consistent with both continuous deformations along arc systems (Figures 3a, 3f) and with stepwise jumps in location (Figures 3g and 3h). There are two characteristics of auroral bulges which are dominant: a well defined maximum in the MLT of the westward leading edge between 20 and 22 MLT, and a direct connection between evening arc system(s) and auroral bulges. These evening arcs must be intimately related to the bulge and hence to the substorm process.

Evening Arcs

As indicated above, one of the more prominent features of the auroral bulge is the emanation of one or more arc systems from its westward leading edge towards earlier local times. Such a feature and its direct connection to the westward traveling surge implies that the substorm process is not confined to the location of the surge but in fact occurs over many hours of local time simultaneously (see for example Figure 2a). This is consistent with the observation of substorm - intensified westward electrojets at earlier local times than the expanding bulge (Hughes et al., 1979). Rostoker et al., (1975) have shown that the extension of this electrojet is due to conductivity enhancements as might be expected from particle precipitation which would also result in optical aurora.

In the previous section we have discussed the characteristics of these evening arcs in relation to auroral bulges. In Figure 5 are shown 4 passes which illustrate the typical features of these long arc systems. On the left are the 3914Å intensities. Figure 5a (740101 at 1222 UT) is an interesting example of the continuity of intensity of these arc systems even though morphologically

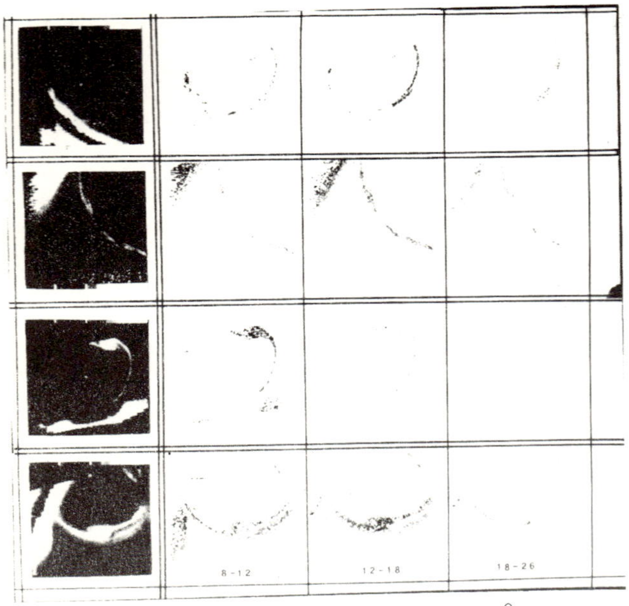

Figure 5: Transformations (all at 3914Å) for 4 passes (on left) along with the distributions of 3914Å intensities in three ratio ranges: .8-1.2, 1.2-1.8, 1.8-2.6 left to right.
(a) 740101 at 1222 UT 4-15 kR
(b) 751202 at 0528 UT 2-10 kR
(c) 741216 at 0811 UT 4-15 kR
(d) 740226 at 0928 UT 2-10 kR
The summation of the 3914Å intensity points in the 3 ratio frames results in the intensity distributions on the left.

they appear to have major discontinuities in them. The latitudinal shift in the arc systems on this pass occurs at 17-19 MLT. The direction of the shift is equatorward, going toward earlier MLT, consistent with the finding of Akasofu and Kan (1980). However, their conclusion that such a discontinuity implies two distinct arc systems and therefore separate dynamos to produce them appears unnecessary for several reasons. First, there is no major intensity variation between these features over the latitudinal shift. Second, such shifts are very numerous as illustrated in Figure 5b. The entire arc system is continuous throughout the evening sector (as is consistent with the results of Snyder et al., 1975) on a macroscopic scale, but at least 2 and possibly 3 discontinuities are evident along its extent. Furthermore, if separate dynamos were involved in establishing these arc systems one might expect different relationships with IMR or AE. While evening arcs intruding into the afternoon sector are strongly dependent on the westward electrojet (Murphree et al., 1980), the occurrence of discontinuities along these arcs does not affect this dependence.

As all 4 examples in Figure 5 show, one of the most striking features of these evening arc systems is their continuity over large regions of MLT. This is obvious in terms of intensity but it can also be demonstrated by calculating the ratio of 5577Å/3914Å along these arcs. This ratio is an index of the characteristic energy of the precipitating particles (Rees and Luckey, 1974) even though the dynamic range of the ratio is small (Mende and Eather, 1975). Nonetheless, significant ratio variations do occur as is illustrated in the right hand frames of Figure 5. Here we present the results of the ratio calculations for the 4 passes discussed above. The three right hand frames for each pass represent the 3914Å intensity values (now shown as black) for three ratio ranges: .8-1.2, 1.2-1.8 and 1.8-2.6. Thus a given 3914Å intensity point will appear in only one of the 3 rightmost frames and the sum of the 3 rightmost frames yields the 3914Å distributions of the left (assuming no ratios < .8 or > 2.6).

The top row in Figure 5 (740101 at 1222 UT) shows clearly that the majority of 3914Å data falls in the 1.2-1.8 ratio bin and that no ratio difference is apparent along the arc system. Similar conclusions apply to the second row (751202 at 0523 UT). The third row (741216 at 0811 UT) shows explicitly the connection of the arc to an auroral bulge and also illustrates that for this pass the majority of 3914Å data occurs in the lowest ratio bin (.8-1.2).

The final pass (740226 at 0929 - see Figure 3a) is shown with a slightly lower intensity range (2 kR-10 kR) in order to include the diffuse aurora. The 3914Å data along the arc systems with the spiral are predominantly in the 1.2-1.8 ratio bin although some are distributed at lower ratios. Most of the diffuse aurora in the .8-1.2 ratio bin is eastward and equatorward of the spiral. However, toward earlier local time the diffuse aurora is predominantly in the 1.2-1.8 bin. Thus local time variations in the ratio exist in the diffuse aurora but are not apparent in the long evening arc system which are observed.

In summary, long evening arc systems are apparently a direct consequence of substorm auroral activity and when their eastward extent can be observed they connect to an auroral bulge. They can extend well past the 18 MLT meridian into the afternoon sector and typically exhibit discontinuities or abrupt shifts in latitude. The continuity of ratio and intensity across these discontinuities as well as along the entire arc, together with the smoothness and continuity of the diffuse aurora at the points where any discontinuities occur, leads us to believe that such an arc system represents a single topological entity with a single source mechanism.

Location of Discrete Features with Respect to the Diffuse Aurora

Since the discovery of the diffuse aurora the question has been raised as to whether discrete features are distinct phenomena poleward of the diffuse aurora or whether they are in fact imbedded in the poleward portions. Since the introduction of the distinction between the Central Plasma Sheet (CPS) and Boundary Plasma Sheet (BPS) (Winningham et al., 1975) and their identification with the diffuse and discrete aurora respectively, it has usually been accepted that the discrete features lie poleward of the diffuse aurora. For instance, in the context of auroral substorms it is the BPS region which is affected primarily (Akasofu, 1977) as the auroral bulge evolves. Part of the difficulty in resolving this question is due to typically low intensity of the diffuse aurora in the evening sector where well-defined arc systems most often occur. Two examples of a clear distinction between the discrete and diffuse aurora are shown in Figure 6. In the top of the figure is the 3914Å data for the 0321 UT pass on 720110. The equatorward boundary of the diffuse aurora is at 67.5° CGL and is characterized by a well-defined edge with a narrow intensification just poleward, especially toward later local times. A minimum in intensity (∿ 5 kR) occurs at 71° CGL. Poleward of this a series of discrete arcs systems is encountered which obviously forms the poleward boundary of the auroral distribution at 75° CGL. In the bottom of the figure on the left the pass ∿ 24 hours later at 0409 UT is shown. The same general characteristics of diffuse and discrete

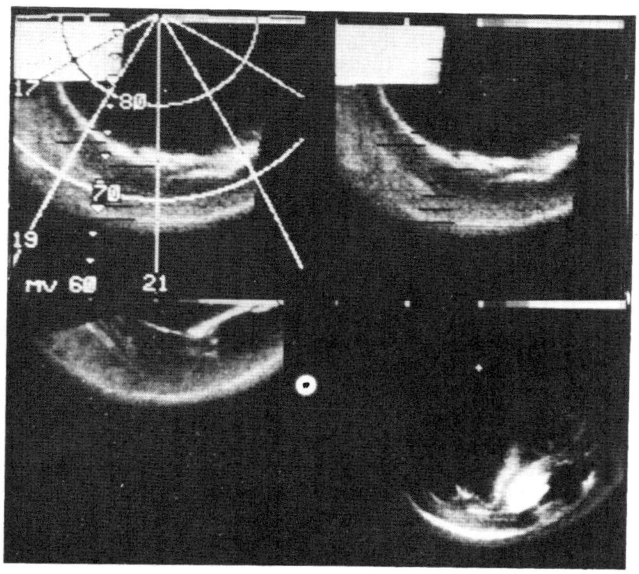

Figure 6: Transformations for three passes illustrating the relationship between discrete and diffuse aurora. Top: 720110 at 0321 UT (3914Å). Intensity grey scale is .5-10 kR and 1-10 kR on left and right respectively. Bottom left: 720111 at 0409 UT (5577Å). Intensity is 2-10 kR on left. Bottom right: 740128 at 1009 UT also with intensity range of 2-10 kR at 5577Å.

aurora apply. It is more evident in this pass, however, that an intensification in the diffuse aurora (CPS) exists which could very well be considered as a discrete arc system. Its characteristics are entirely different from the poleward discrete forms (i.e., its morphology is very smooth with MLT in contrast to the poleward feature) and thus may be a different category of discrete feature such as has been suggested by De La Beaujardiere et al., (1977). It is clear from the 3914Å intensity for the pass in Figure 5d that differences in the relation between discrete arc systems and diffuse aurora as a function of MLT can occur: westward of the spiral the evening arc is apparently imbedded in diffuse aurora (subject to some contamination by ground scattering). Eastward however, the arc clearly forms the poleward boundary.

Because discrete features are normally much more intense than the diffuse aurora it is reasonable to ask what is the relative location of the most intense features along a MLT meridian with respect to the equatorward boundary of the diffuse aurora. This obviously requires the ability to determine quantitatively the equatorward and poleward boundaries of the aurora, taking into account any effects due to ground scattering contamination. Accordingly an analysis was performed for passes along the 22 MLT meridian. Latitude profiles of 3914Å intensities were corrected for ground scattering using the technique of Hays and Anger (1978). A total of 46 passes were selected for which the peak corrected 3914Å intensity was greater than 5 kR within the auroral distribution at that local time. Use of this criterion eliminated passes where only diffuse aurora occurred. The location of the peak 3914Å intensity was then expressed as a percentage of the distance between equatorward and poleward auroral boundaries, and this quantity is plotted in Figure 7 as a function of AE.

The parameter used to define the location of the most intense discrete arc system is subject to several assumptions. First, the latitude resolution employed in determining the profiles is .25° CGL which could result in the smearing out of some discrete arc systems. Second the use of a fraction which is required due to the variations in auroral width, coupled with the above spatial resolution, may lead to slightly different results for similar physical conditions. For instance, if the most intense discrete arc is located at .5° CGL from the poleward boundary, the fraction is 87% for an auroral oval 4° wide and 97% for an auroral oval 15° wide. Finally, there is no requirement that the most intense feature also be the most poleward one. That more than one evening arc system can exist along a meridian is obvious from Figure 1 and some of the examples in Figure 3. In addition, the occurrence of a bulge (points indicated by B in Figure 7) clearly biases the results.

The most obvious feature of this plot is that almost all arcs lie in the poleward half of the auroral distribution. At first glance this may

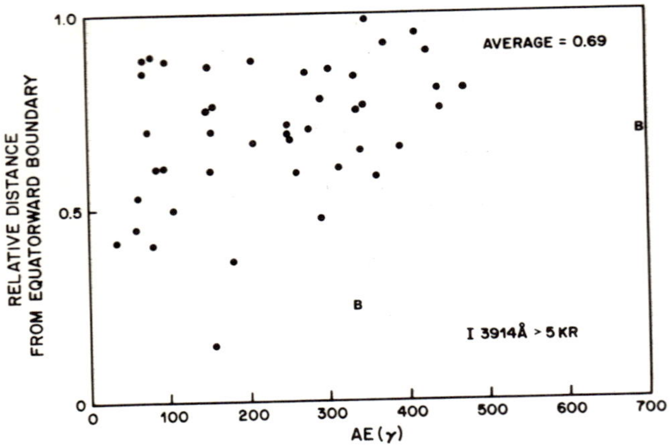

Figure 7: Scatter plot of the relative distance (along the 22 MLT meridian) from the equatorward boundary of the diffuse aurora to where the most intense (3914Å) discrete feature occurs (1.0 is the poleward boundary of the diffuse aurora).

appear to be entirely consistent with the idea that two distinct regions exist, with the discrete aurora confined to the poleward region. However, it does not necessarily mean that the diffuse aurora is confined to the lower region, and on an instantaneous pass basis there are clear examples of discrete arc systems occurring within the diffuse aurora. Figure 3b is a particularly good example of this. The arc system is located almost exactly in the middle of the auroral distribution at early MLT. If one identifies this feature as the brightened arc at the equatorward limit of discrete auroras prior to poleward expansion (Akasofu, 1968) and if substorm initiation occurs at the BPS/CPS interface (Akasofu, 1977) then clearly there is significant diffuse aurora in the BPS.

The above example is somewhat unusual in that normally it is difficult to ascertain whether or not evening arcs that connect to bulges are within the diffuse aurora. Since they predominantly occurnear the poleward edge of the distribution it is difficult, based on either particle or optical measurements, to separate the two phenomena; however, Mende and Eather (1976) report that visible auroral arcs are superimposed on diffuse soft electron precipitation in the evening sector. For the passes shown in Figure 7 the average 3914Å intensity (corrected for ground scattering) at the poleward and equatorward boundaries is 460 and 600R respectively. Thus the limits of the auroral distribution occur at lower intensities than is normally associated with discrete arc systems.

The pass for which the relative location is the smallest value (14%) in Figure 7 is shown as a transformation in the bottom right hand corner of Figure 6. It is clearly atypical in its morphology (see Figure 3) and in the intensity of the discrete feature at 66° CGL, although it is similar to the weaker intensifications in the diffuse aurora in the other two passes shown in Figure 6. This feature is located well within the diffuse aurora and, although longitudinally confined to the 20-24 MLT sector, is clearly the dominant discrete feature. At the very least therefore, one must accept the fact that different classes of discrete features exist and that even the evening arcs connecting to auroral bulges may be embedded in thepoleward portion of the diffuse aurora.

In summary, there seems to be evidence for two types of discrete evening arc systems. The most common type is normally associated with the BPS and connects to auroral bulges. They are generally the most intense features along any MLT meridian and occur in the poleward half of the auroral distribution, most probably within a uniform background of diffuse aurora. The second type, normally weaker in intensity, occurs in the equatorward portion of the diffuse aurora. It is much less distorted in appearance and does not connect to auroral bulges.

Summary

The discrete auroral morphology in the evening/midnight sector as observed by the ISIS-2 satellite consists primarily of long evening arc systems near the poleward boundary of the diffuse aurora which at their eastward end connect to a large scale auroral bulge. These evening arcs are remarkably constant in intensity and 5577Å/3914Å ratio along their length implying an identical particle source over many hours of local time. The connection of intense evening arc systems to auroral bulges associated with substorms implies that these arc systems are an intimate part of the substorm process and illustrate the necessity of interpreting substorms from a magnetospheric rather than a local perspective.

The auroral bulges at which these features terminate can normally be identified as westward traveling surges and their distribution peaks in the 20-22 MLT region rather than at midnight. From the snapshot view available it is difficult to determine whether the bulges form in a stepwise fashion (Wiens and Rostoker, 1975) or whether continuous propagation along an arc is the dominant mode (Akasofu, 1968). Examples consistent with both types can be found.

Typically the connecting evening arcs lie at the poleward edge of the diffuse aurora, but apparently within it, although the determination of a diffuse auroral component is difficult when the arcs are close to the poleward boundary and/or the diffuse auroral intensity is low. It is clear, however, that discrete aurora may exist within the equatorward regions of the auroral distribution. The morphology of these features is much more regular than the convoluted forms generally associated with the BPS.

Acknowledgements

The authors would like to thank D.D. Wallis for helping to develop the two-dimensional coordinate transformation process used in this paper and C. Marcellus for assistance in data processing. This work was supported by National Research Council grants A-7 and A-6762.

References

Akasofu, S.-I., The development of the auroral substorm, Planet. Space Sci., 12, 273, 1964.

Akasofu, S.-I., Polar and Magnetospheric Substorms, D. Reidel, Dordrecht, Netherlands, 1968.

Akasofu, S.-I., Discrete, continuous and diffuse auroras, Planet. Space Sci., 22, 1723, 1974.

Akasofu, S.-I., Recent progress in studies of DMSP auroral photographs, Space Sci. Rev., 19, 169, 1976.

Akasofu, S.-I., Physics of Magnetospheric Substorms, D. Reidel, Dordrecht Netherlands, 1977.

Akasofu, S.-I. and J.R. Kan, Dayside and nightside auroral arc systems, in press, Geophys. Res. Lett., 1980.

Anger, C.D., A.T.Y. Lui and S.-I. Akasofu, Observations of the auroral oval and a westward traveling surge from the ISIS-2 satellite and the Alaskan meridian all-sky cameras, J. Geophys. Res., 78, 3020, 1973.

De La Beaujardiere O., R. Vondark and M. Baron, Radar observations of electric fields and currents associated with auroral arcs, J. Geophys. Res., 82, 5051, 1977.

Davis,T.N. and T.J. Hallinan, Auroral spirals 1. observations, J. Geophys. Res., 81, 3953, 1976.

Feldstein, Y.I., Some problems concerning the morphology of auroras and magnetic disturbances at high latitudes, Geomagn. and Aeronomy, 3, 183, 1963.

Hallinan, T.J., Auroral spirals 2, theory, J. Geophys. Res., 81, 3959, 1976.

Hays, P.B. and C.D. Anger, Influence of ground scattering on satellite auroral observations, App. Optics, 17, 1898, 1978.

Hughes, T.J., D.Y. Oldenburg and G. Rostoker, Interpretation of auroral oval current near dusk using inversion techniques, J. Geophys. Res., 84, 450, 1979.

Kamide, Y. and S.-I. Akasofu, The auroral electrojet and global auroral features, J. Geophys. Res., 80, 3585, 1975.

Kamide, Y. and S.-I. Akasofu and E.P. Rieger, Coexistence of two substorms in the midnight sector, J. Geophys. Res., 82, 1620, 1977.

McPherron, R.L., Magnetospheric substorms, Rev. Geophys. Space Phys., 17, 657, 1979.

Mende, S.B. and R.H. Eather, Spectroscopic determination of the characteristics of precipitating auroral particles, J. Geophys. Res., 80, 3211, 1975.

Mende, S.B. and R.H. Eather, Monochromatic all-sky observations and auroral precipitation patterns, J. Geophys. Res., 81, 3771, 1976.

Murphree, J.S. and C.D. Anger, An observations of the instantaneous optical auroral distribution, Can. J. Phys., 58, 214, 1980.

Murphree, J.S., L.L. Cogger and C.D. Anger, Characteristics of the instantaneous auroral oval in the 12-18 MLT sector, submitted to J. Geophys. Res., 1980.

Rees, M.H. and D. Luckey, Auroral electron energy derived from ratio of spectroscopic emissions 1. model computations, J. Geophys. Res., 79, 5181, 1974.

Rostoker, G., J.C. Armstrong and A.J. Zmuda, Field-aligned current flow associated with intrusion of the substorm-intensified westward electrojet into the evening sector, J. Geophys. Res., 80, 3571, 1975.

Rostoker, G., S.-I. Akasofu, J. Foster, R.A. Greenwald, Y. Kamide, K. Kawasaki, A.T.Y. Lui, R.L. McPherron and C.T. Russell, Magnetospheric substorms - definition and signatures, J. Geophys. Res., 85, 1663, 1980.

Sergeev, V.A. and A.G. Yahnin, The features of auroral bulge expansion, Planet. Space Sci., 17, 1429, 1979.

Snyder, A.L., S.-I. Akasofu and D.S. Kimball, The continuity of the auroral oval in the afternoon sector, Planet. Space Sci., 23, 225, 1975.

Wiens, R.G. and G. Rostoker, Characteristics of the development of the westward electrojet during the expansive phase of magnetospheric substorms, J. Geophys. Res., 80, 2109, 1975.

Winningham, J.D., F. Yashuhara, S.-I. Akasofu and H.J. Heikkila, The latitudinal morphology of 10-ev to 10-Kev electron fluxes during magnetically quiet and disturbed times in the 2100-0300 MLT sector, J. Geophys. Res., 80, 3148, 1975.

SYNOPTIC FEATURES OF AURORAL SYSTEM AND CORRESPONDING ELECTRON PRECIPITATION OBSERVED BY KYOKKO

Eisuke Kaneda

Geophysics Research Laboratory, University of Tokyo, Bunkyo-ku, Tokyo, Japan

Toshifumi Mukai and Kunio Hirao

Institute of Space and Aeronautical Science, University of Tokyo, Meguro-ku, Tokyo, Japan

Introduction

For a study of the synoptic features of auroral display, the effective observational method used during the 1950's was to deploy ground networks of all-sky cameras. The study was supplemented by an extensive air-borne survey in the 1960's. Finally, in the 1970's, imaging from a satellite was realized. The concept of auroral substorm [Akasofu, 1964] has been advocated on the basis of observations in the 1950's and has been confirmed by various observational and theoretical contributions (e.g. Akasofu [1977] and References included therein). However, some fundamental problems concerning substorm generation have remained unsolved due to lack of decisive observational data. The growth-phase controversy is a typical example of such difficulties. The imaging data from satellites launched in the early 70's were, however, not adequate to settle problems related to dynamical features of substorms, owing to methods in image forming. These problems would be suitable observational targets for the snap-shot imaging by KYOKKO which is equipped with a newly developed TV camera capable of VUV aurora imaging every 128 seconds [Kaneda, 1979a]. KYOKKO was launched into the semipolar orbit of 65.3° in inclination, the apogee and perigee heights being 3978 and 641 km, respectively. In this paper, we concentrate on topics concerning some synoptic features of the auroral substorm development and relationships between auroral displays and simultaneously measured precipitating electron fluxes [Mukai and Hirao, 1978].

Eastward auroral motion in the evening hour sector

It was noted by Akasofu [1964] that the auroral substorm manifests its essential part in the poleward expansion of auroras in the midnight hour region. KYOKKO has observed the progressive change of auroral substorm features for a large number of events through continual snap-shot imaging by VUV TV camera installed with a wide-angle mirror lens. First of all, auroral substorm features are shown in an example in which the auroral substorm is proceeding. Figure 1 presents auroral images with the grid in geomagnetic coordinates. These are the photographs (every other shot) taken between 05h35m38s and 05h52m32s (U.T.) on 4th October, 1978. Some of them have already been published in the form of preliminary versions in image processing (Kaneda, 1979a, b). Zig-zag curves in some images of Fig. 1 are caused by noises which jammed signals on telemetry data reception. Figure 1 delineates the development and collapse of an arc system in the nightside region in the course of substorm progress under moderately disturbed conditions (geomagnetic index Kp = 4₋). It can be seen also in Fig. 1 that an arc in the afternoon hours region is extending its one end into the nightside region. This arc along the latitude circle around 67° merges its another end with the daylight region beneath the cusp. We can clearly recognize in Fig. 1c that it get between two latitudinally aligned nightside auroras, the one of these is an active discrete form around 70° in geomag. lat., the other a rather diffuse profile one at about 60°. The same configuration as in Fig. 1c is not always observed, but similar ones are not rare. Figure 2 shows, as another example of such auroral configuration, the auroral images observed between 00h41m59s and 00h50m31s on 7th October 1978 in the same format and time interval (256 sec.) as those in Fig. 1. The auroral display in Fig. 2 indicates also features in the early stage of the substorm expansion phase under weakly disturbed conditions (Kp = 2₋). But in this case, the above discussed arc, one end of which is embedded in the daylight region under the cusp, has already reached the nightside auroral region. In Fig. 2c, we can see that this arc takes too unnatural configuration to connect its another end with the dusk end of the developing nightside aurora. Auroral images, taken thereafter, denote clearly separation of both forms, though they are not included in Fig. 2 for reason

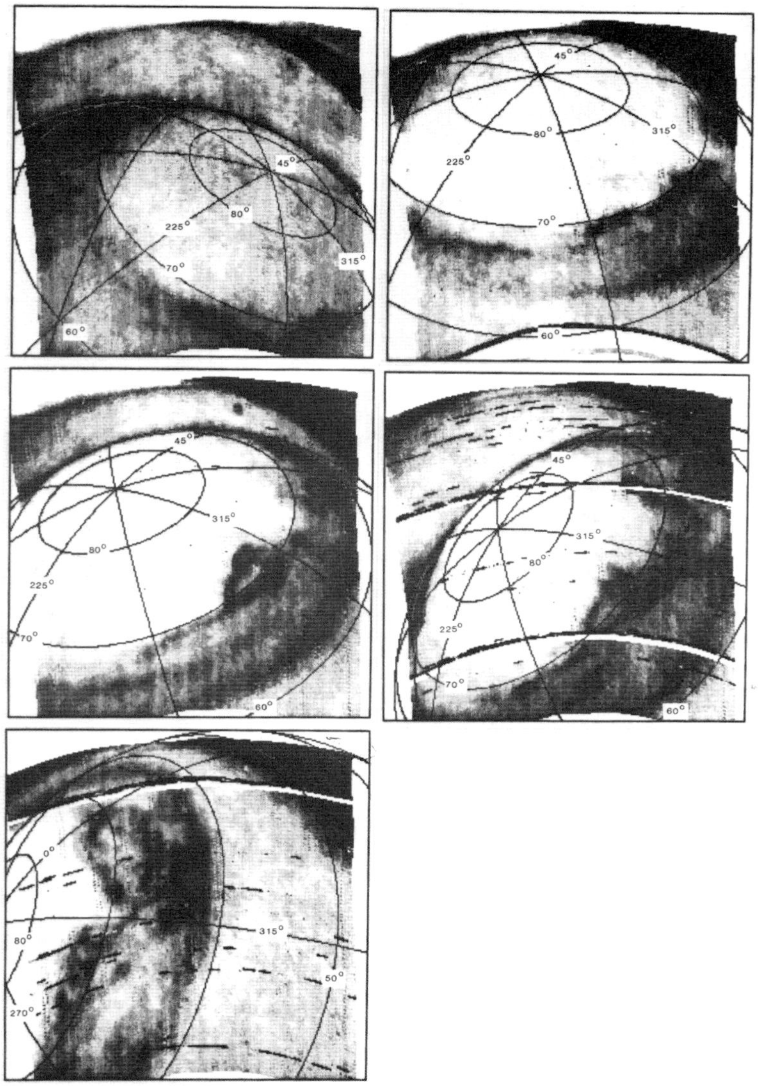

Fig. 1 Auroral images observed in revolution No. 2595 on Oct. 4, 1978;
(a: top-left) Shot time 05h 35m 28s, (b: top-right) Shot time 05h 39m 44s,
(c: middle-left) Shot time 05h 44m 00s, (d: middle-right) Shot time 05h 48m 16s,
(e: bottom-left) Shot time 05h 52m 32s.

of blurring by the effects of the sun which shifted to the view field boundary of the imaging device. Hence, we can recognize this arc without any afternoon branch of the nightside aurora, but with the same kind as discussed in the above. Besides the previous case, there seems to exist other types of non-nightside origin auroras. We prepare Fig. 3 for illustration of another type of auroral display having non-nightside origin. In Fig. 3, panels delineate auroral images taken between 00h21m23s and 00h34m11s on 10th October 1978 by 256 second interval. It can be seen in these panels that a discrete auroral form in the dusk hour sector extends eastward and expands into diffuse forms. It is, in contrast to examples in Figs. 1 and 2, notable that this auroral development does not seem to be associated with auroral substorm expansion around midnight.

Electron precipitation and corresponding aurora

The latitudinal precipitation pattern of electrons in the nightside auroral region is characterized by a band structure in which the electron energy for peak differential intensities shows an inverted-V shaped variation in the energy-time

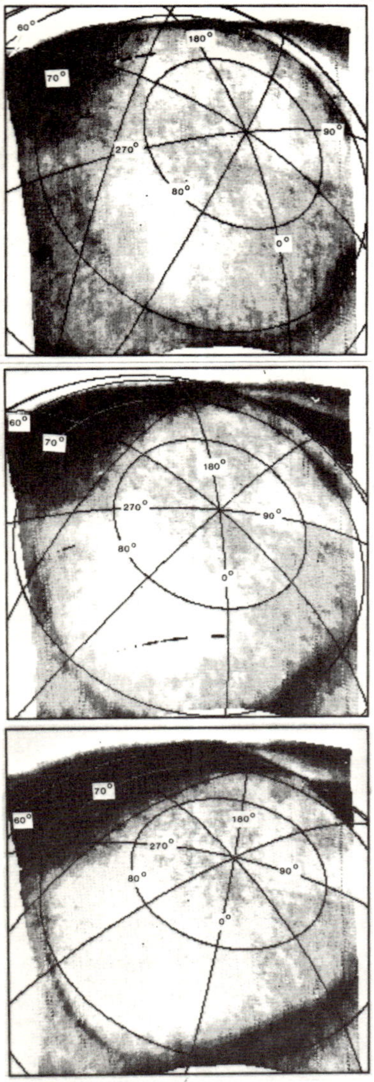

Fig. 2 Auroral images observed in revolution No. 2625 on Oct. 7, 1978:
(a: top) Shot time 00h 41m 59s,
(b: middle) Shot time 00h 46m 15s,
(c: bottom) Shot time 00h 50m 31s.

spectrogram [Heikkila, 1970; Frank and Ackerson, 1971]. An example of the typical inverted-V event can be seen in Fig. 4 which shows the energy-time spectrograms with the energy fluxes in the revolution No. 2360. The data of sensor 1 and sensor 2 correspond to the downward (precipitating) and upward electron fluxes, respectively. One can see an inverted-V precipitation that occurres around 1950 MLT at an invariant latitude of about 69° when the satellite traverses the auroral oval toward higher latitudes with small change of magnetic local times (1930 - 2000 MLT). Hence, Fig. 4 delineates a latitudinal profile of auroral electrons, similar to the earlier observations by the polar-orbitting satellite [e.g. Frank and Ackerson, 1971]. The monoenergetic peak seen as a narrow black band located at higher latitudes than the diffuse precipitation region in the sensor 1 energy-time spectrogram is the typical characteristics of an inverted-V event. The latitudinal width of this event is about 1°. Thus the characteristics of the inverted-V event observed by KYOKKO are similar to those observed by other polar orbitting satellites [Frank and Ackerson, 1972; Winningham et al., 1975]. In other revolutions, the inverted-V event is frequently recognized as an "inverted-U" event that results merely from slant traverse of the sub-satellite track over precipitating regions. However, there exist cases in which the precipitating electron flux denotes very complex changes in their features. One of these will be illustrated in the following.

In the previous section, we have shown rather different auroral substorm features from those obtained by ISIS-2 [Lui et al., 1977]. It would be worthwhile to investigate relations between the aurora and the associated electron precipitation observed by KYOKKO. Because of the continual snap-shot by the VUV camera, we can study the simultaneous correspondence between the both phenomena at image shot time. Even at other times, the interpolation of auroral image data makes it possible to pursue this correspondence with considerable accuracy. This is most useful in studying the aurora in active conditions such as the substorm expansion phase. Auroral images in Fig. 1 and the simultaneously observed precipitating electron energy spectrum are compiled in Fig. 5 as an example of correspondence between the both during high auroral activities. Every auroral image in Fig. 1 is transformed into one expressed in the polar-stereo projection and is respectively delineated in panels from A to E. In these panels, geomagnetic longitude is converted to geomagnetic local time (MLT), as shown in panel A. A sequence of hollowed circle in each panel indicates the locus of the geomagnetic sub-satellite point (projection of satellite position to the ionospheric E-layer height along the geomagnetic field line) in 1 minute intervals. An X-mark pointed by an arrow in each panel denotes respectively the geomagnetic sub-satellite (gm s-s) point at the shot time of auroral image in the same panel. The precipitating electron flux is illustrated by an energy-time spectrogram in the middle of Fig. 5. The shot times of auroral images in panels from A to E are indicated by arrows with the corresponding letters on the top of the spectrogram.

In the revolution No. 2595, as shown in Fig. 5, the gm s-s point crosses the geomag. lat. circle of 60° around 15h MLT zone. It is slantly traversing the arc, which has been identified as of non-nightside origin in the previous section, at 05h38m28s, the shot time of auroral image in

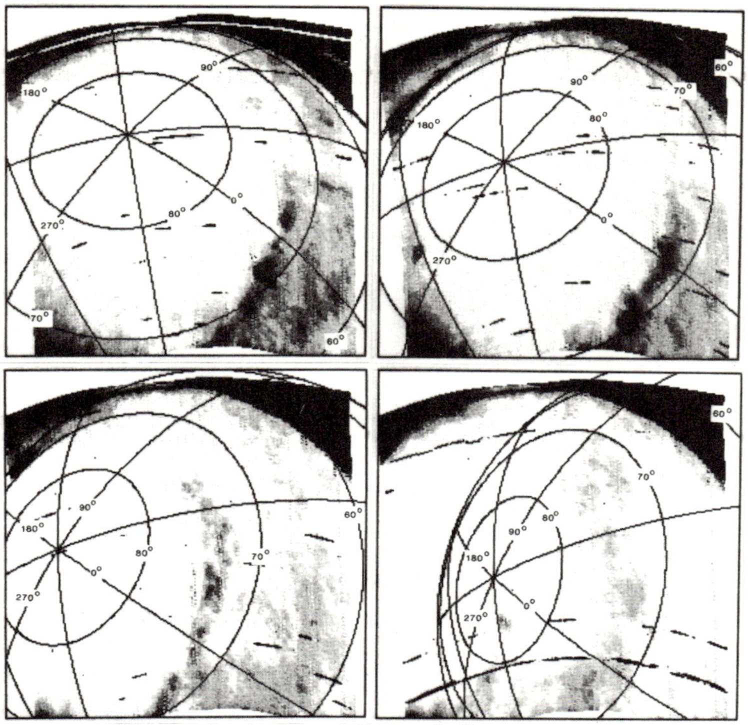

Fig. 3 Auroral images observed in revolution No. 2657 On Oct. 10, 1978; (a: upper-left) Shot time 00h 21m 23s, (b: upper-right) Shot time 00h 25m 39s, (c: lower-left) Shot time 00h 29m 55s, (d: lower-right) Shot time 00h 34m 11s.

panel A. We can consider the gm s-s point to enter, almost along the latitude circle, into the nightside aurora just after the shot time shown in panel B. It can be interpolated from auroral images in panels C and D that the gm s-s point crosses an active form, located equatorward of the 70° latitude circle, between shots C and D. After shot time D, it goes through auroras which collapsed and transformed into turbulent forms.

The corresponding precipitating electron fluxes in E-t diagram can be roughly divided into four main intervals by the aid of this brief description about the movement of gm s-s point accompanied with satellite orbital motion. Namely, the first interval of precipitation is designated as one containing fluxes observed around shot time A, and corresponds to the aurora located in the afternoon hour region. The second one is assigned to the interval between shots B and C. In this interval, the gm s-s point goes almost along the high latitude edge of a westward extending auroral arc. The third interval corresponds to precipitations marked off by shots C and D. We can consider that they are correlative with low latitude portions of an intense loop-shaped aurora around 21h MLT. The fourth terminated by shots D and E is observed during the time interval when the gm s-s point goes equatorward and moves into turbulent auroral forms. The characteristic differences can be recognized among energy spectra in every interval, each of which has respective correspondence to specific auroral forms. To investigate precisely features of the precipitating electron flux, we prepare Fig. 6 which illustrates profiles of its own energy spectrum.

Each panel in Fig. 6 shows energy spectra of precipitating electron flux in every 4 seconds during a 28 second interval which is selected as a typical one in each of the four main intervals. Energy spectra in panel (a) correspond to fluxes observed prior to shot A. We can see no significant flux above 3 keV with them, which show a peak at about 1 keV with a certain scatter. Panel (b) delineates energy spectra of precipitating fluxes just before shot B. There can be seen a knee frofile extending from 100 eV up to about 4 keV. It is remarkable with panel (c) that energy spectra trail the high energy tail exceeding the detection limit. Panel (c) expresses energy spectra of electron fluxes detected in the middle of the interval terminated by shots C and D. Fluxes, measured just prior to shot E, are compiled into panel (d). They show, in energy spectrum, a peak at about 2 keV.

Discussion

The electron precipitation with peaked energy spectra has been found to correspond with a

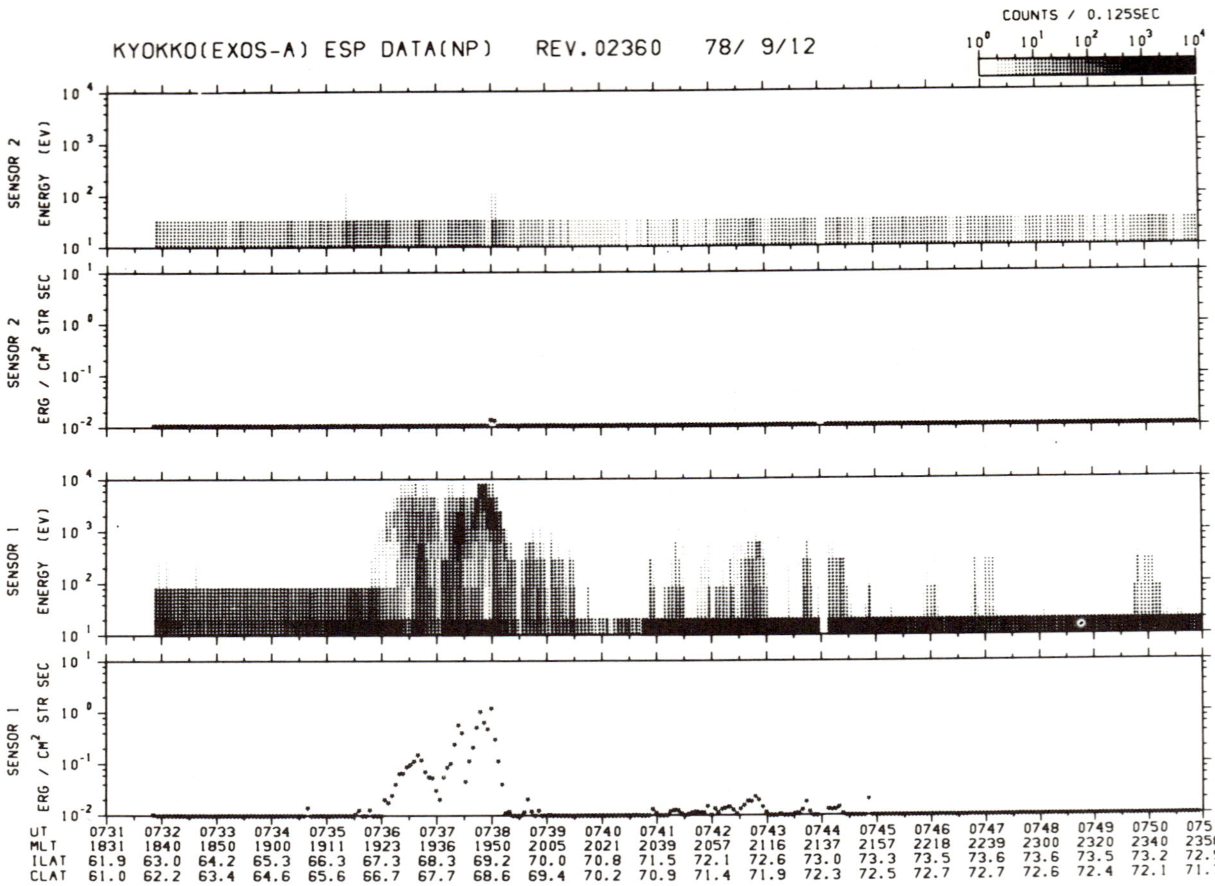

Fig. 4 Energy-time spectrogram and energy flux of electrons in revolution No. 2360; The data of sensor 1 and 2 correspond to the precipitating and upward electron fluxes, respectively. In the energy-time spectrogram the ordinate is the electron energy, and the abcissa is U.T. The detector response is represented by intensity. The bottom three rows are the dipole magnetic local time, the invariant latitude and the conjugate geomagnetic latitude.

"discrete aurora" [Lui et al., 1977; Meng, 1978], and is suggested to be consistent with particle acceleration by the field-aligned electrostatic drops at altitudes of ∿ 1 Re above the auroral ionosphere, as suggested from the S^3-3 observations [Torbert and Mozer, 1978; Mizera and Fennell, 1977]. We have seen in Figs. 6b - 6d that precipitating electron fluxes corresponding to a specific type of discrete auroral forms can be presented respectively by a characteristic profile in energy spectrum. Fluxes showing peaked energy spectra are delineated in Fig. 6d and correspond to turbulent auroral forms collapsed from discrete ones. On the contrary, it can be seen in Figs. 6b and 6c that energy spectra with a knee structure or a high-energy tail are observed with precipitating electrons producing discrete auroral forms before collapse. The latter two types of energy spectrum are not peculiar ones and have been reported in inverted-V events observed by low altitude satellites [e.g. Frank and Ackerson, 1971]. Based upon simultaneous observations of VUV aurora in snapshot imaging and electron precipitation, further studies will be necessary before drawing conclusions about correspondence between auroral forms and types of precipitating electron energy spectrum.

The identification of the "diffuse aurora" [Lui and Anger, 1973] is one of the prominent results from the ISIS-2 imaging. Further, it has been confirmed on the basis of this morphological classification that the "discrete aurora" is embedded in the high latitude boundary of the "diffuse aurora" in the region from midnight to the evening hours [Lui el al., 1977]. However, as shown in Figs. 1 to 3, the configuration of auroral forms is not so simple in VUV-light imaging by KYOKKO as in visual light imaging by ISIS-2. To elucidate this discrepancy, comparisons of the simultaneous imaging data from the both satellites seem effective.

During substorm progress, until the early state of the expansion phase, it is shown by KYOKKO

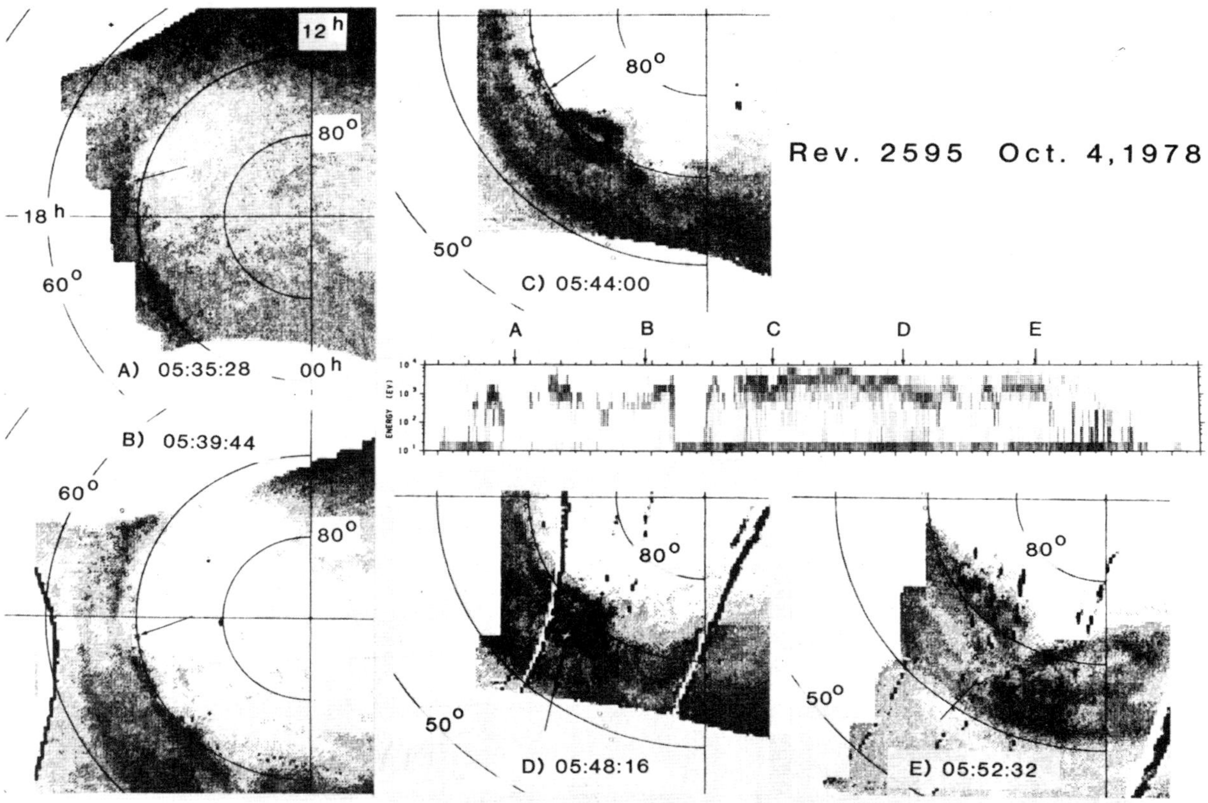

Fig. 5 Auroral images and precipitating electron fluxes in revolution No. 2595, electron fluxes represented by energy-time spectrogram in the middle of figure.

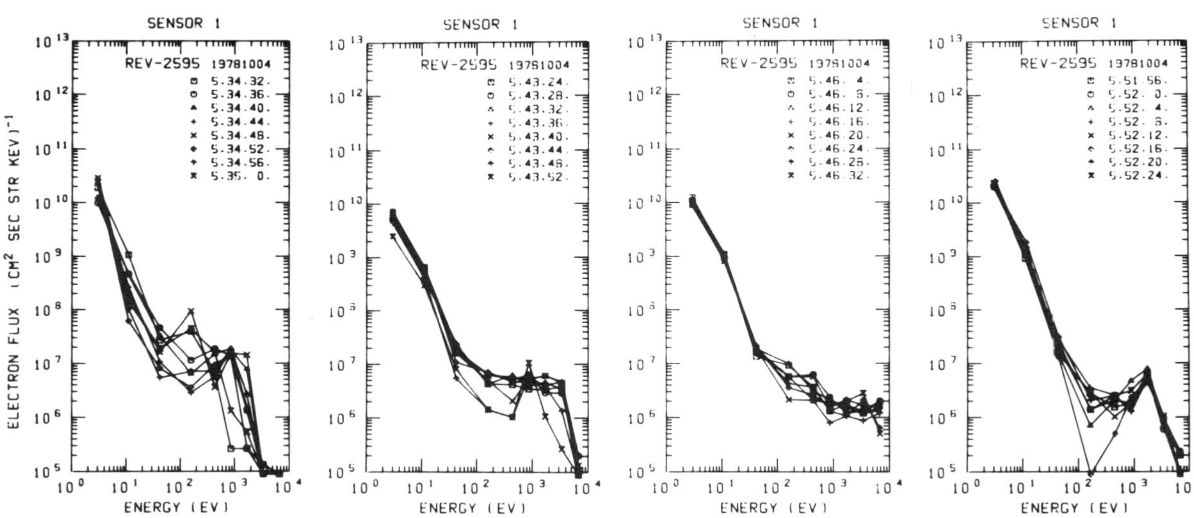

Fig. 6 Energy spectrum profiles of precipitating electron fluxes in revolution No. 2595; (a: leftmost) 05h 34m 32s - 05h 35m 00s, (b: middle-left) 05h 43m 24s - 05h 43m 52s, (c: middle-right) 05h 46m 04s - 05h 46m 32s, (d: rightmost) 05h 51m 56s - 05h 52m 24s.

imaging data that ther exists, in the dusk to late evening hour region of the auroral oval, an arc which merges its one end with the dayside region beneath the cusp. The direction of its extending is reverse to that of the westward auroral motion associated with the substorm expansion in this local time zone. In relation to this observation, it is apparent from Fig. 6a, which represents the energy spectra of the electrons exciting the arc extending towards the nightside, that there is a sharp cut-off of flux at about 3 keV. This cut-off cannot be seen in other panels of Fig. 6 derived from precipitations corresponding to the nightside aurora in discrete forms or collapsed turbulent ones. The difference in flux cut-off energy shown in Fig. 6 can be an evidence for identification of this arc origin as non-nightside. The present result coincides with the conclusion of an analytical study [Kaneda, 1973], based upon the data of ground observations in the neighbourhood of the region under the cusp. It has also been pointed out in his analysis that the discussed arc changes its features, from a rather diffuse profile to a distinct one, at the time of an abrupt decrease in the geomagnetic AL-index, which is understood as a signature of substorm expansion onset. This change will be the same as seen between the arc shown in Figs. 2a and 2c. KYOKKO data indicate this change to take place almost simultaneously over a wide range. In addition to this arc, it is shown by Fig. 3 that there exist in the dusk hour sector eastward moving auroral activities, without any association with obvious auroral substorm expansion. Summarizing all these observational facts, there seems to exist in the magnetosphere different mechanisms which feed separately the particle precipitations, corresponding to the non-nightside and nightside auroras. To work out ambiguities concerning the substorm generation mechanism, it will be necessary in the next step to clarify relations between present observational facts and substorm activities.

References

Akasofu, S.-I., The development of the auroral substorm, Planet. Space Sci., 12, 273, 1964.

Akasofu, S.-I., Physics of magnetospheric substorms, D. Reidel, Hingham, Mass., 1977.

Frank, L.A., and K.L. Ackerson, Observations of charged particle precipitation into the auroral zone, J. Geophys. Res., 76, 3612, 1971.

Frank, L.A., and K.L. Ackerson, Local time survey of plasma at low altitudes over the auroral zone, J. Geophys. Res., 77, 4116, 1972.

Heikkila, W.J., Satellite observation of soft particle fluxes in the auroral zone, Nature, 225, 369, 1970.

Kaneda, E., Dayside auroral activity and its relation to substorm, Rep. Ionosph. Space Res. Japan, 27, 209, 1973.

Kaneda, E., Auroral TV observations by KYOKKO, Proc. Japanese IMS Symp., 146, 1979a.

Kaneda, E., Auroral TV observations by KYOKKO, Proc. Internat. Workshop Select. Topics Magnetosph. Phys., 15, 1979b.

Lui, A.T.Y., and C.D. Anger, A uniform belt of diffuse auroral emission seen by the ISIS-2 scanning photometer, Planet. Space Sci., 21, 799, 1973.

Lui, A.T.Y., D. Venkatesan, C.D. Anger, S.-I. Akasofu, W.J. Heikkila, J.D. Winningham, and J.R. Barrows, Simultaneous observations of particle precipitations and auroral emissions by the ISIS-2 satellite in the 19 - 24 MLT sector, J. Geophys. Res., 82, 2210, 1977.

Meng, C.-I., Electron precipitations and polar aurora, Space Sci. Rev., 22, 223, 1978.

Mizera, P.F., and J.F. Fennell, Signatures of electric fields from high and low altitude particles distributions, Geophys. Res. Lett., 4, 311, 1977.

Mukai, T., and K. Hirao, Characteristics of auroral electron observed in the morning-side from "KYOKKO", Bull. Inst. Space Aeronaut. Sci., 14, 1179, 1978.

Torbert, R.B., and F.S. Mozer, Electrostatic shocks as the source of discrete auroral arcs, Geophys. Res. Lett., 5, 135, 1978.

Winningham, J.D., F. Yasuhara, S.-I. Akasofu, and W.J. Heikkila, The latitudinal morphology of 10-eV to 10-keV electron fluxes during magnetically quiet and disturbed times in the 2100-0300 MLT sector, J. Geophys. Res., 80, 3148, 1975.

TV Observations of Auroral Arcs

Takasi Oguti

Geophysics Research Laboratory, University of Tokyo, Tokyo 113, Japan

Abstract. From the dynamical view point of aurora, it is shown that the discrete aurora so far called, especially in the dusk to the midnight sectors, is a clockwise (viewed along the magnetic field) vortex chain or a shear, including splitting, and that the diffuse aurora, though called so higherto, actually consists of pulsating structures with a diffuse background, especially in the dawn sectors. It is also shown that the transient auroral forms, such as swirling smoke like auroras and highly striated auroras that show both rotational motions and pulsations are often seen on equatorward expanding branches of aurora during an auroral expansion in the midnight sector.

Introduction

It is well known that auroral motions consist of various modes of deformations. Each mode may correspond to a different physical process in the magnetosphere, such as the accelerations and/or pitch angle scatterings of electrons. The temporal characteristics of photometric observation alone are not enough to distinguish these processes, because the deformations in different modes overlap each other in space and in time. It was naturally not until the introduction of a high sensitivity TV camera, using an image orthicon tube in Alaska [Davis, 1966], and later the use of an image intensifier-vidicon system in Canada [Scourfield and Parsons, 1969], that the classification of auroral deformation modes into meaningful categories that gave rise to objects of theoretical studies started.

Quite a lot of observations of auroral deformation modes have been carried out since then. For example, Cresswell and Davis [1966] identified a number of types of auroral deformation modes, Beach et al. [1968] observed the flickering aurora, Hallinan and Davis [1970] reported the "auroral curl", as one of the characteristic deformations. Rapidly moving auroral waves, first noted by Cresswell and Belon [1966] and by Cresswell [1969], by separated field photometer data have been confirmed in TV imaging by Scourfield and Parsons [1971], and the "moving mode" reported by Campbell [1970], also using separated photometers, was found to be identical to the poleward propagation mode on TV images by Oguti and Watanabe [1976].

In spite of all these observations and in spite of several morphological studies such as those by Cresswell [1972], Davis [1972], Thomas et al., [1973], the author [Oguti, 1975a], Royrvik and Davis [1977] and Johnstone [1978], using both photometric and TV observations, an agreed classification for deformation modes and distribution of them along the auroral oval, relating them to physical processes in the magnetosphere is still missing. In this paper, the author proposes a revised classification of auroral deformations after re-examining the early results [Oguti, 1978] and adding new observations in Canada. In addition, a model distribution of auroral deformations with respect to the global auroral pattern is presented that could lead to a consideration of possible mechanisms for the acceleration and/or pitch angle scattering of auroral electrons in the magnetosphere.

Classification of Deformation Modes

The auroral metamorphoses consist of too many varieties to discuss all their features. Consequently, the author intends here to deduce several basic characteristic features from the various deformation modes, and to catagorize them into several groups with some references to the morphological relations between them. In doing so, it is worth noting the specific auroral activity regions, because the aurora usually shows different conditions in different activity regions at one time. For example, the condition of aurora in the remnants behind the poleward expansion is different from that in the poleward expansion front itself, and the aurora that expands equatorwards differes from both of them. This suggests that the terms "expansion" and "recovery" phase define the nature of auroras in distinct regions as well as that of the temporal phases as a whole. The aurora left in the remnants and the aurora that expands towards lower latitudes could be in the recovery phase, while at the same time the aurora in the poleward expansion front is still in the expansion phase.

Considering this, the aurora has been first categorized into three types according to three typical regions, 1) region along and near the front of the poleward expansion, 2) in the remnants left behind the front and expanding equatorward, both mainly in the dusk to the midnight sectors, and 3) in the equatorward region, especially in the midnight to the dawn sectors. Thus the tentative classification of auroral deformation modes is proposed as follows:

1) In and near the discrete auroral forms (Expansion Region)
   Auroral type: sheet like
   a) Splitting or peeling of an arc sometimes followed by rotational unfolding
   b) Clockwise rotations or shear motions across an arc viewed along the magnetic field
2) In the remnant behind the poleward expansion and in the equatorward expanding branch (Transient Region)
   Auroral type: swirling smoke like, and highly striated later
   a) Flickering
   b) Swirling smoke (rotational) variations
   c) Formation of complex striations
   d) Initiation of partial pulsations
3) In the equatorward-expanded branches (Recovery Region)
   Auroral type: pulsating forms with a diffuse background
   a) Pulsation (including expansion)
   b) Poleward propagation of brightening of successive patches
   c) Streaming (propagation of brightening along striation structures)
   d) Flooding (propagation of brightening in arbitrary directions)
   e) Equatorward propagation of faint arc fragments (auroral wave)
   f) Flash (flash of faint, diffuse patches)
   g) Flaming (propagation of brightening along magnetic field lines)

Some examples of these are shown below along with descriptions of motional characteristics.

Category 1

Category 1 deformations are specific motional features of so called discrete aurora in and around the vicinity of the poleward boundary of auroral activities mainly in the dusk to the midnight sectors, and consist of two modes. The one is splitting or peeling of discrete, sheet type aurora, with protrusions toward right hand side (westward and poleward, eastward and equatorward) of the principal arc, and the other is clockwise rotation of bright structures or shear motion of structures across a principal arc, viewed along the magnetic field. The two modes occur in every discrete, sheet type auroras. Enhancements of these modes in motional speed and in luminosities of the structures usually take place in the poleward expanding branch of discrete auroral forms during a course of an auroral expansion.

Splitting followed by rotational unfolding

The splitting and the clockwise rotation usually occurs in combination. For example, Fig. 1 shows a typical deformation of a discrete arc. The left half of the figure shows, in a sequence of photographs, an example of splitting followed by rotational unfolding of a discrete arc; the right half of the figure is a schematic illustration showing how the protrusion on the upper side (high-latitude) of the arc propagates towards the east and the opposite motion is seen on the lower side (low-latitude) during the course of a splitting-unfolding. The rotational sense of the central unfolding part is clockwise viewed along the magnetic field (upward in the southern hemisphere).

Splitting followed by reconnection

The unfolding process is a little modified during an active phase. The split parts (or protrusions) often do not unfold as in Fig. 1, but they tend to fold over further both toward higher and lower latitudes. The fold over branch eventually reconnect to other auroral branches, split from other parts of the principal arc, accordingly forming a new front of poleward expansion. Fig. 2 shows such an example of splitting - foldover - reconnection. The left half of the figure again shows sequential pictures reproduced from video tape and the right half is a schematic illustration representing the fundamental deformation mode seen in the left. The unfolding in the first example apparently looks quite different from the reconnection in the second example, but the two modes are essentially identical. Because the peeled off end in low latitude in the second example either westward retreats into the principal arc as in the unfolding in the first example, eventually forming a new poleward front, or rotates clockwise forming N-S directed arcs. This is often seen in the process of poleward expansion of aurora during an auroral expansion. Thus we can understand that the poleward expansion of aurora usually consists of many steps due to the splitting-reconnection processes of the discrete aurora with various sizes, leaving behind many branches that expand equatorward.

Shear propagation of peeled off ends

The splitting and the rotation can occur in a variety of scale sizes [Oguti, 1975b], forming and releasing typical S-like patterns of discrete aurora. These modes of the deformations in smallest size (5 - 10 km) are usually recognized as a rapid appearance and disappearance of rayed structures along a discrete arc. Small branches appear in succession as if they are peeled off

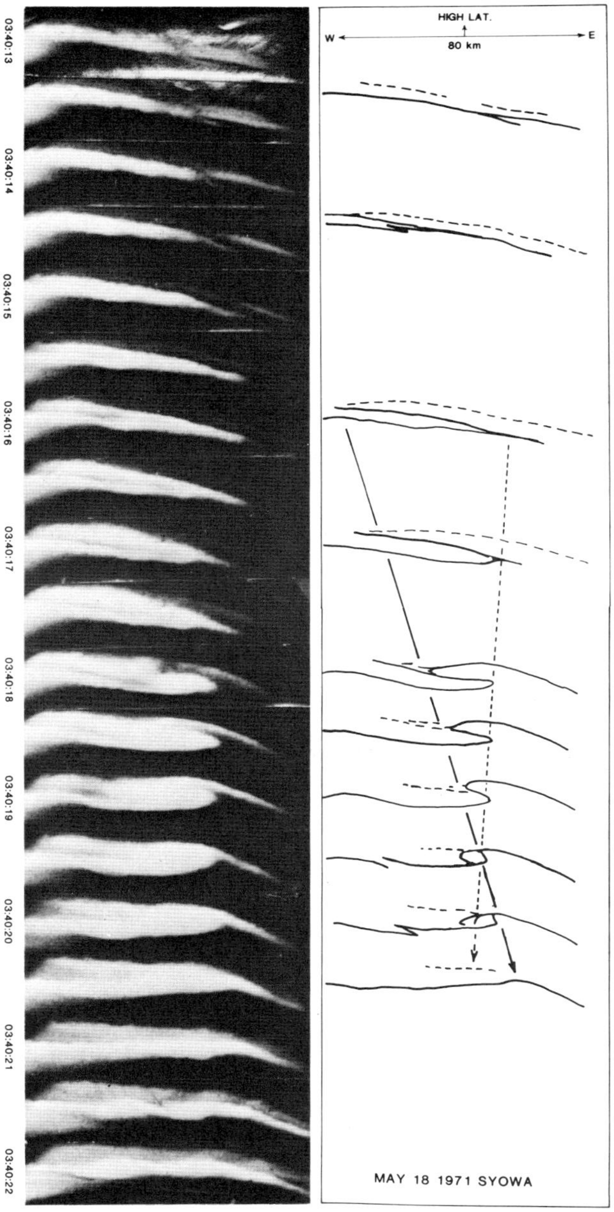

Fig. 1. A typical example of splitting followed by rotational unfolding of a discrete arc. Left panel shows sequential pictures of aurora with a time interval of 1/2 sec and with exposure of 1/60 sec each. The right panel schematically illustrates the main feature of the deformation, how the splitting-unfolding proceeds along this arc. Protrusions on each side of the arc propagate in reverse directions (clockwise viewed along the magnetic field) during the course of the rotational unfolding, as shown by solid and dotted arrows respectively.

from the main arc at one end, and subsequently they often appear to propagate along the arc sometimes as rapidly as 100 km/sec. The peeled off ends propagate eastward when they protrude on the high latitude side of the main band and westward when they are on the low latitude side. An example of these propagating small protrusions is seen in Fig. 3, where the sequence of pictures on the left hand side, reproduced from video-tape at the rate of 60 frames per second, indicates deformations of a very thin arc. The right half of the figure shows how the peeled off ends (protrusions) propagate along the arc.

In contrast to these splittings or peelings in small sizes, giving rise to ray structures, the same mode in a large scale size is seen as the formation of a westward travelling surge or an auroral bulge. Important point is that the mode of deformations hold in spite of the wide variety of scale sizes. Now we can claim that the essential dynamic feature of the discrete aurora is characterized by the splitting and the clockwise rotation of bright portions or shear motions across the principal arc. The discrete arc of aurora, in this sense, can be defined as a vortex-chain consisting of clockwise (viewed along the magnetic field) vortices of various scale sizes, from the dynamical view point of aurora.

Category 2

Category 2 deformations may be understood as transient modes from discrete auroral deformations in the previous section to the pulsating auroras in the next section. Of the four modes in category 2, the flickering and the swirling smoke-like, rotational variations such as auroral curl [Hallinan and Davis, 1970] are often seen in the remnant aurora just behind the poleward expansion front of an expansion aurora and show similar characteristics as those of the category 1.

Flickering

Flickering aurora consists of a number of auroral spot, 5 - 10 km in size, periodically swinging and changing their luminosities viewed along the magnetic field [Cresswell, 1968] while it appears as bundles of periodically swaying rays when viewed in a direction parpendicular to the magnetic field [Oguti, 1975a, 1978].

Many similarities are found between the periodic splitting-unfolding and flickering. The stepwise shift of brightening from one spot to the next in flickering is similar to the sequential occurrence of splitting-unfolding in samll sizes (5 - 10 km) along a discrete band as shown in the previous section and their frequencies are roughly in the same range (3 - 20 Hz).

In addition, the motions of the spots viewed along the magnetic field are often quite similar to the splitting or peeling of discrete arc. The motion of flickering spot often appears to be a

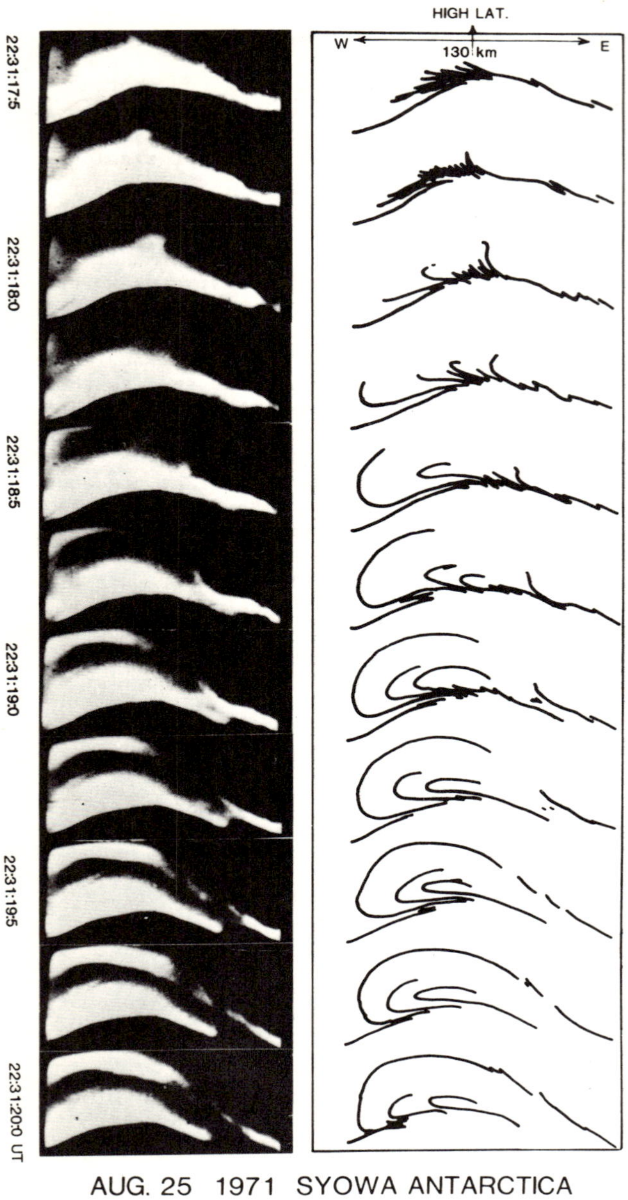

Fig. 2. A typical example of splitting followed by reconnection, forming a new poleward expansion front of a discrete arc. Left panel shows sequential pictures with a time interval of 1/4 sec and with exposure of 1/60 sec each. The right panel schematically illustrates the main feature of the splitting - reconnection process. A protrusion in the low latitude retreats toward left (toward the principal arc) after the reconnection is completed.

rounding motion similar to the rotational unfolding or rotational peeling, and the array of spots is similar to the array of ray structures along a discrete arc. These facts suggest that the physical causes that bring forth the splitting-unfolding process and flickering spots are similar and that the difference in appearance of the modes may be due to differences in the background structures; probably the difference in background precipitation, as in a single sheet for producing ray structures or in multiple sheets for producing flickering ray bundles. Thus a flickering spot could be a vertically extended column that appears in a process of splitting-unfolding of vertically extended sheets of aurora, the sheets themselves being subvisual.

Swirling smoke-like (rotational) varieties

Swirling smoke-like (rotational) varieties such as auroral curl also show clockwide rotations around bright structures as in the discrete arc, revealing themselves to be relatives of the discrete arc.

These two modes in catagory 2 are simialr in their essential point to the modes in category 1 although they appear in the transient region and although they appear apparently different from the modes in category 1.

Formation of complex striations

In contrast to the two modes mentioned above, the formation of complex striations is clearly a transient mode from the rotational modes in the discrete arc to pulsating modes in a lower latitude arc.

During a course of an auroral expansion, many splittings usually occur on the poleward expanding discrete auroral arc. The split arcs usually have eastward and equatorward protrusions. These protrusions shift or expand equatorward while the discrete arc front of the expansion aurora shift poleward. Flickering and swirling smole like (rotational) modes, which appear in the protrusions just after the protrusions split from the poleward discrete arc, gradually decay as the protrusions expand equatorward. The swirling smoke like features in these protrusions tend to change into complicated highly striated forms. Eventually, the protrusions consist of complex striations of aurora. Fig. 4 (bottom) shows the formation of such an equatorward and eastward protrusion that consists of complex striations on sequential pictures reproduced from an all sky video tape. The left picture shows an expansion of a discrete aurora a little south of La Ronge (geomag. lat. 64.8° long. 311.0°) while pulsating auroras are seen in the southern half of the sky, and the second picture shows a split protrusion that extends eastward and equatorward. The split protrusion gradually expands southward over Park Site (geomag. lat. 61.5°, long. 309.8°) as seen in the 3rd and 4th pictures. The 4th and the 5th pictures evidently indicate that the protrusion consists of complex striations.

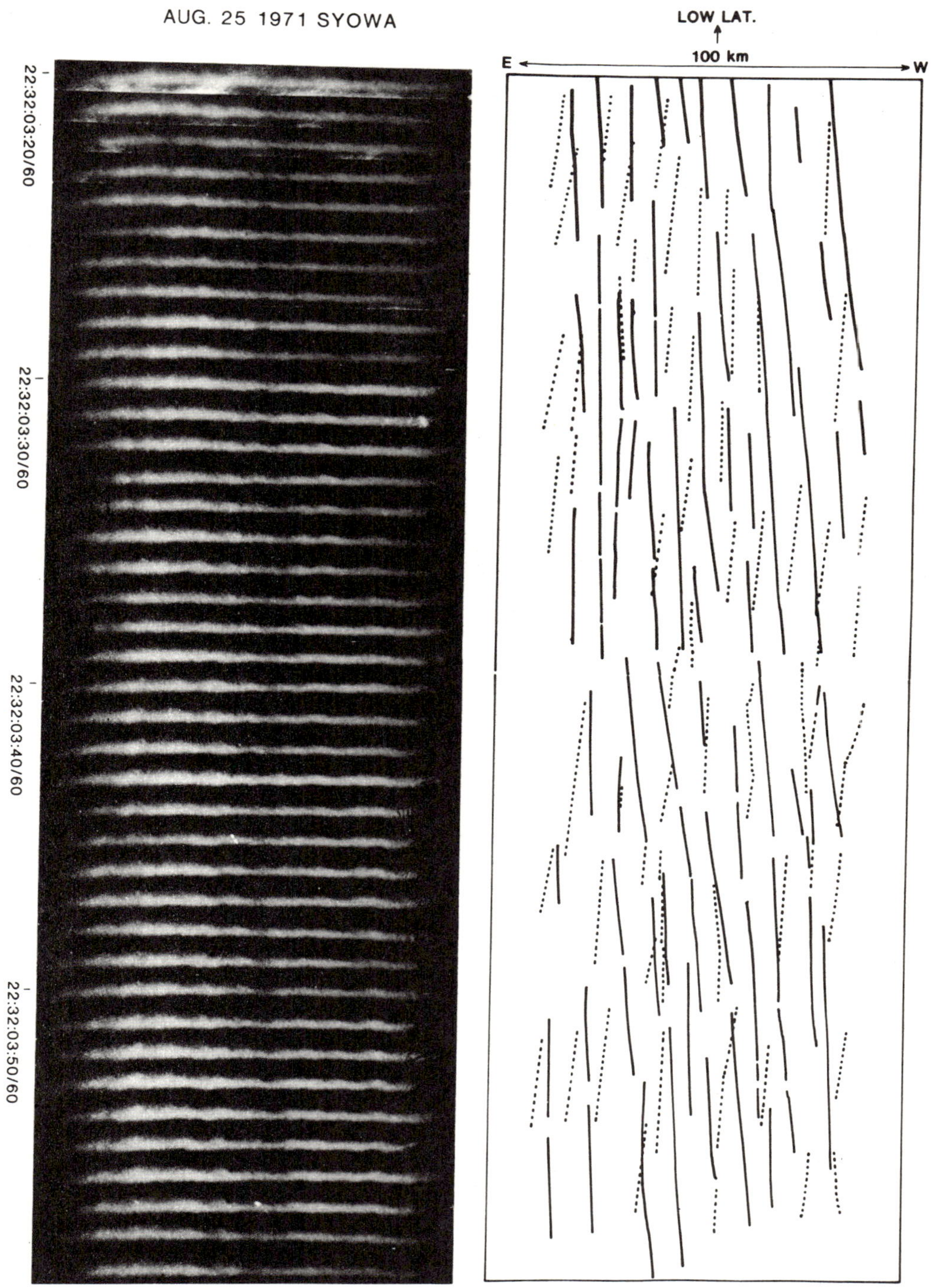

Fig. 3. An example of splitting (or peeling) - fold over deformation of a discrete arc in small sizes (5-10 km), corresponding to the appearance and disappearance of ray structures. The left panel shows sequential pictures of a thin arc with time interval of 1/60 sec. Small protrusions appear along the arc in a sequence to both higher and lower latitudes. The right panel indicates the traces of the propagation of protrusions (peeled off ends) along the arc identified in the left pictures on both poleward (lower) and equatorward (upper) sides, shown by solid and dotted lines respectively. The direction of the propagation is evidently reversed across the arc, showing a clockwise shear of motion viewed along the magnetic field.

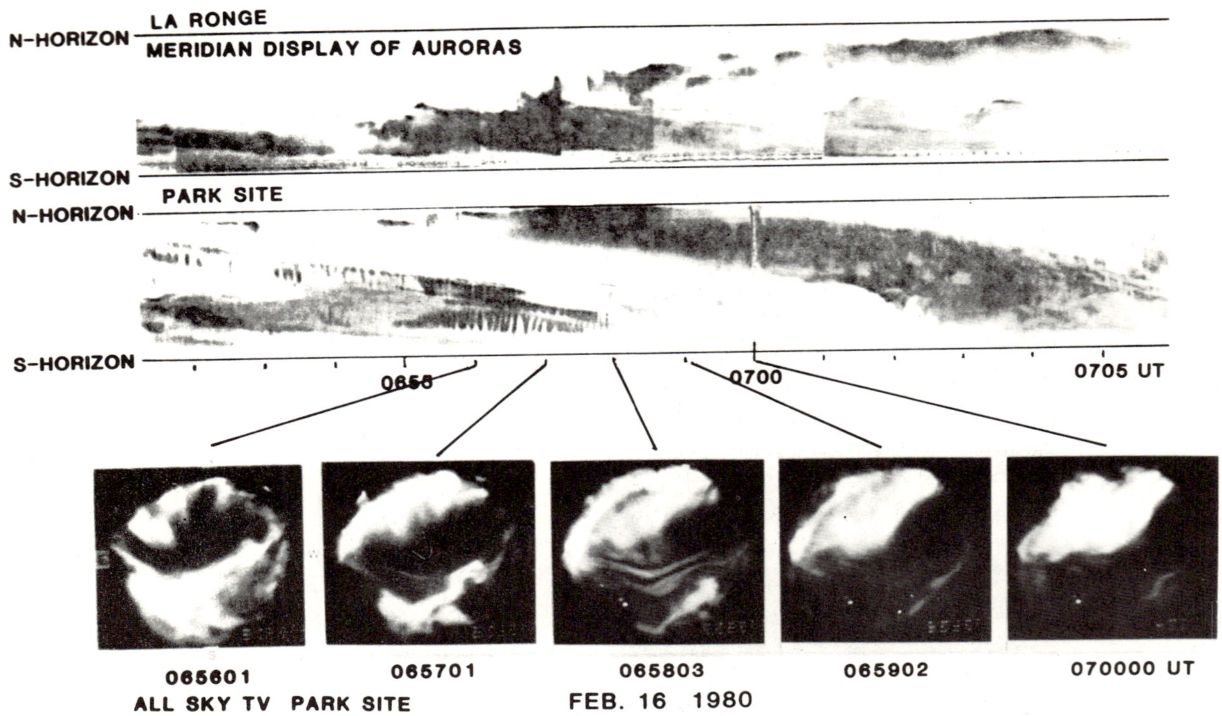

Fig. 4. An example of equatorward expanding branch of aurora during the course of an auroral expansion. Top panel shows meridian display of aurora at La Ronge (geomag. lat. 64.8° long. 311.0°) and Part Site (geomag. lat. 61.5° long. 309.8°). Dark region indicates aurora. The poleward expansion that broke out at 0654 UT is seen at La Ronge, while the equatorward expansion is seen over Park Site at the same time. It is in the bottom pictures of all sky TV from Park Site, that the equatorward expanding branch, initially split from the poleward discrete arc, gradually changes from a swirling smoke like aurora to a highly striated aurora as it expands toward low latitudes. Pulsation starts in this transient aurora, as seen in the top panel. Pulsations before 0659 UT is due to a formerly split arc.

Initiation of partial pulsations

An important point is that 2 modes in category 2, namely swirling smoke and complex striations, sometimes concur with partial pulsations in category 3. Swirling smoke like auroras sometimes pulsate as they swirl, and some parts of the complex striations also pulsate as they fold and meander. Pulsation begins partially in the protrusion, and gradually expands into whole structures. Top panel of Fig. 4 shows temporal variation of aurora from 2 stations along the magnetic meridian. The top is northern horizon, the bottom is southern horizon, and the darkness represents the auroral liminosity. A down-going trace indicates the equatorward expansion of the protrusion, and small dots on the trace represent pulsations. The pulsating aurora seen before 0659 UT is due to a formerly split branch. The aurora in the last picture of Fig. 4 is already pulsating in considerable parts of the structures. This is a reason why these modes are classified into the transient mode. Hence, most of the auroral deformations in category 2 are understood as transient modes from a vortex chain type in category 1 to pulsating auroras in category 3.

Category 3

Category 3 deformations belong to a family of pulsations in the broad sense of the word, consisting of pure pulsations including expansions, streaming along striations or elongated structures, flooding in arbitrary directions, poleward propagation of brightening, equatorward propagating waves, flash, and flaming along the geomagnetic field. The transition, from discrete aurora that is characterized by clockwide vortex-chain to the pulsating aurora that is characterized by the pulsations, usually takes place as follows; The rotational motions, along with complex meandering and folding, gradually quieten down as the split branch in Fig. 4 shifts further equatorward, leaving complex striations that do not much change their configurations, but change their luminosities in pulsative fashion. The auroral arc consisting of category 3 modes is thus formed in lower latitudes than the discrete forms, especially in the down sectors.

Pure pulsations, including expansions

Pure pulsation here defines the change in auroral luminosity with a certain pattern that lasts unchanged for several repetitions of brightening [Oguti, 1976]. This mode can include expansion mode, since the pulsation has a tendency, more or less, to expand from some portions of a patch or a striation into a whole pattern. This mode is frequently seen either at the very beginning of the pulsating auroras especially in the swirling smoke like auroras and complex striated aurora mentioned in the previous section, or in the central region of a well developed torch structure of aurora during a period of high activity. In the former case, the pure pulsation is seen as the pulsation of complex striations and swirling structures, and in the latter the pulsation is mostly seen as the pulsation of complicate shaped patches of various sizes. The boundary of the striations and complicate shaped patches in both cases can be rather clearly defined, and it reveals sometimes a narrow dark zone across which the pulsation is independent on either side of it. The pulsation of striations in the first stage of pulsations in the protrusions shown in Fig. 4, appear to be pure pulsation, whereas the pulsation of the complicated patches in a torch structural appears to be more expanding. Sometimes, when poleward propagation of brightening and equatorward propagation wave concur with the pulsation, the pulsating patches look as if they are illuminated by the propagation modes as the propagation modes pass through the patches.

Poleward propagation of brightening

Poleward propagation of brightening can be seen in almost every part of pulsating aurora except in the lower latitudes where the equatorward propagation wave dominates. The propagation direction can be quite irregular but the poleward propagation is definitely dominant. For example, the streaming along striations or elongated structures shows a strong tendency to shift poleward as each streaming of brightening occurs, and pulsation of complex auroral patches also shows a tendency to propagate poleward during a cycle of pulsation. The most remarkable example of this mode can be seen near the central region of auroral torch structures as a growing mushroom cap pattern, that is a striation or an arc fragment laterally propagating poleward [Oguti and Watanabe, 1976].

The poleward propagation mode is seen on radar aurora too. Kaneda et al. [1964] have shown an example of pc-5 period range in the data from College, Alaska; On the basis of radar aurora observed from Ottawa, McDiarmid and McNamara [1972] have shown that this mode exists in the dawn and in the afternoon sectors as well, with a wide range of period from 10 sec to a few hundreds of seconds.

Streaming along striations or elongated structures

As has been shown by Royrvik and Davis [1977], the streaming condition is very common in the aurora in the dawn sector. It occurs in any direction depending upon the arc orientation. For example, when the aurora is striated roughly parallel to the oval, the streaming usually occurs parallel to the oval, during a slightly active period. On the other hand, the poleward boundary of the aurora tends to deform into torch or omega structures during an active period. The striations on this occasion, usually appear along and roughly parallel to the poleward boundary and streaming occurs also along the torch shaped striations. An example of the streaming along a torch structure is shown in Fig. 5. The top panel shows the temporal variations of auroral luminosities at the zenith, indicating that the streaming causes quasi-periodic pulsation of auroral luminoisty at a fixed point. This figure further shows, that a non pulsating diffuse arc appears to envelop the poleward boundary of the torch (or omega) structure. This is understood by the fact that the pulsation in luminosity recorded by the photometer at the zenith (top) interrupts when the poleward boundary part of the torch passes over the zenith, in spite of the luminosity there being as high as that in the streaming parts. On the other hand, active, mushroom shaped, poleward propagating auroras or complex, pulsating patches usually appear at the center region of the torch structure, as shown in the pictures and in the temporal variations in luminosity at the zenith in Fig. 5, when the central region passes over the zenith. The torch structure of aurora, in this sense, is an activity center of pulsating aurora.

Flooding

The streaming along the striations mentioned above appear to gradually alters into flooding, in the lower latitude of the torch structure. Flooding here means that the brightening is no more limited within a striation but floods beyond the boundaries of striations or elongated patches laterally and longitudinally, involving a number of striations or elongated patches in a sequential and repetitive brightening. The direction of the flooding roughly parallel to the oval during the period of a small activity, but it tends to be in arbitrary directions especially in the central region of a torch structure during an active period.

Equatorward propagation wave

This mode was called the "fast auroral wave" by Cresswell and Belon [1966]. As stated by

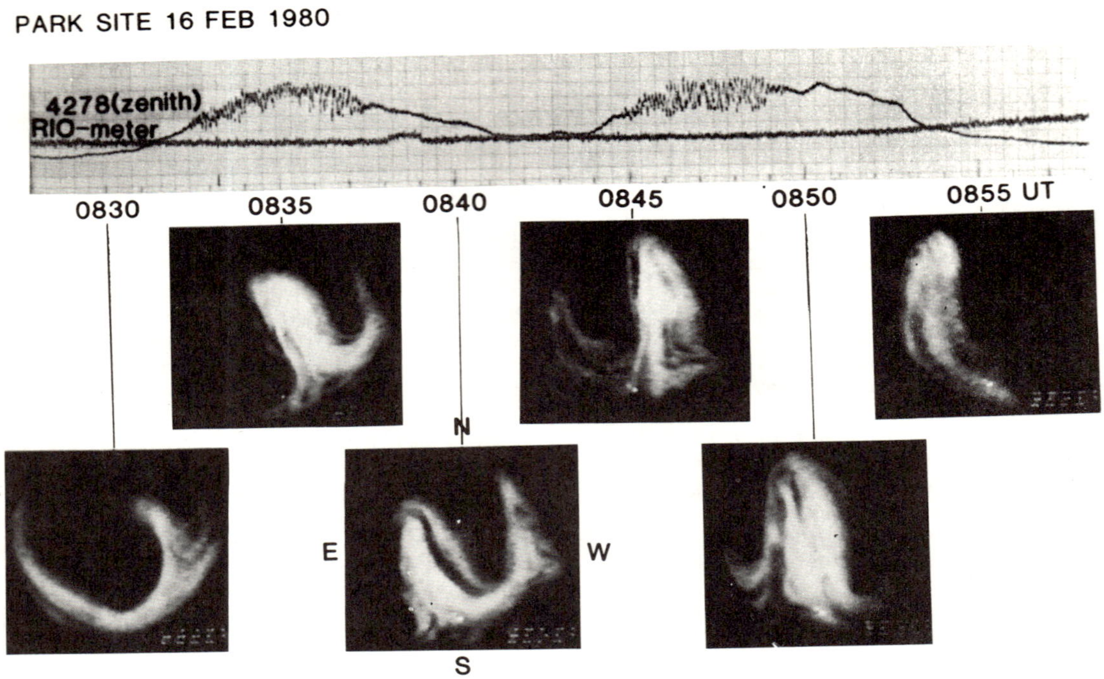

Fig. 5. A typical example of torch structure of aurora, that occurred over Park Site (geomag. lat. 61.5° long. 309.8°) on Feb. 16, 1980. As seen in the top panel, the central region of the torch structure is an activity center of pulsation. There is a non-pulsating diffuse peripheral which appears to envelop the poleward boundary of the torch structure. Striations inside the torch are roughly parallel to the poleward boundary of it, and streaming occurs along the striations. The drift of the torch structure in this particular example is eastward with velocity of about 300 m/sec.

Cresswell [1968], this mode is really a fast wave-like propagation of faint arc fragments usually elongated in an east-west direction, about several tens of km long and about 10 km in width. The mode of propagation of the wave is usually smooth, at almost a constant speed, in contrast to the stepwise propagation of the poleward propagation mode. The occurrence of this mode is limited to the lowest latitude part of auroral oval in the dawn sectors, and possibly to the noon sectors.

Flash

Flash mode [Oguti, 1978] is really flashing appearance and disappearance of faint, diffuse patches with rather large scale sizes 50 to 200 km. This mode usually appears in the lowest latitude part of the auroral oval in late morning hours. Flash mode may be a variation of the equatorward propagation mode that propagate at an extremely high speed as 200-300 km/sec.

Flaming

Flaming along the magnetic field is quite hard to find when aurora consist of diffuse patches. Flaming can be seen only either in the central region of torch structures where complex discrete patches pulsate, or in the equatorward part of the oval where discrete pulsating patches sometimes co-exist with equatorward propagating waves. Although the flaming often appears to be related to the equatorward propagation wave, it will need further observation to confirm the relationship.

Distribution of the modes

On the basis of the facts mentioned in the previous sections, the distribution of various modes along the oval is summarized in Fig. 6 for two magnetic activity ranges. During the period of Kp as low as $0_0 \leq Kp \leq 1_-$, the auroral oval is apt to be recognized to consist only of discrete aurora. However, it is actually not true. The pulsating auroras are seen in a zone between 68° to 70° in geomagnetic latitude in the midnight sectors [Oguti et al., 1980], extending poleward in the dawn, even for the period as quiet as $0_0 \leq Kp \leq 1_-$.

As the activity increases, the range in local time and latitudes where pulsating aurora occurs expands to lower latitudes and toward evening sectors. The pulsating aurora can be seen well before 17h magnetic local time when K_p is greater than 3_- [Oguti et al. 1980]. These facts, in conjunction with the results of the radar aurora

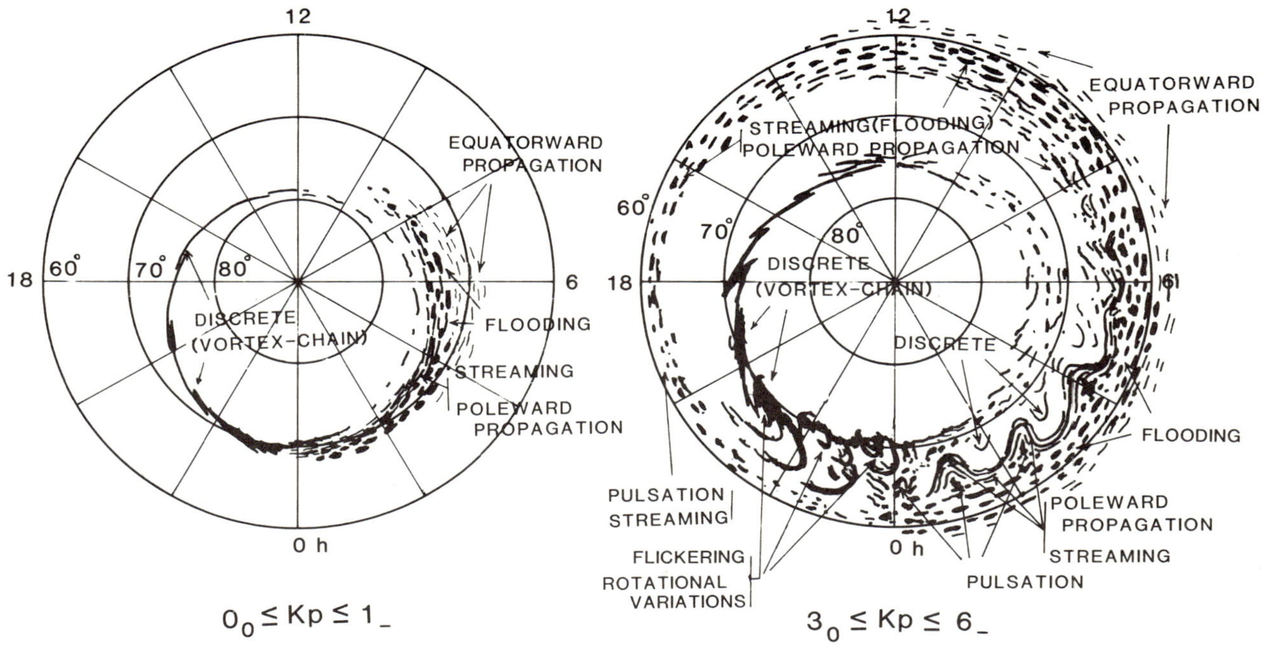

Fig. 6. A model distribution of auroras along the auroral ovals for 2 ranges of activity; $0_0 \leq Kp \leq 1_-$ and $3_0 \leq Kp \leq 6_-$. For the period of low activity, the oval appears to consist of a discrete (vortex-chain) aurora along the dusk half and a pulsating aurora array with a diffuse background along the dawn half of the oval. For the period of high activity the enhanced splitting and rotations occur along the discrete arc and the poleward boundary of the pulsating auroral region deforms into torch structures. The poleward propagation mode and streaming mode are often seen in the poleward part, flooding and pulsation are in the middle zone along with some poleward propagations, and the equatorward propagation mode is limited in the equatorward part of the pulsating auroral region.

[McDiarmid and McNamara, 1973] that showed similar pulsations through the daytime, strongly suggest that the pulsating aurora in the evening hours during high activity are probably the eastward extension of the pulsating auroral region in the dawn sector beyond the dayside.

Drift of Pulsating Auroral Patches

Fig. 7 indicates the temporal variation of the drift direction and speed of pulsating auroral patches that appeared over La Ronge (geomag. lat. 64.8° long. 311.0°) and Park Site (geomag. lat. 61.5° long. 309.8°) on Feb. 16, 1980. The K_p indices for the first half were 6_- untill 12^h UT, and pulsating auroras were seen, over Park Site in the evening hours. As seen in the figure, the direction of drift switches across the magnetic midnight from westward to eastward, being consistent with the general pattern of electric field in this region. This fact indicates that the pulsating auroral patches drift mostly by the electric field, rather than by magnetic effects, and accordingly that the pulsating auroral structure is defined by the colder plasma distributions in the magnetosphere, colder than the precipitations responsible to the pulsating aurora, as suggested by the author [Oguti 1976] and by Johnstone [1978].

Conclusion and Discussion

In conclusion on the TV observations of auroral dynamics, six points are summarized as follows;
1) The discrete arc (so far called) must be defined as a vortex (clockwise viewed along the magnetic field) chain, consisting of vortices of various scale sizes. The splitting, another characteristic mode of discrete aurora, may be understood as a variation of a shear motion.
2) The diffuse arc, called hitherto, must be classified into two at least; the diffuse background of pulsating auroras and the poleward peripherals of torch or omega structures. The latter may be called real diffuse aurora. The diffuse aurora as the background of the pulsating auroras, which has long been called diffuse arc (sometimes the pulsating auroral arc itself has been called diffuse arc due to poor time resolution) is more preferably called pulsating auroral background.
3) The auroral oval in the dawn sectors consists mostly of the pulsating auroras with a background diffuse aurora. It may not be too much to say

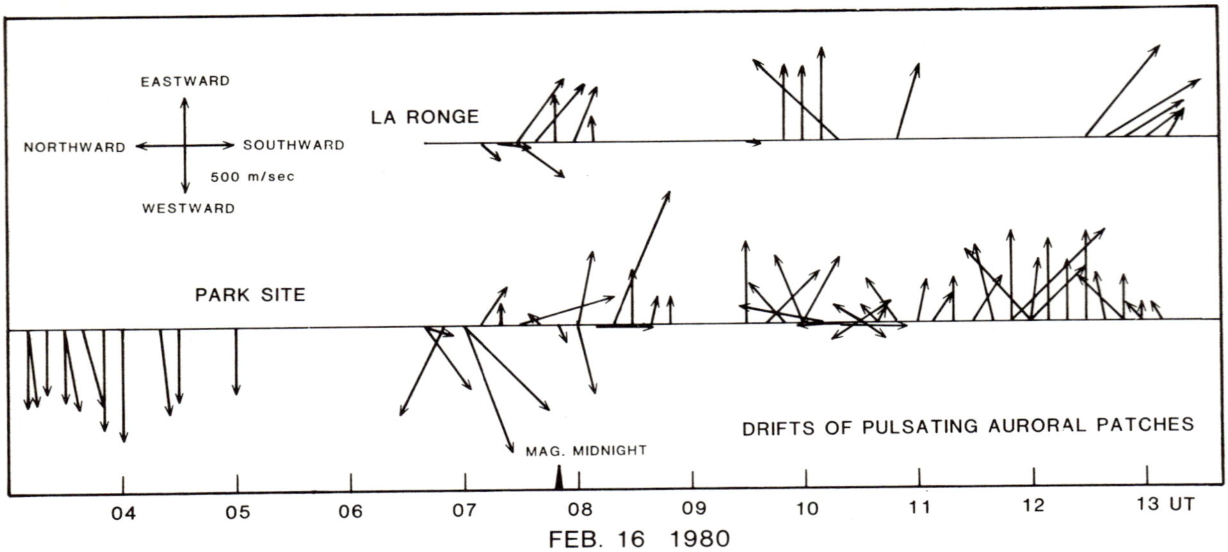

Fig. 7. Drift directions and speeds of pulsating auroral patches over La Ronge (geomag. lat. 61.5° long 311.0°) and Park Site (geomag. lat. 64.8° long. 309.8°). Drift direction is reversed from westward to eastward across the magnetic midnight, suggesting the drift to be due to electric field.

the auroral oval is a combination of vortex chain along the dusk half, and the pulsating structures array along the dawn half of it.

4) The pulsation initiates on the auroral branch that splits from the discrete aurora as the branch drifts or expands equatorward in the midnight sectors.

5) The central region of an auroral torch or omega structure is an activity center of pulsating auroras.

6) Pulsating auroras probably exist encircling the polar cap from the dawn to the dusk sectors through the dayside for the period of high activity as $3_o \leq K_p$, although the dayside part of the pulsating auroral arc is invisible.

7) Drift of pulsating auroral patches are mostly due to electric field, rather than due to the magnetic effects.

On the basis of the fact 1), the precipitation in the discrete aurora must be related to the electric field in the magnetosphere. Probably some mechanism of electric charge separation is always acting in the vicinity along the open-close field boundary in the dusk magnetosphere and the negative charge excess, continuously being produced, would be the source that maintains both the field aligned electric current and the electric potential structure along the discrete auroral magnetic field. The enhancement of the vortex-chain feature of the discrete aurora during an expansion is possibly due to the rapid increase in the rate of electric charge separation, causing the steepening in potential gradient in both directions parallel and perpendicular to the magnetic field, in conjunction with increase in precipitation energies and fluxes.

In contrast to this, there is little doubt that the precipitation of energetic electrons in the pulsating aurora is due to pitch angle scattering by the interaction with electromagnetic or electrostatic VLF waves. Most of pulsation modes may be interpreted by the trapping or ducting of VLF waves in the cold plasma irregularities in which the wave interacts with trapped energetic electrons. When the ducting or trapping of VLF waves is limited in a irregularity, the corresponding pulsation is pure pulsation or expansion. Interactions between oblique propagation wave and energetic particles in an elongated flat irregularity may cause streaming, and when the wave can propagate beyond the relatively smeared boundary of multiple irregularities, the propagation of brightening beyond the patch boundaries can be expected. The poleward and equatorward propagation modes may be related to a compressional MHD wave in addition to the VLF waves, or creeping of VLF waves.

The dominance of pulsating auroras in the dawn sectors(fact 3) is probably due to the relative abundance both of energetic electrons and cold plasma irregularities. The facts 3), 4) and 7) suggest that the cold plasma irregularities responsible to precipitations in the pulsating auroras are detached columns of plasma from the plasma sheet that drift inward, and pile up at the inner most boundary of injection. The low latitude border of the pulsating aurora is possibly identical to the inner most boundary of injection, and the torch or omega structure 5) may be a local pile up of the detached plasma sheet plasma just outside this boundary.

Another possibility of cold plasma irregularities is the plasma blown up from the ionosphere, during the course of an auroral expansion. In any case, the complex striation and complex patches at the beginning of pulsation may indicate the remnant spatial modulation of cold plasma distribution produced by the local growth and decay of small scale electric fields during the period of the discrete auroral activity prior to the pulsations.

Acknowledgements. This study was supported by Grant 40421 from The Science Research Division, Ministry of Education, Science and Culture, Japan, and by Grant 80-18 from The Japanese Society of Promotion of Sciences.

References

Beach, R., G.R. Cresswell, T.N. Davis, T.J. Hallinan and L.R. Sweet, Planet. Space Sci., 16, 1525, 1968.
Campbell, W.H., J. Geophys. Res., 75, 6182, 1970.
Cresswell, G.R. and T.N. Davis, J. Geophys. Res., 71, 3155, 1966.
Cresswell G.R. and A.E. Belon, Planet. Space Sci., 14, 299, 1966.
Cresswell G.R., Planet Space Sci., 16, 1453, 1968.
Cresswell, G.R., J. Atmos. Terr. phys., 34, 549, 1972.
Davis, T.N., Space Sci., Rev., 6, 222, 1966.
Davis, T.N., Planet. Space Sci., 20, 1369, 1972.
Hallinan, T.J. and T.N. Davis, Planet. Space Sci., 18, 1735, 1970.
Johnstone, A.D., Nature, 274,119, 1978.
Kaneda, E., S. Kokubun, T. Oguti and T. Nagata, Rept. Ionosph. Space Res. Japan, 18, 165, 1964.
McDiarmid D.R. and A.G. McNamara, Ann. Geophys., 28, 433, 1972.
Oguti, T., Mem. Nat'l Inst. Polar Res., Ser. A. 12, 1, 1975a.
Oguti, T., J. Atmos. Terr. Phys., 37, 1413, 1975b.
Oguti, T., J. Geophys. Res., 81, 1782, 1976.
Oguti, T. and T. Watanabe, J. Atmos. Terr. Phys., 38, 543, 1976.
Oguti, T., J. Geomag. Geoelectr., 30, 299, 1978.
Oguit, T., S. Kokubun, K. Hayashi, K. Tsuruda, S. Machida, T. Kitamura, O. Saka and T. Watanabe, Can. J. Phys., 59, in press.
Royrvik, O. and T.N. Davis, J. Geophys. Res., 82, 4720, 1977.
Scourfield, M.W.J. and N.R. Parsons, Planet. Space Sci., 17, 75, 1969.
Scourfield, M.W.J. and N.R. Parsons, Planet. Space Sci., 19, 437, 1971.
Thomas, I.L., M.W.J. Scourfield and N.R. Parsons, Can. J. Phys. 51, 2209, 1973.

THE DISTRIBUTION OF VORTICITY IN AURORAL ARCS

T. J. Hallinan

Geophysical Institute, University of Alaska, Fairbanks, Alaska 99701

Abstract. In a magnetized plasma there is a direct proportionality between space charge and vorticity (shear). Because of this, a field-aligned sheet of negative space charge is subject to the Kelvin-Helmholtz instability and rapidly forms an array of vortices. These drift at the $\vec{E} \times \vec{B}$ velocity. In an auroral arc, electric fields parallel to the magnetic field may prevent the charge sheet from penetrating into the ionosphere. However, since energetic electrons reach the ionosphere after traversing this region, the vorticity maps to the ionosphere as an apparent vorticity in the visible auroral structures. The apparent velocity of auroral rays (vortices observed from the side) is the ionospheric manifestation of the $\vec{E} \times \vec{B}$ drift above the region of parallel electric fields. The perpendicular electric field (referred to the 100 km level) and the field-aligned potential are calculated from the observed vorticity in a typical rayed arc. The potential exceeds 2 kv. The field configuration and magnitude (1000 mv/m) are in excellent agreement with the "paired electrostatic shocks" observed from the S3-3 satellite. Despite this agreement, however, there is cause for concern. Most auroral arcs, including some that are quite bright, exhibit active ray motions only sporadically. This suggests that large electric fields are not a necessary feature of auroral arcs. Hence the paired shocks, although highly significant, may not be the basic cause of discrete auroral arcs.

Introduction

It is now generally recognized that large electric fields (1000 mv/m referred to the 100 km level), perpendicular to the magnetic field, commonly occur in association with auroras. Although these electric fields had been postulated on the basis of auroral motions (Hallinan and Davis, 1970; Carlqvist and Bostrom, 1970, Swift et al. 1976), and had been observed on two occasions with high-altitude barium ion tracers (Wescott et al. 1975, 1976), their regular occurrence was first confirmed by in situ measurements by Mozer et al. (1977). A particularly distinctive electric field signature, is the "paired shock" consisting of a pair of oppositely directed electric field spikes. Assuming that these correspond to the passage of the satellite through a stationary structure of electrostatic potential associated with an auroral arc, they imply the existence of a field-aligned sheet of excess negative charge above the aurora. This agrees with the earlier conclusions based on auroral motions.

Since electric fields of corresponding magnitude are rarely if ever observed at altitudes of less than 1000 km, one can assume that these large electric fields do not map to the ionosphere. The S3-3 data confirm that the potential -structures are seen more often at 8000 km altitude than at 2000 km (Mozer, 1980). Confirmation of an abrupt vertical gradient in $E_\perp$ was provided by the skylab shaped-charge barium injection. Large perpendicular electric fields were detected in the region above 5000 km and were non-existent below 5000 km. The observations above and below 5000 km were simultaneous and on the same field line -- a field line through a discrete auroral arc in the ionosphere below (Wescott et al. 1976).

Assuming that the fields are electrostatic, the parallel gradient in $E_\perp$ implies a parallel electric field:

$$\frac{\partial E_y}{\partial z} = \frac{\partial E_z}{\partial y} \qquad (1)$$

where the magnetic field is parallel to z and the measured $E_\perp$ is in the y direction. This suggests U-shaped (Carlqvist and Bostrom, 1970) or V-shaped (Swift et al. 1976) equipotential surfaces. In the center of the structure there is a potential drop above the ionosphere having the proper sign to accelerate electrons downward. Carlqvist and Bostrom (1970) proposed this as a two-dimensional extension of the one-dimensional double-layer described by Block (1972).

This general class of electrostatic potential structures has been discussed widely and includes a variety of specific designations, e.g. V-shock, double layer, oblique shock, paired electrostatic shock, etc. These designations sometimes imply specific geometrical configurations or specific mechanisms for sustaining the field-aligned electric field or both. This paper refers to potential-structures that are either V or U shaped and does not specifically address the question of how the electric fields are maintained. For simplicity of discussion, a U shape will be assumed and the structure will be referred to simply as the U.

Since the electric field is conservative, the potential difference ϕ between the center of the structure and its outer edge is also the potential drop along the field line. Hence the change in kinetic energy (ΔW) of the precipitating electrons passing through the structure is.

$$\Delta W = e\phi = e \int_{\text{Edge}}^{\text{center}} E_y \, dy \qquad (2)$$

(This assumes that the ionospheric electric field is small in comparison to the high-altitude field and therefore makes only a negligible contribution).

Several authors have suggested that this gain in energy (ΔW) may be the key acceleration mechanism for auroral arcs (Block, 1972; Carlqvist and Bostrom, 1970; Swift et al. 1976; Mozer et al. 1977, Kan and Akasofu, 1979). It is also widely felt that the parallel electric fields are somehow related to upward field-aligned currents. Kan and Akasofu (1979) proposed a selfconsistent model relating the current and potential structures.

Vorticity

Perpendicular gradients in the perpendicular electric field can equally well be viewed as shear in the $\vec{E} \times \vec{B}$ plasma flow. This shear can be characterized by the parallel (z) component of the vorticity (ω_s):

$$\omega_s = \nabla \times \vec{v}_\perp \qquad (3)$$

We can express this in terms of the electric field by substituting for v :

$$\omega_s = \nabla \times \left(\frac{E \times B}{B^2}\right) \qquad (4)$$

For simplicity we assume a locally uniform magnetic field to obtain

$$\omega_s = \frac{1}{B^2} [E\ \nabla\cdot B + (B\cdot\nabla)\ E - B(\nabla\cdot E) - (E\cdot\nabla)B] \qquad (5)$$

Since $\vec{B}$ is uniform and the variations in $\vec{E}$ are perpendicular to B, this reduces to

$$\omega_s = -\frac{1}{B} \nabla\cdot\vec{E} \qquad (6)$$

$$\omega_s = -\frac{e\rho}{\varepsilon_0 B} \qquad (7)$$

Thus vorticity is inherently linked to space charge. The potential structures described in the introduction are regions of negative space charge and are therefore regions of strong shear. Since the vorticity (charge) is conserved and the divergence of the plasma flow is zero, the plasma behaves as a fluid. As in any other fluid, vorticity leads to the formation of eddies. In the special case of a field aligned sheet of excess charge, as envisioned in the models described in the introduction, the sheet can be expected to deform into an array of individual vortices (vortex street). This has been demonstrated in laboratory experiments (Fig. 1b) (Kyhl and Webster, 1956, Cutler, 1956) and in computer simulations (Fig. 1a) (Levy and Linson, 1968; Wagner (this volume)). The possibiity of observing these vortex streest in the visible aurora had been discussed by several authors (Webster, 1957; Gartlein, 1959; Kern and Vestine, 1961; Akasofu, 1962); but their actual observation (Hallinan and Davis, 1970) resulted from the introduction of low-light-level television cameras to record the rapid auroral motions (Fig. 1c). Vortex streets (curl systems) are now known to be a common feature of active auroral arcs. The rapidly moving rays in rayed arcs are vortices (curls) observed from the side.

The dispersion relation governing the growth of the vortices is that derived by Rayleigh for the Kelvin-Helmholtz instability (Buneman et al. 1966). In terms of the vorticity, the dispersion relation can be written as (Hallinan, 1976)

$$\omega - \beta v_z - kU_x = \frac{\omega_s}{2} [(1-2ka)^2 - e^{-4ka}]^{1/2} \qquad (8)$$

where ω is the growth rate; k is the wave number in the x direction

$$k = 2\pi/\lambda \qquad (9)$$

λ is the wavelength (distance between vortices); 2a is the thickness of the charge sheet; and β is the spatial growth constant parallel to B. U is the average plasma velocity perpendicular to B (if any) and v_z is the flow parallel to B.

A significant feature of eq. 8 is that the right hand side is either purely real or purely imaginary; but never complex. Real values occur for very short wavelengths and correspond to surface waves drifting at the local $\vec{E} \times \vec{B}$ velocity. Longer wavelengths, involving coupling between the two surfaces and distorting the entire charge sheet, produce imaginary values. These are the wavelengths of interest for vortex formation.

For such wavelengths, ω can have a real component only if there is a fluid velocity U_x. In the rest frame of the plasma ($U_x = 0$), ω is purely imaginary. There is exponential growth with no oscillations. It follows that the vortices are stationary with respect to the rest

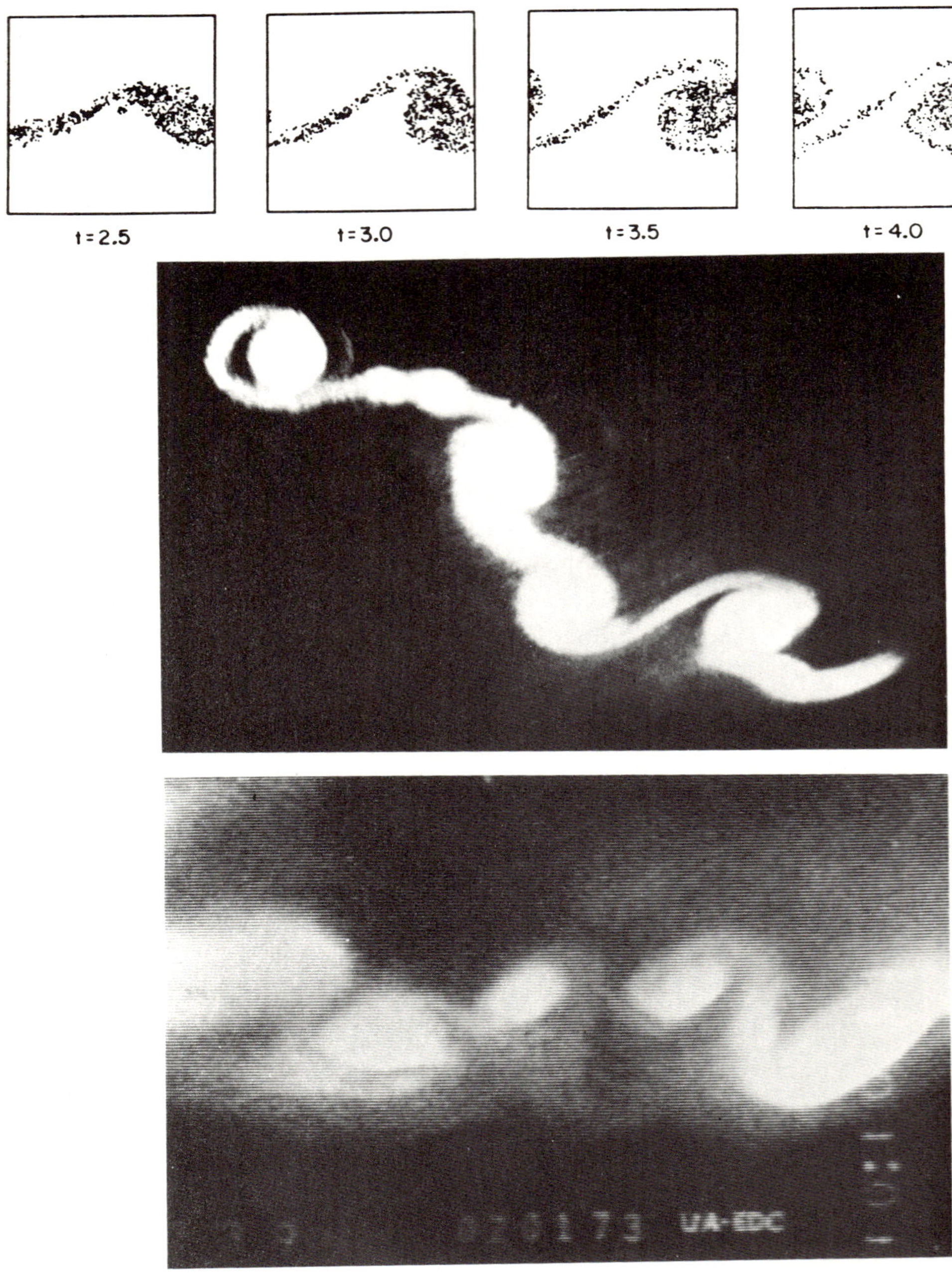

Figure 1a: Computer simulation of the develpment of the Kelvin-Helmholtz instability in a band of negatively charged rods aligned along a magnetic field. (After Levy and Hockney, 1968) B is into the paper. b: Breakup of sheet electron beam in a magnetic field. (after Webster, 1957). B is into the paper. c: Auroral rayed arc viewed in magnetic zenith. B is out of the paper.

frame of the plasma, i.e. the translational velocity of the vortices is the $\vec{E} \times \vec{B}$ plasma velocity. This is simply a special case of the fluid dynamics theorem that vorticity flows with the fluid elements.

Boundary conditions and mapping

Field-aligned charge sheets such as those thought to exist above auroral arcs are clearly unstable to the Kelvin-Helmholtz instability.

They can reasonably be expected to evolve into vortex arrays similar to the observed auroral curl systems (Fig. 1c). The sign of the observed vorticity is proper for negative space charge and the observed velocities (up to 20 km/sec in the ionosphere) are consistent with the measured high altitude electric fields.

But the visible auroral forms and motions are observed in the ionosphere where the Pedersen conductivity is too great to support large electric fields. What relation is there between the observed auroral vorticity in the ionosphere and the plasma vorticity inferred from spatial gradients in electric fields measured at altitudes greater than 2000 km? Is it really reasonable to equate the two phenomena? To answer this, we must look at the boundary conditions to be imposed on eq. 8 and also examine what happens in the distance between the high-altitude space charge region and the ionosphere.

There are two limiting cases of the Kelvin-Helmholtz instability: $\omega = 0$ and $\beta = 0$. Either can be imposed by the appropriate boundary conditions. Both are illustrated in Fig. 1. The computer simulation in 1a imposes the condition that there be no variations along B ($\beta = 0$). The laboratory experiment, (1b) in which the beam begins at one end of a vacuum tube at a fixed cathode is a steady-state phenomena ($\omega = 0$).

If the space charge region above the aurora has boundary conditions similar to those of the laboratory electron beam, one would expect vortices at the lower end of the charge sheet. The only temporal changes would result from temporal variations in the charge density and hence in the vorticity (ω_s). Depending on whether ω_s increases or decreases with time the vortices would wind or unwind.

However, this is a most unrealistic situation. It is hardly likely that there is a fixed rigid "source" comparable to the cathode in the laboratory. There is no obvious reason that ω should be zero. Moreover, there is no average plasma flow (U_z) along the magnetic field. Individual particles are moving in both directions along the field and at a wide variety of speeds. Since there is no directed flow along B, there is no correlation between the altitude of a particle and the time that it has spent in the E-field region. Hence distortions in the sheet do not grow systematically with distance. But they do grow in time. Hence we can set $\beta v_z = 0$ in eq. 8. The instability is absolute rather than convective. Static vortices or unwinding vortices should not be observed. The "charged rod" model used by Levy and Hockney (1968) in their computer simulation (Fig. 1a) resembles the auroral situation more closely than does the laboratory electron beam.

An additional complication in the magnetosphere is the non-uniformity of the magnetic field. e.g. At a distance of 20,000 km along the magnetic field, the magnitude of B is only 1% of the ionospheric field strength. Field lines that are quite close in the ionsophere are separated by large distances at 20,000 km. Perpendicular velocities (E X B) are quite large while perpendicular electric fields are small.

Surprisingly, this large variation has little effect on the Kelvin-Helmholtz instability. What effect there is, results from the fact that north-south dimensions expand somewhat more than longitudinal dimensions. Because of this, the value of ka (eq. 8) between 100 km and 20,000 km along the field-line through College, Alaska varies by 11% from its value at 10,000 km. Where magnetic field lines are equipotentials, even this variation is largely offset by an inverse variation in the value of ω_s. In the long wavelength limit (small ka) the dispersion relation can be approximated by

$$\omega_s = \pm' i \, \omega_s \, ka \qquad (10)$$

As the dimensions change with altitude, ω_s changes as x/y while ka changes as y/x. The product ω_s ka is independent of the distance along the magnetic field. Hence, in this limit, ω_s is independent of distance.

If we ignore the distinction between the mapping of north-south distances and east-west distances and simply use a common symbol (d) to represent distances between field lines, we obtain the relationships shown in Table I. For distances of less than 20,000 km (keeping in mind the tendency of the errors to cancel) these relationships may be used with little resultant error.

In the preceding discussion, it was explicitly assumed that the magnetic field lines are equipotentials. As discussed in the introduction, however, it is now commonly thought that the field lines through auroral arcs are not equipotentials. Indeed, there may well be potential drops of up to a few kilovolts along these field lines.

In order to clarify the effect of the potential drops on auroral vortices, let us consider the relatively simple geometry of U-shaped equipotentials. The U defines three regions: a space-charge region above the U in which the field

TABLE I. "Mapping" along geomagnetic field

Quantity	Approximately Proportional to
$E_\perp$	$\sqrt{B}$
$v_\perp$	$1/\sqrt{B}$
d	$1/\sqrt{B}$
ρ	$\sqrt{B}$
ω_s	constant

lines are equipotentials but in which there are large perpendicular electric fields; a transition region at the base of the U in which there is a parallel electric field; and a region free of electric fields below the U.

In the upper region, inside the U, there is a net negative space charge (vorticity). The charge-sheet is subject to the Kelvin-Helmholtz instability according to the dispersion relation given in eq. 8. Vortices will form in the sheet and they will drift with the average plasma velocity. The velocities, electric fields, etc. will map from one altitude to another approximately as given in Table I.

The transition region at the base of the U is fairly thin -- probably less than a few 100 km in the case of the transition observed in the Skylab barium release. Since the electrons have a large parallel velocity, they spend very little time in this transition region. They acquire kinetic energy from the parallel electric field, but have insufficient time to drift across field lines. They exit this region on essentially the same field lines on which they enter.

Below the U, the field lines are again equipotentials and $E_\perp$ is now zero. Energetic electrons, entering this region from above, flow along the field until they either are mirrored or collide with atmospheric constituents. The spatial pattern of precipitation (visible aurora) reflects the pattern of the charge sheet above the U. The dimensions map as in Table I. Below the U there is no $v_\perp$. However, because there is a continually changing pattern of injection of energetic electrons through the double layer, there is an apparent velocity $v_\perp'$. This apparent velocity maps along the magnetic field (in accordance with Table I) as if it were a real velocity (provided that velocity dispersion below the U does not measureably alter the pattern).

By this means, the vortices in the charge sheets at high altitude produce visible vortices in the aurora in the ionosphere. The apparent auroral ray motions faithfully reflect the true plasma motions at high altitude. The value of $E_\perp$ above or within the U is given by

$$E_\perp = v_\perp' \sqrt{B_o B} \qquad (11)$$

where $v_\perp'$ is the apparent auroral ray velocity in the ionosphere, B_o is the ionospheric magnetic field strength and B is the field strength at the altitude of interest. This is the same relationship that would apply if there were no parallel electric field and $v_\perp'$ were a true plasma velocity. i.e. The auroral velocity maps to the ionosphere even though the perpendicular electric field does not.

A difficulty in comparing electric fields deduced from different types of measurements is that they often pertain to different altitudes. The same equipotential structure can be characterized by a wide range of electric field amplitudes, depending on altitude. One way of providing some degree of standardization is to refer all $E_\perp$ measurements to an altitude of 100 km, using the mapping as given by Table I. This is somewhat fictional since, when parallel electron fields exist, the perpendicular field does not actually map to the ionosphere. Nonetheless, it is a convenient convention. It is particularly convenient for describing fields deduced from auroral motions since the 100 km representation of the electric field is given directly by $v_\perp'$ B where $v_\perp'$ is the observed auroral velocity. At altitudes above the double layer, the real field is then given by

$$E_\perp = E_o \sqrt{B/B_o} \qquad (12)$$

where $E_\perp$ and B are the local values, E_o the field referred to the 100 km level, and B_o the ionospheric magnetic field. Since the altitude of the U is usually unknown, it is not certain above what altitude eq. 12 applies.

Electric fields measured from satellites are usually given as measured, i.e. without being referenced to a standard datum altitude. However, eq. 12 can be used to refer these measurements to the ionosphere. Again, it is necessary to bear in mind that if there is a parallel electric field below the satellite, the "referred" field is an artificial concept and may be two orders of magnitude greater than the actual ionospheric electric field.

The "referred electric field" may seem somewhat artificial and yet it is difficult to keep track of fields measured at different altitudes without some standardization. Much of this difficulty can be avoided if we characterize a V or U-shaped potential-structure by its electrostatic potential (ϕ) given by eq. 2. Although not directly measureable, ϕ can be inferred from other measurements. It has the advantage of being independent of altitude above the double layer. Furthermore, $e\phi$ is also equal to Δw, the maximum energy increase of precipitating electrons from the parallel electric field.

By definition, the electric field referred to 100 km is the electric field that would exist there if the magnetic field lines were equipotentials. It follows that the integral of this fictitious electric field between two magnetic field lines in the ionosphere yields the true potential difference between these field lines above the U. Thus we can evaluate ϕ from the motions of auroral vortices

$$\phi = \int v_\perp' B_o \, dy$$

where the integration is from outside the auroral arc to its center. The "referred" electric field takes on additional meaning since it can be used directly to compute the potential of the V-shock.

Observations

In the previous section it was shown that a negative field-aligned charge sheet, as envisioned in models of V- or U-shaped equipotential surfaces, is subject to the Kelvin-Helmholtz instability. Even though the charge sheet itself (and its associated perpendicular electric field) terminate well above the ionosphere, its temporally varying configuration is imprinted on the ionosphere by the accelerated electrons that originate within the charge sheet. The apparent motions of auroral vortices (curls, rays) reflect the physical motion of the plasma at altitudes above the U. The ray velocity is the velocity that the ionospheric plasma would assume if the high-altitude electric fields mapped to the ionosphere according to the rules for "frozen-in" fields.

One could reasonably expect, based on the configuration of electrostatic pairs observed from the S3-3, satellite that auroral arcs would evolve into pairs of vortex streets having the following properties.

1) Counterclockwise shapes and motions (viewed anti-parallel to B)
2) Wavelengths of less than 10 km
3) Apparent velocities of tens of kilometers per second
4) Relative motion between the two vortex streets that is counter clockwise relative to a point between the streets
5) Vortices that are never stable
6) Vortices that never unwind

These six characteristics of auroral arc distributions were described from observations by Hallinan and Davis (1970). This led to the inference of large high altitude electric fields by arguments somewhat the reverse of those used in this paper. Ten years of subsequent observation have reinforced our confidence in these six characteristics. In particular, we have seen no exceptions to points 1, 4, 5 and 6. Points 1 and 4 identify the sign of the space charge (negative) while 5 and 6 demonstrate that the instability is absolute rather than convective.

These six observations do not apply to the large-scale auroral spirals (Davis and Hallinan, k976). The spirals have been attributed to distortions in the geomagnetic field in the presence of a Birkeland current (Hallinan, 1976).

There seems to be a generally good agreement between the observations of auroral curls and of paired electrostatic shocks. To examine this more closely, let us consider a model auroral curl system that incorporates several typically observed features. The system (auroral arc) has a horizontal thickness of 10 km and is bounded by two arc elements having thicknesses of 1 km. These arc elements, somewhat brighter than the arc as a whole, are distorted into arrays of counterclockwise vortices. The vortices in the two systems drift in opposite directions (eastward in the poleward element and westward in the equatorward form). The velocity in each case is 10 km/ sec.

The electric field and the electrostatic potential (referred to 100 km) for this model arc are shown in Fig. 2. The two dots are given by the ±10 km/sec velocities of the curls. The solid lines have slopes given by the vorticity (ω_s) inferred from the observed growth rate of the curls. The spatial dimensions are those of the auroral form. Outside the visible forms, the observational technique is useless, so we have to guess at the electric field in these regions (dashed curves). The potential curve (Fig. 2c) is derived by integrating Fig. 2b.

It is encouraging to note the similarity between Fig 2b. and the paired electrostatic shocks observed from the S3-3 satellite. The agreement includes both magnitude (as referred to 100 km) and direction.

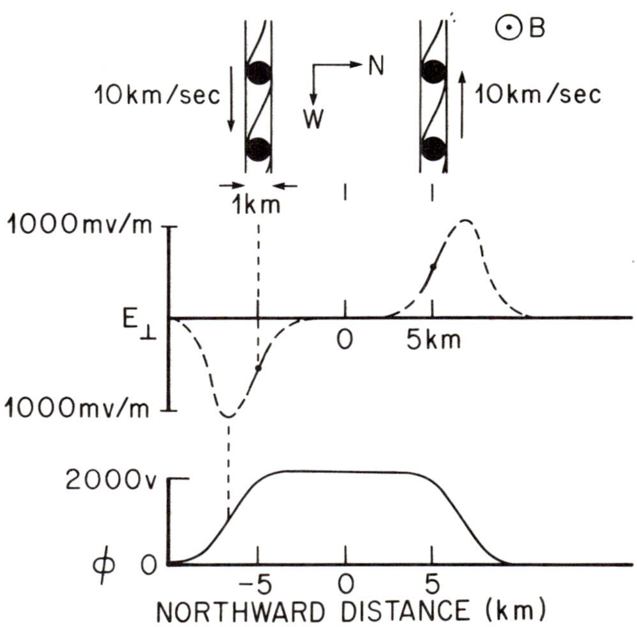

Figure 2: Schematic illustration of model auroral arc bounded by two rayed arcelements (curl systems) moving in opposite directions (top). Middle: The southward electric field above the region of parallel electric fields (but referred to the 100 km level) is determined from the ray velocities (dots) and the rotational velocity of the vortices (solid lines). The dashed lines are a reasonable extension of the solid lines. Bottom: The electrostatic potential is found by integrating across the electric field curve. This is also the potential available for accelerating electrons toward the ionosphere.

Discussion

Mozer (this volume) reported that the paired electrostatic shocks are a common auroral zone feature, being observed on approximately one out of three satellite passes. Since it is likely that most active rayed arcs are curl systems, the curl systems are also common. (Good observations are less common since the observations can be made only when the arc is close to the observer's magnetic zenith).

The "paired electrostatic shocks" are the high-altitude manifestation of U or V shaped potential-structure. It is very difficult to see how such a potential structure could exist without producing a curl system in the visible aurora below. It is equally difficult to see how an auroral curl system could arise without a corresponding fluid shear layer somewhere on the flux tube above. Such a layer would have an electric field configuration strikingly similar to those observed at high altitude.

The observed auroral ray motions imply perpendicular electric fields (referred to the 100 km level) that commonly exceed 1v/m. This is consistent with the satellite measurements. The observations of ray velocities also imply that the electric field is directed inward toward the arc (negative space charge). Again this is consistent with the satellite measurements and with the Oosik high altitude barium tracer (Wescott et al. 1975).

Intriguingly, even though the large perpendicular electric fields do not penetrate into the ionosphere, the auroral ray motions observed in the ionosphere are an accurate reflection of these fields. It is for this reason, not understood at the time, that the existence of the electric fields was recognized from the auroral shapes and motions nearly a decade before their in situ measurement.

It seems clear that high-altitude field-aligned charge sheets play a key role in the morphology of auroral arcs. They are responsible both for the formation of auroral rays (vortices) and for the large apparent velocities of the rays. They also provide a potential drop along some field lines that accelerates electrons toward the ionosphere. This may account for many features of particle distributions measured above auroral arcs.

In the example shown in Fig. 2, the potential through which the electrons are accelerated is somewhat over 2 kv. Since this example assumed a ray velocity of 10 km/sec while velocities of over 20 km/sec are not uncommon, potentials of 4-5 kev would be plausible. It is also interesting to note that the potential exists across the entire arc, not just in the two narrow charge sheets on the edges.

However, some caution is appropriate. In Fig. 2 it is only the two dots (ray velocities) and the two solid line segments (observed vorticity) that are derived directly from the auroral observations. The dashed curves are plausible extensions, but there are other ways they could be drawn that would lead to greater or smaller calculated potentials. Computer simulations such as those reported by Wagner (this volume) may clarify the quantitative relationship between the electrostatic potential and the visible vortex structures by showing how the electric field behaves outside the vortex street.

A more serious problem, however, is that auroral arcs exhibit the formation of rays and fast internal motions only sporadically during their lifetimes. The charge-sheet model serves admirably well as an explanation for rayed arcs --so well that it cannot easily be pressed into service as an explanation for non-rayed arcs. Yet non-rayed arcs (or non-rayed portions of arcs) may be the rule rather than the exception.

It is particularly disturbing that even large-scale bright auroral spirals do not always include rays and fast internal motions. The spirals are thought to result from strong upward field-aligned currents (Webster and Hallinan, 1973; Hallinan, 1976). It has often been suggested that double layers or similar phenomena, result from field-aligned currents. It would seem that spirals and their connecting links would be the best place to look for indications of space-charge regions. Yet these indications are often lacking.

Perhaps these apparent difficulties can be resolved. Meanwhile, even if V or U-shaped potential structures are not the total explanation for auroral arcs, they are evidently a common occurrence and are apparently responsible for much of the fascinating detailed morphology of active rayed arcs.

Acknowledgements. This work was supported by the National Science Foundation - Division of Atmospheric Sciences.

References

Akasofu, S.-I., The auroral rays, J. Atmos. Terr. Phys., 25, 163-165 1962.

Block, L. P., Potential double layers in the ionosphere, Cosmic Electrodynamics, 3, 349-376, 1972.

Buneman, O., R. H. Levy and L. Linson, The stability of crossed-field electron beams, J. Applied. Phys. 37, 3203-3222, 1966.

Carlqvist, P., and R. Bostrom, Space-charge regions above the aurora, J. Geophys. Res., 75, 71407145, 1970.

Cutler, C. C., Instability in hollow and strip electron beams, J. Appl. Phys. 27, 1028-1029, 1956.

Davis, T. N. and T. J. Hallinan, Auroral spirals, I: Observations, J. Geophys. Res., 81, 3953-3958, 1976.

Gartlein, C. W., Visual observation of the aurora, Annal de Geophysique, 15, 31-38, 1959.

Hallinan, T. J., and T. N. Davis, Small-scale auroral arc distortions, Planet. Space Sci., 18, 17351744, 1970.

Hallinan, T. J., Auroral spirals, II: Theory, J. Geophys. Res. 81, 39593965 1976.

Kan, J. R. and S.-I. Akasofu, A model of the auroral electric field, J. Geophys. Res., 84, 507-512, 1979.

Kern, J. W. and E. H. Vestine, Theory of auroral morphology, J. Geophys. Res., 66, 713-723, 1961.

Kyhl, R. L. and H. F. Webster, Breakup of hollow cylindrical electron beams, IRE Trans. Electron Devices, 3, 172-183, 1956.

Levy, R. H. and R. W. Hockney, Computer experiments on low density crossedfield electron beams, Phys. Fluids, 11, 766-771-1968.

Mozer, F. S., C. W. Carlson, M. K. Hudson, R. B. Torbert, B. Parady, J. Yatteau and M.C. Kelley, Observations of paired electrostatic shocks in the polar magnetosphere, Phys. Rev. Lett. 38, 292-298, 1977.

Mozer, F. S., C. A. Cattell, M. K. Hudson, R. L. Lysak, M. Temerin and R. B. Torbert, Satellite measurements and theories of low altitude auroral particle acceleration, Space Sci. Rev., 27, 155-213, 1980.

Swift, D. W., H. C. Stenbaek-Nielsen and T. J. Hallinan, An equipotential model for auroral arcs, J. Geophys. Res., 81, 3931-

Wagner, J., Simulation study of V-potential double layers and auroral arc formation, this volume. Webster, H. F., Structure in magnetically confined electron beams. J. Appl. Phys., 28, 1395-1397, 1957.

Webster, H. F. and T. J. Hallinan, Instabilities in charge sheets and current sheets and their possible occurrence in the aurora, Radio Science, 8, 475482, 1973.

Wescott, E. M., H. C. Stenbaek-Nielsen, T. N. Davis, W. B. Murcray, H. M. Peek and P. J. Bottoms, The L = 6.6 OOSIK barium plasma injection experiment and magnetic storm of March 7, 1972, J. Geophys. Res., 80, 951-967, 1975.

Wescott, E. M., H. C. Stenbaek-Nielsen, T. J. Hallinan, T. N. Davis and H. M. Peek, The Skylab barium plasma injection experiment, II. Evidence for a double layer, J. Geophys. Res., 81, 4495-4502, 1976. 3934, 1976.

SPECTROSCOPIC DIAGNOSTICS OF THE FORMATION OF AURORAL ARCS

M. H. Rees

Geophysical Institute, University of Alaska, Fairbanks, Alaska 99701

Introduction

Spectroscopic observations of auroral arcs have been used principally to study auroral excitation mechanisms, aeronomy in the auroral atmosphere and ionosphere, and the interaction of energetic charged particles with the ambient neutral and ionized gases. Our understanding of these processes has reached a level sufficient to allow us to use auroral spectroscopy to help identify some characteristics associated with the formation of auroral arcs: (1) the species of energetic particles, (2) the energy input rate, (3) variation in the characteristic energy of particles, (4) the time history of buildup and decay of an arc, (5) development of fine-scale structure, and (6) the background conditions or precursors to the formation of an arc.

The Basic Concepts

The brightness of an optical radiation is a measure of the energy input rate into the atmosphere. Specific atomic lines and molecular bands identify the emitting species. If the transition that produces an emission feature is optically allowed, or if the effective lifetime of an excited state is short the temporal evolution of the emission reflects variations in the source of the excitation, i.e. changes in the particle flux incident on the atmosphere. Emission rate ratios of various spectroscopic features can be used to infer the average energy of the exciting auroral particles, as will be illustrated below.

Bombardment of the atmosphere by energetic protons is the only important source of observed emissions of atomic hydrogen, Balmer alpha ($H\alpha$) and Balmer beta ($H\beta$) from the ground and the Lyman series by rocket-borne detectors. (Electron impact excitation of ambient neutral hydrogen is negligible, by comparison.) The required excitation cross sections for the relevant reactions, electron capture

$$H^+ + M \rightarrow H(n = 4,3,2) + M^+ \qquad (1)$$

and excitation

$$H + M \rightarrow H(n = 4,3,2) + M \qquad (2)$$

have recently been measured in the laboratory (Van Zyl and Neumann, 1980). The symbols M and M^+ represent any of the neutral and ionized atmospheric molecules and atoms, N_2, N_2^+, O, O^+, etc. The proton stream that initially enters the atmosphere is efficiently neutralized by electron capture so that reaction (2) can be the dominant excitation source, depending on the energy of the protons. The absolute emission rate of $H\alpha$ or $H\beta$ yields the proton energy deposition rate while the ratio $H\alpha/H\beta$ indicates the average energy of the proton flux. At a spectral resolution of an Ångstrom or higher the Doppler profile of the hydrogen emission yields information on the velocity distribution of the radiating atoms. Figure 1 shows the "blue-shifted" and broadened $H\alpha$ line viewed in the magnetic zenith and the unshifted but broadened line viewed on the magnetic horizon, perpendicular to the field lines. The difference shows that radiating H atoms are traveling at high velocity toward the observer along the magnetic field.

In addition to reaction (2) energetic hydrogen atoms may cause stripping,

$$H + N_2 \rightarrow H^+ + M + e \qquad (3)$$

and ionization,

$$H + N_2 \rightarrow H + M^+ + e \qquad (4)$$

These reactions are sources of ejected electrons that have an initial energy in excess of the thermal energy of ambient electrons and are capable of producing additional ionization and excitation,

$$e + M \rightarrow e + M^+ + e \qquad (5)$$

$$e + M \rightarrow e + M^* \qquad (6)$$

The asterisk in equation (6) signifies an excited state. Indeed, a fraction of the product M^+'s or M's in all the reactions (1) to (6) will be in an excited state, the efficiency depending on the energy of the impacting hydrogen atom or ion. If M^{+*} represents the excited $N_2^+(B^2\Sigma_u)$ state reactions (1) and (4) become sources of

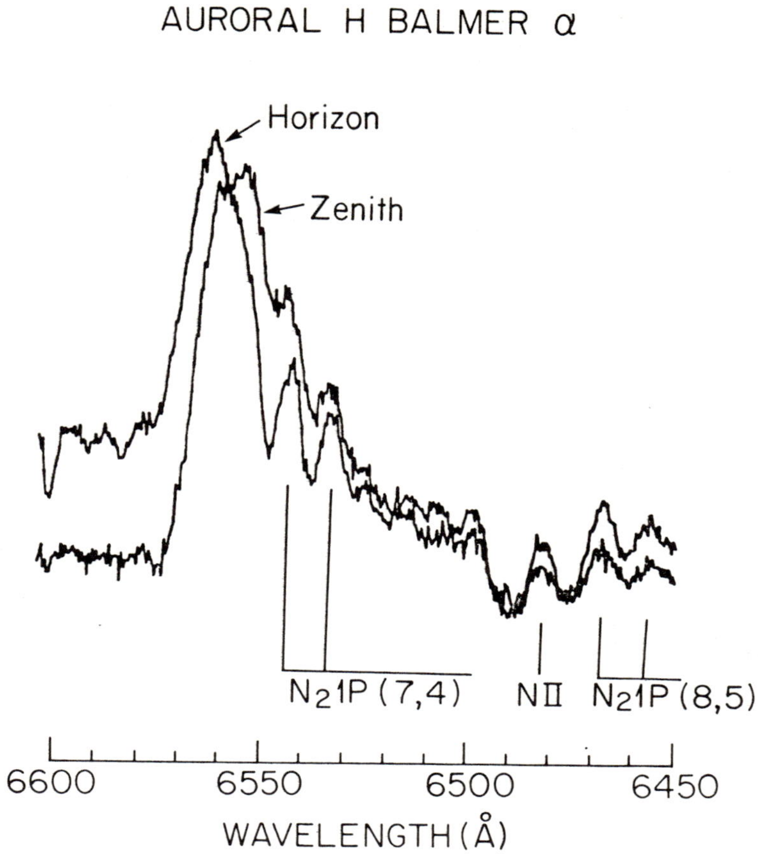

Fig. 1. Doppler profiles of the hydrogen Balmer alpha line viewed along (zenith) and perpendicular (horizon) to the magnetic field. Some nonhydrogenic emission features are identified. The observations were obtained at the Poker Flat, Alaska, optical observatory on 20 January 1977 at about 7 p.m. local time using a one meter Ebert-Fastie scanning spectrophotometer.

the First Negative bands of the nitrogen ion (3914 Å, 4278 Å, etc.). The ejected electrons in reactions (3) and (4) are characterized by a spectrum that falls off sharply with increasing energy, favoring excitation processes (6) over ionization (5), since ionization cross sections have maxima at energies ten times higher (or more) than excitation cross sections.

Reactions (5) and (6) also describe correctly the ionization and excitation of atmospheric gases by energetic auroral electrons, primaries producing mostly ionization and secondaries causing excitation. Electrons, however, do not have an optical signature of their own, as do protons, and their presence and characteristics must be inferred by measuring emissions from atmospheric species (Roble and Rees, 1977; Rees and Luckey, 1974).

Two particularly useful optical emission features produced by auroral electrons are the First Negative bands of molecular nitrogen ions,

$$e + N_2 \rightarrow N_2^+(B^2\Sigma_u) + 2e$$

$$N_2^+(B^2\Sigma_u) \rightarrow N_2^+(X^2\Sigma) + h\nu(3914\ \text{Å},\ 4278\ \text{Å},\ \text{etc.})$$

and the ground configuration transition in singly ionized atomic oxygen

$$e + O \rightarrow O^+(^2P) + 2e$$

$$O^+(^2P) \rightarrow O^+(^2D) + h\nu(7320\ \text{Å},\ 7330\ \text{Å})$$

Subtracting the 4278 Å emission due to proton precipitation from the total observed emission rate yields the contribution due to electron precipitation (Kasting and Hays, 1977) from which the energy input rate associated with primary auroral electron bombardment can be computed (Rees and Luckey, 1974). The First Negative bands originate from an optically allowed transition and the excited state is not collisionally deactivated in the thermosphere. The altitude profile of the

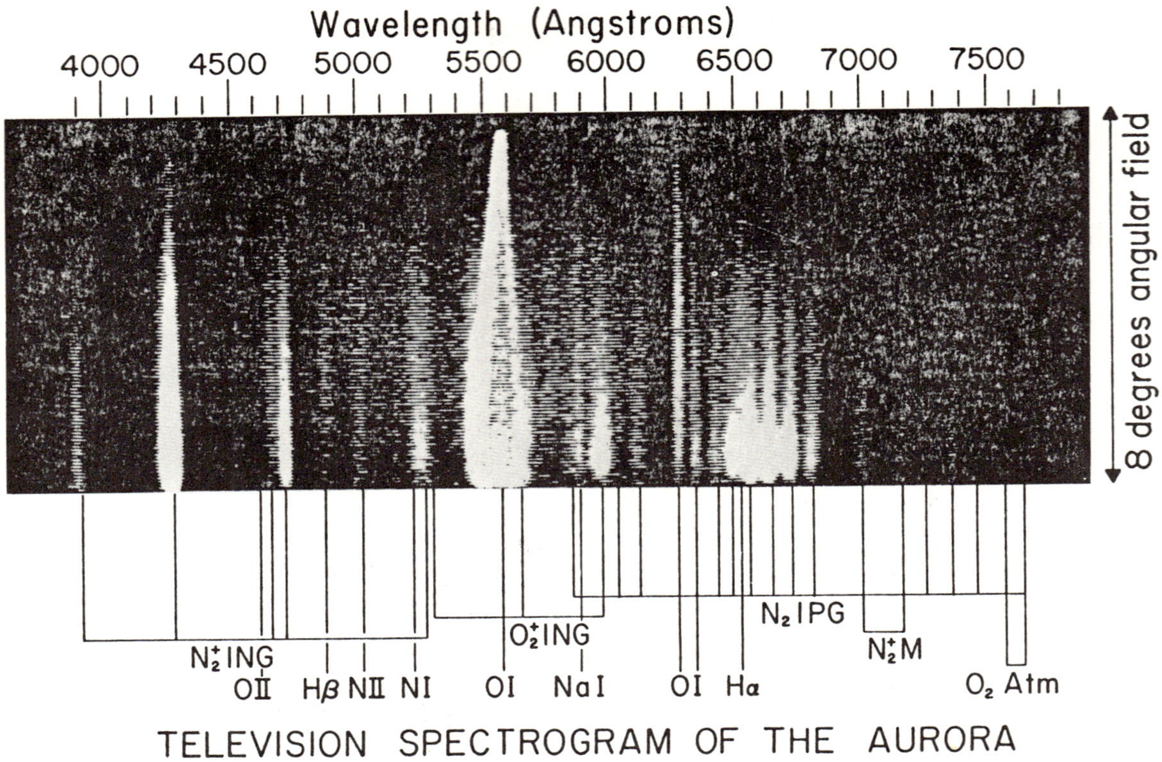

Fig. 2. Television spectrogram obtained by C. S. Deehr at Ester Dome, Alaska, on 20 February 1974. This instrument was pointed at an auroral arc located at 20° elevation in the southern sky. The exposure was one second. Principal auroral emission features are identified.

emission rate therefore accurately reflects the energy deposition rate profile. By contrast, the $O^+(^2P)$ state is metastable (5 second radiative lifetime) and collisional deactivation by N_2 produces an effective height profile of the 7320 Å emission that has a maximum at about 250 km, while the nitrogen emission in normal nighttime aurora peaks between 100 and 130 km. This difference is, of course, also due, in part, to the changing composition of the neutral atmosphere with altitude. The emission rate ratio can therefore be used to infer a mean energy of the bombarding electron flux.

Energy deposition by auroral electrons is usually much larger than the auroral proton energy input but under some circumstances the contributions from the two species can be comparable. Thus, auroral arcs are generally associated with energetic electrons but a possible role for the protons should not be ignored by hypotheses on the formation of arcs. We also note that while regions of hydrogen emission extend over a latitude range that is frequently broader than an identifiable auroral arc this spreading effect is a consequence of the interaction of the proton stream with the atmosphere and does not exclude the existence of narrow beams of protons above the scattering medium comparable in scale to electron beams.

The Observational Aspects

The ideal instrument for spectroscopic studies bearing on the formation of auroral arcs has the following characteristics. Adequate spectral resolution and sensitivity to detect and map the Doppler profile of the weak hydrogen radiations. A spectral range that includes the N_2^+(1NG) band at 4278 Å at the short wavelength end and the $O^+(^2P-^2D)$ transition at 7320 Å at the long wavelength limit. In this wavelength interval lie the H α and H β lines at 6563 Å and 4864 Å respectively. The spectral resolution must be sufficient to isolate the O^+ line from vibration-rotation emission lines of the OH radical in the airglow and rotational lines of a First Positive vibrational band of N_2. Good spatial resolution is required, at least in a meridian plane in order to study spectroscopic differences across an auroral arc. A "fast" instrument (high sensitivity and signal-to-noise ratio) is required to study the temporal evolution of an arc. Together with high spatial resolution, the

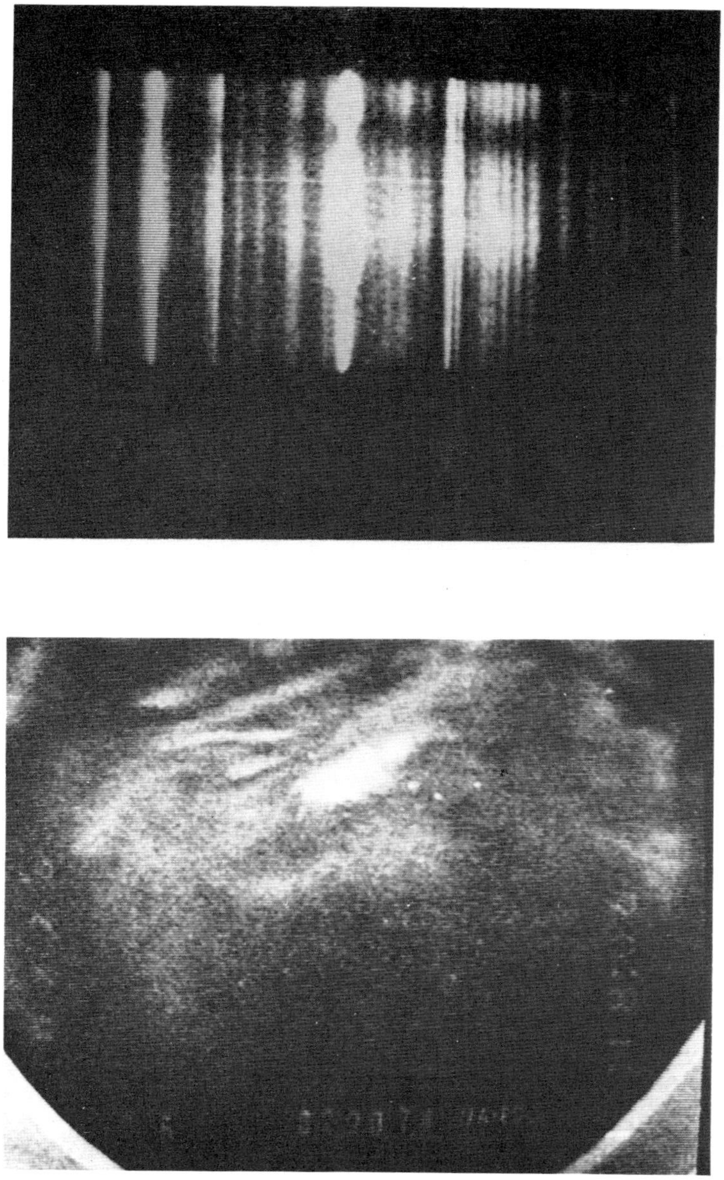

Fig. 3. Auroral corona in the magnetic zenith at Ester Dome, Alaska, obtained on 20 February 1974. The upper portion of the figure shows a spectrogram of a narrow slice across the center of the two-dimensional image at the bottom, showing various segments of the corona.

inception, buildup, motion, and decay of an arc can be followed.

The technology exists to implement the required optical instrument but none has yet been built, to my knowledge. A large aperture spectrometer coupled to an image intensified array detector would be one possible design. There are probably others. As early as 1974, C. S. Deehr at the Geophysical Institute produced television spectrograms of the aurora; an example is shown in Figure 2. A fast objective lens and a transmission grating were used to produce the auroral spectrum recorded by an image orthicon television camera. The wavelength resolution was only 30 Å, insufficient to satisfy the criteria outlined above, but the spatial resolution and the sensitivity were good. The televison image did not yield quantitative measurements. The imaging technique yields both wavelength and spatial information, the latter in one plane only. In order to study the two-dimensional morphology of the aurora and its spectral characteristics a

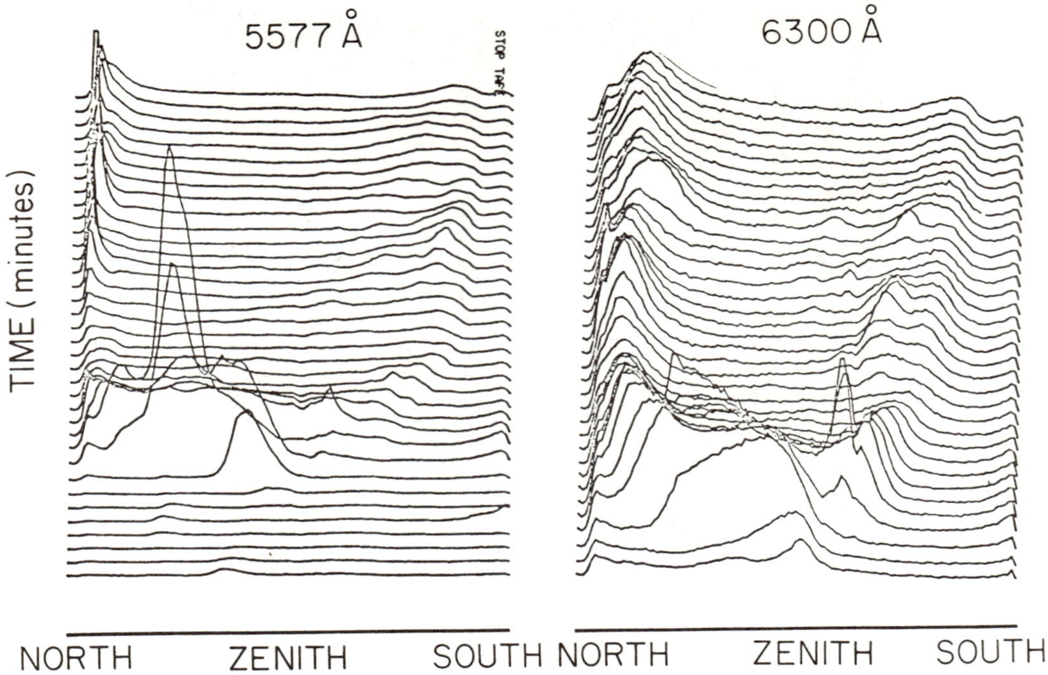

Fig. 4. Sequence of meridian scanning photometer traces obtained at foursecond intervals. The morphology and spatial extent of auroral arcs is shown in the two oxygen emissions, 5577 Å and 6300 Å, the latter originating at a higher level in the atmosphere. Differences between the two optical features are interpreted (see text) in terms of the particle energy.

television image and a television spectrogram were obtained simultaneously, as shown in Figure 3. Individual segments of the corona are well resolved in the spectrogram. The figure shows a single set of frames extracted from a continuous recording of the auroral event, at sixty frames per second for the white light image and at one frame every four seconds for the spectrogram. In my view, much could be learned about the origin of auroral arcs by using instruments that yield absolute emission rates of auroral spectroscopic features in the context of the preceding discussion.

Meridian scanning photometers that were operated during the International Magnetospheric Study (IMS) have yielded spatial/spectral data of selected auroral emission features. The photometers incorporate a narrow passband interference filter capable of isolating prominent auroral radiations: the atomic oxygen lines at 6300 Å and 5577 Å, the 4278 Å N_2^+(1NG) band and the hydrogen H β line. The field-of-view of the photometers is one degree and the telescopes are mounted to view the same region of the sky. A mirror system scans the geomagnetic meridian in four seconds from horizon to horizon. Thusly, aurora in an eight to ten degree latitudinal segment of the oval is viewed and recorded once every four seconds. The photometers are absolutely calibrated to yield the column emission rate or surface brightness. Figure 4 illustrates the capability of this instrument. Auroral arcs are manifested by the increased emission rate above the background (diffuse) aurora. Periodic waxing and waning of two arcs, as well as latitudinal motion, occur in the 5577 Å emission just prior to a rapid increase in the brightness of one of the arcs. The 6300 Å emission, however, while also associated with both arcs initially extends over a wider latitude range. The difference is attributed to a hardening of the electron spectrum in the center of the arc as it becomes brighter, which translates into energetic particle language as an "invertedV" event. Spectroscopic measurements can be used to study the inception, development, lifetime, and decay of "inverted-V" events by continuous ground-based observations.

Discussion and Conclusion

While <u>in situ</u> satellite observations are essential in identifying electric fields associated with auroral arcs, and any electromagnetic radiations that do not propagate to the ground, several characteristics of the energetic particle population can be studied by ground-based spectroscopy (the term is used in a broad context). As theories for the formation of auroral arcs evolve we may look forward to time-dependent models. Ground-based observations of temporal effects may provide input for the development of such models as well as serve to test theoretical predictions. Do protons have a role in the formation of arcs? What is the topology of the potential field across the arc as manifested by the resulting particle energy? Are there identifiable precursors or background conditions or threshold effects necessary for the formation of arcs? Spectroscopic observations have the potential for providing partial answers to these and other questions related to the formation of auroral arcs.

Acknowledgments. My thanks to C. S. Deehr for the use of Figures 2 and 3 and to G. J. Romick for Figure 4. Financial support through NSF grant ATM 79-23266 and NASA grant NSG 5219 is gratefully acknowledged.

References

Kasting, J. F. and P. B. Hays, A comparison between the N_2^+ 4278 emission electron flux in the auroral zone, J. Geophys. Res. 82, 3319, 1977.

Rees, M. H. and D. Luckey, Auroral electron energy derived from ratio of spectroscopic emissions, 1. model computations, J. Geophys. Res. 79, 5181, 1974.

Roble, R. G. and M. H. Rees, Time dependent studies of the aurora: effects of particle precipitation on the dynamic morphology of ionospheric and atmospheric properties, Planet. Spa. Sci. 28, 991, 1977.

Van Zyl, B. and H. Neumann, H and H emission cross sections for low-energy H and H^+ collisions with N_2 and O_2, J. Geophys. Res. 85, 11 6006, 1980.

REVIEW OF AURORAL PARTICLE PRECIPITATION

R. L. Arnoldy

Space Science Center, University of New Hampshire, Durham, N. H. 03824

Abstract. Auroral electron and ion precipitation measured by sounding rockets and satellites will be reviewed. The two observing platforms provide a different perspective of the aurora. An attempt will be made to relate the two data sets to establish the precipitation signature of an auroral arc. Energy spectra and pitch angle distributions of precipitating particles have provided the basis for most auroral acceleration mechanisms. The altitude and particle mass dependence of these parameters, as well as their spatial relationship to the visual aurora will be reviewed. Although optical data have shown the fine structure (temporal and spatial) of an arc, improved response of particle detectors is just beginning to reveal this structure in particle precipitation. For example, intense fluxes of field-aligned electrons are often observed having a duration on the order of a second as seen by rocket detectors. Similar field-aligned bursts fluctuate at a few Hertz in satellite data. The intent of this paper is to summarize the observations that may aid in the formulation of an auroral arc theory.

Introduction

Particle precipitation associated with discrete auroral arcs will be reviewed in this paper emphasizing those observations which are pertinent to theories of auroral arc formation. Because of the large quantity of auroral particle precipitation data available, only observations by rockets and satellites generally made in the evening sector will be considered. Recent reviews of hot magnetospheric ions and electrons associated with optical auroras have been made by Hultqvist (1979, 1974). DMSP data have made is possible to examine on a regular basis optical auroral features as they are associated with electron precipitation (Meng et. al., 1978). This work will be discussed in detail in an accompanying paper (Meng, 1981).

This review will be organized according to observations made in three regions of space: Region 1 accessible to rockets and low altitude satellites (few hundred to a few thousand kilometers); Region 2 extending from 4000 → 8000 km (S3-3 satellite range); and Region 3 near the equatorial plane (geosynchronous satellite measurements).

Region 1

Inverted V Precipitation

The signature of an auroral arc in electron precipitation data is an inverted V structure in an electron spectrogram (Heikkila, 1970; Frank and Ackerson, 1971). In an inverted V the peak in the electron spectrum (monoenergetic peak) increases in energy toward the arc center resulting in an inverted V structure in an energy versus time spectrogram. Arcs or inverted V's are typically a few hundred km in latitudinal width when detected aboard satellites, whereas rocket observations, capable of better spatial resolution, reveal inverted V structures down to several km in width. There is some confusion in terminology in the literature concerning the identification of arcs with inverted V structures. Satellite papers refer to arcs imbedded in inverted V's, whereas from the perspective of rocket measurements any auroral form steady enough in time to traverse with a rocket payload has an inverted V structure. The difference appears to be one of measurement resolution. In fact, rockets have yet to identify the small scale structure (tens of meters) associated with optical arcs (Maggs and Davis, 1968). Finally, the relationship of electrostatic shocks seen by the S3-3 satellite (Mozer et. al., 1977) to inverted V's or auroral arcs also remains unclear at the present time.

Inverted V's occur on open as well as closed field lines. Figure 1 gives the latitudinal and longitudinal distribution of inverted V's seen by the AE-D satellite (Lin and Hoffman, 1979). They primarily occur in the evening sector auroral oval but are also present in the polar cap where field lines are normally considered to be open. Using the definition that field lines are closed if they can support trapped radiation, inverted V's detected at satellite altitudes occur on both open and closed field lines (Frank and Ackerson, 1972; Mizera et. al., 1976; Eliasson et. al., 1978; Lin and Hoffman, 1979). At rocket altitudes, however, inverted V's are generally seen coinci-

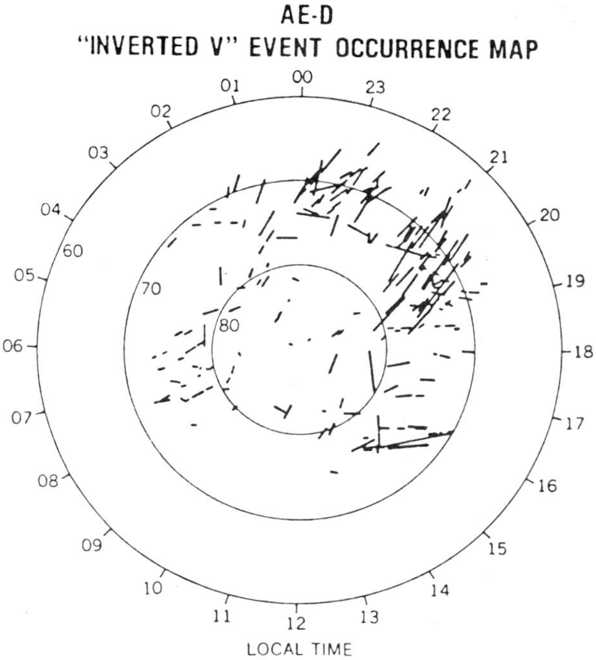

Fig. 1. Spatial occurrence map of 280 inverted V events. Local time and circles of constant invariant latitude are shown as reference (from Lin and Hoffman, 1979).

dent with a mirroring population of energetic ions and/or electrons suggesting closed field lines (Venkataragan et. al., 1975; Whalen and Daley, 1979).

The monoenergetic peak in the inverted V electron spectrum is generally taken as evidence for electron acceleration by a quasi-static electric potential. There are a number of observations, however, that suggest the inverted V electrons have passed through a turbulent region. The pitch angle distribution of the monoenergetic peak electrons is isotropic minus the loss cone at low altitudes (Arnoldy et. al., 1974; Lin and Hoffman, 1979). The temperature of precipitating electrons with energies above the spectral peak is proportional to the peak energy suggesting that they are heated in the acceleration region (Lin and Hoffman, 1979). The monoenergetic peak of inverted V's represents a stable electron distribution when observed at rocket altitudes (Kaufmann et. al., 1978) as the example in Figure 2 shows. Finally, the V-shaped pitch angle distributions observed by the AE-D satellite for energies below the monoenergetic peak has been explained as due to a field-aligned component combined with isotropic backscattered electrons and a significant number of electrons with energies slightly below the peak energy trapped between their mirror points and an electric potential structure (Lin and Hoffman, 1979).

The field-aligned electrons in an inverted V occur at energies below the monoenergetic peak producing double peaked spectra as seen in Figure 3 (Arnoldy et. al., 1974; Lin and Hoffman, 1979; Kaufmann, 1980). The 5 keV peak in the spectra of Figure 3 is isotropic in pitch angle, whereas the 2 → 3 keV peak is field-aligned and observed only at small pitch angles. This field-aligned component will be discussed in more detail later in this review.

Ion Precipitation

The ESRO I data showed that ions and electrons simultaneously precipitated on the same field line, but keV ion fluxes and spectra were less

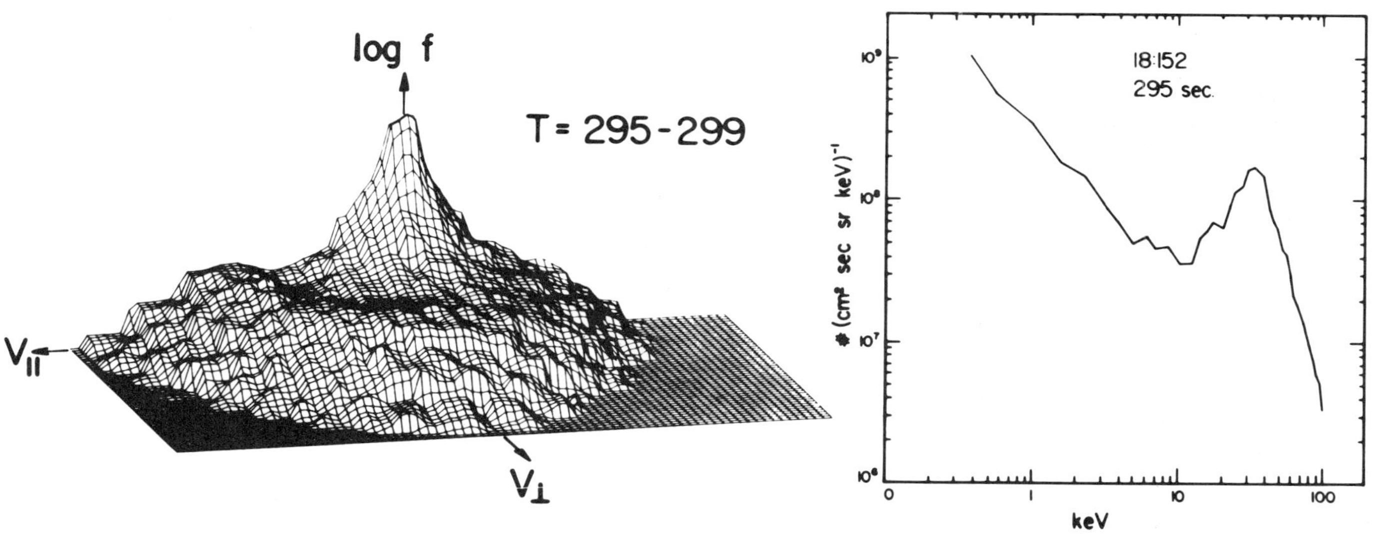

Fig. 2. Example to show how a peak in the differential energy spectrum is not necessarily an unstable structure in the distribution function f.

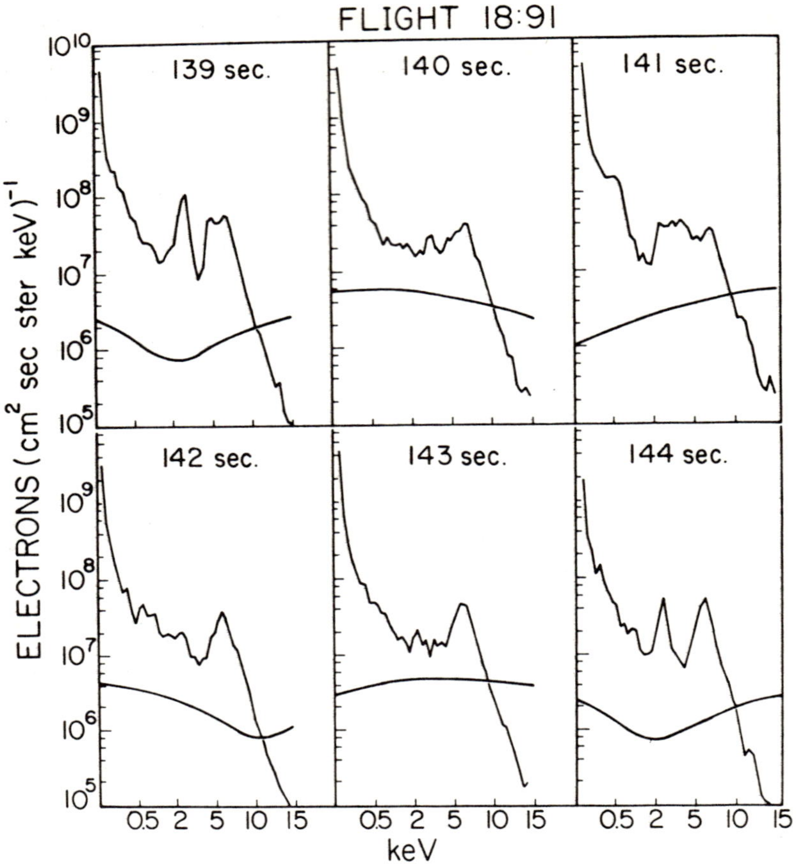

Fig. 3. Sequential electron spectra from flight 18:91. Pitch angle of the electrons is given by the smooth curve. The lower two decades of the flux scale are used to give pitch angles from 1° to 100° (from Arnoldy et. al., 1974).

variable than those of the electrons in the same energy range (Hultqvist, 1979). The ESRO I data also showed that keV electrons were frequently field-aligned in the same region as ions, however, more sporadic and narrower in space/time than the ion field-aligned fluxes. Field-aligned keV ions occur at the edge of inverted V's with spectral softening at the center and a tendency toward isotropy (Hultqvist, 1971). In contradiction to the ESRO I results cited above, the ESRO 4 and INJUN 5 data show no field-aligned ions and rarely any hot ions in inverted V's, respectively (Frank and Ackerson, 1972; Eliasson et. al., 1979). Hultqvist (1971) has interpreted the different results in terms of a solar cycle effect. ESRO I, operating during solar maximum, observed both ions electrons accelerated by a potential above the satellite. The potential accelerating the ions occurred below the electron potential and was created by the precipitating electrons interacting with an elevated neutral atmosphere during solar maximum conditions.

Rocket measurements of ion precipitation are more limited because of the low altitude obtained by many of the flights. Spectral and intensity changes are generally anticorrelated when ions and electrons precipitate simultaneously (Bryant et. al., 1977; Edwards et. al., 1978). A notable exception is the case of general ion and electron anticorrelation until the center of an electron inverted V is reached where suddenly a band of 6 keV ions appeared simultaneously with the electrons (Bryant et. al., 1977). Ion data from a number of Canadian rocket flights (Miller and Whalen, 1976) show relative independence of energy spectra from auroral activity. Pitch angle distributions were observed to be either isotropic over the upper hemisphere or trapped, and there was, in general, little correlation with electron precipitation.

Field-Aligned Precipitation

As discussed above, the monoenergetic peak in the electron spectrum is generally isotropic in pitch angle. The electrons discussed here are sporadic bursts of intense field-aligned precipitation. The field-aligned electrons are usually less than a few keV in energy and are below the monoenergetic peak in the spectrum (Arnoldy et. al., 1974; Lin and Hoffman, 1979). Figure 4 shows a flux ratio spectrogram of a satellite inverted V event. The heavy line tracing the spectral peak lies in a region of general pitch angle isotropy whereas the field-aligned fluxes are transient and of lower energy. The field-aligned fluxes generally occur over a range of energies (Arnoldy et. al., 1974; Raitt and Sojka, 1977; Lin and Hoffman, 1979).

Field-aligned electrons are generally seen during active auroras (McEwen, 1972; Arnoldy et. al., 1974; Bosqued et. al., 1974) and are localized on edges of arcs in regions between different plasmas (Whalen and McDiarmid, 1972; Bryant et. al., 1973; Maehlum and Moestue, 1973; Arnoldy, 1974; Lundin, 1974; Pulliam et. al., 1980). Figure 5 gives two energy channels of a rocket electron detector along with ground photometer measurements in the lower panel. The rapid oscillations in the 1.4 keV detector at 195 and 310 seconds are a spin modulation as the detector swept through the field-aligned precipitation localized near the northern and southern borders of the aurora.

The field-aligned precipitation occurs in bursts or fluctuates at a few Hertz (McEwen, 1971; Arnoldy et. al., 1974; Bosqued et. al., 1974; Boyd and Davis, 1977; Raitt and Sojka, 1977) If fluctuating, all energies field-aligned fluctuate in phase (Lin and Hoffman, 1979). This later observation rules out an energy dependent process for producing the field-alignment. Finally, the field-alignment is generally within 20° of the field line at rocket altitudes and on occasion less than 10° (Bryant et. al., 1973; Arnoldy et. al., 1974; Bosqued et. al., 1974). To maintain such a degree of field-alignment the mechanism could not be too far removed from the measurements. This combined with their short-lived nature and field-alignment over a range of energies suggests that the phenomena is the result of the acceleration of ambient electrons in a large scale transient electric field. The relationship of the field-aligned fluxes to the inverted V acceleration region remains unclear at the present time.

Low Energy Electron Precipitation

The electron differential directional intensity below the monoenergetic peak in the precipitating electron spectrum can be fit to a power law as $E^{-\gamma}$ where γ varies between 1 and 2 (Reasoner and Chappell, 1973; Arnoldy and Choy, 1973. Over an inverted V or a discrete auroral arc these low energy electrons are generally isotropic over the full 180° range of pitch angles. Evans (1974) has interpreted this result in terms of an electric field trapping secondaries and degraded primaries between the potential region and the atmosphere. Recent efforts by Pulliam et. al., (1980) to fit this model to rocket data have been quite successful.

Recent measurements over a diffuse aurora have revealed similar isotropic low energy electrons (Evans and Moore, 1979). Since the diffuse aurora electron precipitation has no peak in the spectrum the electrons apparently have not fallen through a potential region. There is, therefore, no mechanism to reflect the backscattered electrons at 1 → 2 earth radii above the aurora. To achieve the measured isotropic distributions, Evans and Moore (1979) have postulated that the backscattered electrons fill the entire flux tube and as a result of pitch angle scattering could be a source for the plasma sheet electron population. Evans and Moore (1979) have modeled the plasma sheet spectrum and presumably the spectrum precipitated during diffuse aurora are due to a primordial Maxwellian (Figure 6) and the atmospheric

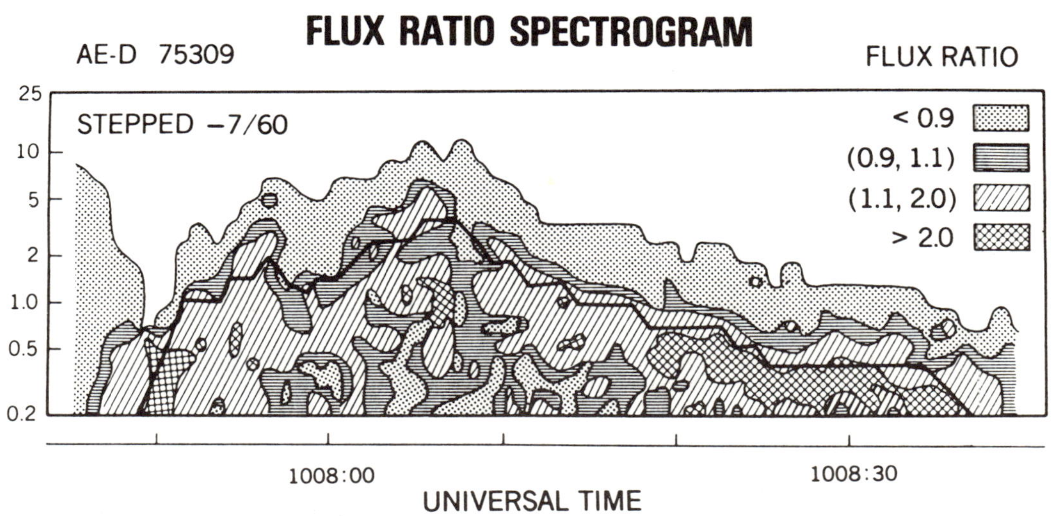

Fig. 4. Spectrogram displaying the flux ratios J(−7°)/J(60°) of an inverted V event. The flux ratios are separated into four categories as shown in the upper right corner (from Lin and Hoffman, 1979).

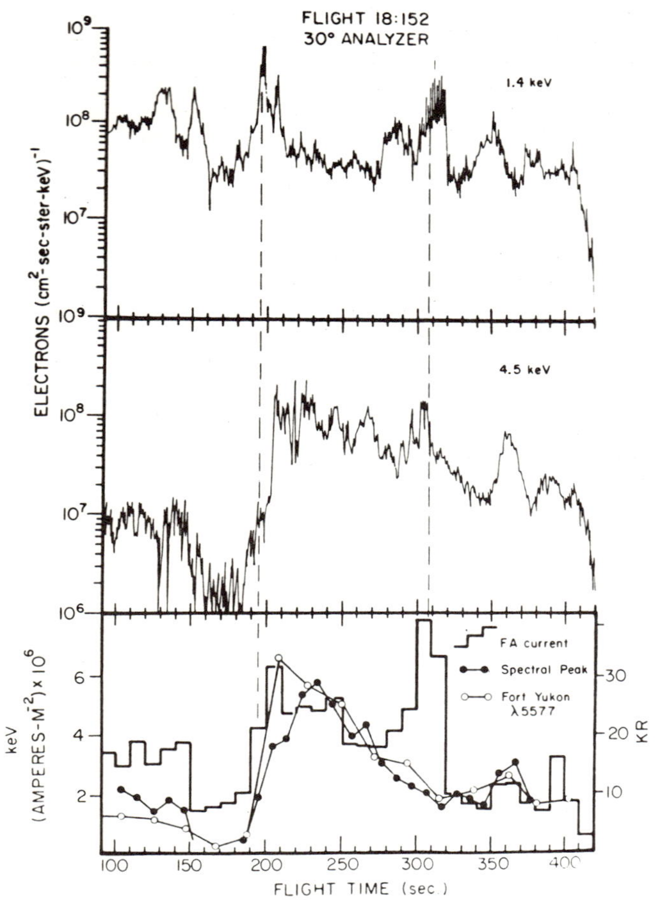

Fig. 5. Top two panels, directional intensities 1.4 and 4.5 keV electrons. Bottom panel, ground meridian-scanning photometer looking at foot of rocket field line, energy of peak current (from Arnoldy, 1977).

backscattered electrons that are in equilibrium with this Maxwellian. In the model spectrum the Maxwellian electrons dominate at high energies and the atmospheric electrons dominate at low energies giving a spectrum which approximates that measured quite well. This topic will be discussed further under Region 3.

When not over visible aurora the low energy precipitated electrons often remain but are not isotropic in pitch angle. Figure 7 gives down and up-going electron energy spectra measured by a sounding rocket in the region between two visible auroras. There is a definite downward streaming population of suprathermal electrons between 100 and 500 eV energy which are not secondarily related to any higher energy component. Figure 8 gives data from another rocket flight where a very narrow arc (spatially coincident with the 3 keV fluxes) was imbedded in widespread low energy precipitation. Above the arc, at 145 seconds flight time, the electrons below 1 keV were isotropic in pitch angle while outside the arc, at 200 seconds there was a net downward streaming of the less than 1 keV electrons. This net precipitation of less than 1 keV electrons carries significant current out of the atmosphere and could represent the high energy tail of the precipitation responsible for the large scale evening sector Region 1 currents except when over an arc where the Region 1 current is carried predominantly by the accelerated energetic electrons.

There is a growing body of evidence that suprathermal electrons are rather easily heated by wave interactions. Measurements from both the Polar V and Echo III electron accelerator payloads give an intense background of suprathermal electrons that remained after the energetic beam injections. These electrons could not be interpreted in terms of beam secondaries nor as scattered and degraded beam primary electrons (Maehlum et. al., 1979; Arnoldy and Winckler, 1980). The accelerator payloads are known, however, to be intense sources of wave turbulence when operated in the upper ionosphere (Cartwright and Kellogg, 1974). Such turbulence could heat the particle population near the payload. Johnson and Sojka (1980) cite a case

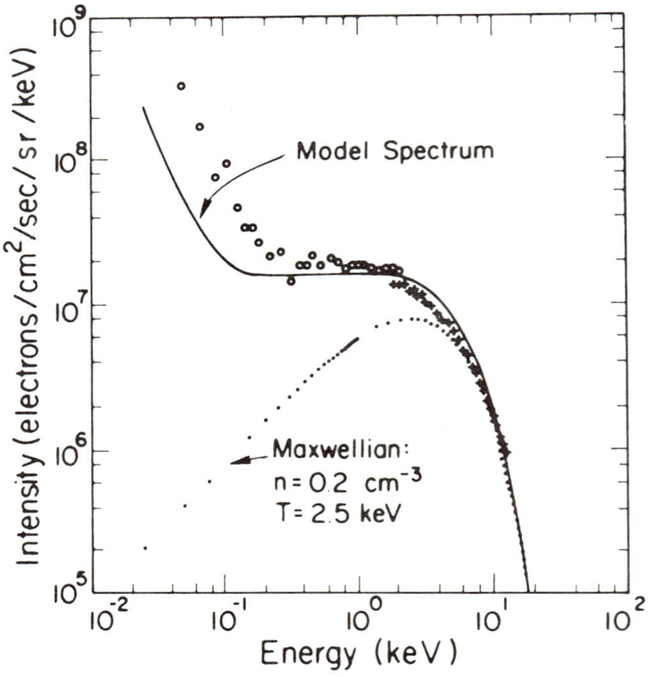

Fig. 6. Measured electron spectrum during a diffuse aurora together with a model spectrum which is the sum of a Maxwellian and the distribution of atmospherically produced backscattered and secondary electrons that is in equilibrium with this Maxwellian. The energy spectrum for the Maxwellian of temperature 2.5 keV and density 0.2 cm^{-3} is also plotted for comparison (from Evans and Moore, 1979).

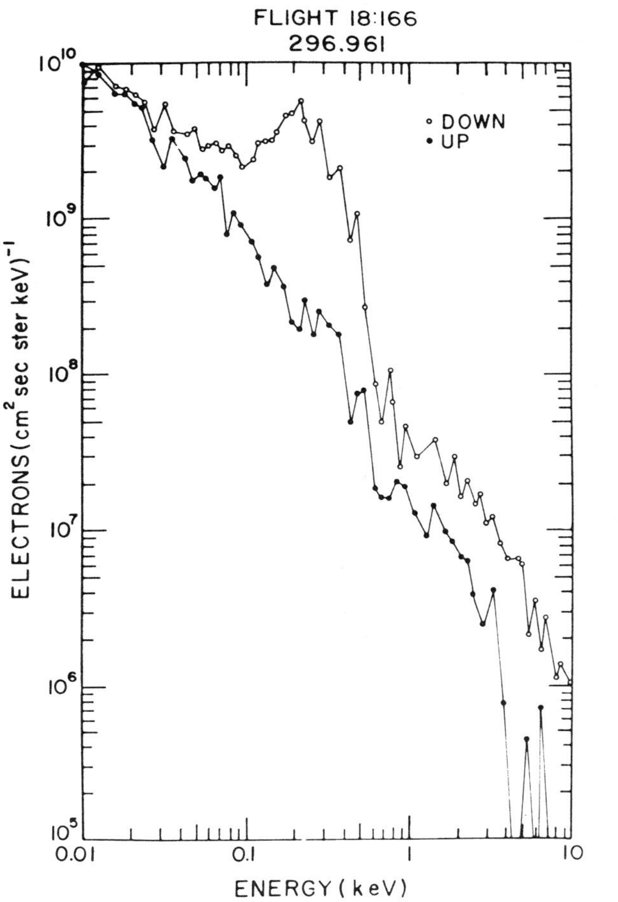

Fig. 7. Spectra of electrons moving up and down the magnetic field line measured simultaneously by rocket detectors.

of ambient thermal electrons being heated to temperatures as high as 10^5 K by the passage of a field-aligned beam of a few hundred eV through the ionosphere at altitudes over 660 km. A beam/plasma interaction is cited as the heating mechanism. Finally, recent results we have obtained from the release of an argon plasma at 200 km altitude show considerable modification of the naturally occurring precipitation below 1 keV. Figure 9 gives some preliminary electron measurements from this experiment. An isotropic population of electrons at a few hundred eV is introduced as well as a component below 100 eV streaming down the field lines. These electrons were closely correlated with plasma injections when wave instruments aboard the payload also measured greatly enhanced intensities. Such modifications of the low energy component can alter the light emission and electrodynamic properties of the upper ionophere from that expected on the basis of atmospheric collisions above.

Region 2

The S3-3 spacecraft surveying the region of space from 4000 → 8000 km above the auroral arc has provided considerable evidence for large scale electric fields along the magnetic field from pitch angle and energy distributions of electrons and ions. The satellite is generally found to be immersed in an electric field ∿ 1 volt/km extending over a few thousand kilometers (Chiu and Schulz, 1978; Cladis and Sharp, 1979). We now briefly review the different types of particle population measured by the S3-3 mission, not all of which are readily interpretable in terms of quasi-steady field-aligned electric field.

UFI (Upward Flowing Ions)

UFI with a directional intensity $J > 2 \times 10^6$ $(cm^2 \text{ sec ster keV})^{-1}$ are a common feature seen on approximately 60% of the S3-3 orbits. They coincide with the statistical auroral oval and have a strong dawn-dusk asymmetry with a maximum in the dusk sector (Ghielmetti et. al., 1978). The predominant species in the UFI are H^+ and O^+ ions (Shelley et. al., 1976). UFI and electrostatic hydrogen cyclotron waves coincide in over 90% of the events studied (Kintner et. al., 1979). Finally, UFI are observed to have a beam pitch angle distribution of ($\Theta \sim 180°$) or a conic pitch angle distribution ($90° < \Theta < 180°$).

UFI Beams

UFI beams have a maximum frequency of occurrence in the dusk sector and are observed primarily above 5000 km altitude (Gorney et.al., (1980). Beams are observed primarily in the region of upward field-aligned current and coincide with electrostatic shocks in the evening sector (Cattell et. al., 1979). It would appear that the beams are a phenomena associated with discrete auroral forms.

UFI Conics

Conics are observed at lower altitudes than beams with a tendency to prefer the higher altitudes during disturbed times (Gorney et. al., 1980). During magnetically quiet times conics tend to be a daytime cusp phenomena becoming more uniformly spread in local time as the activity increases (Ghielmetti et. al., 1979). Conics are observed in all field-aligned current regions (Cattell et. al., 1979) hence appear to be less associated with inverted V's than beams. Conics are apparently the same ions as the transversely accelerated ions (TAI) seen at lower altitudes (Whalen et. al., 1978; Klumpar, 1979; Ungstrup et. al., 1979).

DFI (Downward Flowing Ions)

DFI occur much less frequently than do UFI in the auroral region and do not display a dawn-dusk

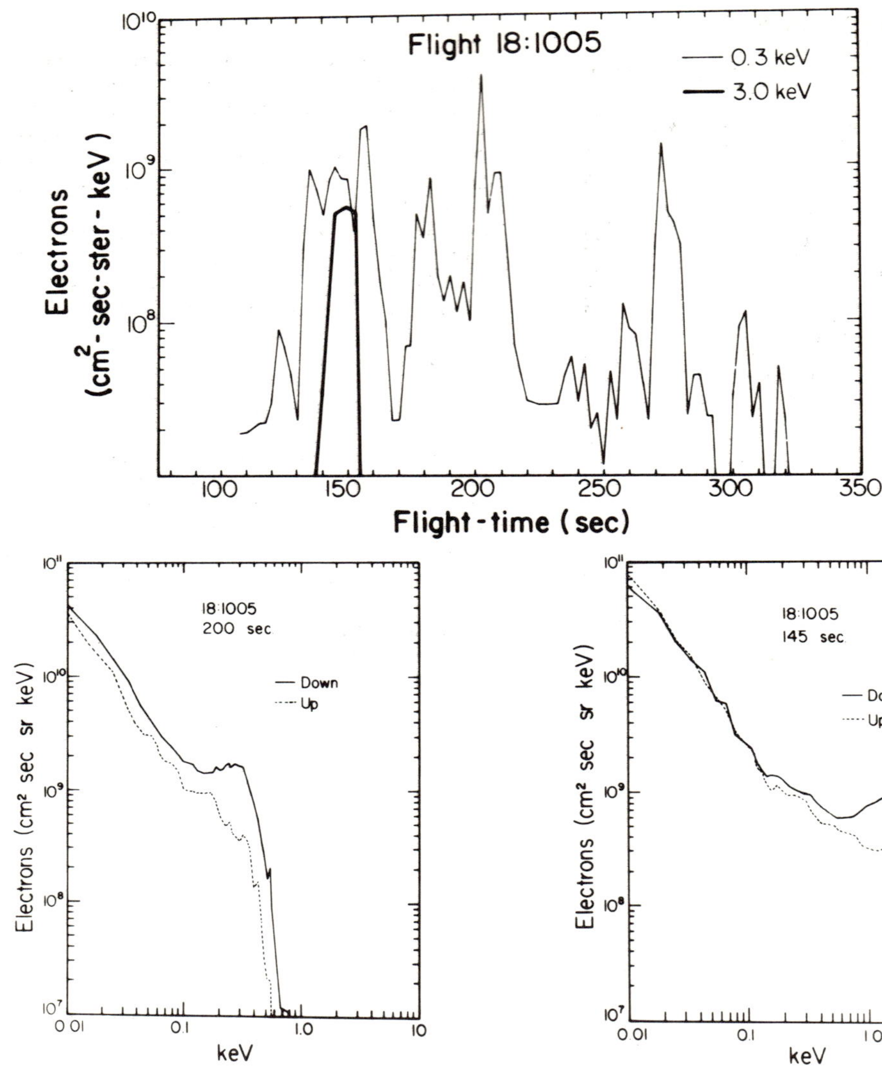

Fig. 8. Rocket measurements of the precipitation of 0.3 and 3 keV electrons and simultaneous up and down spectra taken at two times during the flight. Visible aurora coincided only with the 3 keV precipitation bursts centered on 150 seconds flight time.

asymmetry (Ghielmetti et. al., 1979). DFI occur equatorward of UFI and in regions of inward (evening) and outward (morning) field-aligned currents (Cattell et. al., 1979), hence appear to be associated with the diffuse aurora. DFI are correlated with trapped keV ions and occur preferentially in spatially localized regions of enhanced hot keV plasma ions. Because of this, Ghielmetti, et al. (1979) have associated DFI with the injection of ring current ions in the equatorial plane. Some DFI are associated with retarded electron precipitation suggesting an electric field acceleration opposite to that in inverted V's (Ghielmetti et. al., 1979). The first example of an ion inverted V was presented at this conference by Burgess and Reiff (1980).

Counterstreaming Electron Beams

Counterstreaming electron beams have energies in the keV range. They are often spatially associated with regions of UFI conics (Fennell et. al, 1979). They are field-aligned to within several degrees and are generally comparable in width for both upward and downward beams (Sharp et.al.,1980). The energy spectra for counterstreaming electron beams are broad and generally unstructured. In the region of counterstreaming electron beams, the electron beams, the loss cone of background electrons is not affected, ruling out a quasi-static parallel electric field. Sharp et. al. (1980) have suggested that the counterstreaming electron beams are locally accelerated ambient

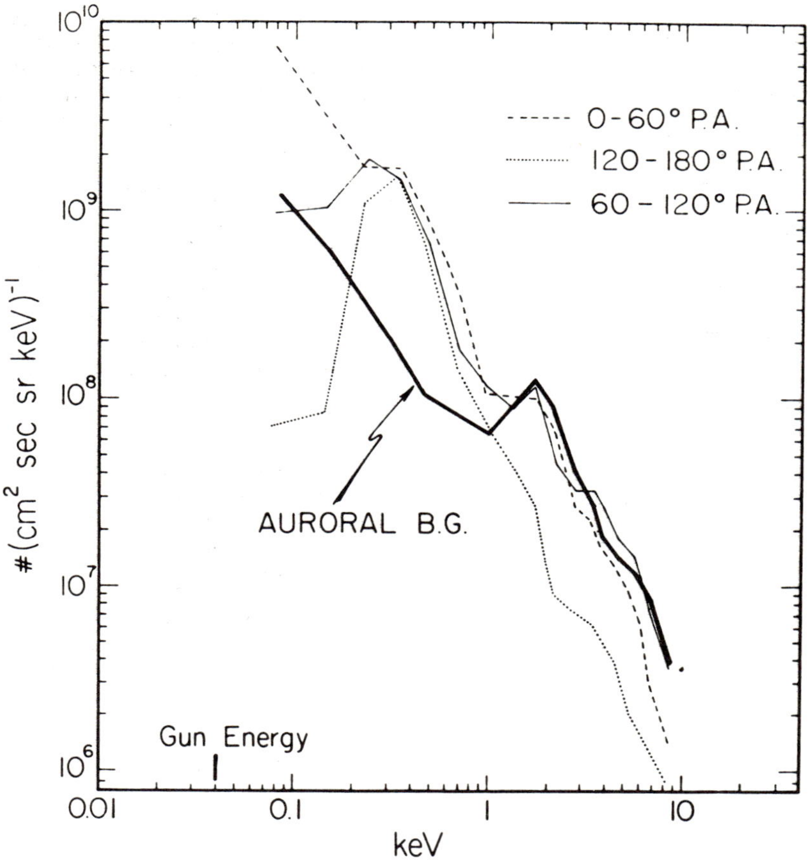

Fig. 9. Background auroral spectrum (heavy line) and three simultaneous spectra taken by detectors scanning the pitch angle ranges indicated during an argon plasma release. The auroral spectrum is dramatically altered below 1 keV.

electrons by flickering double layers. Counter-streaming electron beams appear to be similar in many respects to the field-aligned electron bursts seen by low altitude satellites and rockets discussed above.

Region 3

Plasma injections at synchronous orbit are the equatorial plane signature of any auroral substorm. Although much correlative work between synchronous satellite observations and ground (Akasofu, 1977) and low altitude satellite measurements (Sharp et. al., 1971; Meng et. al., 1979; Zanetti et. al., 1980) has been done, how one region maps to the other is still unclear. Moreover, the physics of the injection of particles at synchronous orbit remains to be defined. That there is an exchange of particles between the equatorial plane and the ionosphere is certain, however, the degree to which the ionosphere provides plasma to the equatorial plane (plasma sheet) is yet to be fully explored. As we have seen, measurements at $\sim 1\ R_E$ (Region 2) reveal many particles leaving the ionosphere. Presumably these particles are those that have been detected at synchronous altitude as "source cone" ions and electrons (McIlwain, 1974; Mauk and McIlwain, 1975). It is to be noted, however, that the source cone particles observed have a wider pitch angle spread than the actual loss cone at 6.6 R_E. If the particles were accelerated in Region 2 they must have been diffused in pitch angle before reaching the equator.

An example of an electron injection at synchronous orbit is given in Figure 10. At 0412 UT the intensity of the higher energy electrons increased by one or two orders of magnitude while the low energy electrons (< few keV) decreased by about as much. The injection represented a dramatic change in spectrum from a monotonically decreasing one to nearly a Maxwellian. Following the injection, the spectrum then gradually returned to approximately its old shape and to pre-injection intensity levels. If an aurora was initiated as a result of this injection, then the gradual return of low energy electrons might represent the filling of the plasma sheet with backscattered

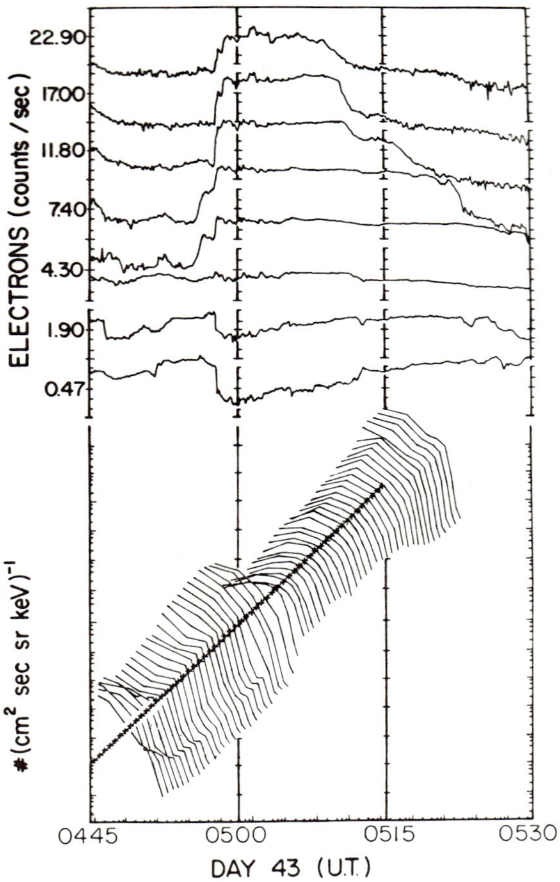

Fig. 10. ATS-6 electron spectra as a function of time during a substorm injection event. The top eight traces give count rates for eight energy channels as a function of time. The energy in keV is given to the right of each trace.

and secondary electrons from the auroral ionosphere as suggested by Evans and Moore (1979).

Concluding Remarks

This review of auroral particle precipitation raises three questions about auroral arc formation. First, what is the role of AC phenomena in auroral particle acceleration? Examples of bursts of field-aligned electrons, simultaneous ion and electron precipitation, counterstreaming electrons and conics (transversely accelerated ions) all suggest AC phenomena. Second, what is the importance of electrons < 1 keV in the auroral beam? Clearly they are not all secondaries. Inverted V's appear to be imbedded in and possibly develop as a result of this precipitation. Finally what role does the lower ionosphere have on the formation of arcs? Recent controlled experiments with sounding rockets which have altered the lower ionospheric conductivity often have resulted in some form of anomalous particle precipitation. In addition, there is a growing evidence for fairly low altitude acceleration of particles occurring naturally, such as bursts of low energy field-aligned electrons and transversely accelerated ions.

Acknowledgments. This work was supported by the National Aeronautics and Space Administration under Grant NSG-6022 and by the Division of Atmospheric Sciences, National Science Foundation under Grant ATM-7920484. Many helpful discussions the author has had with Professor Richard Kaufmann and Dr. Thomas Moore at the University of New Hampshire, is gratefully acknowledged.

References

Akasofu, S.-I., Physics of the magnetospheric substorm, D. Reidel, Boston, 1977.
Arnoldy, R.L., and L.W. Choy, Auroral electrons of energy less than 1 keV observed at rocket altitudes, J. Geophys. Res., 78, 2187, 1973.
Arnoldy, R.L., P.B. Lewis, and P.O. Isaacson, Field-aligned auroral electron fluxes, J. Geophys. Res., 79, 4208, 1974.
Arnoldy, R.L., The relationship between field-aligned current, carried by suprathermal electrons, and the auroral arc, J. Geophys. Res., Ltrs., 4, 407, 1977.
Arnoldy, R.L., and J. Winckler, The hot plasma environment and floating potentials of an electron beam-emitting rocket in the ionosphere, accepted, J. Geophys. Res., 84, 1980.
Bosqued, J.M., G. Cardona, and H. Réme, Auroral electron fluxes parallel to the geomagnetic field lines, J. Geophys. Res., 79, 98, 1974.
Boyd, J.S., and T.N. Davis, Rocket measurements of electrons in a system of multiple auroral arcs, J. Geophys. Res., 82, 1177, 1977.
Bryant, D.A., G.M. Courtier, and G. Bennett, Electron intensities over two auroral arcs, Planet. Space Sci., 21, 165, 1973.
Bryant, D.A., D.S. Hill, D.R. Lepine, and R.W.N. Mason, Electrons and positive ions in an auroral arc, Nature, 266, 148, 1977.
Burgess, G.A., and P.H. Reiff, Particle acceleration of an ion inverted V, Program of AGU Chapman Conference on the Formation of Auroral Arcs, 1980.
Cartwright, D.G., P.J. Kellogg, Observations of radiation from an electron beam artificially injected into the ionosphere, J. Geophys. Res., 79, 1439, 1974.
Cattell, C., R. Lynch, R.B. Torbert, and F.S. Mozer, Observations of differences between regions of current flowing into and out of the ionosphere, Geophys. Res. Ltrs., 6, 621, 1979.
Chiu, Y.T., and M. Schulz, Self-consistent particle and parallel electrostatic field distributions in the magnetosphere--ionosphere auroral region, J. Geophys. Res., 83, 629, 1978.
Cladis, J.B., and R.D. Sharp, Scale of electric field along magnetic field in an inverted V

event, J. Geophys. Res., 84, 6564, 1979.

Edwards, T., D.A. Bryant, and M.J. Smith, Preliminary results of particle measurements from Skylark SL 1421 and Fulra F1, paper presented at MIST, Southampton, 1978.

Eliasson, L., L.A. Holmgren, and K. Ronnmark, Pitch angle and energy distributions of auroral electrons measured by the ESRO 4 satellite, KGI Reprint, 1978.

Eliasson, L., L.A. Holmgren, and D. Hultqvist, Observations of keV ions in inverted V events by means of the ESRO 4 satellite, KGI Report, 1979.

Evans, D.S., Precipitating electron fluxes formed by a magnetic field-aligned potential difference, J. Geophys. Res., 39, 2853, 1974.

Evans, D.S., and T.E. Moore, Precipitating electrons associated with the diffuse aurora: evidence for electrons of atmospheric origin in the plasma sheet, J. Geophys. Res., 84, 6451, 1979.

Fennell, J.F., P.F. Mizera and D.R. Croley, Jr., Observations of ion and electron distribution during the July 29 and July 30, 1977 storm period, Proceeding of Magnetospheric Boundary Layer Conference, Alpach, 1979.

Frank, L.A., and K.L. Ackerson, Observations of charged particle precipitation into the auroral zone, J. Geophys. Res., 76, 3612, 1971.

Frank, L.A. and K.L. Ackerson, Local time survey of plasma at low altitudes near the auroral zone, J. Geophys. Res., 77, 4116, 1972.

Ghielmetti, A.G., R.G. Johnson, R.D. Sharp, and E.G. Shelley, The latitudinal diurnal, and altitudinal distributions of upward flowing energetic ions of ionospheric origin, Geophys. Res. Ltrs., 5, 59, 1978.

Ghielmetti, A.G., R.D. Sharp, E.G. Shelley, and R.G. Johnson, Downward fluxing ions and evidence for injections of ionospheric ions into the plasma sheet, J. Geophys. Res., 84, 5781, 1979.

Gorney, D.J., A. Clarke, D.R. Croley, J.F. Fennell, J.M. Luhmann, and P.F. Mizera, The distribution of ion beams and conics below 8000 km, Aerospace Report # ATR-80 (7744)-1, 1980.

Heikkila, W.J., Satellite observations of soft particle fluxes in the auroral zone, Nature, 225, 369, 1970.

Hultqvist, B., On the production of magnetic field-aligned electric field by the interaction between the hot magnetospheric plasma and the cold ionosphere, Planet. Space Sci., 19, 749, 1971.

Hultqvist, B., Rocket and satellite observations of energetic particle precipitation in relation to optical aurora, Ann. Geophys., 30, 223, 1974.

Hultqvist, B., The hot ion component of the magnetospheric plasma and some relations to the electron component--observations and physical implications, Space Science Review, 23, 581, 1979.

Johnson, A.D., and J.J. Sojka, A beam/plasma interaction in the high-altitude auroral ionosphere, accepted, Planet. Space Sci., 26, 1980.

Kaufmann, R.L., P.E. Dusenbery, and B.J. Thomas, Stability of auroral plasma: parallel and perpendicular propagation of electrostatic wave, J. Geophys. Res., 83, 5663, 1978.

Kaufmann, R.L., Electrostatic wave growth: secondary peaks in the measured auroral distribution function, J. Geophys. Res., 84, 1713, 1980.

Kintner, P.M., M.C. Kelley, R.D. Sharp, A.G. Ghielmetti, M. Temerin, C. Cattell, P.F. Mizera, and J.F. Fennell, Simultaneous observations of energetic (keV) upstreaming and electrostatic hyhydrogen cyclotron waves, J. Geophys. Res., 84, 7201, 1979.

Klumpar, D.M., Transversely accelerated ions: an ionospheric source of hot magnetospheric ions, J. Geophys. Res., 84, 4229, 1979.

Lin, C.S., and R.A. Hoffman, Characteristics of the inverted-V event, J. Geophys. Res., 84, 1514, 1979.

Lundin, R., Rocket observations of electron spectral and angular characteristics of an inverted V event, Planet. Space Sci., 24, 499, 1976.

Maehlum, B.N. and H. Moestue, High temporal and spatial resolution observations by low energy electrons by mother-daughter rocket in the vicinity of two quiescent auroral arcs, Planet. Space Sci., 21, 1957, 1973.

Maehlum, B.N., B. Grandal, T.A. Jacobsen and J. Troim, Polar 5--an electron accelerator experiment within an aurora 2, scattering of an artificially produced electron beam in the atmosphere, Planet. Space Sci., 28, 279, 1979.

Maggs, J.E., and T.N. Davis, Measurements of the thickness of auroral structures, Planet. Space Sci., 16, 205, 1968.

Mauk, B.H., and C.E. McIllwain, USCD auroral particle experiment, IEEE Trans. on Aerospace and Electronic Systems, AES-11, 1125, 1975.

Meng, C.-I., A.L. Synder, Jr., and H.W. Kroehl, Observations of auroral westward trailing surges and electron precipitation, J. Geophys. Res. 83, 575, 1978.

Meng, C.-I., B. Mauk, and C.E. McIllwain, Electron precipitation of evening diffuse auroral and its conjugate electron fluxes near the magnetospheric equator, J. Geophys. Res., 84, 2545, 1979. .

Meng, C.-I., Precipitation characteristics of auroral arcs detected by DMSP satellites, Proceedings of Formation of Auroral Arcs Chapman Conference, Fairbanks, Alaska, 1980, Akasofu ed., American Geophysical Union, Washington, D.C., 1981.

McEwen, D.J., Rocket measurements of low energy electrons during auroral events, Space Research XII, Bowhill, Jaffee and Rycoft, editors, Akademie-Verlag, Berlin, 1972.

McIllwain, C.E., Auroral electron beams near the magnetic equator, Physics of the hot plasma in the magnetosphere, editors, Hultqvist and Stenflo, Plenum Press, New York, 1975.

Miller, J.R., and B.A. Whalen, Characteristics of auroral proton precipitation observed from sounding rockets, J. Geophys. Res., 81, 147,1979.

Mizera, P.F., D.R. Croley, Jr., and J. Fennell, Electron pitch angle distributions in an inverted V structure, Geophys. Res., Ltrs., 3, 149, 1976.

Mozer, F.S., C.W. Carlson, M.K. Hudson, R.B. Torbert, B. Parady, J. Yatteau and M.C. Kelley, Observations of paired electrostatic shocks in the polar magnetosphere, Phys. Res. Ltrs., 38, 1977.

Pulliam, D.M., H.R. Anderson, K. Stamnes, and M. H. Rees, Auroral electron acceleration and atmospheric interactions, Rice Preprint, 1980.

Raitt, W.J., J.J. Sojka, Field-aligned suprathermal electron fluxes below 270 km in the auroral zone, Planet. Space Sci., 25, 5, 1977.

Reasoner, D.L. and C.R. Chappell, Twin payload observations of incident backscattered auroral electrons, J. Geophys. Res., 78, 2176, 1973.

Sharp, R.D., D.L. Carr, R.G. Johnson, and E.G. Shelley, Co-ordinated auroral-electron observations from a synchronous and a polar satellite, J. Geophys. Res., 76, 7669, 1971.

Sharp, R.D., E.G. Shelley, R.G. Johnson, and A.G. Ghielmetti, Counterstreaming electron beams at altitudes of $\sim 1 R_E$ over the auroral zone, J. Geophys. Res., 84, 92, 1980.

Shelley, E.G., R.D. Sharp, and R.G. Johnson, Satellite observations of an ionospheric acceleration mechanism, Geophys. Res. Ltrs., 3, 654, 1976.

Ungstrup, E., D.M. Klumpar, and W.J. Heikkila, Heating of ions to suprathermal energies in the topside ionosphere by electrostatic ion cyclotron wave, J. Geophys. Res., 84, 4289, 1979.

Venkatarangan, P., J.R. Burrows, and I.B. McDiarmid, On the angular distributions of electrons in inverted V substructures, J. Geophys. Res., 80, 1975.

Whalen, B.A., and I.B. McDiarmid, Observations of magnetic field-aligned auroral electron precipitation, J. Geophys. Res., 77, 191, 1972.

Whalen, B.A., W. Bernstein, and P.W. Daley, Low altitude acceleration of ionospheric ions, Geophys. Res., Ltrs., 5, 55, 1978.

Whalen, B.A., and P.W. Daley, Do field-aligned auroral particle distributions imply acceleration by quasi-static parallel electric fields? J. Geophys. Res., 84, 4175, 1979.

Zanetti, L.J., T.A. Potemra, J.P. Doering, J.S. Yee, and R.A. Hoffman, Coincident particle observations from AE-C and ATS-6 during the October 28, 1977, geomagnetic storm, J. Geophys. Res., 83, 4563, 1980.

AURORAL ARCS OBSERVED BY DMSP SATELLITES

Ching-I. Meng

Applied Physics Laboratory, The Johns Hopkins University,
Laurel, Maryland 20810

Abstract. This paper reports the electron precipitation characteristics of discrete auroras distributed over various parts of the polar region, based on the simultaneous optical auroral display observations and the electron precipitation measurements made by the DMSP series of satellites. It was found that all discrete arcs have a common feature, namely the existence of a peak in the electron differential energy spectra. The location of this spectral peak and the precipitated energy flux differs with the auroral brightness and geomagnetic activity. The precipitated energy flux varies from ~ 10 erg $cm^{-2}sec^{-1}$-sr^{-1} for bright active arcs, ~ 1-2 erg $cm^{-2}sec^{-1}sr^{-1}$ for normal quiet arcs to a fraction of 1 erg $cm^{-2}sec^{-1}sr^{-1}$ for faint arcs. The spectral peak is at $\gtrsim$ 8 keV for bright active arcs, ~ 3 keV for normal arcs and $\gtrsim$ 1 keV for faint arcs.

Introduction

Simultaneous observations of auroral displays and the precipitating particles are essential in understanding the physics of auroras. Measurements of precipitating charged particles from low-altitude polar-orbiting satellites were not commonly made with the simultaneous observations of optical auroras from ground-based and airborne photometers or all-sky cameras (see Arnoldy, 1974, for reviews and a complete list of references). Only recently have coordinated particle precipitation and auroral optical emissions made by a single satellite become possible; these observations have been reported by Shepherd et al. (1973) based on an ISIS-2 pass and Deehr et al. (1973) using the ESRO 1/Aurorae observations. Since 1972, the auroral imageries from satellites of the USAF Defense Meteorological Satellite Program (DMSP) with extensive coverage and excellent spatial resolution (~ 3.7 km) have been made available to the scientific community and these data have advanced our understandings of the auroral morphology (Akasofu, 1974). Some of the DMSP satellites also carried low-energy electron detectors. Therefore, it is possible for us to simultaneously measure electron precipitations and the produced visual auroral display in the polar atmosphere. Simultaneous auroral and precipitation observations by DMSP satellites have been reported (Mizera et al., 1975; Meng, 1976). This paper illustrates simultaneous discrete aurora and electron precipitation observations made by DMSP satellites at different locations in the polar region.

Analysis

Satellite observations of polar auroras have revealed two distinct types of auroras, discrete and diffuse (Akasofu, 1974). Discrete auroras occur along all parts of the auroral oval as well as over the polar cap, and they appear as single faint arc or bright strands with a width of about a few tens of kilometers. Recent study of the DMSP data shows that electrons which produce discrete auroras have, as a common feature, a monoenergetic peak in their differential electron energy spectra. Typical examples of the observations at different localities are shown here. Each of these examples consists of three diagrams (1) the auroral photograph taken by a DMSP satellite from about 830 km altitude, (2) the precipitation profile across the auroral region detected by the same satellite along the satellite path which is indicated by a straight line near the middle of the auroral photograph, and (3) the variations of electron differential energy spectra along the subtrack over the main auroral display of interest discussed in these examples. The examples discussed in the following were observed by the DMSP-32 and 33 satellites with a six-channel electrostatic analyzer pointed in the anti-nadir direction (i.e., within about 20° of the magnetic zenith) measuring the precipitated electrons between 200 eV and 20 keV (see Meng, 1976 for a detailed description).

Evening Oval Arcs

The first example is a faint evening arc observed on 23 October 1974 by DMSP-32 satellite (pass 1065). In the DMSP auroral imagery, a faint and geometrically simple discrete auroral

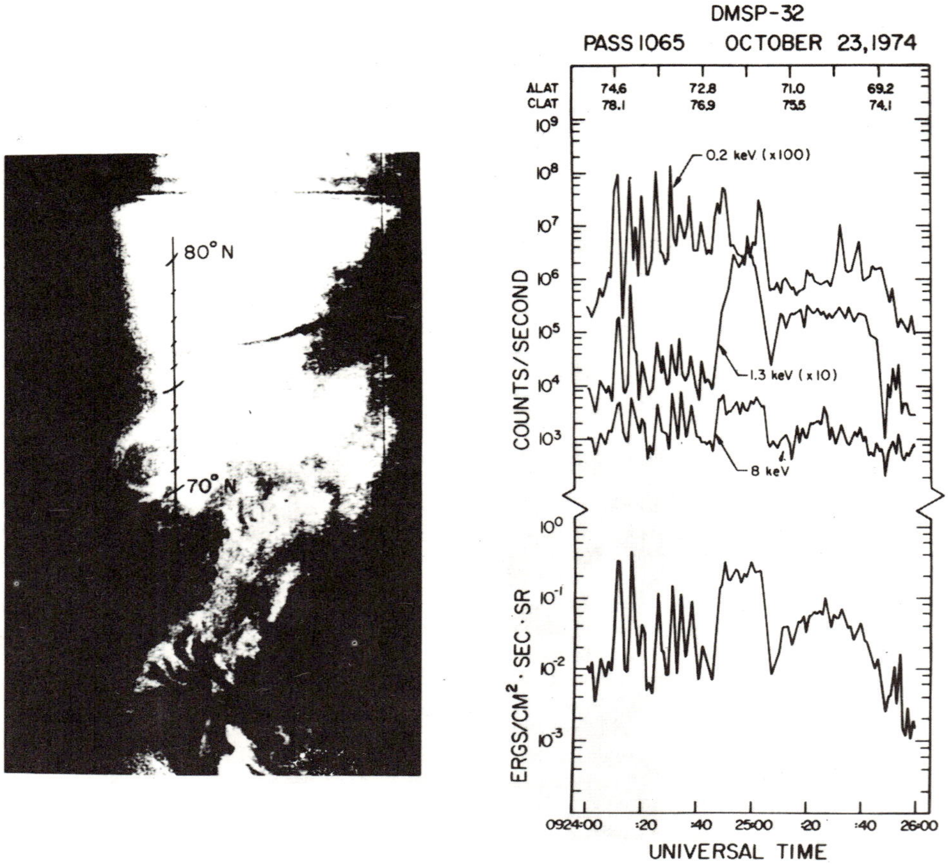

Fig. 1a. Quiet evening faint discrete arcs observed by DMSP-32 on 23 October 1974 (pass 1065). This figure consists of three diagrams: (1) the auroral photograph (in negative), (2) the precipitation profile across the auroral region detected by the same satellite along the satellite path near the middle of the auroral imagery, and (3) the variation of electron differential energy spectra. The satellite location in the corrected geomagnetic latitude (Λ Lat) and the geographic latitude (C Lat) are also illustrated. The enhanced electron precipitation of ~ 0.2 erg cm^{2}sec^{-1}sr^{-1} at ~ 0925:00 UT corresponds to the crossing of the faint discrete arc. The electron spectra were characterized by a spectral peak while the satellite was conjugately over the arc.

arc along the evening oval located at ~ 76.3° geographic latitude (72° geomagnetic latitude) was traversed by DMSP-32 satellite from the polar cap into the lower latitudes at about 0925 UT near 2020 MLT meridian. The geomagnetic condition was very quiet with Kp = 1 (ΣKp = 10) and no substorm activity was recorded by the AE (11) index between 07 and 18 UT. The subtrack of the satellite with geographical latitude marks runs vertically through the middle of the auroral photograph, and the auroral electron precipitation was observed along this line. The electron detector registered enhanced precipitations at all six energy channels (0.2, 0.5, 1.3, 3.2, 8, and 20 keV) during the evening auroral oval crossing. Three selected counting rate profiles and the total precipitated energy flux between 0.2 and 20 keV are shown in the right hand part of Figure 1a, which consists of the auroral imagery and the diagram of the electron precipitation profile along the satellite subtrack. Associated with this faint discrete aurora, the energy flux profile showed an enhanced precipitation for about 15 sec with an intensity of about 0.2 erg cm^{-2}sec^{-1}sr^{-1} over an ~ 0.6° latitudinal width region. The spectral variations across this faint arc from 0924:40 UT to 0925:20 UT are illustrated in Figure 1b; each spectrum was made at the rate of one per second. Two reference lines correspond to differential fluxes of 10^{10} electrons cm^{-2}sec^{-1}sr^{-1}keV^{-1} at 0.2 keV channel and 10^3 electrons cm^{-2}sec^{-1}sr^{-1}keV^{-1} at 20 keV. While the satellite was conjugately over the discrete arc (i.e., the satellite was connected to the arc by the same magnetic field line) from about 0924:54 UT to 0925:00 UT, the

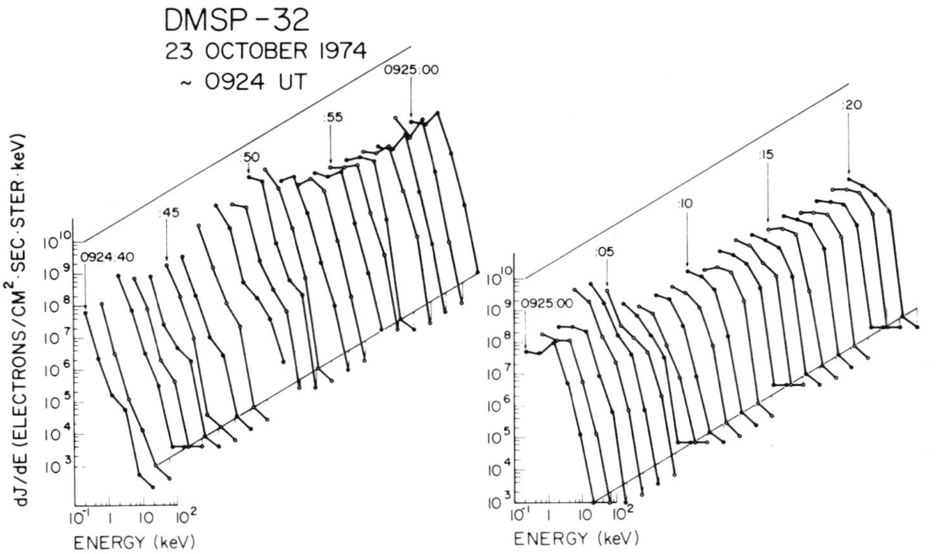

Fig. 1b. Same as Figure 1a.

electron spectra were characterized by a flux peak at the 1.3 keV channel. On both poleward and equatorward boundaries of this faint auroral arc, the spectral peak shifted toward the lower energy indicating a spectral softening outside the arc. Counting rate profiles of 0.2 keV and 1.3 keV also reveal that the most intense fluxes at 0.2 keV were located at the edges of this precipitation band while the 1.3 keV flux peaked in the middle. This feature is known as the inverted 'V' structure on the E-t spectrogram reported previously by Frank and Ackerson (1971).

Figure 2 is an example of bright multiple

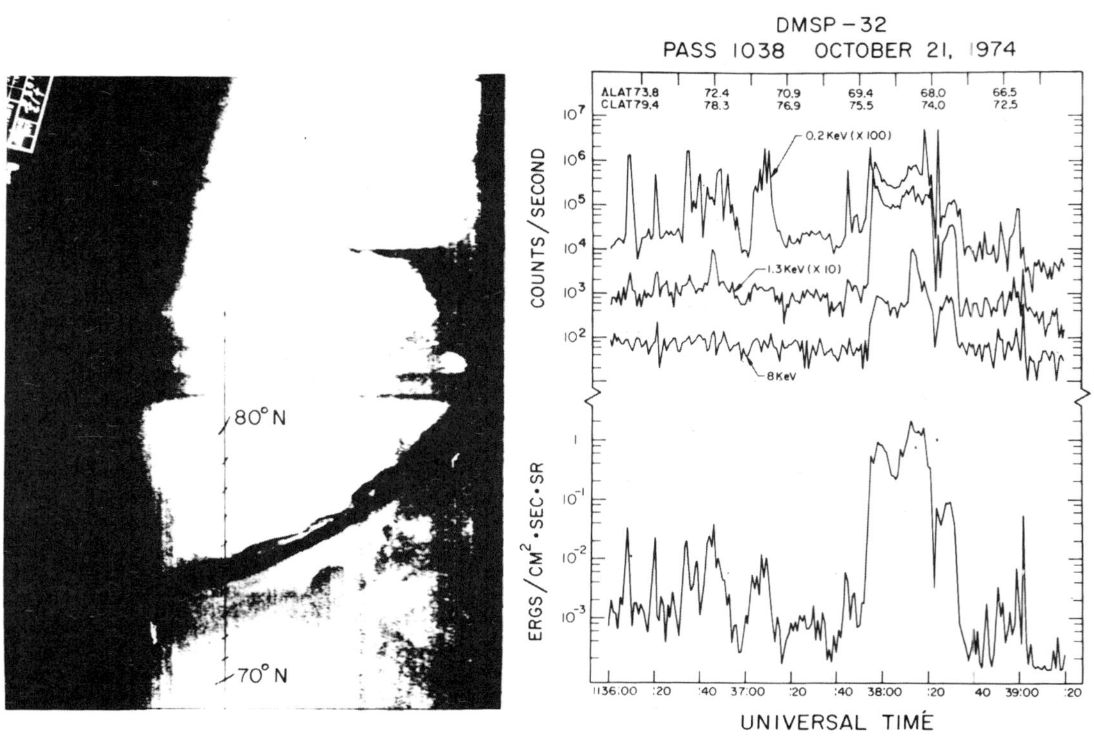

Fig. 2a. Similar to Figure 1, but observed on 21 October 1974 by DMSP-32 (pass 1038) over bright multiple arcs in the evening part of the auroral oval.

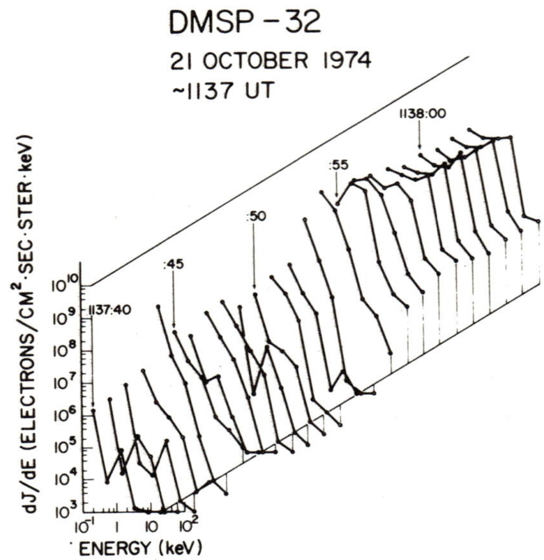

Fig. 2b. Same as Figure 2a.

discrete arcs observed on 21 October 1974 by DMSP-32 (pass 1038) during a slightly active geomagnetic condition when Kp was 2. The magnitude of magnetic disturbance at a midnight auroral zone station (College, Alaska) was less than 100γ and the AE index was between 100 and 200γ during this time span. The auroral photograph shows that two bright auroral arcs, a wider, brighter one at the equatorward side and a narrower, less bright one at the poleward side, with a total width of about 1.1° were

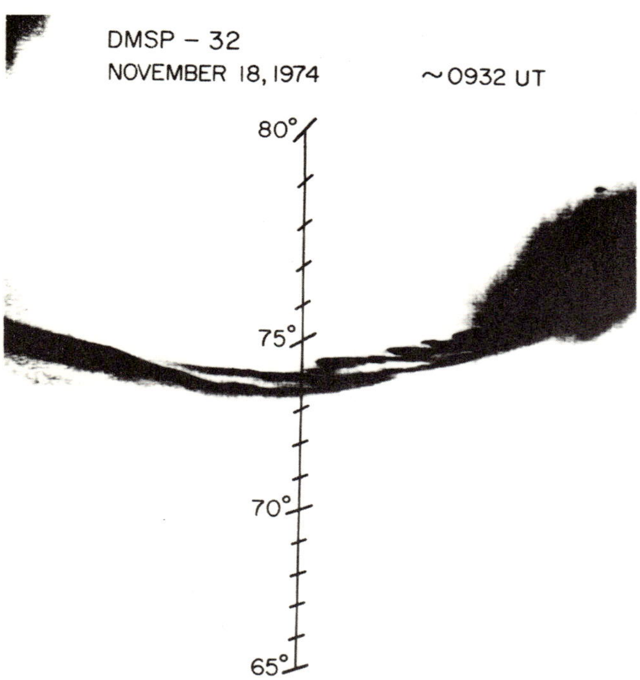

Fig. 3a. Similar to Figure 1, but observed on 18 November 1974 by DMSP-32 (pass 1433) across very bright active arcs ahead of a westward travelling singe.

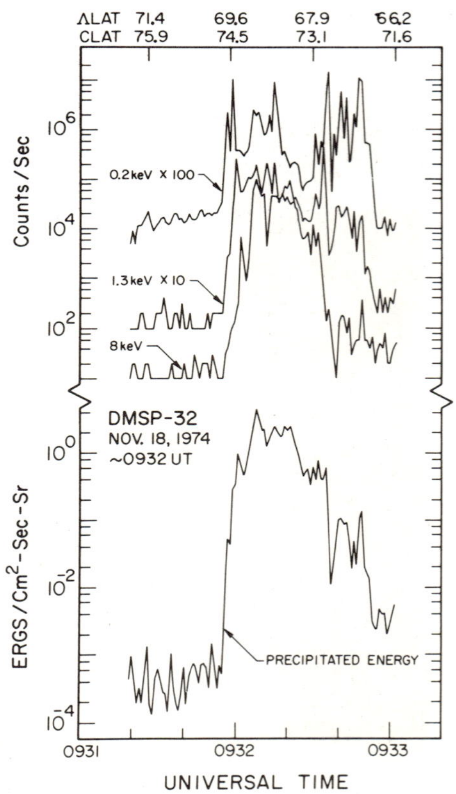

Fig. 3b. Same as Figure 3a.

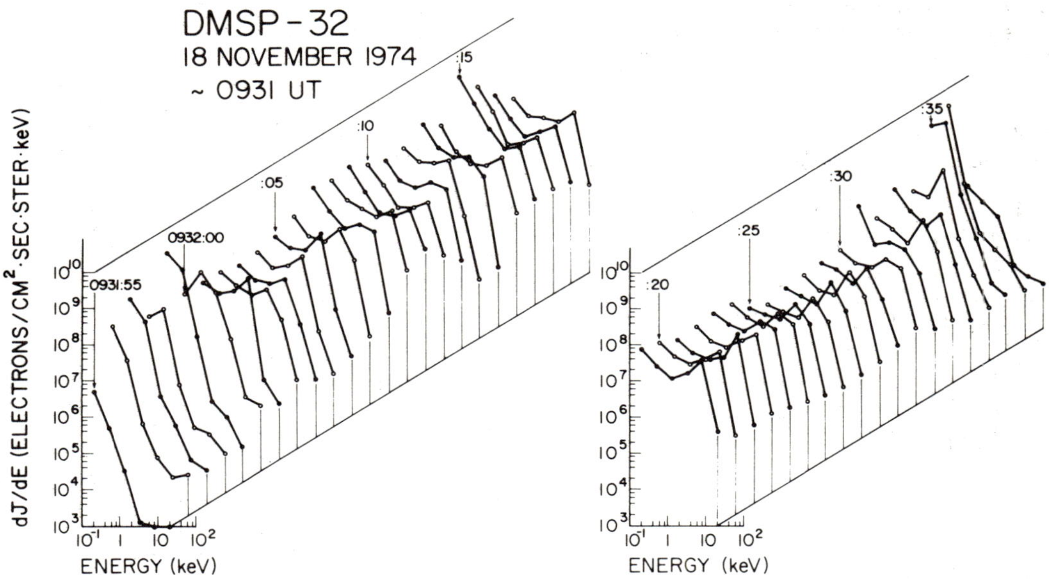

Fig. 3c. Same as Figures 3a and 3b.

observed along the oval near 75°N (69° gm lat). The satellite traversed the auroral region at approximately 1138 UT and detected intense electron precipitations up to about 2 erg $cm^{-2}sec^{-1}sr^{-1}$ with sharp boundaries as shown in the diagram of the precipitation profile. A distinct precipitation peak was associated with each discrete auroral arc. The equatorward peak was more intense and broader than the poleward peak, consistent with the optical features of these two arcs. The latitudinal width of the intense precipitations (> 10^{-1} erg $cm^{-2}sec^{-1}sr^{-1}$) was about 1.1° also in agreement with that of the visual region. The spectral variations (i.e., Figure 2b) revealed two periods (1137:58 to 1138:04 and 1138:11 to 1138:18 UT) of continuous peaked spectra with the peak at the 3.2 keV channel; they were detected when the satellite was conjugately above the bright discrete arcs. No spectral peak appeared in the spectra

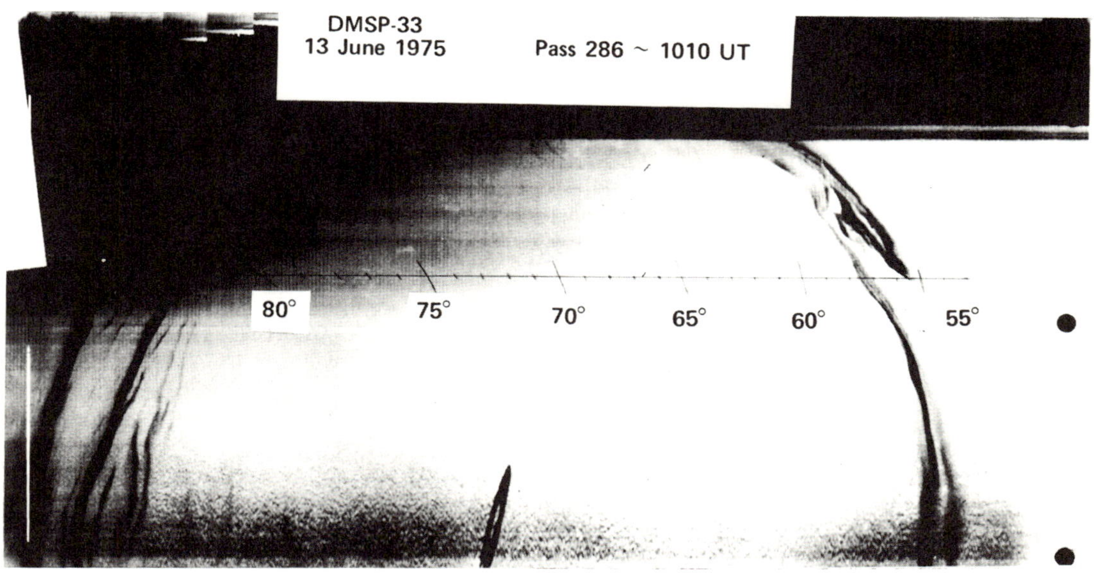

Fig. 4a. Similar to Figure 1 but observed in the southern polar region by DMSP-33 on 13 June 1975 (pass 286) over bright morning discrete arcs (on the left hand side of the auroral photo).

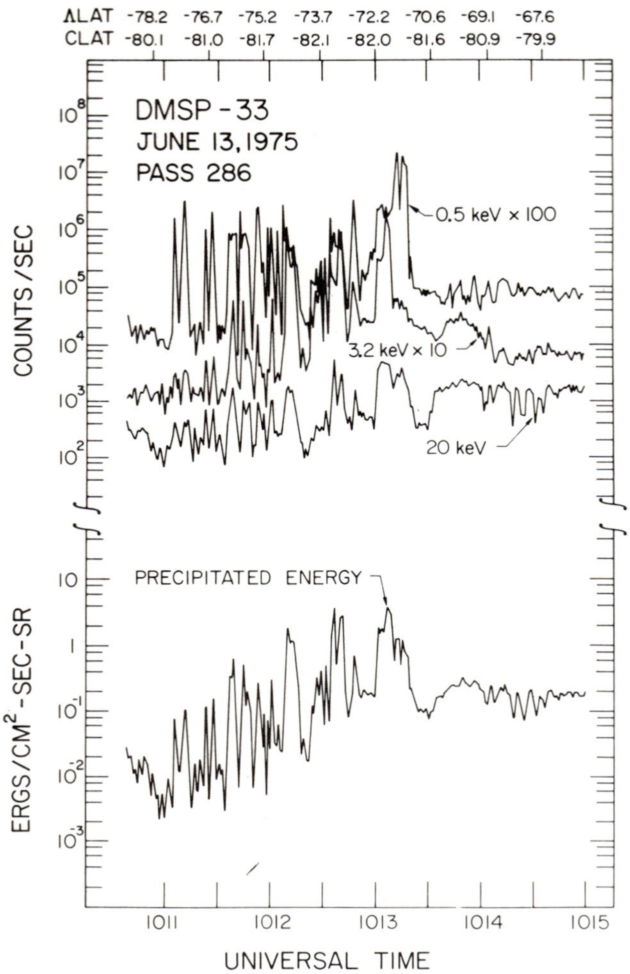

Fig. 4b. Same as Figure 4a.

taken in the region between the two discrete arcs. Spectral hardening and softening were observed as the satellite entered and exited each of the two auroral arcs; this pattern corresponds to the two closely spaced inverted 'V' structures. It is important to point out that the precipitation features are very similar between Figures 1 and 2. The energy of the peak spectra and the energy flux are higher for the bright arcs than those of the faint arc.

As geomagnetic activity increases, auroral displays become more complex. As a result, it is no longer possible to find an extended quiet homogeneous arc along the nightside auroral oval. The auroral substorm originates in the midnight sector of the auroral oval and its effects spread violently in all directions causing various displays in different local time sectors (Akasofu, 1964, 1968). The active discrete auroras were mostly observed in the evening and midnight sectors. Figure 3 is an example of active discrete auroral arcs immediately ahead of a westward travelling surge observed on 18 November 1974 by DMSP-32 (pass 1433) when Kp was 3- (ΣKp = 16) in the expansive phase of a moderate substorm. The surge was located to the east of the satellite subtrack near the midnight sector at ~ 0930 UT. Bright discrete arcs observed from 75.5°N to 73°N (or about 69° gm lat) were crossed by the satellite along 2020 magnetic local time meridian. Just a few minutes before this evening oval traversal, the AE index showed a sharp increase to ~ 800γ, indicating a sudden enhancement of the auroral electrojet intensity, namely, the onset of an auroral substorm. Along the satellite subtrack, enhanced precipitations, with energy fluxes of about 1 erg $cm^{-2}sec^{-1}sr^{-1}$ to 4.5 erg $cm^{-2}sec^{-1}sr^{-1}$ were detected while the satellite was conjugately above two bright discrete arcs. There were two groups of peaked spectra associated with two bright arcs. The first group from 0932:01 to 0932:10 UT showed the spectral peaks at 3.2 keV channel, whereas the second group from 0932:15 to 0932:29 UT showed the spectral peaks ranging from 8 keV to 3.2 keV. The peaked spectrum inside the active discrete evening arcs bears similar characteristics to those of quiet evening arcs but at higher energies and at higher precipitated energy flux.

The above three evening oval discrete arcs clearly reveal a few of the salient features of the particles associated with auroral arcs: (1) the existence of peaked electron differential energy spectra in the particle precipitation region conjugate to the arcs, (2) the spectral peak location increasing with the auroral brightness and activity, and (3) the intensity of the precipitated energy flux increasing with the auroral brightness and geomagnetic activity.

Discrete Arcs at Other Locations

The discrete auroral arcs also occur at other parts of the polar region, even though most of the discrete auroras are observed along the evening and midnight part of the auroral oval. In this section, the precipitation characteristics of discrete arcs along the morning oval, the dayside oval as well as over the polar cap region will be discussed.

Although discrete auroral forms are not very frequently observed along the morning part of the auroral oval, they still have characteristics of discrete arcs. Figure 4 illustrates a few discrete arcs extending along the morning oval observed over the southern polar region on 13 June 1975 by DMSP-33 (pass 286) when Kp was 3- in a recovery phase of a series of weak substorms (~ 400γ). The satellite crossed the morning oval at approximately 0840 LT and detected intense precipitations from 75° gm lat to 71.5° gm lat. Four groups of precipitations correspond to four discrete auroras, two bright ones near the poleward and equatorward boundaries and two faint ones in between. The energy

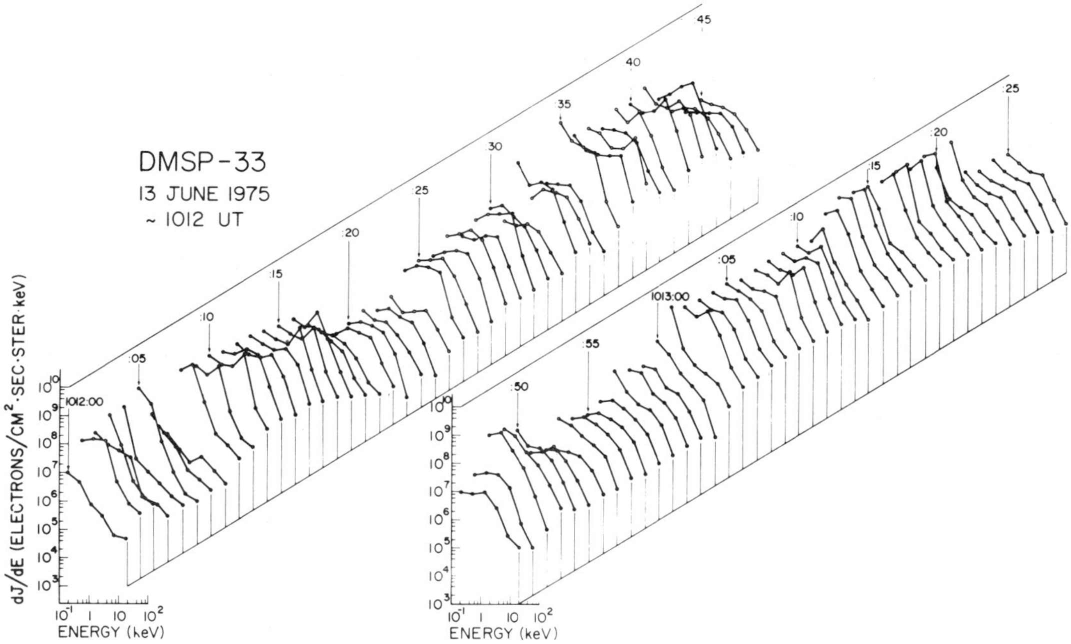

Fig. 4c. Same as Figures 4a and 4b.

fluxes were about a few erg $cm^{-2}sec^{-1}sr^{-1}$. Associated with the two bright auroras near the two boundaries were two groups of peaked spectra, one at 3.2 keV conjugate to the poleward arc and the other ranged from 3.2 keV to 0.5 keV over the equatorward arc observed near 1012:15 and 1013:10 UT.

Discrete arcs are frequently seen in the dayside of the auroral oval particularly near the midday gap of the oval (Snyder and Akasofu, 1976; Dandekar and Pike, 1978; Meng, 1981). The following example illustrates the arcs in the midday auroral oval detected by DMSP-33 (pass 244) at ~ 1040 UT (Kp = 1-) on 10 June 1975

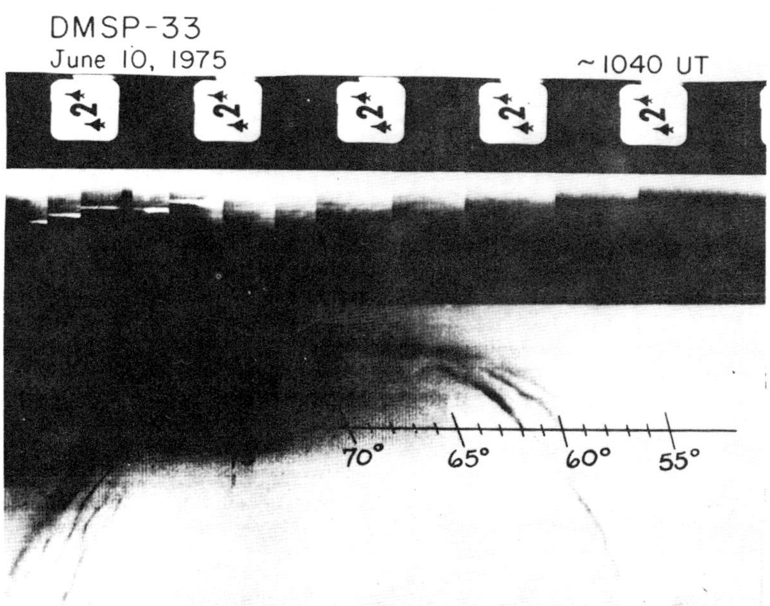

Fig. 5a. Similar to Figure 1 but observed in the southern polar region by DMSP-33 on 10 June 1975 over the midday auroral oval. Discrete arcs were detected over both late afternoon and morning parts of the oval.

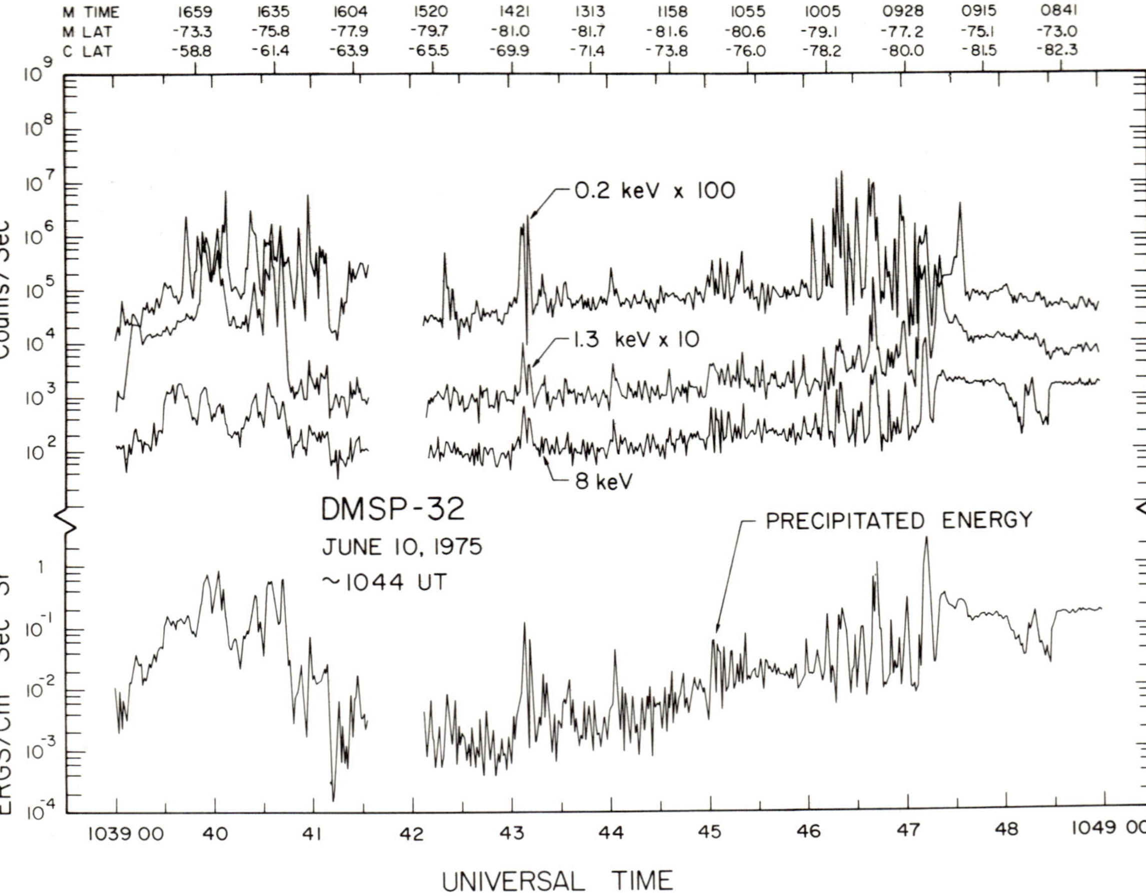

Fig. 5b. Same as Figure 5a.

during a relatively quiet day (ΣKp = 7+) with a low level of continuous AE activity of about 200γ. As shown in Figure 5a, the satellite traversed the afternoon oval between 59°S and 62°S (-73.5° to -76° gm lat) along the 1630 LT meridian and the morning oval at about 79°S (-78° gm lat) near the 0930 LT meridian. Two well-separated discrete auroras were observed in the afternoon oval, and two arcs in the morning oval. The precipitated energy flux was approximately one erg $cm^{-2} sec^{-1} sr^{-1}$ associated with the afternoon arcs and a few erg $cm^{-2} sec^{-1} sr^{-1}$ with morning discrete auroras (Figure 5b). Along the afternoon oval, the spectra with peak near 1.3 keV from 1039:54 to 1040:04 UT and from 1040:35 to 1040:44 UT were detected over the equatorward and poleward discrete arcs, respectively (Figure 5c). The spectral hardening and softening (i.e., the inverted 'V') were observed while the satellite traversed these arcs. Over the morning oval, the spectral variation was more complicated, soft precipitations were measured before 1047:10 UT (i.e., between -79° gm lat and -76.8° gm lat) and two periods of peaked spectra were measured at about 1047:13 UT and 1047:25 UT, respectively. The spectral peak was at the 3.2 keV channel for the poleward discrete arc and varied from 3.2 keV to 0.5 keV over the equatorward discrete aurora.

The morphological studies of the polar auroral display revealed that discrete auroras can be divided into two types according to their spatial locations (Davis, 1960; Denholm and Bond, 1961; Akasofu, 1965). The auroral oval arcs have an orientation approximately along the geomagnetic latitude circles as illustrated in the above examples. The polar cap arcs with a sun-aligned direction occur at high geomagnetic latitudes poleward of the auroral oval. Sun-Earth aligned polar arcs are most commonly observed over the morning half of the polar cap as shown in this example observed by DMSP-32 (pass 1535) on 25 November 1974. The geomagnetic Kp index was 2 (ΣKp = 26+) and the AE index showed that the observation of the polar cap arcs occurred during a ~ 3 hours quiet period (AE at background values) between substorm activities. Several discrete auroral arcs were located between 75°N and 79°N along the subtrack (Figure 6a). Associated with these polar cap

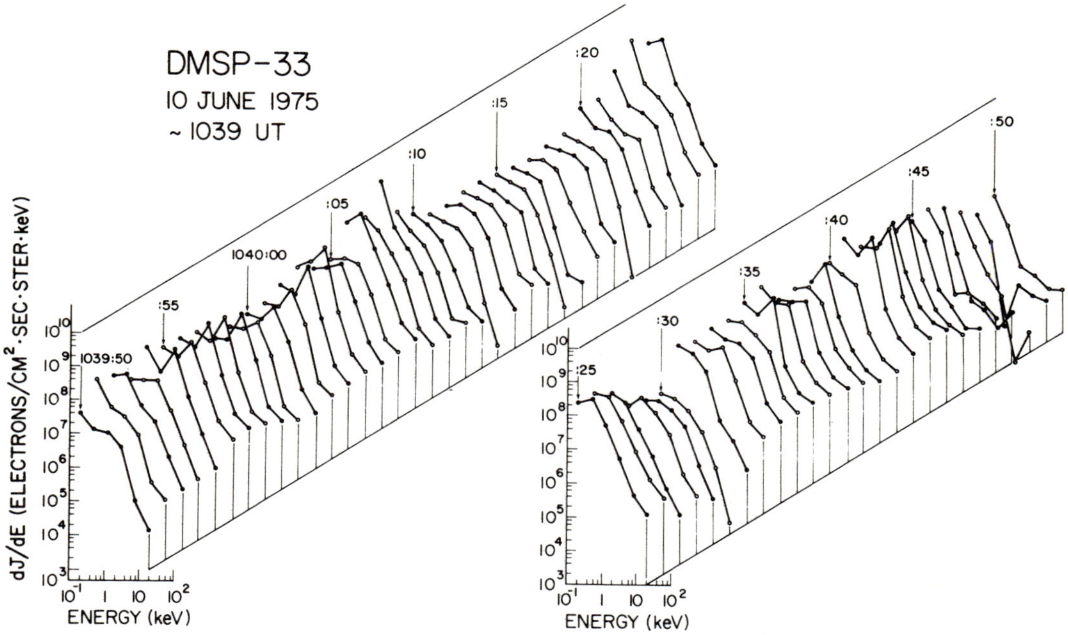

Fig. 5c.1 Same as Figures 5a and 5b.

arcs, intense energy fluxes up to about 2 erg $cm^{-2}sec^{-1}sr^{-1}$ were detected (Figure 6b). The brightest arc near 77°N (~ 76.2° gm lat) corresponds to the most intense precipitation near 1424:10 UT. The spectral variations across only two polar cap arcs, one faint and the other bright, are shown in Figure 6c. It revealed two groups of peaked spectra from 1423:53 to 1423:59 and from 1424:05 to 1424:15 UT, respectively. The spectral peak was at the 1.3 keV channel for the faint arc and near 1.3 or 3.2 keV for the brighter polar cap arc. This example demonstrates that the polar cap sun-aligned discrete auroral arcs have the same electron precipitation characteristics as those of the auroral oval discrete arcs.

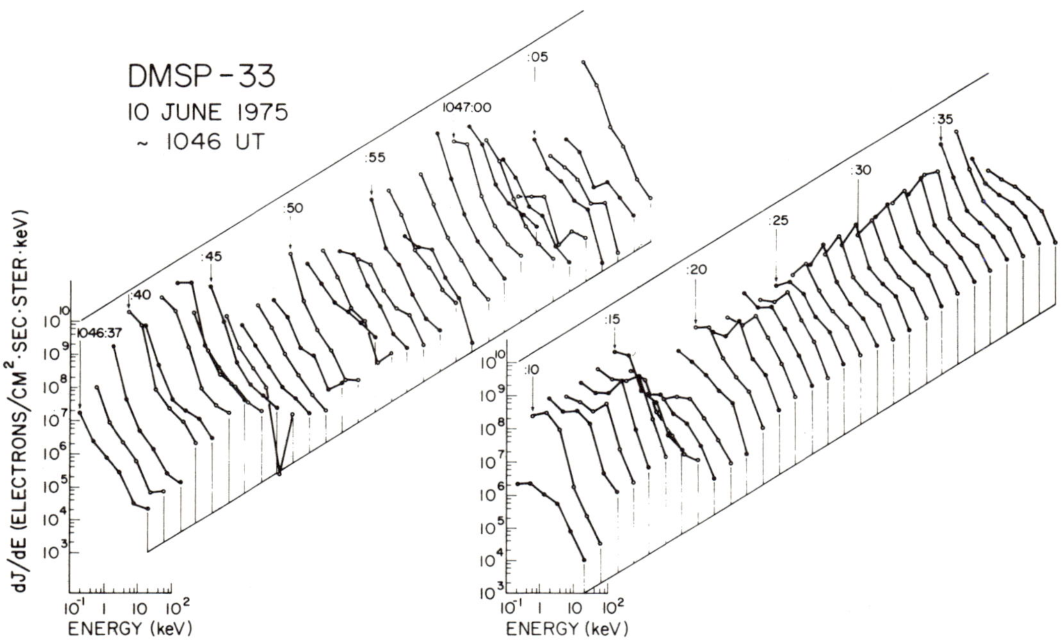

Fig. 5c.2 Same as Figures 5a, 5b and 5c.1.

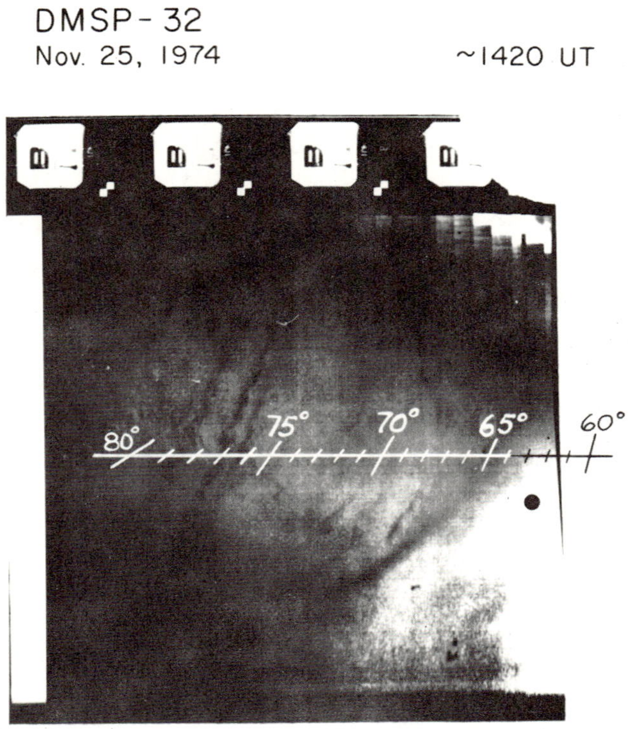

Fig. 6a. Similar to Figure 1 but observed in the northern polar region on 25 November 1974 over the polar cap discrete arcs (76° to 79°N) and the morning oval.

Conclusions and Discussions

The above examples reveal that the discrete auroral arcs all have a common feature which is the existence of a peak in the electron differential energy spectra. The location of this spectral peak and the precipitated energy flux differ with the brightness of auroral arcs and the geomagnetic activity. The precipitated energy flux varies from ~ 10 ergs $cm^{-2}sec^{-1}sr^{-1}$ for bright, active arcs, ~ 1-2 erg $cm^{-2}sec^{-1}sr^{-1}$ for normal quiet arcs to a fraction of one erg $cm^{-2}sec^{-1}sr^{-1}$ for faint quiet arcs. The spectral peak is at $\gtrsim$ 8 keV for bright active arcs, about 3 keV for normal quiet arcs and at $\lesssim$ 1 keV for faint discrete arcs. Based on this observation, it is interesting to discuss the source region of auroral electrons and the formation of discrete arcs.

It has been suggested that the auroral oval is the projection of the magnetospheric plasma sheet onto the polar ionosphere and thus that the electrons in the plasma sheet are likely to be a source of auroral electrons. However, by comparing electron spectra in the aurora with those measured in the plasma sheet (Chase, 1969; Hones et al. 1971; Eather and mende, 1972; Sharp et al. 1971; Rearwin and Hones, 1974), contradictory results were reported. From the latitudinal dependence of spectral characteristics of auroral electrons observed by the ISIS-1 and -2 satellites in the midnight sector during different phases of substorms, Winningham et al. (1975) identified two sources of auroral electrons. One of them is called the boundary layer of the plasma sheet which produce the discrete and variable structures (such as the inverted 'V') in the poleward part of the auroral precipitation region. The other is the near-Earth plasma sheet or the central part of the plasma sheet and sometimes even from the outer Van Allen belt, which produce the diffuse smooth precipitation on the equatorward side. In terms of visual auroral morphology, these identifications implied that the discrete auroras were produced by electrons from the boundary of the plasma sheet and diffuse auroras by electrons from the central plasma sheet and/or the outer radiation zone.

The relationship between polar auroras and the magnetospheric particles has also been studied by many groups using the particle observations of the geosynchronous satellite ATS-5 with the auroral phenomena near the field line conjugate. Akasofu et al. (1974) found that the appearance of active auroras at the foot of ATS-5 geomagnetic field line corresponded to the encounter of ATS-5 with hot plasma. Sharp et al. (1975) found that the trapped electron fluxes at ATS-5 were closely correlated with the auroral electrojet intensity measured at the ATS-5 conjugate point. However, Mende and Shelley

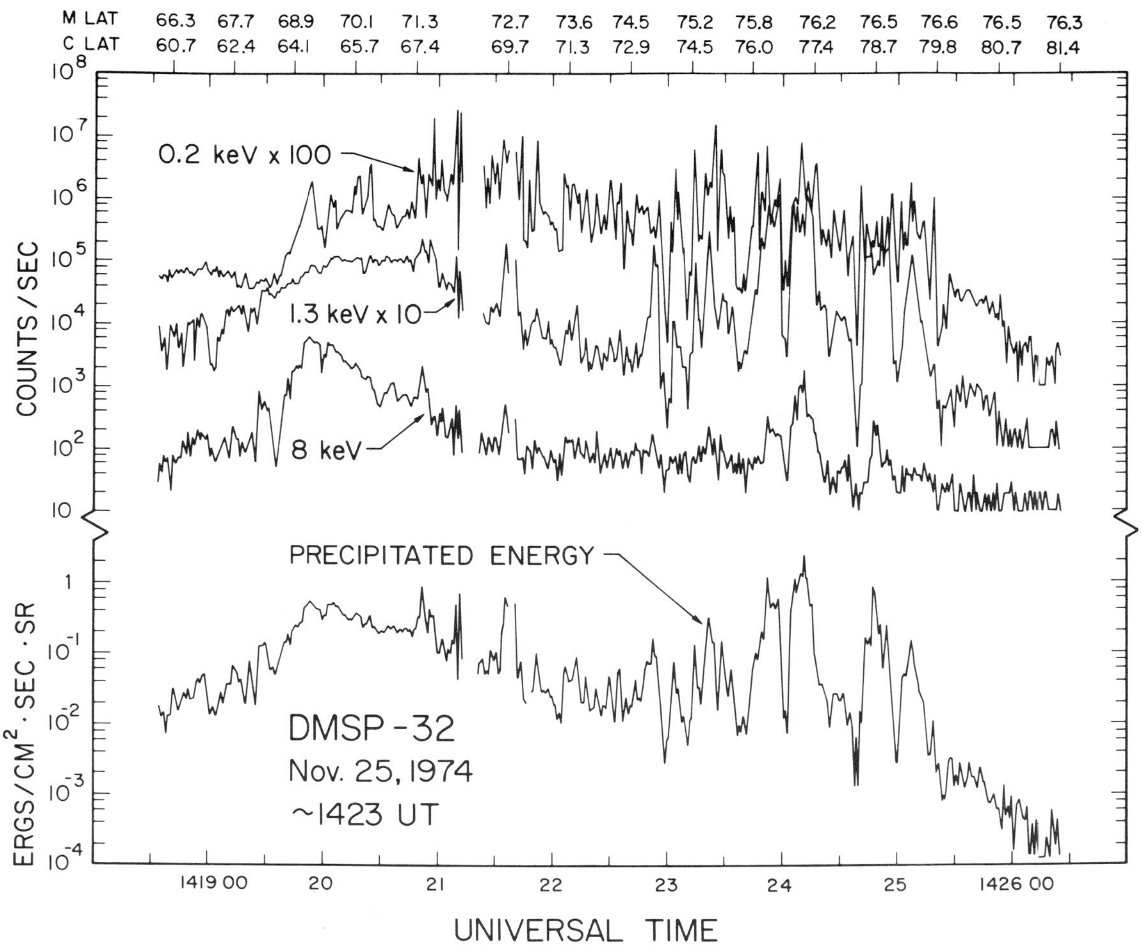

Fig. 6b. Same as Figure 6a.

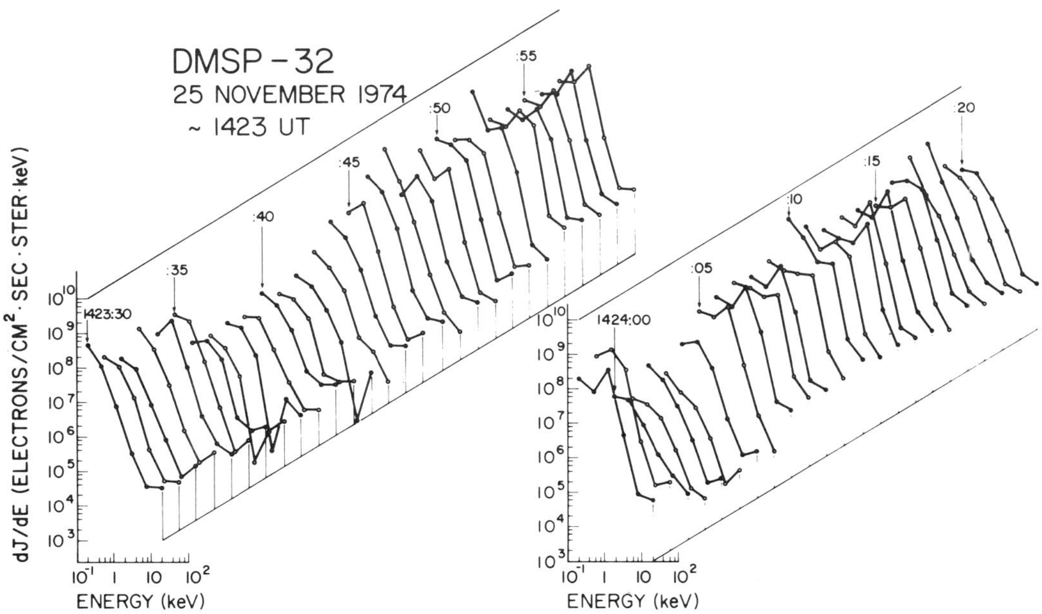

Fig. 6c. Same as Figures 6b and 6c.

(1976) showed that the presence of hot plasma at the synchronous altitude is a necessary but not sufficient condition for the occurrence of conjugate auroras and also that there is a lack of detailed correlation between the fine structures of the trapped plasma fluxes and the small scale auroral features.

The coordinations between plasma observations at synchronous altitude and its conjugate auroral display were also made by DMSP satellites and the ATS-6 satellite. The comparison of the electron component of trapped plasma sheet particles and its conjugate precipitated auroral electrons provides us with the definitive source and mechanisms of auroral electrons. Meng et al. (1979) found that the spectrum and intensity of auroral electrons precipitating into diffuse aurora are identical to those of the trapped electrons in the plasma sheet and concluded that the diffuse auroras are produced by the direct dumping of electrons from the plasma sheet. However, spectral characteristics of electrons producing discrete auroral arcs are found to be distinctly different from the characteristics of the plasma sheet electrons. Therefore, discrete auroral forms are produced by electrons which are accelerated between the magnetospheric equator and the polar ionosphere along the magnetic field line. The observed characteristic monoenergetic peak in the electron spectrum of discrete auroras strongly suggest that a very strong electric field must be involved. The potential drop of an upward pointing parallel electric field which produces the observed monoenergetic spectral peak generally in the range of a few keV is believed to exist from recent satellite observations reported in this meeting. However, how this parallel electric field is created above the auroral region is not clear yet and several different approaches to create this electric field can be found in this monograph.

Acknowledgements. The author is indebted to Drs. S. Ismail and L. J. Zanetti for helpful discussions. The research is supported in part by the Division of Atmospheric Sciences, National Science Foundation, under Grant ATM-79-23240 and by the Air Force Office of Scientific Research under Grant AFOSR-79-0010.

References

Akasofu, S. I., The development of the auroral substorm, Planet. Space Sci., 12, 273, 1964.

Akasofu, S. I., Dynamical morphology of auroras, Space Sci. Rev., 4, 498, 1965.

Akasofu, S. I., Polar and Magnetospheric Substorms, D. Reidel Publ. Co., Dordrecht, Holland, 1968.

Akasofu, S. I., A study of auroral displays photographed from the DMSP-2 satellite and from the Alaska meridian chain of stations, Space Sci. Rev., 16, 617, 1974.

Akasofu, S. I., S. DeForest and C. McIlwain, Auroral displays near the foot of the field line of the ATS-5 satellite, Planet. Space Sci., 22, 25, 1974.

Arnoldy, R. L., Auroral particle precipitation and Birkeland currents, Rev. Geophys. Space Phys., 12, 217, 1974.

Chase, L. M., Evidence that the plasma sheet is the source of auroral electrons, J. Geophys. Res., 74, 348, 1969.

Dandekar, B. S. and C. P. Pike, The midday discrete auroral gap, J. Geophys. Res., 83, 4227, 1978.

Davis, T. N., The morphology of the polar aurora, J. Geophys. Res., 65, 3497, 1960.

Deehr, C. S., A. Egeland, K. Aarsnes, R. Amundson, H. R. Lindalen, F. Soraas, R. Dalziel, P. A. Smith, G. R. Thomas, P. Stanning, H. Borg, G. Gustafsson, L. A. Holmgasen, W. Riedler, J. Raitt, G. Skovli, T. Wedde and R. Jaeschke, Particle and auroral observations from the ESRO 2/Aurorae satellite, J. Atmo. Terr. Phys., 35, 1979, 1973.

Denholm, J. V. and F. R. Bond, Orientation of polar auroras, Aust. J. Phys., 14, 193, 1961.

Eather, R. H. and S. B. Mende, High latitude particle precipitation and source regions in the magnetosphere, in Magnetosphere-Ionosphere Interactions, edited by K. Folkestad, p. 139, Universitetsforlaget, Osla, Norway, 1972.

Frank, L. A. and K. L. Ackerson, Observations of charged particle precipitation into the auroral zone, J. Geophys. Res., 76, 3612, 1971.

Hones, E. W., Jr., J. R. Asbridge, S. J. Bame and S. Singer, Energy spectra and angular distributions of particles in the plasma sheet and their comparison with rocket measurements over the auroral zone, J. Geophys. Res., 76, 63, 1971.

Mende, S. B. and E. G. Shelley, Coordinated ATS-5 electron flux and simultaneous auroral observations, J. Geophys. Res., 81, 97, 1976.

Meng, C. I., Simultaneous observations of low-energy electron precipitation and optical auroral arcs in the evening sector by the DMSP-32 satellite, J. Geophys. Res., 81, 2771, 1971, 1976.

Meng, C. I., B. Mauk and C. E. McIlwain, Electron precipitation of evening diffuse aurora and its conjugate electron fluxes near the magnetospheric equator, J. Geophys. Res., 84, 2545, 1979.

Meng, C. I., Electron precipitation in the midday auroral oval, J. Geophys. Res., (in press), 1981.

Mizera, P. F., D. R. Croley, Jr., F. A. Morse, and A. L. Vanpola, Electron fluxes and correlations with quiet time auroral arcs, J. Geophys. Res., 80, 2129, 1975.

Rearwin, S. and E. W. Hones, Jr., Near-simultaneous measurement of low-energy electrons by sounding rockets and satellites, J. Geophys. Res., 79, 4322, 1974.

Sharp, R. D., D. L. Carr, R. G. Johnson and

E. G. Shelley, Coordinated auroral electron observations from a synchronous and a polar satellite, J. Geophys. Res., 76, 7669, 1971.

Sharp, R. D., E. G. Shelley and G. Rostoker, A relationship between synchronous altitude electron fluxes and the auroral electrojet, J. Geophys. Res., 80, 2319, 1975.

Shepherd, G. G., C. D. Anger, L. H. Brace, J. R. Burrows, W. J. Heikkila, J. Hoffman, E. J. Maier and J. H. Whitteker, An observation of polar aurora and airglow from the Isis-2 spacecraft, Planet. Space Sci., 21, 819, 1973.

Snyder, A. L. and S. I. Akasofu, Auroral oval photographs from the DMSP 8531 and 10533 satellites, J. Geophys. Res., 81, 1799, 1976.

Winningham, J. D., F. Yasuhara, S. I. Akasofu and W. J. Heikkila, The latitudinal morphology of 10 eV to 10 keV electron fluxes during magnetically quiet and disturbed times in the 2100-0300 MLT sector, J. Geophys. Res., 80, 3148, 1975.

STUDY OF INVERTED-V AURORAL PRECIPITATION EVENTS

R. A. Hoffman

Electrodynamics Branch, Goddard Space Flight Center, Greenbelt, Maryland 20771

C. S. Lin

Institute for Physical Science and Technology, University of Maryland, College Park, Maryland 20742

Abstract. A detailed survey of inverted-V events was made possible with the flight on the Atmosphere Explorer-D satellite of the Low Energy Electron Experiment (LEE), which contained an array of 19 charged particle detectors covering the energy range 200 eV to 25 keV. In this review the macroscopic properties of these events will be described first, including their spatial distributions and the extents of individual events. Statistical studies of microscopic characteristics include the distributions in energy and temperature of the monoenergetic components and systematics of the anisotropics in the pitch angle distributions.

A review of new work performed involves the relation of inverted-V events in the pre-midnight hours to magnetic activity. We conclude that inverted-V events are exceedingly common, so common, in fact, that they occur essentially all the time, independent of the state of magnetic activity. We further conclude that these analyses indicate that the basic mechanism for producing the inverted-V structure is nearly time stationary during the passage of the satellite through the event, and perhaps may not change significantly in location for much more extended periods.

Introduction

The Low Energy Electron Experiment on the Atmosphere Explorer-D spacecraft returned comprehensive measurements related to auroral particle precipitation. In particular, data from this instrument have provided many details of the characteristics of electron inverted-V events measured in a low altitude polar orbit. These results will be reviewed.

The review will be divided into three parts:
1. review of previously published results;
2. presentation in summary of some new and very pertinent results related to auroral arcs; and
3. comments regarding the interpretation of the results in terms of parallel electric field mechanisms.

Instrumentation

The Low Energy Electron Experiment flown on the Atmosphere Explorer-D spacecraft, launched on October 6, 1975, contained the largest known array of detectors for auroral particle measurements on any satellite (Hoffman et al., 1973). It was designed in an attempt to obtain good temporal resolution in both the energy spectrum and pitch angle distribution. The design was driven by the results of the OGO-4 Auroral Particles Experiment, whose data indicated the existence of field-aligned electron distributions, which were interpreted as due to electric fields parallel to the magnetic field (Hoffman and Evans, 1968). The instrument contained an array of 19 detectors, each composed of a cylindrical electrostatic analyzer for species and energy selection, followed by a Spiraltron electron multiplier for particle detection.

Table 1 indicates the angular distribution of the detectors, where the angles are given with respect to the +Y axis of the spacecraft, which was radially away from the earth during the times of interest. Also shown is the center energy of the fixed energy detectors and the energy range of the stepped energy detectors. The sets of detectors were placed on either side of 0° to compensate for the inclination of the magnetic field in the high latitude region. As a result, one of the detectors usually measured particles with a center pitch angle of a few degrees, and the other at about 10°. The fixed energy detectors were interrogated sixteen times each second, while the stepped energy detectors obtained a sixteen point energy spectrum each second.

The primary method of analysis utilized for much of the work reported here was performed by visually scanning electron spectrograms (Figure 1 top) to identify the existence of inverted-V structures, and to obtain their location, length in the satellite time frame, and the peak energy. Statistical studies were then based on such observations. This technique, of course, provides a lower limit to any existence studies, since the

TABLE 1. AE-D Low Energy Electron Experiment Electrostatic Analyzer Configuration

	ENERGY (KEV)	ANGLE -7°	+7°	35°	60°
ELECTRONS	0.7	X	X	X	X
	1.4	X	X	X	X
	2.6	X	X	X	X
	6.9		X	X	
	18.0		X	X	
	0.2 to 25	X			X
IONS	0.2 to 25				X

identification of an inverted-V is dependent upon the ability to see the characteristic "monoenergetic peak" in the energy spectrum on the microfilm plots.

Review of Previous Work

The first properties of inverted-V events to be reviewed will be their gross physical characteristics. Three examples will be given to show that the inverted-V structures can exist in a variety of locations. Figure 1 (top) shows an event in the plasma sheet precipitation region shortly before midnight. Figure 2 contains a series of low energy events in the nominal polar cap in the afternoon hours. Some of the one or two second bursts may also be inverted-Vs but they are too short in spacecraft observing time to identify. The third example, Figure 3, contains a very long-lasting, high energy monoenergetic event extending well over one minute with essentially no ions present.

A summary of locations of 430 inverted-V events in invariant latitude, magnetic local time coordinates appears in Figure 4, using data from November 1, 1975, to January 28, 1976, in the northern hemisphere. This is a larger number of cases than published by Lin and Hoffman (1979a) in a similar figure. There was no attempt to obtain an even sampling distribution for this analysis. Note the following features: 1. Inverted-Vs occur at all local times. 2. Their longest extend is in the pre-midnight region. 3. They occur at higher latitudes on the dayside. 4. There is a decrease in occurence around noon.

A number of passes through the dayside cusp were carefully inspected, and in no case was an inverted-V seen within the cusp precipitation. Inverted-Vs having maximum energy below the threshold of this instrument have also been re-

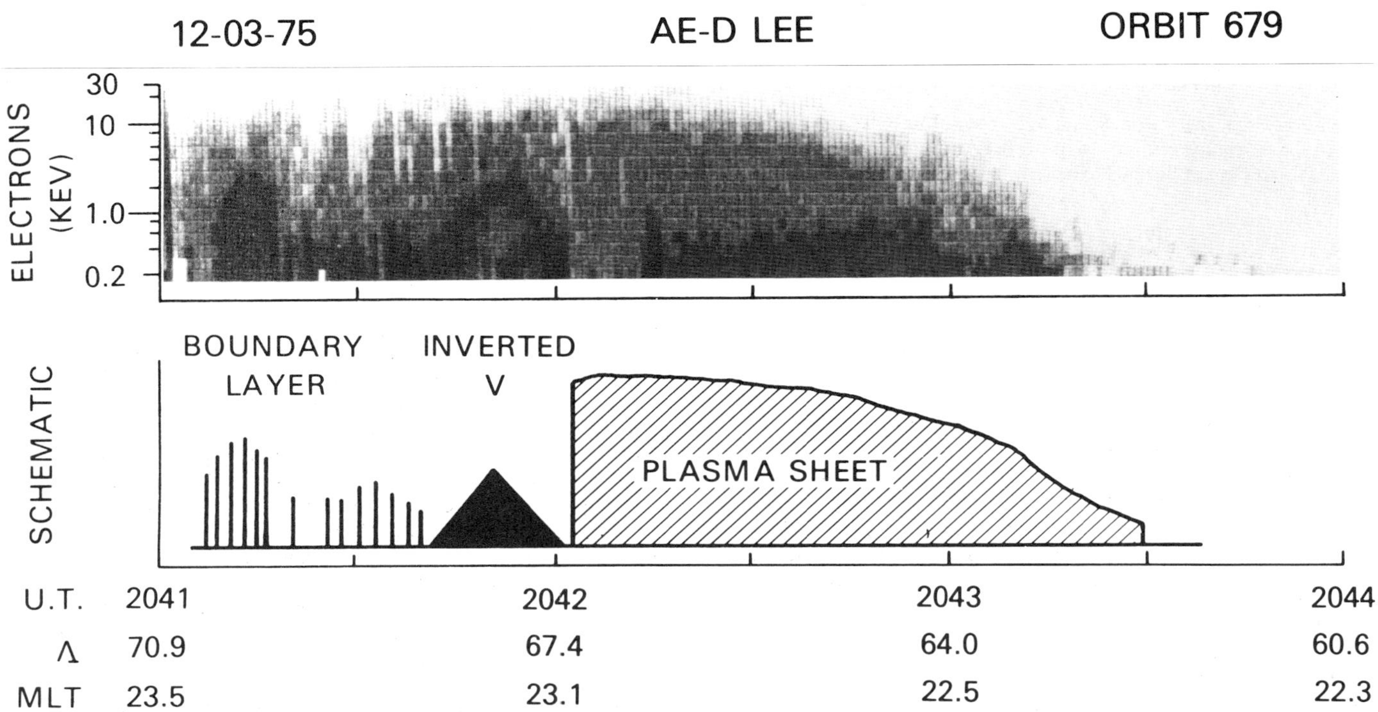

Fig. 1. Top: Energy-time spectrogram for precipitating electron fluxes. Increasing grey is logarithmically related to increasing particle flux. Lower: Schematic of three types of electron precipitation in the nightside auroral zone, used in Figures 11 and 12.

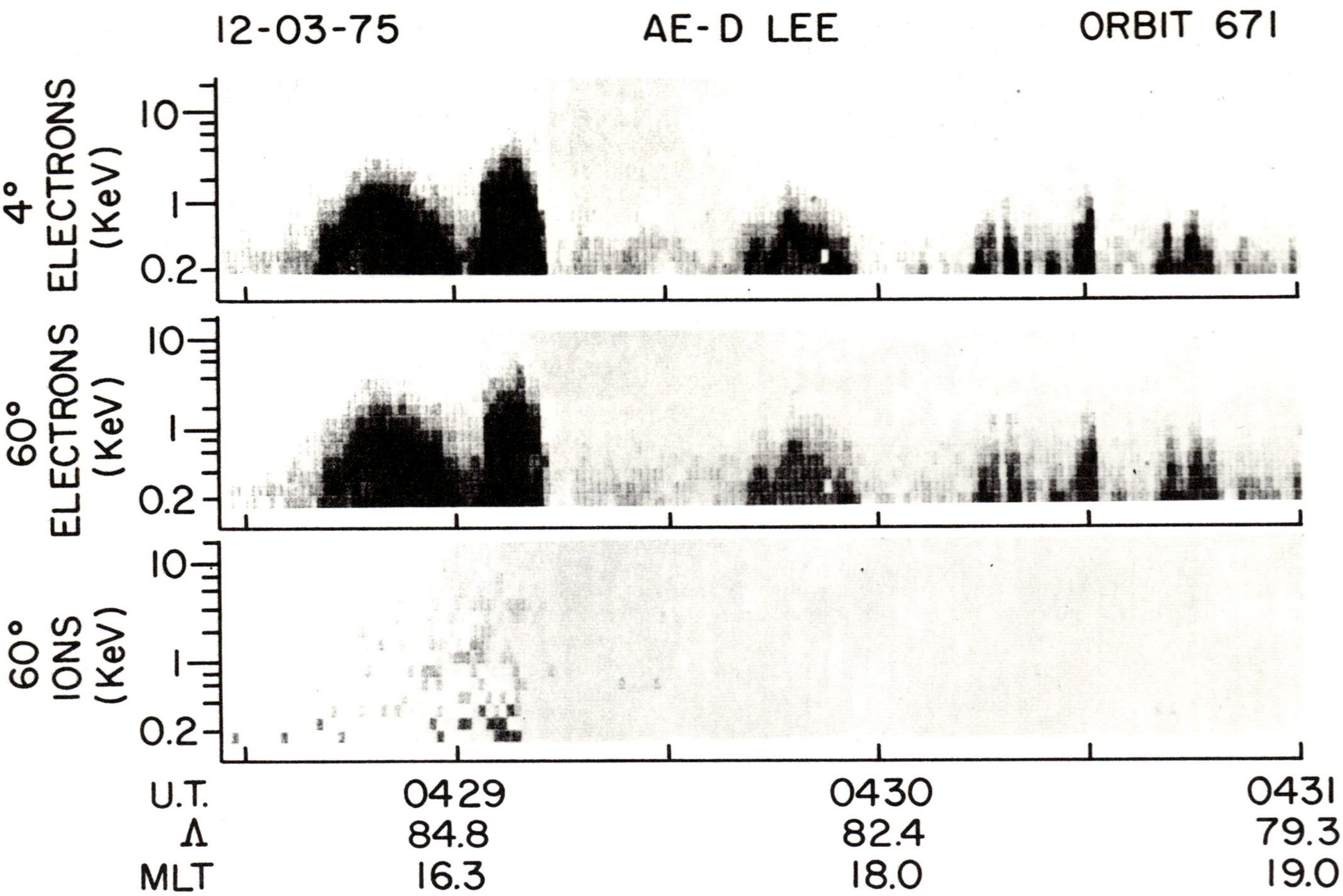

Fig. 2. A series of low energy inverted-V precipitation events in the nominal polar region. The upper two electron spectrograms are from data at pitch angles of about 4° and 60° respectively, and the ion (proton) spectrogram at about 60°.

ported on the dayside, but again near the cusp (Potemra et al., 1977).

The distribution in latitudinal width is shown in Figure 5 for the events in Figure 4. As mentioned in Lin and Hoffman (1979a), the identification of an inverted-V using the stepped detectors requires three or more energy sweeps, so the cutoff at about 0.2° in the distribution is instrumental. By investigating the fixed energy detector responses in very short precipitation events we have identified inverted-Vs as short as 1/2 second, or 4 km in width. Note that while the probability of occurrence rapidaly increases towards small events, there are many of very wide extent.

The satellite orbit of 90° inclination did not provide proper opportunities to measure the dimensions of inverted-Vs in local time or longitude. However, the measured distribution for the 430 events does provide the information that inverted-Vs can have local time widths up to two hours. These length measurements are rather important to note in planning future missions involving multiple spacecraft observations of auroral precipitation, because they indicate the capability of making measurements of the same event from multiple spacecraft slightly out of common magnetic meridian planes.

It is proper and necessary to point out that in the preceeding analyses we have assumed the inverted-Vs to be spatially stable precipitation regions, at least on the scale of the satellite passage time. Certainly with a moving single point of observation, there is yet no proof that this is the case. On the other hand, many events have been seen where the spectrum has remained essentially unchanged for tens of seconds, indicating considerable stability in a source mechanism along the path of the satellite.

We next turn to more detailed properties of inverted-V events. The distribution in maximum peak energy as a function of MLT is shown in Figure 6. Since the lower energy limit of measurement is 200 eV, events can be identified with

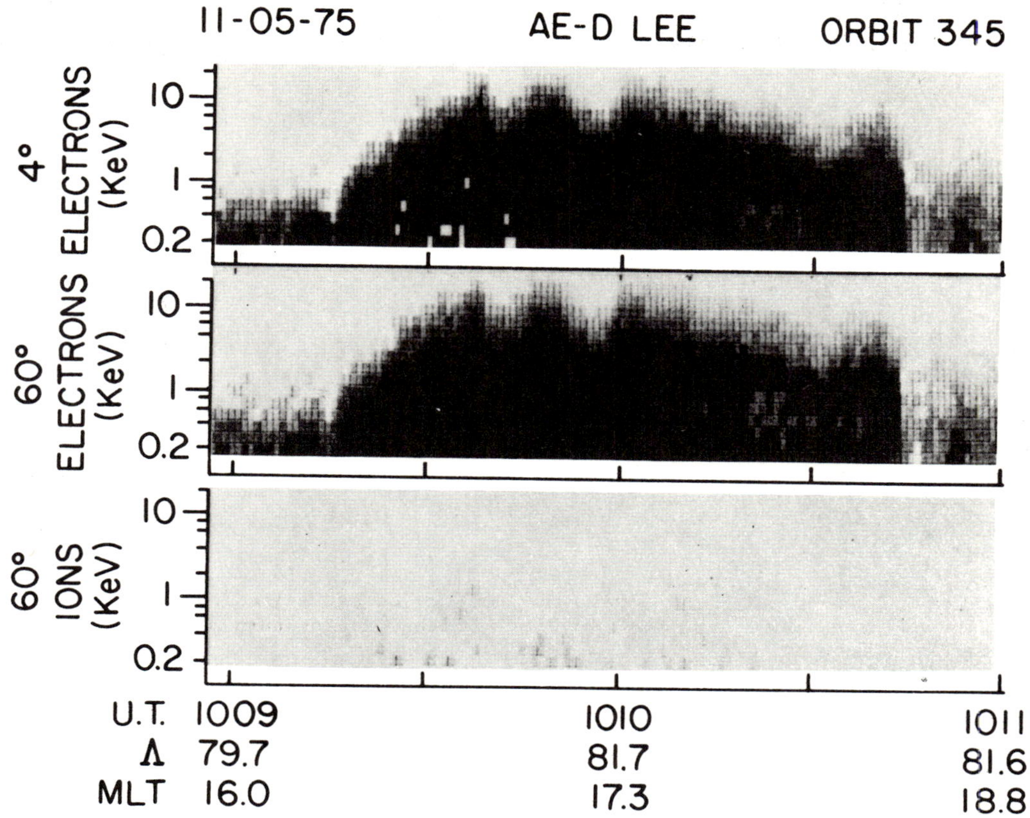

Fig. 3. An extremely extensive inverted-V event at dusk.

peak energies only above about 400 eV. Rather strikingly the more energetic events occured prior to midnight, while the least energetic were seen in the late morning hours. The instrument did observe one event whose maximum energy exceeded the energy band of the highest energy step, which was 29 keV.

The energy spectra as a function of pitch angle across an event can become very complex and varied. However, certain generalities can be presented, but exceptions are always found. Figure 7 contains a sequence of spectra across a long event. Note the following: 1. The spectra progressed through the classical hardening and softening over the event. 2. The "monoenergetic" beam existed at both small and large pitch angles. 3. The intensities at the peak energies were not always the same, nor were the peak energies always identical. However, at no time did the flux at 65° exceed the flux at 3° outside statistical uncertainties. 4. A double peaked spectrum was occassionally observed.

The lower energy peak of a double peaked spectrum is normally very field-aligned, and the electron fluxes show fast fluctuations. Such a situation was first pointed out by Arnoldy et al (1974) on the basis of sounding rocket data, and Lin and Hoffman (1979b) with satellite data.

An example of a very isotropic inverted-V is shown in Figure 8. These data were taken when the spacecraft was spinning. The peak energy as a function of pitch angle for the two stepped energy detectors showed isotropy to about 110°, which included electrons mirroring between the satellite and the atmosphere.

Proceeding to more details, high time resolution distributions can be examined using the fixed energy detectors of the instrument. Four point pitch angle distributions at 0.7, 1.4 and 2.6 keV are shown in Figure 9, with the energy of the monoenergetic beam listed across the top. Note that when the peak energy is at or above the fixed energy detector energies, the distributions from the fixed energy detector are U-shaped, with the minimum intensity lying at about 15° to 20° pitch angle.

As a result of these typical observations, we have deduced generalized characteristics of the pitch angle distributions in inverted-V events: 1. The monoenergetic beam extends to pitch angles including mirroring particles. 2. The pitch angle distributions vary from U-shaped, to isotropic. For the more prevalent U-shaped, the intensity at small pitch angles is slightly more

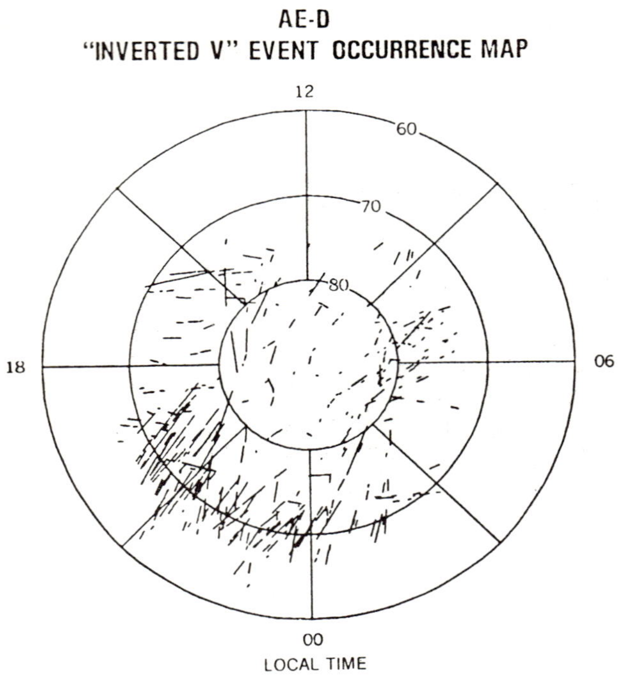

Fig. 4. Spatial occurrence map of 430 inverted-V events in invariant latitude, magnetic local time coordinates.

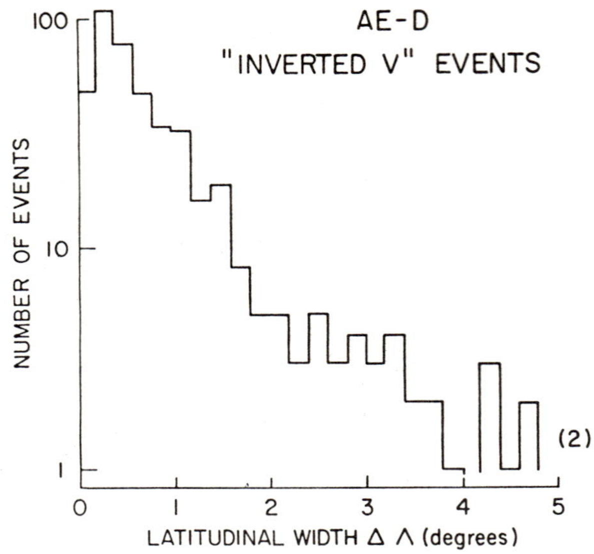

Fig. 5. Distribution of inverted-V events as a function of latitudinal width. Two events had widths greater than 5°.

intense than at mirroring pitch angles.

Finally, we look at the temperature of the monoenergetic component. This is a difficult measurement and can only be obtained from the slope of the spectrum above the peak energy, because of contamination to the beam from backscattered and mirrored particles at energies below the peak. The response of the electrostatic analyzer to this portion of the spectrum can lead to errors in the calculation of the temperature, where the real temperature is probably lower than the calculated temperature. The results from three different inverted-Vs appear in Figure 10, which shows $T_{||}$ to be proportional to the peak energy as it changes during the event. Here the error bars are due to the RMS error of the curve fitting, not to counting statistics. During an event the counting statistics actually improve as the peak energy increases, providing more accurate temperature measurements, since the flux at the peak energy seems to remain quite constant over an event. However, some of T_o may be from the above mentioned errors. The slopes of 32 such analyses cluster around 0.21 (half-width of 0.04), but they vary from one case of no slope to as large as 0.55. We will use this relationship between $T_{||}$ and peak energy when discussing acceleration mechanisms.

The combined pitch angle distributions and tem-

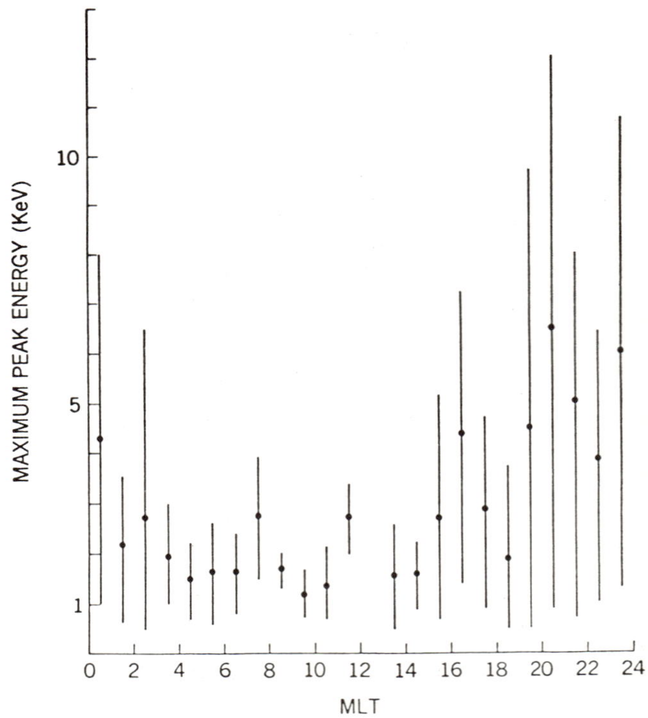

Fig. 6. Distribution of maximum peak energy of inverted-V events as a function of MLT. Solid circles are the average values for each hour interval and the bars give standard deviations.

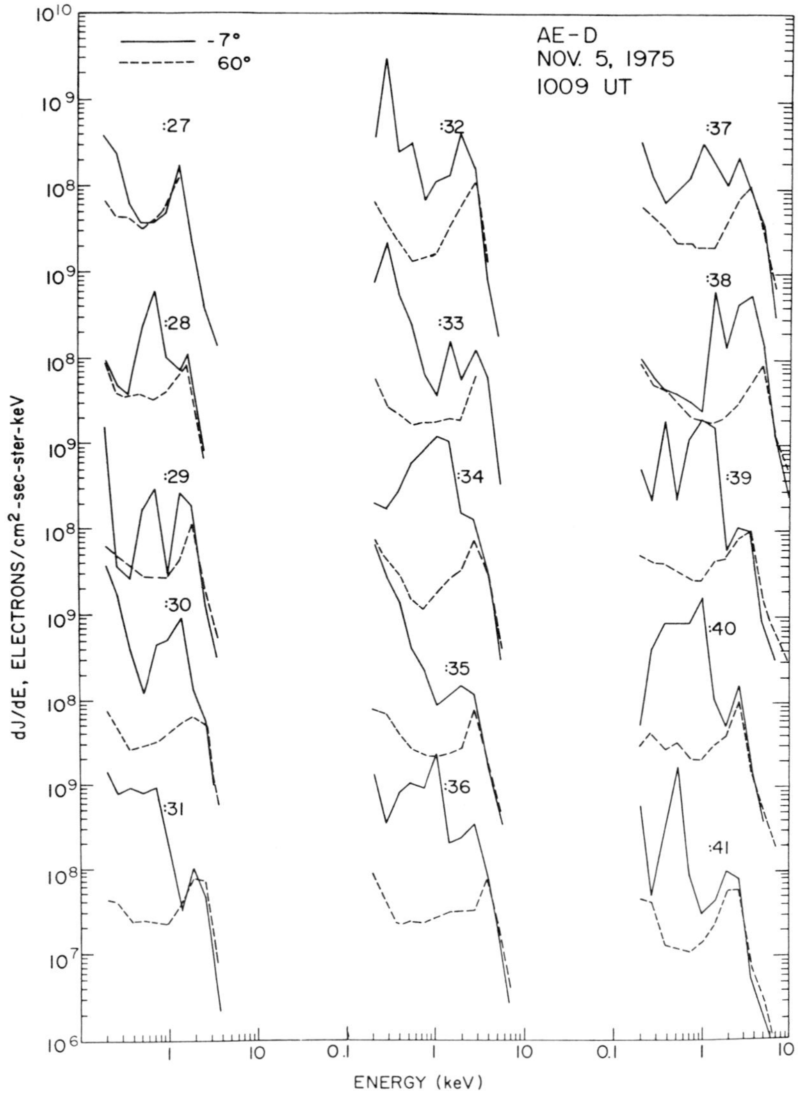

Fig. 7. Electron energy spectra during an inverted-V with a fluctuating lower energy secondary peak. The -7° detector measured electrons with pitch angles around 3°, and the 60° detector around 65°.

perature measurements provide constraints to the parameters of the source. An isotropic pitch angle distribution at the monoenergetic peak can be explained if the source temperature is not an extremely small fraction of the peak energy and the source is not near the ionosphere. The data provide two possible indicators of the source temperature: T_0, the intercept of the fitted curves in Figure 10 and $A \propto T_{\parallel}/E_p$ (since T_0 is fairly small). Both A, the order of 0.2, and the T_0's are small, so if the potential region is at least as high as 2 to 3 R_e, conservation of the first invariant can be applied to show that the fluxes at the peak energy should be isotropic at the satellite. The potential region would have to be exceedingly low in altitude to concentrate the beam to pitch angles smaller than 20°. Thus the small pitch angle portion of the U-shaped distribution is difficult to explain. On the other hand, the enhanced fluxes at mirroring pitch angles can be attributed to a constant source for these particles with no loss mechanism in the absence of scattering. However, should neither T_0 nor A be related in any way to the source temperature, and the source is cold, a highly field-aligned primary beam would result (Hoffman and Evans, 1968).

The situation with the secondaries is different. Since they are continuously created in the atmosphere (Evans, 1974) and are reflected back

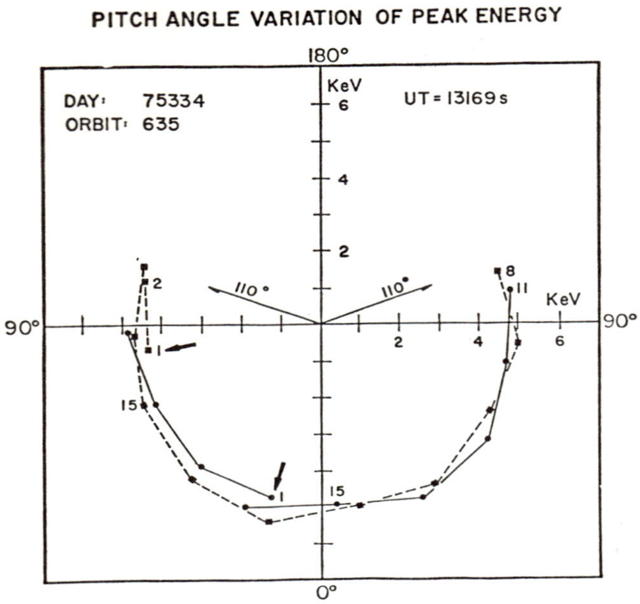

Fig. 8. Peak energy of the monoenergetic component as a function of pitch angle from the two stepped detectors during a pass with the spacecraft spinning. The solid line is from data from the -7° detector, dashed line from the 60° detector. Numbered points were acquired simultaneously.

by the potential, their pitch angle distribution would not include particles in the magnetic mirror pitch angle region. Thus, the U-shaped distributions of the fluxes at energies below the peak energy (Figure 9) are evidence for turbulent scattering in or near the potential region to fill the mirroring pitch angle domain. The presence of field-aligned fluxes below the peak energies further suggests that secondary electrons have experienced additional acceleration or heating along magnetic field lines, since a combination of secondaries and energy-degraded backscattered primaries simply reflected by the electric potential would not maintain the degree of alignment as the primary beam.

New Work

So far we have discussed the properties of inverted-V events without regard to probability of existence or magnetic activity. Recently analyses were performed to investigate such relations in the pre-midnight hours where the most extensive events occurred.

First, a sequence of 49 passes were selected in the local time interval 19 to 01 hours when the satellite was not spinning. The energy-time spectrograms were carefully searched for inverted-V events, and where there was doubt, individual energy spectra were displayed for analysis. It

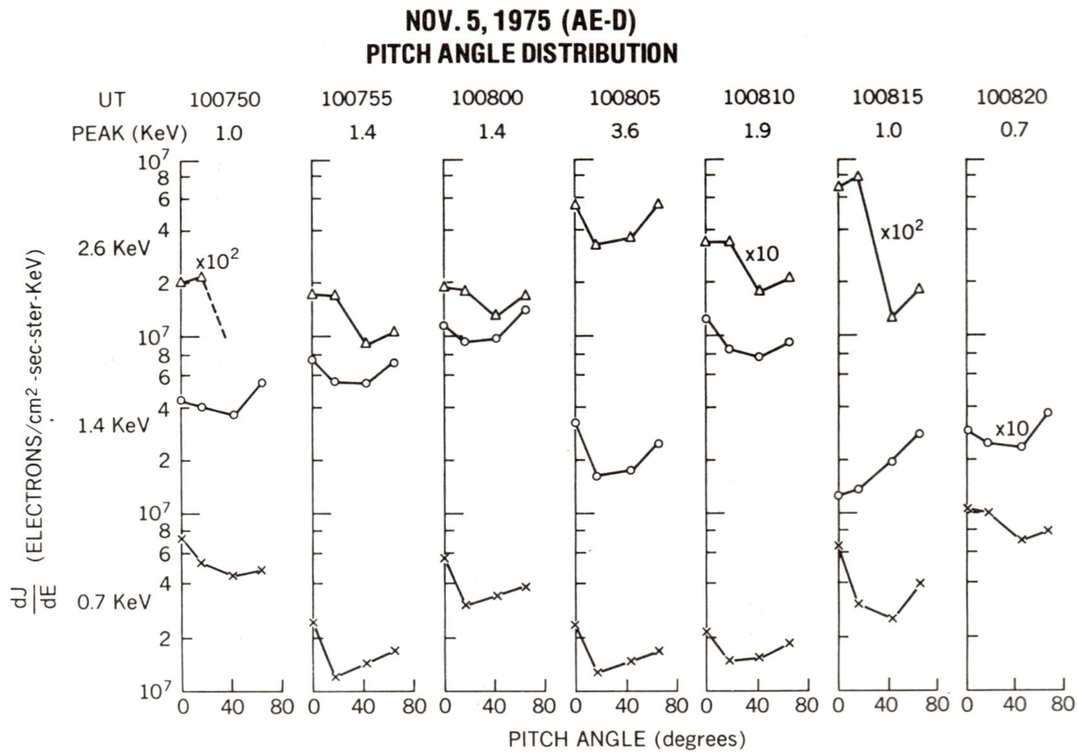

Fig. 9. Pitch angle distribution at energy levels of 0.7, 1.4 and 2.6 keV during an inverted-V event. Time for each data set is given in hours, minutes, seconds.

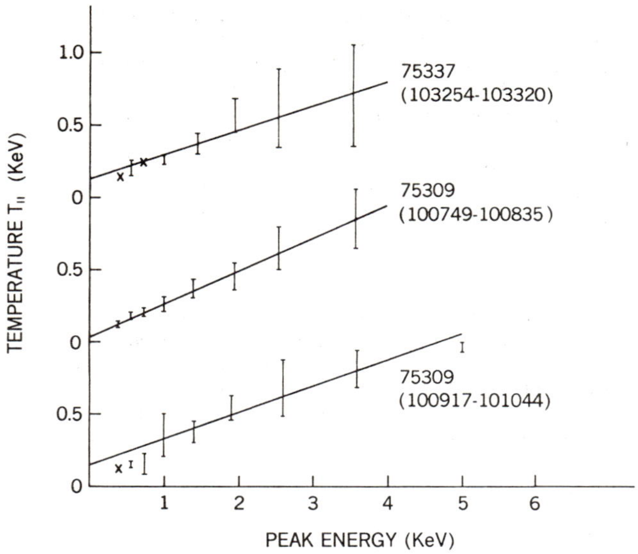

Fig. 10. Electron temperature $T_{\parallel}$ versus peak energy E_p for three inverted-V events. Straight lines are fitted to the data.

was found that on 46 of the 49 passes, at least one inverted-V appeared (see Table 2). If inverted-V structures are assumed to be produced by electric fields parallel to the magnetic field lines, this means that essentially all the time (94%), in this local time region, there exists an area of parallel electric fields. Certainly the location of this area shifts with time, and changes in size and field strength, but it is basically always there. Usually, the region is at or near the poleward boundary of all auroral precipitation (Table 2), and half the time the inverted-V structure is the dominant form of precipitation during a pass. The set of 49 passes covered a wide range of Kp, so the results are not biased towards any magnetic conditions.

For the next analysis, Figure 1 (lower) shows the format of the electron precipitation as it will be schematically illustrated in the following two figures, utilizing the terminology of Winningham et al (1975). Figure 11 contains a sequence of three passes before, during and after a substorm as identified in the magnetograms below the electron data. Kp was moderate. The first pass at 23.6 MLT was over Cape Kemenny, which was quiet, although Dixon Island after midnight and Chelyuskin farther in the morning showed recovery from a substorm. Small inverted-Vs were measured at the boundary of precipitation. The next pass is interesting because it is one of the three which showed no inverted-V, yet it occurred when magnetic activity was extensive at Murmansk and New Aalesund, over which the satellite passed. The third pass occurred well into the recovery phase, and three inverted-Vs were observed.

The next case, (Figure 12) shows a sequence of five passes around two substorms when Kp was quite high. The first pass, during which no inverted-V was found, was near Yellow Knife during a very quite period. The second and fourth passes occurred during substorms and showed inverted-Vs, in contrast to the previous example, while the third and fifth passes were during recovery phases. While there is no temporal history of the inverted-Vs between the passes spaced every two hours, note that for four passes in a sequence an inverted-V was found at a latitude of 67° to 68°.

In another analysis, 77 passes with inverted-V events in the local time interval 20 to 24 hours were separated into four categories based on substorm activity: magnetically quiet, within a half hour of a substorm, during a substorm and during the recovery. It was found that inverted-Vs occurred during all phases of magnetic activity (Table 3).

We conclude from this new work that inverted-V events are exceedingly common, especially in the pre-midnight region, so common, in fact, that they occur essentially all the time, independent of the state of magnetic activity. We further conclude that these analyses indicate that the basic mechanism for producing the inverted-V structure in the electron precipitation is nearly time stationary during the passage of the satellite through the event, and perhaps may not change significantly in location for much more extended periods.

Parallel Electric Field Mechanisms

In this section the observed characteristics of inverted-V events are compared with particle distribution properties predicted from various mechanisms proposed for producing parallel electrostatic fields. Four mechanisms have been suggested: the double layer (Block, 1975; Swift,

TABLE 2. Inverted-V Distributions 19 To 01 Hours MLT

PROBABILITY OF EXISTENCE						
$\frac{46}{49} = 94\%$						
AT OR NEAR POLEWARD EDGE OF PRECIPITATION						
$\frac{33}{46} = 72\%$						
DOMINANT PRECIPITATION FEATURE						
$\frac{23}{46} = 50\%$						
DISTRIBUTION IN Kp						
Kp:	0	1	2	3	4	5
No. Events:	0	6	12	14	12	5

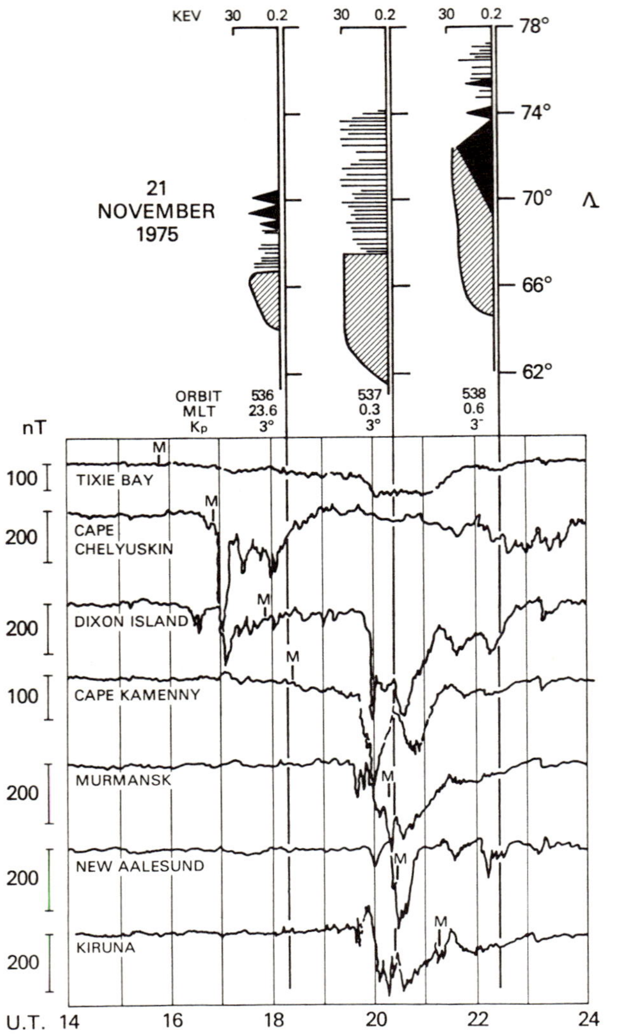

Fig. 11. Schematic illustrations of electron precipitation (see Figure 1) for three passes around a substorm displayed in the magnetograms. The times of the passes are indicated by the heavy vertical lines.

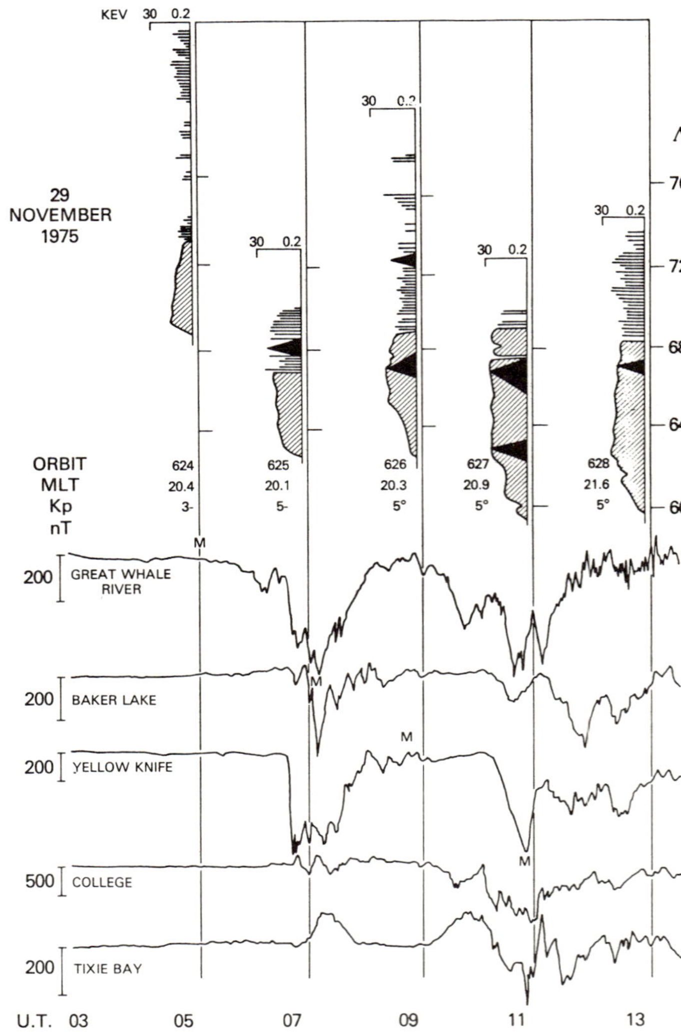

Fig. 12. Schematic illustration of electron precipitation for five passes around two substorms in the format of Figure 11.

Double Layer

In the theory of double layers developed by Block the electron temperature of the electrons emerging from the electric potential is inversely proportional to the monoenergetic beam energy (Lin and Hoffman, 1979a). This is opposite to the relation found in the data, where there was a proportionality constant of about 0.2. Of course, an inverse proportionality could have been produced by electrostatic turbulence in or below the region of the electrostatic field, yielding a temperature which is not characteristic of the primary acceleration mechanism.

Electrostatic Shock

The electrostatic shock in a collisionless plasma is a potential jump caused by thermalization of streaming ions (Tidman and Krall, 1971). The terminology of "electrostatic shock" frequently used in the literature has concepts similar to the double layer (Swift, 1975; Swift, 1975; Kan and Lee, 1980), the electrostatic shock (Tidman and Krall, 1971; Kan, 1975), extended field-aligned potential (Alfven and Falthammar, 1963; Chiu and Schulz, 1978) and anomalous resistivity (Papadopoulos and Coffey, 1974; Papadopoulos, 1977). Recently Lyons (1980) suggested that discontinuities in the magnetospheric convection electric field can generate large-scale regions of parallel potential.

TABLE 3. Inverted-V Events vs Substorm Phase

NUMBER OF PASSES BETWEEN 20 AND 24 HRS MLT CONTAINING INVERTED-V EVENTS			
MAGNETICALLY QUIET	BEFORE BAY <1/2 hr	BAY	RECOVERY
30	8	12	27

1979). In the model of Kan (1975) the energy source of the shock comes from the kinetic energy of streaming ions. No relationship between the electron temperature and the potential jump is expected. The model needs to include electron thermalization in order to conform with observations.

Extended Field-Aligned Potential

In the previously mentioned models, the potential region is limited in length along the field lines. Parallel electric fields extending over large sections of field lines can occur when electrons and ions have unequal pitch angle distributions. The parallel electric field is created in order to inhibit charge separation near the mirroring points. It can be seen from the first invariant that electron fluxes accelerated from the equatorial plane to the satellite altitude by that potential would have isotropic pitch angle distributions, so the temperature of the primary beam would be a function of the initial temperature only.

Anomalous Resistivity

The linear relationship between the temperature $T_{\parallel}$ and the monoenergetic peak energy E_p suggests that the precipitating electrons are heated in the acceleration region. This result supports the idea that the static electric field is created in the region of the anomalous resistivity. In this model the linear relationship simply means that heating is due to the same plasma instability that is also responsible for creating the electrostatic field. Furthermore, the fluctuating fluxes at energies below the primary beam found in some of the inverted-V events support the idea of heating (Lin and Hoffman, 1979b).

It is concluded, then, that by itself, anomalous resistivity is the most favorable mechanism for producing the electron precipitation structures identified in the data as inverted-V events.

Electric Field Divergence

From several points of view the recent suggestion by Lyons (1980) that discontinuities in the magnetospheric convection electric field can generate large-scale regions of parallel currents with associated parallel potential differences is very satisfying. Large-scale fairly constant field regions can be generated. These regions would appear prevalently at the boundary of auroral precipitation in the evening hours, but polar cap fields could also be generated. It provides a U-shaped potential region.

Some problems do remain, especially in the details, as pointed out by Lyons (1980). Perhaps the most serious is the difficulty in applying it to inverted-Vs occurring within the plasma sheet precipitation region, whose field lines most likely map to regions not containing discontinuities in the convection electric field.

Thus one might conclude that more than one mechanism exists for the establishment of parallel electric fields. It should also be pointed out that the effects of electrostatic turbulence on the plasma and the established electrostatic field is not yet clear (Temerin, 1981). It is only by making a complete set of particle and wave measurements in the acceleration region itself that the mechanisms of generating the electrostatic field can be determined. So far, only the S3-3 satellite has made such measurements, but the results are not yet conclusive. Future observations by the Dynamics Explorer satellites should provide important clues to the physical processes responsible for the inverted-V events.

Summary

In summary, we re-emphasize the finding of evidence for the widespread, continuous existence of electric fields parallel to the magnetic field lines. Therefore, because of the common nature of parallel electric fields, mechanisms to produce these fields must not depend upon unusual particle distributions or plasma parameters, but only upon the normal parameters of the plasma distributions and electric field topology.

Acknowledgements. C. S. L. acknowledges the support of the NASA Grant NSG-5404. We thank E. G. Fontheim of the University of Michigan for the use of Figure 8. J. Davidson of Computer Sciences Corporation assisted materially in the data analysis.

References

Alfven, H. and C. G. Falthammar, Cosmic Electrodynamics, pp. 163-167, Clarendon, Oxford, 1963.

Arnoldy, R. L., P. B. Lewis and P. O. Isaacson, Field-aligned auroral electron fluxes, J. Geophys. Res., 79, 4208, 1974.

Block, L. P., Double layers, in Physics of the Hot Plasma in the Magnetosphere, ed. B. Hultqvist and L. Stenflo, Plenum, New York, 1975.

Chiu, Y. T. and M. Schulz, Self-consistent particle and parallel electrostatic distribution in the magnetospheric-ionospheric auroral region, J. Geophys. Res., 83, 629, 1978.

Evans, D. S., Precipitating electron fluxes formed by a magnetic field-aligned potential difference, J. Geophys. Res., 79, 2853, 1974.

Hoffman, R. A. and D. S. Evans, Field-aligned electron bursts at high latitudes observed by Ogo 4, J. Geophys. Res., 73, 6201, 1968.

Hoffman, R. A., J. L. Burch, R. W. Janetzke, J. F. McChesney, S. H. Way and D. S. Evans, Low energy electron experiment for Atmosphere Explorer C and D, Radio Science, 8, 393, 1973.

Kan, J. R., Energization of auroral electrons by electrostatic shock waves, J. Geophys. Res., 80, 2089, 1975.

Kan, J. R. and L. C. Lee, On the auroral double-layer criterion, J. Geophys. Res., 85, 788, 1980.

Lin, C. S. and R. A. Hoffman, Characteristics of inverted-V events, J. Geophys. Res., 84, 1514, 1979a.

Lin. C. S. and R. A. Hoffman, Fluctuations of inverted-V electron fluxes, J. Geophys. Res., 84, 6547, 1979b.

Lyons, L. R., Generation of large-scale regions of auroral currents, electric potential and precipitation by the divergence of the convection electric field, J. Geophys. Res., 85, 17, 1980.

Potemra, T. A., W. K. Peterson, J. P. Doering, C. O. Bostrom, R. W. McEntire and R. A. Hoffman, Low-energy particle observations in the quiet dayside cusp from AE-C and AE-D, J. Geophys. Res., 82, 4765, 1977.

Papadopoulos, K. and T. Coffey, Anomalous resistivity in the auroral plasma, J. Geophys. Res., 79, 1558, 1974.

Papadopoulos, K., A review of anomalous resistivity for the ionosphere, Rev. Geophys. and Sp. Phys., 15, 113, 1977.

Swift, D. W., On the formation of auroral arcs and acceleration of auroral electrons, J. Geophys. Res., 80, 2096, 1975.

Swift, D. W., Auroral mechanisms and morphology, Rev. Geophys. and Sp. Phys., 17, 681, 1979.

Temerin, M., Plasma waves on auroral field lines, this issue, 1981.

Tidman, D. A. and N. A. Krall, Shock Waves in Collisionless Plasmas, Intersciences, New York, 1971.

Winningham, J. D., F. Yasuhara, S. I. Akasofu and W. J. Heikkila, The latitudinal morphology of 10 eV to 10 keV electron fluxes during magnetically quiet and disturbed times in the 2100-0300 MLT sector, J. Geophys. Res., 80, 3148, 1975.

AURORAL PARTICLE DISTRIBUTION FUNCTIONS AND THEIR RELATIONSHIP TO INVERTED Vs AND AURORAL ARCS

J. F. Fennell, D. J. Gorney, and P. F. Mizera

Space Sciences Laboratory, The Aerospace Corporation
P. O. Box 92957, Los Angeles, California 90009

Abstract. Utilizing data from three satellites we show the relationship between the particle energy and angular distributions, the resulting auroral forms, and spatial features such as the inverted V and electrostatic structures. The occurrence of field-aligned electrons over a wide range of energies near and at the edges of inverted V structures in the low altitude local evening auroral region is an outstanding feature in the data. The DMSP pictures often show a brightening of the auroral form in coincidence with similar intense electron precipitation at the edges of inverted V structures. Preliminary examination of the S3-3 data has produced only a few such events with a tendency for them to occur at lower altitudes. As a separate topic, a detailed description of a high altitude inverted V structure is presented using S3-3 data. The region contained strong perpendicular electric field structures embedded therein. The potential drop in the region peaks near 16 kV with 10-12 kV below and 4-6 kV above ~ 8000 km. The modification of the electron distributions due to strong electric field structures is analyzed. The signature of the acceleration in the electrons is consistent with an acceleration region that starts 2000-4000 km above the satellite altitude of 8000 km and extends well below the satellite.

Introduction

There have been available for some time now several comprehensive satellite data sets which contain detailed measurements of the auroral particle spectra (Frank and Ackerson, 1972; Hoffman and Burch, 1973; Winningham et al., 1975; Burch et al., 1975; Meng et al., 1976; Mizera et al., 1976; Mizera and Fennell, 1977; Ghielmetti et al., 1978; and others) and in a few cases auroral imaging (Rogers et al., 1974; Anger and Lui, 1973). The so called inverted V structure has been associated with auroral arcs (Ackerson and Frank, 1972a, b; Frank and Ackerson, 1972; Mizera et al., 1975; Meng, 1976). The particle observations were usually made at low to medium altitudes on satellites and rockets which measured precipitating particles, and the results were compared with photographs of auroral structures. The field aligned nature of the precipitating electrons in the auroral regions was observed by many experimenters. Comparisons of low altitude particle distributions with auroral forms has led to a recognition that the correlation of the inverted V structure and bright forms is more than a crude spatial relationship. There are features in the visual forms which have direct counterparts to features of the particle precipitation. One such feature often occurs near the edges of the inverted V structure and is discussed below.

A second and equally interesting development has been the recent ability to examine the auroral particle distributions within the acceleration region itself (Mizera and Fennell, 1977; Mizera et al., 1980; Mozer et al., 1977; Sharp et al., 1979; Torbert and Mozer, 1978). Of particular interest in this report is the effect that the "electrostatic shocks" (Mozer et al., 1977; Torbert and Mozer, 1978) have on the particle distributions. As is discussed below, the particle distributions can be used to infer the extent of the acceleration region along an auroral field line.

In order to treat these subjects we have brought together data from three different USAF satellites; namely P72-1, DMSP-29, DMSP-F2 and the S3-3. The strengths of each data set are utilized to enhance our understanding of the phenomena. The primary features of the data sets are: 1) the good angular resolution of the P72-1 data at low altitude (~ 800 km circular polar orbit), 2) the combination of simultaneous precipitating electron spectra, auroral X-rays and photographic auroral images (image only on DMSP-29) from the DMSP satellites (~ 800 km altitude, dawn-dusk polar orbits), and 3) the good angular distributions of electrons and ions plus electric field measurements within the auroral acceleration region from the S3-3 satellite (250 x 8200 km altitude polar orbit). The details of the experiments have been described elsewhere (Rogers

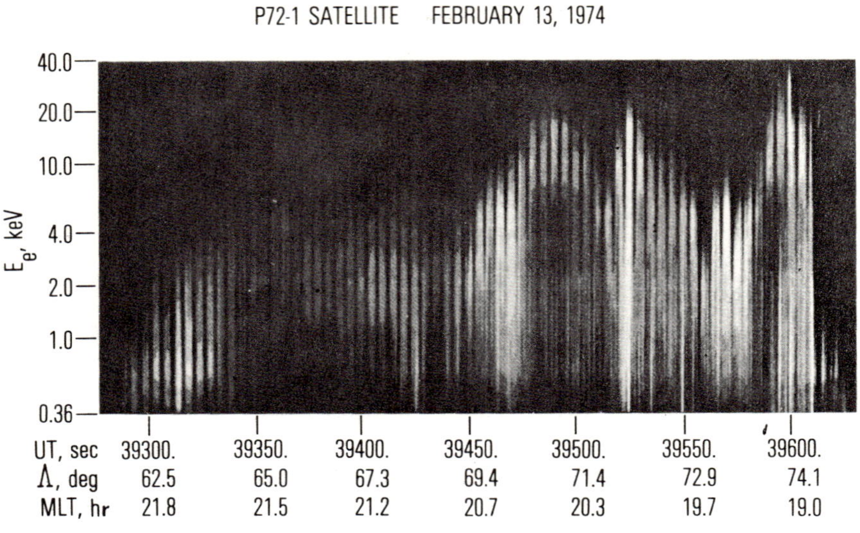

(a)

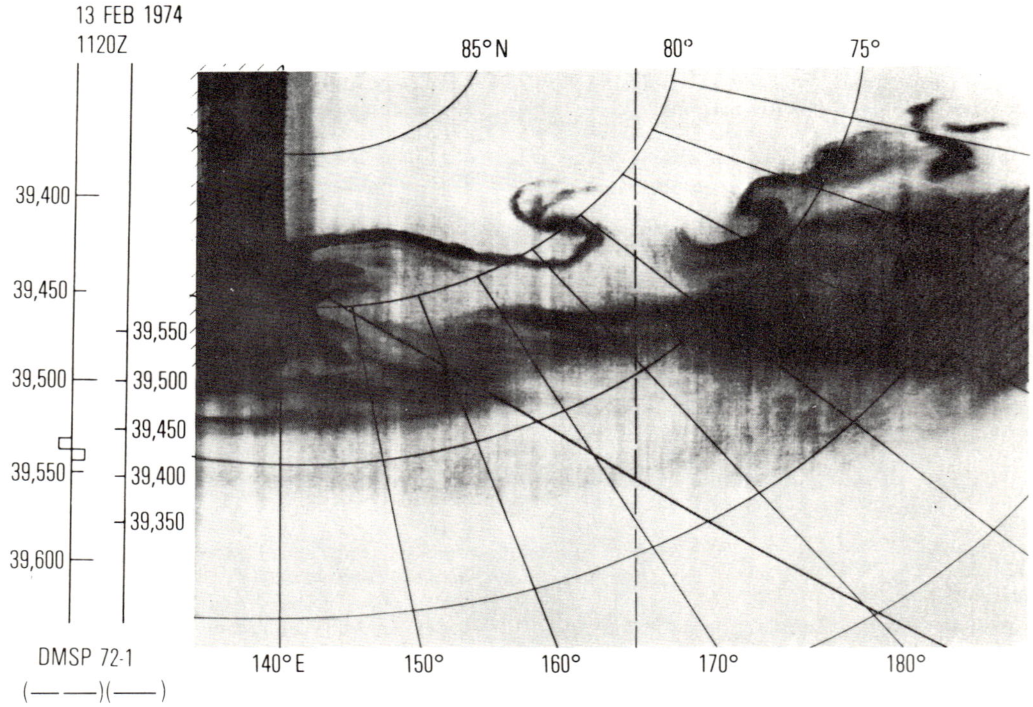

Fig. 1. (a) Spectrogram of late evening auroral electrons taken on February 13, 1974 by the P72-1 satellite. Intensity is proportional to electron energy flux. An inverted V structure is centered near 39480 sec.

(b) Auroral photograph taken by the DMSP satellite at 1058 UT on February 13, 1974. The trajectories of the DMSP (dashed line) and P72-1 (solid line) satellites are shown along with the time (left hand scales). The P72-1 satellite crosses an auroral form near 39480 sec which corresponds to the inverted V structure in Figure 1a.

et al., 1976; Mizera and Fennell, 1977; Mozer et al., 1977; Eather, 1979; Meng, 1979).

By combining the basic results from these sets of measurements we can arrive at an understanding of the auroral processes which is not possible from an individual data set.

Field Aligned Electrons Over Wide Energy Ranges

We will first examine the low altitude inverted V electron signature. The field aligned nature of the precipitating electrons in the inverted V regions has been described by many experimenters on both rockets and satellites. Generally, the angular distributions of the electrons in the inverted V were found to be field aligned for energies corresponding to the peak of the precipitating electron spectrum (Mizera et al., 1976).

Figure 1 shows an electron spectrogram from the P72-1 satellite (top panel) and the corresponding auroral photograph from the DMSP satellite (bottom panel) taken on February 13, 1974. Both satellites are near 800 km altitude in the northern premidnight auroral region. As can be seen in the bottom panel of Figure 1 both satellites crossed the auroral form near 39500 sec. This is near the peak of the inverted V in the top panel. The peak energy of the inverted V can be seen to be near 10 keV on the spectrogram.

DMSP is moving equatorward and P72-1 is moving poleward so that the satellites have simultaneous coverage only at 39500 sec. All of our comments assume that the auroral forms shown in the DMSP image were present when P72-1 traversed the region.

As P72-1 proceeded poleward near 39525 sec UT it encountered a strong precipitation over a range of energies, lasting less than 12 seconds. Careful examination of the auroral photograph shows a very thin auroral "wisp" near the P72-1 ground track (within the errors of the mapping) near 39525-39530 sec. This "wisp" may be an isolated narrow arc. The "wisp" intensity does not appear as great as the main arc although the particle fluxes estimated from the spectrogram are quite intense.

The details of the inverted V for selected satellite spins from P72-1 are shown in the bottom panels of Figure 2. The satellite spin period is ~ 5 sec. Note that the energy at which the electron distributions peak near 0° pitch angle increases from 0.62 to 1 keV at $\Lambda \sim 69.5°$ to ~ 9.7 keV at $\Lambda \sim 70.8$. The energy then rapidly decreases (not shown in complete detail) again at higher latitude. This is the classic inverted V structure as seen by an experiment which measures both electron energy and pitch angle. Such angular distributions and energy dependent field aligned fluxes have been used to evidence the existence of quasi-static parallel electric fields (Evans, 1974; Mizera et al., 1976).

The intense field-aligned fluxes that were observed just poleward of the inverted V in Figure 1 are shown in the bottom right panel of

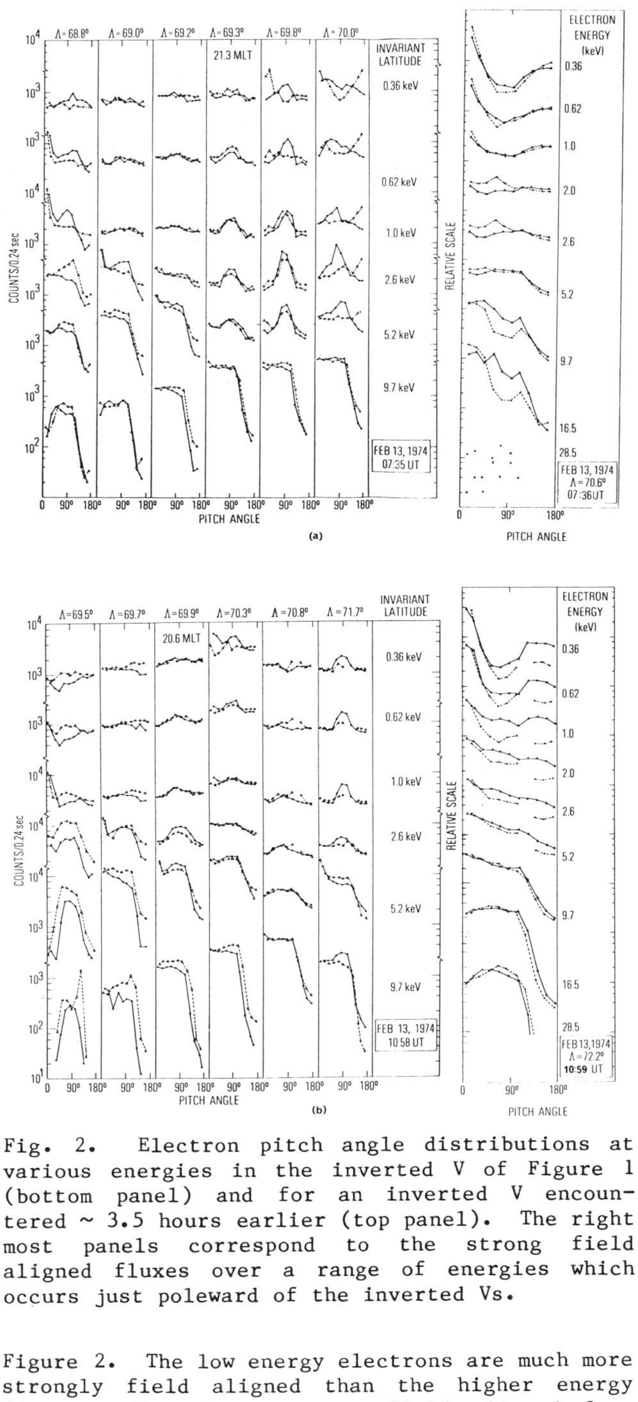

Fig. 2. Electron pitch angle distributions at various energies in the inverted V of Figure 1 (bottom panel) and for an inverted V encountered ~ 3.5 hours earlier (top panel). The right most panels correspond to the strong field aligned fluxes over a range of energies which occurs just poleward of the inverted Vs.

Figure 2. The low energy electrons are much more strongly field aligned than the higher energy fluxes. The electrons are field aligned from 0.36 to 9.7 keV. In this case the field alignment over the wide energy range existed only during one spin period.

This field alignment at the edge of an inverted V structure is not uncommon. Another example is shown at the top of Figure 2. These data were taken ~ 3.5 hours earlier in the same region.

Again, we see the signature of the inverted V structure with the energy of the field aligned electron flux increasing from 0.62-1.0 keV at $\Lambda \sim 68.8°$ to 9.7 keV at $\Lambda \sim 69.3$. The field alignment disappeared and the peak energy decreased at higher altitude. At the poleward edge we once again observed a strong field alignment of electrons over a wide energy range (0.36 to 16.5 keV). The lower energy electrons again showed much stronger field alignments.

As can be seen in Figure 2 the field aligned fluxes are generally well within the atmospheric loss cone (~ 60° for these data) and the backscatter flux is usually much less than the downward fluxes in these edge regions. This is generally not the case at the center of the inverted V where the fluxes below the energy peak are often observed to be isotropic.

It has been argued that such field alignment of the auroral electrons over a wide range of energies is not consistent with a quasi-static parallel electric field acceleration (Whalen and Daly, 1979). We refer the reader to that paper for a discussion of a possible mechanism.

We showed in Figure 1 that the inverted V edge feature (field aligned fluxes) appeared to correspond to a discreet auroral form or "wisp". Do all such features correspond to auroral forms? Do these edge features correspond to a low altitude auroral process? In order to shed more light on this subject we must examine evidence from other observations.

Enhancements in the precipitating electron energy flux over a wide range of energies often appear in the DMSP data at low altitude. Since the DMSP satellite instrument measures only precipitating electron fluxes we cannot tell if these enhanced fluxes are field aligned, but the strong similarities with the P72-1 inverted V profiles is compelling. If we assume the intense precipitating fluxes at the edges of the inverted Vs are the field aligned fluxes we can then examine the auroral forms to see if there are optical features associated with such fluxes.

We present only one example of the many particle spectrogram and auroral pictures we have examined. Figure 3 shows a composite of auroral forms, bremsstrahlung X-rays and precipitating electron fluxes for December 2, 1977. The inverted V structure near 45430 sec UT is reminiscent of Figure 2a. The electron fluxes show strong precipitation at both edges. Examination of the auroral form shows a brightening of the emission at both positions. (The particle fluxes must be mapped from 800 km altitude along the field line to 100 km altitude. This amounts to shifting the electron fluxes poleward in time by 5 - 6 seconds.) Thus, in this instance the "edge" feature corresponds to "discrete" auroral emissions.

Basically, we found that when the maximum energy and intensity of the edge features were comparable to those of the center of the inverted V itself then auroral forms or brightening of the form were observed at that point in the DMSP data. If the edge feature had high energy-flux but low compared to that in the inverted V then no separate feature was distinguishable in the DMSP pictures. Also, the inverted V structure often appears as two half Vs with maximum energy flux at the juncture between the halves and strong flux enhancements over a wide range of energies at the same point. In many such cases the brightest part of the auroral form corresponded best to the position of the juncture.

This short survey implies that field alignment of electron fluxes over a wide range of energies may be a fairly common feature of the edges of the local evening and premidnight inverted V structures (see also Wahlen and Daly, 1979; Craven and Frank, 1975). We have not yet performed a statistical study to determine a probability of occurrance for the phenomenon.

Since all the above and previous observations have occurred at relatively low altitudes (< 1000 km) we used the S3-3 data set to see if this phenomenon was observed at higher altitudes. We have completed a preliminary search of the local evening data starting with low to intermediate altitudes (250-6000 km). The search turned up several candidates most of which were at low altitudes (< 3000 km). Although we have examined large amounts of data in the 6000-8200 km altitude region, we have not found cases of strong field alignment in the electron fluxes over wide ranges of energies. In fact, it is relatively rare to find field aligned electrons at the higher altitudes. A more exhaustive study of the complete S3-3 local evening data base is in progress and may surface a few examples, but it appears that the phenomenon is a relatively low altitude one in this region.

Figure 4 is an example of an S3-3 spectrogram which shows the electron field alignment over a wide range of energies near 2110 UT at ~ 2400 km on April 3, 1978. The right panel shows a plot of the electron angular distributions channel by channel. Note that the maximum energy at which the field alignment occurs is low compared to the lower altitude examples in Fig. 2. The majority of the electron energy flux in the inverted V is carried by the ≥ 2 keV electrons. The field aligned electrons at the edge do not contribute significant energy flux. One would not, in this case, expect an auroral emission to correspond to the edge feature unless the electrons gain much more energy between 2400 km and the top of the atmosphere.

High Altitude Inverted V Structures

The S3-3 data may not contain examples of strong field alignment of the electron fluxes very often but it does contain evidence of strong accelerations. In this section we will examine in some detail the particle distributions in an inverted V structure. The analysis emphasizes the information about the particle accelerations

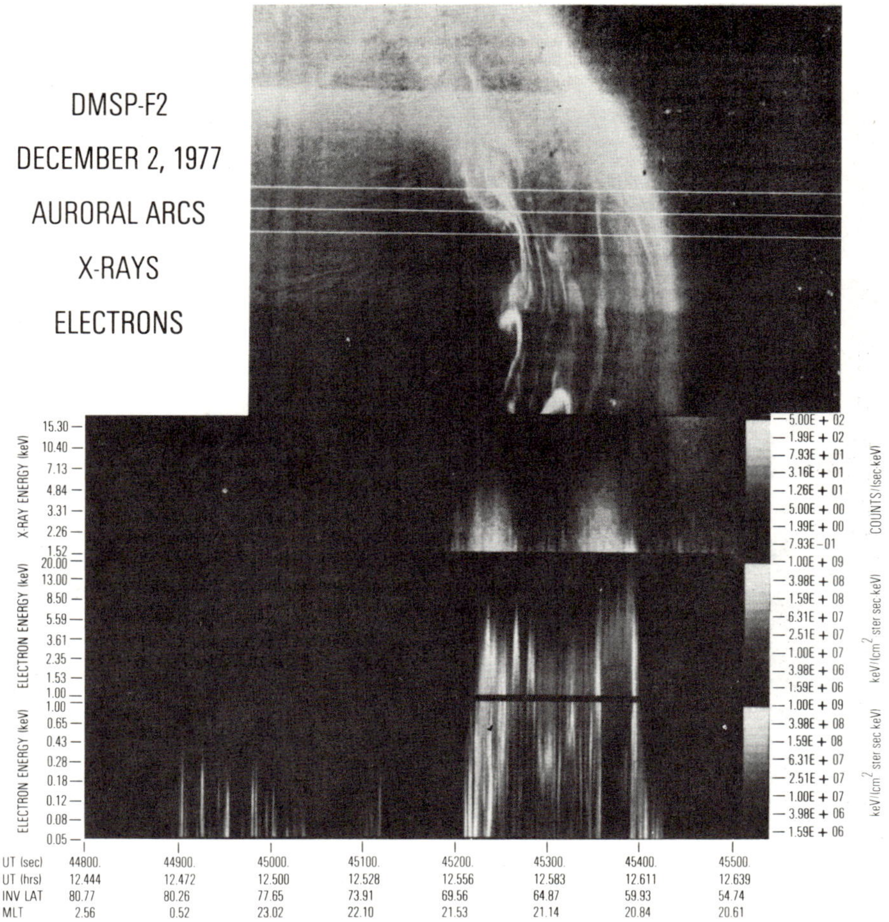

Fig. 3. Example of composite auroral photograph (top panel) and spectrograms of precipitating electron fluxes (bottom two panels) and bremsstrahlung X-rays from the atmosphere (middle panel). The auroral forms, X-rays and > 1 keV electron fluxes start at 45200 sec UT near $\Lambda = 59.9°$.

which is contained in the particle distributions such as the total potential drop and the distribution of the potential above and below the satellite altitude as the structure is traversed. The inverted V in question has been discussed as part of a more inclusive analysis using the ion composition data, AC and DC electric fields and currents by Mizera et al. (1980).

The signatures we examine are impressed on the particles by the acceleration. These signatures are the enhanced electron loss cones and upflowing ions resulting from a potential drop existing below the spacecraft and strong peaks in the precipitating electron spectra evidencing the potential drops above the satellite. As is obvious from the terminology it is assumed that the main acceleration mechanism acting is a field aligned quasi-static electric field.

Figure 5 is a spectrogram showing an inverted V signature in the electrons from 0255 to 0258 UT on September 6, 1976. The S3-3 satellite was at ~ 8000 km altitude near 18 MLT in the auroral zone. The ions corresponding to the inverted V are seen as upflowing beams and the electron loss cones are enhanced at low energies. These features are both indicative of a potential drop below the spacecraft which accelerates ions upward and electrons downward. The electron fluxes have peaked spectra (ref. Figure 7) indicative of a potential drop of the same sense existing above the spacecraft.

These particle signatures can be used to estimate the total potential drop, the potential drop above and the potential drop below the satellite as a function of latitude. This was done for this event and is presented as Figure 6. The potential profile has the appearance of two abutting inverted V structures with peak potential drops of 16-18 kV near 0256:10 UT and 12-14 kV near 0257:05 UT. The ratio of the potential drop

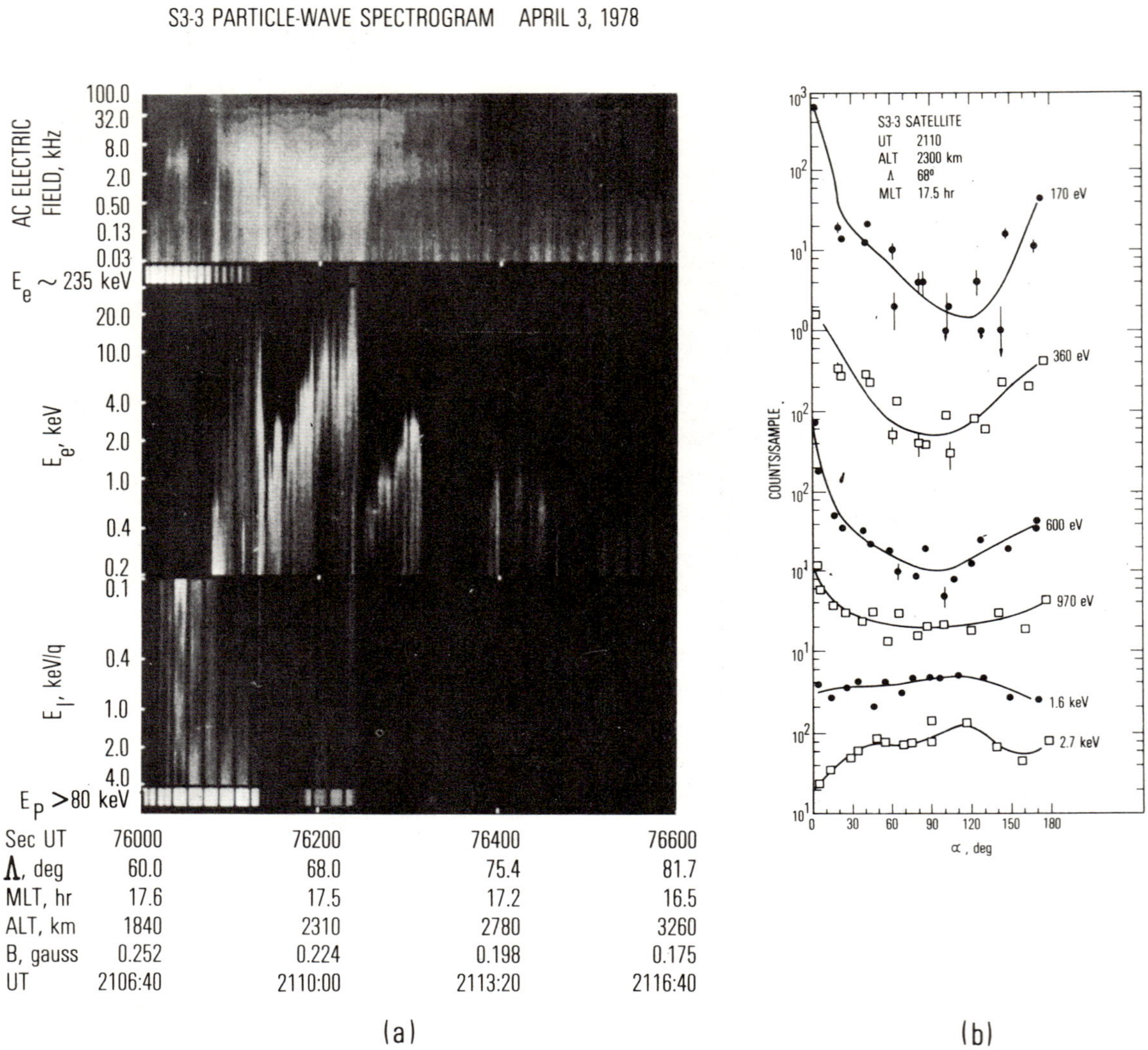

Fig. 4. (a) Spectrogram of S3-3 satellite data on April 3, 1977 near 2300 km.

(b) Pitch angle distribution of intense field aligned fluxes at the edge of the inverted V in panel a.

below the satellite to that above the satellite, V_B/V_A, was not constant in this structure as is shown in Figure 9a. It is clear that prior to the first peak in the total potential and after the second the potential drop was primarily below the spacecraft. At the times of the peaks the potential drop was either equal above and below the satellite or slightly greater above (a factor of 2 or less).

In Figure 7 we show sample electron distribution functions and electron spectra taken at specific pitch angles. Figure 7a shows the plasmasheet electron distribution function measured just outside the inverted V structure. Figures 7b to 7d show the distribution functions measured just inside the inverted V, during the traversal of an electric field structure (see also Fig. 1 of Mizera et al., 1980) and just after traversing the electric field structure, respectively.

The straight lines which intersect at the origin in Figure 7a are the boundaries in velocity space corresponding to the 100 km atmospheric loss cone. We have assumed adiabatic motion of the electrons and that no potential drop existed below the satellite.

It has been shown (Whipple, 1977; Chiu and Schulz, 1978; Croley et al., 1978) that if the potential drop along a magnetic field line is the

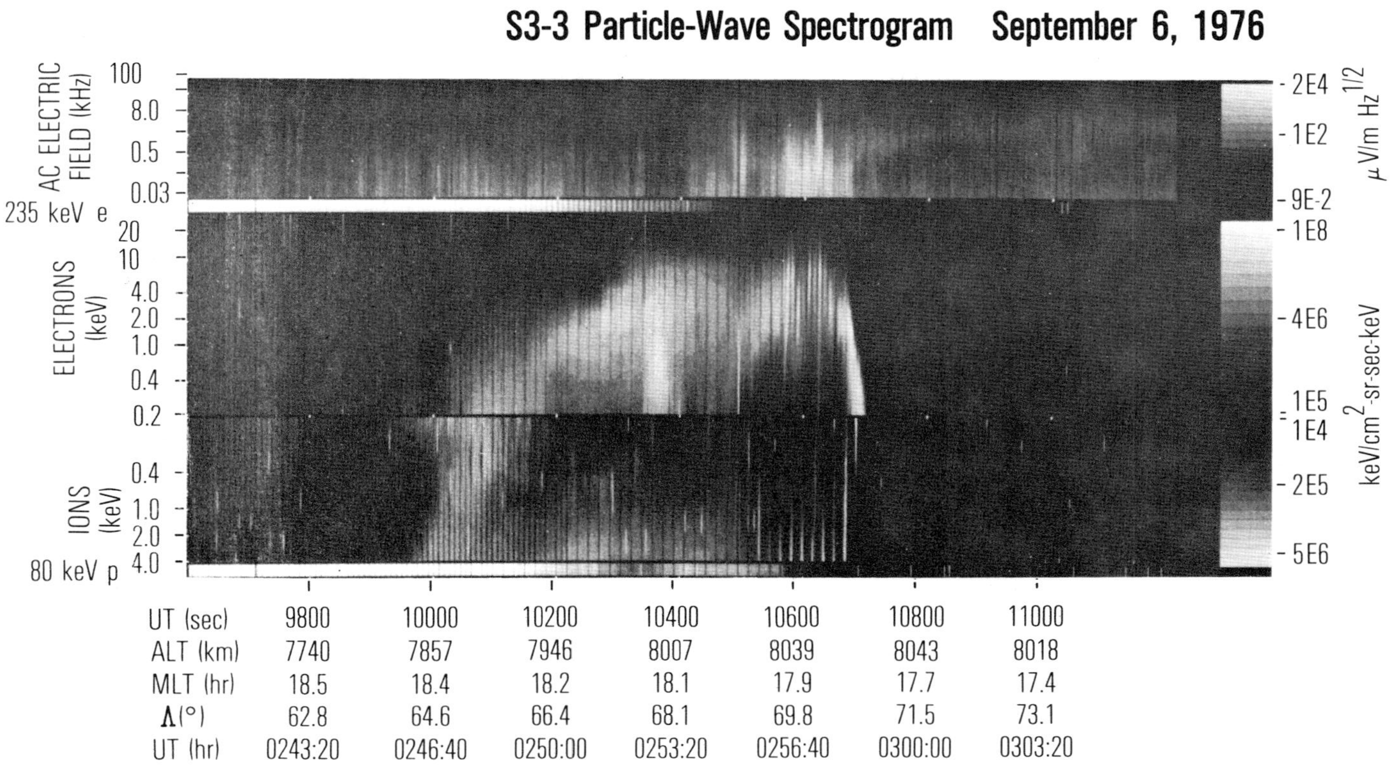

Fig. 5. Spectrogram of S3-3 data taken on September 6, 1976. An inverted V structure was encountered between 10500 and 10700 sec. UT.

result of an electric field of suitable form, and one requires conservation of energy and the first adiabatic invarient, application of Lionville's theorem leads to equations representing particle boundaries in velocity space. The boundary for electrons which come from an arbitrary point below the spacecraft is a hyperbola in velocity space. The region of velocity space to which magnetospheric electrons (i.e., electrons which originate above the satellite) have no adiabatic access is bounded by an ellipse. The shape of the ellipse is determined by the distance to the upper bound of the potential drop. If the majority of that potential drop occurs very far above the satellite the ellipse degenerates to a circle. If the majority of the potential drop above the satellite is local, the ellipse has a very high eccentricity. The region bounded between the ellipse and the hyperbola represents electrons trapped between their magnetic mirror points below the satellite and the potential drop which extends above the satellite (see Croley et al. (1978)). This can be represented by comparing the $V_\parallel$ and $V_\perp$ intercepts ($Vi_\perp$ and $Vi_\parallel$) of the ellipse and constructing a ratio $Vi_\perp/Vi_\parallel$ as a function of the altitude region where the majority of the potential drop would occur. Such a set of curves (one for each satellite altitude) would look like Figure 9b. Note that these curves are very insensitive to ratios that are either very large or very close to unity. In the intermediate range, the ratio $Vi_\perp/Vi_\parallel$ is a sensitive diagnostic of the extent of the potential drop. (For example the ratio region between ~ 1.3 and 2.0.).

In Figures 7b, 7c and 7d we have drawn on the distribution functions the ellipses and hyperbolas corresponding to potential drops of approximately 1.4 kV, 4 kV and 3 kV above the satellite and about 6 kV, 10 to 12 kV and 4 to 8 kV below respectively for these data. The ellipses are represented by circles under the assumption that the majority of the potential drop above the spacecraft occurred far above the satellite. As can be seen, the circle is a pretty good match in Figure 7b but not so in Figures 7c and 7d. The distribution function plots in Figures 7c and 7d are clearly not well represented by a circular trapping boundary, but the most appropriate eccentricity is not clear. The hyperbolic loss cone boundaries represent the data well at high energies, but it is unclear whether they are appropriate for the low energy particles. The following analysis technique has been applied to better define these population boundaries.

The electron distribution function in Figure 7a can be taken as representative of the unaccele-

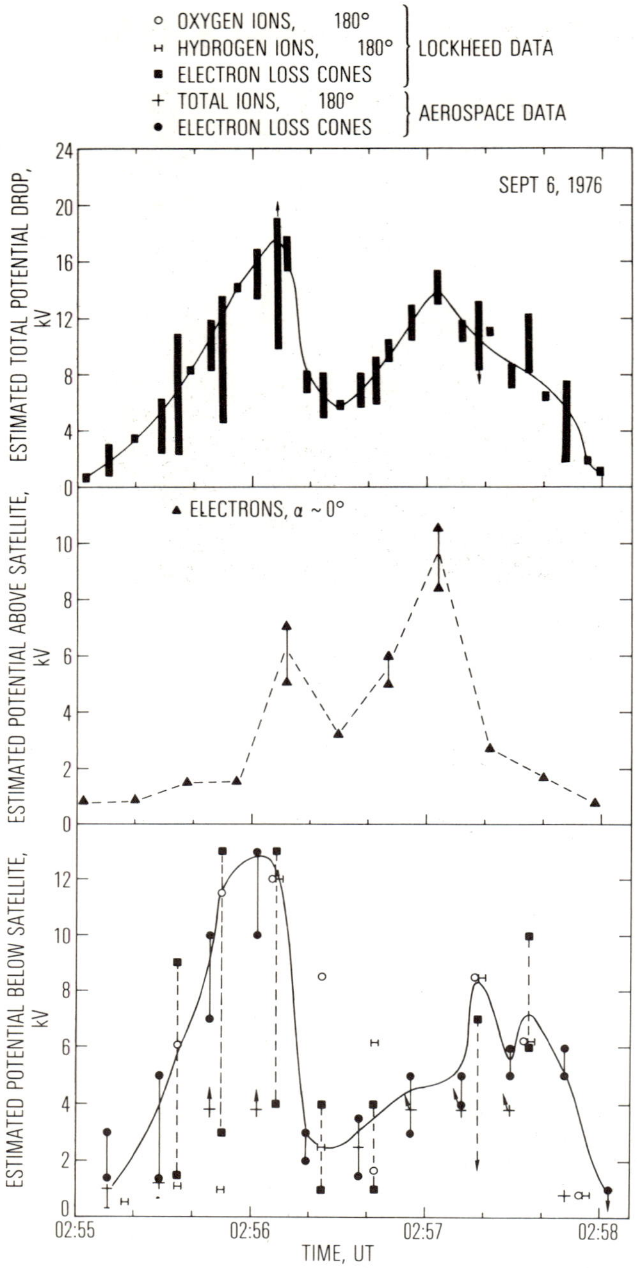

Fig. 6. Potential drops along the magnetic field lines as estimated from the particle data. The total potential drop (top panel), potential drop above the satellite (middle panel) and potential drop below the satellite (bottom panel) are estimated for the inverted V structure in Figure 5 (after Mizera et al., 1980).

rated plasma sheet electrons. If a suitable functional form is fit to the velocity dependence of this plasma sheet distribution, outside the loss cone, one probably has a good representation of the unaccelerated source distribution for the auroral electrons. Subtracting this unaccelerated distribution from the measured distributions produces a display which indicates how particles have been redistributed in velocity space by the auroral potential drop. To obtain reasonable difference plots the high energy tail of the source distribution is normalized to that of the measured distributions. Since the highest energies (10-30 keV) should be relatively unaffected by the potential drop above the spacecraft ($\lesssim$ 7 keV, see Fig. 6 of Mizera et al., 1980) the normalization should be valid and only the shape of the source distribution becomes important.

In Figure 8a the difference plot obtained by subtracting the smoothed source distribution from the measured plasma sheet distribution, of Figure 7a, is shown. Since the source distribution is assumed to be totally isotropic the difference plot should show only the effect of a loss cone which exists in the real distribution. Contour F in the difference plot corresponds to zero difference. Contours G through M correspond to depletions and A through E represent enhancements of the measured distribution over the unaccelerated distribution. The contours are logarithmically spaced with five contour intervals corresponding to one order of magnitude. The straight lines which intersect at the origin, in Figure 7a, are the boundaries in velocity space corresponding to the 100 km atmospheric loss cone as discussed above. The measured loss cone is well represented in this plot, since most of the negative difference contours lie on or inside the region bounded by the straight lines.

This procedure was applied to each of the distributions in Figure 7 and the corresponding difference plots are shown in Figure 8. The same circles and hyperbolas used in Figure 7 are drawn on the difference plots 8b, 8c and 8d. Figure 8b shows a depletion in the loss cone and trapping region with the zero difference contour, F, corresponding very well to both the circle and hyperbola. This would indicate that for Figure 7b the distribution function is consistent with the potential drop above the satellite being predominantly far above the satellite.

Figures 8c and 8d show that a good match is attained between the hyperbola and the zero difference contour at the large parallel speeds (> 0.4×10^5 km/sec and > 0.35×10^5 km/sec respectively). The zero contours extend parallel to the $V_\perp$ axes for these cases. The other contours such as H and I have elliptical shapes and we might estimate the contours cross the $V_\perp$ axis near 7×10^4 to 8×10^4 km/sec in Figure 8c and 4.5×10^4 to 5.5×10^4 km/sec in Figure 8d. The dashed lines show our estimate of the $V_\parallel$ intersection for a 'best' ellipse based on the peak of the precipitating electron spectra. These correspond to ~ 3.6×10^4 km/sec for Figure 8c and ~ 3.0×10^4 km/sec for Figure 8d. The resulting ratios $Vi_\perp/Vi_\parallel$ are ~ 1.9 to 2.2 and ~ 1.5 to 1.8 for Figures 8c and 8d respectively. Com-

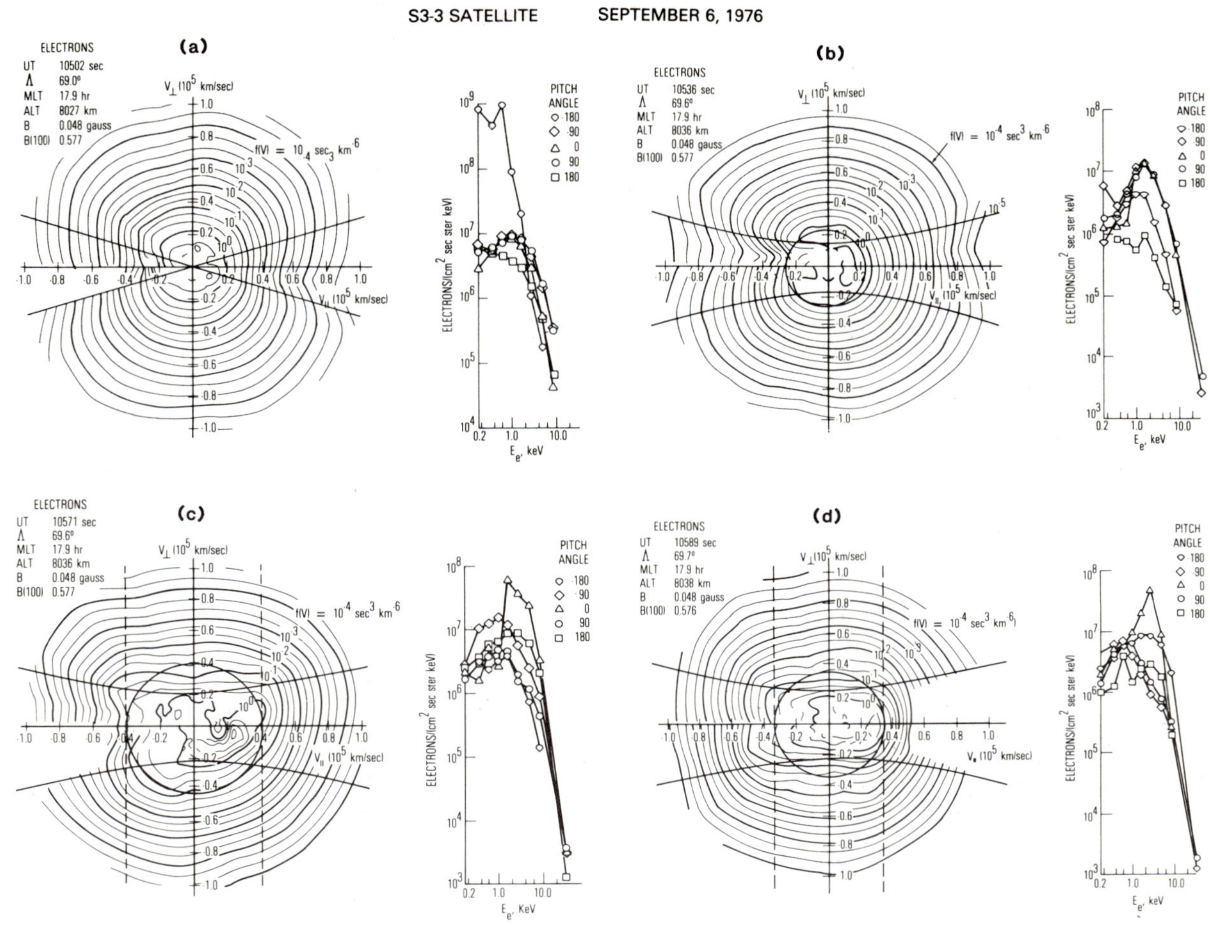

Fig. 7. Electron isodistribution function contours in velocity space and electron energy spectra at specific pitch angles. Data taken in subauroral plasma sheet region (a), near low latitude edge of inverted V (b), at the time of electrostatic 'shock' structure (c) and just after leaving electrostatic structure (d). For details see text.

paring with Figure 9b for a satellite altitude of ~ 8000 km we see that the majority of the potential drop above the spacecraft is roughly within 1500-3000 km of the satellite for Figures 7c and 8c and ~ 3000-5000 km for Figures 7d and 8d.

Thus it appears that the technique of subtracting a source distribution from the measured one highlights the boundaries corresponding to the loss cone and trapping regions as defined above. This clear identification of particle boundaries yields information which allows the particle measurements to be used as diagnostics of the altitude distribution of the auroral potential drop.

Summary and Discussion

We have discussed two separate features of the auroral inverted V structure. The first part of the paper indicates that there is a lower altitude ($\lesssim$ 5000-6000 km) process acting which causes electrons to be precipitated and energized such that the angular distributions are highly field aligned over a wide range of energies. This process appears to be associated with the edges of the local evening inverted V structures. We cannot definitely conclude that the process occurs only in the presence of an inverted V structure. This must await a more systematic study of existing and new data.

These 'edge' features appear to correlate with auroral brightening or arcs when their energy flux is high enough. Why they are seen preferentially in the low altitude data and what mechanism causes them is at this point unknown. It has been argued that they cannot result from quasistatic field aligned electric field acceleration but may be a result of acceleration by

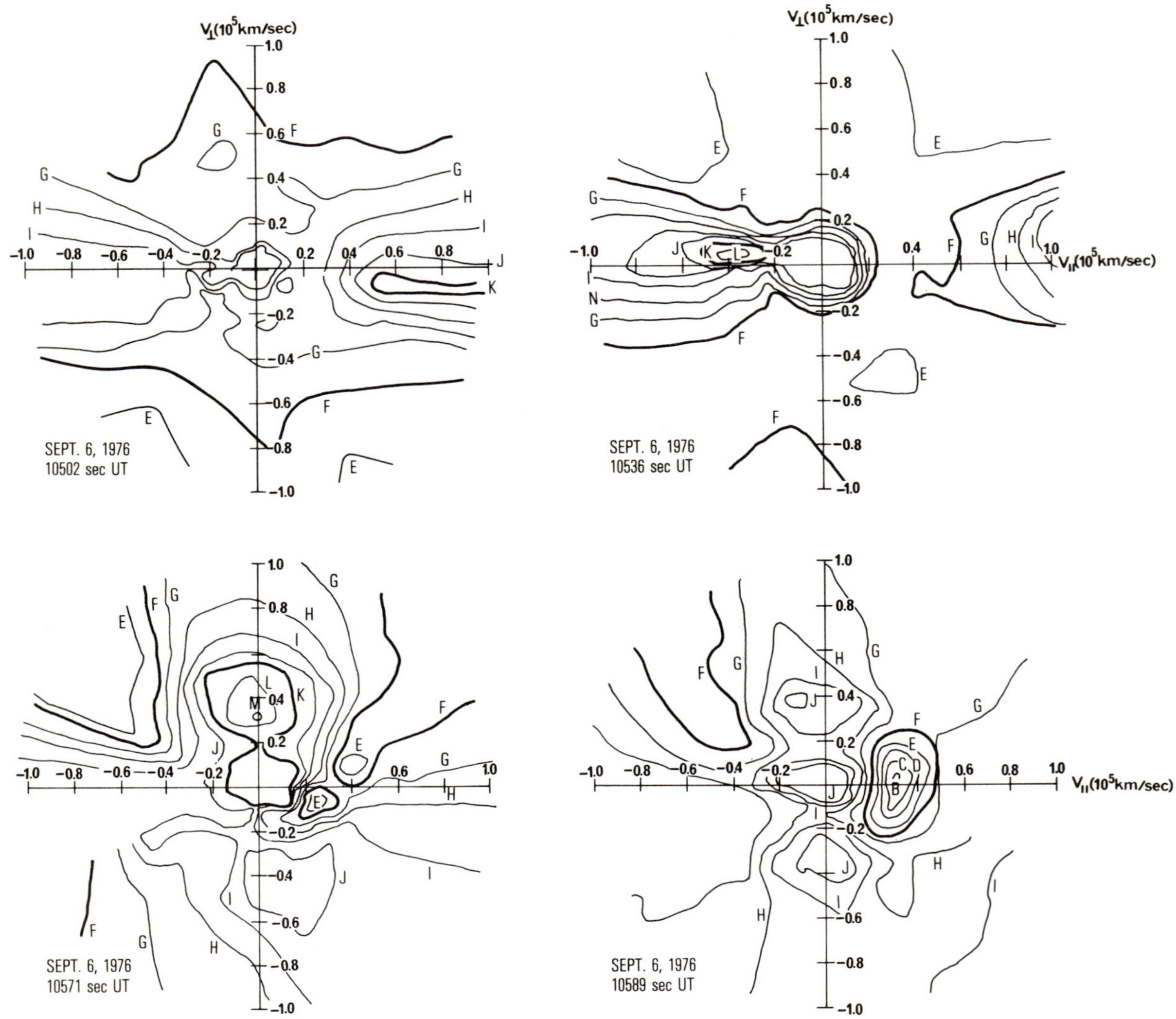

Fig. 8. Difference plots showing the regions of atmospheric loss cone and trapping for the distribution functions of Figure 7. For discussion of boundary curves and differencing procedure see text.

broadband electrostatic waves (Wahlen and Daly, 1979). Such a mechanism does not agree with the observations in detail so at this point we can only emphasize that no complete quantitative description of this feature exists.

The second part of this report shows the kinds of information one can obtain using the particle distribution functions from the S3-3 observations. First, the case study presented showed that the total potential drop and the relative magnitude of the potential drops above and below an observation point can be determined. This determination was based on the assumption that the particle motions were adiabatic in a field aligned quasistatic electric field. A previous report (Mizera et al., 1980) showed that the ion beam energies and electron loss cone enhancements were consistent with those assumptions for the potential drop below the satellite.

The same assumptions were used in analyzing the particle distribution function features to determine the potential drop above the satellite and to make estimates of the altitude distribution of the potential drop.

In order to interpret the distribution function plots that were obtained during the occurrance of a large electrostatic structure (Mizera et al., 1980; Mozer et al., 1977; Torbert and Mozer, 1978) we had to generate a transformation procedure which produced a clearer display of the particle boundaries in velocity space. The boundaries themselves are based on the assumptions of

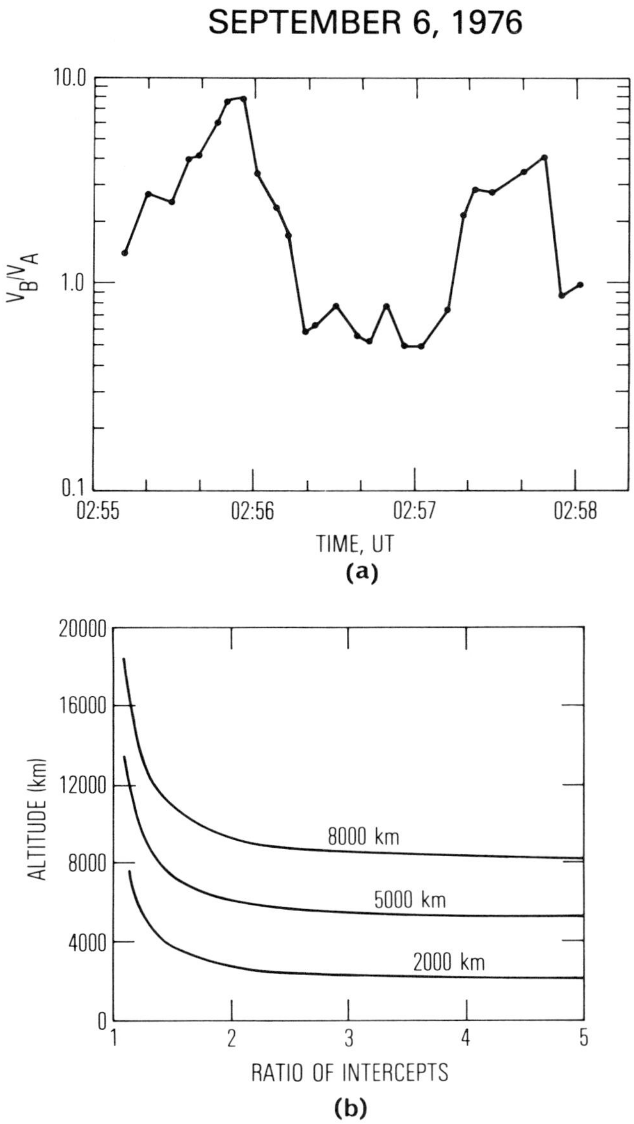

Fig. 9. (a) Ratio of potential drop below to the potential drop above the satellite in the inverted V structure.

(b) Nomogram relating the ratio of the 'trapping' region boundary along $Vi_\perp$ and $Vi_\parallel$ in velocity space to the altitude at which the electron acceleration occurred for observations at three different satellite altitudes. For details see text.

adiabaticity, conservation of energy and electrostatic electric fields parallel to B (Chiu and Schulz, 1978; Croley et al., 1978; Whipple, 1977).

As a result, we found that the majority of the potential drop above the satellite at times occurred within a few thousand kilometers. If we combine this with the previous statistical results for field aligned ion beams which shows that the potential drop corresponding to such ions normally operates above 4500 km (Ghielmetti et al., 1978; Gorney et al., 1980) we then conclude that the electrostatic potential drop corresponding to the electric field can assume various distributions in altitude, and at times can be confined to the low altitude (4500 - 100000 km) region.

Acknowledgments. We would like to thank all those people who made this study possible. Special thanks goes to V. Bledsoe, D. Croley and A. Clarke for organizing the data sets and providing computational support. Thanks also goes to D. Croley, Y. Chiu and M. Schulz for many useful discussions concerning these data. The data reduction and analysis was supported partly by the U. S. Air Force under Contract No. F04701-79-C-0080 and partly by the National Science Foundation under Contract No. ATM-8000518.

References

Ackerson, K. L. and L. A. Frank, Correlated satellite measurements of low-energy electron precipitation and ground-based observations of a visible auroral arc, J. Geophys. Res., 77, 1128, 1972a.

Ackerson, K. L. and L. A. Frank, Correction, J. Geophys. Res., 77, 3002, 1972b.

Anger, C. D. and A. T. Y. Lui, A global view at the polar region on 18 December 1971, Planet. Space Sci., 21, 873, 1973.

Burch, J. L., S. A. Fields, W. B. Hansen, R. A. Heelis, R. A. Hoffman and R. W. Janetzke, Characteristics of auroral electron acceleration regions observed by Atmosphere Explorer C, J. Geophys. Res., 81, 2223, 1976.

Chiu, Y. T. and M. Schulz, Self consistent particle and parallel electrostatic field distributions in the magnetospheric-ionospheric auroral region, J. Geophys. Res., 83, 629, 1978.

Craven, J. D. and L. A. Frank, Observations of angular distributions of low energy electron intensities over the auroral zones with Ariel 4, Proc. R. Soc. Land. A., 343, 167, 1975.

Croley, D. R., Jr., P. F. Mizera and J. F. Fennell, Signature of a parallel electric field in ion and electron distributions in velocity space, J. Geophys. Res., 83, 2701, 1978.

Fennell, J. F., P. F. Mizera and D. R. Croley, Jr., Observations of ion and electron distributions during the July 29 and July 30, 1977 storm period, Proc. Magnetospheric Boundary Layers Conf., ESA SP-148, Alpbach, Austria, 1979.

Eather, R. H., DMSP calibration, J. Geophys. Res., 84, 4134, 1979.

Evans, D. S., Precipitating electron fluxes formed by a magnetic field aligned potential difference, J. Geophys. Res., 79, 2853, 1974.

Frank, L. A. and K. L. Ackerson, Local-time survey of plasma at low altitudes over the auroral zones, J. Geophys. Res., 77, 4116, 1972.

Ghielmetti, A. G., R. G. Johnson, R. D. Sharp and E. G. Shelley, The latitudinal, diurnal and altitudinal distributions of upward flowing energetic ions of ionospheric origin, Geophys. Res. Lett., 5, 59, 1978.

Gorney, D. A., A. Clark, D. Croley, J. Fennell, J. Luhmann and P. Mizera, Auroral ion beam distributions below ~ 8000 km, J. Geophys. Res., 85, 83, 1980.

Hoffman, R. A. and J. L. Burch, Electron precipitation patterns and substorm morphology, J. Geophys. Res., 78, 2867, 1973.

Meng, C. I., Simultaneous observations of low energy electron precipitation and optical auroral arcs in the evening sector by the DMSP 32 satellite, J. Geophys. Res., 81, 2772, 1976.

Meng, C. I., Diurnal variation of the auroral oval size, J. Geophys. Res., 84, 5319, 1979.

Mizera, P. F., D. R. Croley, Jr., F. A. Morse and A. L. Vampola, Electron fluxes and correlations with quiet time auroral arcs, J. Geophys. Res., 80, 2129, 1975.

Mizera, P. F., D. R. Croley, Jr. and J. F. Fennell, Electron pitch-angle distributions in an inverted V structure, Geophys. Res. Lett., 3, 149, 1976.

Mizera, P. F. and J. F. Fennell, Signatures of electric fields from high and low altitude particle distributions, Geophys. Res. Lett., 4, 311, 1977.

Mizera, P. F., J. F. Fennell, D. R. Croley, Jr., A. L. Vampola, F. S. Mozer, R. B. Torbert, M. Temerin, R. Lysak, M. K. Hudson, C. A. Cattell, R. G. Johnson, R. D. Sharp, A. Ghielmetti and P. M. Kintner, The aurora inferred from S3-3 particles and fields, to be published, J. Geophys. Res., 85, 1980.

Mozer, F. S., C. W. Carlson, M. K. Hudson, R. B. Torbert, B. Parady, J. Yatteau and M. C. Kelley, Observations of paired electrostatic shocks in the polar magnetosphere, Phys. Rev. Lett., 38, 292, 1977.

Rogers, E. H., D. F. Nelson and R. C. Savage, Auroral photography from a satellite, Science, 183, 951, 1974.

Sharp, R. D., R. G. Johnson and E. G. Shelley, Energetic particle measurements from within ionospheric structures responsible for auroral precipitation processes, J. Geophys. Res., 84, 480, 1979.

Shelley, E. G., R. D. Sharp and R. G. Johnson, Satellite observations of an ionospheric acceleration mechanism, Geophys. Res. Lett., 3, 654, 1976.

Torbert, R. B. and F. S. Mozer, Electrostatic shocks as the source of discrete auroral arcs, Geophys. Res. Lett., 5, 1978.

Whalen, B. A. and P. W. Daly, Do field-aligned auroral particle distributions imply acceleration by quasi-static parallel electric fields?, J. Geophys. Res., 84, 4175, 1979.

Whipple, E. C., Jr., The signature of parallel fields in a collisionless plasma, J. Geophys. Res., 82, 1525, 1977.

Winningham, J. D., F. Yasuhara, S. -I. Akasofu and W. I. Heikkila, The latitudinal morphology of 10 eV to 10 keV electron fluxes during magnetically quiet and disturbed times in the 2100-0300 MLT sector, J. Geophys. Res., 80, 3148, 1975.

ROCKET STUDIES OF PARTICLE STRUCTURE ASSOCIATED WITH AURORAL ARCS

D. A. Bryant

S.R.C., Rutherford and Appleton Laboratories, Ditton Park, Slough, Berkshire, England

Abstract. The properties of electron streams encountered on six rocket flights over auroral arcs are summarised and discussed. The streams producing the arcs are compared with those in surrounding regions and with the accompanying positive ions. On the basis of the results, which are fully consistent with others appearing in the literature, it is suggested that the auroral arc is essentially a plasma boundary phenomenon in which electrons are accelerated, not by an electrostatic field, but by resonance with plasma waves generated at the boundary.

Introduction

The streams of particles that produce auroral arcs have been examined by a group at the Rutherford and Appleton Laboratories using electrostatic analysers and channel multipliers carried on six rocket flights. Four of the flights were from Andøya in Norway, and two were from Kiruna in Sweden. Summarized here are results from these flights, with the clearest and most straightforward example used as the main point of reference. At the outset it can be said that the particle streams are all found to be very similar in their general properties and to differ only in detail. The observations seem fully in accord with results obtained by others, though we find that we are forced towards conclusions rather different from those that are currently generally accepted. In particular, it seems highly questionable whether the acceleration process is, in the absence of compelling evidence in its favour, basically an electrostatic one as is generally believed.

Flight SL 1221

The flight chosen as the reference is that of the Skylark SL 1221, launched from Andøya at 2044 UT on 1 November 1974. An image of the North Atlantic region taken during the rocket flight by a DMSP satellite (Fig. 1) shows that the arc in this case was an isolated single structure extending from west of Greenland to the north of Norway. It had remained stable for about 15 mins before launch then broke up soon after the rocket, travelling approximately northwards, crossed the particle stream. Electron intensities, shown in Fig. 2 exhibit between flight times of 125 and 235 s the familiar sharp, non-Maxwellian, peak in intensity which is one of the indications that the electrons have been accelerated (McIlwain 1960,

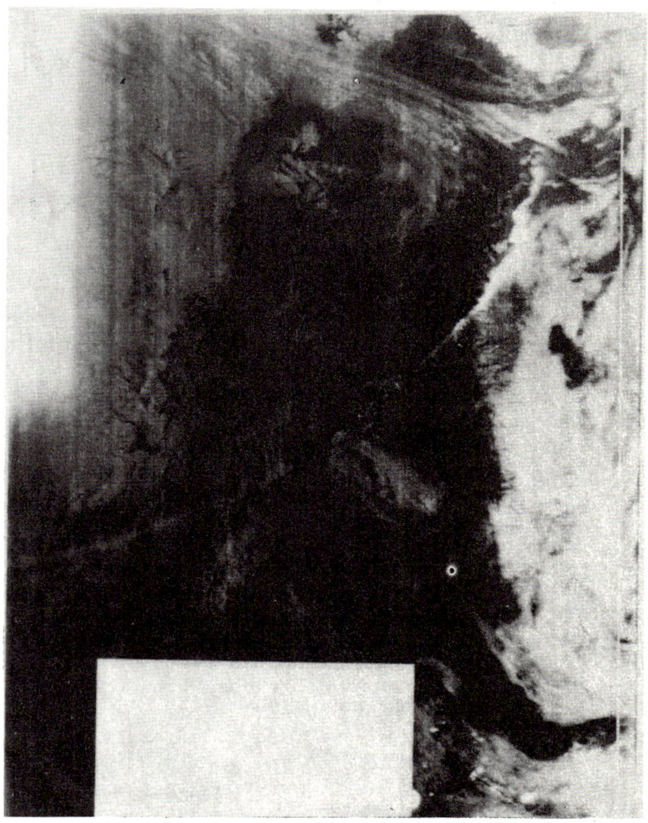

Fig. 1. Image of the north Atlantic taken by the DMSP satellite 9532 showing the auroral arc traversed by the rocket SL 1221 extending from west of Greenland to northern Norway where the rocket was launched.

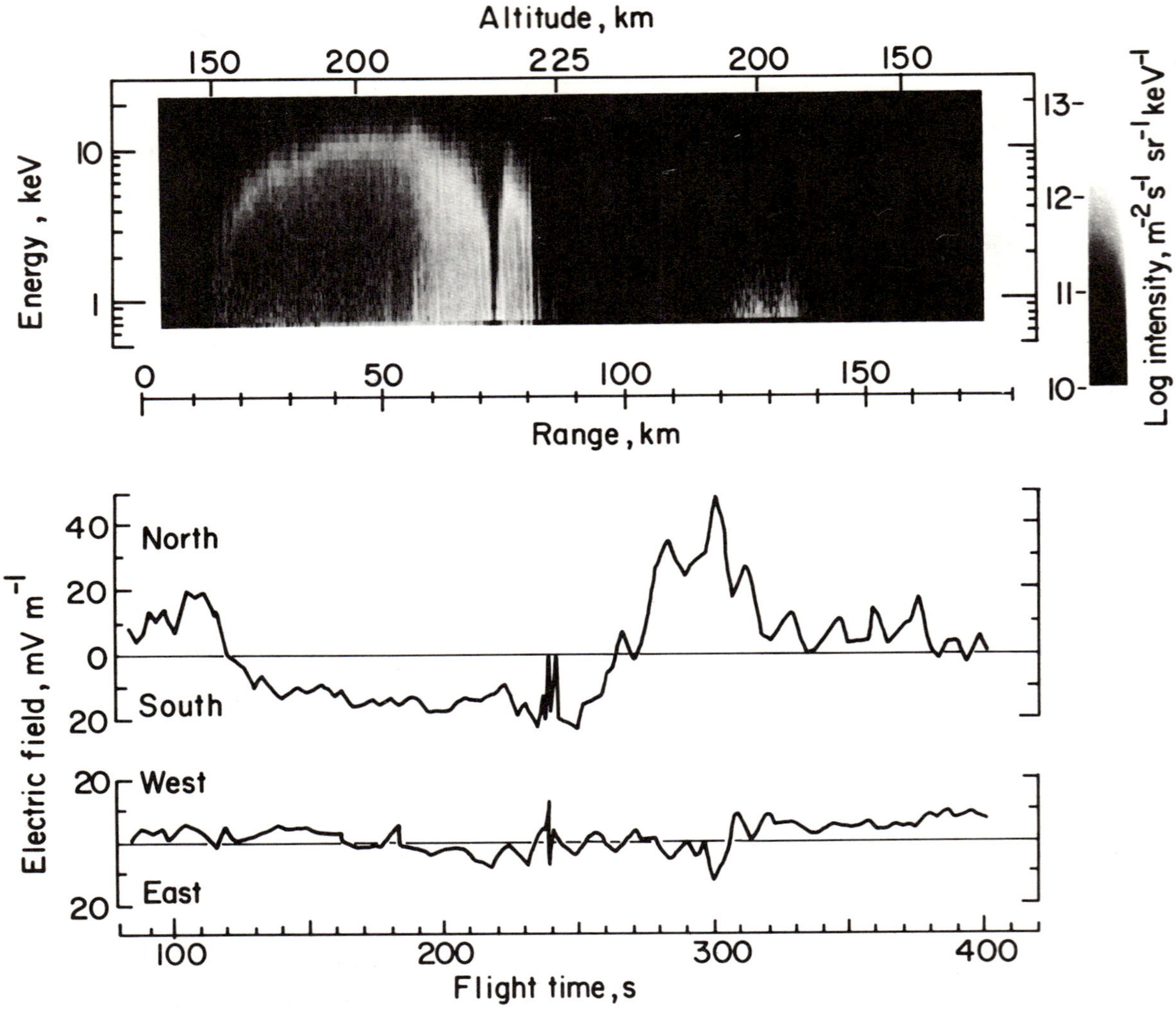

Fig. 2. Electron intensities and electric fields encountered on flight SL 1221. The electron intensities, given by the grey scale, are measured at pitch angles of 35 ± 15°. The altitude of the rocket and its northward horizontal range are indicated.

Albert 1967, Evans 1968). Over a horizontal distance of about 60 km in the north-south direction the energy at which the peak occurs varies systematically, rising to approximately 10 keV near the centre at what might conveniently be termed the core of the particle stream. The spectrum is noticeably broader towards the northern edge (195-220 s) of the main region of precipitation, a fact immediately requiring some versatility of the acceleration mechanism. Intensities encountered in the period 220-240 s are caused by an eastward-moving procession of rays at the northern edge of the particle stream. This interpretation of a television recording made from the ground is supported by additional electron measurements made from a subpayload that had been earlier released from the rocket. The particle measurements shown are in fact those made from the subpayload. The visual arc, which appeared subjectively to have a north-south extent of ~ 10 km, corresponds to a band of electron precipitation where the energy flux is greatest embedded within the 60 km wide overall structure (Lepine et al., 1979). There is some evidence in Fig. 2, to be supported in other flights, that the acceleration region forms a boundary layer between a near-Maxwellian distribution to the south (before 120 s) and a more tenuous and colder plasma to the north (after 240 s).

Electron energy distributions are shown in Fig. 3 in terms of velocity-space density. This quantity, being conserved in a dynamical process, serves as a useful tracer of the evolution

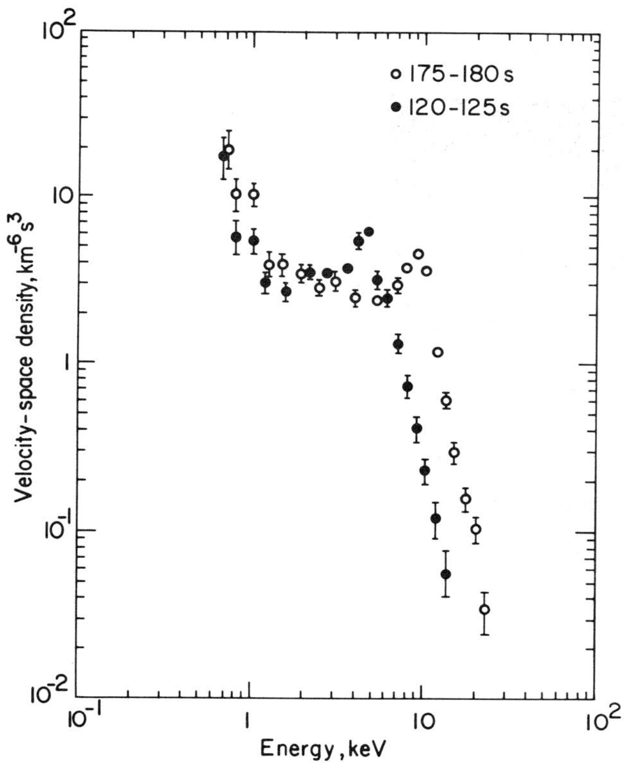

Fig. 3. Electron energy distributions at the core (open circles) and at another representative location (full circles) on flight SL 1221. The pitch angle range is 35 ± 15°.

of a distribution during acceleration. Distributions are shown for the core of the arc where the peak in velocity-space density occurs at its highest energy (10 keV), and for comparison a time when the peak occurs at a lower energy (5 keV). There is little difference between the densities of the lowest energy electrons (below 2 keV) in the two cases, which in view of the marked difference of higher energies, suggests that the low energy electrons are not secondary to the high energy ones.

The same conclusion was reached by Whalen and Daly (1979) and Hall (1980). Ogilvie (1968) concluded that the production of secondaries by ionizing collisions with the atmosphere was too weak to explain the intensities observed below 1 keV. Other secondary processes have been considered (Matthews et al., 1976) but again being secondary they do not lead naturally to constancy of the low energy component.

Another noticeable feature of Fig. 3 is that the peak density is lower in the core distribution than in the other. This effect, which causes the distributions to intersect, is a general one which suggests that the particles forming the different peaks derive from different energies in the source distribution if this is assumed to be common. At energies above the peak the separation in energy for a given density increases from ~ 5 keV near the peak to ~ 9 keV at the highest energy. This broadening in linear terms of the high-energy tails (Burch et al., 1976) appears to represent acceleration by approximately a constant factor, unlike the constant difference expected for straightforward electrostatic acceleration.

Electric fields associated with the arc were

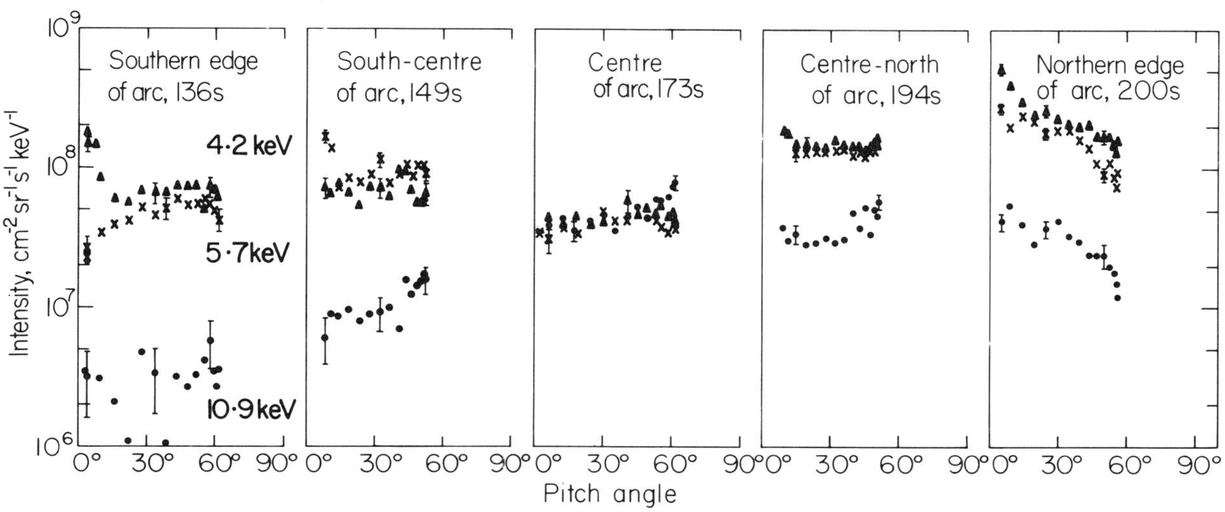

Fig. 4. Electron pitch-angle distributions at various positions within the particle stream encountered on flight SL 1221. Distributions are shown for 3 selected energies as indicated. (From Edwards et al. 1976)

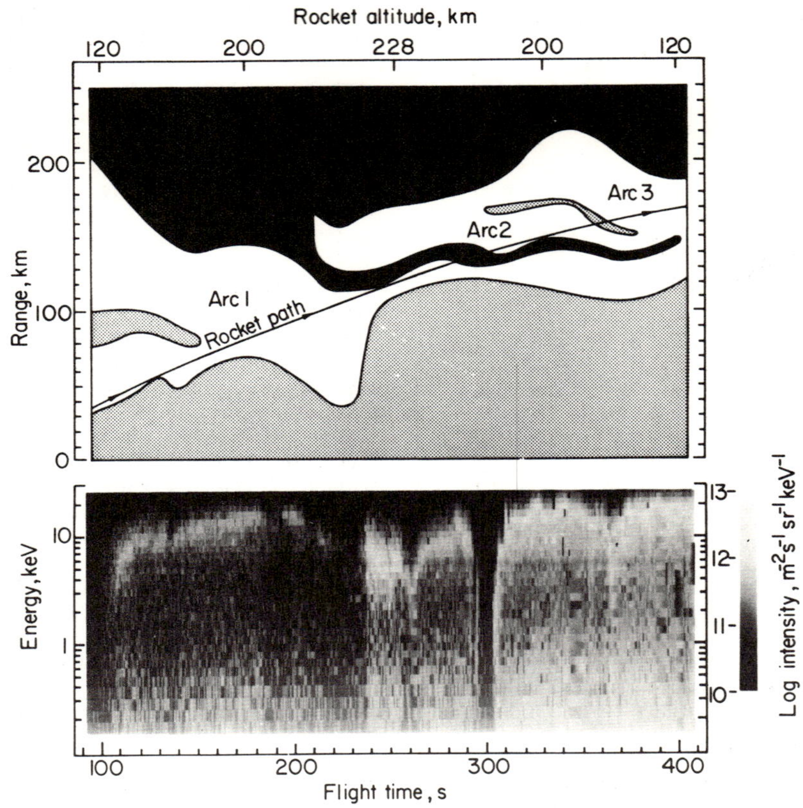

Fig. 5. Auroral arcs traversed by the rocket SL 1122, and electron intensities, given by the grey scale, encountered during the flight. The upper panel reconstructs the approximation northward progress of the rocket across a changing pattern of auroral arcs (white areas) in terms of a sequence of approximately north-south scans. The rocket path and aurora are projected along geomagnetic field lines onto a horizontal plane at an altitude of 100 km. The dark area at the top of the figure to the north of the arcs indicates an absence of visual aurora in this region, and the grey area to the south depicts weak diffuse aurora. The pitch-angle range for the electron measurements (lower panel) is $30 \pm 5^{\circ}$.

measured by the Royal Institute of Technology, Stockholm, and the Space Science Department, ESTEC. As shown in Fig. 2 the field was directed generally northward on the southern side, and southward on the northern side of the main region of precipitation. A northward field is found also to the south of the enhanced precipitation north of the arc but it is not clear in this case whether the measurements relate to a stable structure. Since the fields are directed generally towards regions of electron precipitation, and since potential differences arising from them amount to 1 kV at the most (Edwards et al., 1976), it is possible that the field is a consequence of incomplete neutralization of electron charge by ionospheric and other positive ions. A charge excess of only 1 part in 10^4 distributed non uniformly, could account for the fields observed. Our earlier interpretation of the electron acceleration evident in Figs. 2 and 3 was in terms of a V-shaped electrostatic potential structure. Reservations expressed at the time have now strengthened in view of the fact that it has not been possible to account for the evolution of the distribution function on this basis. The angular distributions, Fig. 4, are also revealing. At and near the core of the stream the electrons are isotropic at all energies. This is generally taken to imply that the acceleration took place far enough away for any collimation to be removed by the converging geomagnetic field experienced during precipitation. At the edges of the particle stream intensities are higher at smaller pitch angles, at least for energies near the spectral peak. This may be understood in terms of a more local acceleration (Hoffman and Evans 1968, Paschmann et al., 1972, Whalen and McDiarmid 1972). While some features of these distributions are explainable in terms of an electrostatic acceleration, the smooth and monotonic decrease of intensity towards larger pitch angles that

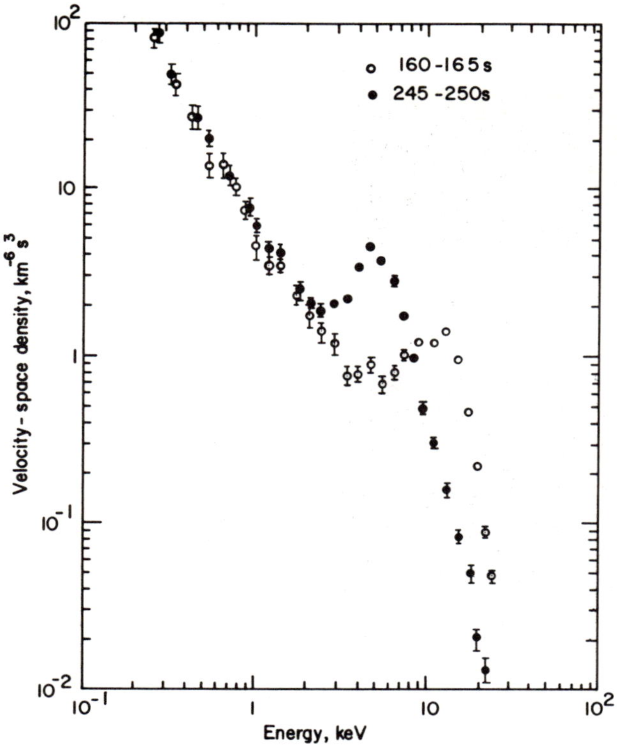

Fig. 6. Electron energy distributions at the core (open circles) and another representative location (full circles) on flight SL 1122. The pitch angle range is 30 ± 5°.

occurs for all energies at the northern edge of the stream is inconsistent with the predicted sharp cut off and systematic change in cut-off angle with energy (Hall and Bryant 1974, Whalen and Daly 1979).

Flight SL 1122

This Skylark 6 rocket was launched from Andøya at 1815 UT on 30 October 1973. Travelling approximately northwards, the rocket traversed a changing pattern of east-west aligned arcs. The progress of the rocket is reconstructed in the upper panel of Fig. 5 using optical measurements from the ground, and employing the electron intensities, shown in the lower panel, to make the reconstruction as accurate as possible (Bryant et al., 1978). As with SL 1221, there is clearly a systematic change in peak energy with position along the trajectory. This trend prevails even through periods of reduced intensity occurring for example from 210-225 s. Electron energy distributions at the core of arc 1 and in a region representative of a lower peak energy are compared in Fig. 6.

In view of the temporal changes taking place in the aurora during the first half of the flight it is not clear that the two distributions result from a common source distribution, as seemed plausible in the more stable aurora, and smooth transitions encountered during flight SL1221. The distributions however still show many features in common with those of Fig. 3. At low energies (again below 2 keV) the distributions are almost indistinguishable, again implying that these electrons are not secondary to the high energy ones whose distributions are very different in the two cases. As with SL 1221, the velocity-space density of the peak is lower when the peak occurs at the higher energy, and the distributions again intersect. At energies above the peak the separation of the distributions increases from ~ 4 keV at a density of 1 km^{-6} s^3 to 6 keV at 10^{-1} km^{-6} s^3. Very similar behaviour is found in arcs 2 and 3. At energies above the peak in the spectrum these distributions can be fitted to the predictions for electrostatic acceleration, but only if the potential difference is assumed to fluctuate (Bryant et al., 1978). It is now clear, though, that in this respect flight SL 1122 is the exception rather than the rule. There is a remarkable geometrical similarity between the core distributions of Figs. 3 and 6. This observation is discussed further in a later section.

Flight SL 1422

In 1976 a 3-stage Skylark 12 rocket was developed in order to reach altitudes high enough for measurements to be made of positive ions, suprathermal electrons, and other related phenomena. The increased horizontal range of the vehicle also permitted the measurements to be seen better in context. The first of these, SL 1422, was launched from Andøya at 2115 UT on 21 November 1976. Figure 7 shows electron and positive-ion intensities. The electrons have a near-Maxwellian distribution with a temperature ~ 3 keV to the south (95-265 s) of a region of acceleration (265-360 s) about 100 km in north-south extent. The arc brightened only a few seconds before the rocket entered the accelerated stream, and there were some temporal fluctuations after this (Rothwell, private communication). However the major intensity changes encountered by the rocket, including the depressions at 285 and 345 s, correspond to a spatial structure in precipitation. Electrons on the poleward side of the accelerated stream exhibit the same low energies and intensities found on other flights. The location of accelerated electrons at a boundary between different types of precipitation, suggested also by the SL 1221 results, is in accord with the fact that auroral arcs commonly appear as a bright boundary layer separating regions of weaker aurora on one side and very much weaker aurora on the other. There is in fact strong evidence from particle and optical measurements

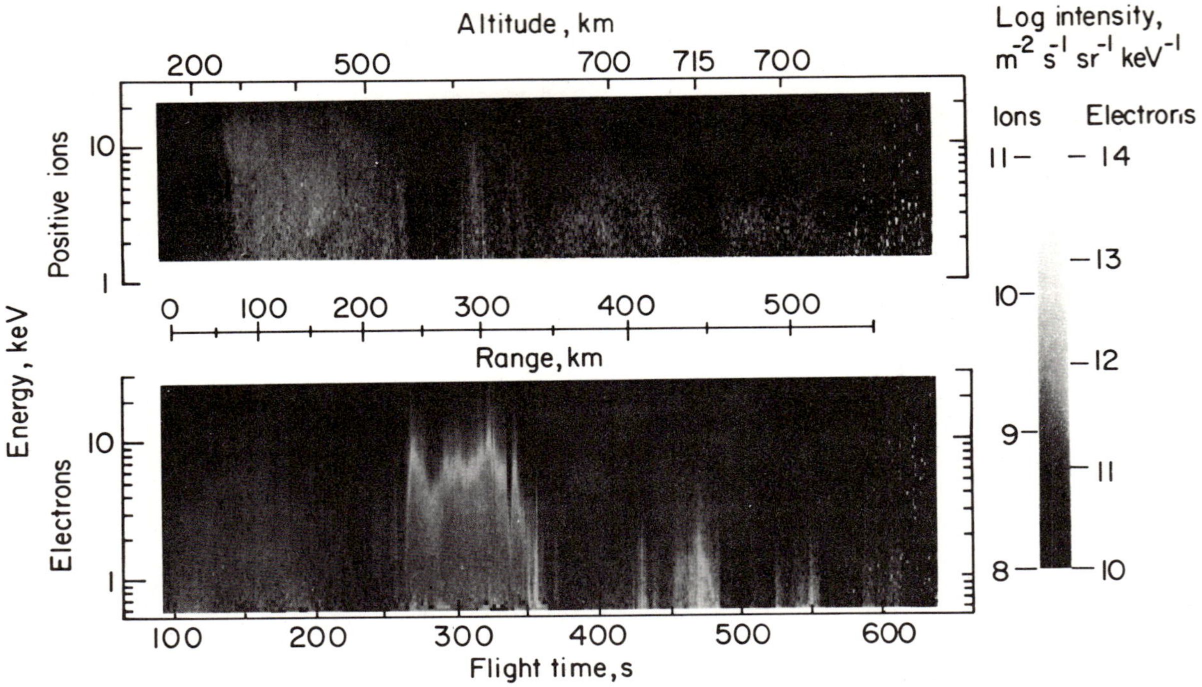

Fig. 7. Electron and positive-ion intensities encountered on flight SL 1422. All species of ion having a given energy per electronic charge are included in the measurements. Intensities are indicated by the grey scales and apply to particles with pitch angles 33 ± 3°. The altitude of the rocket and its horizontal range measured in an approximately northward direction are indicated.

for auroral arcs occurring close to the boundary between different magnetospheric plasmas (Eather and Mende 1971, Winningham et al., 1975, Lui et al., 1977, Meng et al., 1979). From their location and properties these plasmas appear to be the plasma sheet on the equatorward side, and the magnetotail lobe on the poleward side.

Positive-ions have a lower intensity when the accelerated electrons are encountered (265 s). The fall in intensity, though, is considerably greater than that expected for retardation due to a potential difference ~ 10 kV that might be invoked to explain the electron acceleration. The ion intensity increases again near 320 s without any corresponding change in the electrons. Although there is clearly a tendency for ion and electrons to be anticorrelated this behaviour is neither consistent enough nor of the right magnitude to be the direct effect of a potential difference (see also Miller and Whalen 1976).

Representative electron energy distributions (Fig. 8) exhibit the same general features as noted for the other flights. The close geometrical similarity between all core distributions is demonstrated in Fig. 9 which is a composite of the distributions from all flights with suitable scaling of density and energy. It contains also the core distributions from three other flights, the Nike Tomahawk F19 launched from Andøya at 2144 UT on 1 October 1968, and the Skylarks S67-2 and S77-2 launched from Kiruna at 2140 UT on 29 October 1970 and 2150 UT on 24 September 1972, respectively. Flights F19 and S67-2 are discussed in detail by Bryant et al., (1973).

Acceleration by Plasma Waves

There is a striking similarity between the electron distributions observed on these flights with those produced in a laboratory experiment. Franklin et al., (1978) observed the deformation of a near-Maxwellian source distribution by large-amplitude coherent longitudinal electron plasma waves. The evolution of the distribution as the phase velocity of the waves is increased (Fig. 10) follows a pattern which is clearly very similar to that followed by the auroral electrons. It is therefore tempting to consider whether the auroral electrons might have been accelerated by a similar resonant interaction with naturally occurring plasma waves. A broad region of wave turbulence is known to exist on auroral field lines in the magnetosphere (Gurnett and Frank 1977). It is not yet clear, though, what role they play in the electron

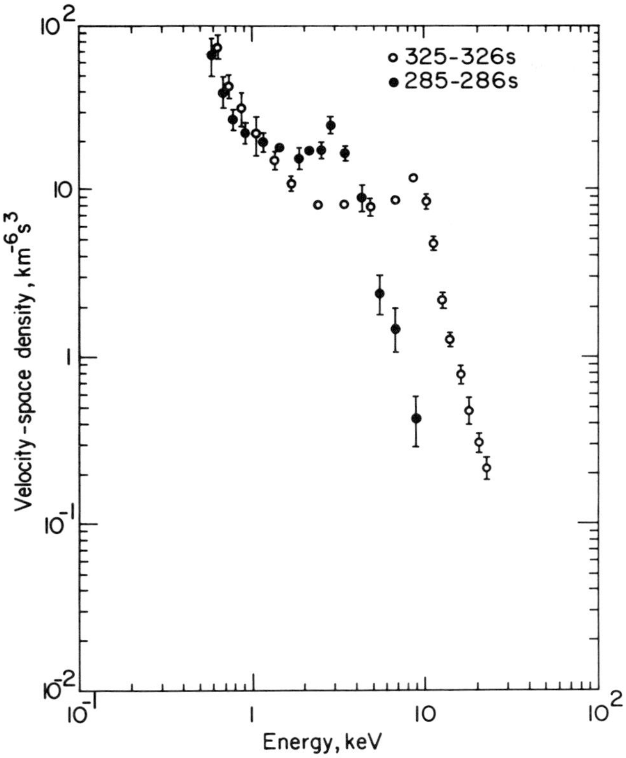

Fig. 8. Electron energy distributions at the core (open circles) and another representative location (full circles) on flight SL 1422. The pitch angle range is 33 $\pm$ 3°.

acceleration process. Arguments for the consideration of a resonant process have been made by Bryant et al., (1978), Whalen and Daly (1979), and Hall (1980).

In such a picture the electron energy distributions would be interpreted as follows. At low energies the electrons are too slow to interact with the waves, so the distribution remains unchanged in this region. The peak in the electron spectrum is produced by a wave spectrum which rises towards higher phase velocities, thereby making electrons with comparable velocities more likely to gain energy than lose energy on interacting with the waves. The sharpness of the peak in the electron spectrum would reflect the steepness of the wave spectrum. The electron velocity-space density at the peak corresponds to that of the lowest energy in the source distribution able to resonate effectively with the waves. The observed relation between the velocity-space density of the peak and the energy at which it occurs suggests that an energy increase of a factor of 5 is the normal limit. The similarity of the core distributions in different arcs suggests further that this distribution may be a limiting one for which the flow of energy between electrons and waves reaches equilibrium. The high energy tail of the electron distribution is formed at velocities where the wave spectrum is flat and produces a stochastic acceleration or heating (Sturrock 1966, Eichler 1978). The flexibility required of the acceleration process can in principle be provided by the wide range of possible wave spectra. Positive ions would, for the same reason as the low-energy electrons, be unaffected by the waves whose phase velocities parallel to the magnetic field match those of 1-20 keV electrons. The general tendency towards anticorrelation between ions and electrons possibly reflects the different populations of the source plasmas either side of the boundary rather than a different response to the acceleration mechanism. Active arcs and their eventual breakup might be seen as a result of firstly a convolution of the boundary and eventually an intermixing of the two plasmas.

It remains necessary of course to identify the waves thought to be responsible for the acceleration and to confirm by measurement that they exist with sufficient intensities.

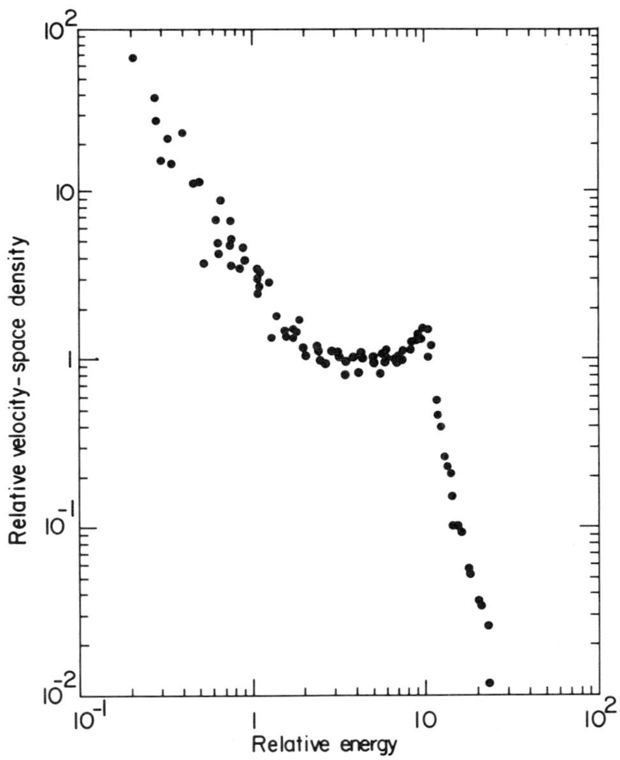

Fig. 9. Composite electron energy distribution combining the scaled distributions from the cores of six auroral arcs.

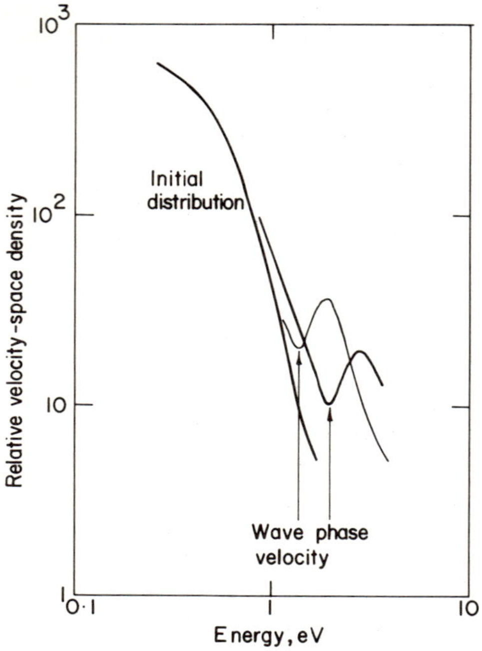

Fig. 10. Evolution of an electron energy distribution in a laboratory experiment produced by plasma waves with phase velocities corresponding to the electron energies indicated. The form of evolution depicted in this figure (derived from Fig. 8 of Franklin et al. 1978) bears a striking resemblance to that occurring in auroral arcs as illustrated in Figs. 3, 6, and 8 of the present paper.

Acknowledgments. The cooperation of the rocket ranges at Andøya and Kiruna is gratefully acknowledged, as is the technical support received within the Norwegian rocket programme (F19), the ESRO rocket programme (S67-2 and S77-2), and the UK rocket programme (SL 1122, SL 1221, and SL 1422).

References

Albert, R. D., Nearly monoenergetic electron fluxes detected during a visible aurora, Phys. Rev. Letters, 18, 369, 1967.

Burch, J. L., S. A. Fields, and W. B. Hanson, Characteristics of auroral electron acceleration regions observed by Atmospheric Explorer C, J. Geophys. Res., 81, 2223, 1976.

Bryant, D. A., G. M. Courtier, and G. Bennett, Electron intensities over two auroral arcs, Planet. Space Sci., 21, 165, 1973.

Bryant, D. A., D. S. Hall, and D. R. Lepine, Electron acceleration in an array of auroral arcs, Planet. Space Sci., 26, 81, 1978.

Eather, R. H., and S. E. Mende, Airborne observations of auroral precipitation patterns, J. Geophys. Res., 76, 1746, 1971.

Edwards, T., D. A. Bryant, M. J. Smith, U. Fahleson, C-G Fälthammar and A. Pedersen, Electric fields and energetic particle precipitation in an auroral arc, Magnetospheric Particles and Fields, p. 285, Ed. McCormac, D. Reidel, Dordrecht, Holland 1976.

Eichler, D., Electron acceleration by strong plasma turbulence, Astrophys. J., 224, 1038, 1978.

Evans, D. S., The observations of a near monoenergetic flux of auroral electrons, J. Geophys. Res., 73, 2315, 1968.

Franklin, R. N., R. R. Mackinlay, P. D. Edgley, and D. N. Wall, Nonlinear behaviour of a finite amplitude electron plasma wave, III The sideband instability, Proc. R. Soc. Lond. A 360, 229, 1978.

Gurnett, D. A., and L. A. Frank, A region of intense plasma wave turbulence on auroral field lines, J. Geophys. Res., 82, 1031, 1977.

Hall, D. S., The influence of energy diffusion on auroral particle distributions, ESA SP-152, 285, 1980.

Hall, D. S., and D. A. Bryant, Collimation of auroral particles by time-varying accelerations, Nature, 251, 402, 1974.

Hoffman, R. A., and D. S. Evans, Field-aligned electron bursts at high latitude observed by OGO-4, J. Geophys. Res., 73, 6201, 1968.

Lepine, D. R., D. A. Bryant, and D. S. Hall, Proceedings of the 7th annual meeting on upper atmosphere studies by optical methods, p. 8, University of Tromsø, 1979.

Lui, A. T. Y., D. Venkatesan, C. D. Anger, S-I. Akasofu, W. J. Heikkila, J. D. Winningham, and J. R. Burrows, Simultaneous observations of particle precipitation and auroral emissions by the ISIS 2 satellite in the 19-24 MLT sector, J. Geophys. Res., 82, 2210, 1977.

Matthews, D. L., M. Pongratz, and K. Papadopoulos, Non-linear production of suprathermal tails in auroral electrons, J. Geophys. Res., 81, 123, 1976.

McIlwain, C. E., Direct measurements of particles producing visible auroras, J. Geophys. Res., 65, 2727, 1960.

Meng, C.-I., B. Mauk, and C. E. McIlwain, Electron precipitation of evening diffuse aurora and its conjugate electron fluxes near the magnetospheric equator, J. Geophys. Res., 84, 2545, 1979.

Miller, J. R. and B. A. Whalen, Characteristics of auroral proton precipitation observed from sounding rockets, J. Geophys. Res., 81, 147, 1976.

Ogilvie, K. W., Auroral electron energy spectra, J. Geophys. Res., 73, 2325, 1968.

Paschmann, G., R. D. Johnson, R. D. Sharp, and E. G. Shelley, Angular distributions of auroral electrons in the energy range 8-16 keV, J. Geophys. Res., 77, 6111, 1972.

Sturrock, P. A., Stochastic acceleration, Phys. Rev., 141, 186, 1966.

Whalen, B. A., and P. W. Daly, Do field-aligned

auroral particle distributions imply acceleration by quasi-static parallel fields? J. Geophys. Res., 84, 4175, 1979.

Whalen, B. A., and I. B. McDiarmid, Observations of magnetic-field-aligned auroral electron precipitation, J. Geophys. Res., 77, 191, 1972.

Winningham, J. D., F. Yaschara, S-I. Akasofu, and W. J. Heikkila, The latitudinal morphology of 10 -eV to 10 -keV electron fluxes during magnetically quiet and disturbed times in the 2100-0300 MLT sector, J. Geophys. Res., 80, 3148, 1975.

POSITIVE ION ACCELERATION IN THE 1 R_E ALTITUDE RANGE

R. D. Sharp

Lockheed Palo Alto Research Laboratory
3251 Hanover Street, Palo Alto, California 94304

Abstract

The signatures of ionospheric acceleration mechanisms operating on auroral field lines at altitudes of about ~1 R_E have been observed with instruments on the S3-3 satellite. The mechanisms accelerate the ambient thermal plasma ions to keV energies and inject them upward into the trapped magnetospheric plasma population. Examples of ions accelerated by quasistatic parallel electric fields and by a primarily transverse acceleration mechanism are presented. The results of a statistical study are used to infer that the upflowing H^+ ions observed with energies $\gtrsim$ 500 eV in the evening local time sector are primarily energized electrostatically, while the O^+ ions in these same ranges receive about half of their energy from electrostatic acceleration and about half from some mass dependent transverse acceleration mechanism.

Introduction

Upward flowing H^+ and O^+ ions with energies in the keV range were frequently observed on auroral field lines at altitudes $\gtrsim$ 4000 km by the energetic ion mass spectrometer on the S3-3 satellite. Their distribution functions indicated that they were accelerated both parallel and transverse to the magnetic field direction. When the parallel acceleration dominated or the energization was at low altitudes compared to the satellite, field aligned pitch angle distributions (ion "beams") resulted. When the transverse acceleration dominated and the energization was at higher altitudes, "conical" pitch angle distributions were observed. The ion beams were often found in association with the signatures of upward directed quasistatic parallel electric fields seen in the electron distributions. In this review we shall show examples of both the ion beams and conics. We will describe the techniques used to estimate the magnitude of the parallel electric potential differences and examine the relationship between the energy of the ion beam and the potential difference below the satellite in individual events in order to estimate the relative contribution of electrostatic acceleration to the ion beam energization process. We will then show some statistical results that allow us to address this question on a broader basis.

Experiment

The measurements to be described were obtained from the S3-3 satellite (1976-65B) which was placed into an elliptical polar orbit with apogee at an altitude of about 8000 km in July 1976. The satellite was spin stabilized at ~3 rpm with its spin axis perpendicular to the orbital plane. The Lockheed instrument consisted of three ion mass spectrometers and four magnetic electron spectrometers. They were mounted with their view directions perpendicular to the spin axis. The electron spectrometers had broad energy bands which are listed in Table 1. The ion spectrometers each acquired a 30 channel mass per unit charge spectrum every 1 second. They each had four possible energy per charge settings (listed in Table 1) that were stepped every 16 seconds. Thus on a single satellite spin a 3 point energy spectrum was acquired for each ion species and ~4 spins were required to obtain a complete 12 point spectrum. In addition to the Lockheed instrument the satellite carried other instruments including an electrostatic analyzer provided by the Aerospace Corporation (Mizera and Fennell, 1977) and an electric field instrument provided by the University of California, Berkeley (Mozer et al., 1977).

Examples of Upward Flowing Ion Events

Figure 1, from Shelley et al. (1976) shows an example of upflowing ion beams observed with the low energy ion spectrometer on 13 July 1976. The top panel shows the pitch angle of the instrument view direction. Below it is indicated the energy per charge setting of the spectrometer. The spectrometer response to O^+ and H^+ ions is illustrated in the two center panels. The narrow field-aligned upward-flowing pitch angle distributions are evident. The flux intensity corresponding to the peak response is 0.3 x 10^8 (cm^2 sec ster keV)$^{-1}$ for the 1.28 keV O^+ ions and

TABLE 1: Detector Characteristics

Detector	Particle	Energy per Charge Range, keV/e			
		STEP 1	STEP 2	STEP 3	STEP 4
CXA-1	ions	0.50	0.68	0.94	1.28
CXA-2	ions	1.76	2.4	3.3	4.5
CXA-3	ions	6.2	8.5	11.6	16.0
CMEA	electrons	0.07 - 0.24			
CMEB	electrons	0.35 - 1.1			
CMEC	electrons	1.6 - 5.0			
CMED	electrons	7.3 - 24			

$0.7 \times 10^8 (cm^2 \text{ sec ster keV})^{-1}$ for the 0.94 keV H^+ ions. The lowest panel shows electron fluxes in the energy range from 0.35 to 1.1 keV. The electron loss cones, and one instance of an anti-loss cone indicative of a closed field line topology are evident. Note how the loss cones are widened and deepened in the region of the upflowing ions relative to the adjacent regions. As will be discussed below, this is the signature of a parallel electric field along the electron's trajectory between the satellite and the atmosphere and suggests that an electrostatic acceleration mechanism is important in the energization of the observed upward flowing ions.

A mass per unit charge spectrum measured at 1.28 keV/e for the upflowing ions at ~0715:47 UT is shown in Figure 2. One sees that the entire flow consisted of O^+ and H^+ ions with no significant He^+ or He^{++} present. This is typical of most of the upflowing ion events although occasionally a measurable He^+ component is encountered.

An example of upflowing O^+ ions exhibiting a conical pitch angle distribution is illustrated in Figure 3, taken from Sharp et al. (1977). The figure shows the angular distribution of the ions observed at 7600 km altitude in the dayside cusp on three consecutive satellite spins with the low energy spectrometer. The energy per charge settings of the instrument during this period are indicated. The observed pitch angle distributions can be interpreted as the result of a transverse acceleration of the ambient thermal O^+ ions in the 4000 - 5000 km altitude range. The pancake-like pitch angle distributions resulting from this transverse acceleration fold inward as the particles move upward adiabatically in the diverging geomagnetic field resulting in the observed conical distribution at the satellite altitude. Similar transverse acceleration phenomena have been observed from the ISIS satellite (Klumpar, 1979) and from a rocket experiment (Whalen et al., 1978) in association with intense electron precipitation events.

Relationship to Parallel Electric Fields

The signatures of parallel electric fields in the particle distribution functions have been described by Kauffman et al. (1976), Whipple (1977) and others. The pitch angle distributions of the trapped and precipitating keV electrons provide useful indicators of the presence of such fields (Sharp et al., 1979; Cladis and Sharp, 1979). This is illustrated by the examples sketched in Figure 4. The lower inset defines the geometry for a situation in which parallel potential drops occur both above (ϕ_A) and below (ϕ_B) the satellite. If one assumes that a

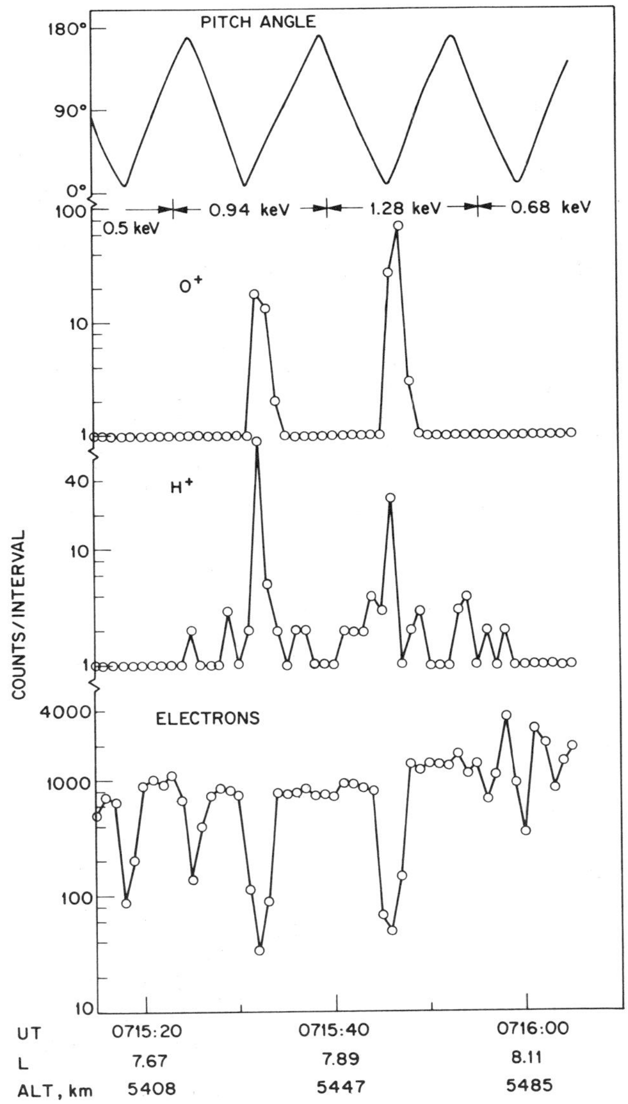

Fig. 1. Upflowing ion beams observed from the S3-3 satellite in association with signatures of a parallel electric field in the electron data (Shelley et al., 1976).

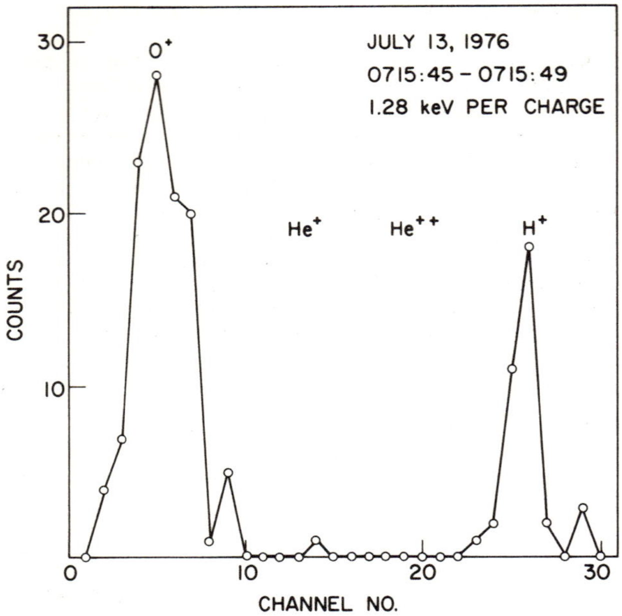

Fig. 2. Ion mass-per-unit-charge spectrum of one of the beams illustrated in Fig. 1 (Shelley et al., 1976).

primary population of isotropically trapped and precipitating plasma sheet electrons exists on the field line and moves under the action of the first adiabatic invariant, then such electrons should exhibit pitch angle distributions as shown in the right hand sketch. They will have widened loss cones of angular width α_L given by the relationship:

$$\sin^2 \alpha_L = \frac{B_S}{B_T} \cdot \frac{E_S + e\phi_B}{E_S}$$

where B and E refer to the geomagnetic field magnitude and electron energy at the altitudes defined by the subscripts and the sketch in the lower inset. The primary electrons with energies only moderately greater than $e\phi_A$ will also exhibit a cutoff in the pitch angle range around 90° given by

$$\sin^2 \alpha_c = \frac{B_S}{B_1} \cdot \frac{E_S - e\phi_A}{E_S}$$

Primary electrons only exist in the energy range $E_S > e\phi_A$. At lower energies will be found secondaries and degraded primaries backscattered from the atmosphere. These exhibit pitch angle distributions with an up-down symmetry such as shown in the shaded portion of the sketch on the left side of Figure 4. This results from the fact that all upcoming electrons with $E < e\phi_A$ will be reflected by the parallel electric field above the satellite. In addition to these short-lived particles topologically connected to the atmosphere, one often sees a population of trapped electrons with energies $< e\phi_A$ that are reflected between the electrostatic mirror above the satellite and the magnetic mirror below. This population can be fed by scattering and fluctuations in the electric fields, and because of intensity fluctuations in the flux levels of the primaries, can build up to levels greater than those observed at a given time in the loss cone. This trapped population is illustrated by the large circles in the sketch on the left side of Figure 4. The flux intensities of the secon-

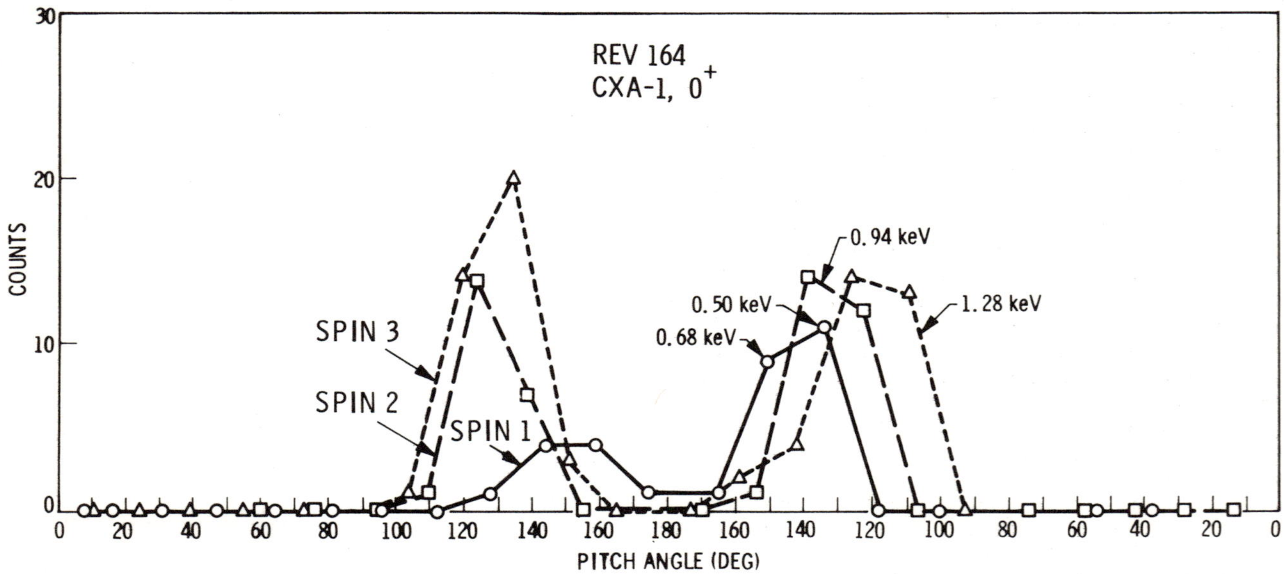

Fig. 3. Oxygen conic observed at an altitude of 7600 km in the dayside cusp (Sharp et al., 1977).

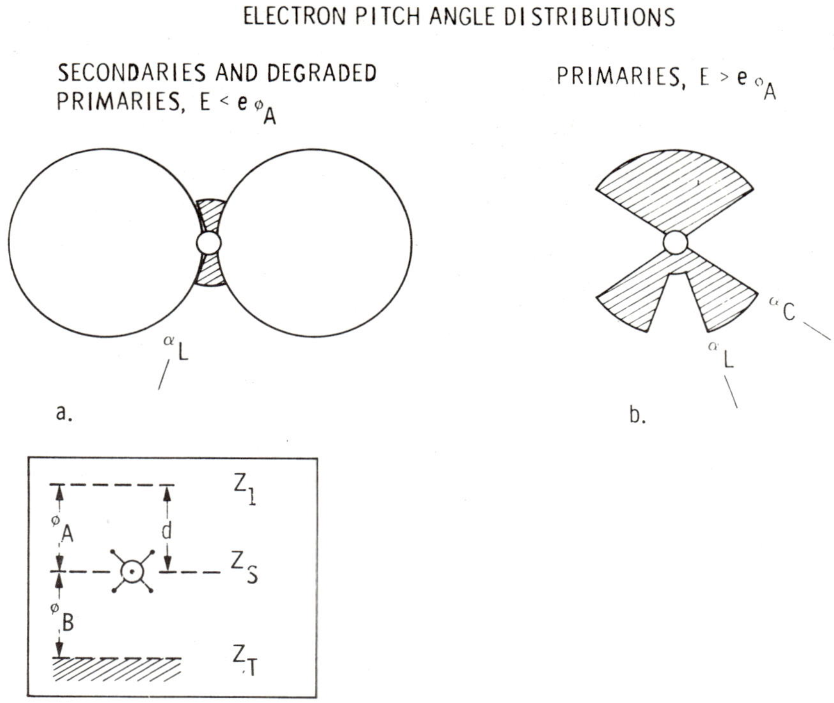

Fig. 4. Signatures of parallel electric fields in electron pitch angle distributions (Sharp et al., 1979).

daries and degraded primaries are also affected by the potential below the satellite which acts both to reflect the low energy secondaries and to reduce the level of the degraded primaries by the factor $E_S/(E_S + e\phi_B)$ which relates the flux at the top of the atmosphere (Z_T) to the flux at the satellite altitude (Z_S) through Liouvilles theorem.

These signatures in the electron fluxes observed during an upward flowing ion event on 15 September 1976 are illustrated in Figure 5. Universal time (SYST), geographic longitude and latitude, altitude (in km) and magnetic local time are indicated on the abscissa. The four lowest panels show the logarithm of the counts per 1/2 second counting interval for the electron spectrometers. The panel labeled pitch shows the pitch angle of the look direction of the spectrometer. The next four panels show the logarithm of the counts from ions with m/q = 1, 2, 4, and 16 respectively, summed once per second from all three of the mass spectrometers giving an approximate measure of the relative number fluxes of the different ion species in the energy range of the experiment. The spin numbers are labeled for reference. The upward flowing ion beams, principally H^+, are evident in spins 2 and 3. The CMEA response in the region of the ion beams shows the up-down symmetry and the 90° maximums characteristic of the trapped secondaries and degraded primaries as discussed above. This establishes that the CMEA energy band is below the energy range of the primaries, and sets a limit of $e\phi_A > 0.24$ keV. A more detailed analysis (Sharp et al., 1979) allows us to infer a peak value of $e\phi_A = 0.54$ keV on spin 3 in this example.

The signatures of the parallel electric field in the angular distributions of primary electrons, i.e. enhanced loss cones and 90° minimums, are evident in the response of CMEB. A more detailed presentation of the CMEB angular distribution in the second half of spin 3 is shown in Figure 6. The observed width of the CMEB loss cone on this spin established that the potential difference between the satellite and the atmosphere was ~1.9 keV (Sharp et al., 1979). At this time the three ion mass spectrometers were set at E/Q = 0.5, 1.8 and 6.2 keV/e, respectively. The peak H^+ flux was observed at 0.5 keV/e, weak H^+ and O^+ fluxes were observed at 1.8 keV/e, and no significant response was found at 6.2 keV/e. Conditions were too variable from spin-to-spin to attempt to construct a detailed spectrum using data from the other energy steps. We conclude that: 1) energetic ions were observed in the energy range corresponding to electrostatic acceleration through the potential difference below the satellite, 2) the ion spectrum was apparently broad and peaked at a lower energy than would be expected for a simple electrostatic acceleration, and 3) higher

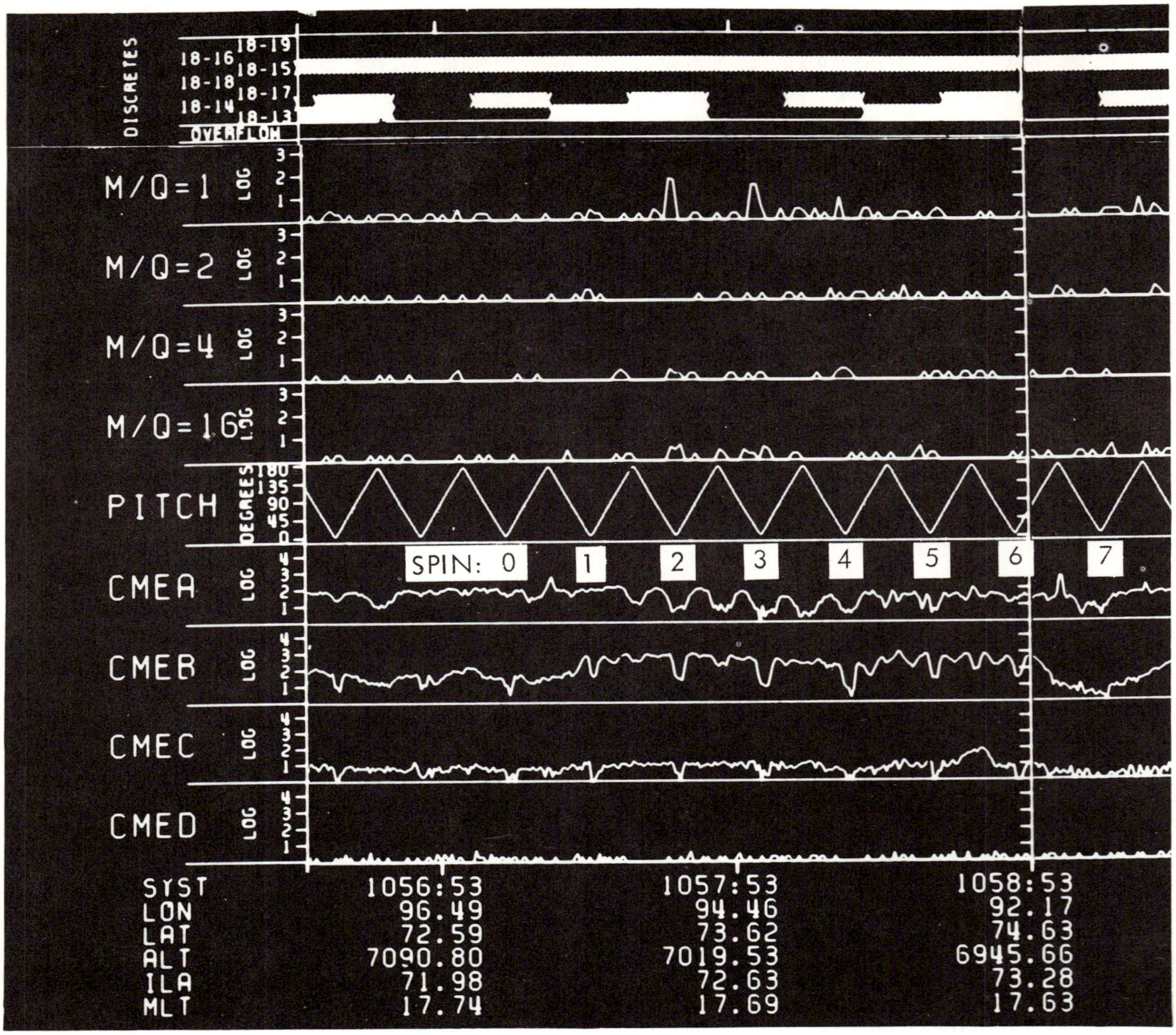

Fig. 5. Survey plots for September 15, 1976 (Sharp et al., 1979).

temporal and energy resolution than is available with the ion mass spectrometer is required to examine this question quantitatively.

If we restrict ourselves to events where the ions peak at energies below 3.9 keV and where we can establish from the mass spectrometer that the ion fluxes are dominated by a single species, we can utilize the higher resolution data available from the Aerospace Corporation's electrostatic analyzer (Mizera and Fennell, 1977) to further investigate the ion acceleration mechanism. Figure 7 shows the data from one such event where the H^+ flux exceeded the O^+ flux by about an order of magnitude. The format is similar to that of Figure 5. One again sees the above described signatures of the parallel electric fields in the electron data in the region of the upward flowing ions (Spins 2, 3, and 4). A detailed analysis of these data (Cladis and Sharp, 1979) results in the estimates of the electrostatic potential differences shown in Figure 8. The circles and triangles denote the potential difference below the satellite determined by two independent techniques from the loss cone widths and depths respectively. The squares denote the total potential difference including the potential difference above the satellite. The H^+ flux measurements from the ion mass spectrometer and the relevant electrostatic analyzer data (courtesy of P. F. Mizera and J. F. Fennell)

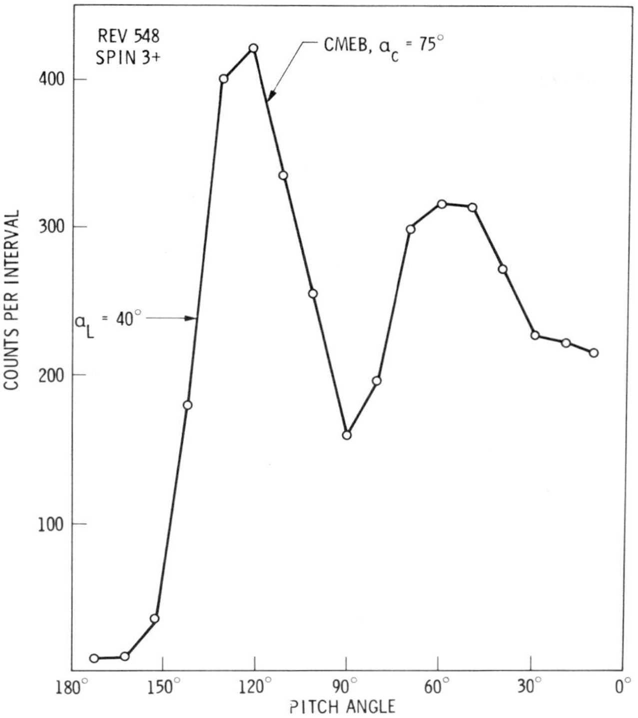

Fig. 6. CMEB angular distribution on Spin 3+, September 15, 1976. See Fig. 5 for definition of spin numbers (Sharp et al., 1979).

are shown in Figure 9. The view direction of the electrostatic analyzer is rotated by 90° with respect to the ion mass spectrometer so it samples a given angular region about 5 seconds earlier. A comparison of the peaks in the spectrums determined by the electrostatic analyzer with the values of the potential differences below the satellite at the appropriate times interpolated from Figure 8 (at points on the abscissa about one quarter of a spin period prior to the numbered spins) indicates that they correspond to roughly comparable energies. The fewer ion mass spectrometer data points are consistent with the inference from Figure 8, that the ion peaks should be at somewhat higher energies on spins 2 and 3 and at a lower energy on spin 4. The data therefore are consistent with the interpretation that the principal ion acceleration mechanism in this event was electrostatic acceleration. We note however that the energy spectrums are broader than would be expected if this was the only mechanism acting to energize the ambient thermal plasma. This suggests that some additional energy transfer process is taking place.

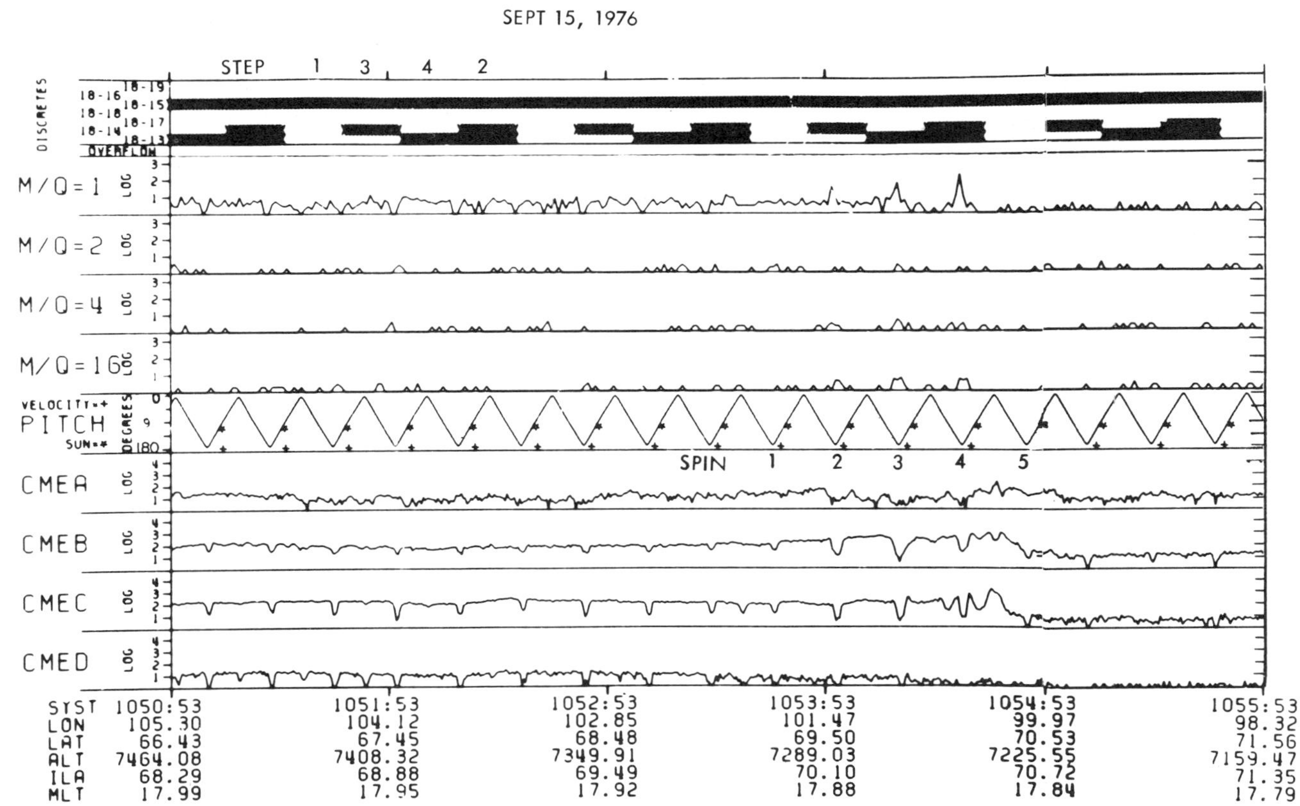

Fig. 7. Survey plots for September 15, 1976 (Cladis and Sharp, 1979).

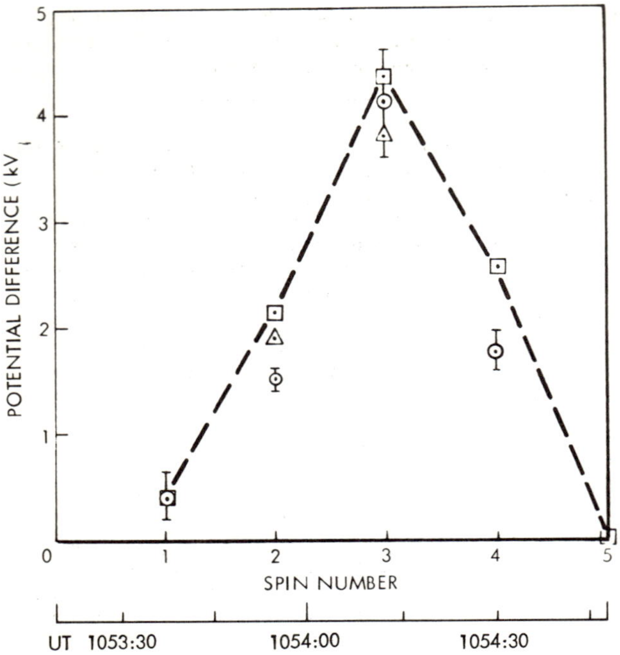

Fig. 8. Potential difference along the magnetic field through the structure of an inverted-V event (Cladis and Sharp, 1979).

Statistical Studies

Collin et al. (1980) have conducted a statistical study using the S3-3 data from the evening local time sector, at altitudes between 6000 and 8000 km, during primarily quiet times. A scatter plot of the average energies of the simultaneously observed upward flowing O^+ and H^+ ions is shown in Figure 10. The O^+ is seen to be systematically more energetic than the H^+. The ratio of their average energies for the data set shown was 1.7. Since electrostatic acceleration is mass independent, this suggests that either some additional process is also energizing the O^+ ions or they entered the electrostatic acceleration region at a systematically lower altitude than did the H^+ ions. An examination of the pitch angle distributions of the two species, however, showed that the O^+ ions had significantly wider beam widths than the H^+. A histogram of the occurrence probability versus half width (at half maximum) of the beams is shown in Figure 11. This is contrary to expectations if the O^+ were electrostatically accelerated to higher energies than the H^+ and suggests that the additional O^+ energy was provided by some mass dependent transverse acceleration mechanism.

From the results of earlier studies by Ghielmetti (1978) and Ghielmetti et al. (1978) we

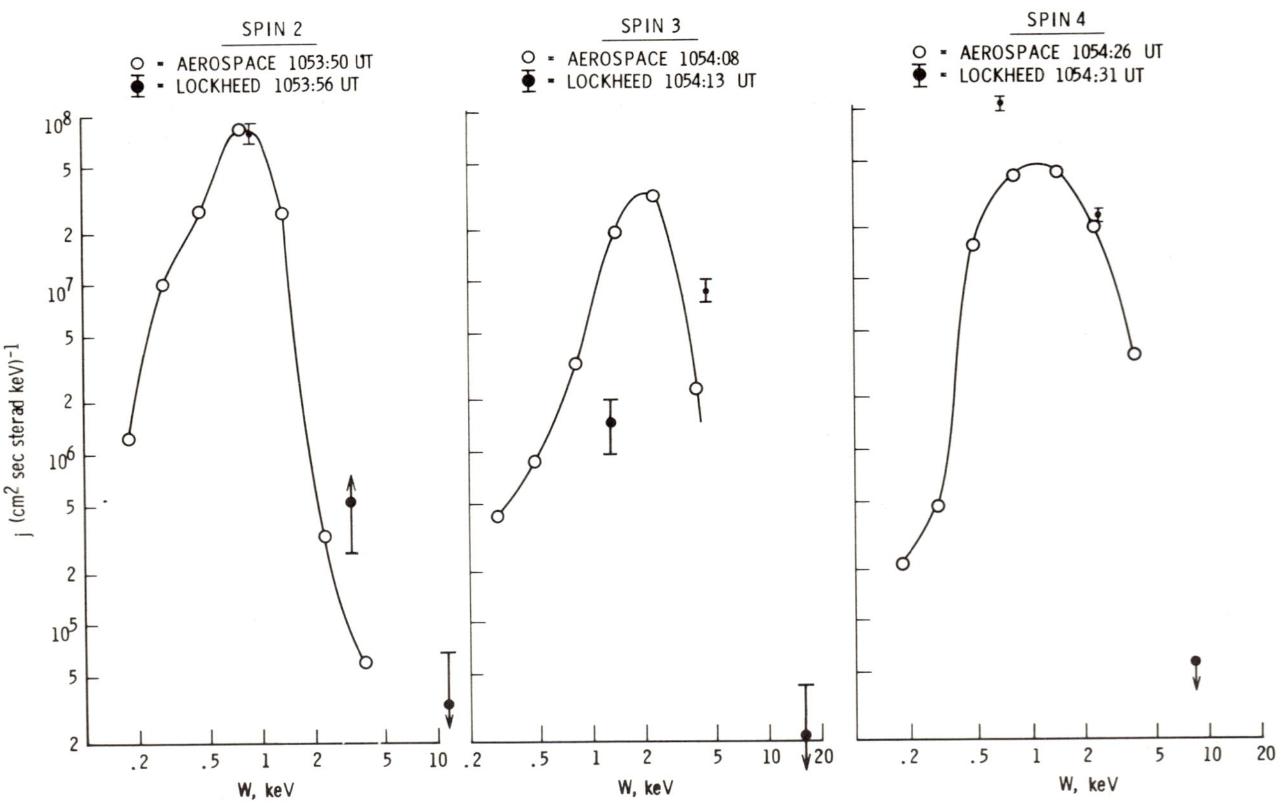

Fig. 9. Energy spectra of upflowing ions within the inverted-V event shown in Fig. 8 (Cladis and Sharp, 1979)

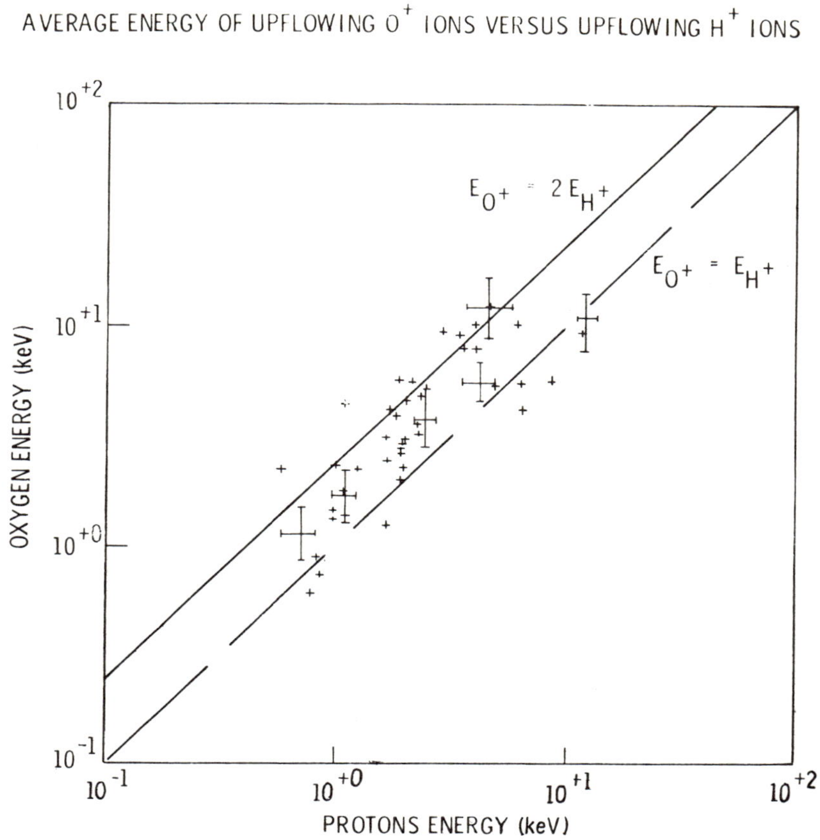

Fig. 10. Scatter plot of average O^+ and H^+ energies in upflowing ion beams (Collin et al., 1980).

know that most of the energy in the upflowing ions observed by the Lockheed spectrometer (E > 500 eV) is acquired above ~4000 km. This can be inferred from Figure 12, which gives the average probability of observing an upflowing ion event as a function of altitude, and Figure 13, which shows the average value of the energy at the peak of the flux intensity for the H^+ and O^+ ions as a function of altitude. We note that at the lowest altitude of observation, the ion energy at the peak flux is already substantially above the energy threshold of the spectrometer (500 eV) and is comparable to the ion energy observed at the highest altitudes.

Returning to the pitch angle distributions (Figure 11) we note that the median width of the O^+ ion peaks was 24° and that of the H^+ ions was 15°. In order to achieve such narrow pitch angles the H^+ ions must have gained less than 15% of their energy from a transverse acceleration at or above 4000 km. O^+ ions on the other hand could derive as much as 45% of their energy from a transverse acceleration at 4000 km or above and still exhibit pitch angles of 24° at 8000 km. Thus both the relative O^+ and H^+ energy distributions (Figure 10) and angular distributions (Figure 11) are consistent with an interpretation in which the bulk of the H^+ energy and roughly half of the O^+ energy is derived electrostatically and the remainder of the O^+ energy is provided by a mass dependent transverse acceleration mechanism which preferentially acts on the O^+ ions. Candidate mechanisms with these characteristics have been proposed by Ungstrup et al. (1979) and Lysak et al. (1980) involving resonant wave-particle interactions, and Lennartsson (1980) involving fluctuating or small-scale transverse electric fields.

Summary and Conclusions

At least two classes of ionospheric acceleration mechanisms are operative in the altitude range above ~4000 km on auroral field lines. Field aligned ion beams are often observed in association with the signatures of parallel electrostatic potential differences and both statistical and case studies suggest that electrostatic acceleration is an important contributor to their energization. Conical pitch angle distributions with minimums along the magnetic field direction are observed which can be interpreted as resulting from a primarily transverse acceleration mechanism. For cases where the transverse

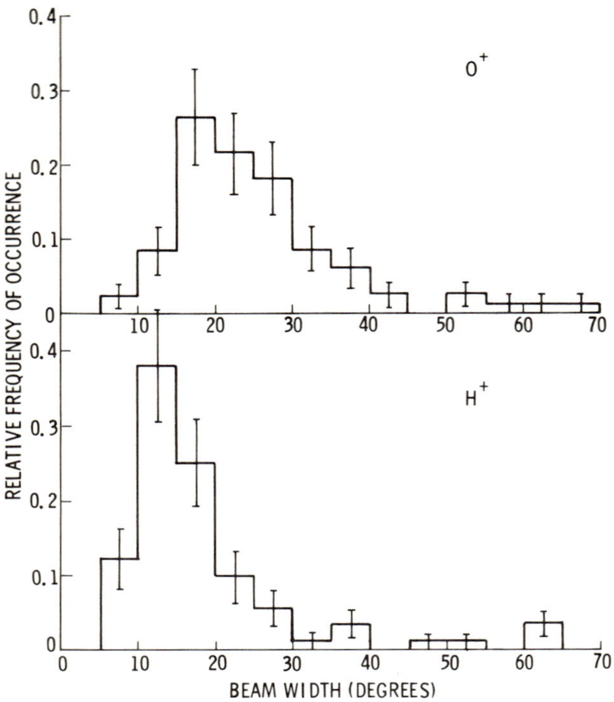

Fig. 11. A comparison of the O^+ and H^+ beams widths from the upflowing ion events shown in Fig. 10 (Collin et al., 1980).

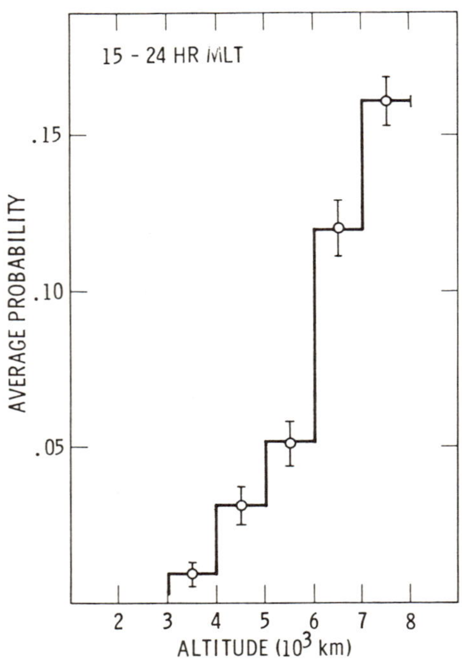

Fig. 12. Average probability of encountering an upward flowing ion event as a function of altitude (Ghielmetti et al., 1978).

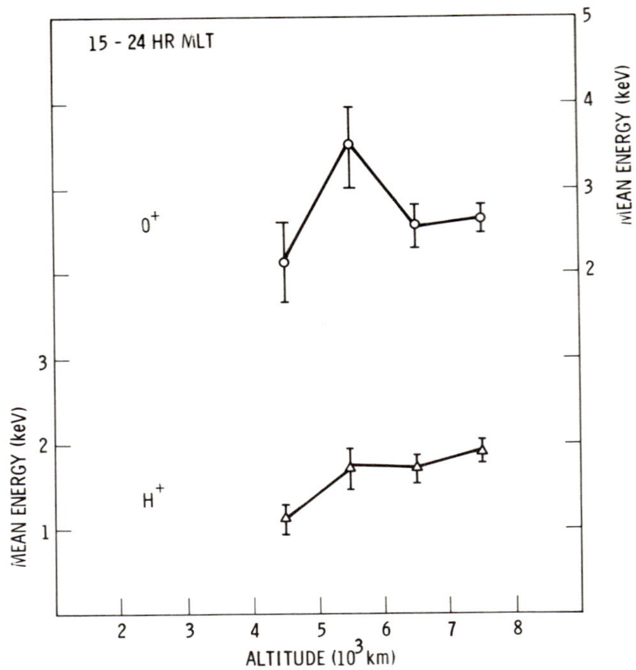

Fig. 13. Mean energy of upflowing ions at peak flux intensity as a function of altitude (Ghielmetti, 1978).

energization occurs at low altitudes with respect to the satellite this minimum subtends a small pitch angle range and cannot always be observed, especially in cases where the pitch angle scan does not completely extend to the magnetic field direction. Thus the observation of an ion beam does not imply that only an electrostatic acceleration is involved. Typically both the pitch angle and energy distributions of the beams are broader than can be understood on the basis of pure electrostatic acceleration and suggest that some additional transverse energization or scattering process is taking place. Similarly the possibility of an electrostatic contribution to the energy of the conics cannot be excluded. Both types of distributions therefore are consistent with the action of combined transverse and parallel acceleration mechanisms. The statistical results suggest that, for the ion beams observed in the evening sector with energies > 500 eV, electrostatic acceleration provides most of the H^+ ion energy and the O^+ ions derive about half of their energy from this mechanism and about half from some mass dependent transverse acceleration mechanism.

Acknowledgements

This work has been supported by the Office of Naval Research and the Atmospheric Sciences section of the National Science Foundation.

References

Cladis, J. B., and R. D. Sharp, Scale of electric field along magnetic field in an inverted-V event, J. Geophys. Res., 84, 6564, 1979.

Collin, H. L., R. D. Sharp, E. G. Shelley, and R. G. Johnson, Some general characteristics of upflowing ion beams over the auroral zone and their relationship to auroral electrons, submitted to J. Geophys. Res., 1980.

Ghielmetti, A. G., R. G. Johnson, R. D. Sharp, and E. G. Shelley, The latitudinal, diurnal, and altitudinal distributions of upflowing energetic ions of ionospheric origin, Geophys. Res. Lett., 5, 59, 1978.

Ghielmetti. A. G., Upward flowing ion characteristics in the high latitude ionospheric acceleration region, EOS Trans. Am. Geophys. U., 59, 1155, 1978.

Kaufmann, R. L., D. N. Walker, and R. L. Arnoldy, Acceleration of auroral electrons in parallel electric fields, J. Geophys. Res., 81, 1673, 1976.

Klumpar, D. M., Transversely accelerated ions: an ionospheric source of hot magnetospheric ions, J. Geophys. Res., 84, 4229, 1979.

Lennartsson, W., On the consequences of the interaction between the auroral plasma and the geomagnetic field, Planet. Space Sci., 58, 135, 1980.

Lysak, R. L., M. K. Hudson, and M. Temerin, Ion heating by strong electrostatic ion cyclotron turbulence, J. Geophys. Res., 85, 678, 1980.

Mizera, P. F., and J. F. Fennell, Signatures of electric fields from high and low altitude particle distributions, Geophys. Res. Lett., 4, 311, 1977.

Mozer, F. S., C. W. Carlson, M. K. Hudson, R. B. Torbert, B. Parady, and J. Yatteau, Observations of paired electrostatic shocks in the polar magnetosphere, Phys. Rev. Lett., 38, 292, 1977.

Sharp, R. D., R. G. Johnson, and E. G. Shelley, Observation of an ionospheric acceleration mechanism producing energetic (keV) ions primarily normal to the geomagnetic field direction, J. Geophys. Res., 82, 3324, 1977.

Sharp, R. D., R. G. Johnson, and E. G. Shelley, Energetic particle measurements from within ionospheric structures responsible for auroral acceleration processes, J. Geophys. Res., 84, 480, 1979.

Shelley, E.G., R. D. Sharp, and R. G. Johnson, Satellite observations of an ionospheric acceleration mechanism, Geophys. Res. Lett., 3, 654, 1976.

Ungstrup, E., D. M. Klumpar, and W. J. Heikkila, Heating of ions to superthermal energies in the topside ionosphere by electrostatic ion cyclotron waves, J. Geophys. Res., 84, 4289, 1979.

Whalen, B. A., W. Bernstein, and P. W. Daly, Low altitude acceleration of ionospheric ions, Geophys. Res. Lett, 5, 55, 1978.

Whipple, E. D., Jr., The signature of parallel electric fields in a collisionless plasma, J. Geophys. Res., 82, 1525, 1977.

TRANSVERSELY ACCELERATED IONS IN AURORAL ARCS

D.M. Klumpar

University of Texas at Dallas, Richardson, Texas 75080

Abstract. Ionospheric ions apparently accelerated transversely to the geomagnetic field in the topside ionosphere are regularly detected by the soft particle spectrometers on the ISIS satellites. Such gyro-accelerated ions are observed in association with precipitating auroral electrons. A detailed study of their relationship with such electrons and with field-aligned currents, together with simultaneous measurements of the local plasma composition and density, reveals the specific conditions present in the topside ionosphere during the generation of such transversely accelerated ions (TAI). A proposed mechanism for generation of TAI involving acceleration by electrostatic ion cyclotron waves is consistent with the present observations.

Introduction

Evidence has accumulated over the past several years that points to the strong role acceleration processes in the ionsophere must play in contributing to the hot plasmas of the magnetosphere (Prange 1978, Shelley 1979, Horwitz and Chappell 1979, Horwitz 1980). There is a constant outflux of ionospheric ions from the earth as a result of a variety of processes. Evidence exists for the direct acceleration of ions to suprathermal energies by apparently different mechanisms that produce two observationally distinct classes of upward flowing ions: One class is evident by the appearance of nearly field-aligned beams of ions that have been accelerated primarily along the magnetic field (Shelley et al. 1976). A second class of ions are observed whose velocity vectors lie along the surface of a cone with an apex half-angle near 90° indicating primarily transverse acceleration with respect to the magnetic field (Klumpar 1975, 1979, Sharp et al. 1977, Whalen et al. 1978). It is this latter class of particles to which this paper is addressed.

Transversely accelerated ions (TAI) are positively charged ionospheric particles that have undergone strong heating preferentially in the guiding center frame. Owing to this acceleration in their cyclotron component of motion, the total energy of the ions rapidly increases and they become subject to the grad B or mirror force by means of which they are ejected into the magnetosphere from the heating region in the ionosphere. The ions are subsequently detected by the soft particle spectrometers on ISIS-1 and ISIS-2 as these satellites pass over the auroral oval. The frequency of detection of TAI indicates that the transverse acceleration of ionospheric ions is a usual and normal process in the high latitude ionosphere, ongoing at all times over a wide area approximately coincident with the auroral oval. Any individual event may be rather localized and is apparently transient on time scales greater than a few minutes. However, a large number of active regions are present in the ionosphere at any one time so that the mechanism represents a significant source of particles to the magnetosphere and furthermore is intimately associated with the process that leads to auroral displays.

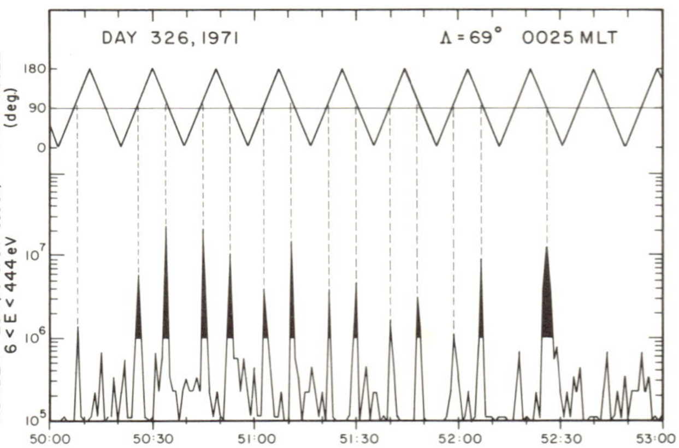

Fig. 1. Low energy ion number flux versus time for a 3-minute interval on day 326, 1971. The transversely accelerated ions are the flux enhancements of more than an order of magnitude observed twice during each satellite spin at pitch angles somewhat greater than 90° (i.e. upward moving).

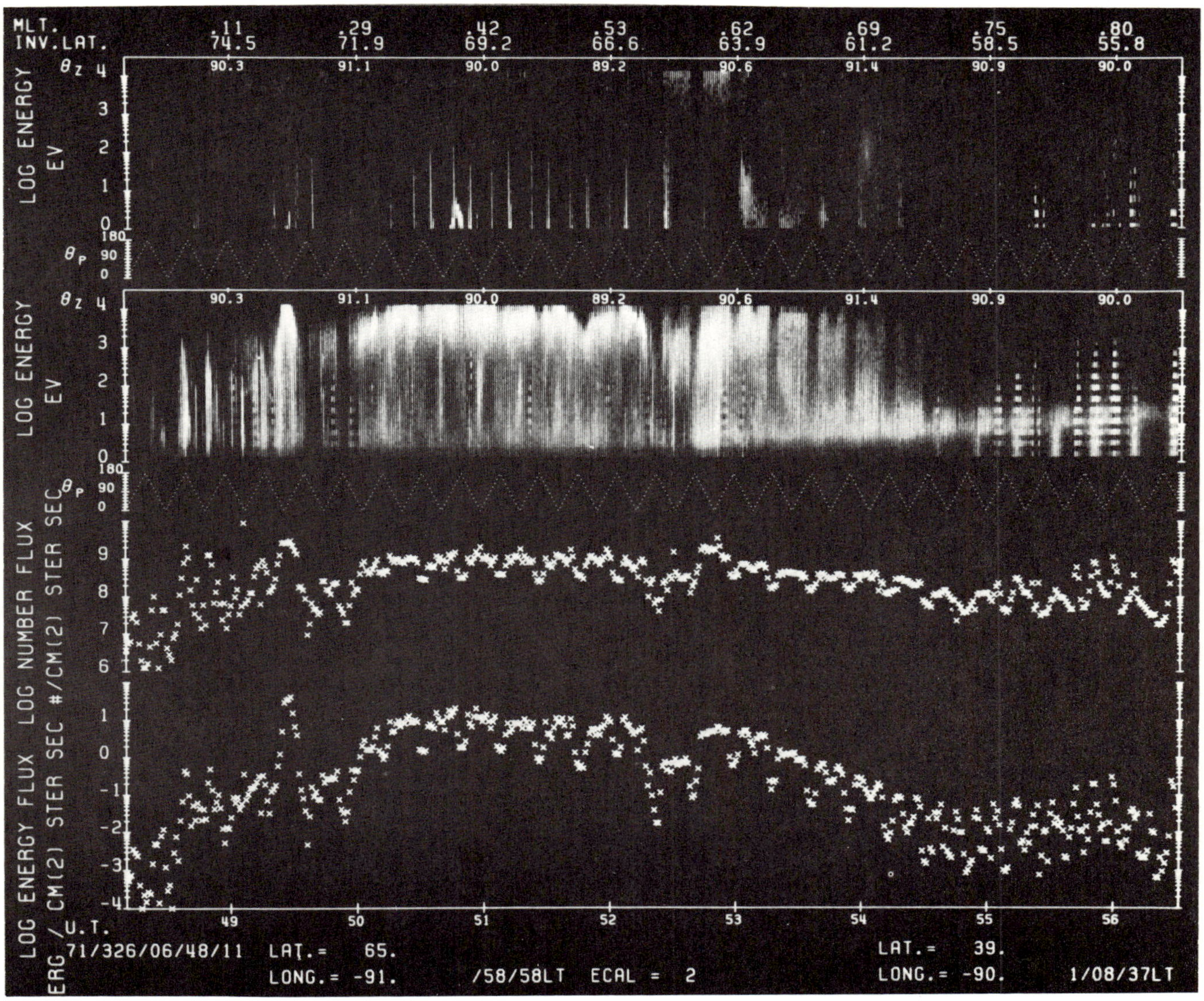

Fig. 2. Ion (upper panel) and electron (lower panel) spectrograms for the auroral oval crossing shown in detail in Figure 1.

It is the purpose of this paper to discuss a multi-parameter set of observations associated with this phenomena. We adopt a working hypothesis for the mechanism responsible for heating the ions and evaluate that model in light of the data. The observations are found to be consistent with the hypothesis proposed by Ungstrup et al. (1979) that electrostatic ion cyclotron waves (EIC) are responsible for the heating. Such waves, generated in the ionosphere by field-aligned currents of sufficient intensity, but of either sign, are capable of interacting with a portion of the local ionospheric plasma and of accelerating some fraction of the ions in a direction primarily transverse to the geomagnetic field. As a result of the local magnetic mirror force these ions are injected into the magnetosphere where they have subsequently been detected from a number of high altitude spacecraft.

Observations

Transversely accelerated ions are readily recognized by their characteristic signature on a flux versus time plot of data from a spinning satellite where the entrance aperture of the particle detector sweeps repeatedly through a large range of pitch angles. Figure 1 displays the flux of ions with energies between 6eV and

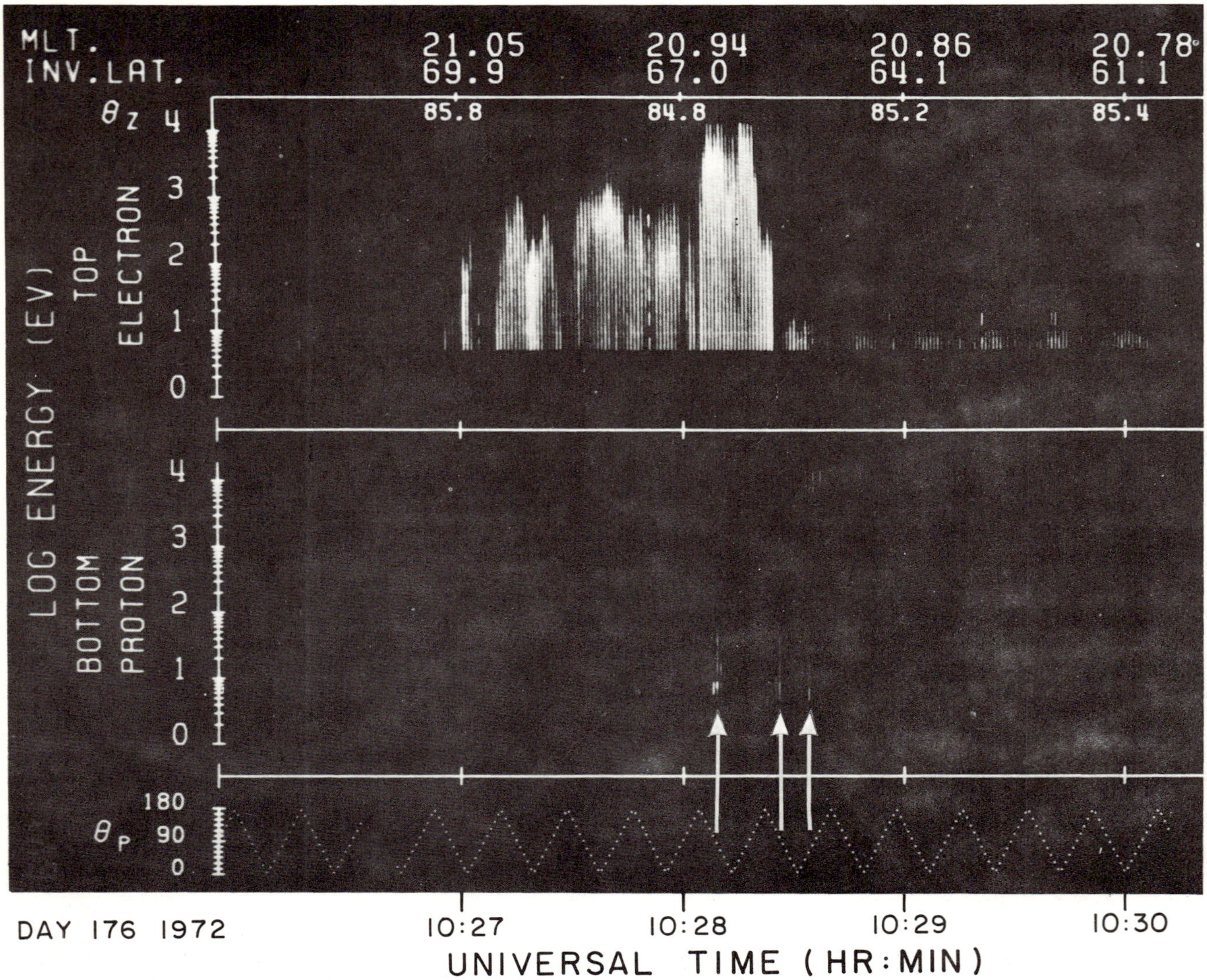

Fig. 3. Expanded electron and ion spectrograms for a pre-midnight auroral oval crossing on day 176, 1972. The transversely accelerated ions (marked by vertical arrows) are associated with the intense auroral electron precipitation but extend ≈100 km equatorward.

~450eV observed as a function of time for a three-minute interval during a pass of ISIS-2 over the midnight auroral oval. The satellite's spin axis is aligned approximately perpendicular to the magnetic field so that the radially looking ion detectors sample the entire range of pitch angles from zero to 180° every one-half spin or every nine seconds. It is apparent from the plot that factor of ten or more enhancements in the intensity of these low energy ions are being detected each time the instrument samples ions with pitch angles near 100°. In the nomenclature used here particles at zero degrees pitch angle are moving along the magnetic field directly downward toward the ionosphere while 180° particles are moving upward toward the magnetosphere. Hence the presence of flux enhancements at 100° with little or no measureable ion fluxes at other pitch angles indicates that the satellite is immersed in a region where virtually all of the measured ions are spiraling outward along the magnetic field with their velocity vectors lying along the surface of a cone with an apex half-angle of 80°.

Figure 2 illustrates in the top panel the signature of these ions in our usual grey-scale spectrogram format. In such spectrograms the grey-shade is modulated as a function of energy

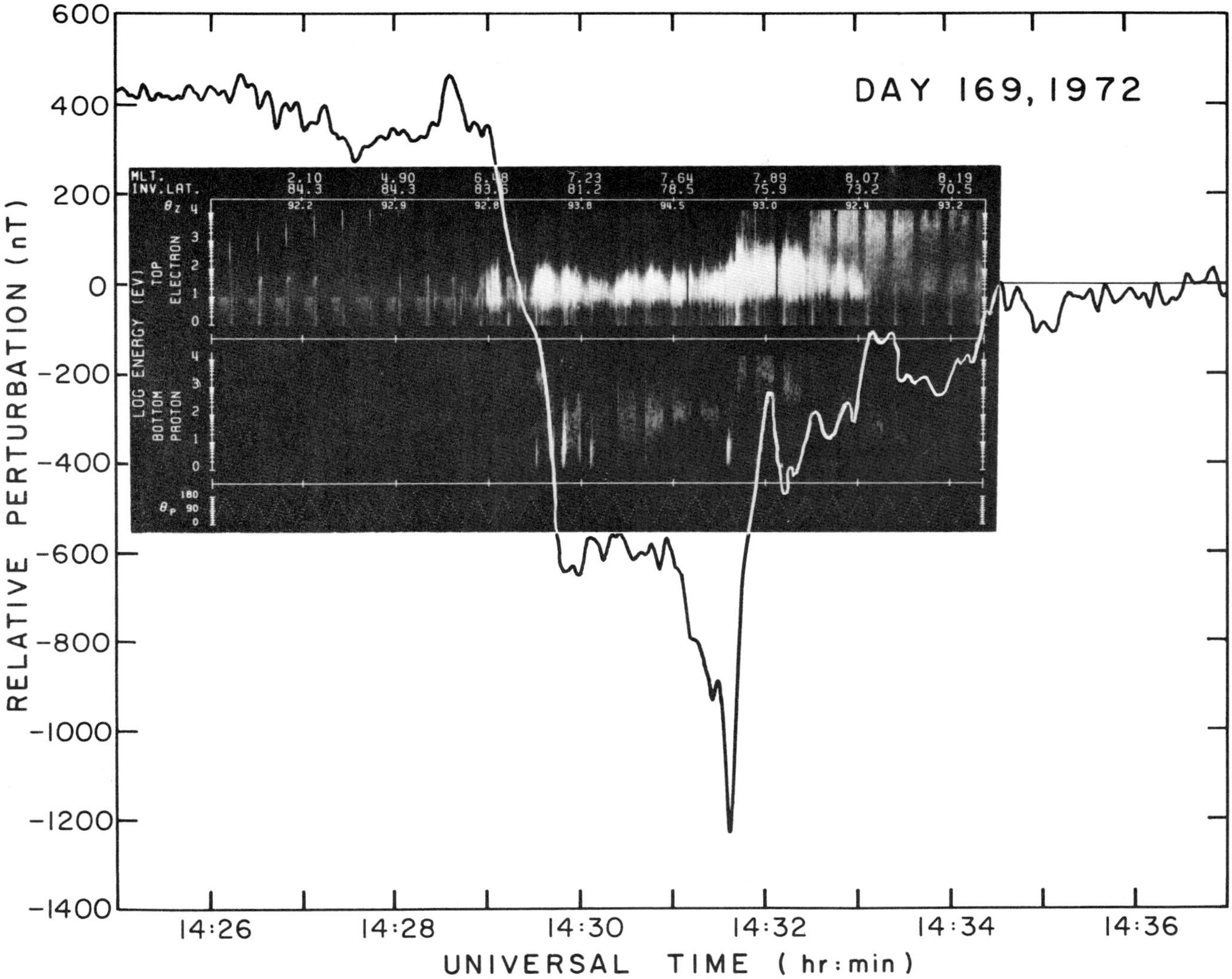

Fig. 4. Concurrent particle spectrograms and E-W magnetic field perturbation. Intense TAI observed from 14:29:20 to 14:30:10 and at 14:31:40 are associated with steep gradients in the magnetic perturbation. Such gradients are the signatures of strong field-aligned currents.

and time by the measured energy flux of incident particles where increasing lightness indicates increased energy flux. Also shown in the figure is the simultaneously obtained electron spectrogram accompanied by the integrated intensity and energy flux of the electrons.

Two pertinent points may be made from this data set. Firstly, the upward moving ions are not accompanied by downcoming fluxes of ions and thus cannot be attributed to a mirrored population of ions originating above the satellite. The observed pitch angle distribution is consistent with energization perpendicular to the magnetic field in a narrow altitude regime a few hundred kilometers below the satellite. Secondly, the ions are associated in a gross sense with strong fluxes of electrons, whose energy flux in this case exceeds 10 ergs/cm^2 sec sr.

In order to examine the association between these ions and auroral electrons in somewhat greater detail, a second example, shown in Figure 3, will be discussed. Again, this is a composite electron and ion spectrogram this time with ions in the lower panel. Three vertical arrows mark the location of transversely accelerated ions observed near 90°. The region of transverse ion acceleration in this example extends up to 100 km equatorward of the low latitude edge of any detectable electron fluxes of auroral energy. This and other similar cases have led us to dismiss any requirement for a direct cause and effect relationship between the auroral electrons and TAI. There must be more to the story. The situation must be more complex than is shown in the particle data alone.

The hypothesized association with electrostatic ion cyclotron waves led to an examination of data from an on-board magnetometer for the indication of field-aligned currents since theory (Kindel and Kennel 1971) predicts EIC to be generated as a result of field-aligned currents. ISIS-2 carried a three-axis magnetometer initially designed to provide satellite attitude information. Data from this magnetometer has been successfully used to detect the magnetic field perturbations arising from local field-aligned currents using techniques described by McDiarmid et al. (1978). Such magnetic perturbation data have been examined during intervals when TAI are present. It was found, as expected under the hypothesized EIC heating mechanism, that TAI are indeed present in regions where strong field-aligned currents are flowing.

An example of this relationship is shown in Figure 4. Here the simultaneous electron and ion spectrograms have been overlaid with the concurrent east-west magnetic field perturbation derived from the ISIS magnetometer measurements. Again, it is apparent that TAI occur in general in association with intense fluxes of electrons, but also that they occur at steep gradients in the magnetic perturbation. These steep slopes in ΔB are usually taken to be the signatures of local field-aligned currents, estimated in this example to be several $\mu A/m^2$. Thus it would appear that TAI are associated with intense field-aligned currents. However, not all field-aligned current regions were found to have associated transversely accelerated ions.

Guided by the hypothesis that TAI may be generated by ion cyclotron waves, we again appeal to the theory of Kindel and Kennel (1971) who show that the current necessary for EIC instability is highly dependent upon the local ion composition, density and temperature. The greater the ionospheric density, the higher should be the field-aligned current required to drive the instability.

Local ion composition and density are measured on ISIS-2 by an ion mass spectrometer. Relative changes in total ion density are measured with very fast temporal resolution by a Langmuir probe. Examination of these data revealed that (1) oxygen is always the dominant ion when TAI are observed from ISIS-2, and (2) that TAI are accompanied by depletions in local ionospheric density.

This association is illustrated in Figure 5, which summarizes the event discussed previously in connection with Figure 3. In this auroral oval crossing the TAI region extended well equatorward of the measured electron precipitation. Indeed, Figure 5 illustrates that a strong downward field-aligned current (estimated at $2\frac{1}{2}$ $\mu A/m$) was present in the transverse ion region and furthermore that the ion mass spectrometer measured the ambient O^+ density to be depleted to <100 ions/cm^3in this region. An

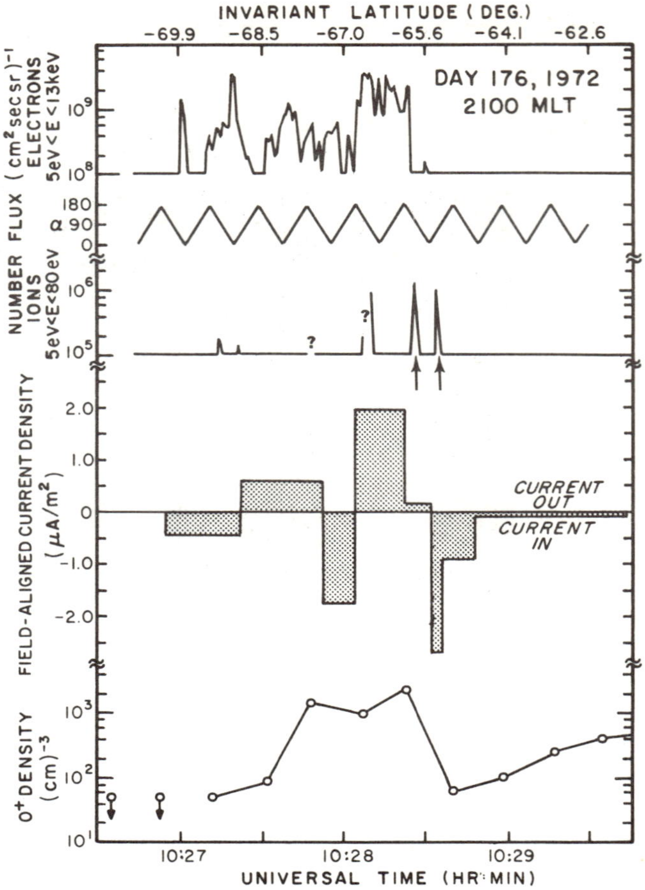

Fig. 5. The event of Figure 3 together with the simultaneously measured field-aligned currents and O^+ ion density. The TAI are associated with large current densities and low plasma density.

examination of the higher resolution Langmuir probe data for this pass revealed that the total ionospheric plasma density actually dropped to <10/$cm^3$and was highly turbulent in this region. The absence of detectable electrons or ions above 5eV sufficient to account for this current seems to point to upflowing ionospheric electrons as the likely current carriers, an interpretation entirely consistent with the deep ionospheric depletion observed at this spot.

The theoretical expectations indicate that EIC instability is generated by field-aligned currents and furthermore that the instability has a higher growth rate for a given field-aligned current when heavy ions are present and when the local plasma density is low. As the plasma density decreases, the current needed to generate ion cyclotron waves is reduced. Thus, if the presence of TAI is indeed an indicator of the ion cyclotron instability, one would expect to find an empirical relationship between plasma density and current intensity that governs the occurrence of TAI. Specifically, TAI should not be present in regions where high ionospheric densities exist and where the field-aligned currents are weak. Conversely, when the ionospheric density is depleted relatively small field-aligned currents should drive the ionosphere unstable. In other words, the parameter space composed of parallel current density and ionospheric plasma density should be composed of a stable region marked by the absence of heated ions and an unstable region containing TAI.

Figure 6 illustrates the preliminary results of an on-going analysis to test this hypothesis. Each point represents a distinct observation of the field-aligned current density and local O^+ density both in regions where TAI are present as well as where they are absent. As expected, there appears to be a region across the upper left portion of the figure in which low field-aligned currents and high ionospheric densities do not result in production of TAI. On the other hand, as the current density increases for a given O^+ number density a threshold for instability is reached and TAI are observed. This is precisely the behavior that was expected from the hypothesis; namely, that transversley accelerated ions result from heating by electrostatic ion cyclotron waves.

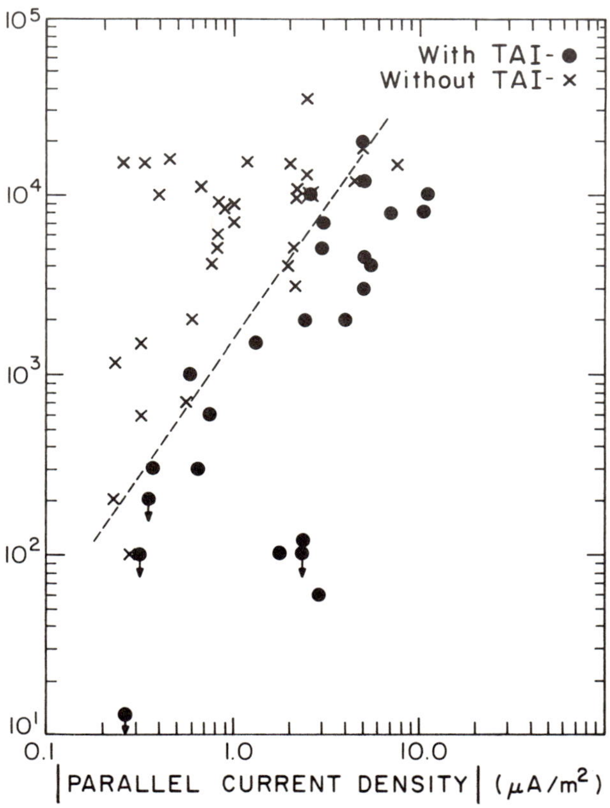

Fig. 6. Scatter plot of oxygen ion number densities and field-aligned current densities for auroral oval crossings both with and without TAI. The absence of TAI in the upper left hand portion of this parameter space indicates ionospheric stability to the TAI acceleration mechanism.

Summary

Transversely accelerated ions are found to be a common feature in the topside ionosphere within the auroral oval. Their association with the auroral acceleration process, their contribution as a source of particles for the magnetosphere, and their role in ionosphere-magnetosphere coupling make it imperative that the acceleration process be fully understood. Observations presented here support the hypothesis that TAI result from primarily perpendicular heating of a portion of the local plasma ions by electrostatic ion cyclotron waves that are generated in an ionosphere unstable to field-aligned current densities exceeding certain thresholds determined by the local ionospheric plasma composition, density, and temperature. Direct measurements by instruments on ISIS-2 permit the conditions present in the acceleration region to be determined.

Acknowledgements. The author is grateful to J.R. Burrows, L. Brace, and J. Hoffman for their generosity by providing access to their magnetic perturbation, Langmuir probe, and ion mass spectrometer data respectively. This material is based upon work supported by the National Science Foundation under Grant Numbers ATM75-03985 and ATM-8002117 and by the National Aeronautics and Space Administration under grant NSG 5085.

References

Horwitz, J.L., Conical Distributions of Low-Energy Ion Fluxes at Synchronous Orbit, J. Geophys. Res., 85, 2057, 1980.

Horwitz, J.L. and C.R. Chappell, Observations of Warm Plasma in the Dayside Plasma Trough at Geosynchronous Orbit, J. Geophys. Res., 84, 7075, 1979.

Kindel, J.M. and C.F. Kennle, Topside Current Instabilities, J. Geophys. Res., 76, 3055, 1971.

Klumpar, D.M., Auroral Zone Non-Thermal Positive Ions Observed with ISIS-2 (abstract), in Program and and Abstracts, XVII General Assembly, IAGA Bull. 36, International Association of Geomagnetism and Aeronomy, Grenoble, 1975.

Klumpar, D.M., Transversely Accelerated Ions: An Ionospheric Source of Hot Magnetospheric Ions, J. Geophys. Res., 84, 4229, 1979.

McDiarmid, I.B., J.R. Burrows, and M.D. Wilson, Comparison of Magnetic Field Perturbations at High Latitudes with Charged Particle and IMF Measurements, J. Geophys. Res., 83, 681, 1978.

Prange, Renee, Energetic (keV) Ions of Ionospheric Origin in the Magnetosphere: A Review, Ann. Geophys., 34, 187, 1978.

Sharp, R.D., R.G. Johnson, and E.G. Shelley, Observations of an Ionospheric Acceleration Mechanism Producing Energetic (keV) Ions Primarily Normal to the Geomagnetic Field Direction, J. Geophys. Res., 82, 3324, 1977.

Shelley, E.G., Heavy Ions in the Magnetosphere, Sp. Sci. Rev., 23, 465, 1979.

Shelley, E.G., R.D. Sharp, and R.G. Johnson, Satellite Observations of an Ionospheric Acceleration Mechanism, Geophys. Res. Lett., 3, 654, 1976.

Ungstrup, E., D.M. Klumpar, and W.J. Heikkila, Heating of Ions to Superthermal Energies in the Topside Ionosphere by Electrostatic Ion Cyclotron Waves, J. Geophys. Res., 84, 4289, 1979.

Whalen, B.A., W. Bernstein, and P.W. Daly, Low Altitude Acceleration of Ionospheric Ions, Geophys. Res. Lett., 5, 55, 1978.

X-RAY IMAGES OF AN AURORAL BREAK-UP

Barry Mauk and George K. Parks

Geophysics AK-50, University of Washington, Seattle, Washington 98195

Abstract. X-ray images have been obtained by a newly developed X-ray pin-hole camera during auroral breakups. A sequence of X-ray images sampled during the initiation of an auroral breakup shows the presence of a 20-km spot-like event that had a relatively steep energy spectrum superimposed on a more energetic component that was spatially uniform. The very intense acceleration which was observed subsequent to the spot event appears to have been an intensification of the harder component. These X-ray characteristics are reminiscent of the behavior of electron distributions on the equator during dispersionless injection events. At the equator, cool beam-like and hot isotropic populations often appear in coordination with each other. We suggest that the energy and spatial structures within the X-ray images reflect the details of the dynamics corresponding to the equatorial particles travelling along the geomagnetic field.

Introduction

It is a characteristic of magnetospheric dynamics to produce multi-component particle distributions (Evans, 1974; McIlwain, 1975). The low energy component (peaking below 10 keV) is believed to be of ionospheric origin and is, possibly, accelerated by double layers or electrostatic shocks (Mozer, 1980). Because of the form of the distribution, the higher energy component (dominating above 10 keV) is believed to be accelerated deeper in the magnetosphere (by processes yet to be identified), and is believed to be derived from the plasmasheet on closed field lines (Parks et al., 1978). Geostationary observations indicate that the multiple components, while not appearing in one-to-one correlation, do occur in close coordination with each other (McIlwain, 1975; Parks et al., 1979). Understanding the relationships among the various components, when and how each is formed, is of fundamental importance in auroral substorm phenomena. It is clear that all components must be understood before the basic acceleration mechanisms can be understood.

Images taken by all-sky cameras (Akasofu, 1968), by TV's (Davis and Hicks, 1964), and more recently by polar orbiting spacecraft (Anger and Lui, 1973), have provided detailed pictures of the visual aurora. The visual images result from an ambiguous mix of the two components. The presence of the high energy components makes the interpretation of the visual pictures more complicated since these pictures cannot be simply interpreted either in terms of "foot-prints" of processes working in more distant regions, or in terms of local processes alone.

Very little information is available on the spatial properties of the more energetic components alone. Some attempts have been made in the past to obtain spatial information of the more energetic component by detecting the atmospheric bremsstrahlung X-rays of the precipitated energetic electrons (Parks, 1967; Yamagami et al., 1978). These instruments divided the sky into essentially a few sectors with a spatial resolution of 20-30 km at ionospheric heights (100 km altitude). Use of several balloons equipped with similar X-ray instruments flown from regions separated by several hundred kilometers has provided large-scale information (Kremser et al., 1973). Also, observations of $\geq$45 keV X-rays from above by polar orbiting spacecraft have provided spatial information on the more energetic component with a spatial resolution $\gtrsim$200 km (Imhoff et al., 1974).

For the present study a new X-ray imaging device (an X-ray pin-hole camera) has been developed to explore in more detail the spatial characteristics of the energetic precipitated components. The X-ray pin-hole camera has been successfully flown on balloons on several occasions, and X-ray pictures have been obtained during auroral breakups. The purpose of this paper is to present some preliminary pictures of electron precipitation regions responsible for the X-rays (a more complete presentation is given by Mauk et al., 1981). In particular, we will show a sequence of pictures that were obtained during a breakup initiation interval. These pictures show that the breakup was initiated by a (20 km) "spot" -like precipitation that had a relatively steep energy spectrum. Along with the "spot" event, a much harder, spatially uniform component was observed. The very intense acceleration which followed the spot event appears to have been an in-

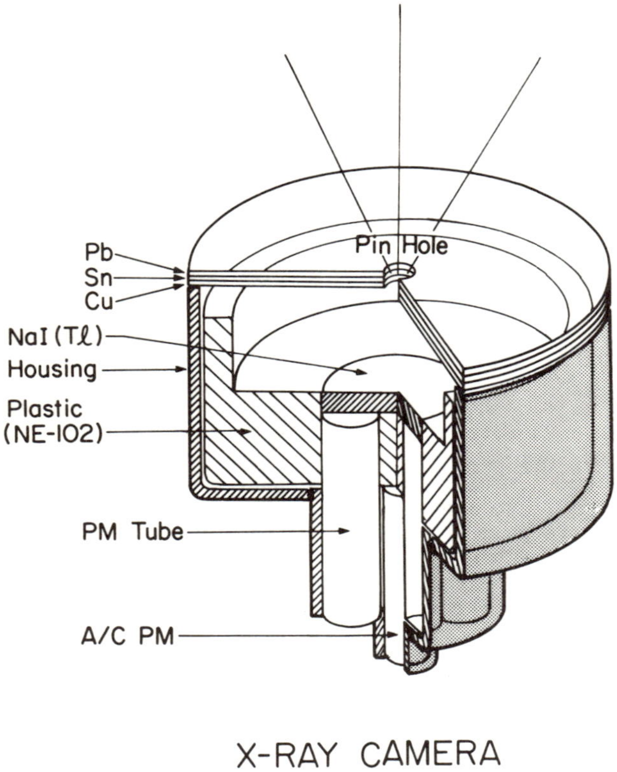

Fig. 1. Cutaway view of the X-ray imaging system.

tensification of the harder component. These energy spectral features are very similar to the behavior of particle distributions observed on the equator during dispersionless injection events (McIlwain, 1975; Parks et al., 1979). We suggest that the X-ray images represent the details of the dynamics corresponding to the equatorial particles with small pitch angles during injection events.

Section 2 briefly describes the workings of the X-ray camera. Examples of the X-ray images are shown in Section 3. Section 4 discusses our observations.

An X-ray Pin-hole Camera

A cutaway drawing of our pin-hole camera is shown in Figure 1. The camera consists essentially of a (1.9 cm) pin-hole (lens) and a 9 cm NaI (Tl) scintillator (film) that is viewed by four photomultiplier tubes. An auroral X-ray which deposits its energy in the crystal generates many secondary photons. The proportion of the secondary photons detected by each photomultiplier tube depends on the relative distance between the photomultiplier tube and the striking position. The positional information can be directly obtained by pulse-height analyzing each X-ray photon simultaneously by the four photomultiplier tubes. Information on the X-ray energy can be obtained by summing the pulse-height data from all of the tubes. The positional information on the crystal is projected to the sky through the pin-hole. X-rays in the interval 22-110 keV arriving from $\sim\pm 30^{\circ}$ from zenith are detected and analyzed. The pin-hole size projects to a 15 km spot at auroral altitudes, and the entire field of view projects to an 80 km disk. To reduce contribution from cosmic rays and their secondaries, the instrument was shielded with lead and anti-coincidence plastic NE-102 scintillator.

X-ray Images

Examples of X-ray images will be shown from data obtained during the August 31, 1978 balloon

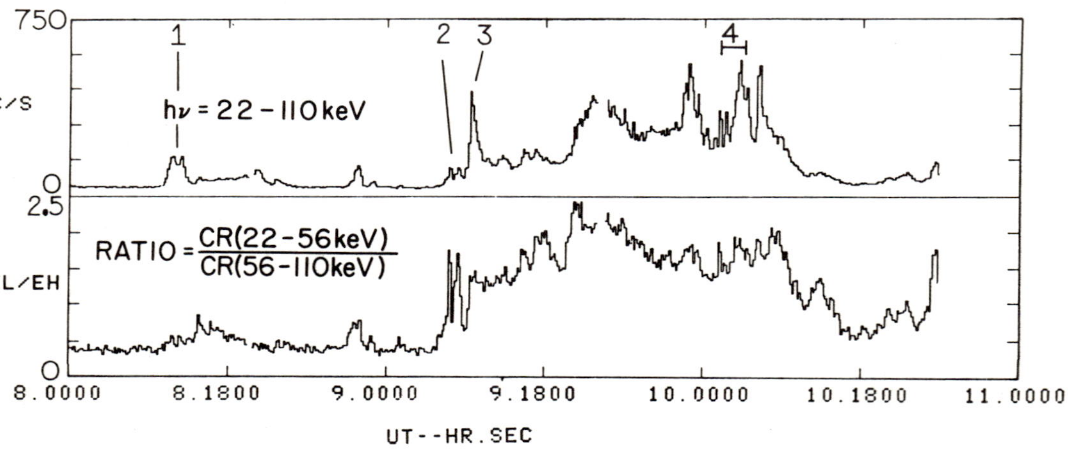

Fig. 2. Three-hour summary plot showing the total (dead time corrected) count-rate and a spectral hardness index for a portion of a balloon flight on August 31, 1979. The scales are linear.

flight. Figure 2 shows a 3-hour summary of a break-up interval. The plot shows the X-ray count-rates (top curve) and a spectral hardness index which consists of the ratio of X-ray counts detected in the energy range 22-56 keV and 56-110 keV. It is appropriate to mention here that this representation of data is a typical one used by instruments that detect X-rays without imaging (see Brown, 1966; Anderson, 1965). Several features indicated here are well known: the impulsiveness of the precipitated energetic electrons and the softening of the energy spectra associated with the breakup (Barcus and Rosenberg, 1960).

A detailed examination of the bottom panel shows that the energy spectral index is not a simple monotonic function of the X-ray count-rate. For example, note the isolated "precursor" count-rate enhancement around 08.1000 UT (hour.seconds), at the point marked "1". This event was not accompanied by softening of the spectrum. In contrast, the series of short bursts beginning around 09.0600 UT (near the point marked "2") was accompanied by drastic softening of the energy spectra, more drastic than even the major burst which followed (marked "3"). The X-ray images which we will examine demonstrate that multiple acceleration and precipitation processes are involved in auroral dynamics.

Figure 3B and 3C are X-ray images taken during the event marked "2" in Figure 2. These figures, as compared to Figure 3A (taken during a "quiet period" and representing background conditions), clearly indicate spatial structure (see Mauk et al., 1981 for a discussion of the detector response). Figure 3B is taken during the first "hump" shown in Figure 2, and much of this precipitation was confined to a 20 km region near the middle of the figure. Figure 3C, taken during the second "hump", shows that the peak X-ray fluxes have somewhat increased and shifted slightly in position. These structures did not result from an increase of precipitation from preexisting structures nor from motions of structures that existed outside the camera's field of view. The structures emerged out of the background.

The spot-like event shown in Figure 3 appears in Figure 2 to have signaled and perhaps initiated a major change in the intensity and energy spectral characteristics of the precipitated electrons. Further study is clearly warranted. The left panel of Figure 4 shows the raw count-rate energy spectra sampled from different regions of an image which averages together Figures 3B and 3C. The upper spectrum (solid line) was sampled near the middle of the image and the middle spectrum (solid circles) was sampled near the very lowest portion of the image. The lowest spectrum (open circles) shows a background spectrum sampled just prior to all of the events depicted in Figure 2. The spectra shown on the right panel represent spectra with the background subtracted.

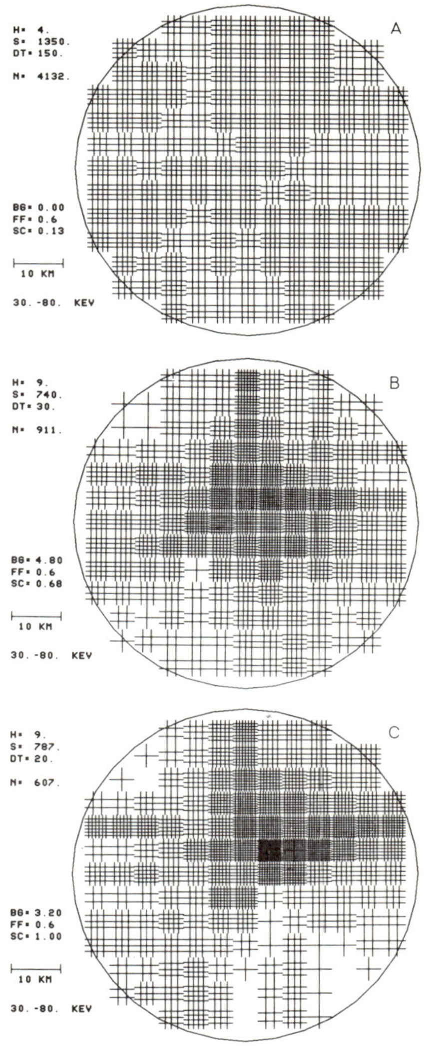

Fig. 3. Grey-scale images sampled during background conditions (A) and during the dramatic softening event labeled "2" on Figure 2 (B and C). The number of horizontal or vertical lines (M) within each "square" is lineraly proportional to the number of photons (n) which struck that square. The key gives the start time (H,S = hour, second), the accumulation time in seconds (DT), the total number of photons after background subtraction (N), the number of background counts subtracted from each square (BC), the scaling factor (M = n·SC), a standard length, and the accumulated energy range. The statistical errors (1σ) are ∿1/2 of the square root of the counts within the square.

There are two important points to be made about the spectra shown in the right panel of Figure 4. The first is that the spectrum sampled near the middle of the spot in Figures 3B and 3C (solid

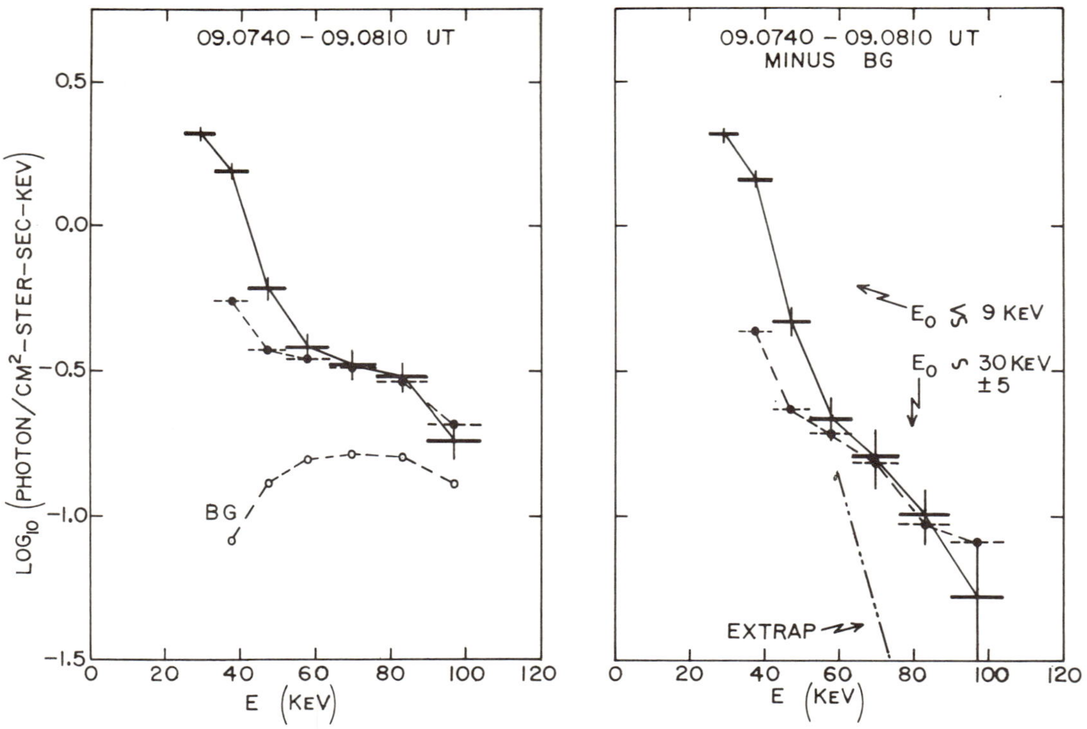

Fig. 4. The left portion shows a background energy spectrum (open circles), a spectrum sampled from the lowest portions of an average of Figures 3A and 3B (closed circles), and a spectrum sampled from the middle portions of an average of Figures 3A and 3B. The right portion shows the spectra shown on the left after background subtraction. The horizontal bars on each point on the spectra are about 1/4 shorter than the actual resolution capabilities of the instrument.

line) shows two distinct components: one has an e-folding energy of ≲9 keV (determination is instrumentally limited) and the other ∿30 ±5 keV. Second, the main difference that exists between the two spectra shown occurs within the low energy component. The spectra shown and others (not shown) indicate that the high energy component is relatively uniform withing the instrument's field of view. The 20 km spot shown in Figures 3B and 3C consisted of a spectrally soft component which appeared on top of a spatially uniform and spectrally hard component. The different spatial scales for the two energy components suggests that the parent particle distributions had different forms and had been generated by separate and distinct mechanisms.

Immediately following the spot-like event discussed above was a very major intensity burst marked "3" in Figure 2, as well as the initiation of a period when the spectral characteristics of the X-rays were relatively stable (see the bottom panel). Images constructed during this burst show that the emissions were quite uniform in space (not shown). Figure 5 shows an energy spectrum sampled during the peak of the intensity (closed circles). This spectrum is compared with the spectrum of the spot event already discussed. It is significant that the spectral characteristics of the new spectrum (e-folding energy ∿27 keV) is very similar to that of the high energy component of the old spectrum (e-folding energy ∿30 keV). Figure 5 suggests, along with the spatial uniformity of the event, that the major burst, and the activity that followed, resulted from an intensification of processes which were initiated during the spot event.

Discussion

The energy spectral details of the spot event and the events that followed are very reminiscent of the behavior of particle distributions observed during dispersionless substorm injection events at synchronous altitudes (McIlwain, 1975; Parks et al., 1979). On the equator, during the very initial stage of local injection events, a cool electron component (kT ∿ several keV) often accompanies a hotter (kT ∿ 10 keV) component. The forms of the two distributions are quite different, with the cool component showing a strongly field-aligned beam-like structure, and the hotter component being relatively isotropic. In addi-

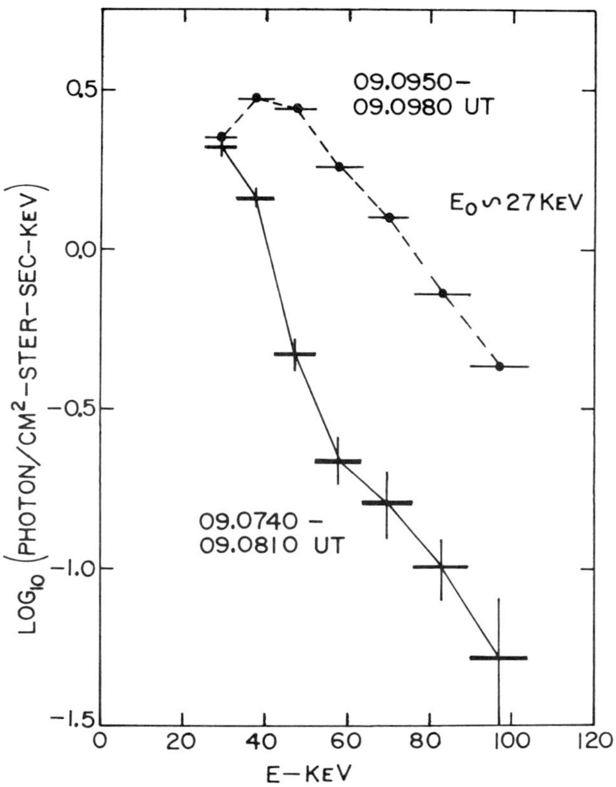

Fig. 5. Spectra sampled from the burst labeled "3" in Figure 2 (closed circles), compared with one of the spectra shown in Figure 4 (solid bars) sampled during the softening event labeled "2" on Figure 2.

tion, the cool component changes form very rapidly (less than 10 minutes) while the hotter component lasts for hours. The two components do not necessarily occur in time coincidence. In fact, much like the X-ray observations, the cool component at times precedes the hotter component by several minutes. All of the qualitative features detected at synchronous altitudes (the multiple components, the timing of their arrivals, and the different distribution forms for the two components) appear very similar to our X-ray observations already described. Note again that the different spatial scales within the X-rays suggest different forms for the parent distributions. Note also that the fact that the hotter geostationary population is most likely derived from tail populations suggests that the hard precipitated component is similarly derived from tail populations and not from local acceleration mechanisms.

The 20 km scale length of the spot-like X-ray emissions cannot be derived from scale lengths observed within the geostationary environment. If indeed the X-ray spot event is a reflection of the high energy tail of an equatorial field-aligned beam, the beams must undergo local ionospheric striating effects as they propagate to the ionosphere. It would not be surprising if the shock-like and other small-scale electric field structures observed above the auroral zone by Mozer et al. (1980) were related to the low-energy component which we observe. If the relationship is direct, mechanisms must be available to generate high-energy tails on the accelerated distributions. This tail generation may occur in more distant regions of the magnetosphere, perhaps near the equator where the distributions are often broad in energy.

The X-ray observations suggest a rather direct association between the auroral and geostationary regions. When comparing the auroral and the geostationary observations it must be noted that the observation of X-rays requires that not only must the particles be present in the magnetosphere, but that they must be unstable in a manner to cause the particles to be moved into the loss cone. For instance, the relative stability of the energy spectral properties of the X-rays beginning at the point marked "3" in Figure 2 suggests that the hotter populations are continuously present after that time, much like geostationary observations of injections. It must be the high variability of the precipitation mechanisms which leads to the variation in the X-ray count-rate.

In presenting the breakup initiation event discussed above, we have left the impression that small-scale spatial structure is a limited feature of energetic precipitation, occurring only in the very low-energy components observed. Figure 6, presented to eliminate this impression, shows images sampled within the region marked "4" in Figure 2. The images were sampled from successive burst features in the count-rate profile. Structure is clearly evident, and considering the time sequence, continuity exists from image to image. Small scale structure is quite prevalent within energetic precipitation. The e-folding energies measured during the event shown were of order 18 keV.

Conclusions

The X-ray images have in general the potential to isolate the processes which contribute to energetic electron precipitation. In this report, we have shown examples that indicate that the initiation of an auroral breakup involves multiple and quite different energization mechanisms which may be occurring in coordination with each other. The different spatial structures we observed in X-rays together with the spectral information can be used to distinguish between different types of energization mechanisms and parent distributions. The parent distributions suggested by our observations are in fact observed at geosynchronous altitudes during substorm initiation. We have also indicated with another event that small-scale spatial structure can be quite prevalent within energetic precipitation.

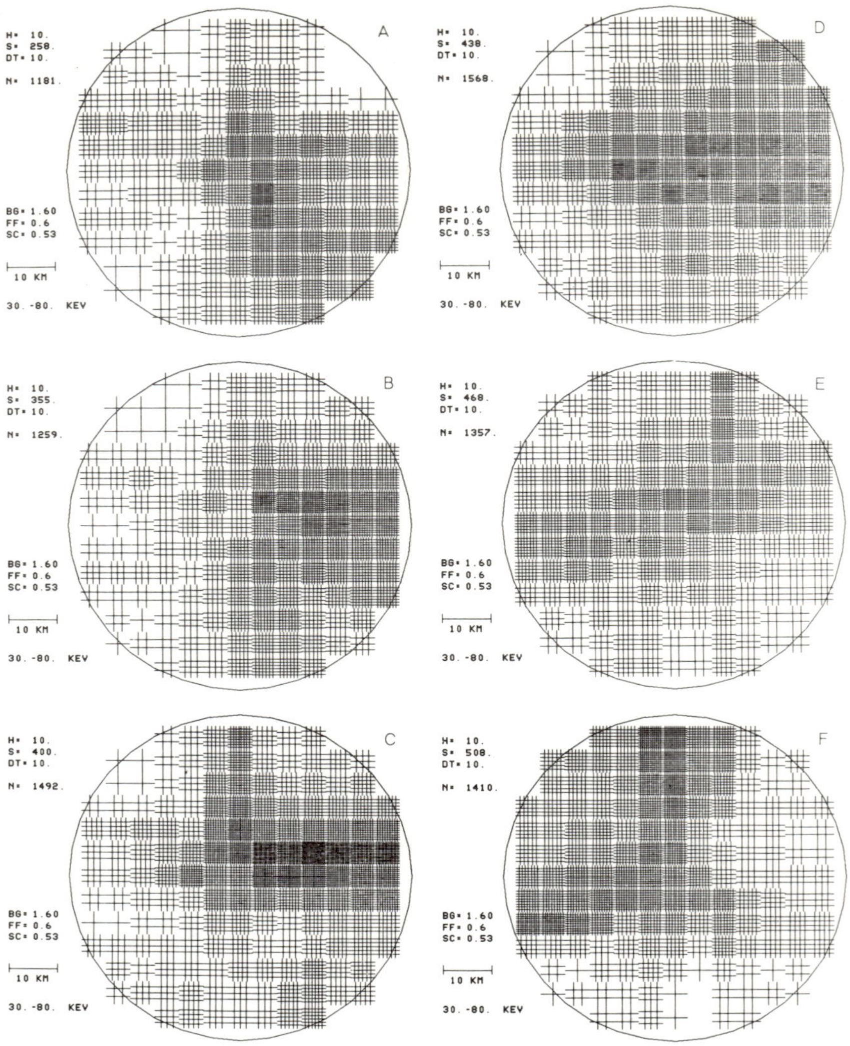

Fig. 6. Images sampled during the burst events marked with the number "4" in Figure 2. The time begins on the upper left and reads vertically.

The observations of small-scale structure in X-rays is important because the characteristic sizes of energetic electron precipitation regions provide additional clues toward unravelling the physics of deep space particle interactions. Characteristic plasma-scale lengths may be involved (for instance, the cyclotron diameter of a 10 keV proton at the equator projects to a 15 km circle within our image). It is significant that small-scale structure is so prevalent.

Acknowledgments. We would like to acknowledge the untiring efforts of John Chin in obtaining the data set here described.

The research at the University of Washington is in part supported by grants received from the National Science Foundation, Atmospheric Sciences Division (ATM77-09080), and the Office of Naval Research, Contract N00014-78-C-0333.

References

Akasofu, S.I., Polar Magnetospheric Substorms, D. Reidel, Dordrecht, Holland, 1968.

Anderson, K.A., Balloon Measurements of X-rays in the auroral zone, in Auroral Phenomena, ed. by M. Walt, Stanford Univ. Press, Stanford, CA, 1965.

Anger, C.O. and A.T. Kui, A global view at the polar region on 18 December 1971, Planet. Space Sci., 21, 873, 1973.

Barcus, J.R. and T. J. Rosenberg, Energy Spectrum for auroral zone X-rays, 1, Diurnal and type effects, J. Geophys. Res., 71, 803, 1966.

Brown, R.R., Electron precipitation in the auroral zone, Space Sci. Rev., 5, 311, 1966.

Davis, T.N. and G.T. Hicks, Television cinemaphotography of auroras and preliminary measurements of auroral velocities, J. Geophys. Res., 69, 1931, 1964.

Evans, D.S., Precipitation electron fluxes formed by a magnetic field-aligned potential difference, J. Geophys. Res., 79, 2853, 1974.

Imhof, W.L., G.H. Nakano, R.G. Johnson, and J.B. Reagen, Satellite observations of bremsstrahlung from widespread energetic electron precipitation events, J. Geophys. Res., 79, 565, 1974.

Kremser, G., K. Wilhehm, W. Riedler, K. Bronotad, H. Trefall, S. Vllalaud, J.P. Legrand, J. Kangas, and P. Tanskanin, On the morphology of auroral zone X-ray events, II. Events during the early morning hours, J. Atmos. Terr. Phys., 35, 713, 1973.

Mauk, B.H., J. Chin, and G. Parks, Auroral X-ray Images, submitted, J. Geophys. Res., 1981.

Mozer, F.S., On the lowest altitude S3-3 observations of electrostatic shocks and parallel electric fields, Geophys. Res. Lett., 7, 1097, 1980.

McIlwain, C.E., Auroral electron beams near the magnetic equator, in The Physics of the Hot Plasma in the Magnetosphere, ed. by Bengt Hultqvist and Lennart Stenflo, Plunum Press, New York and London, 1975, p. 91.

Parks, G.K., Spatial characteristics of auroral zone X-ray microbursts, J. Geophys. Res., 72, 215, 1967.

Parks, G.K., C.S. Lin, B. Mauk, S. DeForest, and C. McIlwain, Characteristics of magnetospheric particle injection deduced from events observed on August 18, 1974, J. Geophys. Res., 82, 5208, 1978.

Parks, G.K., B.H. Mauk, C. Gurgiolo, and C.S. Lin, Observations of Plasma Injection, in Dynamics of the Magnetosphere, ed. by S. Akusofu, D. Reidel Pub. Co., Dordrecht, Holland, 1979, p. 371.

Yamagami, T., M. Fujii, J. Nishimura, H. Morakami, Y. Hirasima, M. Kajiwara, K. Okodiara, and M. Kodama, Balloon Observations of Auroral X-rays in Canada, I. Determination of auroral X-ray illumination regions, J. Geomag. Geoelectr., 30, 663, 1978.

THE LOW ALTITUDE ELECTRIC FIELD STRUCTURE OF DISCRETE AURORAL ARCS

F. S. Mozer

Physics Department and Space Sciences Laboratory
University of California, Berkeley

Abstract. The structure and properties of the latitudinally confined spatial regions of large electric fields seen between altitudes of 1000 kilometers and at least 7 R_e on auroral zone field lines are described. Evidence is presented that discrete auroral arcs are produced by particle acceleration inside these structures.

The major objective of this paper is to summarize the evidence in support of the view that discrete auroral arcs are produced by particle acceleration in spatially confined regions of large electric fields that have been observed on auroral zone field lines at altitudes between 1000 kilometers and at least 7 earth radii. These regions have been called "electrostatic shocks" or "double layers" by different groups that have proposed at least five different (and often contradictory) definitions to distinguish between these terms. To avoid semantic arguments, the regions of large, spatially confined electric fields observed on auroral zone field lines will be called "auroral field structures" in the remainder of this paper.

The observational properties of auroral field structures will be reviewed by reference to Figure 1, which covers a five minute interval of S3-3 satellite data during which the spacecraft traversed the auroral zone at an altitude near 2000 kilometers. The auroral field structures are indicated by the large values of the perpendicular, equatorward component of the electric field near 1514.6 and 1515.5 UT. Five seconds of expanded data near each of these times is presented at the bottom of the figure. Although the field structure may not have been resolved by the measurements made every 0.125 seconds in time or 800 meters of spacecraft motion, there is evidence both that the field was as big as 400 mV/m and that the structure sometimes consisted of paired, oppositely directed electric fields. Since a 400 mV/m field at an altitude of 2100 km would map to a field strength of about 600 mV/m in the ionosphere if there were no parallel electric fields, and since 600 mV/m fields are not observed in the ionosphere, strong parallel electric fields involving total potential drops of several kV must have existed at altitudes below a few thousand kilometers. Thus, the lower ionosphere must have been electrically decoupled from plasmas existing at altitudes of a few thousand kilometers and the concept of "frozen-in" magnetic field lines cannot be universally applied in the magnetospheric plasma.

The auroral field structures of Figure 1 are typical in that the equatorial perpendicular electric field component is larger than either the westward perpendicular component or the parallel component. Thus, electric equipotential contours associated with auroral field structures are approximately field-aligned and elongated in the east-west direction. Further evidence in support of this geometry is the large occurrence frequency of such structures, which requires that their equipotentials cover wide ranges of longitude and altitude while they remain narrow in latitudinal extent.

A meridional section through the equipotentials associated with one possible type of auroral field structure is given in Figure 2 [*Gurnett,* 1972; *Block,* 1975; *Mozer et al.,* 1977]. This figure is oversimplified in that it is not drawn to scale and it does not illustrate the frequent repetition in latitude of such structures. Since the equipotentials of Figure 2 are double sided (henceforth called V-shaped), a poleward moving spacecraft at an altitude of 1-2 earth radii would measure an electric field pattern consisting of paired, oppositely directed electric fields. Such field structures are clearly observed in some events [*Mozer et al.,* 1977] and there is at least a suggestion of such structures in others, such as those of Figure 1. In an attempt to determine the occurrence frequency of V-shaped equipotential patterns in auroral field structures, 207 events observed during the first 6 months of satellite operation have been studied. As illustrated by Figure 1, it is often difficult to determine the potential profile because of the occurrence of many such structures in a confined spatial region, because their small spatial size is poorly resolved by 8 field measurements per second, and because of the possibility of temporal variations during the measurement of a single structure. Even with these difficulties, it is clear that less than 50% (and probably less than 25%) of the observed structures can be explained by V-shaped equipotentials like that depicted in Figure 2. An example of auroral field structures that are not consistent with V-shaped potential patterns is given in Figure 3. Thus, the understanding of the most typical potential patterns associated with auroral field structures requires consideration of S-shaped contours, such as that illustrated in Figure 4a, rather than the more popular V-shaped patterns of Figures 4b and 4c.

Individual auroral field structures, such as those of Figure 1, typically span an invariant latitudinal width of 0.01° to 0.1°. These small-scale structures are immersed in typically few-degrees-wide spatial regions of low frequency wave turbulence, field-aligned currents, ion cyclotron wave emission, and upgoing, locally accelerated ions. Such a turbulent spatial region can most easily be seen in the example of Figure 1 by examination of the raw data plots of the two center panels. Each raw data plot is the potential difference between oppositely oriented pairs of spheres that are on the ends of 18-meter wire booms and which rotate about the satellite spin axis with a spin period of about 16 seconds during this orbit. In a constant external field, the two

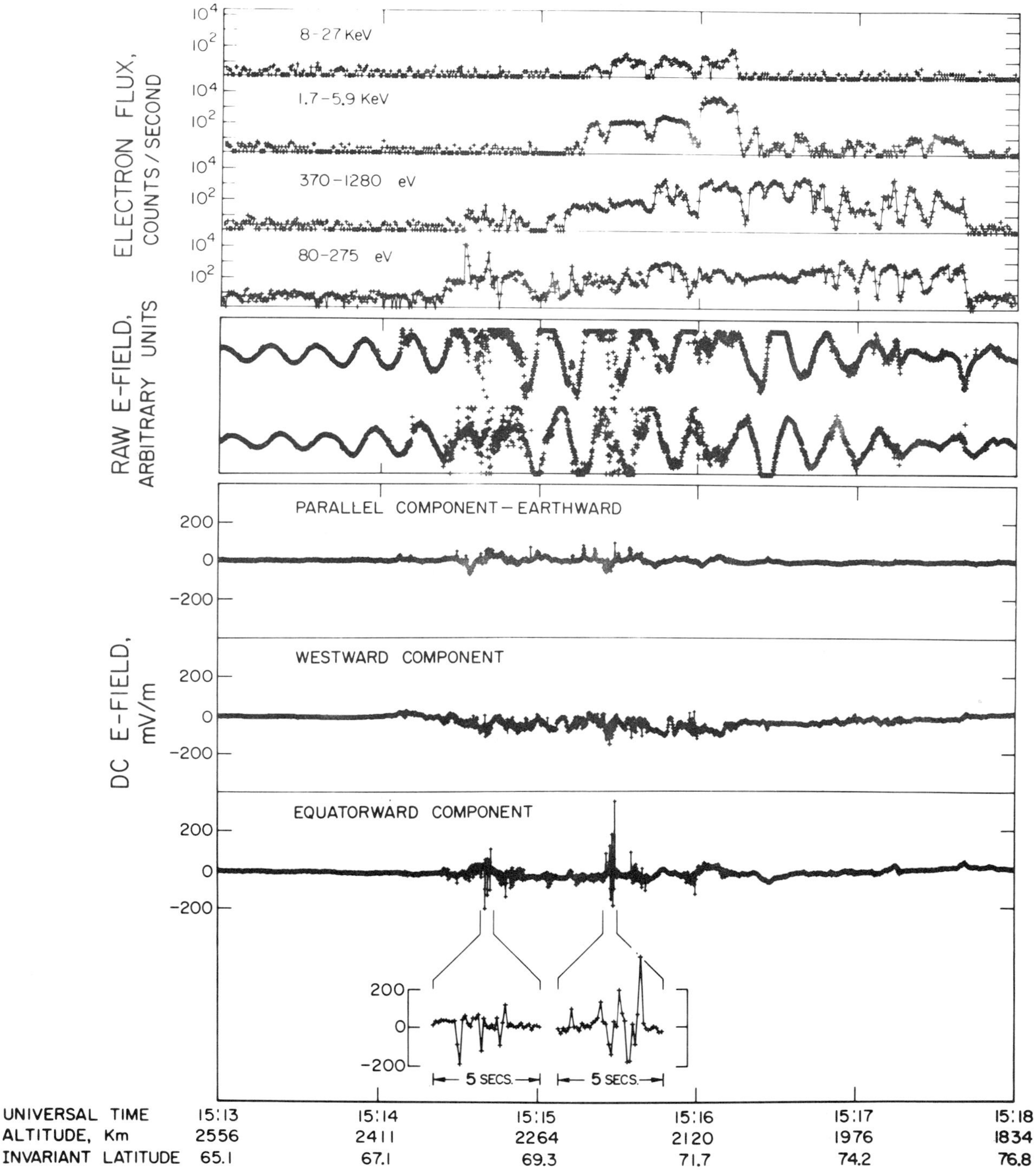

Fig. 1. Examples of analyzed and raw electric field data during passage through several auroral field structures.

raw data curves would be sinusoidal at the spin frequency, with equal amplitudes that are proportional to the electric field component in the spin plane, and with a 90° phase difference that arises from the 90° orientation of one sphere pair with respect to the other. Between universal times of about 1514.4 and 1516.2, the character of the sine waves is distorted by the <30 Hz turbulence superimposed on them. This turbulence spans an invariant latitude width of about 4° and is also evident in the analyzed DC data of the lower curves in Figure 1.

The histogram of Figure 5 presents the invariant latitudinal

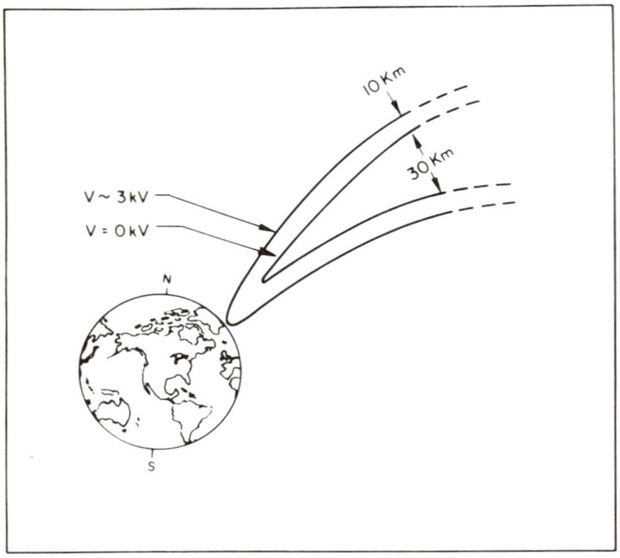

Fig. 2. Meridional plane view of a possible equipotential structure associated with paired auroral field structures.

width of the low frequency turbulent regions surrounding 207 auroral field structures observed during the first six months of the satellite operation. The typical 1° or 2° width of the low frequency turbulence is to be contrasted with the fact that the broadest single auroral field structure observed in these 207 events had an invariant latitude width of about 0.1° and many were an order of magnitude thinner. This evidence confirms the existence of two spatial scales in the auroral zone.

In the following discussion, the types of phenomena that exist on each of the two spatial scales will be illustrated with reference to the example of Figure 6 in which the particle, field, wave, and

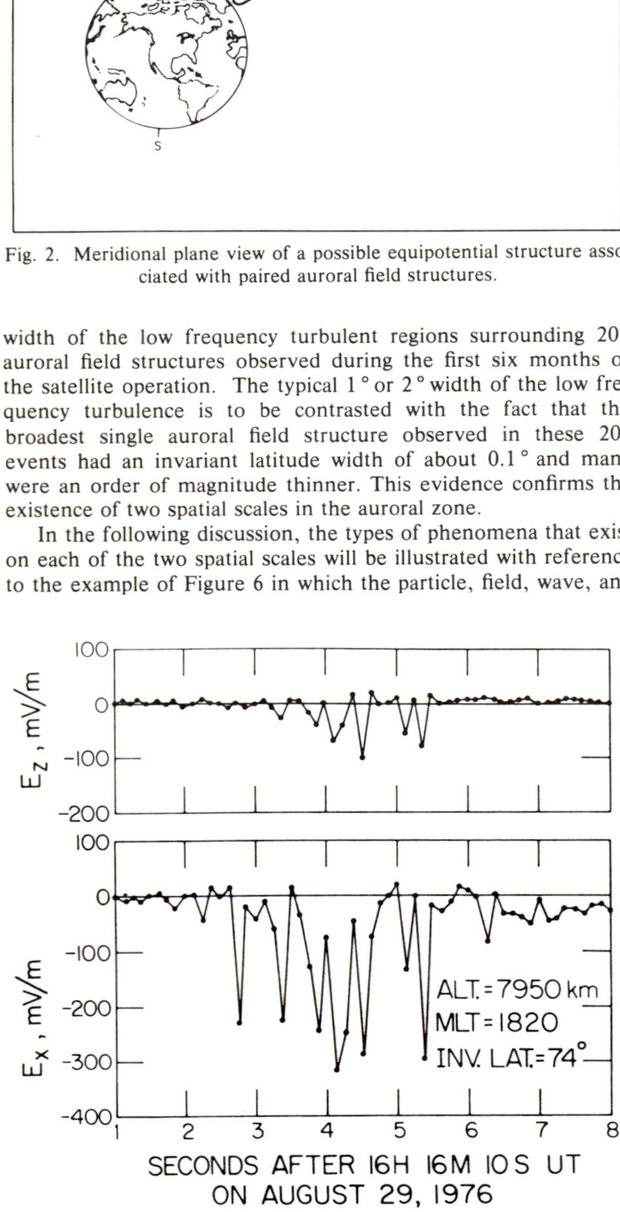

Fig. 3. The perpendicular and parallel electric fields in a series of narrow auroral field structures.

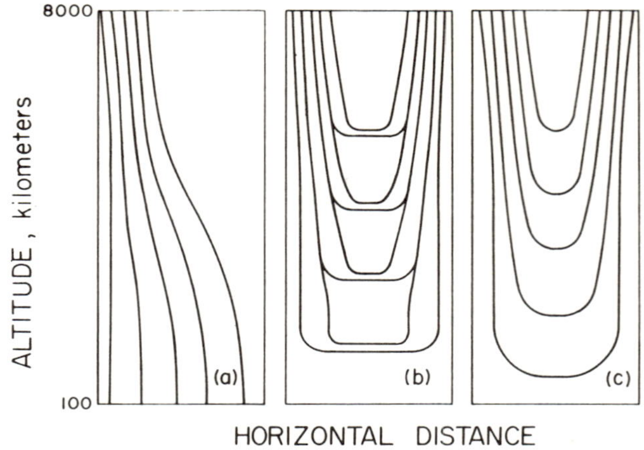

Fig. 4. Possible equipotential contours of auroral field structures.

field-aligned current data obtained during a single satellite pass from the plasmasphere to the polar cap are presented. During the 45 minute data interval of this figure, the spacecraft was at an altitude of about 7500 km at a local time near dusk. The plasmapause was crossed near 0242 UT, as determined by the reversal of the 17 second average of the x-(meridional) component of the electric field (panel d) from the small positive values associated with (anti-sunward) corotation in the plasmasphere to the larger negative values associated with sunward convection in the auroral zone. The boundary of the polar cap was crossed at about 0317 UT, as determined from the electric field reversal to the positive values in the polar cap that are associated with anti-sunward polar cap convection. The line integral of the electric field along the spacecraft trajectory (panel c) shows that there was a 65 kV potential drop between the plasmapause and the polar cap in this example.

The auroral field structures that were observed on this orbit are illustrated in the top two panels of Figure 6, which show the existence of several $< 0.1°$ wide structures having x-components of their electric fields that are larger than the y-components. (None of these structures appear in the 17 second

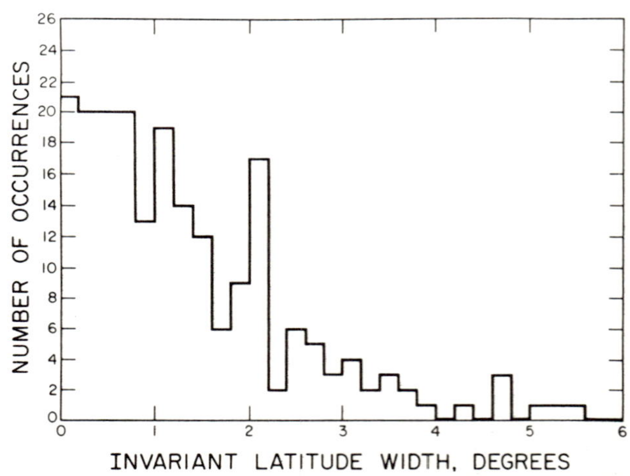

Fig. 5. Histogram of the invariant latitudinal widths of the low frequency turbulent regions surrounding 207 auroral field structures.

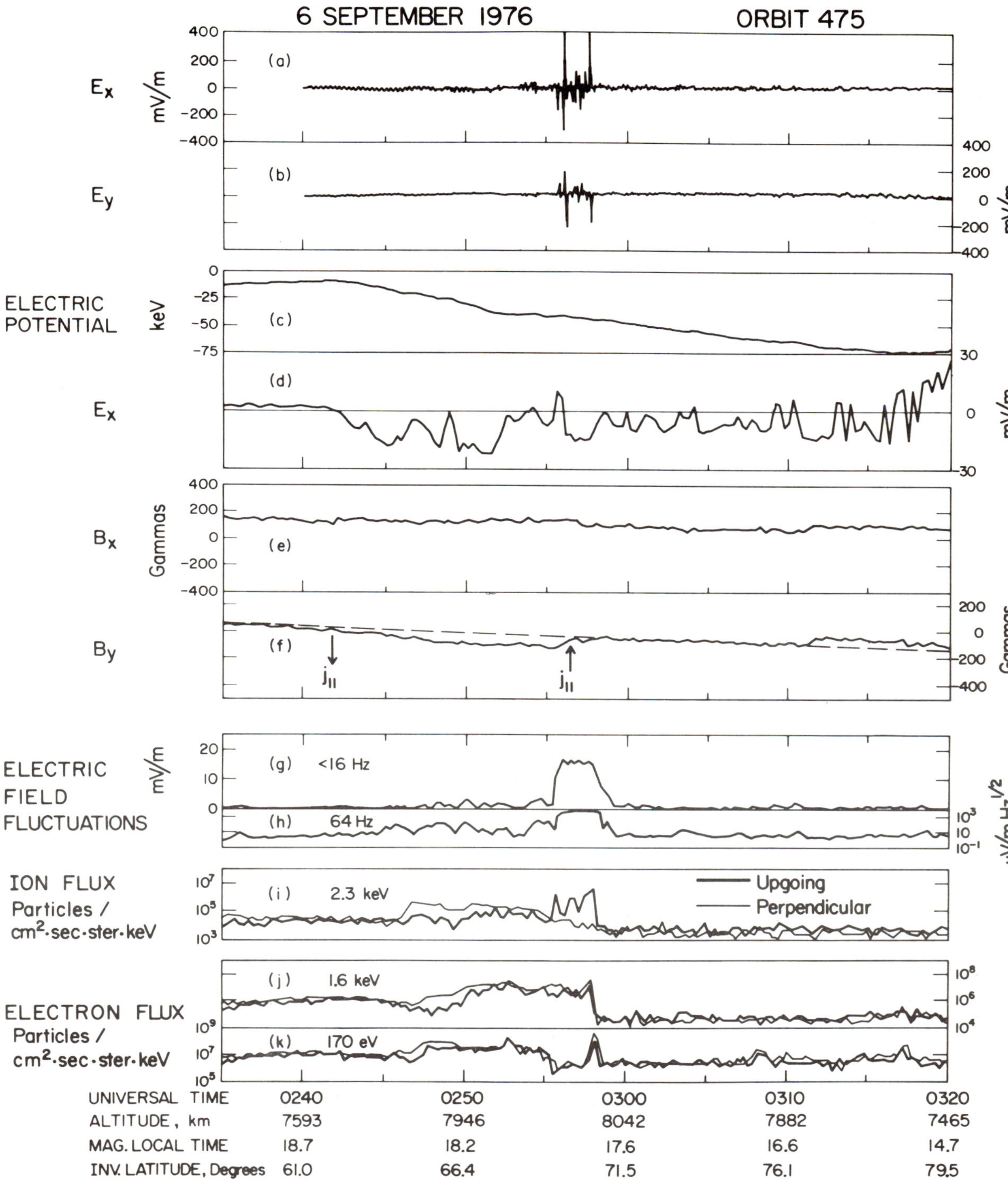

Fig. 6. Auroral field structures, the large scale electric field, field-aligned currents, waves and particles measured during a typical S3-3 satellite passage from the plasmasphere to the polar cap.

average electric field of panel d because of their short durations). These small scale structures are immersed in the low frequency wave turbulence that is illustrated in panel g and that spans a latitudinal interval of 2° to 3°. Panel f presents observations of the magnetic field component that is most strongly affected by longitudinally extensive sheets of field-aligned currents. The data suggest a downward field-aligned current in the low latitude auroral zone and an upgoing current in the vicinity of the low frequency turbulence. For a plasma density of 10 cm^{-3} and for ion and electron temperatures of about 1 eV, this upgoing current exceeds the threshold of the electrostatic ion cyclotron instability (which corresponds to a 30γ magnetic field variation in the 17 second spin period) over most of the upward current region. Thus, the factor of more than 100 enhancement of wave electric field power near the local hydrogen ion cyclotron frequency at this time (panel h) can be understood as electrostatic ion cyclotron emissions. This identification is supported by measurements of density fluctuations, by observations of harmonics of the gyro-frequency, and by detailed analyses [*Mozer et al.*, 1977; *Kintner et al.*, 1978].

The upflowing and perpendicular fluxes of 2.3 keV ions are presented in panel i, in which the upflowing flux exceeded the perpendicular flux by at least a factor of 100 over an $\approx 2°$ latitudinal interval included in the ion cyclotron emissions and low frequency turbulence. Thus, in this and in most examples, the $< 0.1°$ wide individual auroral field structures are immersed in a $> 1°$ wide region of low frequency turbulence, field-aligned currents, ion cyclotron waves, and upward moving ions. Several points in favor of associating the $< 0.1°$ region of large electric fields with discrete auroral arcs will next be discussed.

Point 1. *The electric potential across auroral field structures is adequate to explain peaks in auroral arc electron energy spectra.*

Considerable evidence exists to show that auroral electron beams are produced by acceleration through electrostatic potentials of a few kilovolts. The integrated potential difference across auroral field structures is also typically a few kV. This potential difference is measured to be nearly perpendicular to the local magnetic field. If, however, this potential exists along the magnetic field line somewhere, as it must in either V-shaped or S-shaped electric potential models, it would accelerate electrons to the energies observed in discrete auroral arcs.

Point 2. *Auroral field structures are observed at the same latitudes as auroral arcs.*

In Figure 7, the mean invariant latitudes of the auroral field structures are plotted as a function of the magnetic local times of observation. The average of these data exhibits a diurnal variation having properties similar to that of the auroral oval. That is, auroral field structures and auroral arcs are both observed at higher latitudes near noon than midnight. Figure 7 also shows a relative paucity of events near local dawn and a relative excess near dusk. These behaviors may be explained by non-uniform sampling in local time due to the orientation of the satellite orbit during the data collection period. Auroral arcs and auroral field structures also occur at similar latitudes on an individual basis [*Torbert and Mozer*, 1978].

Point 3. *The geometry of auroral field structures is like that of auroral arcs.*

Auroral field structures are observed at all local times (see Figure 7), as are arcs. They are extensive in longitude (since their westward electric field component is typically smaller than their equatorward component and because the large occurrence frequency is inconsistent with any other geometry) and they are repetitive in latitude (see Figure 1), as are arcs. Their typical thickness, extrapolated to the ionosphere, ranges from the measurement resolution of a few hundred meters to ~10 kilometers, which is similar to scale sizes observed in arcs [*Maggs and Davis*, 1968].

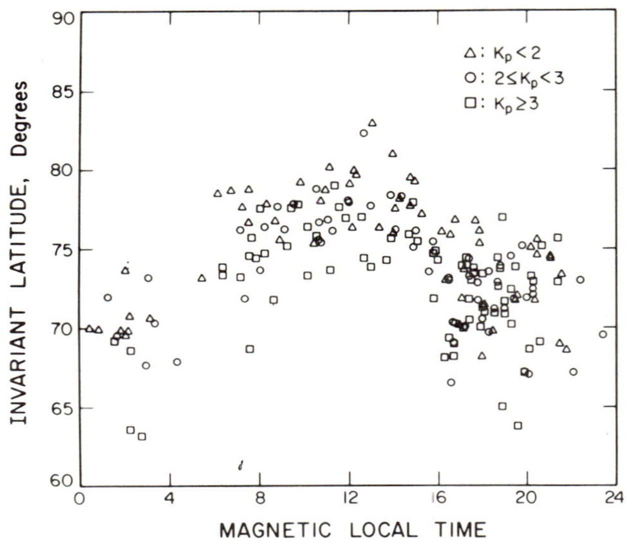

Fig. 7. The distribution in magnetic local time and invariant latitude of the locations of 207 auroral field structures.

Point 4. *Auroral field structures and auroral arcs have similar high occurrence frequencies.*

Auroral field structures having fields greater than or equal to 120 millivolts/meter were observed on the average of once per orbit of data collected during the first six months of satellite operation, near solar minimum. The observation frequency would be greatly enhanced if the threshold for definition of an event were reduced below 120 millivolts/meter or if the satellite apogee had been higher than 8000 kilometers since nearly half of the events had a peak field less than 180 millivolts/meter and one-third of the events occurred above an altitude of 7000 kilometers. Thus, auroral field structures are an ordinary occurrence in the polar region, as are auroral arcs in the ionosphere.

Point 5. *The dependence of auroral field structures on magnetic activity is similar to that of auroral arcs.*

In Figure 7, events are divided into three regimes of K_p, each containing one-third of the observations. Since the high K_p observations of this figure typically lie equatorward of those at low K_p, the magnetic activity dependence of the typical latitude of auroral field structures is similar to that of auroral arcs.

The occurrence frequency of auroral field structures is also positively correlated with K_p [*Torbert and Mozer*, 1978].

Point 6. *Auroral field structures and auroral arcs both occur near field-aligned current regions, the trapping boundary, and VLF saucers.*

Discussions of these correlations are given above and in other publications [*Mozer et al.*, 1977; *Temerin et al.*, 1977; *Mozer et al.*, 1980].

Point 7. *Downward accelerated electrons are often observed inside auroral field structures.*

The prior six points are statistical in nature, while this seventh point is direct evidence that auroral electrons are accelerated in auroral field structures. One example of this evidence is given in Figure 8, which is an expanded plot of the

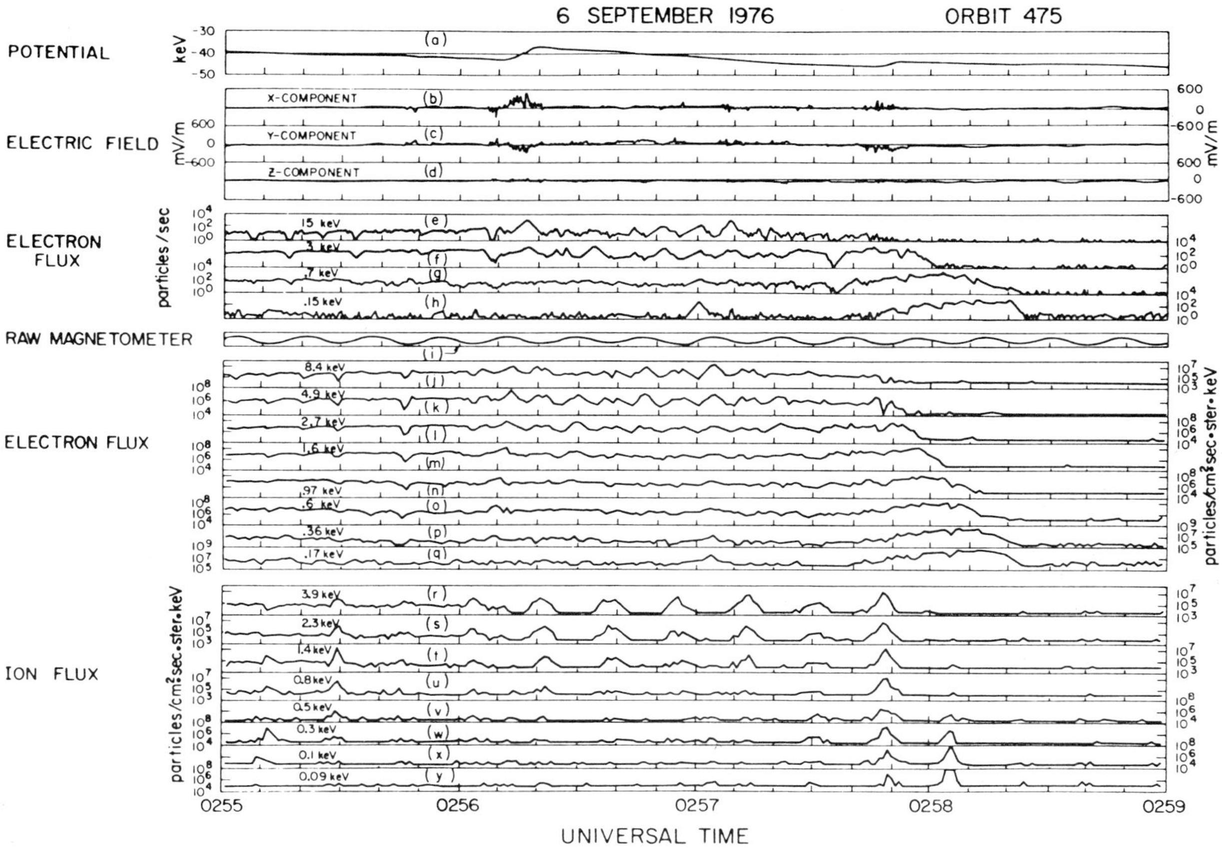

Fig. 8. An example showing electron and ion acceleration in auroral field structures.

acceleration region shown in Figure 6 and described above. Panels b, c, and d show the existence of large electric fields near 0256:15 UT, 0257:10 UT, and 0257:45 UT. Panels e through h are electron count rates from the Lockheed detectors which sample particles coming up the field at the positive going zero-crossings of the sine wave in panel i. Panels j through y are the electron and ion fluxes from the Aerospace detectors which measure particles moving up the field line at minima of the sine wave in panel i. There are upflowing ion beams (primarily at 1.7 keV $< E <$ 3.9 keV) throughout the two minute interval containing the auroral field structures, indicating a parallel potential drop below the satellite. The beams are broad in pitch angle, which implies perpendicular heating as well as parallel acceleration. At the time of the electric field structure near 0256:15 UT, the Lockheed detector, which was looking at particles coming down the magnetic field line, saw a strong beam in the 15 keV channel (panel e). At the same time the Aerospace detectors were looking perpendicular to the magnetic field, and saw much lower fluxes. This event shows that electrons were accelerated to parallel energies of at least 15 keV inside an auroral field structure. By chance, the same geometry and results occurred in the smaller structure at 0257:50 UT. The second large auroral field structure near 0257:50 UT occurred when the Aerospace detectors were looking at ions and electrons coming up the magnetic field line. At this time the largest upflowing ion beam of the entire event was observed, which suggests that ions are also accelerated inside the confined spatial regions of large electric fields. In virtually every event where particle detectors were observing field-aligned fluxes inside auroral field structures, large field aligned beams have been observed [*Mizera et al.,* 1979].

Given that discrete auroras are produced by particle acceleration inside auroral field structures, the next question concerns the acceleration mechanism. The experimental evidence suggests that acceleration is achieved by both wave-particle interactions and by the parallel electric fields inside the auroral field structures [*Mozer et al.,* 1980]. Measurements of such parallel electric fields can give important information on both the mechanism of formation of the auroral field structures and on the particle acceleration within them. Previously published examples [*Mozer et al.,* 1977; *Torbert and Mozer,* 1978] as well as that of Figure 3, indicate that the magnitudes of the parallel fields may sometimes be reasonable fractions of the perpendicular field magnitude. In some of these cases [*Mozer et al.,* 1977], this conclusion may be erroneous because the electric field detector was saturated by the input signal. In other cases, the apparent parallel field component could be erroneous because:

(a) The sensor's short antenna pair, whose field measurement is always questionable due to spacecraft perturbations, may not have been oriented predominantly perpendicular to B, and may, therefore, have contributed to a spurious parallel field measurement.

(b) The sensor's long antenna pairs may not have produced a clean measurement due to asymmetries between sphere pairs caused by spacecraft perturbations, spacecraft shadowing of one or the other sphere, or spacecraft photoemission. All such asymmetries are absent when the satellite is in a dawn-dusk orbit, since the spin axis points toward the sun at these times.

(c) One component of the measured electric field is obtained from the potential difference between non-identically oriented spheres when the instrument is in the density mode, and one sphere is used as a Langmuir probe [*Mozer et al.*, 1979].

For these reasons, the 207 shock events have been culled to select only those occurring in dawn-dusk orbits, when the instrument was not in the density mode, when the magnetic field line was within a few degrees of the satellite spin plane, and when the observed electric field was smaller than that required to saturate the instrument. One of the several events satisfying these criteria is presented in Figure 3. For this event, the angle between the satellite spin plane and the magnetic field was 1° ± 1°, while the spin plane and the plane containing the model magnetic field line coincided to the same accuracy. Hence, the relatively poorer measurement made by the short antenna pair cannot have influenced the magnetic equatorward and parallel components illustrated in the figure. Because this example satisfies stringent criteria aimed at eliminating spurious apparent parallel field measurements, there is a high level of confidence that the observed 50 - 100 mV/m parallel fields are real. Because they are negative, the parallel fields of Figure 3 act to accelerate electrons downward, in agreement with the hypothesis that discrete auroral arcs are produced by electron acceleration in electrostatic shocks.

The observed parallel field, being less than but the order of the perpendicular field, cannot be that of an ion acoustic shock (also called "double layer") since the parallel electric field in such structures should be orders of magnitude larger than the perpendicular field [*Hudson and Mozer,* 1978]. The observed parallel field is at least an order of magnitude too large to be explained by anomalous resistivity [*Hudson et al.*, 1978; *Shawhan et al.*, 1978]. Thus, the most plausible interpretation of Figure 3

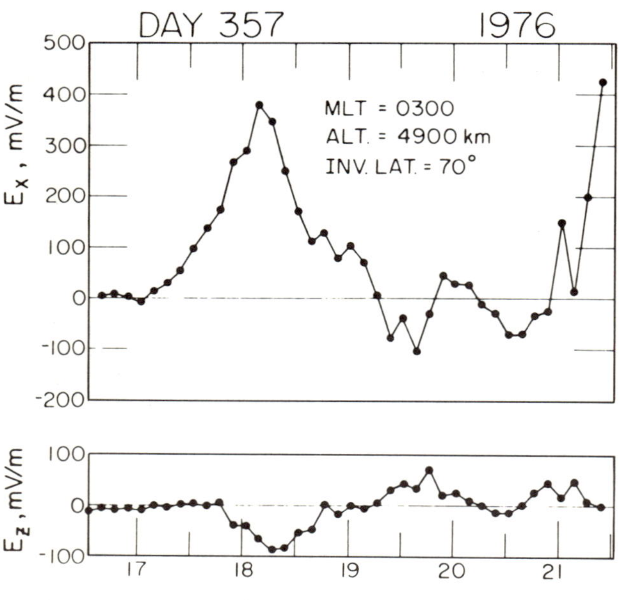

Fig. 9. Perpendicular and parallel fields in an auroral field structure.

is that the parallel field is produced by the obliqueness of the potential structure relative to the magnetic field line [*Swift,* 1976; *Hudson and Mozer,* 1978]. The observed field ratio suggests that the equipotentials were tilted about 20° off the magnetic field line. In this picture the parallel electric field on a given magnetic field line extends over a distance equal to the perpendicular thickness of the structure divided by tan 20°, and the total parallel potential is equal to the perpendicular potential across the event.

Another, more complex, example of parallel fields inside auroral field structures is given in Figure 9. In this example the parallel electric field points in opposite directions within a one second interval or about 5 kilometers of spacecraft motion. Thus, parallel fields and, by inference, field-aligned currents, particle fluxes, and wave fields, have complex variations on short spatial (or temporal) scales.

In summary, experimental evidence has been presented on the large perpendicular and parallel electric fields that exist within latitudinally confined regions in the auroral zone at altitudes between 1000 kilometers and at least 7 R_e. Both direct and statistical evidence has been presented in support of the view that discrete auroral arcs are produced by particle acceleration at altitudes below a few R_e inside these structures.

Acknowledgments. This work was performed under Office of Naval Research Contract N00014-75-C-0294.

References

Block, L. P., "Double layers," *Physics of the Hot Plasma in the Magnetosphere,* edited by Hultquist and Stenflo, Plenum Pub. Co., New York, 1975.

Gurnett, D. A., "Electric field and plasma observations in the magnetosphere," in *Critical Problems of Magnetospheric Physics,* edited by E. R. Dyer, National Academy of Sciences, 1972.

Hudson, M. K., R. L. Lysak, and F. S. Mozer, "Magnetic field-aligned potential drops due to electrostatic ion cyclotron turbulence," *Geophys. Res. Lett.*, **5**, 143, 1978.

Hudson M. K. and F. S. Mozer, "Electric shocks, double layers, and anomalous resistivity in the magnetosphere," *Geophys. Res. Lett.*, **5**, 131, 1978.

Maggs, J. E., and T. N. Davis, "Measurements of the thickness of auroral structures," *Planet. Space Sci.*, **16**, 205, 1968.

Mizera, P. F., J. F. Fennell, D. R. Croley, Jr., A. L. Vampola, F. S. Mozer, R. B. Torbert, M. Temerin, R. L. Lysak, M. K. Hudson, C. A. Cattell, R. G. Johnson, R. D. Sharp, P. M. Kintner, and M. C. Kelley, "The aurora inferred from S3-3 particles and fields," in preparation, 1979.

Mozer, F. S., C. W. Carlson, M. K. Hudson, R. B. Torbert, B. Parady, J. Yatteau, and M. C. Kelley, "Observations of paired electrostatic shocks in the polar magnetosphere," *Phys. Rev. Lett.*, **38**, 292, 1977.

Mozer, F. S., C. A. Cattell, M. Temerin, R. B. Torbert, S. Von Glinski, M. Woldorff, and J.Wygant, "The DC and AC electric Field, plasma density, plasma temperature, and field-aligned current experiments on the S3-3 satellite," *J. Geophys. Res.*, **84**, 5875, 1979.

Mozer, F. S., C. A. Cattell, M. K. Hudson, R. L. Lysak, M. Temerin, and R. B. Torbert, "Satellite measurements and theories of low altitude auroral particle acceleration," *Space Sci. Rev.*, in press 1980.

Shawhan, S. D., C. -G. Fälthammar, and L. P. Block, "On the nature of large auroral zone electric fields at 1 R_e altitude," *J. Geophys. Res.*, **83**, 1049, 1978.

Swift, D. W., "An equipotential model for auroral arcs," *J. Geophys. Res.*, **81**, 3935, 1976.

Temerin, M., R. B. Torbert, and F. S. Mozer, "Creation of VLF saucers and V-shaped emissions by electrostatic shocks," Space Sciences Laboratory, University of California, preprint, 1977.

Torbert, R. B., and F. S. Mozer, "Electrostatic shocks as the source of discrete auroral arcs," *Geophys. Res. Lett.*, **5**, 135, 1978.

The Boundary of the Polar Cap and its Relation to Electric Fields, Field-Aligned Currents, and Auroral Particle Precipitation

R. B. Torbert, C. A. Cattell, F. S. Mozer*

Space Sciences Laboratory, University of California, Berkeley, California 94720

C. -I. Meng

Applied Physics Laboratory, The Johns Hopkins University, Laurel, Maryland 20810

Abstract. Simultaneous observations of electric fields, field-aligned currents, and auroral particles on the S3-3 spacecraft over the earth's polar regions have been used to illustrate how the low-altitude extent of field lines open to interplanetary space may be unambiguously identified. We define the polar cap to be the region of open field lines, which are only those containing plasma flowing anti-sunward and possessing precipitating low-energy electron distributions characteristic of the magnetosheath. Field-aligned current signatures are found near the polar cap boundary, but not always coincident with it. Within this polar cap, gradients in the observed soft particle flux are explained as a result of the density decrease previously observed in the magnetosheath as one progresses anti-sunward along the magnetotail. Presumably depending upon the interplanetary magnetic field orientation, the location of the polar cap is occasionally drastically shifted. At these times, there are high-latitude regions, extending even up to the magnetic pole, which are closed and convecting towards the sun. Since boundaries, identified by either the poleward extent of plasma sheet particle populations or the 33 keV electron intensity cutoff or isotropy boundary, are well-separated at these times from the region of open field lines, we conclude that there is no physical connection between these boundaries and the edge of the polar cap; but, rather, that they are topologically often found near each other. The location of field-aligned current sheets appears to depend, in a more complex way, on the gradients in the electric field and ionospheric conductivity, as indicated by auroral precipitation.

One of the long unresolved, irrepressible, and irritating questions in magnetospheric physics is "Are these field lines open or closed?" Nevertheless, it is a central question in understanding the dynamical interaction of the solar wind and the magnetosphere. In particular, the existence of open field lines plays a pivotal role in the theory of reconnection [*Dungey,* 1961; *Vasyliunas,* 1975], which has been used to explain many magnetospheric processes, including the general convection of plasma around the earth [*Axford,* 1969].

This question arose with the study of cosmic ray cut-off rigidities [*Störmer,* 1955; *Lüst,* 1957; *McCracken,* 1962], which were, at first, interpreted assuming dipole fields lines penetrating deep into interplanetary space, but were later used to explore the details of magnetospheric topology [*Freon and McCracken,* 1962]. Observations of solar energetic protons and electrons (100 keV up to several meV) in the polar regions have been modeled as diffusive entry into an extensive and perhaps closed magnetotail. (For a review, see *Morfill and Scholer,* 1973.) The case of several hundred keV solar flare electrons observed over the pole may indicate direct illumination by these particles [*Vämpola,* 1971].

Lower energy precipitating electrons (< 1 keV) observed over the pole were named "polar rain" by *Heikkila and Winningham* [1974], who concluded that they were of magnetosheath origin. Subsequently *Mizera et al.,* [1974]; *Fennel et al.,* [1975]; and *Meng and Kroehl,* [1977], showed that the polar-rain flux tracked the interplanetary electron intensity and further revealed that the north-south polar asymmetry in the precipitating flux could be explained by the anisotropy of the solar-wind electrons, and was controlled by the x-component of the IMF. *Meng et al.,* [1977] reported a dawn-dusk gradient in the "polar rain" whose direction is opposite in opposite hemispheres and is controlled by the sign of interplanetary B_y: there is a dawn-directed gradient in the northern hemisphere when $B_y > 0$.

The anti-sunward $\mathbf{E} \times \mathbf{B}$ plasma convection across the polar cap has been observed by balloons and satellites [*Mozer et al.,* 1974; *Heppner,* 1972; *Mozer and Torbert,* 1980]. This flow is the return of magnetic flux previously convected towards the sun in the general convection pattern of the magnetosphere, whether this pattern is driven by reconnection (and thus is open) or by viscous diffusion (and may be closed). It is known that B_y affects the distribution of the dawn-dusk electric field in the polar cap. [*Heppner,* 1972]. This would imply that some connection, perhaps only indirect, exists between the cap and interplanetary space.

Triad measurements identified a poleward region of field-aligned currents (labeled "Region I"), closely identified in space with the assumed location of the polar cap [*Armstrong and Zmuda,* 1973; *Iijima and Potemra,* 1976]. This association was confirmed by ISIS 2 particle measurements [*Burrows et al.,* 1976; *McDiarmid et al.,* 1978], and by particle and electric field data from S3-3 [*Mozer et al.,* 1980; *Cattell,* 1980]. *Cattell,* [1980], showed (and we will further illustrate) that the location of the poleward field-aligned current sheets is only an approximate, and sometimes poor, indicator of the electric field reversal region,

*also Physics Department

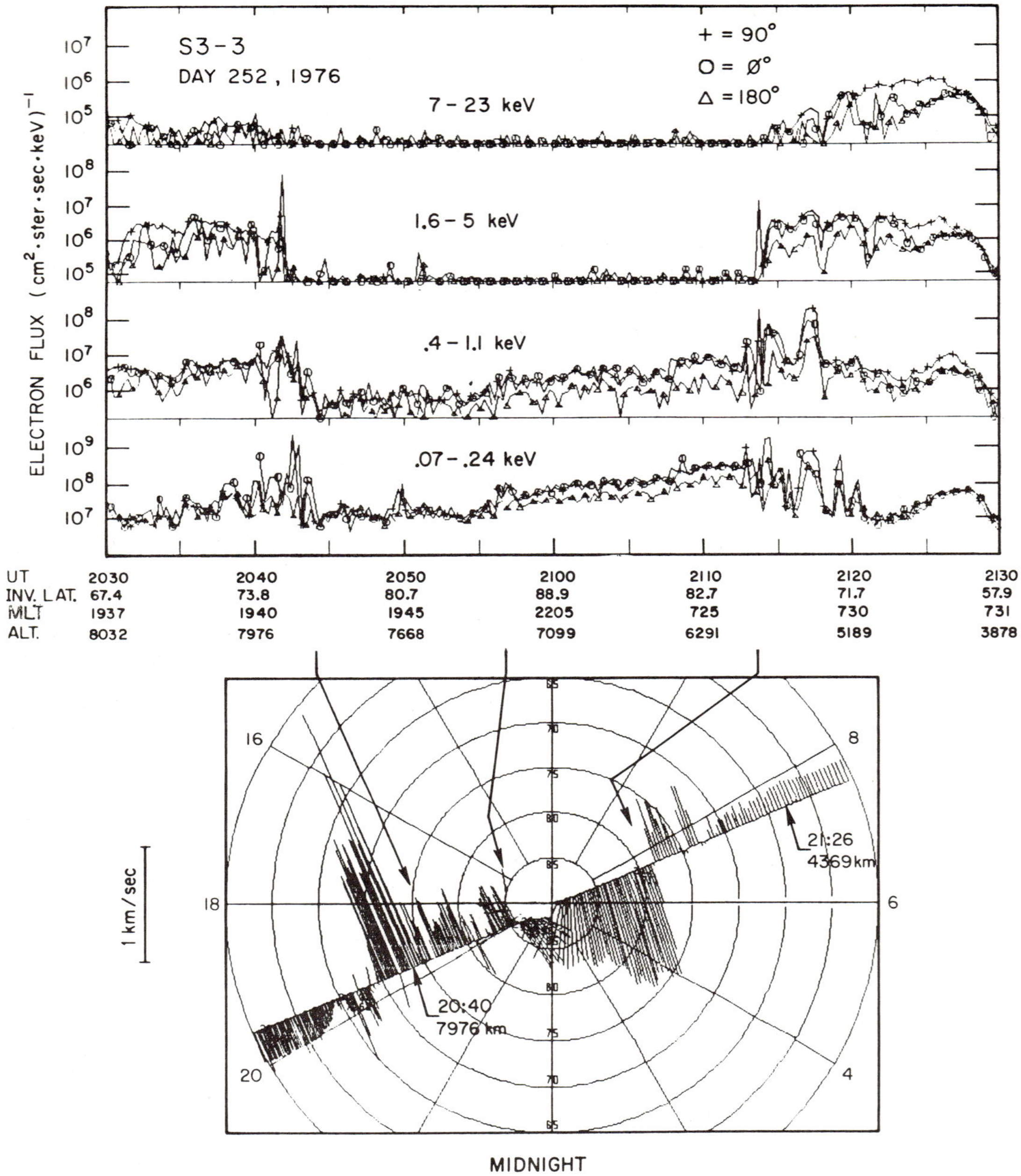

Fig. 1. Convection flow vectors derived from the E_y component in one pass over the northern hemisphere. The vectors have been normalized to a reference altitude of 100 km. The electron flux data is sectored into three pitch angle bins once each spin period.

since the currents are related to gradients in both the electric field and the ionospheric conductivity, as has been discussed by *Smiddy et al.,* [1980].

The following simultaneous observations of particles and the electric field over the polar region show that: 1) the contiguous portion of the anti-sunward flow, found nearly always somewhere in the polar regions, corresponds precisely to that region exposed to soft precipitating electrons of interplanetary origin; 2) the gradient in these electrons results from known thermodynamic processes in the magnetosheath; and, therefore, 3) that a well-defined "polar cap" of open field lines can be identified.

Instruments

The Air Force S3-3 satellite was launched into an elliptical polar orbit with an initial inclination of 97.5°, an apogee of 8040 km altitude and a perigee of 240 km. We present here measurements of electric fields deduced from the potential difference between spherical probes separated by 37.5 meters along extended wire boom elements. These probes lay in the spin plane of the satellite, which lay very nearly in the orbital plane. Information on field-aligned currents was deduced from differences in magnetic perturbations, as measured by the on-board magnetometer. The resolution of the high gain state of this instrument, some 80 nT per bit, was improved to 10 nT by carefully averaging the data over one spin period, thereby reducing spatial resolution to ~80 km per measurement. For further experimental details, see *Mozer et al.,* [1979].

Particle flux data used here are mainly from the Aerospace Corporation electrostatic analyzer covering the range 0.17 to 33 keV for electrons and .09 to 3.9 keV/q for ions (see *Mizera and Fennell,* 1977). In addition, we have used electron flux data with better time resolution for broad energy channels from the electron spectrometer of the Lockheed Corporation. (See *Shelley et al.,* 1976).

Polar Cap As Defined By Electric Fields and Particles

An example of the electric field and electron data over the polar region is given in Figure 1. In the lower half of this figure, the **E × B** flow, in the noncorotating reference frame and deduced from the spin-period average of the measured electric field, is plotted in corrected magnetic local time (MLT) and invariant latitude (ILA) co-ordinates. All the flow vectors are nearly perpendicular to the trajectory of the satellite, since only the electric field in the spin plane was used and this plane contained the velocity vector of the spacecraft. The flow vectors are normalized to a constant altitude of 100 km, under the assumption that magnetic field lines are equipotentials. As has been shown by *Mozer and Torbert,* [1980], this is not a very good assumption, but has been used in the present study only as a zeroth order correction for altitude dependent effects. In this pass over polar region, the satellite leaves the co-rotating plasmasphere at ~ 2022 UT near 62° ILA in the dusk sector and reenters on the morning side somewhere equatorward at 65° ILA. In the top panel, there are electron fluxes in four energy channels from 70 eV to 23 keV measured by the Lockheed Corporation electron spectrometer. These fluxes have been sectored into three pitch angle bins: 0° - 45*deg*; 45° - 135*deg*; 135° - 180°.

Just at the equatorward edge of the dusk plasmasheet as seen in the 1.6 - 5 keV electron channel, at 2031 UT near 69° invariant latitude, the satellite encountered strong sunward convection, which persists far poleward of the signatures of discrete auroral precipitation that are observed near 73°. This is a somewhat higher latitude than is normally found for the discrete auroral zone. Although there is a strong shear in the plasma flow at 74° invariant latitude (characteristically found at the poleward boundary of discrete aurora), the convection remains strongly sunward up to 86° ILA. In this region from 2044 to 2057 UT, low electron fluxes, previously associated with the low density polar cap, are observed at all energies. There is a slight enhancement in the lowest energy electron channel at 2050 UT, which is associated with the perturbation seen in the flow at ~81° ILA. Such a signature is again seen just at the reversal boundary. This precipitation might be associated with weak sun-aligned, so-called "polar cap" arcs.

After the reversal to steady anti-sunward flow, the satellite encounters enhanced soft electron fluxes which uniformly increase up until the flow reversal in the morning at 78° ILA. Simultaneously, discrete auroral fluxes (note the peak in the 1.6 - 5 keV channel) are observed. The sunward flow persists below an invariant latitude of 60° in a manner consistent with co-rotation.

The soft electron fluxes between 2057 and 2113 are precipitating into the ionosphere with an isotropic pitch angle distribution, except for the depleted loss cone, as is indicated in Figure 1 by the lower fluxes in the 180° sector. As the spectrum of these particles is similar in magnitude and slope to those measured in the magnetosheath, and ranges in magnitude between those measured in the magnetosheath and those in the solar wind (see Figure 2), we conclude that these electrons have direct access to the polar cap from the magnetosheath, as has been discussed previously [*Winningham and Heikkila,* 1974; *Mizera et al.,* 1974; *Fennel et al.,* 1975; *Meng et al.,* 1977]. The name "polar rain" has been used to describe this soft, steady precipitation.

The definite gradient in the soft electron flux towards the dawn on day 252, 1976, corresponds with a gradient of the plasma flow in the same direction. Such gradients in both the soft electron flux and the electric field have been separately correlated with interplanetary B_y [*Heppner,* 1972; *Meng et al.,* 1977]. Although there is no interplanetary field data available for this particular day, we have examined many other cases of S3-3 data where unambiguous gradients exist. In most of these, we find agreement with the conclusions of previous authors: i.e.,

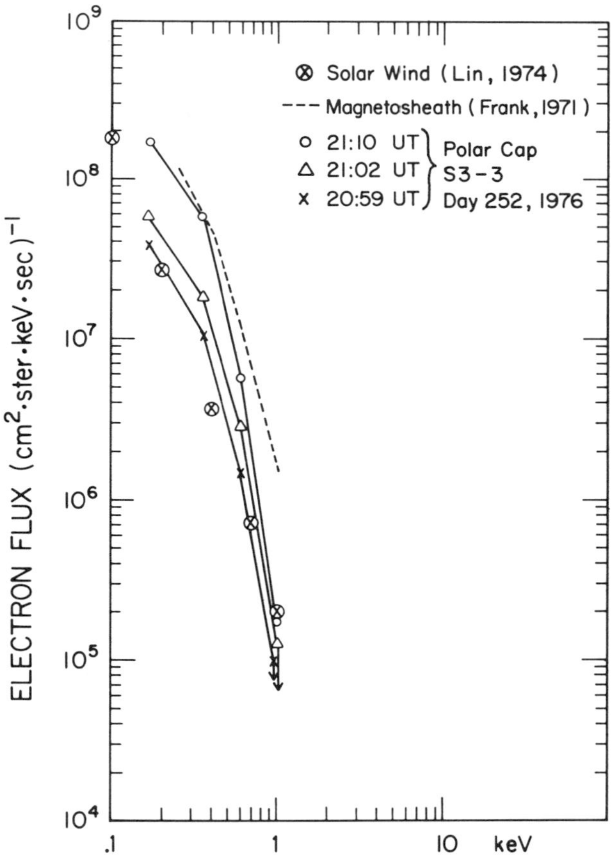

Fig. 2. Electron spectra at 0° pitch angle taken at three different points along the gradient in the soft precipitating polar cap flux.

for a dawn-directed gradient (of either particles or fields), $B_y > 0$; and vice-versa for either $B_y < 0$ or the southern hemisphere. Thus, we would infer that probably $B_y > 0$ on day 252. This is in agreement with the shift towards the morning of the poleward boundaries of auroral precipitation seen here as computed by *Gonzales and Mozer* [1974], and observed by *Lassen and Danielsen* [1978].

What is significant in this pass is that the region of soft electron precipitation out of the magnetosheath seen from 2057 to 2113 UT corresponds exactly with the region of anti-sunward flow; and therefore, the region of open field lines extends only from 86° in the evening to ~ 78 ° in the morning. In the evening, the polar cap is not delineated in any way by the lack of plasma sheet or discrete auroral particles. In fact, in the evening the large region of low fluxes from 2044 to 2057 UT presumably is connected to parts of the low density magnetotail which are closed (no magnetosheath particles visible) and definitely flowing swiftly sunward.

The next example is shown in Figure 3 in the same format as Figure 1. On day 246 (2 September), 1976, the interplanetary B_y ranged between +3.3 and 1.5 nT in the time period shown. B_Z was approximately −0.5 nT. Again, the definite flow gradient towards the morning coincides with the gradient seen from 21:40:30 UT to the onset of occasional low energy auroral electron precipitation at 2200. This "polar rain" can be seen in the faster time resolution spectrogram of the same event presented in Figure 4. (Recall that the data in Figures 1 and 3 are spin averaged). In the middle panel of this figure, the electron differential energy flux between .2 and .4 keV/q shows that the low energy component is present, among evidence of low-energy discrete auroral arcs, right up until 2209 at ~74.5°, when the electric field makes the sharp reversal to sunward convection. Equatorward of this reversal are the common auroral zone signatures of plasma sheet and trapped energetic electrons, together with ring current protons and low-energy upgoing conical proton distributions often seen in discrete aurora [*Mozer et al.*, 1977; *Cattell et al.*, 1978; *Gorney et al.*, 1980]. The presence of soft discrete arcs on field lines poleward of 74.5° implies that the processes responsible for low-altitude acceleration can occur on both open and closed field lines. For instance, they may occur wherever there are sufficiently strong field-aligned currents. This result agrees with those of *Heelis et al.*, [1980], who found that the flow reversal occurs within the "boundary plasma sheet" — poleward of the "central plasma sheet." The poleward boundary of the "central plasma sheet" occurs at ~73° in the morning on this pass.

The magnitude of the precipitating polar flux is much less than that seen in Figure 1, reflecting presumably a lower density in the magnetosheath and interplanetary space, as has been reported by *Fennell et al.*, [1975]. Thus, it may be that such fluxes could always be seen over that part of the polar magnetosphere which is open to interplanetary space, but is present at flux levels below the sensitivity of the detectors on board S3-3.

The dusk-side boundary of the open field region is uncertain because of the low flux levels seen on this day. At 21:40:30 there is an enhancement in the lowest energy electrons, indicating the presence of a weak auroral arc, possibly sun-aligned. At this time, the flow reverses to the anti-sunward direction until nearly 85° ILA and then continues for a short period sunward. Thus, one might place the polar cap boundary at ~82° ILA in the evening. In either case, there exists a large region of sunward flowing plasma well poleward of the intense plasma sheet with energetic electrons.

In the past, it has been assumed that polar regions of no observable electron flux (such as between 2125 and 2140 UT in Figure 3) were on open field lines. With the agreement between both the electric field and particle data presented here, however, the boundaries of the polar cap are clear. Although at any one time the polar cap may be considerably off-centered from the pole, it is still clearly evident. In fact, an examination of some 100 orbits reveals only three cases where a continuous region of sustained anti-solar flow is not observed somewhere over the polar regions. Thus, the conventional convection pattern seems to be a nearly ever-present feature of the magnetosphere. Reports of sustained sunward flow within the polar cap [*Mozer et al.*, 1974] perhaps were deduced from measurements made not in the polar cap as defined here, but within these closed, sunward flowing portions of the high-latitude ionosphere.

Field-Aligned Current Correlations

The top panel of Figure 5 plots the y (westward) component of the magnetic field along the satellite trajectory on day 246, 1976, in the same MLT-ILA polar plot as the convection flow data in the bottom panel, which is identical to that in Figures 3 and 4. Since $\hat{z}$ in this system is along the model magnetic field (the IGRF (10-68) model), any non-zero component in $\hat{x}$-(equatorward) or $\hat{y}$-direction represents a difference between the model and the actual field. These vectors are again normalized to a reference altitude of 100 km, just as in the electric field case. Gradients in the y-component can be interpreted as sheet currents aligned in the east-west direction. In particular, a change from sunward to anti-sunward as the satellite progresses from dusk to dawn in the northern hemisphere here corresponds to a field-aligned current into the ionosphere.

The morning region of strong downward current (nearly 1 ampere/meter) between 73° and 75° invariant latitude overlaps the convection reversal point seen in the bottom panel and with the onset of keV auroral particles seen at 2208 UT in Figures 3 and 4. This coincidence is a regular, but far from universal, feature in our data. In fact, there are many passes where the reversal region is as unambiguous as in Figure 5 and, nevertheless, the current sheet and the convection reversal are displayed 2° to 4° ILA from each other.

This uncertain correlation can be seen in the evening side of this figure. The large upward current ("Region I" of Iijima and Potemra) is found between 69° and 71° at 2000 MLT, corresponding to electron precipitation seen mainly in the .4 - 1.1 and 1.6 - 5 keV channels of Figure 3. The largest electric field shear occurs at 72°. Likewise the downward sheet between 74° and 75° occurs in the middle of the sunward flowing region from 73° to 77.5° ILA. This would be a consistent result of current patterns set up in the ionosphere by the electric fields on either side of this current sheet. The next upward current sheet appears to be a weak one, distributed over the range 74.5° to 82° ILA, within which there are several field reversals.

A survey of the location of field-aligned currents relative to that of the convection reversal is given in Figure 6a and Figure 6b, previously published in *Mozer et al.*, [1980]. The normalized extents, magnitudes, and directions of the field-aligned current sheets and their relationships to convection boundaries are shown where the abscissa is normalized in each event such that the distance (in invariant latitude) between the plasmapause and the polar cap boundary is one unit. The morning and evening polar cap boundaries and the evening plasmapause are defined by the zero-crossings of the $\hat{x}$ component of **E** in a non-corotating frame. The morning plasmapause is defined as the position where this flow deviates from the corotation value. The average evening plasmapause and polar cap boundaries [*Mozer et al.*, 1980] are

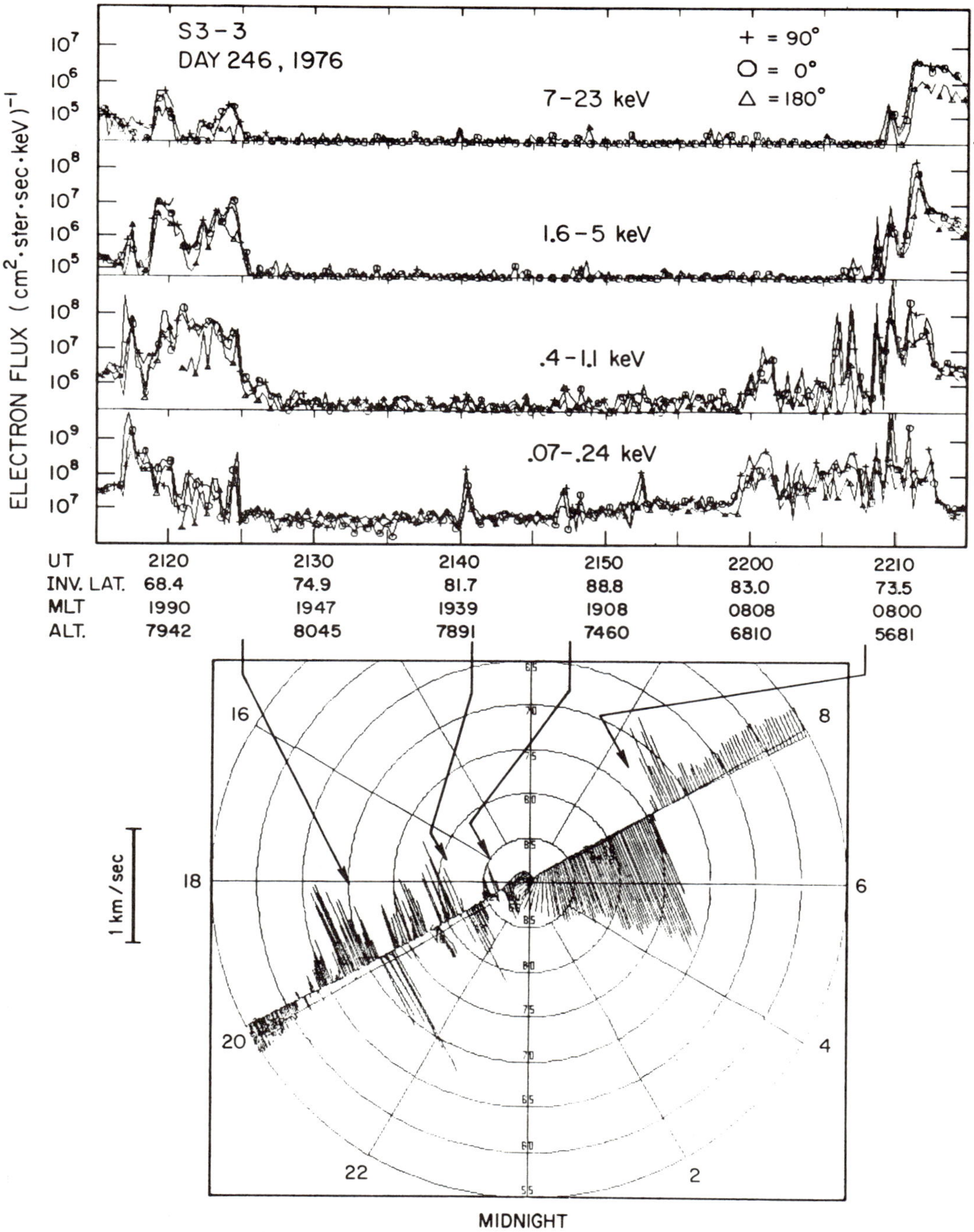

Fig. 3. The convection flow and electron spectra for day 246, 1976, just as in Figure 1.

at latitudes of ~65° and ~74° respectively; in the morning, the average positions are ~62° and ~77°. (The morning average values are less accurate than those at dusk because the number of events is smaller and because the plasmapause is more difficult to identify in the morning.) Negative values of the ordinate of Figure 6 correspond to currents flowing into the ionosphere. The relevant portions of this figure are the top two panels (poleward and middle sheets) taken together, but excluding the small collection of upward current sheets (positive) in the top panel of 6a.

Examination of these figures, together with the relevant particle data, has led us to the following conclusions:

1) *In the evening,* typically the polar cap convection reversal boundary occurs at (less frequently, just inside of) the poleward edge of the region of upward field-aligned current and

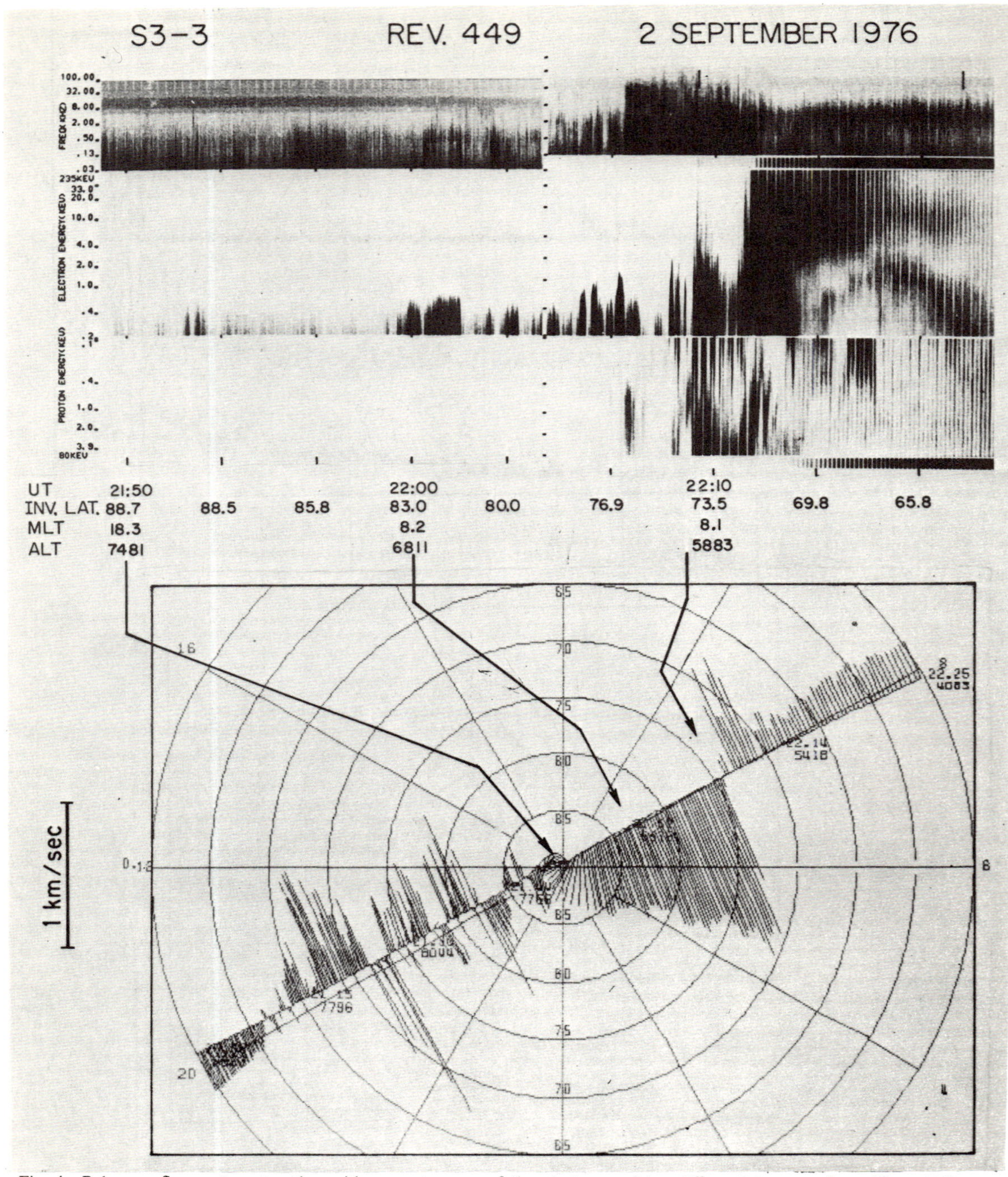

Fig. 4. Polar cap flow pattern together with a spectrogram of the electron and ion differential energy flux. The middle panel exhibits electrons from .2 to 33 keV. Below that, with increasing energy going down, are ion data between .1 and 3.9 keV/q. Above the electrons and below the ions are narrow strips for single-energy channels of 235 keV electrons and 80 keV protons, respectively. The top spectrogram is of AC-electric power spectra from 30 Hz to 100 KHz, not discussed here. Note the dotted line down the middle indicating an unavoidable scale change and sensitivity change in the case of the wave spectra.

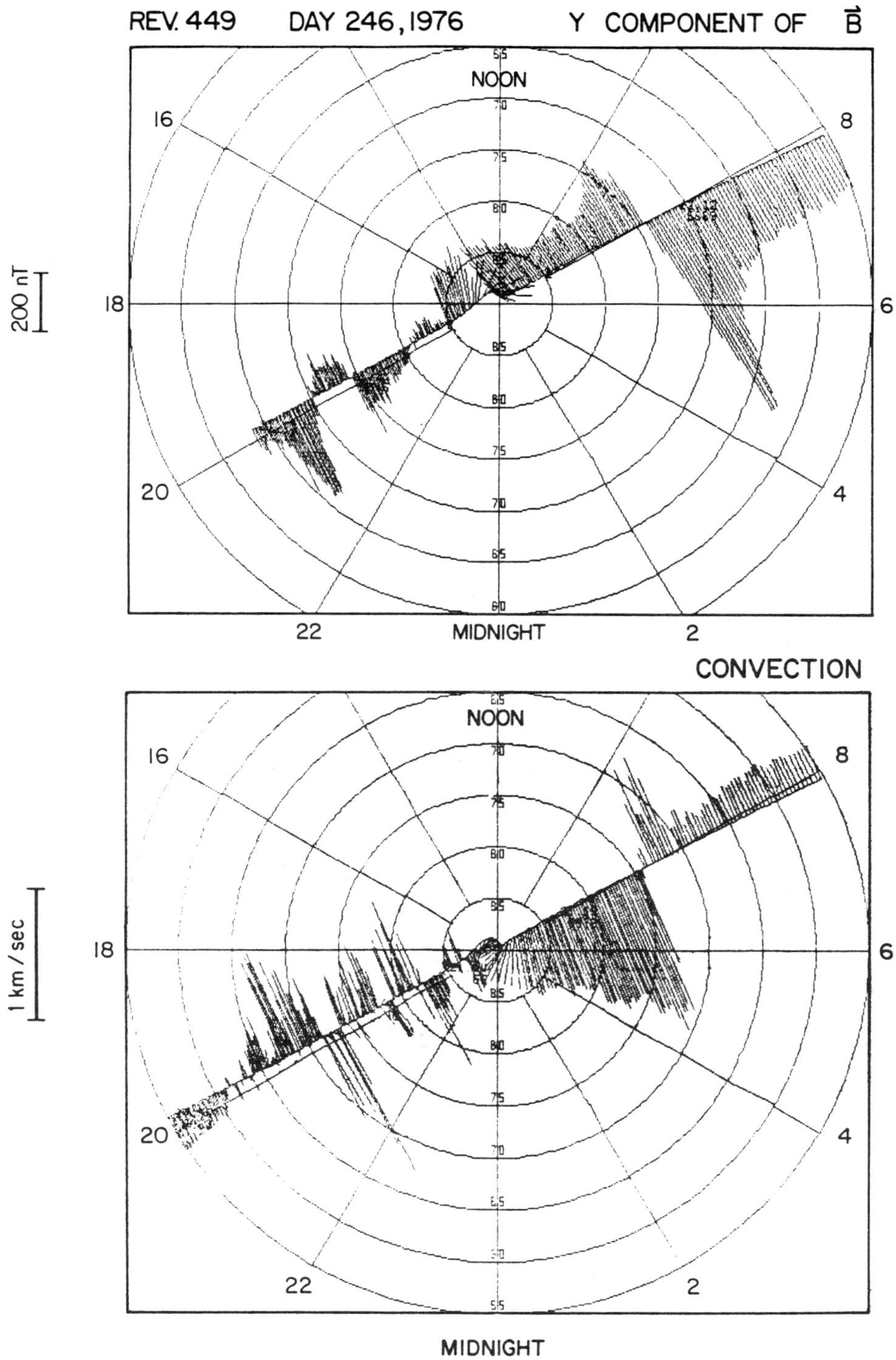

Fig. 5. Magnetic perturbation vectors along the y-axis for day 246, 1976, together with convection flow.

the structured (arc-like) electron precipitation of the plasma sheet. In terms used by *Winningham et al.*, [1975] and *Heelis et al.*, [1980], the convection reversal occurs at or just inside of the poleward boundary of the "boundary plasma sheet." The reversal always occurs poleward of the "central plasma sheet."

2) *In the morning*, this reversal occurs typically within the structured plasma sheet electrons and the region of downward

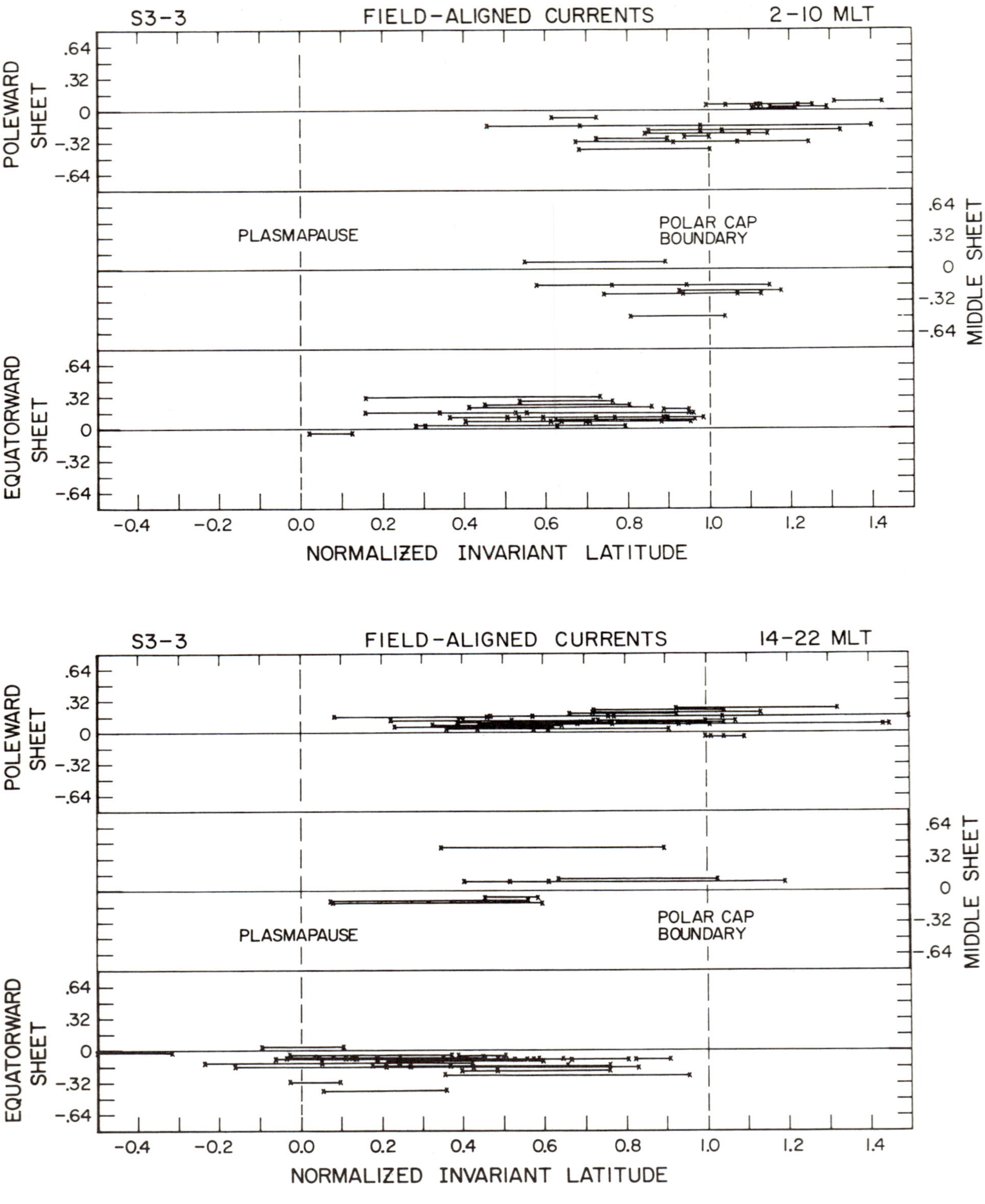

Fig. 6. (a, b) The locations and magnitudes (in amperes/meter) of field-aligned currents relative to the plasmapause and polar cap boundaries for 2 - 10 MLT (a) and 14 - 22 MLT (b).

field-aligned current and less frequently, but regularly, just poleward of these plasma sheet electrons.

Although statistically the poleward field-aligned currents are thus found near the reversal, our discussion above makes clear that one cannot deduce the location of the polar cap by examination of the currents in any one pass, as assumed by *McDiarmid et al.*, [1978, 1980]. For example, it would be impossible to determine the evening boundary of the polar cap in Figure 5 from the top panel alone. Since the location of parallel currents is tied to the gradient of not only the electric field, but also the ionospheric conductivity, which in turn is intimately related to plasma sheet precipitation, it may be that, just as for the particles, the location of strongest field-aligned currents has no direct physical relation to the location of the polar cap. It is simply found near it.

Explanation of Gradients

That both the electric field and soft-particle flux over the polar cap have gradients in the same direction, either to the dawn or to the dusk, could be inferred from previously published results on their separate dependences with respect to the interplanetary B_y component. [*Heppner*, 1972; *Meng et al.*, 1977]. Here, we have shown explicitly that this is the case. We believe these gradients have a simple explanation.

It is well-known from both theoretical studies [*Spreiter and Alksne*, 1969] and *in-situ* measurements [*Siscoe*, 1967] that the number density in the magnetosheath decreases, even below that of the solar wind, as one progresses further back in the tail. This arises from the fact that the plasma is heated and slowed as it crosses the bow shock (thus increasing its density) and then adiabatically expands as if into a vacuum around the magnetosphere, picking up speed and therefore decreasing its density. At some point down the tail, the magnetosheath (MS) flow becomes nearly parallel with the solar wind (SW) outside. Then the pressure must balance,

$$n_{ms}\, T_{ms} = n_{sw}\, T_{sw}$$

But the entropy is much greater in the magnetosheath,

$$S_{ms} = \frac{T_{ms}}{n_{ms}^{2/3}} >> \frac{T_{sw}}{n_{sw}^{2/3}}$$

implying that $n_{ms} < n_{sw}$. Eventually, the density must, of course, again reach values equal to the solar wind as the magnetosheath cools progressively down along the tail. Thus, the further any open field line extends down the magnetotail, the lower the density (and therefore also the lower the temperature, as the entropy is preserved) of plasma penetrating into the magnetosphere. For example, in the sketch in Figure 7, which attempts to reconstruct the situation that might resemble Figure 1, field line A is near the dawn polar cap boundary and exits the magnetosphere sunward of field line B. Since both the density and the temperature are lower at field line B, the flux, which is proportional to $nT^{1/2}$, is even further reduced. According to this explanation, on any orbit from noon towards midnight, one should encounter progressively less flux. We have examined data on both S3-3 and the DMSP satellite and find this supposition to be true. Therefore, we believe that the major component of the gradient in the "polar rain" is from midnight towards noon across the polar cap. We will present this result in a later paper.

The coincidence of the gradients in **E** and the soft particle flux has been established, both by us and by others [*Heppner*, 1972] in primarily dawn-dusk orbits. Typically it is the case for

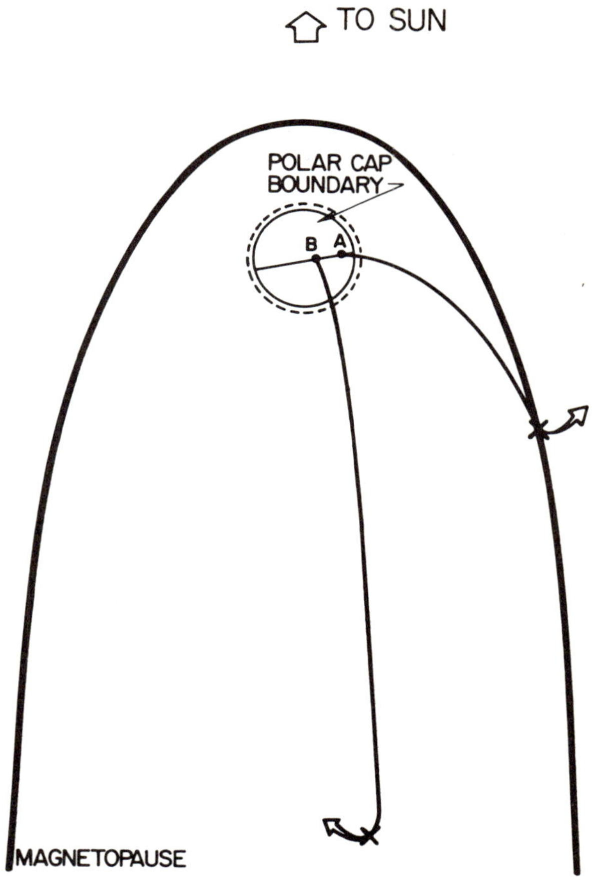

Fig. 7. A pictorial view of the loci of the projection into the equatorial plane of two field lines in the northern dawn polar cap. Field line A exits the magnetosphere at the "x" along the dawn flank. Field line B exits through the top of northern tail lobe.

the Northern hemisphere that, whenever the polar cap is substantially shifted to the dawn (dusk) due to $B_y > 0$ ($B_y < 0$), the electric field in the polar cap is strongest at the dawn (dusk) boundary. Apparently, in these cases, maximum flux transport towards the tail occurs along the flanks. Although the gross topological shift of the outer magnetosphere and polar cap with interplanetary B_y can be explained with geometric reconnection models [*Gonzales and Mozer*, 1974], the enhanced flow to one side must be intimately related either to reconnection processes of the magnetopause or to details of flow of open field lines within the magnetosphere. Whatever the reason, this pattern must imply that field lines near the dawn (dusk) polar cap boundary when $B_y > 0$ ($B_y < 0$), have more recently entered the magnetosphere and are therefore connected to the solar wind sunward of those points along the satellite trajectory that are farther away from this boundary. That is, the direction of the "polar rain" *gradient* in the polar cap shifts around from noon to the dawn (dusk) as B_y increases (decreases). Thus, we believe that along a nearly dawn (dusk) trajectory of the satellite, an electric field gradient to the dawn due to $B_y > 0$ must imply that field lines more towards the dawn on this trajectory enter the magnetosheath more sunward and therefore intercept higher electron flux, as described in the preceding paragraph. Thus the two gradients should always be in the same direction.

Summary

In the light of these data, the following conclusions should be kept in mind when studying phenomena near or depending upon the boundary between open and closed field lines:

1. A contiguous polar region, both where the electric field is dawn to dusk (convecting anti-sunward) and over which "polar rain" of magnetosheath origin is precipitating, may be consistently considered to be "open." From satellite data such as S3-3, the boundary of these open field lines can be determined to within 1° or 2°.
2. The energetic (30 keV) particle intensity or isotropy boundary is *not* a consistent indicator of the polar cap boundary, especially when the magnitude of B_y is significant.
3. The location of the large poleward parallel currents is also not a reliable criterion.
4. Ground-based observers using only the electric field or the absence of energetic particle precipitation may be severely misled about the location of the polar cap, because,
5. There may be significant portions of the high-latitude ionosphere which are closed; yet, briefly or within a small area, they may be flowing anti-sunward or sunward, depending upon auroral arc activity.
6. The connection between dawn-dusk gradients in the polar cap of both the electric field and the soft "polar rain" arises out of their common dependence on the interplanetary magnetic field. In the case of the particles, the gradient can be explained using the topology of high-latitude field lines together with known thermodynamic effects in the magnetosheath.

Acknowledgments. We are indebted to our collaborators at the Aerospace Corporation, Drs. P. F. Mizera, J. F. Fennel, and others, and those at the Lockheed Palo Alto Research Laboratory, Drs. Sharp, Shelley, and Johnson for the liberal use of their particle data from S3-3. Helpful comments from V. Vasyliunas and C. W. Carlson are appreciated. This work has been supported by ONR grant N00014-75-C-0294 and by NSF grant ATM-8000518 to the University of California; and by AFOSR grant 79-OO10 and by NSF grant ATM-79-23240 to the Johns Hopkins University.

References

Armstrong, J. C., and A. J. Zmuda, "Three axis magnetic measurements and field-aligned currents in the auroral region: initial results," *J. Geophys. Res., 78*, 6802, 1973.

Axford, W. I., "Magnetospheric convection," *Revs. Geophys., 7*, 421, 1969.

Cattell C. A., R. L. Lysak, R. B. Torbert, and F. S. Mozer, "Observations of differences between regions of current flowing into and out of the ionosphere," *Geophys. Res. Lett., 6*, 621, 1979.

Cattell, C. A., "Magnetic field-aligned currents in the earth's magnetosphere," Ph.D. thesis, University of California, Berkeley, 1980.

Dungey, J. W., "Interplanetary magnetic field and the auroral zone," *Phys. Rev. Lett., 6*, 47, 1961.

Fennell, J. F., P. F. Mizera, and D. R. Croley, "Low energy polar cap electrons during quiet times," *Proc. Int. Conf. Cosmic Rays, 14th,, 4*, 1267, 1975.

Freon, A., and K. G. McCracken, "A note on the vertical cutoff rigidities of cosmic rays in the geomagnetic field," *J. Geophys. Res., 67*, 888, 1962.

Gonzalez, W. D., and F. S. Mozer, "A quantitative model for the potential resulting from reconnection with an arbitrary interplanetary magnetic field," *J. Geophys. Res., 79*, 4186, 1974.

Gorney, D. J., A. Clarke, D. Croley, J. Fennell, J. Luhman, and P. Mizera, "The distribution of ion beams and conics below 8000 km," *J. Geophys. Res.*, in press 1980.

Heikkila, W. J., and J. D. Winningham, "Penetration of magnetosheath plasma to low altitudes through the daysides magnetospheric cusps," *J. Geophys. Res., 76*, 883, 1971.

Heelis, R. A., J. D. Winningham, W. B. Hanson, and J. L. Burch, "The relationships between high-latitude convection reversals and the energetic particle morphology observed by Atmosphere Explorer," *J. Geophys. Res., 85*, 3315, 1980.

Heppner, J. P., "Polar cap electric field distributions related to the interplanetary magnetic field direction," *J. Geophys. Res., 77*, 4877, 1972.

Iijima, T., and T. A. Potemra, "The amplitude of field-aligned currents at northern high latitudes observed by Triad," *J. Geophys. Res., 81*, 2165, 1976.

Lassen, K., and C. Danielsen, "Quiet time pattern of auroral arcs for different directions of the interplanetary magnetic field in the y - z plane," *J. Geophys. Res., 83*, 5277, 1978.

Lüst, R., "Impact zones for solar cosmic-ray particles," *Phys. Rev., 105*, 1827, 1957.

McCracken, K. G., "The cosmic ray flare effect," *J. Geophys. Res., 67*, 447, 1962.

McDiarmid, I. B., J. R. Burrows, and M. D. Wilson, "Magnetic field perturbations in the dayside cleft and their relationship to IMF," *J. Geophys. Res., 83*, 5753, 1978.

McDiarmid, I. B., J. R. Burrows, and M. D. Wilson, "Comparison of magnetic field perturbations at high latitudes with charged particle and IMF measurements," *J. Geophys. Res., 83*, 681, 1978.

McDiarmid, I. B., J. R. Burrows, and Margaret D. Wilson, "Comparison of magnetic field perturbations and solar electron profiles in the polar cap," *J. Geophys. Res., 83*, 1163, 1980.

Meng, C. -I., S. -I. Akasofu, and K. A. Anderson, "Dawn-dusk gradient of the precipitation of low-energy electrons over the polar cap and its relation to the interplanetary magnetic field," *J. Geophys. Res., 78*, 5271, 1977.

Meng, C. -I., and H. W. Kroehl, "Intense uniform precipitation of low-energy electrons over the polar cap," *J. Geophys. Res., 82*, 2305, 1977.

Mizera, P. F., J. F. Fennell, and D. R. Croley, Jr., "Quiet-time polar cap electrons," *EOS Trans. AGU, 56*, 1175, 1974.

Mizera, P. F., and J. F. Fennell, "Signatures of electric fields from high and low altitude particle distributions," *Geophys. Res. Lett., 4*, 331, 1977.

Morfill, G., and M. Scholer, "Study of the magnetosphere using energetic solar particles," *Space Sci. Rev., 15*, 267, 1973.

Mozer, F. S., C. W. Carlson, M. K. Hudson, R. B. Torbert, B. Parady, J. Yatteau, and M. C. Kelley, "Observations of paired electrostatic shocks in the polar magnetosphere," *Phys. Rev. Lett., 38*, 292, 1977.

Mozer, F. S., C. A. Cattell, M. Temerin, R. B. Torbert, S. Von Glinski, M. Woldorff, and J. Wygant, "The DC and AC electric field, plasma density, plasma temperatures and field-aligned current experiments on the S3-3 satellite," *J. Geophys. Res., 84*, 5875, 1979.

Mozer, F. S., C. A. Cattell, M. K. Hudson, R. L. Lysak, R. B. Torbert, and M. Temerin, "Satellite measurements and theories of low-altitude auroral particle acceleration," *Space Sci. Rev., 27*, 155, 1980.

Mozer, F. S., R. B. Torbert, U. V. Fahleson, C. G. Fälthammar, A. Gonfalone, A. Pederson, and C. T. Russell, "Direct observation of a continuous tangential electric field at the magnetopause," *Geophys. Res. Lett., 6*, 305, 1979.

Mozer, F. S., and R. B. Torbert, "An average parallel electric field deduced from the perpendicular electric field below 8000 kilometers," *Geophys. Res. Lett., 3*, 219, 1980.

Mozer, F. S., W. D. Gonzalez, F. Bogott, M. C. Kelley, and S. Schutz, "High latitude electric fields and the three-dimensional interaction between the interplanetary and terrestrial magnetic fields," *J. Geophys. Res., 79*, 56, 1974.

Mozer, F. S., and P. Lucht, "The average auroral zone electric field," *J. Geophys. Res., 79*, 1001, 1974.

Siscoe, G. L., L. Davis, E. J. Smith, P. J. Coleman, and D. E. Jones, "Magnetic fluctuations in the magnetosheath," *J. Geophys. Res., 72*, 1, 1967.

Shelley, E. G., R. D. Sharp, and R. G. Johnson, "Satellite observations of an ionospheric acceleration mechanism," *Geophys. Res. Lett., 3*, 654, 1976.

Smiddy, M., W. J. Burke, M. C. Kelley, N. A. Saflekos, M. S. Gussenhaven, D. A. Hardy, and F. J. Rich, "Effects of high-latitude conductivity on observed convection electric fields and Birkeland Currents," *J. Geophys. Res., 85*, 6811, 1980.

Southwood, D. J., and R. A. Wolf, "An assessment of the role of precipitation in magnetospheric convection," *J. Geophys. Res., 83*, 5227, 1978.

Störmer, C., *The Polar Aurora,* Oxford University Press, 1955.

Vämpola, A. L., "Access of solar electrons to closed field lines," *J. Geophys. Res., 76*, 36, 1971.

Vasyliunas, V. M., "Theoretical models of magnetic field line merging," *Rev. Geophys. Space Phys., 13*, 303, 1975.

Winningham, J. S., F. Yasuhara, S.-I. Akasofu, and W. J. Heikkila, "The latitudinal morphology of 10 eV to 10 keV electron fluxes during magnetically quiet and disturbed times in the 2100 to 03:00 MLT sector," *J. Geophys. Res., 76*, 36, 1971.

AE-C OBSERVATIONS OF ELECTRIC FIELDS AROUND AURORAL ARCS

R. A. Heelis and W. B. Hanson, Univ. of Texas at Dallas, Richardson, TX 75080

J. L. Burch, Southwest Research Institute, San Antonio, TX 78284

Introduction

The large-scale features of the ionospheric convection pattern at high latitudes are quite well documented and understood. There are, however, significant variations in the observed ion drift in the auroral zone that are associated with discrete events in the energetic particle precipitation. Measurements from rockets (Evans et al., 1977) and ground-based equipment have shown a great deal of variability in the spatial scales of these structures, and in general a satellite is able to resolve discrete structures of order 25 km in spatial extent. Gurnett (1972) and Gurnett and Frank (1973) first reported the existence of structured electric fields associated with inverted V events on satellites. In particular they noted that events that occur as oppositely directed electric field spikes adjacent to the inverted V may be interpreted in terms of field-aligned electrostatic potential gradients. Similar properties of the energetic particle spectrum and the electric field have been reported from rockets (Anderson and Vondrak, 1975) and from ground-based radars (de la Beaujardiere et al., 1977). The Atmosphere Explorer C satellite was launched into a 68° inclination orbit in December 1973 and sampled the high latitude ionosphere for more than half of its operational lifetime. The instrument complement included the capability of measuring the total ion drift velocity vector (Hanson et al., 1973) with a resolution of 1½ to 6 seconds (10 km to 50 km) and the energetic electron spectrum (Hoffman et al., 1973) with a resolution of 1 second (8 km) along the satellite track. This data has been used to investigate the spectral characteristics of auroral arcs (Burch et al., 1976a) which show that a discrete arc is well characterized by an accelerated maxwellian distribution. Typical field-aligned potentials were found to be in the region 3 kev to 10 kev near dusk, with somewhat smaller values occurring closer to local noon. The effect of substorms on the electric field and particle morphology has been investigated by Burch et al. (1976b), where they found that a region of discrete inverted V events embedded in the rapid equatorward ion drift velocity near midnight is the most dominant feature in the high latitude region. Studies of the electric field configuration around polar cap discrete arcs have also been investigated (Burch et al., 1979) and will be addressed further here. Throughout this discussion we use the terms "discrete arc, inverted V and auroral arc" somewhat synonymously realizing that they frequently have different meanings. First the spatial scales of the events to which we refer are of order 100 km, while similar phenomena are observed from rockets with spatial scales of order 10 km. From the ground discrete visual auroral forms have spatial scales of 1 km and these are also termed auroral arcs. The relationships between these phenomena are not well known, although there is evidence to show that visual forms are associated with inverted V events observed from satellites. It should be noted that the electric field signatures associated with "satellite" arcs will not be the same as those associated with smaller scale arcs.

It is the purpose of this paper to discuss the variety of ion drift (or equivalently electric field) signatures that are observed by AE-C during its passage through the auroral zone. Only a few examples will be used to illustrate the features seen in many orbits and described in the previously referenced material. The appropriateness of models to interpret the various signatures will be discussed. A complete discussion of the Atmosphere Explorer mission and its instrumentation can be found in the April 1973 edition of Radio Science.

Instrumentation

The RPA and Ion Drift Meter on Atmosphere Explorer are used to derive three mutually perpendicular components of ion drift velocity as well as the ion concentration and temperature. The techniques involved are described in detail by Hanson and Heelis (1975). We note here that the drift velocity along the satellite track is derived from the RPA by measuring the ion energy distribution in this direction, and the horizontal ion velocity across the spacecraft track is determined from the ion drift meter by measuring the incoming ion arrival angle. These two drift components are labelled "ram drift" and "cross drift," respectively, in the figures. The electrostatic potential distribution along the satellite track is obtained by calculating the electric field with knowledge of a model magnetic field and integrating the component parallel to the orbit along the satellite path. The electron energy spectrum is determined by a single detector that steps through 16 logarithmically spaced energies between 200 eV and 25 keV. A complete spectrum is determined every second. When the spacecraft is despun as is the case in all examples shown, the sensor pitch angle is essentially constant and indicated on each figure.

Observations

We will describe here the electric field signatures observed in the vicinity of auroral arcs in the nighttime ionosphere. An auroral arc is identified from the energetic electron detector as an accelerated particle spectrum with peak energy exceeding 1 kv. The total ion velocity vector with a resolution of 6 sec along the satellite track is used to locate the features within the global convection pattern and establish that the spacecraft is moving approximately radially through the auroral zone. The single drift component transverse to the satellite orbit is then used to examine the electric field structure perpendicular to the arc in more detail. The vector ion drift velocity perpendicular to B is used to calculate the electrostatic potential distribution along the satellite track, which proves to be useful for interpretation in terms of current models of magnetosphere-ionosphere coupling.

Figure 1 shows the vector ion drift velocity data obtained from a southern hemisphere pass of AE-C across the nighttime high latitude region. The signature of a two-cell convection pattern is quite clearly seen, but of more interest on this occasion is the structure of the flow velocity both in magnitude and direction near the global convection reversals.

Figure 2 shows the simultaneously measured energetic particle spectrum, the ion drift velocity component perpendicular to the satellite track, and the derived electrostatic potential distribution for the evening side of this pass. Since the satellite is passing almost radially through the auroral zone positive velocities indicate a poleward directed electric field, while negative velocities denote an equatorward directed field. The global convection reversal is located at about 19:11:47 UT with velocity structure associated with auroral arcs both equatorward and poleward of the reversal.

We note that only very occasionally do discrete auroral arcs on spatial scales of 100 km exist as singular structures so that usually the electric field signature is the result of several auroral forms. Nevertheless, several consistent features in the electric field signature can be seen in this pass.

There are three inverted V structures that we wish to consider, which span the time periods from 19:11:05 to 19:11:30, from 19:11:45 to 19:12:05, and from 19:12:17 to 19:12:30. First note that the electric field appears to decrease inside the auroral arc. In fact the electric field returns to a reasonably smooth background level within the arcs, as can be easily seen in

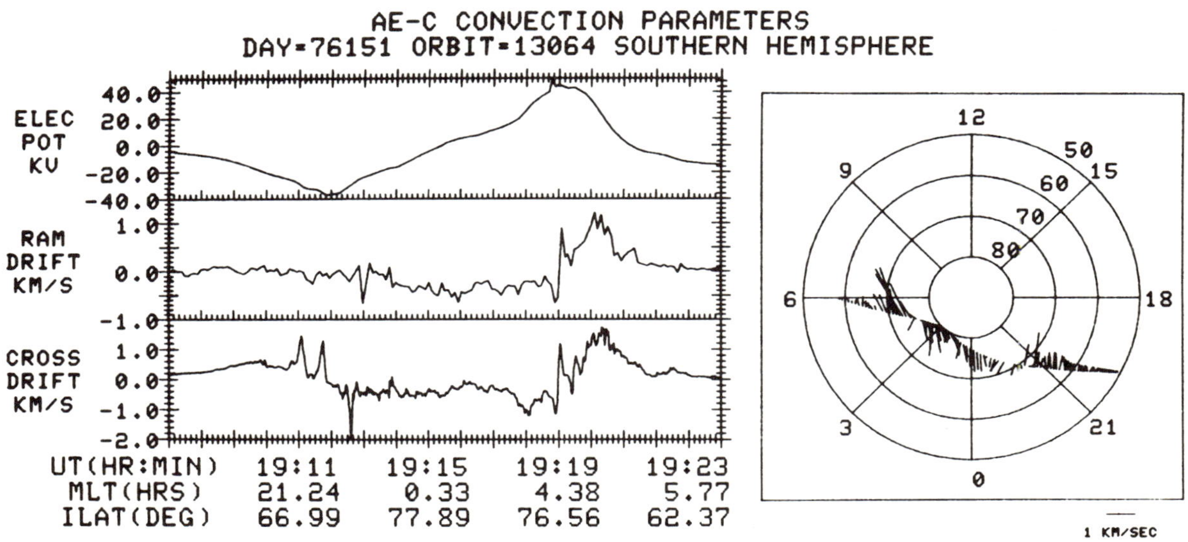

Fig. 1. Ion drift velocity plotted against UT and invariant latitude and magnetic local time for the southern hemisphere pass 13064.

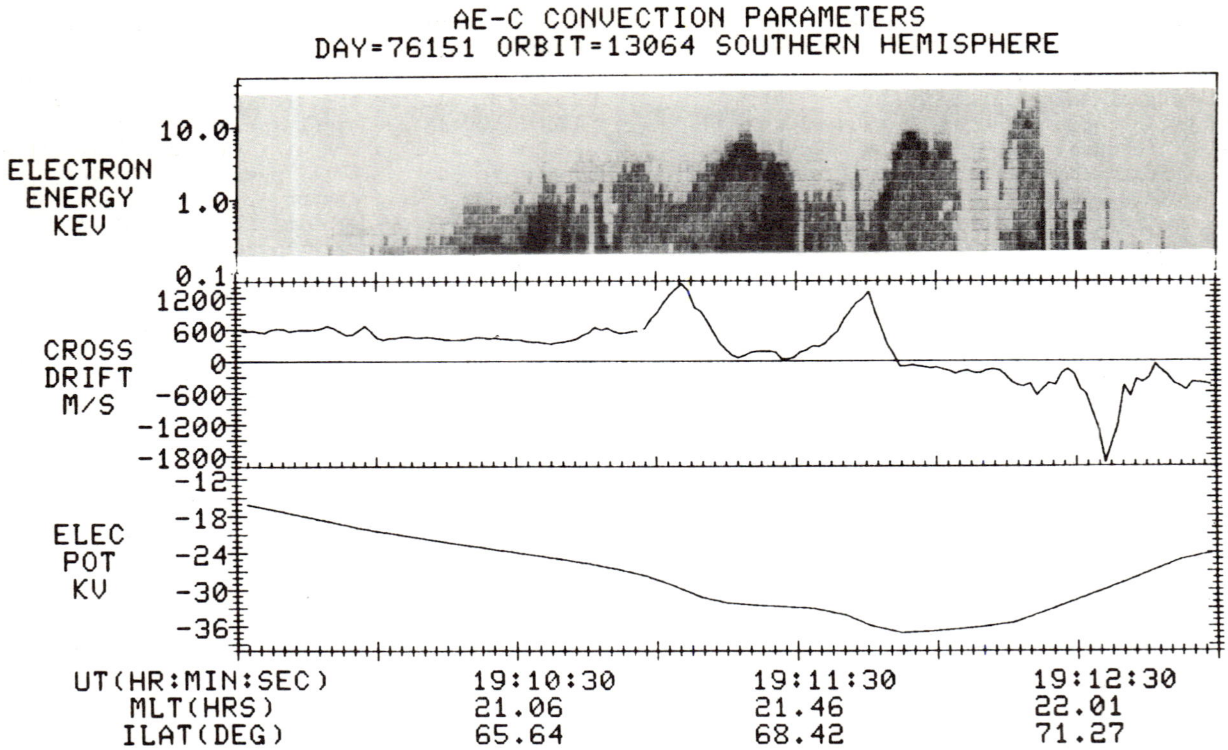

Fig. 2. Expanded view of dusk side data shown in Figure 1. The particle sensor pitch angle for this pass is about 60°.

Figure 1. The apparent decrease in the field strength is in fact due to the electric field enhancements at the edges of the auroral arc. In this regard a critical examination of the individual events shows that while the electric field appears to be enhanced on both sides of the arc, it is in fact more reasonable to note that the field is enhanced only on one side. For the case of the arcs on the equatorward side of the reversal the electric field is enhanced on the equatorward side of the arc and is directed poleward. The arc that appears poleward of the reversal shows an enhanced equatorward field on its poleward side. Further emphasis on this feature will be given later.

It should also be pointed out in most cases the increase in electric field at the edge of the arc takes place just outside the location of particle acceleration in excess of 1-2 kv, so that observationally the electric field increase appears to span the arc. Of course it is difficult to be precise on the location of the arc since it is accompanied by smaller but significant particle fluxes on each side. In addition it can be seen that the total potential difference in the arc is comparable to or greater than the acceleration energy of the electrons seen in this event. Note that in all cases the electric field is consistent with a net horizontal current being driven toward the arc and presumably closed by an upward field aligned current within the arc. Finally, we note that at the dusk side convection reversal, which is clearly of a shear nature, the global electric field is directed toward the reversal both poleward and equatorward of it. Horizontal current is then driven toward the reversal, and it is therefore not surprising to observe auroral arcs at the dusk side reversal as is the case here. Notice that the electric field reduces to quite small values as the convection reversal is traversed on the dusk side.

Figure 3 shows the observed electric field structure from the dawn side crossing of orbit 13064 shown in Figure 1. Since we have crossed over the pole, negative velocities now denote a poleward directed electric field and positive velocities denote an equatorward electric field. Observation of Figure 1 shows clearly that the global convection reversal occurs at 19:18:57 and that auroral arcs occur both equatorward and poleward of this reversal. There are two discrete features on which we wish to focus in this pass. One from 19:18:20 to 19:18:50 is on the poleward side of the reversal and one from 19;19:07 to 19:19:25 is on the equatorward side of the reversal. We see that the poleward-most arc is characterized by an enhanced flow veloc-

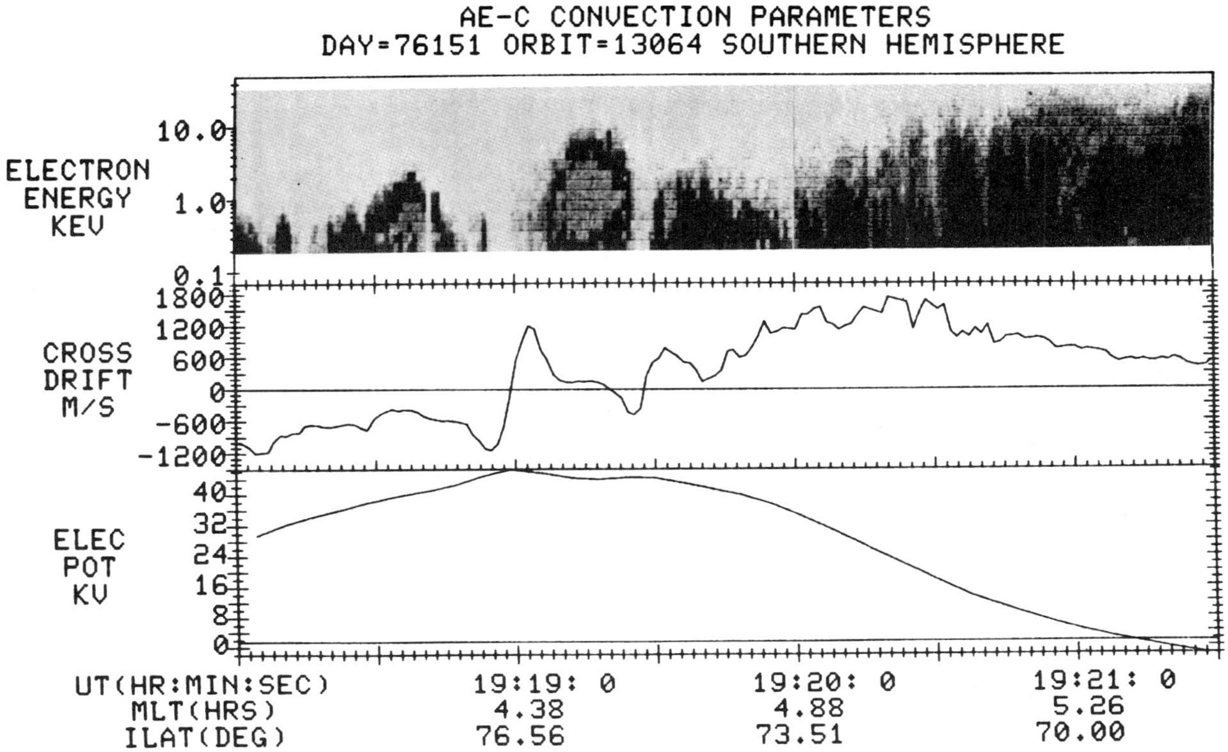

Fig. 3. Expanded view of dawn side data shown in Figure 1. The particle sensor pitch angle for this pass is about 45°.

ity on one side and that this time the flow is enhanced on the equatorward side and directed poleward. This is again consistent with a horizontal current driven into the arc and closed by upward field-aligned current in the arc. Notice also that the horizontal potential difference across this arc is again similar to the energy of the precipitating particles. The arc on the equatorward side of the convection reversal is of a different nature than those discussed previously. While it has an enhanced electric field on its poleward edge, consistent with horizontal current being directed toward it, it also has a poleward directed electric field on its equatorward edge. Investigation of the occurrence of these so-called spike-like fields associated with auroral arcs has been carried out by Gurnett (1972) and Burch et al. (1976). Burch et al. (1976) have shown their existence both equatorward and poleward of the global convection reversal. They also showed that it is not necessary to pass through the field-aligned potential difference as suggested by Gurnett (1972) but that this signature can also be seen if the spacecraft passes below the potential difference region. Note that the electric field always returns to some smooth background level within the arc whenever it is accompanied by enhanced fields at its edges. On this occasion the potential difference in this arc is only about 2 or 3 kv, which is much smaller than the energy of the precipitating particles.

In contrast to the global electric field configuration at the dusk convection reversal, the dawn side reversal has the electric field directed away from it on both the poleward and equatorward side. It should be noticed that at the dawnside reversal there is a distinct lack of energetic particles presumably because the electric field configuration requires a downward field aligned current at this location. This is a reproducible feature when the dawn side convection reversal is a shear and is analogous with the anti inverted V event reported by Burch et al. (1976). We also point out that in the event of multiple reversals as in this case there is also an absence of energetic particles at 19:19:27 where again the electric field is directed away from this point on both sides. It is also important to notice that the electric field does not return to a smooth background level in these cases but simply changes sign abruptly. Thus the dawn side convection reversal shear is much more dramatic than a shear reversal on the dusk side. We emphasize here that the convection reversal does not delineate any obvious boundary in the particle data since discrete arcs are observed on both sides of it. This may indicate that the

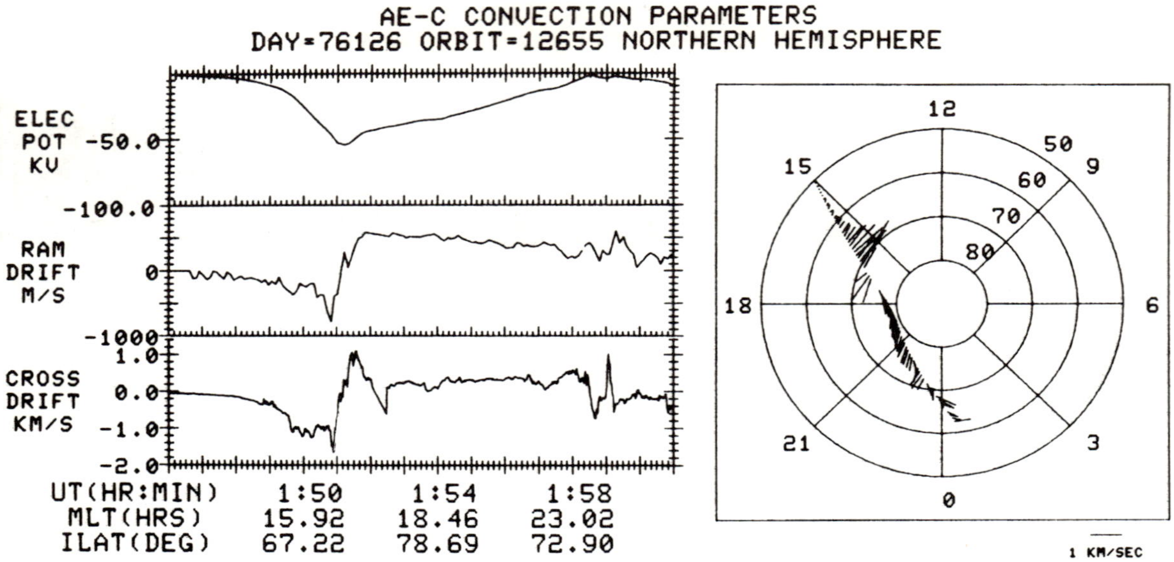

Fig. 4. Ion drift velocity plotted against UT and invariant latitude and magnetic local time for the northern hemisphere pass 12655.

convection reversal also does not delineate an obvious boundary at the magnetosphere (Heelis et al., 1980).

Figure 4 shows the data obtained during a pass across the high latitude northern hemisphere. The convection reversal near 01:51 UT is clearly a shear in nature, while a somewhat more complex flow geometry is seen near midnight. Closer examination of flow characteristics near the afternoon reversal in Figure 5 show that variations in the auroral particle flux, which may appear as discrete features, may simply be associated with a changing gradient in the electric field. We notice that in all cases the field behaves in a manner that maintains current continuity. That is, auroral arcs appear in the region where the electric field gradient is positive and tend not to appear where the gradient is negative. This trend is followed on quite small scales by satellite definitions as can be seen using the time lines drawn in this figure. Finally we note that near midnight, Figure 6 shows that discrete auroral precipitation such as that near 01:58:20 can occur within the antisunward convecting region and not be associated with electric field gradients. In addition, the potential difference across this arc is comparable with that of the mean energy of the precipitating particles. The convection reversal is almost shear in nature on this occasion, and the electric field is directed away from the reversal on both sides. Again we see the absence of energetic electrons in such a configuration. The spike-like electric field associated with the auroral precipitation at 01:58:50 lies within the sunward convecting plasma region and has a spatial scale similar to that shown in Figure 3 (i.e. ∿200 km). The characteristics of electric fields around auroral arcs discussed here are also seen in isolated precipitation regions within the polar cap (Burch et al., 1979).

The following characteristics of the electric field configuration around auroral arcs are observed in the AE-C data.

1) Auroral arcs occur both equatorward and poleward of the global convection reversal.
2) Within an auroral arc the electric field perpendicular to the arc reaches a smooth background level or is reduced below this level.
3) Arcs can appear with no unusual electric field signature inside or around the arc.
4) Significant increases in the electric field from a smooth background level are always accompanied by discrete energetic particle fluxes.
5) Arcs occurring completely within a given convection zone usually have enhanced fields of the same sign as the background field and only on one side of the arc. Table 1 shows the configuration of electric field and arc position.
6) Arcs with oppositely-directed electric fields on both sides can occur within the sunward or antisunward convecting regions or at the dusk side convection reversal.
7) Arcs with oppositely-directed fields always have the field directed toward them on both sides.
8) Arcs are distinctly absent in regions where neighboring electric field enhancements are directed away from each other.

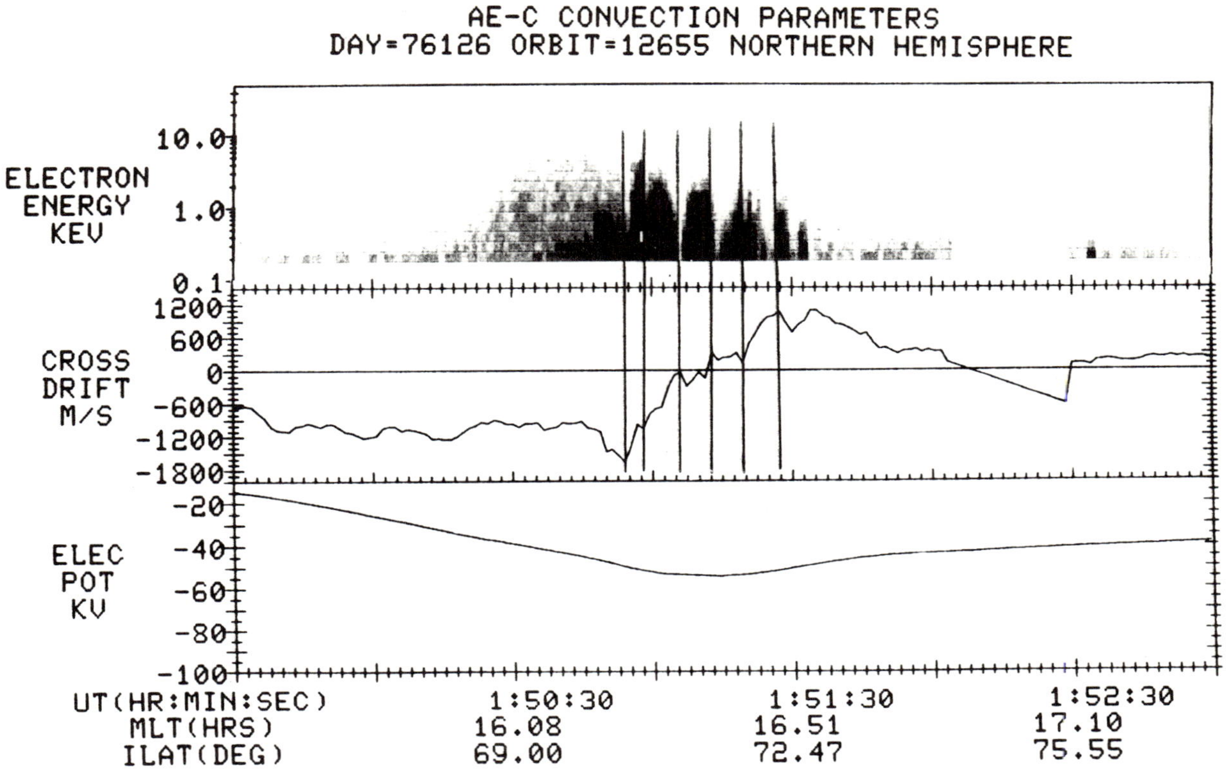

Fig. 5. Expanded view of the afternoon side data shown in Figure 4. The particle sensor pitch angle for this pass is about 40°.

In particular, arcs are always absent at a shear convection reversal on the dawn side.

9) A reduction of the electric field to background levels within an auroral arc usually leads to much less dramatic shear convection reversals on the dusk side than on the dawn side.

Discussion

It is not our intent to use the observations here to justify the presence of parallel electric fields but rather to discuss the implications of the observations in the context of present models of the ionosphere-magnetosphere system. In this regard, it is important to note that observations of the electric field configuration in the ionosphere show the combined effects of the applied magnetospheric potential distribution modified by field-aligned potential differences and ionospheric polarization fields. It is not possible to separate these effects without independent knowledge of two of the phenomena in addition to their sum. Recently Lyons (1980, 1981) has developed a model of the ionosphere-magnetosphere coupled system that allows a self-consistent determination of the applied magnetosphere potential distribution from knowledge of the precipitating electron energy spectrum and ionospheric electrostatic potential distribution. Viewed very simply and to first order, this model predicts that the magnetospheric potential distribution is simply the observed ionospheric distribution minus a potential equal to the acceleration energy of the precipitating electron spectrum. Such a model predicts that discrete particle accelerations will appear at places where there exist large changes in the gradient of the magnetospheric potential. Such a model does not address how the field-aligned potential is distributed along the fields but does indicate quite clearly that it is not unreasonable to expect potential drops of 5 to 10 kv associated with field-aligned currents of 10^{-6} A m^{-2}. The theory of magnetosphere-ionosphere coupling developed by Kan and Lee (1980) also predicts similar features.

In order to place our findings within the context of such models, it is necessary to delineate quite precisely the location of the electric field variations and the electron precipitation regions associated with auroral arcs. In this regard we note that when electric field enhancements occur on one side of an auroral arc, particle precipitation is in fact present on both sides, even though the flux

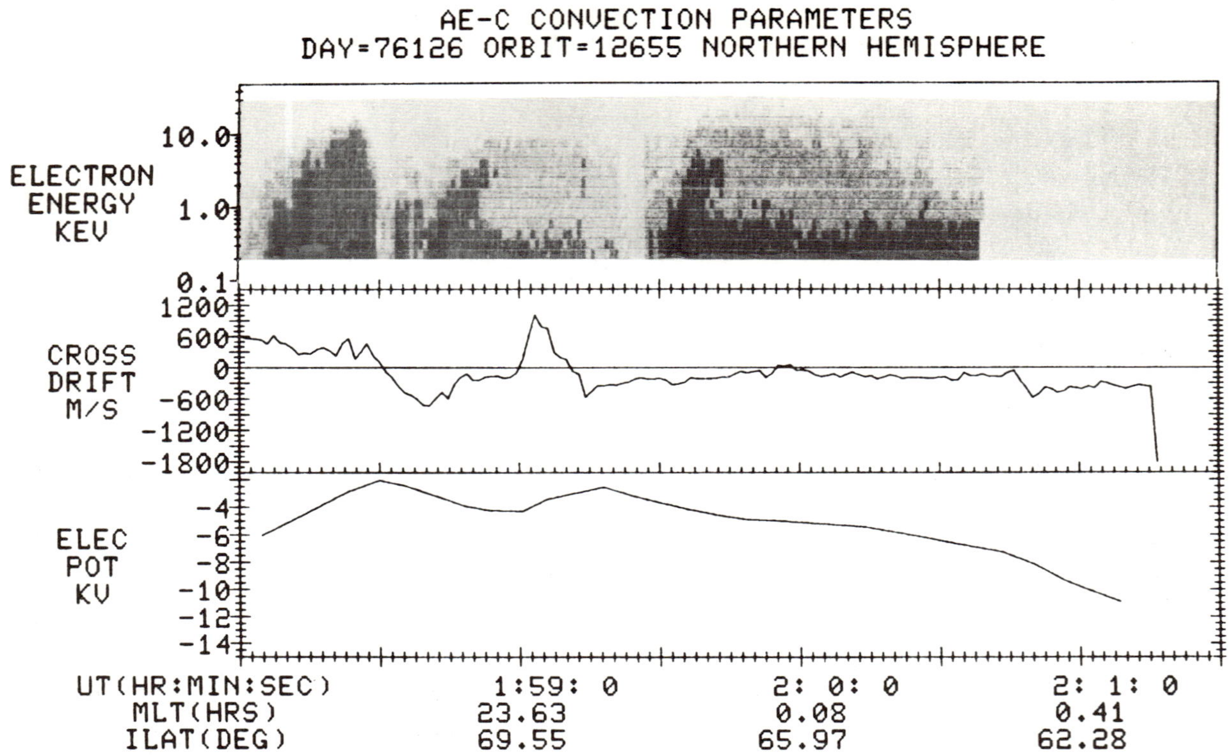

Fig. 6. Expanded view of the midnight crossing of the data shown in Figure 4. The particle sensor pitch angle for this pass is about 60°.

variations are sufficient to identify a discrete feature. However, in general, the field-aligned potential drop associated with the accompanying particles is not as high as that required in the discrete arc. The models would thus predict that at least some portion of the increased ionospheric potential gradient associated with the electric field enhancement is applied from the magnetosphere. In most cases we have noted that the electric field is small inside the arc and that typical flow velocities within the arc required latitudinal potential differences across the arc that are in some cases smaller than and in most cases comparable to the field-aligned potential difference determined by the energy of the particles. In principle, it is the total energy flux rather than the peak energy that depends upon the field-aligned potential difference, but certainly we would expect to observe particles in the incident spectrum of an energy equivalent to the field-aligned potential. The observations suggest

TABLE 1. Relationships Between Electric Fields and Auroral Arcs

Location Relative to Arc	Position of Arc Relative to Reversal	Location of Reversal
	Enhanced Poleward Field	
Equatorward	Equatorward	Dusk
Equatorward	Poleward	Dawn
	Enhanced Equatorward Field	
Poleward	Poleward	Dusk
Poleward	Equatorward	Dawn

and indeed a case study by Lyons (1981) shows that over many discrete arc features the magnetospheric potential gradient must change sign even though the observed ionospheric potential, which is modified by the parallel electric fields, does not. This situation is shown schematically in Figure 7 and illustrates quite clearly that such a configuration would be consistent with the observed location of enhanced fields with respect to the arc. Here a field-aligned potential drop of maximum value 5 kv is assumed and electrostatic equipotential lines are drawn to illustrate that both "S" shape and "U" shape potential contours can be present in one event (cf. Kan and Lee, 1980). We note that in the case that the potential drop across the arc is greater than the particle acceleration energy, it would not be necessary to have a change in sign of the magnetospheric potential gradient and thus only "S" shape equipotential lines would result. The situation is very similar for the spike-like fields associated with discrete arcs that are not coincident with the convection reversal. In this case a change in sign of the potential gradient is present both in the magnetosphere and the ionosphere, but the magnetospheric potential gradient change is more pronounced. These properties have some interesting implications about the nature of the flow regimes in the magnetosphere that are connected by the earth's magnetic field to the auroral arcs observed in the ionosphere.

Observations from all-sky cameras (Davis, 1978) and from satellite auroral images (Meng et al., 1977) have shown that the discrete arcs often extend over many hours of local time. In addition, recent investigation of the local time or longitudinal variations in the electric field

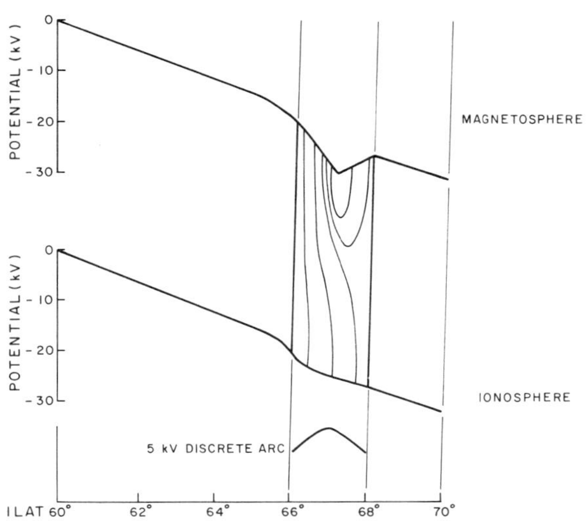

Fig. 7. Representative electrostatic potential distribution in the magnetosphere and ionosphere associated with a 5 kv discrete arc in the dusk side sunward convection region.

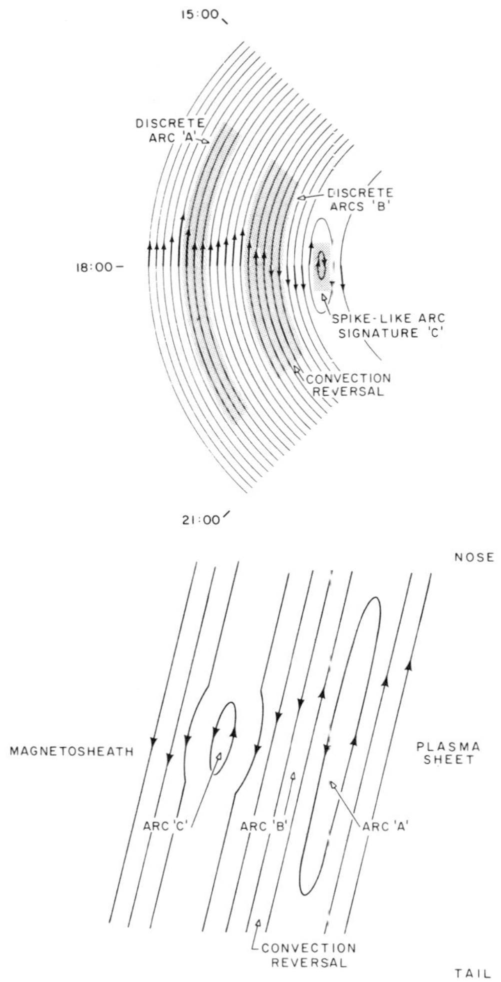

Fig. 8. Schematic flow trajectories in the ionosphere and magnetosphere associated with the variety of discrete arcs observed by the AE-C satellite. Flow configuration is shown in the dusk side only for

A arcs with enhanced fields on one side
B arcs associated with the global convection reversal
C arcs with spike-like electric field signatures

signature of an arc show that it does not change significantly (de la Beaujardiere et al., 1980). This implies that significant regions of the magnetosphere consist of three or more adjacent regions where the plasma flow is oppositely directed. This may well be expected in a magnetosphere boundary layer. The observation of oppositely directed spike-like flows places a further implication on the flow characteristics. From a statistical examination of electric field data at high latitudes (Heppner, 1977) and convection velocity data (Heelis and Hanson, 1980) it can be shown that a single global convection reversal is always observed and that multiple reversals produced by the presence of spike-like features around auroral arcs do not occur in any predictable fashion. It is therefore not reasonable to expect the spike-like signature in the electric field to persist over large longitudinal distances either in the ionosphere or the magnetosphere. In this case the ionospheric and magnetospheric signatures are more reasonably like small eddies of the type expected in the presence of a point source charge. This is in contrast to the case of discrete arcs where the electric field is enhanced on only one side. The different flow regions that we anticipate in the ionosphere and magnetosphere are shown in Figure 8. Here the arrows in the ionosphere denote the magnitude and direction of the flow and the flow lines are not joined for large convective cells.

Finally, it is worthwhile to note some restrictions on the distribution of field-aligned potential drops that result from observations of the horizontal electrostatic potential distribution. In a region of predominantly sunward or antisunward convection the potential distribution is either monotonically increasing or decreasing. In such cases the potential distribution below the magnetosphere can only be of the "S" shape described by Kan and Lee (1980). Similarly a "U" shaped potential distribution below the magnetosphere can only occur if the magnetospheric potential gradient changes sign. This is always the case at the global convection reversal but as we have seen here, may also be true for many discrete arcs poleward and equatorward of the reversal.

Conclusions

We have used a very limited data set to describe features observed by a much larger data base from Atmosphere Explorer. These features described here and in more detail elsewhere indicate that the electric field signature around auroral arcs in the ionosphere is quite reproducible. We have found that in terms of currently acceptable models the region of the magnetosphere that is connected to the discrete arcs by the earth's magnetic field must have both large and small areas in which adjacent flow regions are opposed to one another and that several such regions must exist in moving from the plasma sheet to the magnetosheath.

Acknowledgments. This work was supported by the National Aeronautics and Space Administration under contract NAS5-11407 and grant NGR44-004-120.

References

Anderson, H. R., and R. R. Vondrak, Observations of Birkeland currents at auroral latitudes, Rev. Geophys. Space Phys., 13, 243, 1975.

Burch, J. L., S. A. Fields, W. B. Hanson, R. A. Heelis, R. A. Hoffman and R. W. Janetzke, Characteristics of auroral electron acceleration regions observed by Atmosphere Explorer -C, J. Geophys. Res., 81, 2223, 1976a.

Burch, J. L., W. Lennartson, W. B. Hanson, R. A. Heelis, J. H. Hoffman, and R. A. Hoffman, Properties of spike-like shear flow reversals observed in the auroral plasma by Atmosphere Explorer-C, J. Geophys. Res., 81, 3886, 1976b.

Burch, J. L., S. A. Fields and R. A. Heelis, Polar cap acceleration regions, J. Geophys. Res., 84, 5803, 1979.

Davis, T. N., Observed characteristics of auroral forms, Space Sci. Rev., 22, 77, 1978.

de la Beaujardiere, O., R. R. Vondrak and M. Baron, Radar observations of electric fields and currents associated with auroral arcs, J. Geophys. Res., 82, 5051, 1977.

de la Beaujardiere, O., R. Vondrak, R. Heelis, W. B. Hanson and R. A. Hoffman, Auroral arc electrodynamic parameters measured by AE-C and the Chatanika radar, submitted to J. Geophys. Res., 1980.

Evans, D. S., N. C. Maynard, J. Troim, T. Jacobsen and A. Egland, Auroral vector electric field and particle comparisons, 2, Electrodynamics of an arc, J. Geophys. Res., 82, 2235, 1977.

Gurnett, D. A., Electric fields and plasma observations in the magnetosphere, in Critical Problems of Magnetospheric Physics, publ. by IVCSTP Secretariat, Washington, D.C., 1972.

Gurnett, D. C. and L. A. Frank, Observed relationships between electric fields and auroral particle precipitation, J. Geophys. Res., 78, 145, 1973.

Hanson, W. B., D. R. Zuccaro, C. R. Lippincott and S. Sanatani, The retarding potential analyzer on Atmosphere Explorer, Radio Sci., 8, 333, 1973.

Hanson, W. B. and R. A. Heelis, Techniques for measuring bulk gas motions from satellites, Space Sci. Instr., 1, 493, 1975.

Heelis, R. A. and W. B. Hanson, High latitude ion convection in the nighttime F-region, J. Geophys. Res., 85, 1995, 1980.

Heelis, R. A., J. D. Winningham, W. B. Hanson and J. L. Burch, The relationship between high latitude convection reversals and the energetic particle morphology observed by Atmosphere Explorer-C, J. Geophys. Res., 85, 3315, 1980.

Heppner, J. P., Empirical models of high latitude electric fields, J. Geophys. Res., 82, 1115, 1977.

Hoffman, R. A., J. L. Burch, R. W. Janetzke, J. F. McChesnay, S. H. Way and D. S. Evans, Low energy electron experiment for Atmosphere Explorer C and D, Radio Sci., 9, 393, 1973.

Kan, J. R. and L. C. Lee, Theory of imperfect magnetosphere-ionosphere coupling, Geophys. Res. Lett., accepted for publ., 1980.

Lyons, L. R., Generation of large scale regions

of auroral currents, electric potentials and precipitation by the divergence of the convection electric field, J. Geophys. Res., 85, 17, 1980.

Lyons, L. R., The field aligned current versus electric potential relation and auroral electrodynamics, AGU, Geophysical Monograph No. 25 - Auroral Arcs, ed. by S.-I. Akasofu, 1981.

Meng, C-I., R. H. Holzworth and S.-I. Akasofu, Auroral circle - delineating the poleward boundary of the quiet auroral belt, J. Geophys. Res., 82, 164, 1977.

ELECTRIC FIELDS, BIRKELAND CURRENTS AND ELECTRON PRECIPITATION IN THE VICINITY OF DISCRETE AURORAL ARCS

William J. Burke
Air Force Geophysics Laboratory, Hanscom AFB, MA 01731

Abstract. Simultaneous measurements of electric fields, Birkeland currents and electron fluxes are presented from one winter, polar-cap and two auroral zone passes, of the USAF satellite S3-2. At the time of the polar-cap pass the IMF had a 7.1 nT northward component. The large-scale electric field was highly irregular. Five upward Birkeland current sheets accompanied by enhanced electron precipitation are taken as signatures of sun-aligned arcs in the polar cap. These current sheets were all embedded in regions of negative space charge. The precipitating electrons had average energies $\leq$ 2keV, and carried energy fluxes up to 3 ergs/(cm^2-sec-ster).

During a substorm period "electrostatic shock" like structures were observed at altitudes of 1050 and 1350 km, on the dawn and dusk sides of the auroral oval. In the dawn (dusk) side event the transverse magnetic field component deflected by 100 (300) nT and recovered in 4 (7) seconds. In both cases the satellite traversed an upward current sheet in 1/4 seconds corresponding to average current densities of 45 and 135 μ A/m^2 contained in latitudinal widths of < 2 km. Between the current sheets where $j_\parallel \approx 0$ the electric field rotated and returned to its original configuration. Spatial aliasing of electron measurements precludes a positive identification of the energy distribution of the current carriers. Return current sheets were more spatially spread out, having maximum intensities of 10 to 15 μ A/m^2.

As a whole, the data support a magnetospheric rather than an ionospheric generator model of discrete arcs.

Introduction

Until recently there was a generally conceded division of labor between rocket and satellite borne instrumentation. Instruments on rockets are clearly capable of delineating small-scale ($\leq$ 10 km) characteristics of discrete auroral arcs. The high speeds of satellites tend to restrict their measuring capabilities to structures of larger scales. The much publicized inverted-V structures regularly observed by polar-orbiting satellites in the topside ionosphere can extend for more than a full degree in latitude. High-time resolution experiments on board USAF and NASA satellites launched in the mid-1970's opened the possibility of investigating characteristics of discrete auroral arcs within their large-scale, environmental contexts. The observation of rapidly varying electrostatic structures in the high latitude ionosphere by means of instruments on the S3-3 satellite constitutes the most exciting measurement of small-scale auroral features reported to date (Mozer et al., 1977). These structures, commonly referred to as "electrostatic shocks" appear as latitudinally narrow reversals of the components of the convective electic field. They are frequently found embedded in regions of plasma turbulence, at altitudes greater than 4000 km and are associated with the acceleration of positive ions out of, and keV electrons into the ionosphere.

Depending on the regions of the ionosphere in which they are found, two broad classes of discrete arcs are distinguished in the literature. The first class of discrete arcs is found in the auroral oval and tend to be aligned in the magnetic east-west direction. Their luminosity is produced by the precipitation of electrons with energies of several keV. They originate on closed magnetic field lines that map into the boundary plasma sheet. A second class of arcs, found in the polar cap, have a tendency to be sun-aligned. They appear most frequently during periods of magnetic quieting while the IMF has a northward component. The electrons responsible for these arcs typically have average energies of $\leq$ 1 keV.

The purpose of this paper is to present detailed observations taken in places where electric and magnetic field variations and electron fluxes suggest that S3-2 was passing over discrete arcs. In the following section the complement of S3-2 scientific instruments and the methods for data reduction are described. The observation section is made up of two major subsections. In the first subsection we consider particle and field measurements taken during a cross polar cap pass while the IMF had a northward component. In the second subsection we present measurements taken in the auroral oval during a substorm period. Special attention is directed to the Birkeland current distributions associated with electric field stuctures similar to those reported by Mozer et al., (1977). The structures reported here were encountered at altitudes between 1000 and 1400 km. Some implication of these new measurements for understanding the development of discrete arcs are discussed in the final section.

Instrumentation

The S3-2 satellite was launched into polar orbit during December 1975 with an initial apogee, perigee and inclination of 1557 km, 240 km and 96.3°, respectively. It is spin-stabilized with a nominal spin period of 20 sec. The spin axis is nearly perpendicular to the orbital plane. The scientific package on S3-2 includes: (1) an electric field experiment consisting of two dipoles, (2) a triaxial fluxgate magnetometer, and (3) an energetic electron spectrometer. Electric and magnetic field components are usually calculated in a spacecraft centered coordinate system with X positive along the satellite velocity, Z is positive toward local nadir and Y completes the right hand system. When the orbital plane is near the dawn-dusk meridian, as in the cases studied here, Y is positive in the antisunward direction (Smiddy et al. 1980).

The instrumentation on S3-2 and the methods used to reduce data have been described in detail by Burke et al. (1980). Briefly, one of the dipoles has a length of 29.72 m. It lies in the satellite's spin-plane and measures the electric field component along the satellite trajectory (E_x). The second dipole has a length of 11.18 m and lies close to the satellite's spin axis. Because the spin-axis aligned dipole frequently has an unknown dc offset potential, only the spin-plane data are normally presented. The dc offsets vary on time scales much longer than the 10 sec required for the satellite to traverse "electrostatic shock" structures. In such instances the spin axis component of the electric field (E_y) can be used. Potential differences between the ends of the dipole are sampled 32 times per second. Data are usually presented as 5 second averages of E_x which are calculated assuming $\underline{E} \cdot \underline{B} = 0$. To resolve structures such as polar cap arcs, in this paper, we use all of the measurements to calculate values of E_x with a much higher resolution. Magnetic field measurements were taken and are presented at a rate of 32 samples per second, with a one bit resolution of 5nT, as differences between the measured and IGRF 75 model components. The electron spectrometer has an aperture mounted in the spin plane of the satellite and measures electrons with energies between 80 eV and 17 KeV in 32 energy channels, compiling a full spectrum every second.

Observations

Electric and magnetic field signatures of a quiet-time, discrete arc in the auroral oval as simultaneously observed by S3-2 and a DMSP optical imager have been discussed by Burke et al., (1980). The observations were consistent with many previously reported rocket, and radar measurements over discrete arcs in the auroral oval. However, the simultaneous electric and magnetic signatures of polar cap arcs and small-scale "electrostatic shock" structures have not been reported previously. To fill this deficiency, we present data taken during three, high-latitude passes of S3-2 when the orbital plane was close to the dawn-dusk meridian. The first pass is across the winter polar-cap during a period of northward IMF. Characteristics of three small-scale structures that are, most probably associated with discrete polar cap arcs are examined in detail. The second and third orbits were consecutive passes during a substorm period when small-scale structures, similar to those reported by Mozer et al., (1977) were observed at S3-3 altitudes. It is unusual to find these structures at altitudes <1500 km. Thus, these events provide a unique opportunity to study the spatial distribution of Birkeland currents in the vicinity of electrostatic shock structures.

A. Discrete Arcs in the Polar Cap

In this subsection we concentrate on measurements taken during a single north polar pass of S3-2 on 8 December 1976. The IMF had an hourly averaged Z component of 7.1 nT. Several hours before this pass, it had been strongly southward. Figure 1 gives plots of E_x, ΔB_y and the directional flux of energetic electrons and their pitch angles during Rev 5231 as functions of UT, invariant latitude (Λ), magnetic local time (MLT) and altitude. During Rev 5231, the satellite passed within 1° of the magnetic pole along the MLT dawn-dusk meridian. Although imagery from DMSP is not available for this time, the orbital and IMF conditions make it highly likely that we will detect signatures of sun-aligned arcs in the polar cap. Values of E_x, which are very complex in structure are 5 second averages obtained by fitting the measured signal to a sine function. Data interruptions marked by the letter C indicate periods when the electric field experiment went through calibration sequences. Positive values of E_x in the northern hemisphere correspond to a dusk to dawn (sunward convection) component of the electric field. Regions of negative slope in ΔB_y have field-aligned currents out of the ionosphere. Eight narrow regions of current out of the ionosphere marked in Figure 1 are discussed in detail below. Designation of the boundary between the auroral oval and the polar cap by means of large scale E_x and ΔB_y signatures during periods of a northward IMF is ambiguous (Smiddy et al., 1980). In the evening sector the electric field has several reversals. ΔB_y signatures for a simple region 1/region 2 large-scale auroral current system are frequently absent in the winter auroral zone during periods

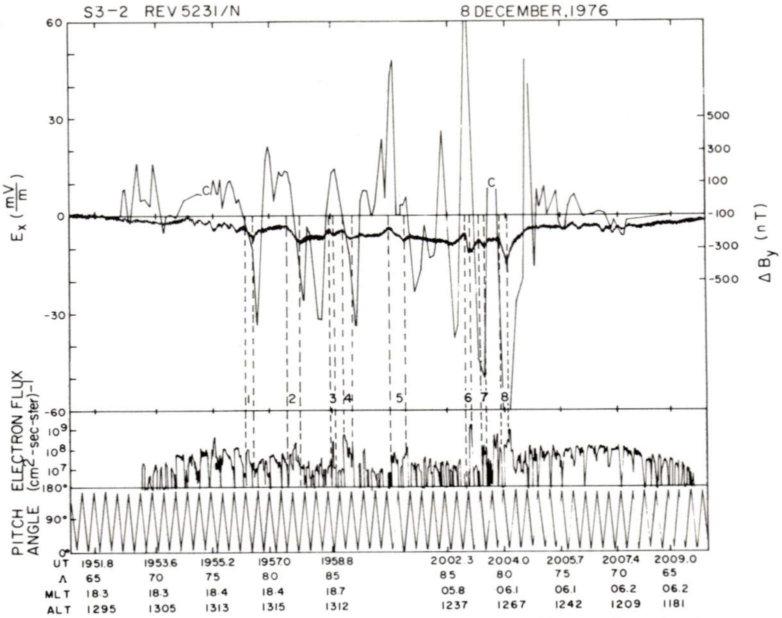

Figure 1. Electric field, magnetic field components, the directional electron flux and pitch angle as functions of UT, invariant latitude, magnetic local time and altitude from S3-2 Rev 5231 N. Eight events marked by field aligned currents out of the ionosphere are denoted by dashed lines. The hourly averaged values of the IMF X, Y, and Z solar magnetospheric components were -3.7, 4.4 and 7.1 nT.

of northward IMF in S3-2 data. We note that significant reversals in E_x occur at 1956:06 UT ($\Lambda = 77°$) and 1957:48 UT ($\Lambda = 82°$). In both instances (events 1 and 2) the E_x reversal are accompanied by currents out of the ionosphere and locally enhanced electron fluxes. Equatorward of the event 2 the electron fluxes range between 5×10^7 and 3×10^8 $(cm^2\text{-sec-ster})^{-1}$ and are fairly isotropic over the downcoming hemisphere. Poleward of event 2, the electron flux decreases abruptly to values near the lower limit of sensitivity for the detector, typical of polar-rain situations. The morning side E_x reversal at 2003:30 UT ($\Lambda = 81.6$) also marks a transition in electron fluxes from polar rain-like to more intense and nearly isotropic over the downcoming hemisphere. For this reason we conservatively set the evening and morning side polar-cap boundaries at $\Lambda = 82°$ and $81.6°$, respectively. Thus, events 3 through 7 which are marked by enhanced electron fluxes and currents out of the ionosphere lie entirely within the polar cap. The 5 second averaged values of E_x show that all of the events with the possible exceptions of events 3 and 6 lie in a region of negative slope in E_x. We now consider some of these events on smaller time scales.

In Figures 2, we have plotted the potential measured between the ends of the spinning dipole between 1958:00 and 2000:30 UT (events 3 and 4) and between 2002:30 and 2005:00 UT (events 6,7, and 8), respectively. ΔB_y, the electron flux and electron pitch angles are provided for reference. Measurements from the electric field experiment have had potential differences induced by satellite motion across magnetic field lines and a constant dc bias removed. The dashed sine wave is a reference signal that gives the potential that would be measured in the presence of a constant electric field of 12 mV/m that is directed from dusk to dawn and has no component along $\underline{B}$. Thus, when the reference and measured signals are in phase (180° out of phase) the ambient electric field has a dusk to dawn (dawn to dusk) component. The + and - signs at the top of the figure designate periods when the measured and reference signals are in and out of phase, respectively. We note that, despite rapid variations in the ambient electric field, the measured and reference signals pass through zero at the same times. This indicates that there is no measureable electric field component along B at these times. It is possible that components of E along $\underline{B}$ exist during periods away from nodal points of the reference signal.

Twelve seconds prior to the beginning of event 3 the measured signal changes from out of phase to in phase with the reference signal. During event 3, the potential measured along the dipole decreased to zero then returned to a value in phase with the reference signal. At the time of event 6 the measured value charged from a high value in phase with the reference signal to a small value out of phase, i.e. event 6 is in a

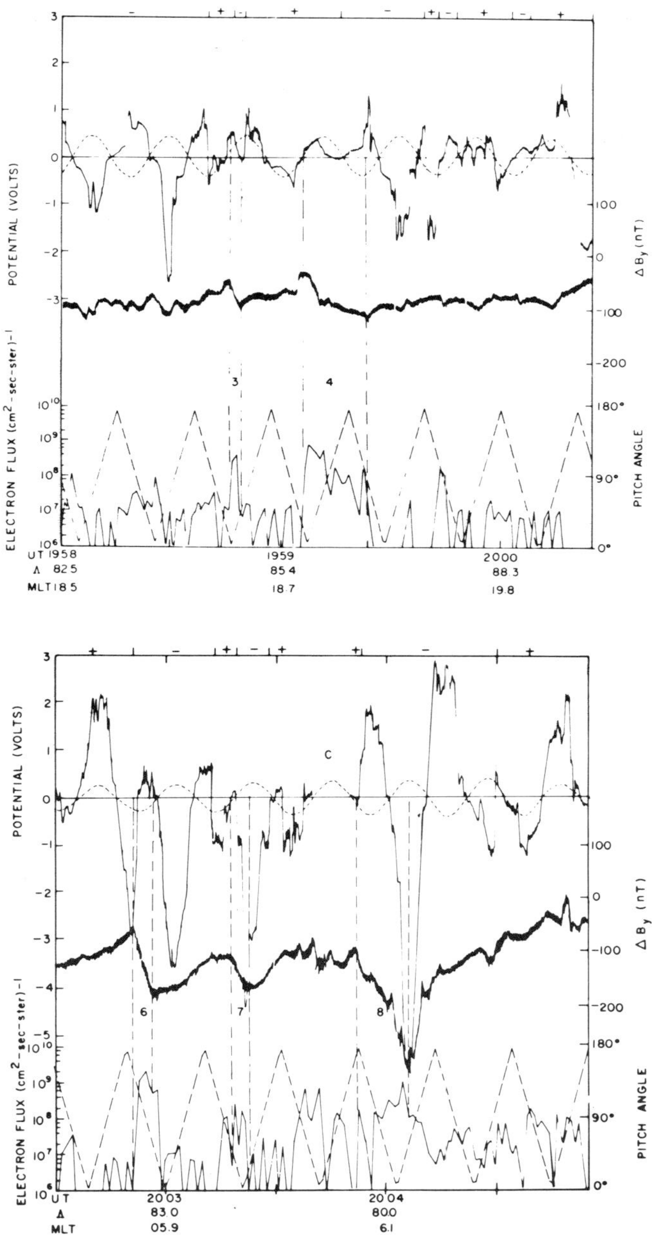

Figure 2. Potential measured by spinning dipole (solid line) and the potential that would be measured in the presence of a constant field of 12 mV/m directed from dusk to dawn (dashed line) from periods near events 3 and 4, (top) and events 6, 7. and 8 (bottom).

region of decreasing E_x. Plots of 1 second average values of E_x based on the potentials measured near the time of events 3, 4, 6, and 7 (Figure 3) clearly show that the regions of upward Birkeland currents and enhanced electron fluxes are also regions of decreasing E_x.

At the time of event 3, the integral flux inreased abruptly from background to 3 x 10^8 $(cm^2\text{-sec-ster})^{-1}$ for two seconds, then returned to back-ground. The peak directional differential flux was between 1 and 2 x 10^9 $(cm^2\text{-sec-ster-keV})^{-1}$ and appeared in < 1 keV energy channels. Between 1958:46 and 1958:49 UT, E_x and ΔB_y decreased by 50 mV/m and 47 nT, respectively. If we assume that most of the electric and magnetic field variations occur along the satellite trajectory, then we can use the infinite sheet approximation to Maxwell's equations to calculate the excess space charge density (ρ) and the field aligned current density ($j_\parallel$) associated with event 3. In this approximation

$$\rho\ (\text{charges/cm}^3) \approx 7.9\text{x}10^{-6}\ \frac{\partial\ E_x\ (\text{mV/m})}{\partial t\ (\text{sec})} \qquad (1)$$

and

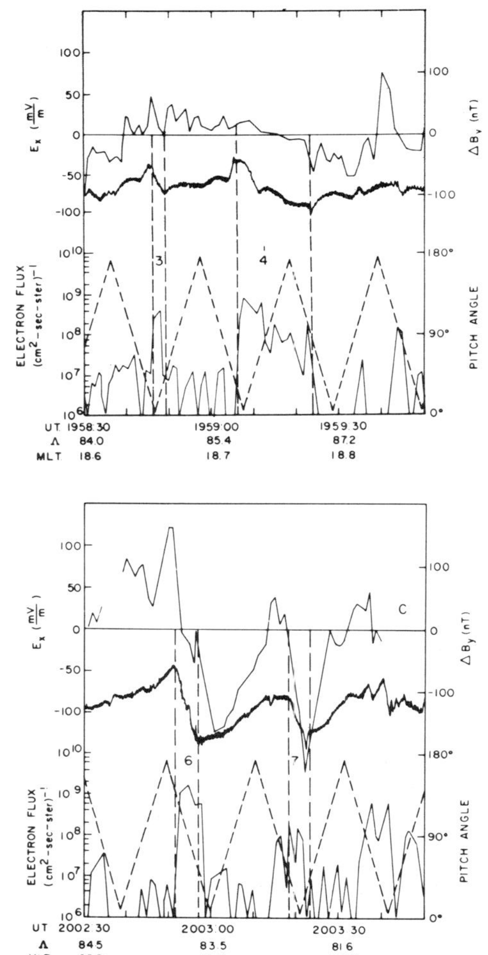

Figure 3. One second averages of of E_x deduced from potential measurements in Figure (2) from period of events 3 and 4, (top) and events 6 and 7 (bottom).

$$j_{\parallel}\ (\mu\ A/m^2) \approx 0.113\ \frac{\partial \Delta\ B_y\ (nT)}{\partial\ t\ (sec)} \qquad (2)$$

where 1 charge = 1.6×10^{-19} Coulomb. The net charge density in event 3 is 1.3×10^{-4} electrons/cm^3. A thermal electron probe on S3-2 indicates that at satellite altitude (~ 1300 km) the thermal plasma density ranged between 0.3 and 3×10^3cm^{-3}. Thus, the excess charge is about 1 part in 3 to 30 million. The current density calculated from equation (2) is 1.8 μ A/m^2. This current density corresponds to an electron flux of ~ 10^9 (cm^2-sec)$^{-1}$. For this to be consistent with a measured directional flux of 3×10^8 (cm^{2}sec-ster)$^{-1}$ the electrons associated with event 3 must be nearly isotropic in their distribution over the downcoming hemisphere. Both spectra which had peak fluxes in the 200 and 500 eV channels, were taken while the detector was looking close to the magnetic field direction. The energy flux associated with event 3 is a few tenths of an erg/cm^2-sec-ster, and is only sufficient for producing a subvisual arc.

During event 4 Δ B_y decreased uniformly by 47 nT between 1959:07 and 1959 :09 UT. A current density of 1.4 μ A/m^2 is consistent with the uniform flux of 4×10^8 (cm^2-sec-ster)$^{-1}$ if the electron flux is weakly field aligned. An examination of individual spectra (not shown) indicate that the event has the characteristics of an inverted-V structure. The peak flux appeared in the 2 keV channel at 1959:09 and in lesser energy channels at earlier and later times. A maximum directional differential flux of 10^9 (cm^2-sec-ster-keV)$^{-1}$ was measured measured in the 1 keV channel when the detector was looking closest to $\underline{B}$.

Between 2002:52 and :57 UT (event 6) E_x and Δ B_y decreased by 160 mV/m and 127 nT indicating that $\rho = 2.5 \times 10^{-4}$ electrons/cm^3 and $j_{\parallel} = 2.8$ μ A/m^2. The latter value is consistent with measured electron flux of 1.2×10^9 (cm^2-sec -ster)$^{-1}$. Spectra taken at this time show strongly peaked fluxes in the 1 keV channel. Most of the spectra were measured while the detector was looking up, but at relatively large angles to $\underline{B}$. When the spectra were converted to distribution functions primary and secondary populations were clearly distinguishable (Burch et al., 1979).
The primary electrons had temperatures of several hundred eV and had been accelerated though potential drops of ~ 1 kV. The energy fluxes associated with event 6 ranged between 2 and 3 ergs/cm^2-sec-ster. Thus, event 6 was probably associated with a visible, polar-cap arc.

B. Field-Aligned Currents near two Small-Scale Shock-like Structures

In this subsection, we consider the morphology of two "electrostatic shock" like structures. Events of this type are rarely detected at S3-2 altitudes. They were observed in widely separated places during a substorm period on 19 September 1976. Attention was first drawn to these events by their magnetic rather than electric field signatures. Figure 4 contains plots of Δ B_y measured on the dawn and dusk side, southern hemisphere auroral oval passes of Rev 4079A and 4079B. The data are plotted as functions of UT (in seconds of the day and in hours-minutes), invariant latitude, magnetic local time and altitude. The first event is at 0953:06 UT (Λ = 68.8°, MLT=03.4), and appears as a 100 nT excursion of Δ B_y that lasts for 4 seconds. The second event is a 300 nT excursion of Δ B_y beginning at 1148:16 UT (Λ = 66.9° MLT = 19.5) and lasting for 7 seconds. The altitudes of observation were 1050 and 1350 km, respectively. The Birkeland current densities associated with these Δ B_y excursions greatly exceed those normally found near auroral arcs.

A detailed analysis of the events requires that the individual electric and magnetic field measurements across the events be examined. Rapid fluctuations (not shown) are found in the potential differences measured across both the spinning and axial dipoles. The fluctuations are similar to those measured at higher altitudes near small-scale electrostatic structures (Mozer et al., 1977). The satellite was sufficiently high in altitude that variations in the axial dipole measurements may be attributed to changing E_y rather than to chemically induced changing dc offsets on the active elements of the electric

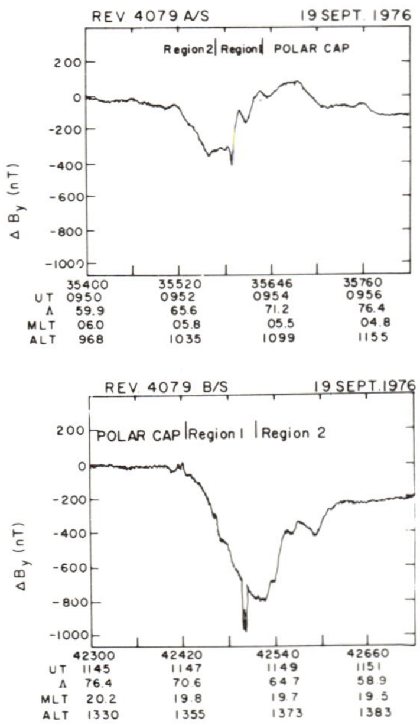

Figure 4. Measurements of Δ B_y for S3-2 pass over the dawn-side, southernhemisphere, auroral oval during Rev. 4079A/S, (top) and Rev 4079B/S (bottom). Rough estimates of region 1, region 2 and polar cap positions are provided for reference.

field sensors. Values of E_y were calculated by subtracting the electric field component due to satellite motion across magnetic field lines. There may still remain a small, unknown dc bias in the measurements. However, relative variations in E_y are accurate. The electric field components were transformed from satellite to geomagnetic north-south (E_{NS}) and east-west (E_{EW}) coordinates by assuming that $\underline{E} \cdot \underline{B} \approx 0$ everywhere. The three components of $\underline{\Delta B}$ were also computed in geomagnetic coordinates. For the remainder of this subsection we consider the highest possible time resolution for S3-2 measurements of $\underline{E}$ and $\underline{\Delta B}$ for the dawn and dusk side small-scale structures.

(i) Dawnside Structures

Figure 5 is a plot of four and one half seconds of electric field measurements (geomagnetic coordinates) starting at 35586.2 (0953:06.2) UT. Current densities derived from the slope of ΔB_{EW} are represented in histogram form with positive currents out of the ionosphere. ΔB_{UD} and ΔB_{NS} show no systematic variations across the region of interest. In the quarter of a second after 35586.9 UT ΔB_{EW} increased from 350 to 450 nT. Over the next 3.5 seconds it returned to a nearly constant value of 300 nT. The most intense current density is 45 $\mu A/m^2$ directed out of the ionosphere. The return current is more dispersed over the event. The latitudinal thickness of the upward current sheet projected to an altitude of 100 km is ~ 1.5 km, a typical width of optical arcs when viewed from the ground in the direction of magnetic zenith.

Vector representations in the bottom panel show that the electric field was originally oriented to the west of north and remained that way into the intense current sheet. Between 35586.2 and 35587.0 UT E_{EW} had a nearly constant westward value of ~ 20 mV/m. During the same interval the northward component increased from 50 to 200 mV/m. In the dawnside, southern hemisphere an increasing northward electrostatic field component in satellite coordinates corresponds to a region of negative electric field divergence. Thus, this upward current sheet is also a region of negative space charge. After 35587.2 the electric field rotated through east to southeast then back through east to northwest. It is important to note that E_{EW} and E_{NS} underwent major reversals in a region where ΔB_y was essentially constant.

Because a full electron spectrum is compiled in 1 second and the current sheet was crossed in one quarter of a second, it is not possible to identify the current carriers. An examination of electron spectrometer measurements shows an intense burst of 3keV electrons at pitch angles of

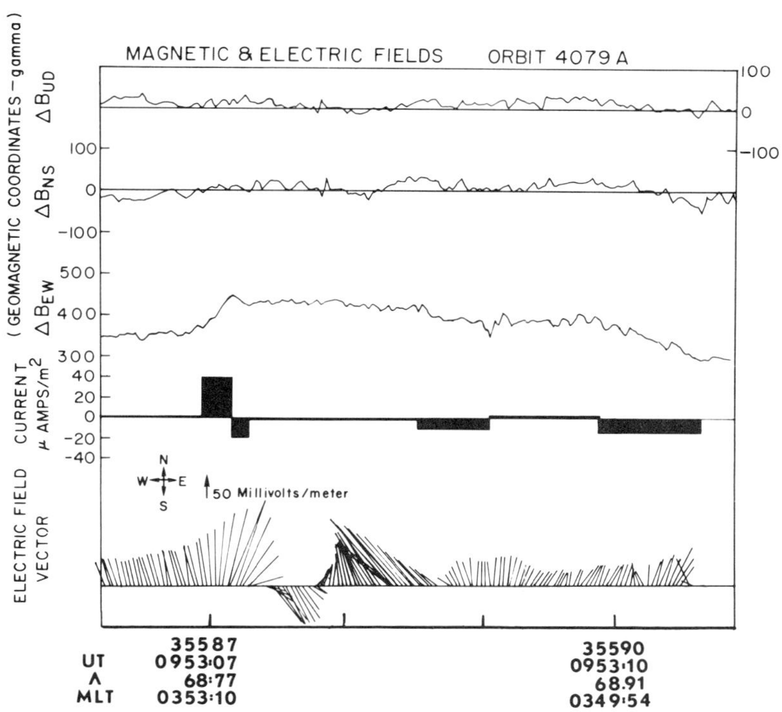

Figure 5. The up-down (UD), north-south (NS) and east-west components of Δ B in geomagnetic coordinates across dawn side event. Values of $j_{\parallel}$ are positive for currents out of the ionosphere. Vector representations of the electric field components are also in geomagnetic coordinates.

~ 90° in the upward current sheet and background counts elsewhere.

(ii) Duskside Structure

The electric and magnetic field components plotted in Figure 6 show that the duskside structure had the more complex internal structure. Near the poleward boundary of the structure, E_{NS} decreased from 75 to ~ 0 mV/m between 42496.8 and 42497.6 UT while E_{EW} maintained a nearly constant, low value. After this time, both components increased in magnitude then reversed directions. The magnitude of E after the reversal exceeded 300 mV/m. The components then returned to their original values. After 42498.0 UT E_{NS} underwent large variations in magnitude, but generally remained southerly in orientation. E_{EW} underwent large variations in both magnitude and direction before returning to a near zero value at 42504.8 UT.

Magnetic field variations across the evening side structure show patterns that on a large scale are similar to those of the morning side structure. In the quarter of a second prior to 42497.8 UT, ΔB_{EW} decreased by 300 nT, corresponding to an average current density of 135 $\mu A/m^2$. ΔB_{EW} subsequently recoverd to its preevent value during the four seconds after 42500.8 UT. Again, the electric field underwent a rotation, this time of 360°, at the interior edge of the intense upward current sheet. However, the evening side event is much more complex in its internal structure. This is reflected in the "multiple current sheets" found between 42497.8 and 49500.8 UT. The intensities of these current sheets were calculated from the slope of ΔB_{EW} i.e. assuming that the satellite crossed a longitudinally extensive structure at normal incidence. The presence of significant variations in ΔB_{NS} as well as ΔB_{EW} indicates that the "multiple current sheet" level of complexity is itself too simple. If ΔB_{EW} and ΔB_{NS} varied in the same sense across the structure, then the measurements would indicate that the satellite trajectory crossed a series of current sheets at an angle other than 90°. Indeed similar variations of the two components near the the poleward edge of the structure indicate that the arc was crossed at an angle of 9° away from normal incidence. The anticorrelated variations in ΔB_{NS} and ΔB_{EW} near 42500.8 UT suggest that S3-2 passed close to a line current embedded within an extended sheet current.

Another level of complexity is found in the measurements taken between 42498.8 and 42499.8 UT. Variations in $\underline{\Delta B}$ coincide with variations in the magnitude, but not direction, of $\underline{E}$. The measurements are consistent with the satellite having detected large amplitude Alfven waves. The relationship between the amplitudes of the electric and magnetic fields of Alfven waves comes directly from Maxwell's equation $\delta E = V_A \; \delta B$. The Alfven speed, $V_A = B_o / \sqrt{\mu_o \rho}$, where B_o is the main field of the earth, and ρ the mass density of the plasma. In the centered dipole approximation, at an altitude of 1350 km along the Λ = 67° field line $B_o = 3.3 \times 10^{-5}$nT. For a dominant ion species of O^+ and a number density between 10^{10} and 10^9 m^{-3}, V_A is between 1.8 and 5.7×10^6 m/sec. Thus, for variations in δB of 75 nT, as seen in the period of interest, δE should vary between 135 and 420 mV/m. The variation in $\underline{E}$ during this time is ~ 300 mV/m. Ten seconds prior to encountering the structure a planar thermal ion sensor on S3-2 measured a plasma density of 3×10^9 m^{-3}. Note, however, that an Alfven wave can not explain the variations in ΔB_{EW} near 42497.8 UT. A 300 nT magnetic field amplitude would require a δE between 540 and 1700 mV/m mV/m, far greater than the actually measured electric field. Finally, we note that even if the variations of $\underline{E}$ and $\underline{\Delta B}$ between 42498.8 and 42499.8 UT are due to Alfven waves, it is not possible to specify directions of propagation uniquely. Since $\underline{E}$ varied in magnitude, but not direction, $\underline{\delta E}$ lies approximately in the NE-SW direction. Unfortunately, establishing an "unperturbed baseline" from which we measure δB demands an unacceptably high level of subjectivity.

Limit of space do not allow a detailed presentation of electron measurements at this times. Because of the spatial aliasing the intense upward current carriers cannot be identified. A sharp "turn on" was observed 3 and 5 keV energy channels just prior to entering the upward current sheet. Inside and just interior to the current sheet a large flux of downcoming, low-energy (<1keV) electrons was detected suggesting the presence of electrostatic barrier at an altitude > 1350 km. A spectrum compiled in the second starting at 42498.8 UT (the region of possible Alfven waves) has a broad thermal shape between 0.1 and 1 keV. The absence of spike-like fluxes at this time is more consistent with an Alfven wave, rather than multiple current sheet interpretation of the magnetometer data. In the region of the return current no downcoming electrons were detected.

Summary and Discussion

Observations presented in the previous section provide the first simultaneous measurements of small-scale electric and magnetic field variations and electron fluxes in the vicinity of discrete arcs in the polar cap and "electrostatic shocks" in the auroral oval.

During the winter polar cap pass studied here, the IMF had a strong northward component; the satellite trajectory was close to the dawn-dusk meridian and passed within 1° of the magnetic pole. These are ideal conditions for detecting the characteristics of sun-aligned, polar-cap arcs. The electric field had a highly irregular dawn-dusk component. Five field-aligned current events were identified in the polar cap which are believed to be associated with discrete arcs embedded in regions of negative space charge. Local charge imbalances were in the range one part

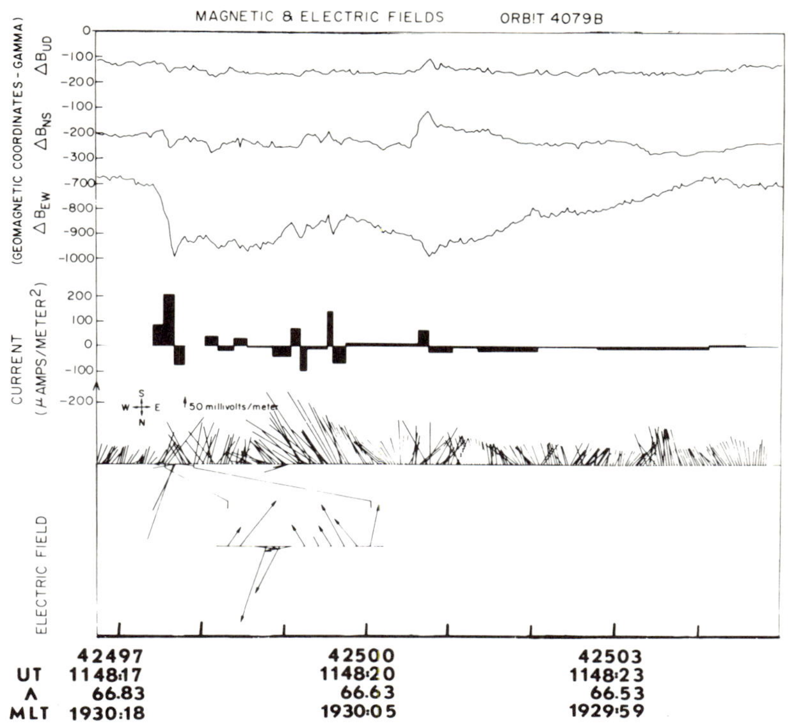

Figure 6. Magnetic field deflections, current intensities and electric field vectors for evening side structure in the same format as for Figure 5.

in three to thirty million. During several of these field-aligned current events, the electron spectrometer was looking upwards allowing an inspection of individual spectra. Comparisons of spectra with measured Birkeland current densities and and with each other as a function of pitch angle suggest that the primary electrons are nearly isotropic or weakly field-aligned over the downcoming hemisphere (Burch et al., 1979). Energy fluxes derived from the spectra ranged from a few tenths of an erg to 3 erg per (cm^2-sec-ster). At this upper (lower) limit of energy flux, a discrete arc would be visible (subvisual). Finally, we note that despite the fact that the primary electrons were accelerated by ~ 1 keV and there were many reversals in E_x, there was no direct evidence of field-aligned potential drops at the satellite altitude.

Two small-scale, field-aligned current events, the first near dawn the second near dusk, were observed at auroral latitudes during a substorm period. In the dawn side (dusk side) event the transverse magnetic field component decreased by 100 nT (300 nT) and recovered in 4(7) seconds while the satellite was at an altitude of 1050 (1350) km. In both cases upward Birkeland current sheets, which had average intensities of 45 and 135 $\mu A/m^2$, were traversed in 1/4 sec. These upward current sheets were also embedded in regions where $\partial E_x/\partial x < 0$. At the interior edges of the upward current sheets the electric field underwent rotations. These events would be identified in data from a satellite equipped only with electric field probes as "electrostatic shocks". It appears also that $j_\parallel \approx 0$ in the region where the electric field components reversed directions. The return currents were more spread out having maximum values of $\sim 10\ \mu A/m^2$. Interior to the dawn side event, no field-aligned currents were detected. Much more complex electric and magnetic field variations were found within the dusk side structure. The variations are consistent with either the presence of Alfven waves or multiple current sheets. Because of spatial aliasing it is not possible to make a positive identification of the average energy of the main current carriers in the intense upward current sheets. Available evidence suggests that they are at least in the several keV range.

A detailed discussion of auroral theory is well beyond the scope of this essentially observational report. However, the electric and magnetic field measurements presented here do allow some comments concerning the generator of auroral arcs. Currently, there are two major theoretical concepts for auroral generation which we call the "ionospheric generator" (Miura and Sato, 1980 and references therein) and the magnetospheric generator, (Lyons, 1981 and references therein) models.

Both models are highly developed and each, granting its initial assumption predicts the growth of arc-like structures. There does not appear to be anything specific to these theories which makes them inapplicable to cases of polar-cap arcs. These theories have diametrically opposite predictions concerning the relationship between the electric fields and Birkeland currents above arcs. The "ionospheric generator" model (cf Figure 1 of Miura and Sato, 1980) requires that arcs develop in regions of positive space charge ($\nabla \cdot E > 0$). The "magnetospheric generator" theory makes the opposite prediction, i.e. arcs should appear in regions of negative space charge ($\nabla \cdot E > 0$). The data presented here all show that the upward Birkeland current sheets are embedded in regions of negative space charge.

A further requirement of the ionospheric (magnetospheric) generator model is that the Poynting flux ($\underline{\delta E \times \delta \Delta B}$) be directed out of (into) the ionosphere (Mallinckrodt and Carlson, 1978) From the cases presented in Figure 3, it is clear that the Poynting flux is directed into the ionosphere where presumably, it is dissipated as heat energy.

Finally, from the intensity of the upward current sheets associated with the "electrostatic shock" events, it can be argued that the generator must be contained in the magnetosphere. Mallinckrodt and Carlson (1978) showed that in the wave model there is an upper limit of 0.1 A/m that can be driven by an ionospheric generator. Deflections of 100 and 300 nT correspond to integrated currents of 0.08 and 0.24 A/m, respectively. While the dawn-side current sheet is within the upper limit, for the ionospheric generator, the dusk side sheet exceeds the limit predicted by the Alfven wave model by more than a factor of two. The calculated upper limit for a magnetospheric generator is 1 A/m. Note that where the magnetospheric generator can maintain very large current densities using hot magnetospheric electrons as carriers, there seems to be an upper limit of 10 to 15 μ A/m^2 when cold, ionospheric electrons are the carriers. Burke et al., (1980a) have found that in the altitude range 1000-1500 km, 10 μ A/m^2 approximates the critical level for the onset of ion-cyclotron turbulence (Kindel and Kennel, 1971) within the return current sheet.

Acknowledgement. The author wishes to thank R. C. Sagalyn, M. Smiddy, B. Shuman and R. P. Vancour of AFGL for their work in designing the instrumental package of S3-2. Special thanks is extended to D. A. Hardy of AFGL, F. J. Rich of Regis College and M. C. Kelley of Cornell University without whose long term help and cooperation this contribution would not be possible. The author also expresses gratitude to Arthur Gentile, and Karen Leccese of AFGL for drafting the figures and typing the text.

References

Burch, J.L., S.A. Fields and R.A. Heelis, Polar cap electron acceleration regions, J. Geophys. Res., 84, 5863, 1979.

Burke, W.J., D.A. Hardy, F.J. Rich, M.C. Kelley, M. Smiddy, B. Shuman, R.C. Sagalyn, R.P. Vancour, P.J.L. Wildman, and S.T. Lai, Electrodynamic structure of the late evening sector of the auroral zone, J. Geophys. Res., 85, 1179, 1980.

Kindel, J.M., and C.F. Kennel, Topside current instabilities, J. Geophys. Res., 76, 3065, 1971.

Lyons, L., Descrete auroras as the direct result of an inferred high-altitude generating potential, J. Geophys. Res., 86, 1, 1981.

Mallinckrodt, A.J., and C.W. Carlson, Relations between transverse electric fields and field-aligned currents, J. Geophys. Res., 83, 1426, 1978.

Miura, A., and T. Sato, Numerical simulation of global formation of auroral arcs, J. Geophys. Res., 85, 73, 1980.

Mozer, F.S., C.W. Carlson, M.K. Hudson, R.B. Torbert, B. Parady, J. Yetteau, and M.C. Kelley, Observations of paired electrostatic shocks in the polar magnetosphere, Phys. Rev. Lett., 38, 292, 1977.

Smiddy, M., W.J. Burke, M.C. Kelley, N.A. Saflekos, M.S. Gussenhoven, D.A. Hardy, and F.J. Rich, Effects of high-latitude conductivity on observed convection electric fields and Birkeland currents, J. Geophys. Res., 85, 6811, 1980.

COLD PLASMA BOUNDARIES AND AURORAL ARCS

C. E. McIlwain

Department of Physics, and Center for Astrophysics and Space Sciences
University of California, San Diego
La Jolla, California 92093

Abstract. Auroral arcs often extend for more than a thousand kilometers with little deviation of their relative position within the auroral oval. At high altitudes, the outer limits of the plasmasphere are usually marked by sharp decreases in the cold plasma densities. It is suggested that some auroral arcs follow the ionospheric trace of these boundary shells.

Why Thin Arcs?

Why do auroral displays often take the form of thin east-west oriented arcs? The answer to this question has long been sought for the insight that it should provide into the underlying processes responsible for creating auroras. The extention of arcs over thousands of kilometers (Akasofu, 1963) might be explained in the fashion suggested by Störmer (1955, p 299), i.e., that the arcs follow approximately constant magnetic latitudes and thus correspond to a constant magnetic selection process along their length. A modern version of this idea would include the effect of magnetospheric convection, but even this would not overcome a difficulty pointed out by Störmer: the arcs can be very thin, thus requiring very narrow velocity distributions in the parent particles.

Possibilities

Some auroral forms are likely caused by the precipitation of existing particle populations while they are transported by large scale magnetic fields and by perpendicular electric fields. It is now widely believed, however that most bright forms are associated with the acceleration of particles by parallel electric fields somewhere on the local magnetic field lines. It is not known how thin auroral arcs relate to the spatial distribution of parallel electric fields, so let us consider three possibilities:

1) Arcs reflect the presence of shells of extra wave activity which causes a localized increase in the precipitation of a broadly distributed particle population.

2) Arcs directly reflect the presence of parallel electric fields, and the absence of such fields on neighboring shells (or at least of the correct sign to accelerate electrons into the ionosphere).

3) Arcs reflect the overlap of a region with parallel electric fields and the boundary of an appropriate source of particles waiting to be fed into the acceleration region.

In the case of the first two possibilities, there would have to be something special about the arc's shell to explain that particular localization of waves or parallel electric fields. Any sharp change in conditions is a possible candidate. The only change in the magnetic field which might be sufficiently sharp is the transition between open and closed field lines. It is probable that some dayside arcs do correspond to this transition, i.e., that they are ionospheric traces of the magnetopause. On the nightside, it is generally believed that some other explanation must be sought. About the only other thing left, to sharply mark a particular shell, as special, is the spatial distribution of plasma.

Carpenter (1963) discovered a 'knee' in the radial profile of plasma density which he later (Carpenter, 1966) showed to be local time dependent. This dependence was summarized in his classic sketch of the plasmapause. In-situ measurements (Chappell et al. 1970a) confirmed this picture in the 3 to 5 earth radii region, but also revealed the existence of further structure at greater distances. They (Chappell et al. 1970b) developed a picture in which the outer regions are filled during quiet magnetic periods and peeled away during active periods.

Injection Boundaries

The determination of a dynamic three dimensional structure with measurements from a single spacecraft is, of course, impossible, but various statistical studies seem to lead to similar results. Vasyliumas (1968) showed that the inner edge of the plasma sheet approximately follows the auroral oval in the evening to midnight sector, and that it moves inward during active periods. McIlwain (1974) showed that fresh hot plasma appears up to a limiting 'injection boundary' during each substorm, and that the plasma inside this boundary is left relatively undisturbed. Mauk and McIlwain (1974) showed that the shape of this boundary follows the auroral oval in the evening to midnight sector in agreement with Vasyliunas (as it should be, because the observed 'fresh hot plasma' constitutes the inner edge of the plasma sheet).

Energy dependent magnetic gradient and curvature drifts cause the hot plasma particles to disperse, but the cooler plasma particles convect together. This means that each substorm leaves a 'cold plasma boundary' which remains intact until substorm activity increases, or the slow refilling and loss processes cause it to fade away. In this picture, the early observations of the plasmapause correspond to the cold plasma boundary as left by the last big substorm (which could have occurred days or even weeks before). The standard idea of the plasmapause occurring in the 3 to 5 earth radii region is thus correct only in the sense that that is where the largest sharp decrease in density is often found. It is not correct to think of this inner boundary as the one which plays an important role in determining the electric fields and currents associated with auroral activity. During periods of decreasing activity, the boundaries left by successive substorms may be left relatively undisturbed as the magnetosphere expands. The primary shielding of externally applied electric fields might then occur at one of the outer boundaries, rather than an old inner one.

Examination of energy-time spectrograms, such as those made from ATS-6 data, reveals that the coexistence of the residue of a number of injection boundaries is in fact quite common. The dispersion curves can be used to obtain a measure of the distance between the observing spacecraft and the cold plasma boundaries (McIlwain, 1974). Preliminary studies indicate that new injection boundaries tend to follow the cold plasma boundaries left by previous injections, but this needs to be verified by multiple spacecraft studies.

Density Gradients

Returning to the possibilities mentioned earlier, it seems possible that all three could be related to sharp gradients in the plasma density:

1) All of the conditions for sustained wave generation may be met only within those regions spanned by the plasma gradient. Kurth et al. (1979) report intense wave activity just outside the plasmapause, but it is not yet known how these waves may relate to auroral arcs.

2) Long-lived parallel electric fields may also require specific conditions only met within a boundary region, or they might also simply tend to stabilize along gradients.

3) The very concept, that externally imposed electric fields are shielded from the inner magnetosphere, implies an interaction region where there is a penetration of the field into a denser plasma. Even if the accelerating field is much smaller (due to partial shielding), the greater number of particles being accelerated might cause a brighter aurora along the boundary.

Plasma Removal and Replacement

The notion that auroral processes establish long-lived structures in the plasma distributions, which then determine where future auroras will occur, raises the question of how plasma is removed and replaced. Observations now show (Mauk and McIlwain, 1975; Baugher et al. 1980) that these processes are much more complicated than previously believed. In particular, the observed plasma distributions exhibit strong composition and pitch angle dependences as well as complex dependences upon time, radial distance, and local time.

The quest for an understanding of the interplay between plasma distributions and auroral processes is only in the beginning stages, but perhaps it will soon yield an answer to the auroral arc question.

Acknowledgements. The assistance of J. Quinn and C. Freeman is acknowledged. Partial funding was provided by NASA Grant NGL 05-005-007.

References

Akasofu, S. -I., The dynamical morphology of the aurora polaris, J. Geophys. Res. 68, 1667-1673, 1963.

Baugher, C. R., C. R. Chappell, J. L. Horwitz, E. G. Shelley, and D. T. Young, Initial thermal plasma observations from ISEE-1, Geophys. Res. (Letters), 7, 657-660, 1980.

Carpenter, D. L., Whistler evidence of a 'knee' in the magnetospheric ionisation density profile, J. Geophys. Res. 68, 1675-1682, 1963.

Carpenter, D. L., Whistler studies of the plasmapause in the magnetosphere, J. Geophys. Res. 71, 693-709, 1966.

Chappell, C. R., K. K. Harris, and G. W. Sharp, A Study of the influence of magnetic activity on the location of the plasmapause as measured by OGO 5, J. Geophys. Res. 75, 50-56, 1970.

Chappell, C. R., K. K. Harris, and G. W. Sharp, The morphology of the bulge region of the plasmasphere, J. Geophys. Res. 75, 3848-61, 1970.

Kurth, W. S., J. D. Craven, L. A. Frank, and D. A. Gurnett, J. Geophys. Res. 84, 4154-4164, 1979.

Mauk, B. H., and C. E. McIlwain, Correlation of Kp with the substorm-injected plasma boundary, J. Geophys. Res. 79, 3193-3196, 1974.

Mauk, B. H., and C. E. McIlwain, ATS-6 UCSD auroral particles experiment, IEEE Trans. J. AES-11, 6, 1125-1130, 1975.

McIlwain, C. E., Substorm injection boundaries, Magnetospheric Physics, 143-154. B. McCormac, ed., L. Reidel Pub. Co., Dordrecht, Holland, 1974.

Störmer, C., The Polar Aurora, Clarendon Press, Oxford, England, 1955.

THE ELECTRIC FIELD STRUCTURE OF AURORAL ARCS AS DETERMINED FROM BARIUM PLASMA INJECTION EXPERIMENTS

E. M. Wescott

Geophysical Institute, University of Alaska, Fairbanks, Alaska 99701

Abstract. Barium plasma injection experiments have revealed a number of features of electric fields in and near auroral forms extending from a few hundred to many thousands of km in altitude. There is evidence for V-type potential structures over some auroras, but not in others. For some auroral arcs, large E fields are found at ionospheric altitudes outside the arc but the E field inside the arc is near zero. In a few other auroras, most recently one investigated in an experiment conducted from Poker Flat on March 22, 1980, large, rapidly fluctuating E fields were detected by barium plasma near 600 km altitude. These E fields suggest that the motion of auroral rays can be an effect of low-altitude electric fields, or that V-type potential structures may be found at low altitudes.

Introduction

The use of barium as an ionospheric-magnetospheric tracer is analogous to injecting fluorescein dye into a body of water to measure the current. In the case of barium, the ions fluoresce in sunlight and can be observed from the ground if the solar angle is sufficiently low to reduce the sky brightness. The motion of the barium ions is controlled by their initial velocity, their collisions with the neutral and ionized components of the ambient atmosphere, the electric and magnetic fields, and the gravitational field.

The electric field component perpendicular to the magnetic field is deduced from the transverse velocity of the plasma (Haerendel et al., 1967). In all the experiments described in this paper $E_{\perp} = 1/c(B \times v_{\perp})$ because the releases were at sufficient altitude where collisional terms are negligible and the size of the releases were small enough not to significantly perturb the medium.

The early experiments created ionospheric and also magnetospheric barium clouds from rockets by burning a mixture of barium and copper oxide (barium thermite). The barium vapor had initial velocities ranging up to about 1 km/sec. A later technique, using a high-explosive shaped charge with a conical liner of barium metal, produces a jet of barium vapor with velocities up to 13 km/sec. If the jet is aimed nearly up along the magnetic field direction, the ions will travel out along the field line just like beads on a wire. Perpendicular fields can be detected by measuring the horizontal drift of the field-aligned streak, but parallel electric fields may also be deduced from the ion motion along the field. The fast ions have an initial kinetic energy of about 100 eV, and are quite sensitive to small parallel electric fields which may either retard or accelerate the streaming motion.

High Electric Fields Near and Perpendicular to Auroral Forms

In some of the earliest auroral zone barium thermite releases, electric fields were measured in the vicinity of auroras in the altitude range 200-320 km. In one classic situation electric fields on the order of 100 mV/m were found to be nearly perpendicular to discrete arcs [Wescott et al., 1969]. The close parallel of the barium ion cloud motion to the alignment of the arcs (during that event can be seen in Figure 1, which is derived from a series of four releases no a Nike-Tomahawk flight, Sept. 2, 1967, from Andøya, Norway. Electric field probe measurements on an instrumented rocket flown into the same display revealed that the electric field direction was the same on both sides of the arc, and the field magnitude was very low inside the arc (Aggson, 1969). The low perpendicular electric field on the magnetic field lines threading through discrete auroral arcs was also seen from barium thermite releases from Andøya, Norway, Sept. 20, 1968. During the interval of observing the electric drift of three ion clouds, discrete auroral forms moved back and forth across the location of the ion clouds. Figure 2 shows the total perpendicular electric field deduced from the barium ion motion and the intervals in black when the barium and auroras were on the same magnetic field line. With few exceptions the field was low above the auroral forms. In the case of the drift motions of ion clouds 1 and 2 (I1 and I2), the electric field was in the same direction on both sides of the arc.

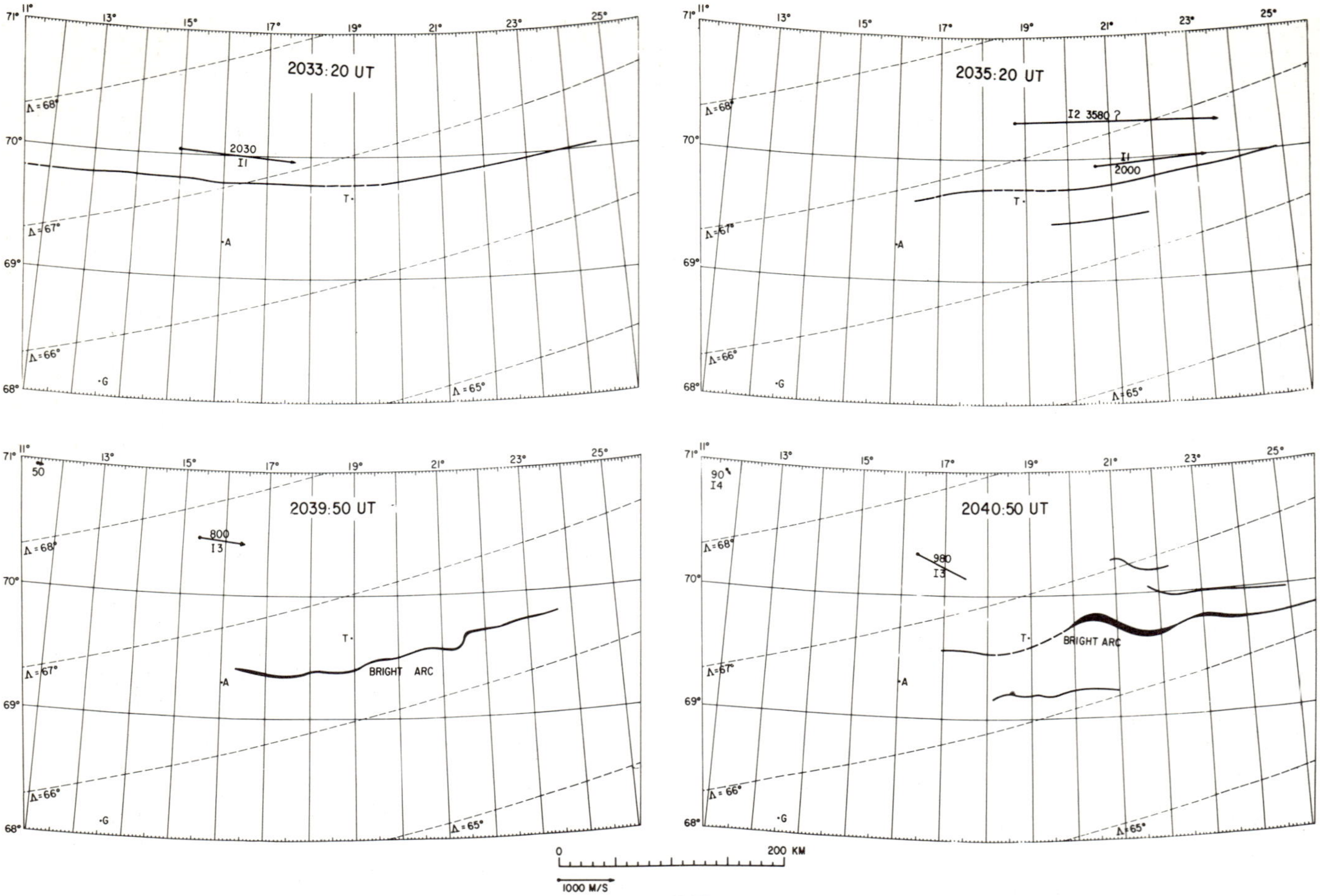

Fig. 1. Time sequence of typical auroral situations and ionized barium cloud positions and velocities (m/sec) of three barium clouds released from a rocket launched from Andøya, Norway, Sept. 2, 1967. Cloud positions have been projected down magnetic field lines to 100 km-datum altitude. E is 90° clockwise of v and approximately = v/20 (mV/m). From Wescott et al. (1969).

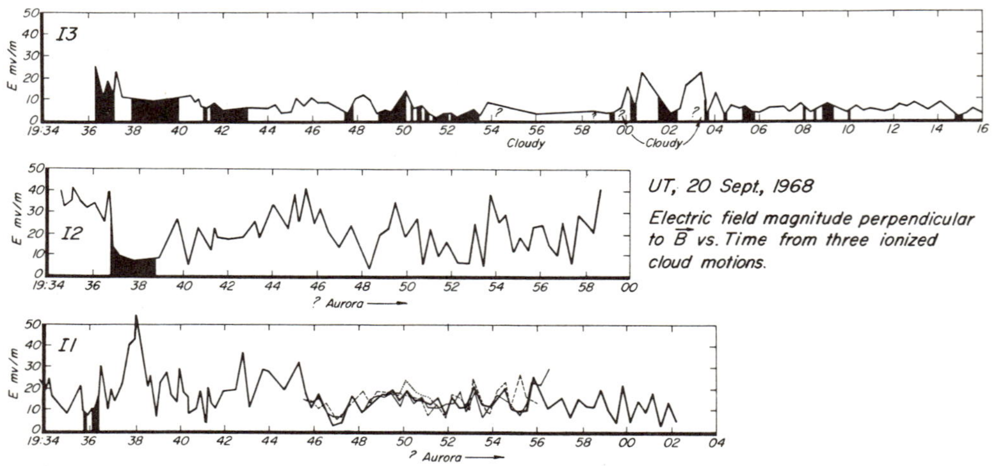

Fig. 2. Magnitude of $E_{\perp}$ vs. time from three barium ion clouds released from a rocket flown from Andøya, Norway, Sept. 20, 1968. The solid black areas are times when discrete auroral arcs were located on the same magnetic field lines as the barium cloud. Question marks indicate times when clouds obscured the auroral position or the barium was too far towards the horizon to locate the aurora.

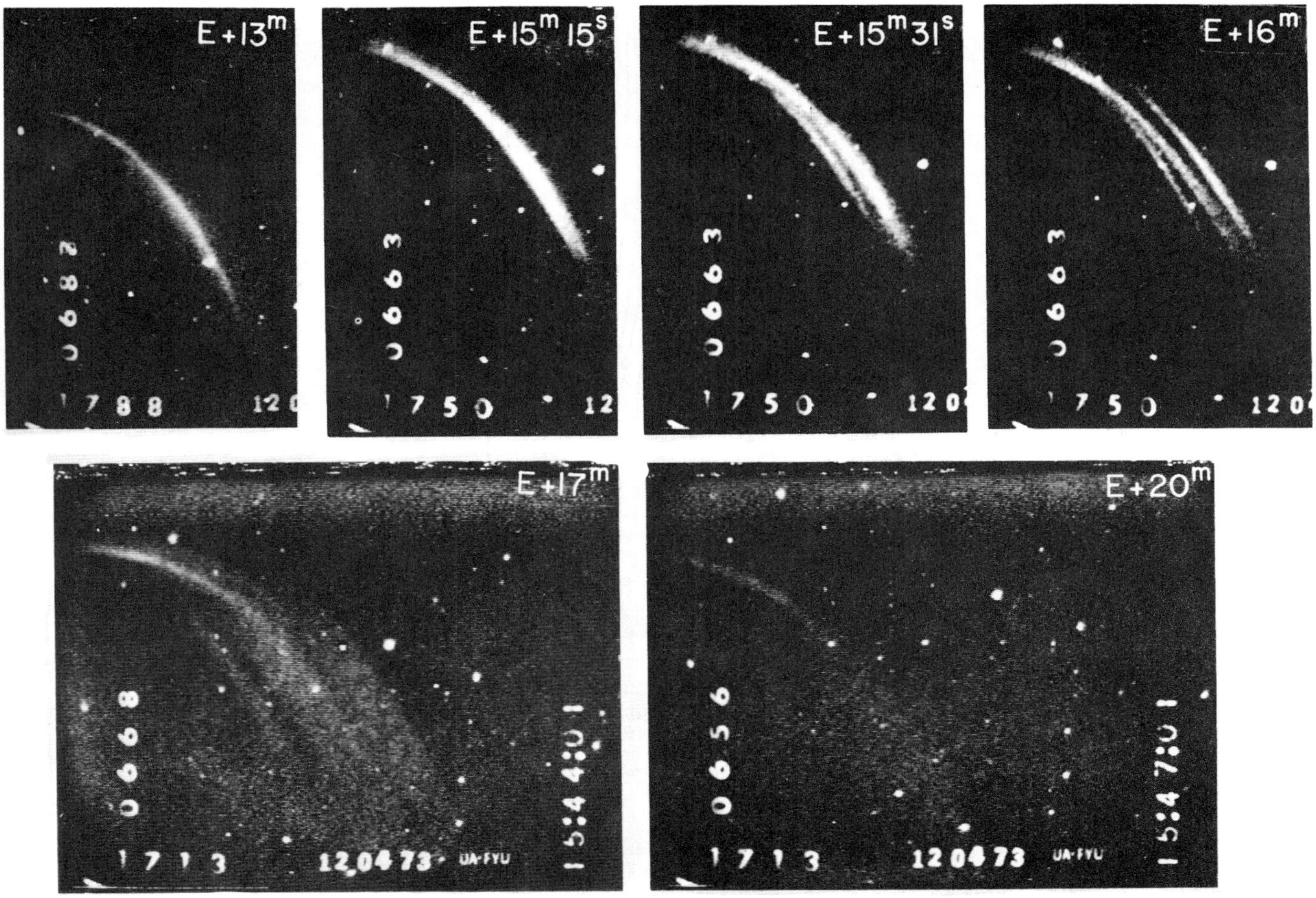

Fig. 3. Parts or complete TV frames of the anomalous brightening of the barium plasma flux tube of Skylab Beta, Dec. 4, 1973. In the upper left frame, at 13 minutes after injection, the pre-brightening streak can be seen as a single flux tube extending from 4500 km altitude at upper left to near 10,500 km altitude near center right. From E + 15^m15^s the streak brightened rapidly to 3x its original value as seen in the second frame. (This is due to rapid motion in the cloudsun direction, causing the absorbed light to be Doppler shifted out of the solar Fraunhoffer line at 4554 Å). The third frame shows the evolution of the first of several discrete streaks which separated at high velocity from the original streak above 5700 km altitude. The E + 16^m frame shows the lack of low altitude continuation of the separate streaks (cut-off altitude near 5700 km). The bottom two full TV frames show the continued rapid diffusion of the upper barium streaks to undetectability at E + 20^m. The last frame shows the original single streak truncated near 7600 km altitude. From Wescott et al. (1976).

We cannot conclude, however, that all discrete arcs have the same electric field configuration. Barium ion streaks extending to much higher altitudes have encountered large electric fields correlated with auroral forms, as will be discussed later. It is also clear that appreciable east-west electric fields are observed with some auroras. The fact that barium clouds may drift across auroral forms demonstrates that the electric field is not always perpendicular to the arc. Also the north south motion of auroral forms crossing barium clouds suggests that such motion is not due to ionospheric electric fields or to electric fields in the altitude range of the barium.

V-Shaped Electric Potential Structures Above Auroral Forms

The barium shaped charge technique for generating escape-velocity plasma traveling up auroral zone magnetic field lines is a powerful tool for investigating the electric field structure over auroral arcs. Due to the initial velocity disper-

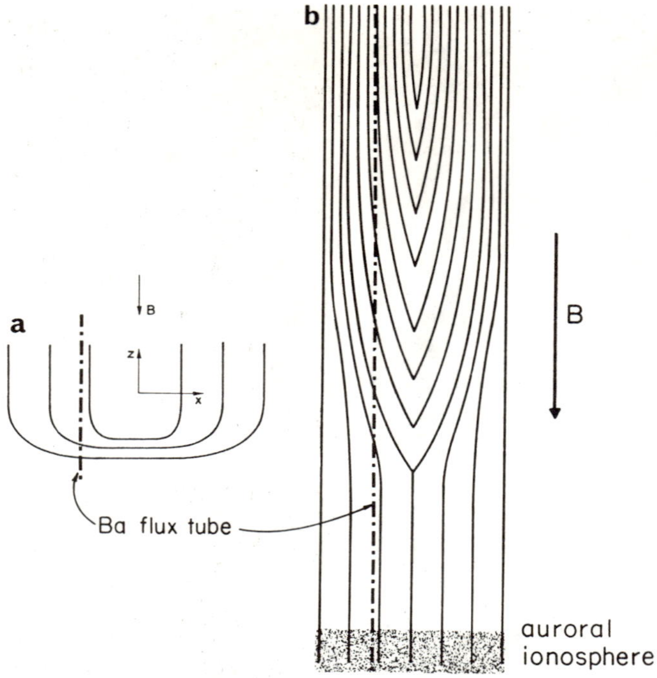

Fig. 4. Left, equipotentials from a double layer turning upwards around a field-aligned current sheet (after Block, 1972). The dashed line represents the position of the barium streak. Right, V-shaped equipotential model for an auroral arc. Lines represent equipotential surfaces; the potential gradient is away from the center axis of the figure. After Swift et al. (1976).

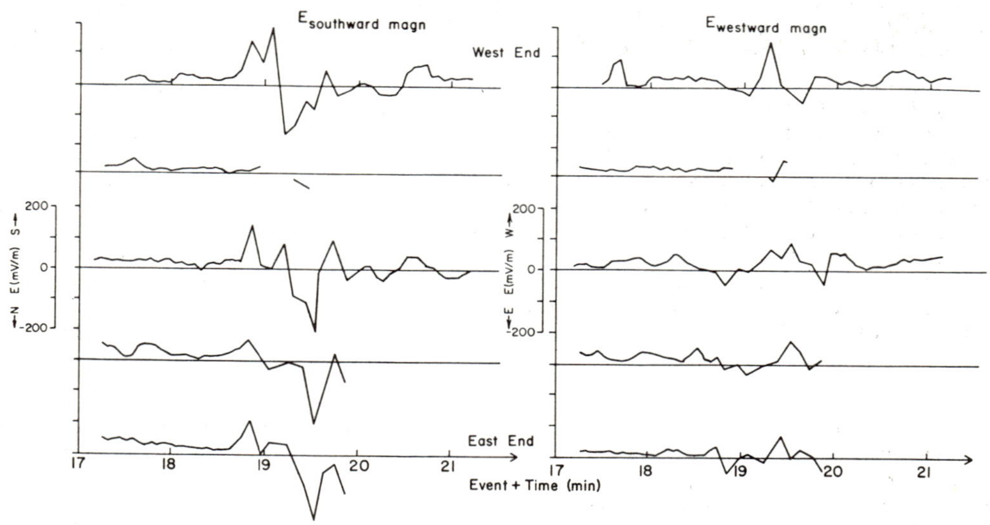

Fig. 5. Magnetic south and west components of the electric field deduced from drift motion of the five brightest barium ion streaks in the Oosik experiment near the time the aurora expanded poleward through the barium. Each trace represents one of the five barium jets. The five streaks were nearly aligned in a magnetic eastwest array. High north-south components are seen near the aurora; the reversal in polarity coincides with the crossing of the most poleward auroral arc. The field is directed inward toward the auroral arc, as could be expected from the V-shaped potential model shown in Figure 4 (after Wescott et al., 1975).

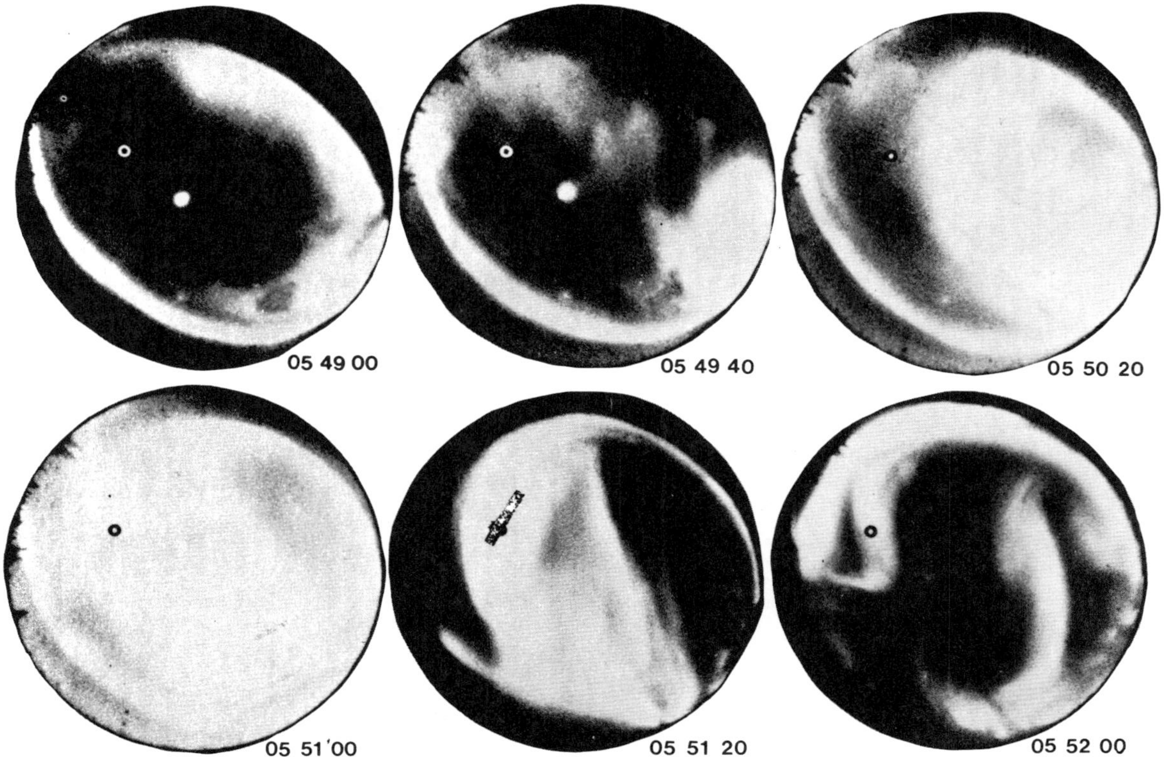

Fig. 6. All-sky camera frames from Ft. Yukon, Alaska, during the first three minutes of "The Duchess" experiment, Feb. 27, 1978. Frames are printed so that geographic north is up, east is to the right, etc. The near-zero velocity barium vapor and explosive debris can be seen in several frames, and the projected position down the magnetic field to 100 km is shown as an open circle. Nothing happened to the barium as the aurora passed through; the electric field was low.

sion, low-light-level TV systems may observe plasma extending thousands of km along a field line. The observed plasma streak can be compared with an appropriate magnetic field model that includes terms for the external sources, such as the Mead-Fairfield K_p dependent terms (Mead and Fairfield, 1973).

Two important factors may cause the observed streak to depart from the model: parallel Birkeland currents which distort the local magnetic field lines, and anomalous electric potential structures which cause the plasma to drift away from field alignment. The purpose of many shaped charge injection rocket flights into auroras has been to search for anomalous electric potentials. In some situations it is not possible to differentiate between distortion caused by parallel currents and that caused by potential structures, particularly in cases where only a short segment of flux tube without much curvature is observed.

However, in several cases unequivocal evidence of V-shaped electric potential structures was found from observation of extended plasma streaks. Figure 3 shows a sequence of TV pictures of a shaped charge injected plasma, streak "Skylab Beta" of December 4, 1973 (Wescott et al., 1976). The streak at E (event) plus 13 minutes extends from 4500 km in the upper left to near 10,500 km altitude near center right. In the following frames the streak can be seen to brighten due to Doppler shifting o·t of the Fraunhofer line, split into multiple streaks above 5700 km altitude (the shear zone), and eventually convect away rapidly and diffuse until at the end only the single flux tube originally below the shear zone is left. The cause of the flux tube splitting above the shear zone is not known, but the subsequent convection is due to the large perpendicular E field and gradients above the shear zone. The perpendicular field below the shear zone is small, so the barium flux tube is unaffected. Figure 4 shows two possible potential structures which could explain the phenomenon.

A second example of a V-shaped potential structure over a discrete aurora was observed in the "Oosik" experiment of March 7, 1972 (Wescott et al., 1975). A barium shaped charge was launched on a rocket from Poker Flat, Alaska, to the north of existing auroral forms. For nearly 18 minutes the barium drifted southward and

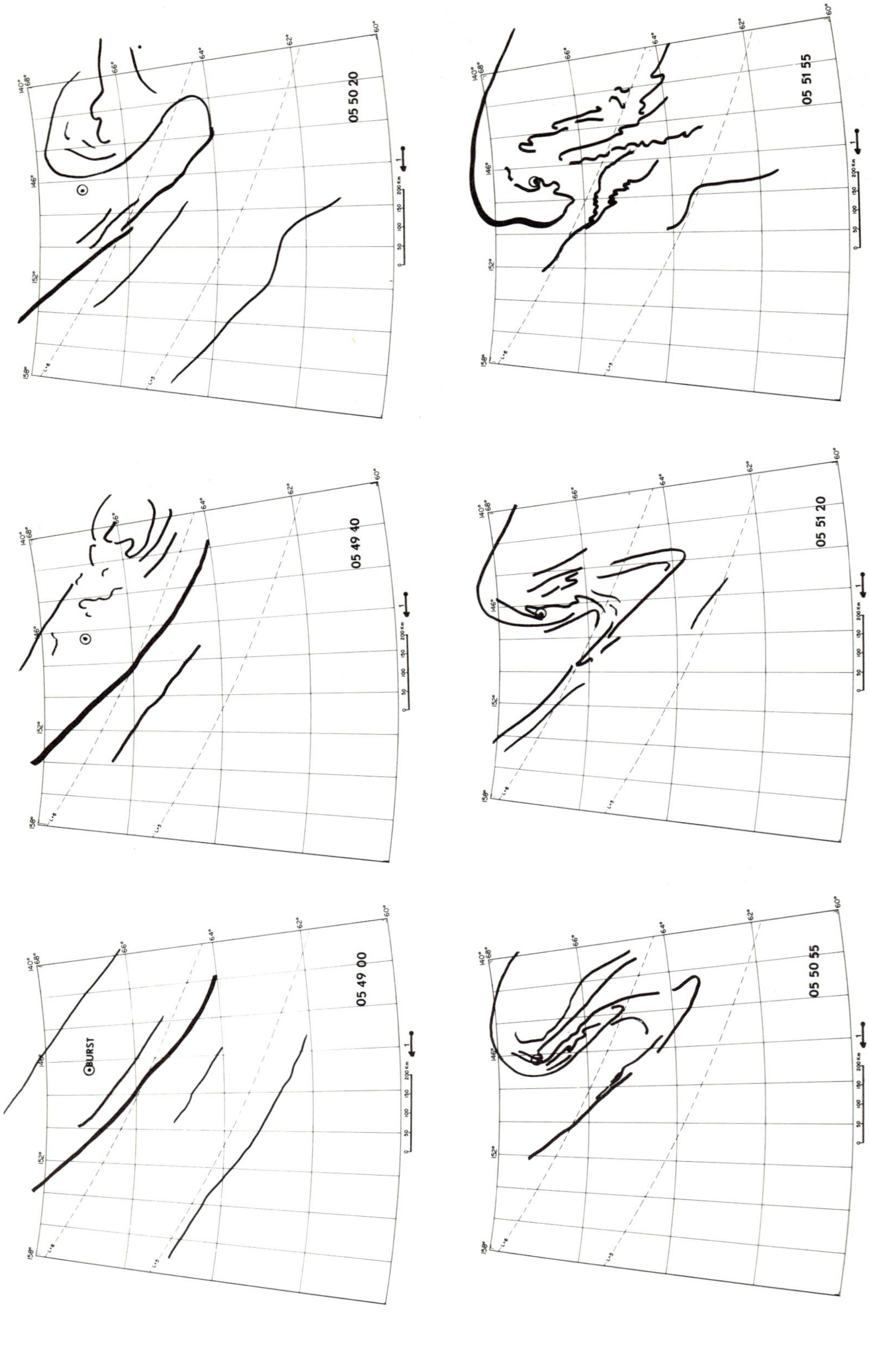

Fig. 7. Mapped auroral lower borders and the 100 km projected position of the barium for the interval corresponding to Figure 6. The arrow at the bottom indicates an ion drift velocity perpendicular to B of 1 km/sec. All drifts during this interval were too small to show as a vector.

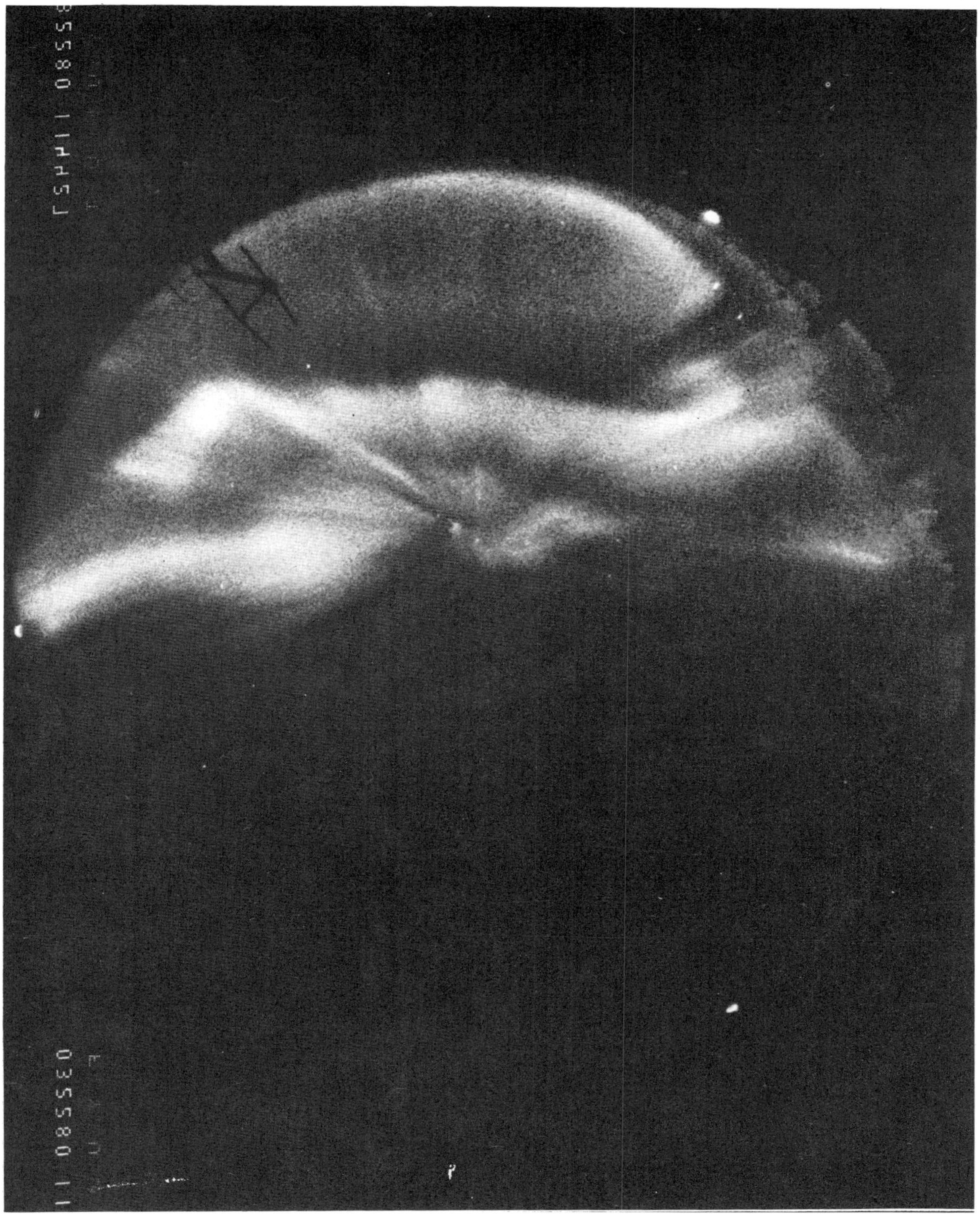

Fig. 8. All-sky camera frame from Ft. Yukon, Alaska, during an interval of rapid ray motions in the aurora and rapid fluctuating motion of the barium above the ionosphere on field lines within the aurora.

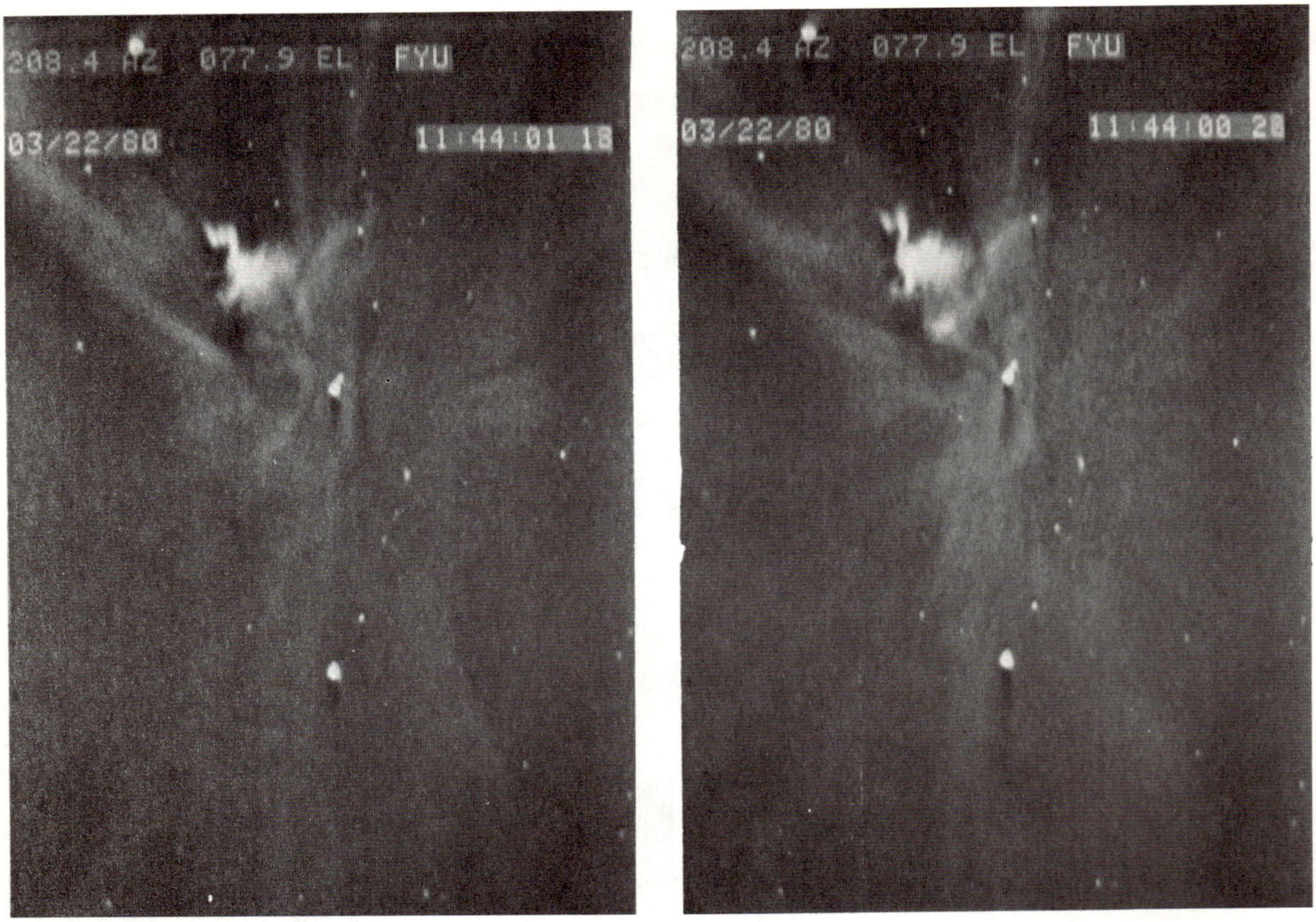

Fig. 9. A pair of TV frames in white light from experiment "Miss Peggy" Mar. 22, 1980, one second apart (56 and 57 sec after injection). Rapid fluctuating distortion of the barium (bright crab-like form center left) can be seen as well as coronal auroral rays. The barium and auroral motions appear to be similar.

also striated into about a dozen separate flux tubes separated in an eastwest direction by several hundred km. A large westward-moving spiral form passed through the barium, and as it did, very rapid motions of the flux tubes from the apparent lower end at 2000 km upwards to 10,000 km were observed. Figure 5 shows the calculated electric field from the motion of several prominent streaks. The sharp reversal of polarity is characteristic of the "laminar V-shocks" reported from satellite data by Mozer et al. (1977). In this case the shear zone of the V shock would be at a lower altitude than 2000 km.

Most recently, however, Wescott et al. (1981) have reported a similar auroral surge which passed through the location of a barium flux tube without any indication of high electric fields. Figs. 6 and 7 show all sky camera frames and mapped lower border positions respectively for the interval of the surge passage. The 100 km projected position of the barium flux tube is shown. The barium motion perpendicular to B was small as the aurora passed through, indicating small E field from about 500 km altitude to the tip of the barium. The barium had reached an altitude of only 1670 km at the time of the first auroral encounter, so we cannot rule out the presence of a V-type potential structure above the tip of the barium.

There is evidence, though, that the aurora did not show any of the shear-type fine-scale motions which were taken as the early evidence for V-type structures to exist above auroras. From this, Wescott et al. (1981) proposed that the V-type potential structure did not exist above the aurora in that particular experiment.

Fine Structure in Auroral Electric Fields

On several occasions, barium plasma clouds and auroral forms have drifted into coalesence on the same magnetic field lines when the auroral fine structure was undergoing rapid motions. On these occasions similar transitory motion has been observed in the barium with TV systems. This suggests that the rapid motion of auroral fine structures can be an effect of low altitude electric fields, or that V-type potential structures may be found at low altitudes sometimes.

The most recent observation of such phenomenon occurred following a barium shaped charge detonation at 617 km altitude over Ft. Yukon, Alaska, on March 22, 1980. By 45 seconds after the detonation the field-aligned upward-streaming plasma was well within a band of active rayed aurora. Figure 8 shows the Ft. Yukon 35 mm all-sky camera frame at one minute 18 seconds after the injection. The aurora and the barium are close to the magnetic zenith. For about two minutes there were rapid auroral ray motions with typical velocities of a few km/sec. During this period the barium striated along field lines and sheared horizontally, and various portions of the assembly of flux tubes exhibited rapid fluctuating horizontal motions. The typical fluctuation had a period of about two seconds and a velocity of about 2 km/sec. This implies fluctuating electric fields of about 100 mV/m between 617 km altitude and upwards to at least 2000 km. Figure 9 shows the Ft. Yukon TV pictures of the barium (brightest forms) and the auroral rays one second apart, at 57 seconds after injection. Figure 10 also shows a pair of Ft. Yukon TV frames, taken three seconds apart, about one minute and 29 seconds after injection. The rapid distortion of the barium is obvious. The TV recordings also show higher frequency fluttering of the barium, indicating the presence of wavelike electric fields of higher frequency.

Analysis at this time is incomplete, but it seems that the barium plasma motions are similar to and contemporaneous with the motions of auroral fine structures. It is most likely that both are

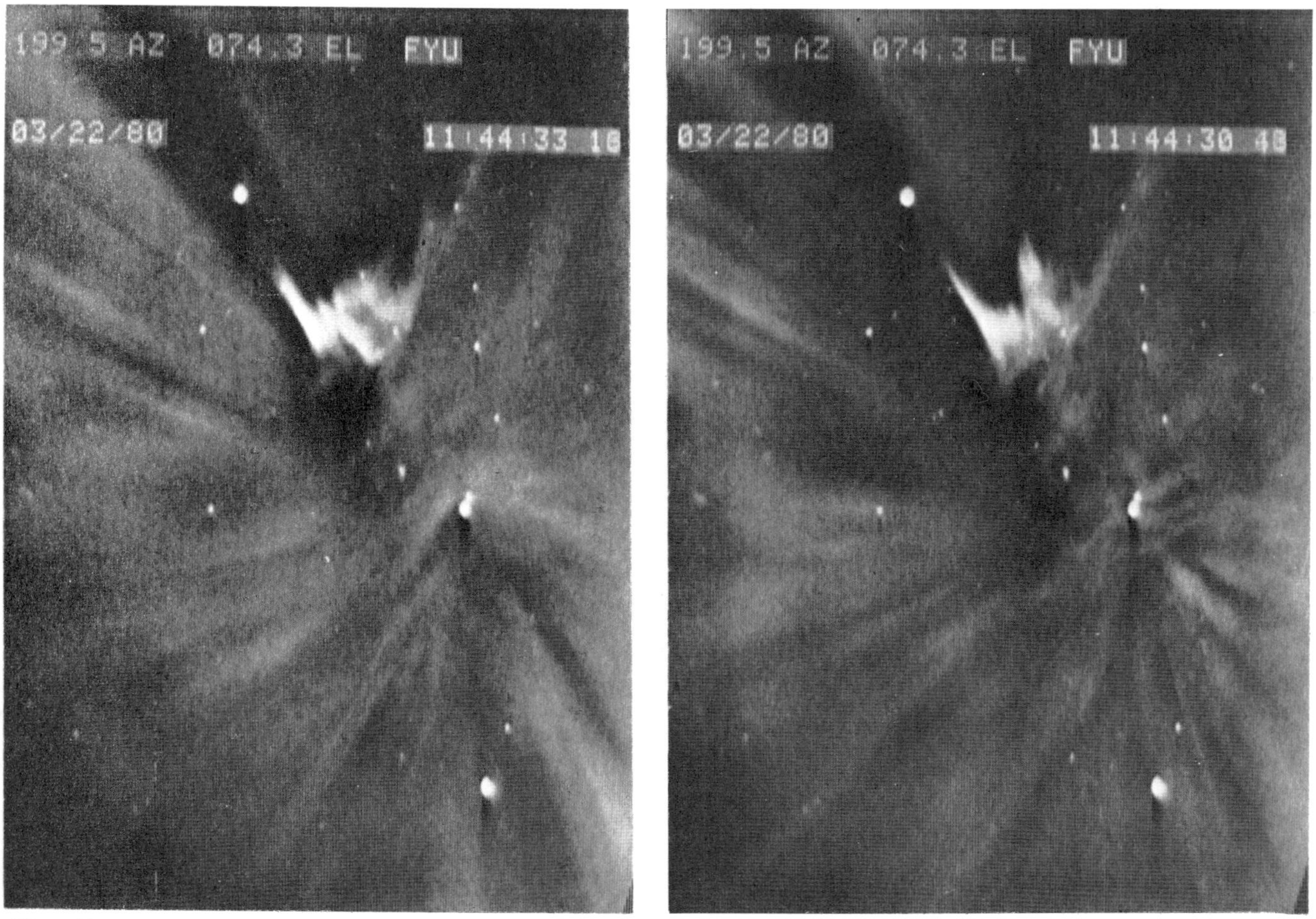

Fig. 10. Pair of TV frames from "Miss Peggy" experiment, taken three seconds apart about 30 seconds later than those shown in Figure 9. The continued rapid fluctuating distortion of the barium can be seen (bright jagged form left center). Rapidly moving coronal auroral rays are also shown. The motion of the barium and auroral rays appear to be similar.

due to electric fields. This implies that auroral fine-structure motions are true indicators of electric fields.

Conclusions

Barium plasma releases allow a different viewpoint of ionospheric and magnetospheric electric fields structures than that provided by probe measurements from rockets or satellites. Since the barium ions drift with magnetospheric convection velocity, we can trace the convection for extended periods of time and cover large spatial distances. Especially useful are shaped charge injection experiments which can provide fluorescent plasma tracers extending for tens of thousands of km along field lines where electric field shears due to V-type potential structures can be observed.

From low-altitude thermite releases it is clear that often auroral arcs have near zero electric fields inside, and large uni-directional fields on either side.

Sometimes, however, where the auroral forms show fine-structure rapid motions, electric fields of high magnitude and fluctuating direction are observed in arcs even at ionospheric altitude. We suggest from our observations, that rapid shear-like motions in the aurora are a necessary condition for V-type potential structures to exist above discrete arcs.

Acknowledgments. I would like to acknowledge the large contributions of my collegues, Tom Hallinan, Hans Stenbaek-Nielsen, and T. Neil Davis in some of the barium release experiments reported in this paper. This work was supported by a grant from NASA; NAG-6 and from Wallops Flight Center for payload and rocket launch support.

References

Aggson, T. L., Probe measurements of electric fields in space, in Atmospheric Emissions, Ed. B. M. McCormac and A. Omholt, Van Nostrand Reinhold Co., New York, 305, 1969.

Block, L. P., Potential double layers in the ionosphere, Cosmic Electrodyn. 3, 349-376, 1972.

Haerendel, G., R. Lüst, and E. Rieger, Motion of artificial ion clouds in the upper atmosphere, Planet. Space Sci. 15, 1, 1967.

Mead, G. D., and D. H. Fairfield, Quantitative magnetospheric models derived from magnetometer data, Goddard Space Flight Center Document X-641-73-363, Greenbelt MD., 1973.

Mozer, F. A., C. W. Carlson, M. K. Hudson, R. B. Torbert, B. Parady, J. Yatteau, and M. C. Kelley, Observations of paired electrostatic shocks in the polar magnetosphere, Phys. Rev. Lett. 38, 292, 1977.

Swift, D. W., H. C. Stenbaek-Nielsen, and T. Hallinan, An equipotential model for auroral arcs, J. Geophys. Res. 81, 3931, 1976.

Wescott, E. M., J. D. Stolarik, and J. P. Heppner, Electric fields in the vicinity of auroral forms from the motions of barium vapor releases, J. Geophys. Res. 74, (14), 3469-3487, 1969.

Wescott, E. M., H. C. Stenbaek-Nielsen, T. N. Davis, W. B. Murcray, H. M. Peek, and P. J. Bottoms, The L = 6.6 Oosik barium plasma injection experiment and magnetic storm of March 7, 1972, J. Geophys. Res. 80, (7), 951, 1975.

Wescott, E. M., H. C. Stenbaek-Nielsen, T. J. Hallinan, T. N. Davis, and H. M. Peek, The Skylab barium plasma experiments, Part II. Evidence for a double layer, J. Geophys. Res. 81, (25), 4495, 1976.

Wescott, E. M., H. C. Stenbaek-Nielsen, T. J. Hallinan, and H. Föppl, Electric fields above a westward travelling surge and post surge deceleration of barium plasma, J. Geophys. Res., in press 1981.

CHATANIKA RADAR MEASUREMENTS OF THE ELECTRICAL PROPERTIES OF AURORAL ARCS

Richard R. Vondrak

Radio Physics Laboratory, SRI International, Menlo Park, California 94025

Abstract. The Chatanika radar has been used to study the electrical properties of the ionosphere in and near auroral arcs. The spatial and temporal variation of electron density, Hall and Pedersen conductivity, horizontal electric fields, electrojet currents, precipitating electron energy flux, and the Joule heating rate can be derived from the incoherent-scatter radar measurements. Discrete auroral arcs are generally imbedded in a region of enhanced conductivity that is several tens of kilometers in latitudinal extent. The maximum ratio of Hall-to-Pedersen conductivity is typically at the center of the arc, although the conductivity gradients are not always symmetric about the arc center. The latitudinal variation in the meridional electric field seems to depend on the location of the arc with respect to the ion convection reversal. In the early evening, a localized reduction in the ambient northward electric field occurs within the arc. Near the ion convection reversal, the northward electric field reverses direction at the arc. In the morning hours when the ionospheric electric field is directed to the southwest, the southward electric field is increased within the arc.

Introduction

To understand the processes responsible for the formation of auroral arcs, we need information about the physical properties of the ionosphere in and near arcs. The incoherent-scatter radar at Chatanika, Alaska can be used to measure most of the electrical and thermal properties of the auroral ionosphere. The basic quantities measured by the radar are the electron density, plasma drift, and the ion and electron temperatures as a function of range along the radar line of sight (see, for example, Leadabrand et al., 1972; Wickwar, 1975; Banks and Doupnik, 1975; Vondrak and Baron, 1976; and references therein). From these quantities other parameters can be derived, including the conductivity, electric currents, Joule heating rate, and the precipitating electron energy distribution. Because the radar is fully steerable and can make continuous measurements for long periods of time, it is able to monitor well the spatial and temporal variation of these ionospheric plasma parameters.

This paper summarizes what has been learned from Chatanika radar measurements of auroral arcs. In particular, these measurements have identified the pattern of the latitudinal variation of the ionization, conductivity, and electric field in and near auroral arcs.

Radar Observations

The Chatanika incoherent-scatter radar is located in Alaska at the geomagnetic latitude of 65° ($L = 5.7$). The radar transmits 3 to 5 MW of power at a frequency of 1290 MHz, a very small fraction of which is radiated back by the ionospheric electrons. The received electromagnetic radiation is analyzed in several ways. The received power provides a measure of the electron density, the doppler shift corresponds to the line-of-sight ionospheric plasma-drift velocity, and the frequency spectrum indicates the electron and ion temperatures. From these basic quantities we can deduce most of the other ionospheric parameters.

The altitude profile of ionization is usually measured with a range resolution of 4.5 km along the radar line of sight. In auroral arcs the E-region electron density exceeds 10^5 cm^{-3}; an integration time of only a few seconds is needed to measure such an ionization concentration with an accuracy within ten percent.

The line-of-sight ion drift velocity is generally measured at eight different range gates. Each gate has an extent in range of about 48 km and the gates are usually spaced at altitude intervals of about 50 km, with the first gate in the E region at an altitude of 110 km. During auroral arc experiments, the integration time required to have a good velocity estimate is typically 15 s to 60 s. Ion drift measurements are possible with a range resolution of 10 to 24 km (Brekke and Rino, 1978), but such an increased spatial resolution requires integration times of several minutes, which are too long for most auroral arc studies.

Two basic types of operating modes have been used to study auroral arcs with the Chatanika

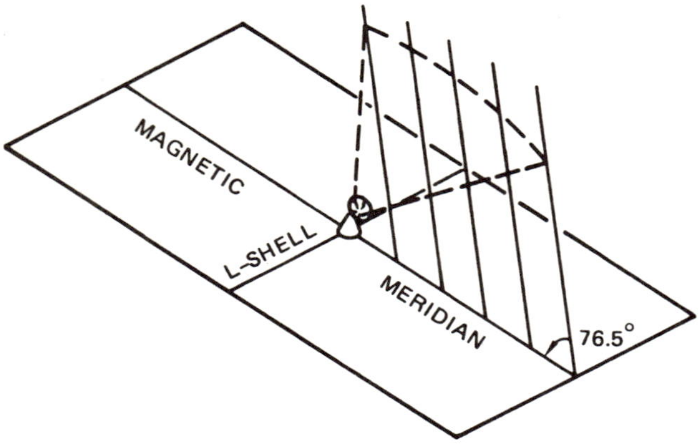

Figure 1. Geometry of an elevation scan in the geomagnetic meridian plane. The radar beamwidth is 0.6°.

radar. The first is shown in Figure 1, which illustrates the elevation-scan technique. The antenna azimuth is kept in the geomagnetic meridian plane. By continuously scanning in elevation, the radar measures the latitude and altitude variation of ionospheric plasma parameters (Vondrak and Baron, 1976). This technique produces a two-dimensional cross-sectional map of the auroral forms, which usually extend in an east-west direction perpendicular to the meridian plane. The angular resolution of the measurements is determined by the product of the scan rate and the integration time (typically 1° to 3°), convolved with the radar beamwidth of 0.6°. This corresponds to a spatial resolution of about 10 km. The region surveyed by the radar has a latitudinal extent in the E region of about 600 km. The length of time needed for a scan depends on the angular range surveyed and the scan rates, but is generally 5 to 15 minutes.

During an elevation scan the latitudinal variation of the east-west component of the electric field, $\vec{E}$, is computed from the north-south ion drift in the F region. At altitudes above about 160 km, the ionospheric plasma is drifting with a velocity equal to $\vec{E} \times \vec{B}/B^2$. The north-south component of the electric field is calculated by using the E-region ion velocity. The effect of ion-neutral collisions in the E region is to reduce and rotate the ion velocity vector toward the direction of $\vec{E}$, and the amount of rotation can be computed from a model atmosphere. Thus, the measurements at the two altitudes are equivalent to measurements of ion drift at two different viewing angles, and the complete vector of the horizontal electric field can be computed (de la Beaujardiere et al., 1977, 1980). The contributions of the neutral wind and vertical drift to the measured line-of-sight ion velocity are usually either neglected or modeled. The accuracy of the vector electric field derived from the elevation-scan measurements depends on the ionospheric conditions. Typically, the total uncertainties are less than 5 mV/m for the zonal electric field and 10 mV/m for the meridional electric field. Such accuracies have been verified by comparisons between radar measurements and simultaneous rocket-borne electric field measurements (Robinson et al., 1981; Mikkelsen et al., 1981).

The finest spatial resolution possible with the radar is obtained in a fixed-position experiment, the second type of experiment used to study auroral arcs. As shown in Figure 2, the radar beam is held fixed (typically in the local L shell) and ionospheric parameters are measured while the auroral forms drift through the radar beam. If the motion of the arc is determined from observations with all-sky cameras, the measurements can be identified with distance from the arc center. The spatial resolution is the sum of the E region beamwidth (1 km) and the arc motion during one integration time (typically 1 to 5 km). As in an elevation scan, the vector electric field is measured by comparing drifts at two altitudes.

Measurements of Ionization and Conductivity

The first measurements of ionization in auroral arcs were made at Chatanika with fixed-position experiments (Baron, 1974; Hunsucker, 1975). Elevation scan experiments now are generally used to identify the ionization morphology of auroral arcs and their relation to other ionospheric features.

Auroral conditions during a scan through a typical evening-sector arc shown in Figure 3. Visible in the all-sky photograph is clear sky to the south, diffuse aurora overhead, and a bright auroral arc in the north. The discrete arc in the north is separated from the diffuse

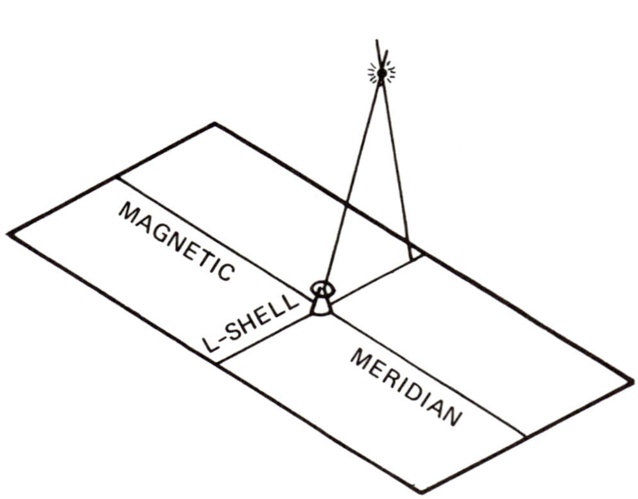

Figure 2. Radar measurements with the antenna pointing in a fixed position in the local L-shell.

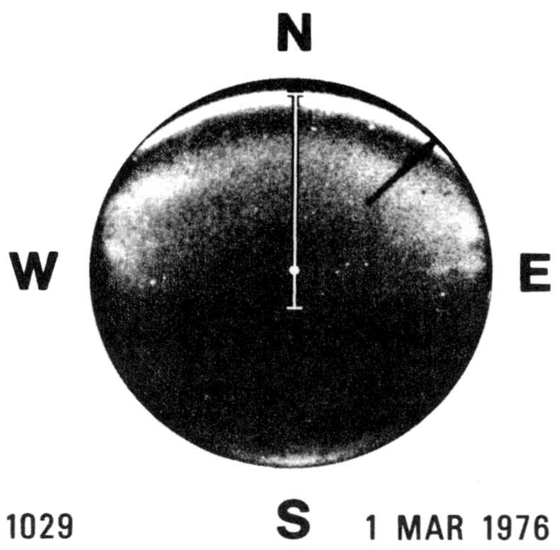

Figure 3. All-sky camera photograph at 1029 UT (0029 local time) on 1 March 1976. The limits of the radar elevation scan are indicated.

aurora by a dark gap. The Chatanika radar measurements of ionization at this time are shown in Figure 4. All of the auroral morphological features present in the all-sky photograph can be identified in the latitude/altitude plot of electron concentration. The diffuse aurora is reasonably uniform, with a maximum ionization of about 2×10^5 cm^{-3} at an altitude of 120 km. The arc has a maximum electron density of 10^6 cm^{-3} at an altitude near 100 km. The arc is imbedded within a region of enhanced ionization approximately 100 km wide, corresponding to 1° of latitude. This region has an asymmetric distribution of ionization with latitude, such that the steepest gradient is on the equatorward side of the

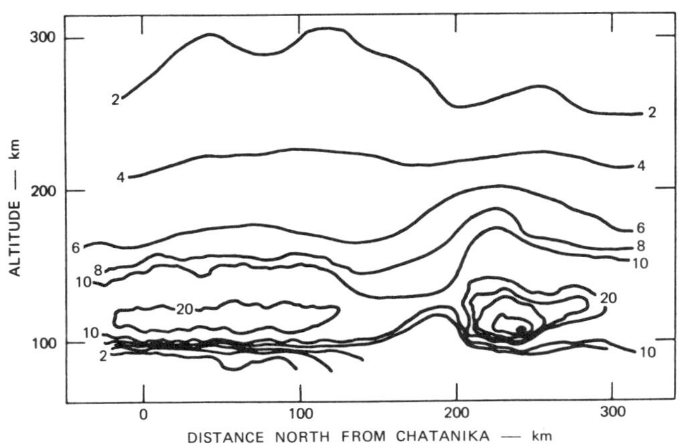

Figure 4. Contour map of the spatial distribution of ionization in the visual aurora shown in Figure 3. Ionization contours are in units of 10^4 cm^{-3}.

arc. This E-region pattern is frequently observed, although sometimes arcs are imbedded within the diffuse aurora. This morphology is in agreement with the latitudinal structure of the auroral oval deduced from satellite observations (Lui et al., 1977). The measurements shown in Figure 4 were made on a winter night near the solar-cycle minimum; little F-region ionization is present above either the diffuse aurora or the discrete arc. More recent observations often show the frequency presence of a dense F layer (5 to 10×10^5 cm^{-3}) above the aurora.

The electron concentration measurements can be converted into volume conductivity by multiplying by the Pedersen and Hall mobilities (Brekke et al., 1974; Vickrey et al., 1981). Figure 5 shows contours of the conductivity distribution. The upper panel shows the latitudinal variation of the height-integrated Hall and Pedersen conductivities. The conductivities are fairly uniform within the diffuse aurora and, of course, increase within the arc. The ratio of the Hall to Pedersen conductivity reaches a maximum at the arc center and decreases in a linear fashion toward the edges of the arc. This pattern results from the latitudinal variation of the energy distribution of the precipitating auroral electrons. This inverted-V hardening of the electron

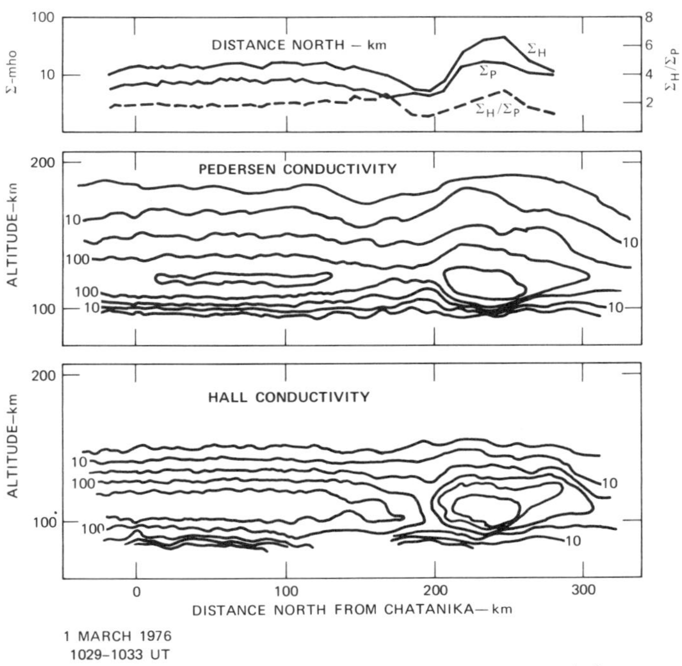

Figure 5. Electrical conductivity computed from the ionization measurements made during the scan shown in Figure 4. The upper panel shows the latitudinal variation in the height-integrated Hall conductivity (Σ_H) and Pedersen conductivity (Σ_P). The lower panels show the volume conductivities at intervals of 3, 10, 100, 300, 1000, and 3000 micromho/m.

energy spectrum results in a similar pattern of conductivity ratio, because a harder energy spectrum preferentially increases the Hall conductivity (Vickrey et al., 1981).

Electric Field Measurements

Detailed measurements of the electric field in and around auroral arcs have been made in several fixed-position experiments (de la Beaujardiere et al., 1977; de la Beaujardiere and Vondrak, 1981; Stiles et al., 1980).

One example of the spatial variation of ionization and electric field around an auroral arc is shown in Figure 6. This is a premidnight arc that has been studied by de la Beaujardiere et al., (1977) and de la Beaujardiere and Vondrak (1981). This arc was aligned in the east-west direction and drifted through the radar beam with a very regular velocity during the time period shown in Figure 6. For these reasons, the radar measurements can be used to identify the variation of ionization, electric field, and related quantities with a spatial resolution of about 2 km. The northward electric field decreases near the arc center and increases at the arc edges where the ionization and conductivity are small. This is the typical pattern of evening arcs.

Evidence for the acceleration of the precipitating auroral electrons within the arc is obtained from the altitude profiles of ionization. Four ionization profiles in and near the arc are shown in Figure 7. Nearer the center of the arc the ionization profile peaks at a lower altitude.

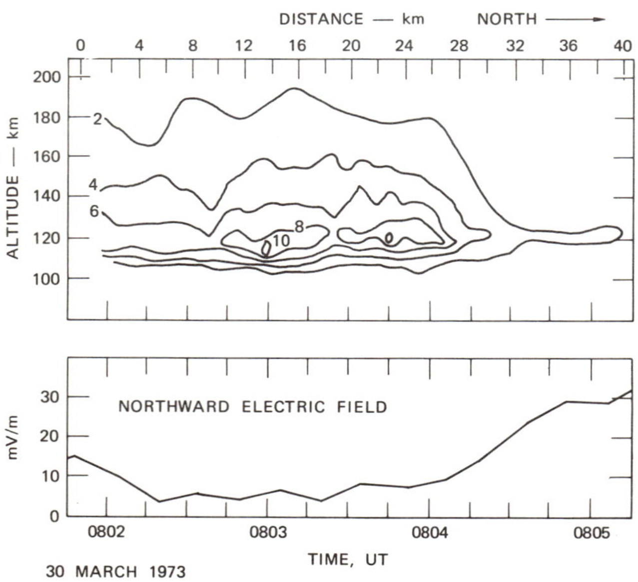

Figure 6. Latitudinal variation of ionization and northward electric field in and near a discrete evening auroral arc. Ionization contours are in units of 10^5 cm^{-3}.

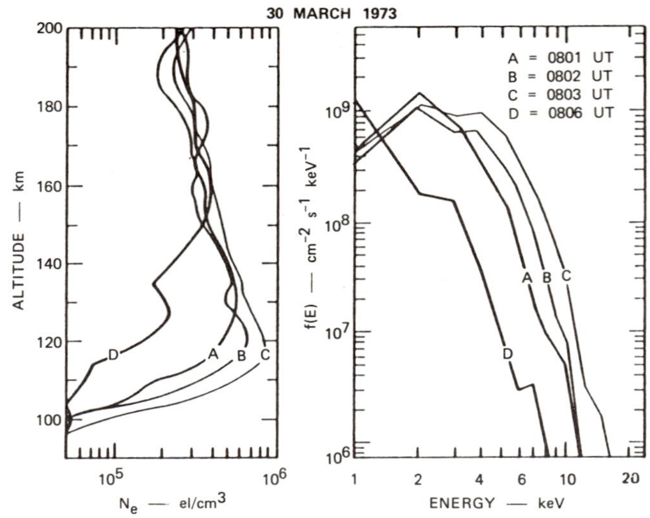

Figure 7. Ionization profiles and corresponding electron energy distributions associated with the auroral arc shown in Figure 6.

Also shown in Figure 7 are the differential flux distributions of auroral primary electrons that will produce these altitude profiles (Vondrak and Baron, 1977). These spectra resemble Maxwellian distributions that have been accelerated through a potential difference at some indeterminate altitude. The potential drop at the arc center is approximately 5 kV. The total horizontal potential difference across the auroral arc is equal to the latitude integral of the northward electric field in the ionosphere (Figure 6). This latitudinal potential difference is only 0.6 kV, a small fraction of the estimated field-aligned potential difference between the ionosphere and the magnetosphere. This inequality between the latitudinal and vertical potential differences indicates that the ionospheric electric field is strongly decoupled from the electric field in the magnetosphere, presumably by the presence of field-aligned electric fields at altitudes above the ionospheric F region.

The large-scale electric field variation with latitude can be surveyed by elevation-scan measurements. Figure 8 is an example that has been described in detail by de la Beaujardiere et al., (1981). Two arcs are present near Chatanika, a bright arc in the south and a weaker arc in the north. The latitudinal variation of the northward electric field is different around each arc. At the southern arc the field is large (90 mV/m) on the equatorward edge and reduced on the poleward side. At the northern arc the electric field reverses direction; it is directed northward on the equatorward side and southward on the poleward side. These and other observations reported by de la Beaujardiere et al., (1981) indicate that, at any given time, the electric field variation around an arc depends upon the latitude of the arc.

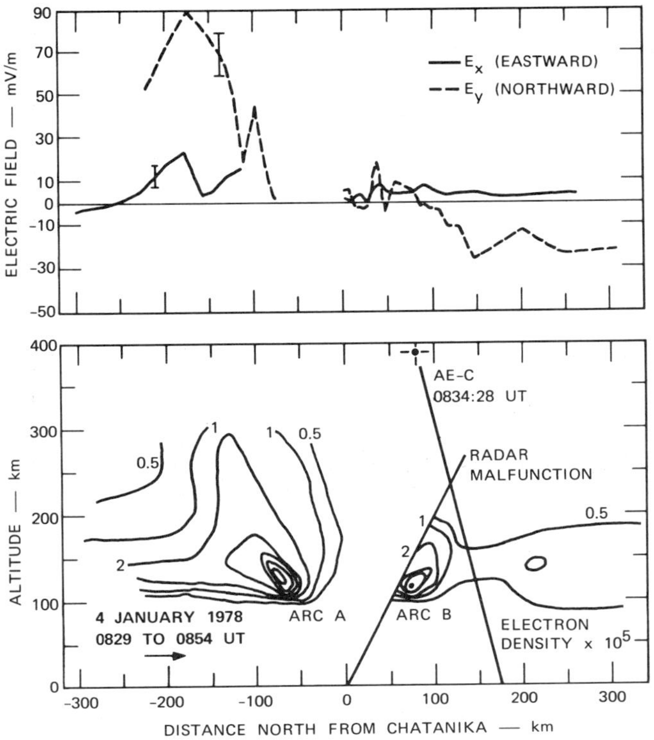

Figure 8. Latitudinal variation of the vector electric field (upper panel) and the spatial distribution of ionization (lower panel) measured near two auroral arcs.

Because the data set shown in Figure 8 coincided with a pass of the Atmosphere Explorer-C (AE-C) spacecraft, it can be used to study the longitudinal variation of the electric field pattern. The AE-C trajectory crossed the southern arc at two longitudes that were separated from the Chatanika meridian by approximately two hours of local magnetic time. The nearly simultaneous measurements of the latitudinal variation of the northward electric field around this arc are shown in Figure 9. The electric field pattern is remarkably similar at all three local times. This pattern also resembles that found on another occasion in a Chatanika fixed position experiment by Stiles et al., (1980) and in a coordinated rocket and radar measurement reported by Robinson et al., (1981). In Figure 9 is shown the latitudinal profile measured by Stiles et al., (1980).

The temporal variation of the electric field pattern around an auroral arc can be determined by repetitive radar measurements. Stiles et al., (1980) and de la Beaujardiere et al., (1981) have made such measurements for time periods of about one hour. Both studies found little change in the electric field pattern, even though the arc location, intensity and width changed substantially.

Summary

The general pattern of ionization and electric field around auroral arcs in the evening sector is shown schematically in Figure 10. This configuration is a composite synthesized from many

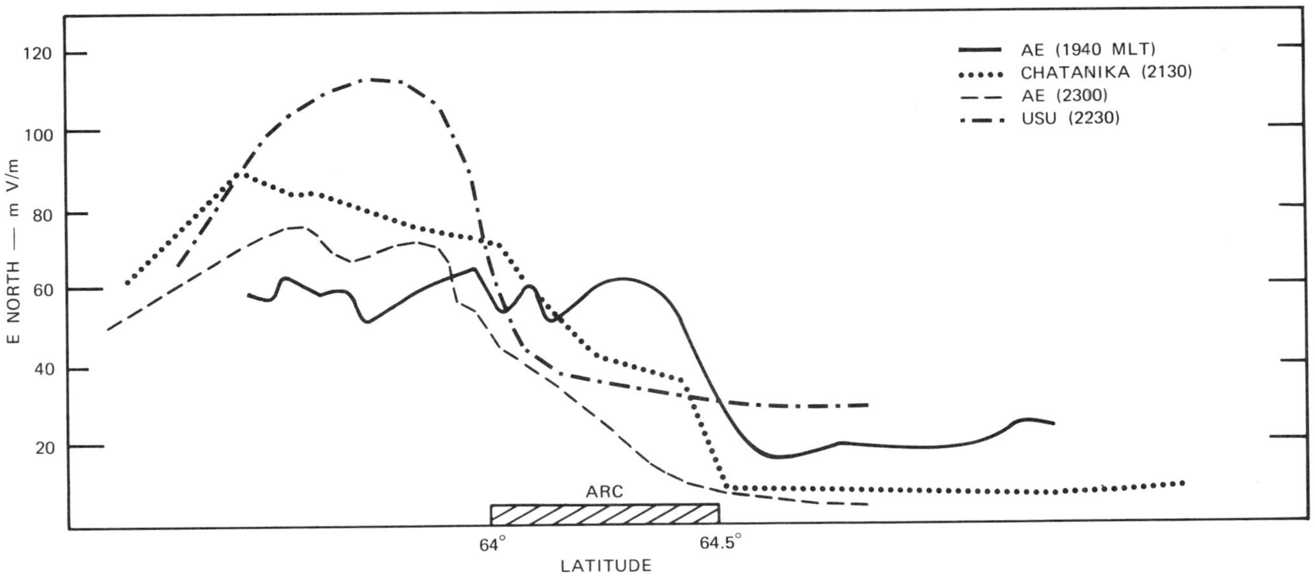

Figure 9. The latitudinal variation of the meridional electric field associated with an auroral arc. Three of the plots are nearly-simultaneous measurements of a single discrete arc near Chatanika that was crossed twice by the AE-C spacecraft (de la Beaujardiere et al., 1980). The other plot (labeled USU) is a measurement of a different arc reported by Stiles et al., (1980).

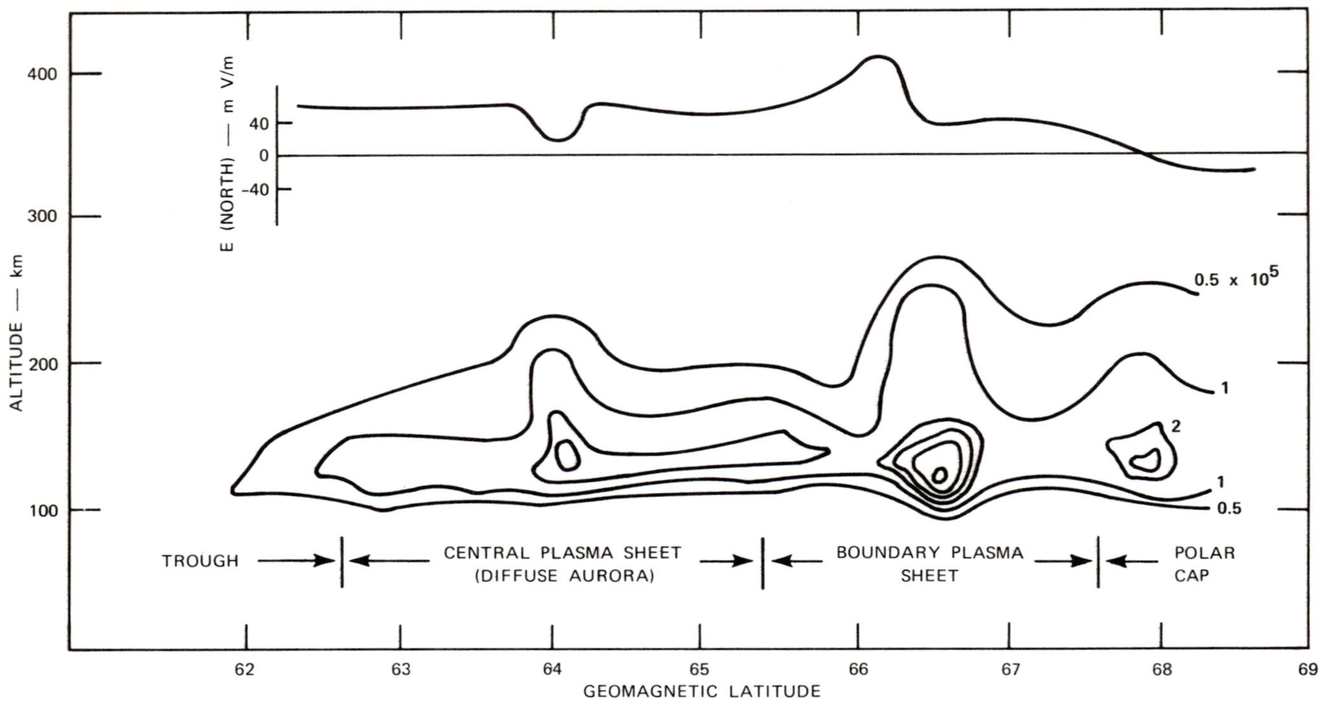

Figure 10. Schematic representation of the latitudinal variation of ionization and meridional electric field at high latitudes in the evening sector.

individual radar measurements, because the radar field of view usually includes only a portion of the auroral oval. The main feature is the different patterns of the northward electric field variations. Arcs located within the diffuse aurora are local conductivity enhancements that generally result in local reductions in the northward electric field. Discrete arcs at higher latitudes corresponding to the boundary plasma sheet have an asymmetric electric field pattern in which the northward electric field is reduced and remains small on the poleward side. At still higher latitudes there is often a weak arc near the ion-convection reversal that corresponds to the polar cap boundary in the early evening and the Harang discontinuity in the midnight sector. This arc seems to represent a boundary where the ionospheric electric field reverses direction from northward to southward.

Few observations have been made in the morning sector. De la Beaujardiere et al., (1977) showed that in the morning hours when the ionospheric electric field was directed to the southwest, the southward electric field was increased within the arc.

These latitudinal variations in the ionospheric electric field are interpreted as the result of a latitudinally varying magnetospheric electric field applied to a nonuniform conducting ionosphere. To preserve current continuity, polarization electric fields develop in the ionosphere. The direction and magnitude of the polarization electric fields depend upon the direction and magnitude of the magnetospheric electric field, the spatial variation of the ionospheric Hall and Pedersen conductivities, and the intensity of field-aligned currents between the ionosphere and magnetosphere.

Acknowledgments. I am grateful for the assistance and useful suggestions of the members of the SRI Radio Physics Laboratory, particularly Odile de la Beaujardiere. This work was supported by Grant ATM-7823658 from the Atmospheric Sciences Section of the National Science Foundation and by NASA Contract NAS5-25875.

References

Banks, P.M. and J.R. Doupnik, A review of auroral zone electrodynamics deduced from incoherent scatter radar observations, J. Atmos. Terr. Phys., 37, 951-972, 1975.

Baron, M.J., Electron densities within aurorae and other auroral E-region characteristics, Radio Sci., 9, 341-348, 1974.

Brekke, A., J.R. Doupnik and P.M. Banks, Incoherent scatter measurements of E-region conductivities and currents in the auroral zone, J. Geophys. Res., 79, 3773, 1974.

Brekke, A. and C.L. Rino, High-resolution altitude profiles of the auroral zone energy dissipation due to ionospheric currents, J. Geophys. Res., 83, 2517-2524, 1978.

de la Beaujardiere, O., R. Vondrak and M. Baron, Radar observations of electric fields and currents associated with auroral arcs, J. Geophys. Res., 82, 5051-5062, 1977.

de la Beaujardiere, O., R. Vondrak, R. Heelis, W. Hanson and R. Hoffman, Auroral arc electrodynamic parameters measured by AE-C and the Chatanika radar, J. Geophys. Res., in press, 1981.

de la Beaujardiere, O. and R. Vondrak, Chatanika radar observations of the electric field, current, and potential distribution in an auroral arc, to be submitted to J. Geophys. Res., 1981.

Hunsucker, R.D., Chatanika radar investigation of high-latitude E-region ionization structure and dynamics, Radio Sci., 10, 277-288, 1975.

Leadabrand, R.L., M.J. Baron, J. Petriceks and H.F. Bates, Chatanika, Alaska, auroral zone incoherent-scatter facility, Radio Sci., 7, 747, 1972.

Lui, A., D. Venkatesan, C. Anger, S. Akasofu, W. Heikkila, J. Winningham and J. Burrows, Simultaneous observations of particle precipitation and auroral emissions by the ISIS-2 satellite in the 19-24 MLT sector, J. Geophys. Res., 82, 2210, 1977.

Mikkelsen, I., T. Jorgensen, M. Kelley, M. Larsen, E. Pereira and J. Vickrey, Neutral winds and electric fields in the dusk auroral oval: 1, Measurements, J. Geophys. Res., 86, 1513-1524, 1981.

Robinson, R., E. Bering, R. Vondrak, H. Anderson and P. Cloutier, Simultaneous rocket and radar measurements of currents in an auroral arc, submitted to J. Geophys. Res., 1981.

Stiles, G.S., J.C. Foster and J.R. Doupnik, Prolonged radar observations of an auroral arc, J. Geophys. Res., 85, 1223-1234, 1980.

Vickrey, J.G., R.R. Vondrak and S. Matthews, The diurnal and latitudinal variation of auroral-zone ionospheric conductivity, J. Geophys. Res., 86, 65, 1981.

Vondrak, R. and M. Baron, Radar measurements of the latitudinal variation of auroral ionization, Radio Sci., 11, 939, 1976.

Vondrak, R.R. and M.J. Baron, A method of obtaining the energy distribution of auroral electrons from incoherent scatter radar measurements, in Radar Probing of the Auroral Plasma (Proceedings of the EISCAT Summer School, Tromsø, Norway, June 5-13, 1975), A. Brekke, ed., pp. 315-330, Universitetsforlaget, Tromsø-Oslo-Bergen, 1977.

Wickwar, V.B., Chatanika radar measurements, Atmosphere of Earth and the Planets, 111-124, B.M. McCormac, ed., D. Reidel Publishing Co., Dordrecht-Holland, 1975.

ON THE EMPIRICAL RELATIONSHIP BETWEEN FIELD-ALIGNED CURRENTS AND AURORAS

Y. Kamide

Kyoto Sangyo University, Kamigamo, Kita-ku, Kyoto 603, Japan

Abstract. The development of several new observational techniques during the last decade has contributed significantly to our understanding of the characteristics of ionospheric and field-aligned currents at high latitudes as well as of the auroral distribution. This paper attempts to point out several pitfalls in earlier studies of the empirical relationship between field-aligned currents and auroras, with special emphasis on the variability of auroral forms.

Introduction

In order to answer the important questions "What are the main charge carriers responsible for field-aligned currents?" and "Where do these particles originate?", it is necessary to make clear how the field-aligned currents relate to different auroral features at various local times under different substorm conditions. During the last few years, new data sets of the auroral distribution have been presented which yield details in auroral structure, permitting their correlation with electric current flow. In this paper, we point out several problems in the past studies of the relationship between field-aligned currents and auroras, and suggest some areas which need to be clarified.

Discrete and Diffuse Auroras

One of the important findings in auroral physics during the last decade on the basis of satellite-viewed auroral imagery is the identification of discrete and diffuse auroras [see reviews by Akasofu, 1974a; Davis 1978]. Earlier studies of the distribution of auroral forms which relied primarily upon all-sky cameras on the earth's surface, have the limitations that an all-sky camera has only a limited field of view and very often fails to detect the diffuse aurora. Winningham et al. [1975] identified two different types of auroral particle precipitations corresponding to the two types of auroras. The characteristics of both types of auroral precipitations represent a complicated function of local time and latitude, as well as of substorm time. Detailed inspection of the Isis 2 electron energy spectra led Lui et al. [1977] to conclude that the precipitation belt can be spatially divided into two regions in the midnight sector. In the poleward half of the belt, both the intensity and energy of electrons are structured highly variable corresponding to discrete arcs. Akasofu [1974a] examined how various auroral patterns, such as rayed arcs, patches, torches and omega bands, etc., in the combined all-sky camera records are represented by the large-scale structure in satellite photographs. Equatorward of this, there is an extensive precipitation region with fairly uniform intensity, in which the pitch angle distribution is generally isotropic. This is the region of the diffuse precipitation.

This distinction of the two types of auroras is quite important, since it is now believed that the first indication of an auroral substorm, viz., the auroral breakup, occurs near an arc located between the discrete and diffuse auroras in the midnight sector. Akasofu [1977] summarized some of the differences in morphological features between the discrete and diffuse auroras. In essence, a discrete aurora appears as a single, bright strand, well separated from other auroras by a dark luminosity with a latitudinal width of, at least, several tens of km.

Correlative studies have been conducted of the discrete and diffuse auroras with other signatures of electric fields and currents as well as of particle precipitation in auroral latitudes. Horwitz et al. [1978] showed using data from the Chatanika radar that the electric field in the evening diffuse aurora is directed generally northward, whereas it is southward in the region of the discrete aurora. Kamide and Akasofu [1975] observed that it is only the evening diffuse aurora region where the eastward electrojet is observed. This is also the region where the diffuse radar aurora is confined [Tsunoda et al., 1976]. In the evening sector, the downward field-aligned current was shown to be co-located with the diffuse aurora [Kamide et al., 1979], while the intense upward current is located in the discrete aurora region [Kamide and Akasofu, 1976]. Deehr et al. [1976] made a simultaneous observation of the two auroral regions, one from above by the Isis 2 satellite and the other from below by the NASA 711 aircraft, which provided the electron spectrogram and all-sky pictures, respectively. An active, discrete aurora located poleward and the relativaly stable, diffuse aurora covering the equatorward half are clearly seen in the two data sets. Similar correlative studies of diffuse and discrete auroras were also made by Mizera et al. [1975], Cavary [1975], Meng [1976] and Kaufmann et al. [1978].

However, such a distinction between the two main auroral types cannot always be made both in ground-based and satellite observations. The equatorward boundary of the diffuse aurora may often be misidentified as a homogeneous arc. Akasofu [1974b] also noted that in the morning sector, the diffuse aurora tends to develop into several discrete auroras and some of the discrete forms spread over the morning half of the polar cap. Meng and Akasofu [1976] observed that there is a particular type of polar cap arc which lies parallel to auroral oval arcs in the evening sector and which bends away from the oval to the usual sun-earth line in the vicinity of the poleward boundary of the auroral oval. Whalen et al. [1977] and Pike et al. [1977] showed that both discrete and diffuse components are contained in a band of enhanced 6300Å emission extended over several degress in latitude in the midnight sector.

One of the most prominent features of the diffuse aurora when viewed from a global perspective is its well-defined equatorward boundary. However, Anger and Murphree [1976] cautioned that the auroral intensity does not actually cut off as sharply as one might expect from looking at satellite auroral imagery, because each satellite picture has a specific intensity threshold. Furthermore, although the identification of the diffuse aurora in evening hours is relatively easy, it is difficult to define it exclusively in the morning sector, where complicated post-breakup auroral signatures are generally observed. Anger and Murphree [1976] and Akasofu [1977] pointed out that the terms mantle aurora [Sandford, 1968] and continuous aurora [Pike and Whalen, 1974] and the present diffuse aurora are essentially interchangeable. Snyder and Akasofu [1974] indicated that the diffuse aurora can also have considerable structure on a small (25-50km) scale especially in the morning sector. Pulsating auroras which make up parts of the diffuse aurora in the morning sector can occur at times on the poleward, leading edge of the midnight sector surge, which is no doubt, the discrete aurora.

We note, by taking all these complications into account, that the classification into the two types of auroras, i.e., discrete and diffuse, is only the first step to understand physical processes in the polar ionosphere occurring in association with auroral particle precipitation. Sometimes, this distinction in auroral records cannot be made unambiguously, depending on coordinate frames and substorm times.

What is an Auroral Arc?

The term 'auroral arc' had originally been used for a long time to describe a special class of band-like forms which appear as a simple, curving arch for visual observations. However, this term has recently been used to denote a discrete curtain-like form of auroral display, departing from the classical definition. The width of auroral arcs ranges from a minimim of 50m (a few electron gyro-radii) to a maximum of order 10km (Davis, 1978).

The distribution of visible auroras can now be observed by ground-based all-sky cameras and TV camera systems, airborne all-sky cameras, and satellite scanners from above the poles. There is no doubt that auroral physics has been advanced to a significant extent by studies of photographs from scanning cameras aboard the Isis 2 and a series of the DMSP satellites [e.g., Anger et al., 1973; Snyder et al., 1974]. These types of cameras can reconstruct a global map of auroral luminosity from horizon-to-horizon scans of the earth by cameras sweeping perpendicular to the orbital path, thus satellite auroral imagery is inherently of a 'snapshot' character, providing of itself virtually no information on the temporal development of auroras. The scanner data have revealed several new auroral characteristics which had not been clearly recognized by all-sky camera studies [Akasofu, 1974a]. These satellite-viewed aurora imagery has modified to a significant degree the terminology in common use [Davis, 1979]. Furthermore, it is now possible to identify the positions of electric currents, such as the auroral electrojets in the ionosphere and field-aligned currents, with respect to the auroral display observed from the satellites.

However, the availability of a number of auroral data obtained from various techniques has resulted in researchers discussing different types of auroras but giving them the common label of auroral arc.

Kan et al. [1980] have pointed out that the discrete auroras can be divided into the auroral arc and the inverted-V aurora on the basis of the latitudinal structure in particle flux, electric field, and auroral luminosity measurements. The discrete aurora associated with the inverted-V events is typically hundreds km in latitidinal width, while the auroral arc is typically much less than that, perhaps hundreds meters. It was shown that different mechanisms must be considered to generate the different auroral structures.

Using data of the auroral luminosity at 3914 and 5577Å obtained by the auroral scanner photometer on board Isis 2, Wallis et al. [1976] constructed a number of latitudinal profiles of the auroral intensity, which were employed to locate the latitudes of discrete arcs. The discrete arcs were identitied by relatively narrow latitudinal features in which emissions rise to, at least, 1.5 times the adjacent intensities. As long as this definition is followed on the basis of the data whose resolution is approximately 10km, the arcs defined by Wallis et al. [1976] represent perhaps regions of enhanced auroral liminosity produced by groups of auroral arcs.

Anger and Murphree [1976] examined a number of Isis 2 passes over the northern polar region including a good view of a large portion of the auroral belt, and found that auroral arcs lie within the diffuse aurora, not outside of it. This contradicts to the previous findings using DMSP imagery that there is a latitudinal separation between the discrete and diffuse auroras [Snyder and Akasofu, 1974]. Note that the term 'auroral arc' was used by Anger and Murphree [1976] to describe any elongated and narrow auroral form, although extended arcs in the evening sector which move to higher latitudes at earlier local times are common.

Using all-sky camera and Triad satellite data, Armstrong et al. [1975] showed that the poleward-most descrete arc marks the poleward boundary of the field-aligned current regime. Kamide and Akasofu [1976] extended the study by examining such simultaneous satellite and ground-based observations and found that in the evening sector, discrete arcs are confined within the latitudinal region of the upward field-aligned current. Evans et al. [1977], Carlson and Kelley [1977], and Kamide and Rostoker [1977] showed that the precipitating keV electrons constitute the upward field-aligned current within auroral arcs. Most of the rocket observations of the field-aligned currents indicated that in the region of individual auroral arcs, the precipitating electrons with energy between approximately 500eV and 20keV can carry a significant fraction of the upward field-aligned current [see regiew paper by Arnoldy, 1974]. However, some rocket data [e.g., Pazich and Anderson, 1975; Spiger and Anderson, 1975] showed that although a total upward current coincides with the main auroral arc, the measured electrons (0.5-20keV) carry only a small fraction (<15%) of the upward current necessary to produce the simultaneously oberved magnetic perturbations in the vicinity of the arc.

We note that this apparent discrepancy may be accounted for, at least, partially by making clear; what we mean by 'auroral arcs', what is the typical thickness, what is the technique to observe the arc, what is the local time, where the arcs are obseved, what types of arcs (isolated or embedded in the diffuse aurora), and longitudinal size? Caution must be exercised that discrete auroras are not simply auroral arcs!

The Harang Discontinuity

Birkeland's initial work developed the concept that most of the high-latitude magnetic perturbations could be explained by latitudinally localized ionospheric currents, called the auroral electojets, which must be connected with field-aligned currents. After the Birkeland' extensive work, many studies of average storm-time magnetic disturbances led to the conclusion that the eastward electrojet was flowing in the evening sector and the westward electrojet was flowing in the morning sector [e.g., Silsbee and Vestine, 1942]. Harang [1946] is the first who carried out a definitive study using a so-called 'meridian chain' of magnetic observatories in which the configuration of the westward electrojet penertrating into the evening sector at the poleward edge of the eastward electrojet is presented. The Harang discontinuity is generally defined as the demarcation line separating the westward electrojet (to the poleward) and the eastward electojet (to the equatorward); see Heppner [1972] for details. Various phenomena reverse or change significantly their characteristics across the discontinuity [Davis, 1962; Maynard, 1974; Burch et al., 1976]. The line crosses at different latitudes at different local times in the premidnight sector, and shows quite dynamic behaviours during substorm activity. Hughes and Rostoker [1979] have shown that the westward electrojet in the evening sector exists even if substorms are not in progress.

Kamide [1978] has suggested that although various phenomena, such as ground magnetic fields, ionospheric currents, auroral features, and electric fields have been proposed to identify the Harang discontinuity, the loci defined by these signatures do not always coincide. It was then suggested that the electric field best expresses the basic characteristics of the Harang discontinuity. As one moves poleward from the eastward electrojet into the westward electrojet regions, the ionospheric electric field undergoes a poleward to equqtorward transition at the Harang discontinuity.

However, it is important to note that the reversal of the electric field does not occur abruptly, from poleward to equatorward, across a discrete boundary line, but that the poleward-directed field would gradually rotate counterclockwise; namely the electric field becomes westward in the discontinuity region [Maynard, 1974; Wedde et al., 1977]. Based on Barium cloud release experiments and the incoherent scatter radar at Chatanika, Wescott et al. [1969] and Banks et al. [1973] also reported such a predominance of the wastward electic field.

The Harang discontinuity may in fact be the 'Harang continuity'! Anything seen like a discontinuity in variations in some types of data at a particular observatory may represent temporal changes (not spatial discontinuity) associated with substorm onset. For example, if an auroral breakup took place when an auroral observatory was located in the vicinity of the Harang discontinuity and had been observing the diffuse aurora, one might incorrectly conclude that auroral luminosity changed from diffuse type in evening hours to active, rayed forms across the Harang discontinuity. However, the fact is that the active aurora just comes to the field of view of the observatory due to the rapid westward expansion of the substorm disturbed region.

Kamide [1978] argued that there is nothing specific about the Harang discontinuity region in terms of field-aligned currents, unless the discontinuity in the electric field takes place along a line in a strictly mathematical sense. As an extreme case it is possible that there is no field-aligned current into or out from the discontinuity region, but the ionospheric currents merely change the direction. This view is contrasted with the earlier current models that an intense upward field-aligned current should flow from the Harang discontinuity [Heppner et al., 1971].

Baumjohann et al. [1980] have recently shown by using two-dimensional observations of ionospheric electric fields and ground magnetic perturbations that during one period, there is no field-aligned current in the Harang discontinuity region, where the eastward electrojet turns northward instead of diverging up along field lines, while on another occasion, the easten-terminating part of the eastward electrojet diverges up the field lines. Vickrey et al. [1981] have used the Chatanika radar data to estimate the field-aligned current distribution. It was found that the upward current in the Harang discontinuity was very weak. It may well be that the occurrence of one of these two different mechanisms depends upon local time.

Quiet Periods

The empirical relationship of field-aligned currents with the auroral distribution during extremely quiet times is also not settled. As is discussed in the previous sections, recent studies have indicated the existence of latitudinally separated zones of auroras and field-aligned currents. The two regions of auroras, diffuse and discrete, coincide generally with the two regions of the field-aligned currents, upward and downward, in the dark sector.

There arises one puzzling problem when quiet-time features are considered. Auroral electron observations have shown that the diffuse auroral particle precipitation on the equatorward side of the nighttime auroral oval is a persistent feature [Winningham et al., 1975], although the precipitation belt tends to shift poleward (up to 70° invariant latitude) during very quiet times [Kamide and Winningham, 1977]. On the other hand, only the poleward portion of the field-aligned current (downward and upward in the morning and evening sectors, respectively) remains during extremely quiet periods, such as K_p=0 [Iijima and Potemra, 1976]. This portion is thought to be associated with the $S_q{}^P$ current system representing quiet-time magnetospheric convection. Kamide and Akasofu [1976] have suggested that even during quiet periods, discrete auroral arcs can often be found within the region of the upward field-aligned currents in the poleward half of the evening oval, indicating the presence of discrete electron precipitation there. It may be, however, that such a discrete precipitation is confined in latitude and the pattern is more stable during extremely quiet times than during substorm times.

Morning Sector Features

Most previous studies of field-aligned currents and their spatial relation to auroral forms and the auroral electrojets have been made only for the evening and midnight sectors. As far as the present author is aware, there are only five published papers which have examined empirically such a relationship for morning hours [Klumpar et al., 1976; Kamide et al., 1976; de la Beaujardiere et al., 1977; Sulzbacher et al., 1978; Hughes and Rostoker, 1978] in contrast to hundreds of papers for the evening sector. Since it has been shown that characteristics of auroras in the morning sector are quite different from those in the evening sector and that the direction of the field-aligned currents is reversed between morning and evening sides of the auroral oval, it is important to extend such correlation studies to the morning sector.

By examining the electron fluxes over the energy range 5 eV<E<13 keV by the Isis 2 soft particle spectrometer [Rostoker et al., 1981], it is found that the latitudinal distribution of the fluxes is, in general, characterized by the following four elements:

A: A region of low energy electrons featuring localized regions of intense fluxes, and having the character of the boundary plasma sheet in the evening sector but lying poleward of the westward electrojet in the morning sector

B: A region featuring energetic electrons which have a power law distribution in the poleward portion of the wastward electrojet.

C: A region of precipitating energetic electrons in the equatorward portion of the westward electrojet. This particle population consists of a hot Maxwellian superposed on a spectrum which has a power-law distribution.

D: A region of trapped energetic electrons below the equatorward boundary of the westward electrojet.

This character in the morning sector seems more complicated than that in the evening sector, where only the two major precipitation patterns, diffuse and discrete, have been reported. This difference between the evening and morning sector features may imply that the development of these electron regions in the morning sector is determined in the context of the acceleration of boundary plasma sheet material as it drifts not only from the dawn but also from the dusk regions toward the center of the magnetotail under the influence of the east-west component of the magnetospheric electric field.

Conclusion

There have already been many excellent review articles on various features of electric field currents as well as of auroras. Further, efforts have been made to combine various complicated findings into a consistent view in terms of basic physical concepts. However, there are still some areas which are not satisfactorily settled. We believe that this difficulty might be resolved, at least partially, by taking into account the variability of auroras with respect to different local times and substorm times.

One uncertainty in the previous studies of field-aligned current system is the relationship between the small-scale current configuration associated with individual auroral arcs and the large-scale current pattern extending over the entire auroral oval. It may be needed to design high-quality meaurements of particle fluxes and magnetic fields in the vicinity of auroral display simultaneously by sounding rockets and polar-orbiting satellites.

Simultaneous observations of DMSP auroral imagery and Triad field-aligned currents have indicated that the auroral electrojets and field-aligned currents are more tightly confined to the region of auroral luminosity than was indicated by data of Isis 2 precipitations and the meridian chains of ground magnetic observatories. The latter noted significant auroral luminosity outside of the boundaries of the auroral electrojets. Furthermore, recent simultaneous measurements of magnetic fields and soft particle distributions by Isis 2 have shown that the region of the field-aligned currents extends on the average 2.4° equatorward of the low-latitude boundary of 1-keV electron precipitation in the evening sector [Klumpar, 1979]. These disagreements involve an important problem of how field-aligned currents and the auroral electrojets are connected, and how these currents can be located to the region of auroral forms and auroral precipitations in the light of the threshold of the DMSP and Isis 2 satellite detectors.

Acknowledgements. I am grateful to many colleagues for their stimulating conversations during the AGU Chapman Conference on the Formation of Auroral Arcs. Portions of this paper were prepared during my stay at the Geophysical Institute of the University of Alaska in 1980 summer. This research was supported in part by National Institute of Polar Research, Tokyo, and in part by the Ministry of Education Japan under a grant 554105.

References

Akasofu, S.-I., A study of auroral displays photographed from the DMSP-2 satellite and from the Alaska meridian chains of stations, Space Sci. Rev., 16, 617, 1974a.

Akasofu, S.-I., Discrete, continuous and diffuse auroras, Planet. Space Sci., 22, 1723, 1974b.

Akasofu, S.-I., Recent progress in studies of DMSP auroral photographs, Space Sci. Rev., 19 169, 1976.

Akasofu, S.-I., Physics of Magnetospheric Substorms, Dordrecht, D. Reidel, 599pp, 1977.

Akasofu, S.-I., Morphology of auroral arcs, this issue, edited by S.-I. Akasofu, American Geophysical Union, 1981.

Anger, C. D., and J. S. Murphree, ISIS-2 satellite imagery and auroral morphology, in Magnetospheric Particles and Fields, edited by B. M. McCormac, Dordrecht, D. Reidel, pp. 223-234, 1976.

Anger, C. D., A. T. Y. Lui, and S.-I. Akasofu, Observations of the auroral oval and Westward traveling surge from the Isis 2 satellite and the Alaska meridian all-sky cameras, J. Geophys. Res., 78, 3020, 1973.

Armstrong, J. C., S.-I. Akasofu, and G. Rostoker, A comparison of satellite observations of Birkeland currents with ground observations of visible aurora and ionospheric currents, J. Geophys. Res., 80, 575, 1975.

Arnoldy, R. L., Auroral particle precipitation and Birkeland currents, Rev. Geophys. Space Phys., 12, 217, 1974.

Banks, P. M., J. R. Doupnik, and S.-I. Akasofu, Electric field observations by incoherent scatter radar in the auroral zone, J. Geophys. Res., 78, 6607, 1973.

Baumjohann, W., J. Untiedt, and R. A. Greenwald, Joint two-dimensional observations of ground magnetic and ionospheric electric fields associated with auroral zone currents, 1. Three-dimensional current flows associated with a substorm-intensified eastward electrojet, J. Geophys. Res., 85, 1963, 1980.

Burch, J. L., W. Lennartsson, W. B. Hanson, R. A. Heelis, J. H. Hoffman, and R. A. Hoffman, Properties of steplike shear flow reversals observed in the auroral plasma by Atmospheric Explorer C, J. Geophys. Res., 81, 3886, 1976.

Carlson, C. W., and M. C. Kelley, Observation and interpretation of particle and electric field measurements inside and adjacent to an active auroral arc, J. Geophys. Res., 82, 2349, 1977.

Cavarly, R. S., Observations from a chain of all-sky cameras coordinated with satellite auroral zone particle precipitation, M. S. Thesis, University of Alaska, Fairbanks, 78pp, 1975.

Davis, T. N., The morphology of the auroral displays of 1978-1958, 2. Detail analysis of Alaska data and analysis of high-latitude data, J. Geophys. Res., 67, 75, 1962.

Davis. T. N., Observed characteristics of auroral forms, Space Sci. Rev., 22, 77, 1978.

Davis. T. N., Chemical releases in the ionosphere, Rep. Prog. Phys., 42, 1565, 1979.

Deehr, C. S., J. D. Winningham, F. Yasuhara, and S.-I. Akasofu, Simultaneous observations of discrete and diffuse auroras by the Isis 2 satellite and airborne instruments, J. Geophys. Res., 81, 5527, 1976.

de la Beaujardiere, O., R. R. Vondrak, and M. Baron, Radar observations of electric fields and currents associated with auroral arcs, J. Geophys. Res., 82, 5051, 1977.

Evans, D. S., N. C. Maynard, J. Trpim, T. Jacobsen, and A. Egeland, Auroral vector electric field and particle comparisons, 2. Electrodynamics of an arc, J. Geophys. Res., 82 2235, 1977.

Harang, L., The mean field of disturbance of polar geomagnetic storms, Terr. Magn. Atmos. Elec., 51, 353, 1946.

Heppner J. P., The Harang discontinuity in auroral belt ionospheric currents, Geofys. Publ., 29, 105, 1972.

Heppner, J. P., J. D. Stolarik, and E. M. Wescott, Electric-field measurements and the identification of aurrents causing magnetic disturbances in the polar cap, J. Geophys. Res., 76, 6028, 1971.

Horwitz, J. L., J. R. Doupnik, P. M. Banks, Y. Kamide, and S.-I. Akasofu, The latitudinal distributions of auroral zone electric fields and ground magnetic perturbations, and their response to variations in the interplanetary magnetic field, J. Geophys. Res., 83, 2071, 1978.

Hughes, T. J., and G. Rostoker, A comprehensive model current system for high-latitude magnetic activity - I. The steady state system, Geophys. J. R. Astro. Soc., 58, 525, 1979.

Iijima, T., and T. A. Potemra, The amplitude distribution of field-aligned aurrents at northern high latitudes observed by Triad, J. Geophys. Res., 81, 2165, 1976.

Kamide, Y., On current continuity at the Harang discontinuity, Planet. Space Sci., 26, 237, 1978.

Kamide, Y., and S.-I. Akasofu, The auroral electrojet and global auroral features, J. Geophys. Res., 80, 3585, 1975.

Kamide, Y., and S.-I. Akasofu, The location of the field-aligned aurrents with respect to discrete auroral arcs, J. Geophys. Res., 81 3999, 1976.

Kamide, Y., and G. Rostoker, The spatial relationship of field-aligned currents and auroral electrojets to the distribution of nightside auroras, J. Geophys. Res., 82, 5589, 1977.

Kamide, Y., and J. D. Winningham, A statistical study of the 'instantaneous' nightside auroral oval: The equatorward boundary of electron precipitation as observed by the Isis 1 and 2 satellites, J. Geophys. Res., 82, 5573, 1977.

Kamide, Y., S.-I. Akasofu, and G. Rostoker, Field-aligned currents and the auroral electrojet in the morning sector, J. Geophys. Res., 81, 6141, 1976.

Kamide, Y., J. S. Murphree, C. D. Anger. F. T. Berkey, and T. A. Potemra, Simultaneous observations of field-aligned currents and visible auroras by the Triad and Isis 2 satellites, J. Geophys. Res., 84, 4425, 1979.

Kan, J. R., S.-I. Akasofu, and L. C. Lee, Physical processes for the onset of magnetospheric substorms, in Dynamics of the Magnetosphere, edited by S.-I. Akasofu, D. Reidel, pp. 357-368, 1980.

Kaufmann, R. L., P. B. Dusenbery, B. J. Thomas, and R. L. Arnoldy, Auroral electron distribution function, J. Geophys. Res., 83, 586, 1978.

Klumpar, D. M., Relationships between auroral particle distributions and magnetic field perturbations associated with field-aligned currents, J. Geophys. Res., 84, 6524, 1979.

Klumpar, D. M., J. R. Burrows, and M. D. Wilson, Simultaneous observations of field-aligned currents and particle fluxes in the post-midnight sector, Geophys. Res. Lett., 3, 395, 1976.

Lui, A. T. Y., D. Venkatesan, C. D. Anger, S.-I. Akasofu, W. J. Heikkila, J. D. Winningham, and J. R. Burrows, Simultaneous observations of particle precipitations and auroral emissions by the Isis-2 satellite, J. Geophys. Res., 82, 2210, 1977.

Maynard, N. C., Electric field measurements across the Harang discontinuity, J. Geophys. Res., 79, 4620, 1974.

Meng, C.-I., Simultaneous observations of low energy electron precipitation and optical auroral arcs in the evening sector by the DMSP 32 satellite, J. Geophys. Res., 81, 2771, 1976.

Meng, C.-I., and S.-I. Akasofu, The relation between the polar cap auroral arc and the auroral oval arc, J. Geophys. Res., 81, 4004, 1976.

Mizera, P. F., D. R. Croley, Jr., F. A. Morse, and A. L. Vampola, Electron fluxes and correlations with quiet time auroral arcs, J. Geophys. Res., 80, 2129, 1975.

Pazich, P. M., and H. R. Anderson, Rocket measurement of auroral electron fluxes associated with field-aligned currents, J. Geophys. Res., 80, 2152, 1975.

Pike, C. P., and J. A. Whalen, Satellite observations of auroral substorms, J. Geophys. Res., 79, 985, 1974.

Pike, C. P., J. A. Whalen, and J. Buchau, A 12-hour case study of auroral phenomena in the midnight sector: Flayer and 6300-A° measurements, J. Geophys. Res., 82, 3547, 1977.

Rostoker, G., Y. Kamide, and J. D. Winningham, Simultaneous observations of the auroral electrojet and energetic particle fluxes in the morning sector, in preparation, 1981.

Sandford, P. B., Variations of auroral emissions with time, magnetic activity and the solar cycle, J. Atmos. Terr. Phys., 30, 1921, 1968.

Silsbee, H. C., and E. H. Vestine, Geomagnetic bays, thier frequency and current-system, J. Geophys. Res., 47, 195, 1942.

Snyder, A. L., Jr., and S.-I. Akasofu, Major auroral substorm features in the dark sector by a USAF DMSP satellite, planet. Space Sci., 22, 1511, 1974.

Snyder, A. L., Jr., S.-I. Akasofu, and T. N. Davis, Auroral substorms observed from above the north polar region by a satellite, J. Geophys. Res., 79, 1393, 1974.

Spiger, R. J., and H. R. Anderson, Electron currents associated with an auroral band, J. Geophys. Res., 80, 2161, 1975.

Sulzbacher, H., W. Baumjohann and T. A. Potemra, Coordinated magnetic observations of morning sector auroral zone currents with Triad and the Scandinavian Magnetometer Array: A case study, J. Geophys., 48, 7, 1980.

Tsunoda, R. T., R. I. Presnell, and T. A. Potemra, The spatial relationship between the evening radar aurora and field-aligned currents, J. Geophys. Res., 81, 3791, 1976.

Vickrey, J., R. Vondrak, and S. Matthews, The diurnal and latitudinal variation of auroral-zone ionospheric conductivity, J. Geophys. Res., 86, 65, 1981.

Wallis, D. D., C. D. Anger, and G. Rostoker, The spatial relationship of auroral electrojets and visible aurora in the evening sector, J. Geophys. Res., 81, 2857, 1976.

Wedde, T., J. R. Doupnik, and P. M. Banks, Chatanika observations of the latitudinal structure of electric fields and particle precipitation on November 21, 1975, J. Geophys. Res., 82, 2743, 1977.

Wescott, E. M., J. D. Stolarik, and J. P. Heppner, Electric fields in the vicinity of auroral forms from motions of Barium vapor releases, J. Geophys. Res., 74, 3469, 1969.

Whalen, J. A., R. A. Wagner, and J. Buchau, A 12-hour case study of auroral phenomena in the midnight sector: Oval, polar cap, and continuous auroras, J. Geophys. Res., 82, 3529, 1977.

Winningham, J. D., F. Yasuhara, S.-I. Akasofu, and W. J. Heikkila, The latitudinal morphology of 10-keV electron fluxes during magnetically quiet and disturbed times in the 2100-0300 MLT sector, J. Geophys. Res., 80, 3148, 1975.

WATERHOLE - AN AURORAL-IONOSPHERE PERTURBATION EXPERIMENT

B. A. Whalen, A. W. Yau, F. Creutzberg

Herzberg Institute of Astrophysics, National Research Council, Ottawa, Canada K1A 0R6

M. B. Pongratz

Los Alamos Scientific Laboratory, University of California, Los Alamos, N.M. 87545

Abstract. A sounding rocket carrying 100 kg of high explosives and plasma diagnostic instrumentation was launched from Churchill Research Range on 6 April 1980 over a premidnight auroral arc. The object of the experiment was to produce an ionospheric hole or plasma density depletion at about 300 km altitude on field lines connected to an auroral arc. The plasma depletion is produced when the explosive by-products (mostly water) charge-exchange with the ambient O^+ ions and then rapidly recombine. It was speculated that the presence of the "hole" would interfere with the field-aligned current systems associated with the arc and would in turn perturb the auroral source mechanism. The release occurred about 10 km poleward of the auroral arc fieldlines. As expected, a large ionospheric hole was detected by rocket-borne plasma sensors. Within a few seconds following the release (a) the energetic electron precipitation observed in the hole dropped to background levels, (b) the luminosity of the auroral arc observed by a ground-based auroral scanning photometer decreased by a factor of two, and (c) the ionospheric E region density below the hole decayed at a rate consistent with a sudden reduction in particle precipitation. The simultaneous onset of these gross changes in electron precipitation coincident with the release strongly suggests a cause and effect relationship. In particular, these results suggest that the ionospheric plasma and the field-aligned current systems play a crucial role in the auroral acceleration process.

Introduction

In this report, preliminary results from a recent rocket flight, in which an attempt was made to perturb the auroral particle acceleration mechanism, are presented. These results indicate that either this experiment had a most profound effect on the auroral arc system or that nature has contrived to produce a remarkable coincidence. It is left to the reader to decide which is the most credible interpretation.

Experimental Concept

In Figure 1, we show the experimental concept. Inherent in many recent models of the auroral acceleration mechanism is the requirement for strong field-aligned current systems to be present on geomagnetic field lines connected to discrete auroral arcs. Current continuity is retained by closure through the ionosphere and magnetosphere. Field-aligned currents are crucial to these mechanisms since it is these currents which cause the formation of the accelerating parallel electric fields either by a current driven instability creating a region of anomalously high resistivity (Kindel and Kennel, 1971), or perhaps by the formation of a double layer (Block, 1978). In any case, it is these currents which are an essential element in the transmission of energy from the deep magnetosphere to the ionosphere.

The perturbation concept relies on the assumption that the thermal plasma supplies some or all of the field-aligned current carriers, presumably the thermal electrons, at least at low altitudes. This assumption seems to be borne out by many observations reported in the literature. If now the plasma density over the auroral arc is artificially reduced, this current system could be interrupted which in turn may perturb the acceleration mechanism.

Let us consider some of the possible consequences of a chemical release experiment which depletes the plasma in the auroral ionosphere. If the plasma in a region of space were completely removed, no current could pass through the "hole" and it would rapidly polarise, inducing a parallel electric field ($E_{\parallel}$) inside the depleted volume. The induced potential drop across the "hole" could approach that of the presumed ionosphere-magnetosphere potential difference. An $E_{\parallel}$ of this magnitude would have a significant effect on the keV electron and ion precipitation in the "hole", and on the current density along the field line.

Perhaps a more probable scenario involves the

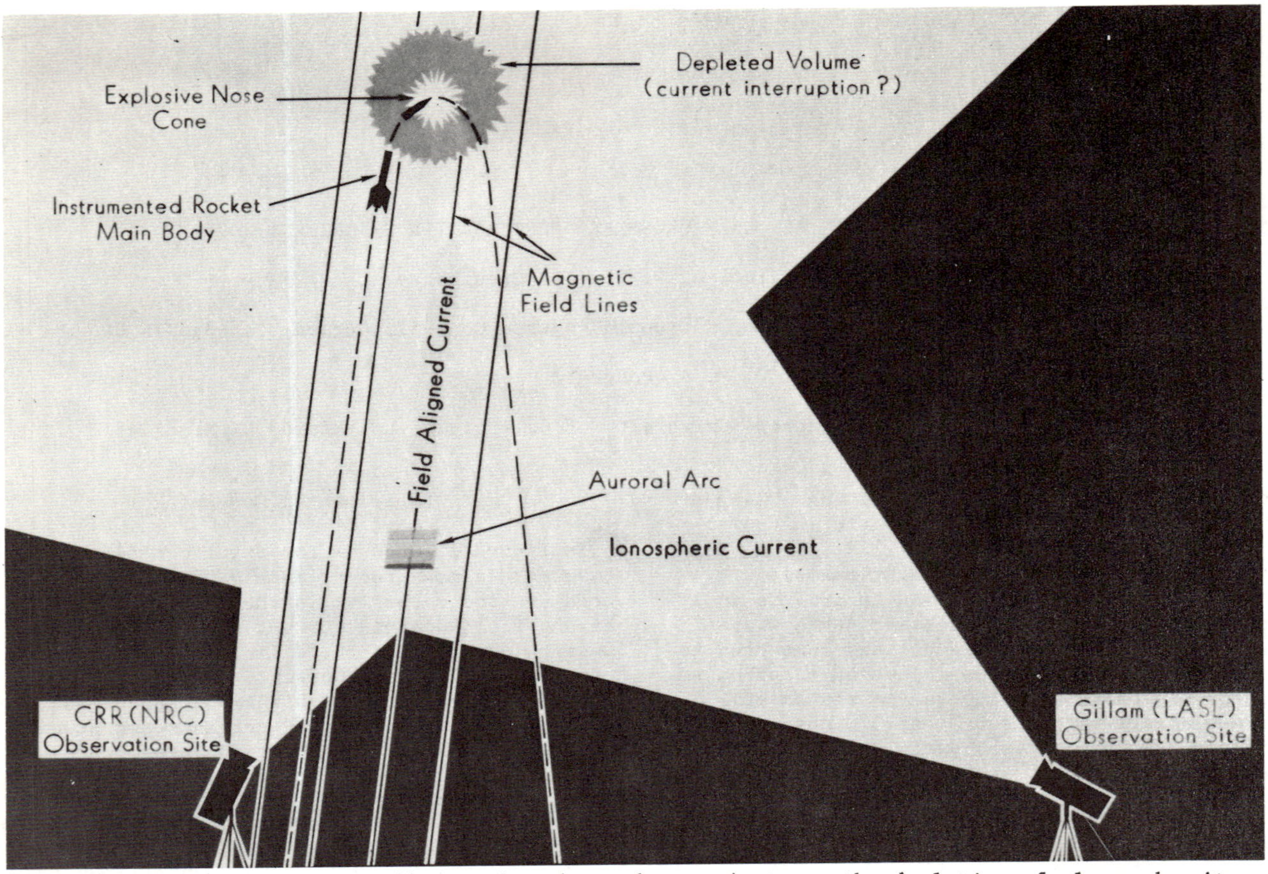

Fig. 1. Schematic diagram illustrating the rocket trajectory, the depletion of plasma density on the magnetic field lines connected to the auroral arc, assumed field-aligned currents, and ground observing stations at Churchill and Gillam.

following considerations. Since the release does not remove all the charge carriers, but simply reduces the density, current conservation requires the remaining carriers (presumably thermal electrons) to increase their directed velocities to keep the current flowing through the "hole" constant. Depending on the magnitude of the plasma depletion and the current density, the electron drift velocity may exceed the critical drift velocity. When this occurs, a number of instabilities may develop in the flow (Kindel and Kennel, 1971). One of the consequences of these instabilities may be the formation of an $E_{\parallel}$ in the "hole" and the modification of the field-aligned current systems. Assuming a nominal field-aligned current density of 0.01 mA/m^2 and an electron temperature of $3000^{o}K$, a depletion of an order of magnitude is required to make the drift velocity exceed the critical drift velocity. Depletions comparable to this were measured in previous experiments at lower latitudes (Operations Lagopedo Uno and Dos, see Pongratz et al., 1978).

Both of these scenarios suggest that the perturbation (the artificial depletion of the plasma density) may modify the particle precipitation and the field-aligned current in the aurora.

Experiment

In this experiment, the reduction in plasma density or depletion was produced by the explosive injection of chemically reactive vapors into the ionospheric F region (Pongratz et al., 1977). A 100 kg mixture of nitromethane and ammonium nitrate was detonated, releasing 1.5 x 10^{27} molecules of H_2O, 6.6 x 10^{26} molecules of N_2, and 3.3 x 10^{26} molecules of CO_2. Stoichiometrically,

$$2\ CH_3NO_2 + 3\ NH_4NO_3 \rightarrow 9\ H_2O + 4\ N_2 + 2\ CO_2. \quad (1)$$

The molecules produced by the release charge-exchanged with the ambient O^+ ions to form molecular ions, which then rapidly recombined, producing the plasma depletion. For the H_2O molecules, for example, the charge-exchange reaction

$$H_2O + O^+ \rightarrow H_2O^+ + O \quad (2)$$

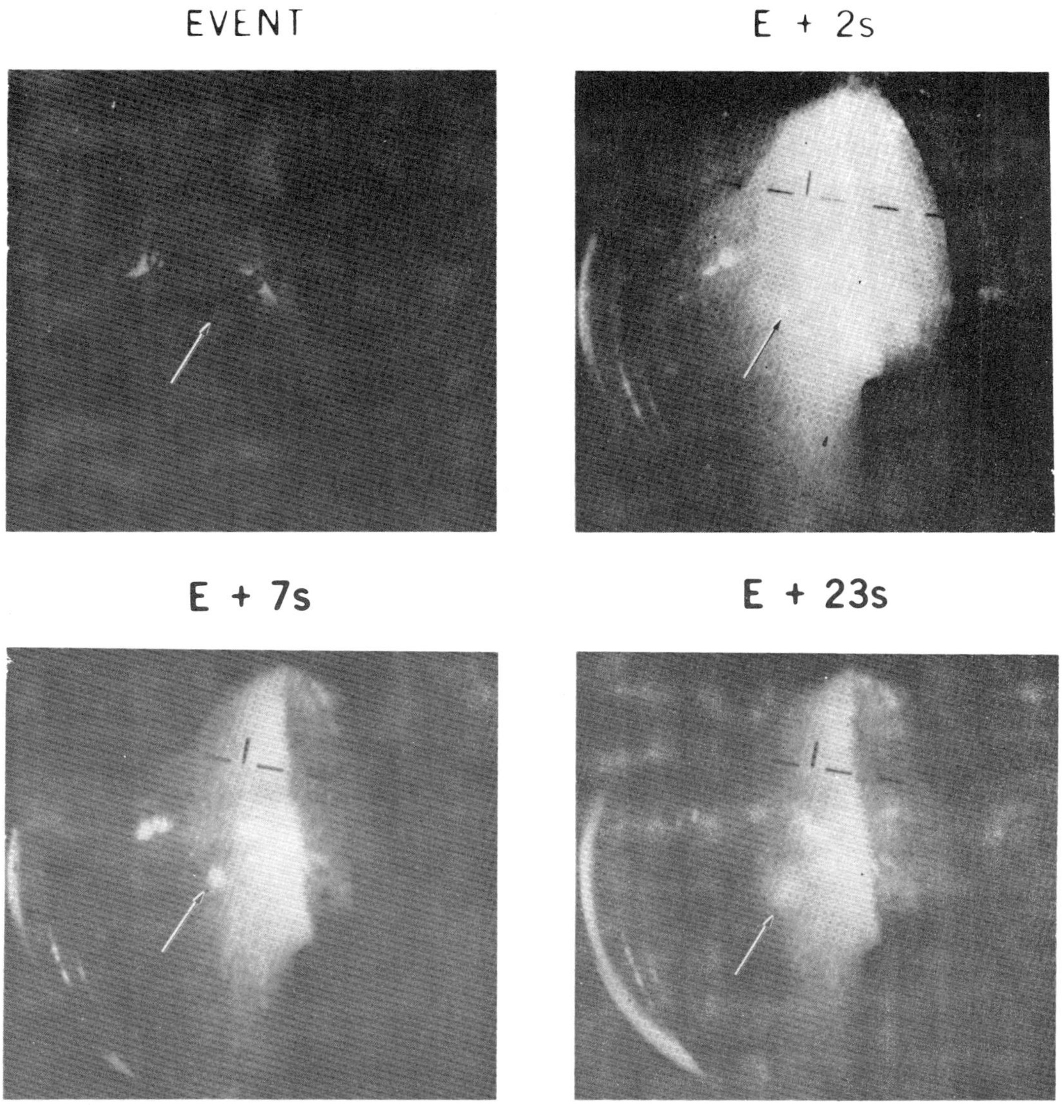

Fig. 2. White-light image intensified photograph of the cloud expansion as viewed from the Gillam site at the time of the event, and at 2, 7 and 23 s after the event.

is followed by the dissociative recombination reaction

$$H_2O^+ + e^- \rightarrow H + OH. \tag{3}$$

The net result of reactions (2) and (3) is the conversion of the $O^+ - e^-$ pair to neutrals:

$$H_2O + O^+ + e^- \rightarrow H + O + OH. \tag{4}$$

The plasma loss rate due to (4) is orders of magnitude faster than the corresponding rate due to direct recombination of ambient ions. It was anticipated that the release would produce a hole approximately 50 km in diameter inside which the plasma would be depleted by at least one order of magnitude.

The sounding rocket payload was a mother-daughter configuration. The ejected forward section carried the explosive release and the aft or "mother" section carried plasma diagnostics, energetic particle detectors and a bank of photometers to measure optical emissions induced by the release.

Three ground stations were used to measure emissions associated with the event as well as to monitor the large scale electron precipitation pattern. Two sites were established at Churchill, one at the range and one 10 km west of the launch site. A third site at Gillam, Manitoba (about

200 km south of CRR) was instrumented by the Los Alamos Scientific Laboratories team.

Results

A Nike-Black Brant V rocket carrying the "Waterhole" payload was launched on April 6, 1980 at 0333 UT into a relatively stable early evening auroral arc. At lift-off the arc was situated such that the pre-timed release would occur at 300 km on field lines connected to the aurora. Soon after launch the arc moved southward; however, the rocket flew on a more southerly azimuth than predicted, correcting somewhat for the unanticipated auroral motion. The net result was that the release occurred approximately 10 km poleward of the arc.

Figure 2 shows a time-lapse white-light image intensified photograph of the cloud expansion as viewed from the Gillam site. Also shown on the figure in the lower portion of each photograph is the auroral arc. The dark, irregularly shaped patches are reflections off the dome housing the camera and should be ignored. This figure demonstrates the rapid and more or less symmetrical expansion of the cloud about the release point to a diameter of approximately 40 km 23 seconds after the release.

Some of the plasma parameters measured by instruments on the "mother" section of the rocket payload and by a ground-based scanning photometer appear in Figure 3. The first panel (a) indicates the rocket altitude as a function of time just previous to and for 4 minutes after the release. Panel b shows Langmuir probe measurements of the local electron density (courtesy of A. G. McNamara, HIA, NRC) as a function of time. At the event time, indicated by the vertical line (0337:14) the electron density decreased abruptly by at least two orders of magnitude then recovered to approximately 1/10 of the original density where it remained for about three minutes. This profile results from a complicated mixture of temporal and spatial effects associated with the

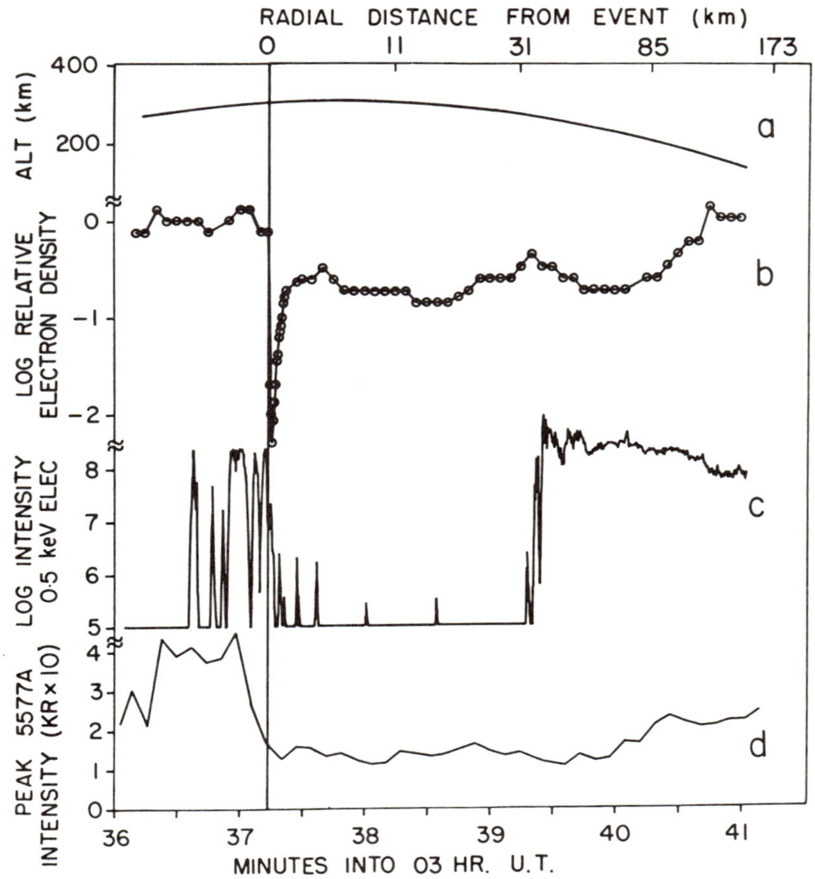

Fig. 3. Summary plots of in-situ particle and ground-based photometric observations before and after the event. Shown here are (a) rocket altitude; (b) relative local electron density; (c) precipitating electron intensity at 0.5 keV in units of $cm^{-2}s^{-1}sr^{-1}keV^{-1}$; (d) peak column emission intensities of the auroral green line measured at CRR. The vertical line indicates the time of the event (0337:14).

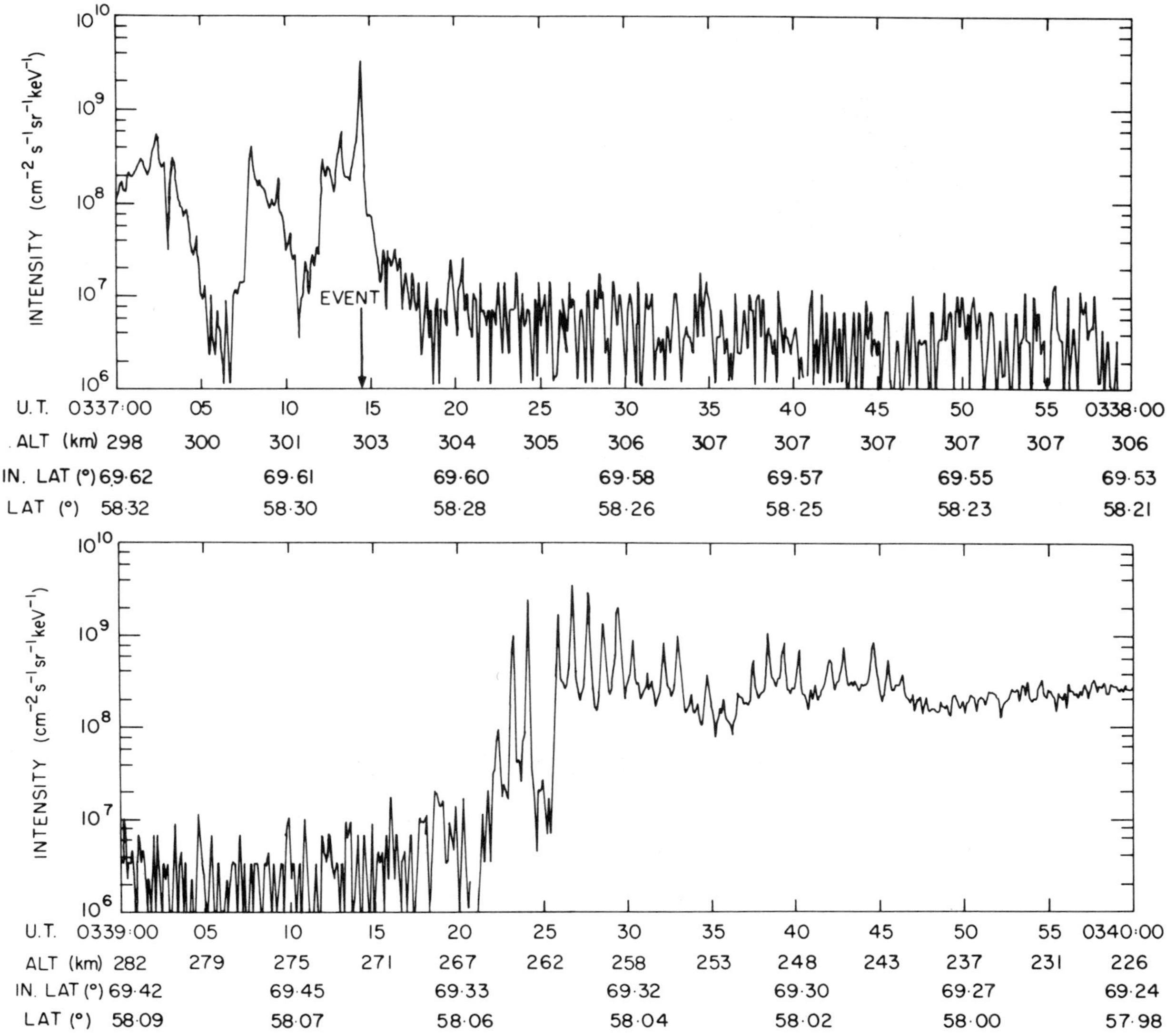

Fig. 4. Measured electron intensity at 0.5 keV from 0336:35 to 0340:05 showing the 2-s decay of the intensity.

cloud expansion about the release point and rocket motion out of the cloud.

Plotted in panel c is the precipitating electron intensity at 0.5 keV which is representative of the spectrum which peaked near this energy at this time. Before the release the intensity was observed to be fluctuating with the average energy and intensity increasing as the rocket approached the arc. Within two seconds of the release the 0.5 keV electron intensity dropped to background levels and remained there for approximately 130 seconds at which time it recovered to slightly in excess of the pre-event level.

The same 0.5 keV electron data are shown in an expanded time scale in Figure 4. The intensity decrease is seen to begin immediately after the event and reaches background level within 2 seconds. This extremely rapid response is both convincing evidence that the release caused the intensity decrease and the most difficult result to explain.

In Figure 5 measurements made by the meridian scanning photometer are displayed. The azimuth of the instrument was preset on the predicted rocket trajectory (135°); therefore the scan line ran from the northwest horizon (top of the figure) through the zenith to the southeast (bottom of the figure), one scan every seven seconds. The grey-scale used to indicate the observed 5577 Å intensity is shown at the bottom of the figure.

These data show that near lift-off (0333 UT) the aurora was near the zenith (90°) and moved

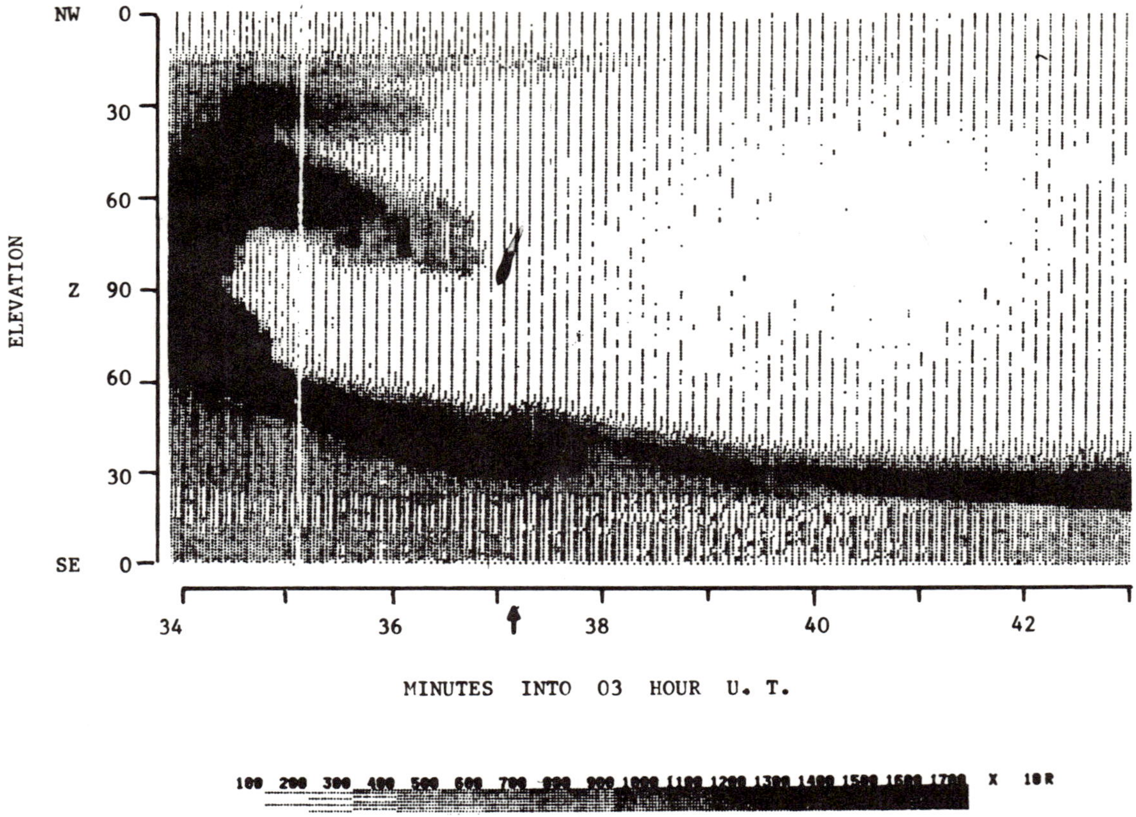

Fig. 5. The column emission intensity of the auroral green line measured by the meridian scanner from 0334 to 0343, as a function of elevation angle. The scan line ran from the northwest horizon (top of the figure) through the zenith to the southeast horizon (bottom of the figure). The grey-scale used to indicate the observed intensity is given at the bottom of the figure.

southward soon after. For a time period of approximately 5 minutes before the event the intensity of the arc on the line viewed by the photometer fluctuated between 20 and 30 kR. These fluctuations were caused by the motion of auroral rays or filaments through the field of view of the instrument. Starting at 0337:10, the average intensity of the arc decreased to around 10 kR and remained there for about three minutes when it recovered to its original state.

Peak intensities in the arc profile were measured for each scan and these data are plotted in Figure 3d. The large amplitude fluctuations observed preceeding the event are clearly shown. The intensity on the scan line was near a minimum at the event and no further enhancements were observed for about three minutes. All-sky camera (ASC) images indicate that near 0337:00, there was a gap in the arc at 135° azimuth, and the apparent intensity decrease observed in the scanner at this time was local in nature. On the other hand, the intensity decrease from 0338 to 0340 was large scale; indeed, the ASC images indicate that in this time period, the morphology of the aurora changed from a highly structured arc with rapidly moving rays to one which was more diffuse, homogeneous and less intense.

A total electron content beacon was also included in the rocket payload. This instrument measured the electron content between the rocket and a ground-based receiving station. Because of the rocket-receiving station geometry this instrument can be regarded as measuring the total electron content in the flux tube below the rocket. Results from this instrument have been reported elsewhere (MacDougall et al., 1980). Early in the flight the normal E and F region density profile, with some fluctuation associated with electron density irregularities, was observed. Coincident with the event, the TEC dropped precipitously and then followed a more gradual decay. The content then recovered to its expected level approximately 3 minutes later. These data can be shown to best fit a model which includes a rapid local depletion of plasma near the rocket followed by a decay in the TEC in the flux tube below, consistent with a complete cutoff of energetic electron precipitation on field lines passing through the "hole". These results are therefore in complete agreement with the electron detectors on the rocket.

Summary

To summarize, it appears that for a short time (130 seconds) following the release, the energetic electron precipitation through the "hole" was completely cut off. During roughly the same time period the local morphology of an auroral arc about 10 km equatorward of the centre of the "hole" changed from an intense rayed structure to a diffuse, less intense form. The mechanism, apparently induced by the release, responsible for altering the local electron precipitation pattern is not understood. A modification in the field-aligned current system could play an important role in this process; however, the near instantaneous response of the electron precipitation is hard to understand.

It is always possible, and perhaps the easiest explanation to accept, that the changes observed coincident with the release were fortuitous; however, the simultaneity of these events strongly suggests a cause and effect relationship. Plans are now underway to repeat this experiment in early 1982.

References

Block, L. P., A double layer review, Astrophys. Space Sci., 55, 59, 1978.

Kindel, J. M. and C. F. Kennel, Topside current instabilities, J. Geophys. Res., 76, 3055, 1971.

MacDougall, J. M., J. A. Fulford, and P. A. Forsyth, Ionospheric electron content measurements during "Waterhole", Can. J. Phys., submitted, 1980.

Pongratz, M. B., G. M. Smith, C. D. Sutherland, J. Zinn, P. W. Daly, B. A. Whalen, and T. E. Eastman, Artificial hole in the F region ionosphere. Theory and experiment, EOS Trans. AGU, 58, 455, 1977.

Pongratz, M. B., G. M. Smith, C. D. Sutherland, and J. Zinn, Lagopedo - two F-region ionospheric depletion experiments, in Effect of the Ionosphere on Space and Terrestrial Systems, ed. J. M. Goodman, US Govt. Printing Office, Washington, D.C., 1978.

FORMATION OF AURORAL ARCS AND INVERTED V PRECIPITATIONS: AN OVERVIEW

J. R. Kan and L. C. Lee

Geophysical Institute, University of Alaska, Fairbanks, Alaska 99701

Abstract. Recent theoretical developments on the formation of discrete auroras are reviewed to assess the progress of our understanding of discrete auroral phenomenon. In spite of the great complexity involved, a unified physical picture begins to emerge in which the roles of various auroral acceleration processes can be identified in the context of magnetosphere-ionosphere coupling. Potential drops for the formation of inverted V precipitations are supported predominantly by a combination of double layer and pitch-angle anisotropy processes. Additional electron energizations for the formation of thin auroral arcs are most likely due to filamentations of electrostatic ion cyclotron turbulence along the inverted V field lines. The discrete auroras can be understood as a manifestation of the imperfect coupling state of the magnetosphere-ionosphere interaction due to enhanced magnetospheric convection.

1. Introduction

The purpose of this paper is to provide a useful guide for those who are interested in the theoretical development of discrete auroral research. To set the stage for our theoretical discussion, it is necessary to briefly summarize what has been learned from observations including auroral morphology, field-aligned currents, electric fields, particles and waves on auroral field lines.

Discrete auroras appear as bright curtain-like structures extending along geomagnetic field lines. The latitudinal width of discrete auroras ranges from ~100 m for active arcs to ~10 km for homogeneous arcs [Davis, 1978], and up to ~100 km for inverted V electron precipitation bands [Frank and Ackerson, 1971; Lin and Hoffman, 1979]. Several thin auroral arcs often appear closely packed in a region believed to coincide with the inverted V precipitation band.

Discrete auroras are associated with precipitating electron fluxes peaked at energies between ~1 to 10 keV. These peaked electron fluxes appear to have been accelerated along geomagnetic field lines by parallel electric fields [Evans, 1972]. Outside the bright auroral forms, the average energy of precipitating electrons is usually around a few 100 eV [Arnoldy, 1974]. Field-aligned current density associated with bright discrete auroras is $\sim 10^{-5}$ A/m^2 [Anderson and Vondrak, 1975] and decreases to $\sim 10^{-6}$ A/m^2 under quiet conditions [Iijima and Potemra, 1978].

Equipotential structures deduced from electric field measurements on auroral field lines can be characterized as "V-shaped" and "S-shaped" [Gurnett, 1972; Cattell et al., 1979] as shown in Figure 1. The scale lengths L_x and L_z represent the latitudinal (i.e., transverse) and the field-aligned dimensions of the equipotential structure. For an auroral arc, $\lambda_D \lesssim L_x \lesssim \rho_i$; for an inverted V, $L_x \gg \rho_i$, where λ_D is the Debye length and ρ_i is the ion gyroradius. Observations indicate that the potential drop along auroral field lines is often extended with $L_z \gg \lambda_D$ [Mizera and Fennell, 1977; Sharp et al., 1979] rather than localized.

Particle observations [Meng, 1978; Hultqvist, 1979] revealed the existence of five particle species along auroral field lines. These are:

(1) Precipitating magnetospheric electrons;
(2) Background magnetospheric ions;
(3) Upstreaming ionospheric ions;
(4) Ionospheric electrons;
(5) Trapped electrons and backscattered electrons produced from precipitating electrons (due to interactions with neutrals and waves). The trajectories of these five particle species along the auroral field lines are schematically shown in Figure 2.

Plasma waves observed on auroral field lines can be classified as electromagnetic emissions and electrostatic turbulence. Electromagnetic emissions include auroral hiss, saucers, ELF noise and auroral kilometric radiation [Gurnett, 1978; Benson and Calvert, 1979]. Intense electrostatic turbulence in the frequency range of ~10 Hz to ~10 kHz has been observed to correlate with field-aligned currents [Gurnett and Frank, 1977; Temerin, 1978; Kintner et al.,

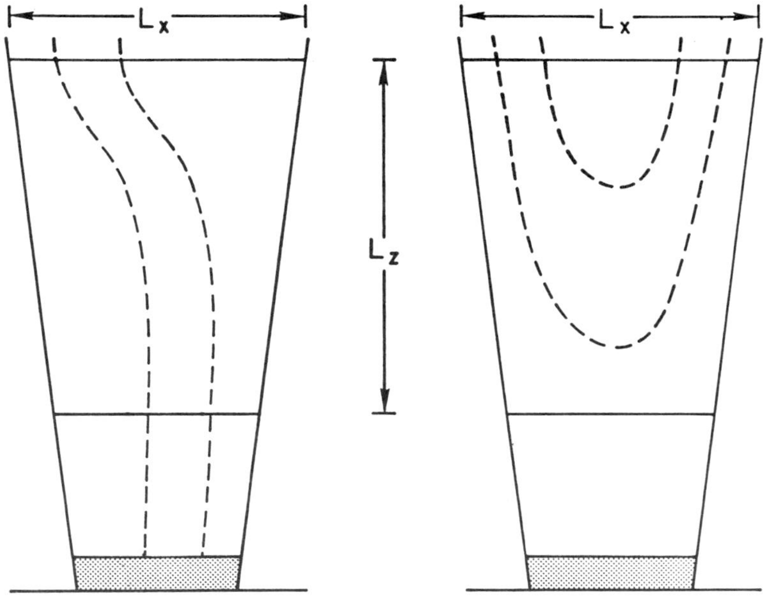

Figure 1. Schematic illustration of the S-shaped and V-shaped equipotential structures (dashed curved) along converging magnetic field lines.

1979]. These electrostatic waves are likely due to current-driven instabilities [Kindel and Kennel, 1971].

In addition to the east-west aligned arcs along auroral oval, there is a distinct class of sun-aligned arcs in the polar cap [Burke et al., 1981]. The polar cap arcs are known to correlate with the northward interplanetary magnetic fields, and hence present a unique opportunity for studying the energy transfer processes from the solar wind to the magnetosphere and the ionosphere under relatively undisturbed conditions.

2. Origin of Potential Drops Along Auroral Field Lines

In this section we summarize recent developments pertaining to the cause of potential drops along auroral field lines. It has been shown that enhanced magnetospheric convections lead to enhanced field-aligned currents and that potential drops are required when the enhanced upward field-aligned current density exceeds the limit set by the atmospheric loss cone. These interrelated processes are integral parts of the magnetosphere-ionosphere coupling system. The coupling can be termed "perfect" if geomagnetic field lines are equipotential with $E_{\parallel} = 0$ and "imperfect" if some field lines are non-equipotential with $E_{\parallel} \neq 0$. In this connection, the formation of discrete auroras can be considered as a manifestation of the imperfect coupling state due to enhanced magnetospheric convections.

2.1 Perfect Coupling State ($E_{\parallel} = 0$)

The importance of the ionosphere in regulating the magnetosphere was first recognized by Axford and Hines [1961], Dungey [1961] and Cole [1961]. Under the assumption of $E_{\parallel} = 0$, the interactions between the magnetosphere and the ionosphere have been studied extensively during the last two decades [Karlson, 1963; Fejer, 1964; Block, 1966; Wolf, 1970; Vasyliunas, 1970, 1972; Jaggi and Wolf, 1973; Mal'tsev, 1974; Wolf, 1974; Volland, 1975; Yasuhara and Akasofu, 1977; Nisbet et al., 1978; Sato, 1978; Nopper and Carovillano, 1978, 1979; Kamide and Matsushita, 1979a and b; Gizler et al., 1979; Miura and Sato, 1980; Harel et al., 1980a and b; Spiro et al., 1980].

A comprehensive effort in modeling the large-scale (~400 km in latitudinal width) magnetosphere-ionosphere coupling is being carried out by the Rice group [Harel et al., 1980a and b; Spiro et al., 1980]. This coupling model takes into account the effects of ionospheric conductivity and field-aligned currents in regulating the convection in the inner magnetosphere, where the convective inertia force and the plasma pressure gradient force are both important. Simulation results obtained from the Rice model can provide quantitative information for understanding the region II field-aligned current [Iijima and Potemra, 1976], the injection phenomenon at the synchronous orbit [Deforest and McIlwain, 1971] and the shielding process of the convection near the inner edge of the plasma sheet [Southwood and Wolf, 1978].

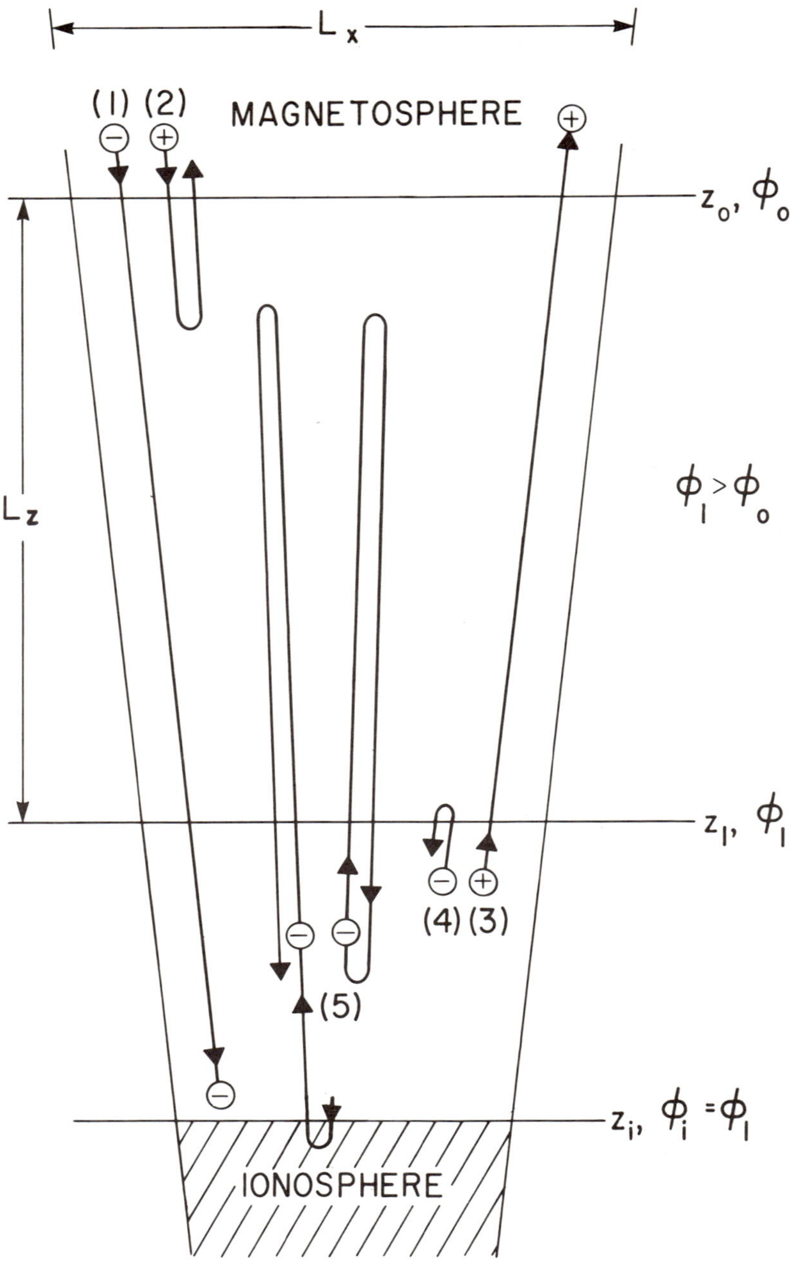

Figure 2. Schematic illustration of particle trajectories of (1) precipitating magnetospheric electrons; (2) magnetospheric ions; (3) ionospheric ions; (4) ionospheric electrons; and (5) trapped and backscattered electrons.

In addition to the large-scale magnetosphere-ionosphere coupling through the large-scale field-aligned current system [Iijima and Potemra, 1976], there are small-scale field-aligned current sheets [Cloutier et al., 1970] imbedded in the large-scale current region. The source mechanism for driving the large-scale current system can be understood in terms of the magnetospheric convection and pressure distribution in the plasma sheet [Wolf, 1970; Vasyliunas, 1972, Boström, 1975; Rostoker and Boström, 1976; Sonnerup, 1980]. The mechanism for driving the small-scale field-aligned current sheets is not so clear at present -- the source could be located either in the ionosphere or in the magnetosphere.

A possible ionospheric source for the small-scale current sheets was proposed by Sato [1978] and further studied by Miura and Sato [1980], according to whom a series of upward and downward current sheets can be produced by a feedback instability driven by the north-south ionospheric current. They found that the ionospheric current is unstable to electrostatic perturbation if the magnetosphere acts as an inductive load in the equivalent electrical circuit. The model is formulated from the linearized MHD equations under the assumptions that $E_{\parallel} = 0$ and that the waves are totally reflected from the equatorial plane in the magnetosphere. These limitations of the model should be removed to see whether or not the model is applicable to the formation of auroral arcs.

There are several possible magnetospheric sources for the small-scale current sheets. One has to do with the nonlinear ion acoustic and ion cyclotron waves driven by the large-scale field-aligned current [Chaturvedi, 1976; Myra and Liu, 1979; Temerin et al., 1979; Böhmer and Fornaca, 1979; Lee and Kan, 1981]. These nonlinear electrostatic waves are accompanied by intense perpendicular electric fields with wave length on the order of the ion gyroradius. These wave electric fields can drive pairs of small-scale upward and downward field-aligned current sheets closing through the ionospheric Pedersen current. It is also possible to generate field-aligned current sheets by the ion tearing instability in the plasma sheet as proposed by Goldstein and Schindler [1978].

2.2 Imperfect Coupling State ($E_{\parallel} \neq 0$)

Alfvén [1958] was the first to point out the importance of the parallel electric field on auroral field lines. Axford and Hines [1961] attributed the geomagnetic activity during disturbed periods to enhanced magnetospheric convection. The connection between magnetospheric convection and parallel electric fields has been studied by Coroniti and Kennel [1972], Boström [1974], Kan and Akasofu [1976], Lennartsson [1977], Goertz and Boswell [1979], Chiu et al. [1980], Lyons [1980] and Sonnerup [1980]. Recently Kan and Lee [1980c] formulated a theory of steady state imperfect magnetosphere-ionosphere coupling ($E_{\parallel} \neq 0$) in which the magnetosphere as well as the ionosphere is allowed to respond to the effects of the parallel electric field.

Perhaps the single most important fact behind the imperfect magnetosphere-ionosphere coupling is the limitation of the upward field-aligned current by the atmospheric loss cone. The presence of upward parallel electric fields is required to relax this current limitation. [Knight, 1973; Lyons et al., 1979; Fridman and Lemaire, 1980]. In other words, an upward parallel electric field is required whenever the upward field-aligned current density exceeds $\sim 10^{-6}$ A/m^2. On the basis of this requirement, the imperfect coupling can be understood as a consequence of an enhanced magnetospheric convection [Kan and Lee, 1980c]. The basic physical processes leading to the imperfect coupling are summarized in the flow chart shown in Figure 3. An enhanced magnetospheric convection ($\underset{\sim}{E}_m$) leads to an increase in the field-aligned current density ($J_{\parallel}$) feeding into the Pedersen current ($\underset{\sim}{I}_p$) in order to speed up ionospheric convection ($\underset{\sim}{E}_i$). If the resulting upward field-aligned current density exceeds a certain value, a parallel potential drop is required, enabling the magnetospheric electrons to carry the current by modifying the atmospheric loss cone. The current-carrying electrons are accelerated by the parallel potential and subsequently enhance the ionospheric conductivity (Σ_p) and reduce the ionospheric electric field. On the other hand, the enhanced field-aligned current leads to enhanced cross-field current in the magnetosphere ($\underset{\sim}{J}_{\perp}$) which increases the loading effect ($\underset{\sim}{J}_{\perp} \cdot \underset{\sim}{E}_m < 0$) on the magnetospheric convection. The above processes have been formulated within the framework of the MHD equations [Kan and Lee, 1980c]. They showed that the equipotential contours can be distorted into V-shaped structures centered on the convection reversal boundary, and S-shaped structures away from the

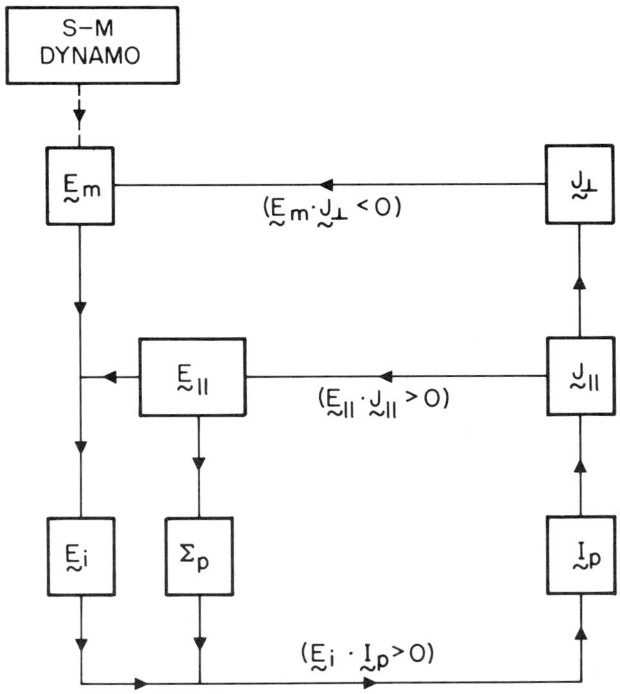

Figure 3. A flow chart for the imperfect magnetosphere-ionosphere coupling.

reversal boundary, as illustrated schematically in Figure 4.

The latitudinal scale length for the imperfect magnetosphere-ionosphere coupling [Kan and Lee, 1980c] is determined by the Pedersen conductivity via the closure of field-aligned currents in the ionosphere [Chiu and Cornwall, 1980]. Under moderately disturbed conditions, the latitudinal scale length is ~50 to 100 km, and increases with increasing Pedersen conductivity. In view of this scale length, we conclude that the inverted-V precipitation is a direct consequence of the imperfect coupling state, in which $E_{||}$ is supported by the auroral double layer process. Additional processes, possibly ion cyclotron turbulence or nonlinear ion cyclotron waves, are requried to generate fine structures ($L_x \lesssim \rho_i$) for the formation of thin auroral arcs inside the inverted V precipitation region.

3. Processes Supporting Parallel Electric Fields on Auroral Field Lines

Several elementary processes have been proposed in the literature to show that parallel electric fields can be supported in collision-

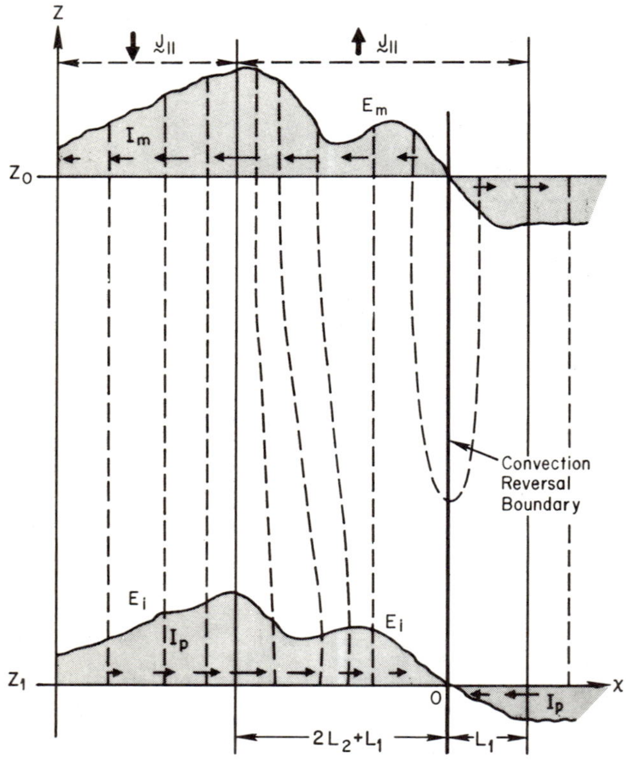

Figure 4. A schematic illustration of the imperfect magnetosphere-ionosphere coupling along a meridianal plane (the x axis is positive northward) in the evening sector.

less plasmas [Fälthammar, 1978]. These elementary processes include: (a) double layers, (b) electrostatic shocks, (c) differential pitch-angle anisotropy, (d) thermoelectric process, (e) anomalous resistivity, and (f) Alfvén waves. The main purpose of this section is to bring forth the concept that the formation of discrete auroras can be better understood in terms of a combination of elementary processes, rather than a single elementary process.

3.1 Elementary Processes Supporting Parallel Electric Fields

(a) Double-Layer Process. Double layer is an electrical discharge phenomenon first studied by Langmuir [1929]. A double layer is a potential structure self-consistently supported in a current-carrying plasma in which oppositely streaming electrons and ions are accelerated and extract energy from the double layer potential. Energy is supplied to the double layer potential by an external electromotive force (emf) in a closed electrical circuit. The potential drop in an unmagnetized double layer is localized in a few Debye lengths. On the other hand, the scale length of a magnetized double layer may depend on the ion gyroradius, the converging field scale length or the turbulent dissipation scale length which can be much greater than the Debye length. The presence of trapped electrons on auroral field lines play an important role in providing charges needed to maintain quasi-charge-neutrality in the extended potential structures along field lines. Magnitude of the double layer potential depends primarily on the applied emf in the circuit.

Double layers have been studied theoretically by Block [1972], Knorr and Goertz [1974], Kan [1975], Swift [1975], Levine and Crawford [1980], Kan and Lee [1980a]; experimentally by Langmuir [1929], Torven and Babic [1975], Quon and Wong [1976], Levine et al. [1978], Coakley and Hershkowitz [1979], Iizuka et al. [1979], and Stenzel et al. [1981]; and numerically by Goertz and Joyce [1975], DeGroot et al. [1977], Joyce and Hubbard [1978], Hubbard and Joyce [1979], Singh and Thiemann [1980], Sato and Okuda [1980] and Wagner et al. [1980].

(b) Electrostatic Shock Process. Electrostatic shocks are potential structures self-consistently supported in a streaming plasma [Moiseev and Sagdeev, 1963; Kennel and Sagdeev, 1967; Montgomery and Joyce, 1969; Forslund and Shonk, 1970; Tidman and Krall, 1971; Sakanaka et al., 1971; Mason, 1972; Biskamp, 1973]. There are no electrical currents in these electrostatic shock models. As the plasma flows through the shock potential, incident ions are decelerated while incident electrons are accelerated. The electron flux and the ion flux are individually conserved through the shock transition. Energy in the shock potential is supplied by the streaming ions and taken up by the streaming electrons. Since the ion flux equals the electron flux, the net energy gain by the shock potential is zero. The potential jump in a laminar electrostatic shock is localized in a few Debye lengths. This constraint on the scale length can be relaxed in the turbulent electrostatiac shocks [Biskamp, 1973]. Magnitude of the shock potential is bounded by the ion streaming energy.

It should be noted that there is widespread confusion in the auroral literature concerning the distinction between the double layer and the electrostatic shock. This confusion arose when the term "electrostatic shock" was first introduced into auroral research by Kan [1975] and Swift [1975]. It turns out that the model considered by Kan [1975] was a hybrid between the double layer and the shock, while the model considered by Swift was an oblique double layer. It will become clear in the next section that the electrostatic shock process can make only a secondary contribution to the acceleration of auroral electrons.

(c) Differential Pitch-Angle Anisotropy Process. Alfvén and Fälthammar [1963] pointed out the possibility of supporting parallel electric fields by differential pitch-angle anisotropy between electrons and ions in the absence of field-aligned currents. This mechanism was further developed by Persson [1963, 1966], Whipple [1977], Ponyavin et al. [1977], Chiu and Schulz [1978], Chiu and Cornwall [1980]. The converging magnetic field plays an important role in this process. The energy source for the potential in this model lies in the thermal energy associated with the anisotropy. The maximum potential difference generated by this process is bounded by the thermal energy of Maxwellian electrons or ions, whichever is smaller [Ponyavin et al., 1977]. This restriction can be relaxed under special conditions [Alfvén and Fälthammar, 1963]. The potential drop in this case is extended, and $L_z >> \lambda_D$.

(d) Thermoelectric Potential Process. Hultqvist [1971, 1972] suggested the thermoelectric contact potential between hot magnetospheric plasma and cold ionospheric plasma as a possible source of the parallel electric field on geomagnetic field lines. This effect can make a contribution to downward-directed electric field due to the higher mobility of the hot magnetospheric electrons, and thus could lead to deceleration rather than acceleration of precipitating electrons.

(e) Anomalous Resistivity Process. Anomalous resistivity results from the scattering of current-carrying electrons by the fields of unstable waves in a collisionless plasma. Steady-state anomalous resistivity depends on the ability of the plasma to sustain the wave turbulence in the nonlinear regime. The existence of steady-state anomalous resistance due to current-driven instability along auroral field lines remains somewhat controversial [Palmadesso et al., 1974; Ionson et al., 1976; Papadopoulos, 1977; Hudson et al., 1978 and references therein]. The most likely candidate for producing anomalous resistivity on auroral field lines appears to be electrostatic ion cyclotron turbulence [Kindel and Kennel, 1971] observed near 1 R_E altitude [Kintner et al., 1979]. A high level ion cyclotron turbulence can be sustained only if the perpendicular ion heating by the instability can be balanced by a heat transfer cooling process. Laboratory experiments on ion cyclotron turbulence [Bohmer and Fornaca, 1979] showed that ion heating results in a low-density, warm core surrounded by a denser hot ion cloud. Therefore, a high level ion cyclotron turbulence can be sustained in a core of ion gyroradius scale. Thus, it is possible for an unstable field-aligned current to produce intense ion cyclotron turbulence in thin sheets of widths comparable to the ion gyroradius. This filamentation process can produce fine structures in an unstable field-aligned current region and lead to the formation of auroral arcs imbedded in the inverted V precipitation region.

(f) Alfvén Wave Process. Parallel electric fields associated with oblique Alfvén waves have been considered as a possible mechanism for accelerating auroral electrons by Fejer and Kan [1969], Hasegawa [1976], and Goertz and Boswell [1980]. This process may contribute to transient or periodic acceleration of electrons along filed lines. It should be noted that Mallinckrodt and Carlson [1978], Sato [1978], Miura and Sato [1980] and others have studied propagations of Alfvén waves without parallel electric fields on auroral field lines. Possible sources for generating Alfvén waves on auroral field lines will be discussed in a later section. Such parallel electric fields arise mainly due to finite Larmor radius corrections to the usual MHD Alfvén mode. These short wavelength ($k_\perp \rho_i \simeq 1$) Alfvén waves are usually called kinetic Alfvén waves.

3.2 Distinction Between Double Layers and Electrostatic Shocks

As is evident in the literature, the double layer and electrostatic shock theoreis were developed as two independent entities until the term "electrostatic shock" was introduced into auroral research by Kan [1975] and Swift [1975]. The resulting confusion was further compounded by Hudson and Mozer [1978]. Goertz [1979] made an attempt to clarify the confusion.

Kan [1980] noted that the basic physical differences between the double layer and the electrostatic shock processes are: (i) the current $J_{||} \neq 0$ in the double layer, while $J_{||} = 0$ in the electrostatic shock; (ii) the immediate energy source for the double layer potential is electrical (through $J_{||} \neq 0$), while for the electrostatic shock it is mechanical (through ion streaming energy); (iii) the streaming ions are accelerated in the double layer while they are decelerated as they flow through the electrostatic shock.

Figure 5 illustrates schematically the differences between the double layer and the electrostatic shock. The average velocities $\langle V_e \rangle$ and $\langle V_i \rangle$ are in opposite directions in the double layer while they are in the same direction in the electrostatic shock. The electron flux and ion flux are individually conserved, which leads to the conservation of current with $J_{||} \neq 0$ in the double layer and $J_{||} = 0$ in the electrostatic shock. The kinetic energy K_e gained by the electrons is positive in both the double layer and the electrostatic shock; the kinetic energy K_i gained by the ions is positive in the double layer and negative in the shock.

From the above discussion, it can be seen that the electrostatic shock has very little to do with the acceleration of auroral electrons since the precipitating electron flux is always much greater than the precipitating ion flux on discrete auroral field lines. The energy supplied to the potential by the electrostatic shock process is proportional to the precipitating ion flux which is only a small fraction of the energy extracted by the precipitating electron flux. Hence, the energy for the potential can be expected to be supplied predominantly by the electromotive force in the circuit as in the double layer process.

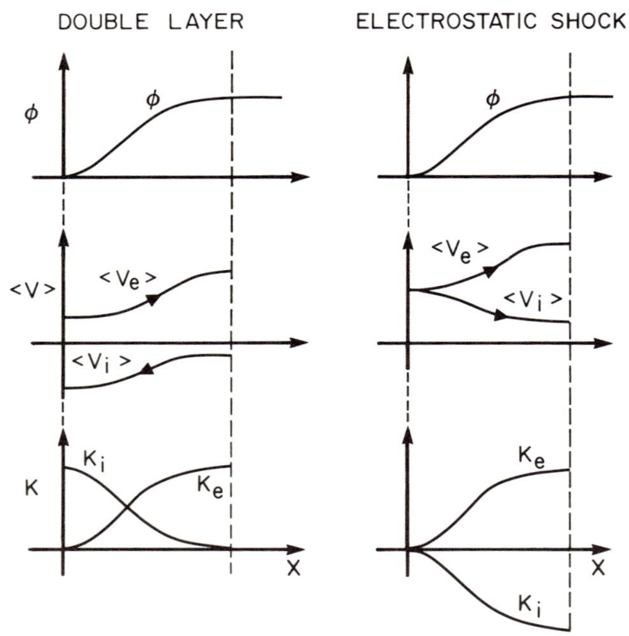

Figure 5. Comparison of the potential φ, the average velocity ⟨V⟩ and the average kinetic energy gain K between the double layer and the electrostatic shock.

3.3 Auroral Acceleration Process

As discussed earlier, auroral observations indicate: (i) the presence of upward field-aligned currents on discrete auroral field lines, $J_{||} \neq 0$; (ii) the presence of both hot magnetospheric plasma and cold ionospheric plasma; (iii) upstreaming ionospheric ions; (iv) trapped and backscattered electrons; and (v) an extended potential drop along field lines ($L_z >> \lambda_D$).

Comparison between the observed characteristics and the elementary processes (discussed in Section 2.1) suggests that the primary auroral acceleration process is a combination of the four elementary processes (a)-(d). Of these four processes, the double layer process and the differential pitch-angle anisotropy process are probably dominant in maintaining the potential drop along auroral field lines. The extended potential structure on auroral field lines ($L_z >> \lambda_D$) is certainly a characteristic of the pitch-angle anisotropy process. However, since upward field-aligned currents are always associated with discrete auroras and the potential drops (a few KV) are often much greater than the thermal energy of the magnetospheric electrons on auroral field lines (a few 100 eV), the double layer process appears to dominate over the pitch-angle anisotropy process at least when the precipitating electrons are more energetic (~5 to 10 keV). Moreover, the main energy source for the potential drop on auroral field lines is the electromotive force (emf) powered by the magneotspheric convection as discussed in Section 2. From the electrical circuit point of view, the potential drops along auroral field lines are maintained by the double layer process, regardless of the scale length of the potential structure. For these reasons, we suggest the term auroral double layer for the observed extended potential structures along auroral field lines.

Models developed in recent years for the auroral acceleration region are slowly converging toward the concept of the auroral double layer [e.g., Swift [1976; 1979], Chiu and Shulz [1978], Kan et al. [1979], Gile [1979], Chiu and Cornwall [1980], Kan and Lee [1980a; 1980b], Wagner et al. [1980]]. From these studies, it is found that the presence of trapped electrons and backscattered electrons is of fundamental importance in supporting the extended potential structure alng congerging auroral field lines. This and other characteristics of the auroral double layer are reviewed in the monogrpah by Lee and Kan [1981].

At this point, it may be useful to comment on the difference among existing two-dimensional auroral double layer models. Swift [1976; 1979] presented a model with $L_x \sim 2\rho_i$ in a uniforma

TABLE (1). AURORAL ACCELERATION PROCESSES

Processes \ Characteristics	$J_\parallel$	Energy Source	Remarks
(a) double layer	$J_\parallel \neq 0$	electrical energy (emf)	Auroral double layer process = (a) + (b) + (c) + (d), biased toward (a) and (c).
(b) electrostatic shock	$J_\parallel = 0$	ion streaming energy (mechanical)	Quasi-static laminar potential structure
(c) pitch-angle anisotropy	$J_\parallel = 0$	thermal energy	$L_x \sim \lambda_D$ to $>> \rho_i$, $L_z >> \lambda_D$
(d) thermoelectric	$J_\parallel = 0$	thermal energy	potential drop along field line $\Psi >$ few kV
(e) anomalous resistivity	$J_\parallel \neq 0$	electron streaming energy	Turbulent potential structure $L_x \lesssim \rho_i$, $L_z >> \lambda_D$
(f) Alfvén wave	$J_\parallel \neq 0$	wave energy	Transient or periodic potential $L_x \gtrsim \rho_i$, $L_z >> \lambda_D$

magnetic field. Since the backscattered and trapped electrons are not included in Swift's model, quasineutrality along field lines can only be maintained by the polarization drift of the upstreaming ionospheric ions. Due to this limitation, Swift [1979] noted that explicit solutions of his model have not yet been obtained. Kan et al. [1979] presented a model with $L_x \sim \lambda_D < \rho_i$ in a converging magnetic field. This model is developed for thin auroral arcs of a few 100 m [Maggs and Davis, 1968]. The upstreaming ions in this model are highly nonadiabatic. These ions oscillate back and forth across field lines while being accelerated upward by the parallel electric field. The highly nonadiabatic motions of these ions cannot be approximated by the polarization drift. For this reason, we choose to describe the number density of the upstreaming ions by treating them as unmagnetized. This could have underestimated the ion number density by a factor of $\sqrt{2}$. The quasineutrality along field lines is achieved by the backscattered and trapped electrons. This model has been further studied by Wagner et al. [1980] using a numerical simualtion technique. The formation of the V-potential double layer is indeed found to depend critically on the back-scattered and trapped electrons.

Chiu and Cornwall [1980] presented a model with $L_x >> \rho_i$ in which the polarization drift in unimportant. In this model, the scale length L_x is determined by the ionospheric conductivity through the closure of the large-scale field-aligned current. Again, the quasineutrality along field lines is maintained by the trapped and backscattered electrons in this model. Finally, it should be noted that Giles [1979] presented a model with $L_x \sim \lambda_D >> \rho_i$. This model is not applicable to the auroral problem due to the assumption of $\lambda_D >> \rho_i$ which cannot be satisfied on auroral field lines.

3.4 Summary

From the above discussion, it seems reasonable to conclude that the primary process for supporting quasi-static parallel electric fields on auroral field lines is the auroral double layer process, a combination of the four elementary processes (a)-(d) but dominated by the elementary double layer processes. Since the anomalous resistivity process (e) can operate in thin current sheets with thickness comparable to the ion gyroradius through filamentation of an unstable current by ion cyclotron turbulence [Böhmer and Fornaca, 1979], it may contribute to the potential drop in thin auroral arcs. The Alfvén wave process (f) may also contribute to the parallel electric fields on auroral field lines, but is unlikely to be a major factor for the acceleration of auroral electrons. An overall summary of this section in given in Table 1.

4. Conclusion

The formation of discrete auroras can be understood as a manifestation of the imperfect coupling state of the magnetosphere-ionosphere interaction due to enhanced magnetospheric convection. Acceleration processes for auroral electrons are reasonably well understood. Potential drops for the formation of inverted V precipitations are supported predominantly by a combination of double layer and pitch-angle anisotropy processes. Formation of thin auroral arcs imbedded in the inverted V region are most likely due to filamentations generated by electrostatic ion cyclotron turbulence.

Acknowledgments. This research was supported in part by the Atmospheric Science Section of NSF under Grant ATM79-16470 and by the U.S. Air Force under contract F19628-79C-0046 to the University of Alaska. We thank C.F. Kennel and J.M. Cornwall for illuminating comments on the manuscript.

References

Alfvén, H., and C.-G. Fälthammar, Cosmical Electrodynamics, Fundamental Principles, p. 162, Clarendon Press, Oxford, 1963.

Anderson, H. R., and R. R. Vondrak, Observations of Birkeland currents at auroral altitudes, Rev. Geophys. Space Phys., 13, 243, 1975.

Arnoldy, R. L., Auroral particle precipitation and Birkeland currents, Rev. Geophys. Space Phys., 12, 217, 1974.

Axford, W. I., and C. O. Hines, A unifying theory of high-latitude geophysical phenomena and geomagnetic storms, Can. J. Phys., 39, 1433, 1961.

Benson, R. F., and W. Calvert, ISIS 1 observations at the source of auroral kilometric radiation, Geophys. Res. Lett., 6, 479, 1979.

Biskamp, D., Collisionless shock waves in plasmas, Nuclear Fusion, 13, 719, 1973.

Block, L. P., On the distribution of electric fields in the magnetosphere, J. Geophys. Res., 71, 855, 1966.

Block, L. P., Potential double layers in the ionosphere, Cosm. Electrodyn., 3, 349, 1972.

Böhmer, H., and S. Fornaca, Experiments on nonlinear effects of strong ion cyclotron wave turbulence, J. Geophys. Res., 84, 5234, 1979.

Boström, R., Ionosphere-magnetosphere coupling in Magnetospheric Physics, ed. by B. M. McCormac, D. Reidel Publ. Co., Dordrecht-Holland, 1974.

Burke, W.J., Gussenhoven, M.S., Kelley, M.C., Hardy, D.A., and F.J. Rich, Electric and magnetic field characteristics of discrete arcs in the polar cap, in Physics of Auroral Arc Formation, AGU Publication, 1981.

Cattell, C. A., M. K. Hudson, R. L. Lysak, M. Temerin, R. B. Torbert, and F. S. Mozer, Measurements of mechanisms associated with low altitude auroral particle acceleration by the S3-3 satellite, EOS, AGU 60, 18, 346, 1979.

Chaturvedi, P. K., Finite amplitude electrostatic ion cyclotron modes, Phys. Fluids, 19, 1064, 9176.

Chiu, Y.T. and M. Schulz, Self-consistent particle and parallel electrostatic field distributions in the magnetosphere-ionosphere auroral region, J. Geophys. Res., 83, 629, 1978.

Chiu, Y. T., and J. M Cornwall, Electrostatic model of a quiet auroral arc, J. Geophys. Res., 85, 543, 1980.

Chiu, Y. T., J. M. Cornwall, and M. Schulz, Auroral magnetosphere-ionosphere coupling: A brief topical review, Solar-Terrestrial Predictions Proceedings, ed. R.F. Donnelly, Vol. 2, p. 494, NOAA, Boulder, Co., 1980.

Cloutier, P. A., H. R. Anderson, R. J. Park, R. R. Vondrak, R. J. Spiger, and B. R. Sandel, Detection of geomagnetically aligned currents associated with an auroral arc, J. Geophys. Res, 75, 2595, 1970.

Coakley, P., and N. Hershkowitz, Laboratory double layer, Phys. Fluids, 22, 1171, 1979.

Cole, K. D., On solar wind generation of polar geomagnetic disturbance, Geophys. J., 6, 103, 1961.

Coroniti, F.V. and C.F. Kennel, Polarization of the auroral electrojet, J. Geophys. Res., 77, 2835, 1972.

Davis, T. N., Observed characteristics of auroral forms, Space Sci. Rev., 12, 77, 1978.

Deforest, S. E., and C. E. McIlwain, Plasma clouds in the magnetosphere, J. Geophys. Res, 76, 3587, 1971.

DeGroot, J. S., C. Barnes, and A. E. Walstead, Localized structures and anomalous dc resistivity, Phys. Rev. Lett., 38, 1283, 1977.

Dungey, J. W., Interplanetary magnetic field and the auroral zones, Phys. Rev. Lett., 6, 47, 1961.

Evans, D. S., Auroral particles and fields, Ann. Geophys., 28, 639, 1972.

Fälthammar, C.-G., Generation mechanisms for magnetic-field-aligned electric fields in the magnetosphere, J. Geomag. Geoelectr., 30, 419, 1978.

Fejer, J. A., Theory of geomagnetic daily disturbance variations, J. Geophys. Res., 69, 123, 1964.

Fejer, J. A., and J. R. Kan, A guiding center Vlasov equation and its application to Alfvën waves, J. Plasma Phys., 3, 331, 1969.

Forslund, D. W., and C. R. Shank, Formation and structure of electrostatic collisionless shocks, Phys. Rev. Lett., 25, 1699, 1970.

Frank, L. A., and K. L. Ackerson, Observations of charged particle precipitation into the auroral zone, J. Geophys. Res., 76, 3612, 1971.

Fridman, M., and J. Lemaire, Relationship between auroral electron fluxes and field-aligned electric potential difference, J. Geophys. Res., 85, 664, 1980.

Giles, M. J., Approximate equipotentials for electrostatic V-shocks, Planet. Space Sci., 27, 1311, 1979.

Gizler, V. A., V. S. Semenov, and O. A. Troshichev, Electric fields and currents in the ionosphere generated by field-aligned currents observed by Triad, Planet. Space Sci., 27, 223, 1979.

Goertz, C. K., and G. Joyce, Numerical simulation of the plasma double layer, Astrophys. Space Sci., 32, 165, 1975.

Goertz, C. K., and R. W. Boswell, Magnetosphere-ionosphere coupling, J. Geophys. Res., 84, 7239, 1979.

Goldstein, H., and K. Schindler, On the role of the ionosphere in substorms: Generation of field-aligned currents, J. Geophys. Res., 83, 2574, 1974.

Gurnett, D. A., Electric field and plasma observations in the magnetosphere, in Critical Problems of Magnetospheric Physics, ed. by E. R. Dyer, 123, Nat. Acad. Sci., Washington, D.C., 1972.

Gurnett, D. A., and L. A. Frank, A region of intense plasma wave turbulence on auroral field lines, J. Geophys. Res., 82, 1031, 1977.

Gurnett, D. A., Electromagnetic plasma wave emissions from the auroral field lines, J. Geomagnet. and Geoelectr., 30, 257, 1978.

Harel, M., R. A. Wolf, P. H. Reiff, R. W. Spiro, W. J. Burke, F. J. Rich, and M. Smiddy, Quantitative simulation of a magnetospheric substorm, 1. Model logic and overview, submitted to J. Geophys. Res., 1980a.

Harel, M., R. A. Wolf, R. W. Spiro, P. H. Reiff, C.-K. Chen, W. J. Burke, F. J. Rich, and M. Smiddy, Quantitative simulation of a magnetospheric substorm, 2. Comparison with observations, submitted to J. Geophys. Res., 1980b.

Hasegawa, A., Particle acceleration by MHD surface wave and formation of aurora, J. Geophys. Res., 81, 5083, 1976.

Hubbard, R. F., and G. Joyce, Simulation of auroral double layers, J. Geophys. Res., 84, 4297, 1979.

Hudson, M. K., and F. S. Mozer, Electrostatic shocks, double layers, and anomalous resistivity in the magnetosphere, Geophys. Res. Lett., 5, 131, 1978.

Hudson, M. K., R. L. Lysak, and F. S. Mozer, Magnetic field-aligned potential drops due to electrostatic ion cyclotron turbulence, Geophys. Res. Lett., 5, 143, 1978.

Hultqvist, B., H. Borg, W. Riedler, and P. Christopherson, Observations of magnetic field-aligned anisotropy for 1 and 6 keV positive ions in the upper ionosphere, Planet. Space Sci., 19, 279, 1971.

Hultqvist, B., On the interaction between the magnetosphere and the ionosphere, in Solar-Terrestrial Physics/Proc. of the International Symp., Leningrad, USSR, May 11-19, 1970, ed. by E. R. Dyer, pp. 176-198, D. Reidel, Dordrecht-Holland, 1972.

Hultqvist, B., The hot ion component of the magnetospheric plasma and some relations to the electron component -- observations and physical implications, Space Sci. Rev., 23, 581, 1979.

Iijima, T., and T. A. Potemra, Large-scale characteristics of field-aligned currents associated with substorms, J. Geophys. Res., 81, 3999, 1976.

Iizuka, S., K. Saeki, N. Sato, and Y. Hatta, Buneman instability, Pierce instability, and double layer formation in a collisionless plasma, Phys. Rev. Lett., 43, 1404, 1979.

Ionson, J. A., R. S. B. Ong, and E. G. Fontheim, Anomalous resistivity of the auroral plasma in the topside ionosphere, Geophys. Res. Lett., 3, 549, 1976.

Jaggi, R. K., and R. A. Wolf, Self-consistent calculation of the motion of a sheet of ions in the magnetosphere, J. Geophys. Res., 78, 2842, 1973.

Joyce, G., and R. F. Hubbard, Numerical simulation of plasma double layers, J. Plasma Phys., 20, 391, 1978.

Kamide, Y., and S. Matsushita, Simulation studies of ionospheric electric fields and currents in relation to field-aligned currents, 1. Quiet periods, J. Geophys. Res., 84, 4083, 1979a.

Kamide, Y., and S. Matsushita, Simulation studies of ionospheric electric fields and currents in relation to field-aligned currents, 2. Substorms, J. Geophys. Res., 84, 4099, 1979b.

Kan, J. R., Energization of auroral electrons by electrostatic shock waves, J. Geophys. Res., 80, 2089, 1975.

Kan, J. R., L. C. Lee, and S.-I. Akasofu, Two-dimensional potential double layers and discrete auroras, J. Geophys. Res., 84, 4305, 1979.

Kan, J. R., and L. C. Lee, On the auroral double layer criterion, J. Geophys. Res., 85, 788, 1980a.

Kan, J. R.,. and L. C. Lee, Double-layer criterion on the altitude of the auroral acceleration region, Geophys. Res. Lett., 7, 429, 1980b.

Kan, J. R., and L. C. Lee, Theory of imperfect magnetosphere-ionosphere coupling, Geophys. Res. Lett., 7, 633, 1980c.

Karlson, E. J., Streaming of plasma through a magnetic dipole field, Phys. Fluids, 6, 798, 1963.

Kennel, C.F. and R.Z. Sagdeev, Collisionless shock waves in high β plasmas, 1, J. Geophys. Res., 72, 3303, 1967.

Kindel, J. M., and C. F. Kennel, Topside current instabilities, J. Geophys. Res., 76, 3055, 1971.

Kintner, P. M., M. C. Kelley, R. D. Sharp, A. G. Ghielmetti, M. Temerin, C. Cattell, P. F. Mizera, and J. F. Fennell, Simultaneous observations of energetic (keV) upstreaming ions and electrostatic hydrogen cyclotron waves, J. Geophys. Res., 84, 7201, 1979.

Knorr, G., and C. K. Goertz, Existence and stability of strong potential double layers, Astrophys. Space Sci., 31, 209, 1974.

Langmuir, I., The interaction of electron and positive ion space charges in cathode sheaths, Phys. Rev., 33, 954, 1929.

Lee, L. C., and J. R. Kan, Nonlinear ion acoustic wave and soliton in a magnetized plasma, Phys. Fluids, March issue, 1981.

Lee, L. C., and J. R. Kan, Auroral double layers in Physics of Auroral Arc Formation, Geophysical Monograph Series, AGU, 1981.

Levine, J. S., D. B. Ilič, and F. W. Crawford, Laboratory observations of plasma double layers, J. Geomag. Geoelectr., 30, 461, 1978.

Levine, J. S., and F. W. Crawford, A fluid description of plasma double-layers, J. Plasma Phys., 23, 223, 1980.

Lin, C. S., and R. A. Hoffman, Characteristics of the inverted V event, J. Geophys. Res., 84, 1514, 1979.

Lyons, L. R., D. S. Evans, and R. Lundin, An observed relation between magnetic field aligned electric fields and downward electron energy fluxes in the vicinity of auroral forms, J. Geophys. Res., 84, 457, 1979.

Lyons, L. R., Generations of large-scale regions of auroral currents, electric potentials, and precipitation by the divergence of the convection electric field, J. Geophys. Res., 85, 17, 1980.

Maggs, J. E., and T. N. Davis, Measurements of the thickness of auroral structures, Planet. Space Sci., 16, 205, 1968.

Mallinckrodt, A. J., and C. W. Carlson, Relations between transverse electric fields and field-aligned currents, J. Geophys. Res., 83, 1426, 1978.

Mal'tsev, Yu. P., The effect of ionospheric conductivity on the convection system in the magnetosphere, Geomag. Aeron., 4, 128, 1974.

Mason, R. J., Computer simulation of ion-acoustic shocks, II. Slug and piston problems, Phys. Fluids, 15, 845, 1972.

Meng, C.-I., Electron precipitations and polar auroras, Space Sci. Rev., 22, 223, 1978.

Mizera, P. F., and Fennell, J. F., Signature of electric fields from high and low altitude particle distributions, Geophys. Res. Lett., 4, 311, 1977.

Miura, A., and T. Sato, Numerical simulation of global formation of auroral arcs, J. Geophys. Res., 85, 73, 1980.

Moiseev, S. S., and R. Z. Sagdeev, Collisionless shock waves in a plasma in a weak magnetic field, Plasma Phys., 5, 43, 1963.

Montgomery, D., and G. Joyce, Shock-like solutions of the electrostatic Vlasov equations, J. Plasma Phys., 3, 1, 1969.

Myra, J. R., and C. S. Liu, Nonlinear stabilization of the ion-beam-cyclotron instability, Phys. Rev. Lett., 43, 861, 1979.

Nisbet, J. S., M. J. Miller, and L. A. Carpenter, Currents and electric fields in the ionosphere due to field-aligned auroral currents, J. Geophys. Res., 83, 2647, 1978.

Nopper, R. W. Jr., and R. L. Carovillano, Polar equatorial coupling during magnetically active periods, Geophys. Res. Lett., 5, 699, 1978.

Nopper, R. W. Jr., and R. L. Carovillano, Ionospheric electric fields driven by field-aligned currents, in Quantitative Modeling of the Magnetospheric Processes, Geophys. Monogr. Soc., 21, ed. by W. P. Olson, p. 557, AGU, Washington, D.C., 1979.

Palmadesso, P. J., T. P. Coffey, S. L. Ossakow, and K. Papadopoulos, Topside ionosphere ion heating due to electrostatic ion cyclotron turbulence, Geophys. Res. Lett., 1, 105, 1974.

Papadopoulos, K., A review of anomalous

resistivity for the ionosphere, Rev. Geophys. Space Phys., 15, 113, 1977.

Persson, H., Electric field along a magnetic line of force in a low-density plasma, Phys. Fluids, 6, 1756, 1963.

Persson, H., Electric field parallel to the magnetic field in a low-density plasma, Phys. Fluids, 9, 1090, 1966.

Ponyavin, D. I., M. I. Pudovkin, and S. S. Sazhin, Self-consistent field-aligned electric field in the earth's magnetosphere, Geomag. and Aeron., 17, 323, 1977.

Quon, B. M., and A. Y. Wong, Formation of potential double layers in plasmas, Phys. Rev. Lett., 37, 1393, 1976.

Rostoker, G., and R. Boström, A mechanism for driving the gross Birkeland current configuration in the auroral oval, J. Geophys. Res., 81, 235, 1976.

Sakanaka, P. H., C. K. Chu, and T. C. Marshall, Formation of ion-acoustic collisionless shocks, Phys. Fluids, 14, 611, 1971.

Sato, T., A theory of quiet auroral arcs, J. Geophys. Res., 83, 1042, 1978.

Sato, T., and H. Okuda, Ion acoustic double layer, Phys. Rev. Lett., 44, 740, 1980.

Sharp, R. D., R. G. Johnson, and E. G. Shelley, Energetic particle measurements from within ionospheric structures responsible for auroral acceleration processes, J. Geophys. Res., 84, 480, 1979.

Singh, N., and H. Thiemann, Some features of inverted V events as seen from simulated double layers, Geophys. Res. Lett., (accepted), 1980.

Sonnerup, B. U. Ö., Theory of the low-latitude boundary layer, J. Geophys. Res., 85, 2017, 1980.

Southwood, D. J., and R. A. Wolf, An assessment of the role of precipitation in magnetospheric convection, J. Geophys. Res., 83, 5227, 1978.

Stenzel, R. L., M. Ooyama, and Y. Nakamura, Potential double layers formed by ion beam reflection in magnetized plasmas, in Physics of Auroral Arc Formation, AGU Publication, 1981.

Swift, D. W., On the formation of auroral arcs and the acceleration of auroral electrons, J. Geophys. Res., 80, 2096, 1975.

Swift, D. W., An equipotential model of auroral arcs, 2. Numerical solutions, J. Geophys. Res., 81, 3935, 1976.

Swift, D. W., An equipotential model for auroral arcs: The theory of two-dimensional laminar electrostatic shocks, J. Geophys. Res., 84, 6427, 1979.

Temerin, M., The polarization, frequency, and wavelengths of high-altitude turbulence, J. Geophys. Res., 83, 2609, 1978.

Temerin, M., M. Woldorff, and F. S. Mozer, Nonlinear steepening of the electrostatic ion cyclotron wave, Phys. Rev. Lett., 43, 1041, 1979.

Tidman, D.A. and N.A. Krall, Shock Waves in Collisionless Plasmas, John Wiley and Sons, Inc., 1971.

Torvén, S. and M. Babic, Current chopping space charge layers in a low pressure arc plasma, Proc. 12th Int. Conf. on Phenomena in Ionized Gases, Eindhoven, Netherlands, Aug. 18-22, 124, North-Holland Publ. Co., 1975.

Volland, H., Models of global electric fields within the magnetosphere, Ann. Geophys., 31, 154, 1975.

Vasyliunas, V. M., Mathematical models of magnetospheric convection and its coupling to the ionosphere, in Particles and Fields in the Magnetosphere, ed. by B. McCormac, p. 60, D. Reidel, Dordrecht-Holland, 1970.

Vasyliunas, V. M., The interrelationship of magnetospheric processes, in Earth's Magnetospheric Processes, ed. by Mccormac, p. 29, D. Reidel, Dordrecht-Holland, 1972.

Wagner, J. S., T. Tajima, J. R. Kan, J. N. Leboeuf, S.-I. Akasofu, and J. M. Dawson, V-potential double layers and the formation of auroral arcs, Phys. Rev. Lett., 45, 803, 1980.

Whipple, E. C., The signature of parallel electric fields in a collisionless plasma, J. Geophys. Res., 82, 1525, 1977.

Wolf, R. A., Effects of ionospheric conductivity on convective flow of plasma in the magnetosphere, J. Geophys. Res., 75, 4677, 1970.

Wolf, R. A., Calculations of magnetospheric electric fields, in Magnetospheric Physics, ed. by B. McCormac, p. 167, D. Reidel, Dordrecht-Holland, 1974.

Yasuhara, F. and S.-I. Akasofu, Field-aligned currents and ionospheric electric fields, J. Geophys. Res., 82, 1279, 1977.

DOUBLE LAYERS IN THE LABORATORY AND ABOVE THE AURORA

Lars P. Block

Department of Plasma Physics, Royal Institute of Technology, S-100 44 Stockholm, Sweden

Abstract. Recent laboratory double layer experiments have simulated, much better than before, the conditions prevailing on auroral field lines at high altitudes. In particular, magnetic fields strong enough to magnetize the electrons (but not quite the ions) have been used. Particle and wave spectra have been measured. Wave-particle interaction has been shown to play a minor role in the only case that has been quantitatively analyzed. The three-dimensional potential distribution has been mapped. The particle budget requires the radial electric field to be outward in the no magnetic field case but inward with magnetic field, in agreement with what is observed above the aurora.

Introduction

Laboratory studies and computer simulations can often contribute greatly to our understanding of plasma processes in space (cf. e.g. Block, 1976). This is eminently true for the double layer phenomenon. The existence of double layers (henceforth abbreviated DL in this paper) would not even have been suggested, had they not been observed earlier in the laboratory.

It is, however, well known that a proper scaling between laboratory and space is difficult and sometimes almost impossible. The early DL experiments suffered from several deficiencies, the most important being:
(1) the chamber walls may have played an important role in the laboratory,
(2) the short Debye lengths in the laboratory made it virtually impossible to study the interior properties of the DL in sufficient detail to discriminate between monotonic and non-monotonic potential variations, to localize excitation of waves, or to measure particle velocity distributions at different locations within the DL,
(3) the laboratory plasmas were unmagnetized,
(4) the three-dimensional (3-D) particle flow and potential distribution were not measured,
(5) no detailed particle and momentum flow analysis was made,
(6) the role of wave-particle interaction was not established although many workers have reported wave emissions,
(7) no attempts were made to inject, on both sides of the DL, particles with velocity distributions that realistically reproduce those injected e.g. below and above DL on auroral geomagnetic field lines.

Recent improvements on the six first points listed above will be discussed here, mainly from the point of view of theoretical understanding of the DL phenomenon, *per se*. Discussion of (7) above on particle velocity distributions is deferred to the second half of the paper, devoted to comparisons with auroral DL.

Experimental Improvements in the Laboratory

Wall Effects

In the earliest experiments DL were observed at a position where the discharge tube cross section changed more or less abruptly (Crawford and Freeston, 1963; Schönhuber, 1963; Andersson *et al.*, 1969). A probable explanation for the existence of a DL at such a position is that the plasma properties depend on the tube cross section. In a narrow tube the rate of ionization has to be higher to cover the greater particle losses to the walls. That implies a higher electron temperature. A matching of two plasmas with different temperatures can most easily be done by means of a weak DL with a potential drop of the order of kT_e/e. No critical minimum current density is required for the existence of this kind of DL.

Unstable, very fast growing DL, where the potential drop can rise to 100 kV or more in a fraction of a millisecond and then collapse, have been shown to be due to a sudden rarefaction of the plasma density at a certain point (Torvén, 1968). Normally the plasma density is maintained by ionization. If, however, the current is so strong that it nearly equals the random electron current, a small decrease in the rate of ionization, due to e.g. an increased adsorption of neutral particles on the walls, will lead to a rarefaction instability and a DL (Carlqvist, 1972; Raadu and Carlqvist, 1979). The subsequent heating will then release gas from the walls with a DL collapse as a result.

The importance of wall effects is obvious in

the above examples. Torvén and Babić (1975) showed, however, that a relatively stable DL can be formed in a plasma tube with uniform cross section provided the current density exceeds a critical value corresponding to an electron drift speed about equal to the electron thermal speed. Still, the tube was narrow (2 cm diameter and 100 cm long) so wall effects could not be excluded. Later experiments in larger plasma chambers and/or axial magnetic fields have considerably reduced the influence of the walls. Other detailed studies to be discussed below have also shown that the walls do not play a decisive role.

Interior DL potential profile

In a large plasma chamber particle losses to the walls play a relatively minor role, due to the decreased area to volume ratio. Hence a more tenuous plasma with correspondingly longer Debye length can be produced. A DL will therefore be thicker, thus facilitating investigation of its interior. In this way DL in plasmas with Debye lengths of the order of millimeters (Quon and Wong, 1976; Coakley et al., 1978) or centimeters (Baker et al., 1981) have been studied and the potential variation within them have been found to be monotonic. Torvén and Lindberg (1980) were able to show the same in a plasma with only 0.4 mm Debye length, by feeding the signals from two narrowly spaced probes to the x- and y-inputs of an oscilloscope. The measured electric fields, potential drops and DL widths as related to Debye length and electron temperature agree reasonably well with theory and computer simulations in all these experiments.

Magnetized layers

The first indication of a DL in a longitudinal magnetic field was made by Lutsenko et al. (1975). However, their experiment was pulsed and the potentials measured with capacitive probes outside the tube. The first measurements with in situ (emissive) probes were made independently by Coakley et al. (1979) and Torvén and Andersson (1979), and later by Baker et al. (1981). Thus, it is now established beyond doubt that magnetized layers can exist under the same general conditions as unmagnetized ones. However, only the electrons have been well magnetized, while the ion gyro radii could not at all, or (Baker et al., 1981) only marginally be made smaller than the plasma column radius.

Some characteristics of DL could be demonstrated more easily in a magnetic field, due to the better beam confinement (Baker et al., 1981), as will be discussed below.

Attempts to magnetize the ions are very desirable and should be tried in the near future.

Two- and three-dimensional potential distribution

To understand the formation and stability of DL it is essential to consider the 2-D and 3-D potential distribution. With a few exceptions (Swift, 1976; Kan et al., 1979; Wagner et al., 1980) self-consistent theories have only been one-dimensional. Unfortunately, the few 3-D theories are not applicable to the laboratory experiments since they all assume magnetized ions as well as electrons. Even so, the laboratory measurements of 3-D potential distributions in non-magnetized and weakly magnetized DL plasmas (Coakley et al., 1979; Torvén and Lindberg, 1980; Baker et al., 1981) represent a considerable progress. Fig. 1 shows the potential distributions measured by Baker et al. (1981), who discuss their results qualitatively in terms of particle production, focussing and losses due to the radial electric field. In the no magnetic field case the radial electric field must be directed outward to maintain quasi-neutrality by confining the electrons (cf. Fig. 1a). With magnetized electrons (Fig. 1b) ion losses must be reduced at the high potential side of the layer, so the radial electric field is partly reversed. Note how the magnetization tends to make the potential distribution similar to that observed above the aurora.

Particle and momentum flow

Since the need for parallel electric fields was recognized, different mechanisms for production of such fields have been proposed. (For a reference see Block and Fälthammar, 1976).

The essential difference between them is the different processes balancing the momentum gain from the electric field for the majority of the electrons. For normal resistivity it is binary collisions, for anomalous resistivity wave-particle interaction, and for DL it is particle inertia. As is discussed in the following section, wave emissions habe been observed from most or all DL. It is, therefore, very important to quantitatively evaluate the relative roles of the three above mentioned processes.

The presence of a well-defined electron beam on the high potential side has been demonstrated by Crawford and Freeston (1963), Andersson et al. (1969), Andersson (1977), Coakley et al. (1978) and Baker et al. (1981). An approximate analysis by Baker et al. (1981) for the simpler case of magnetized electrons, allowing an essentially one-dimensional treatment, showed that inertia balances at least 50% of the momentum input from the electric field, binary collisions about 25%, and wave-particle interaction at most 25%. Beam divergence, which was not measured, could have shifted some of the 25% wave-particle interaction to inertia. Figure 2 shows the electron flux spectra upon which the above numbers are based. There is a clear beam population already on the low potential side of the layer. It is due to a strong potential drop in a sheath at the negative grid where plasma is injected. It can be seen that most of the momentum loss is suffered by this beam, which indicates some beam plasma interaction, but on the other hand beam divergence

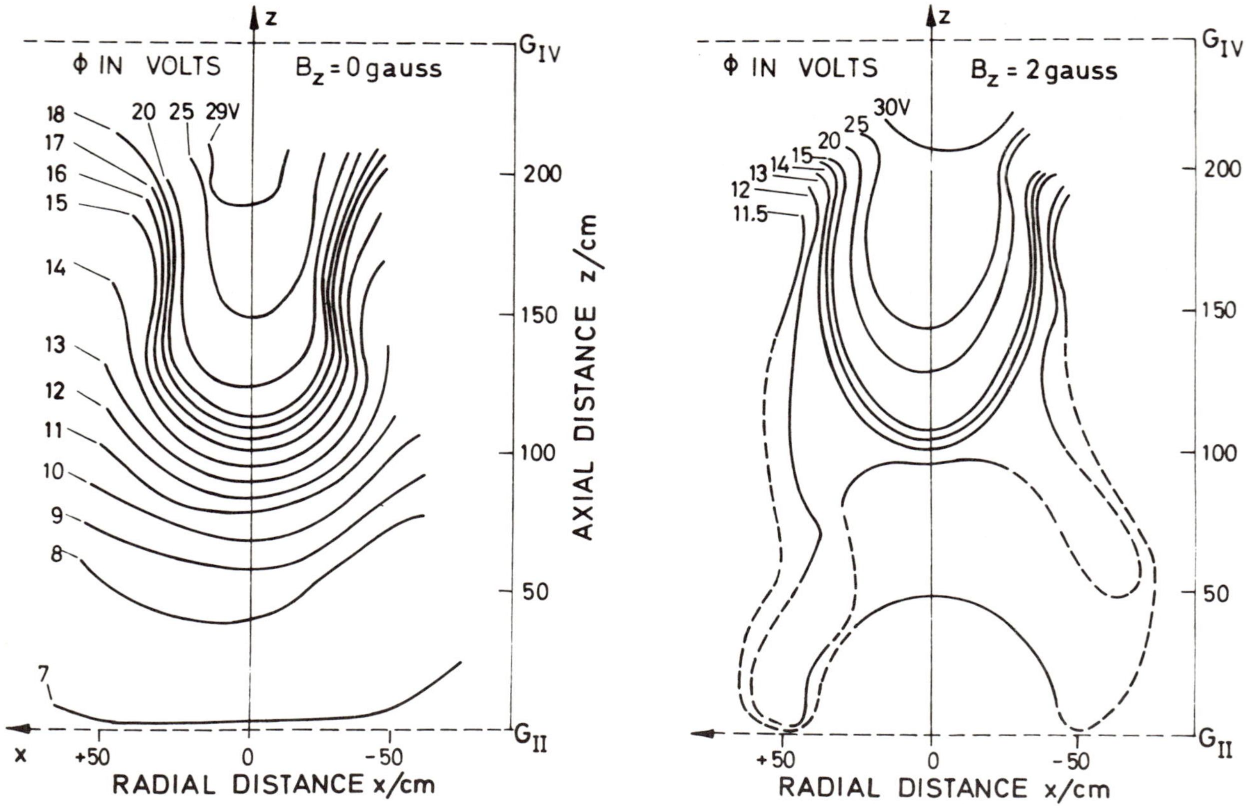

Fig. 1 Two-dimensional potential distributions in experiments by Baker et al. (1981): to the left with zero magnetic field and to the right with B = 2 gauss. Note that mostly the radial component is away from the central z-axis, thus reducing electron losses to the zero volt chamber walls. However, with 2 gauss field the radial E-field can partly turn towards the central axis.

could account for some of the apparent degradation, since the magnetic field confinement is least efficient for the most energetic electrons.

Role of wave-particle interaction

The above mentioned analysis of Baker et al. at least puts an upper limit (< 25%) on the role of wave-particle interaction, i.e. anomalous resistivity, for the momentum balance. We do not know of any other similar quantitative analysis, but the electron beams observed on the high potential side in several DL experiments (for references, see the preceeding section) strongly indicate that wave-particle interaction in any case does not play a major role.

Torvén and Lindberg (1980) observe low frequency fluctuations (~50 kHz << electron gyro frequency but >> ion gyro frequency at B = 50 gauss, cf. Fig. 3) generated and spatially concentrated within the layer due to rather regular fluctuations in DL potential and position. On the high potential side the electron beam is observed to generate a wave spectrum (100 - 1000 MHz) with maximum intensity at the plasma frequency (500 MHz). These waves propagate away from the DL with a nearly frequency independent phase velocity that is 10 - 20% less than the electron beam velocity. They are damped out within a few hundred Debye lengths from the DL without strongly degrading the beam. Apparently, the electron velocity distribution requires only a slight modification to make its wave-inducing power negligible.

More detailed wave measurements in laboratory DL are urgently needed.

Comparisons with Auroral Double Layers

There are both similarities and differences between the laboratory and space conditions that have to be taken into consideration when discussing the chances for DL formation above the aurora. We shall here discuss the following aspects of the problem:

(1) the need for plasma sources on both sides of the DL,
(2) a minimum electron drift velocity on the low potential side (Bohm condition),

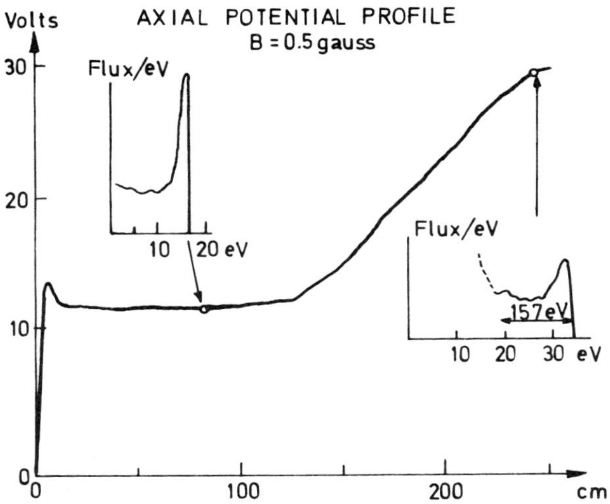

Fig. 2 Axial potential profile and energy distributions measured at indicated positions for the electron flux towards the positive plasma source before and after passage through the DL (Baker et al., 1981). The potential energy difference between the two spectra is 17.5 eV. The full curve part of the "After-spectrum" (17.5 - 34 eV) carries a momentum flux that is ~5.7 nN/m^2 larger than the "Before-spectrum"). Theoretically it should carry ~11.4 nN/m^2 more. Half of the missing 5.7 nN/m^2 can be attributed to binary collisions. The other half is shared in an unknown proportion between beam divergence and wave particle interaction. The dashed part of the "After--spectrum" is due to ionizing collisions within the DL. The ionization potential of Argon is 15.7 eV. The beam peak in the low potential spectrum is due to the sheath potential drop which in space may be equivalent to a DL at higher altitude. The DL at 130 - 250 cm thus proves that an already accelerated beam can induce a second DL.

(3) particle velocity distribution before and after passage through the DL,
(4) wave-emissions,
(5) magnetic field,
(6) potential distribution.

Plasma sources

Quon and Wong (1976), Torvén and Babić (1975) and Torvén and Lindberg (1980) have used a plasma source only on the low potential side of the DL. Torvén and Lindberg show that in their device ionization by the energetic electrons acts as a plasma source on the high potential side. The same mechanism may also work in Quon and Wong's double plasma device. In any case, it is very tricky to obtain a double layer with only one plasma source, but difficult to avoid it with two plasma sources at different potentials. In space there is a vast plasma source in the ionosphere, and the magnetospheric convection supplies plasma from the magnetotail to the auroral field lines, while simultaneously generating the necessary potential difference.

Critical electron drift velocity

It is well known from theories and computer simulation that formation and conservation of a DL requires that the electrons entering the DL on the low potential side have a minimum drift velocity of the order of the thermal electron velocity (see e.g. Block, 1972, 1978; Goertz and Joyce, 1975; Singh, 1980). This was also demonstrated experimentally by Baker et al. (1981). It is called the Bohm criterion (Bohm, 1949).

The exact value for the critical drift velocity depends on the particle velocity distributions on both sides of the layer. Kan and Lee (1980) show that it can be made arbitrarily small if sufficiently many trapped electrons with energies up to the DL potential drop exist

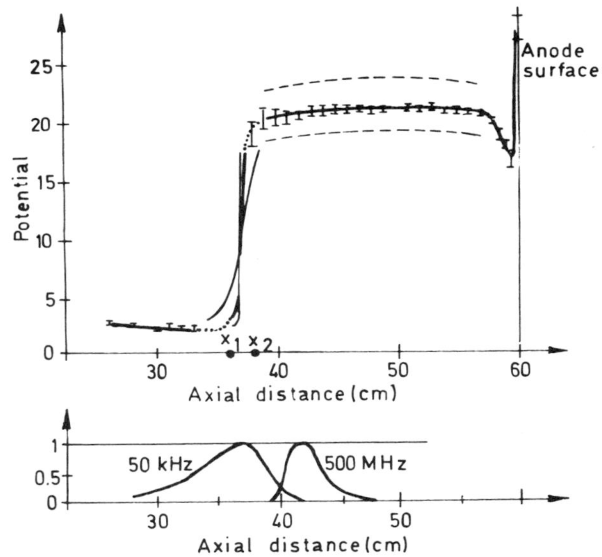

Fig. 3 Axial potential profile at the symmetry axis (upper diagram) with B = 50 gauss, in an experiment by Torvén and Lindberg (1980). The DL is situated between x_1 and x_2. The high potential level fluctuates between the two dashed levels, implying varying steepness of the potential curve within the DL, as indicated. This fluctuation, together with fluctuations back and forth in the x--direction with an amplitude of about the layer thickness, gives rise to a low frequency (50 kHz, bandwidth 8 kHz) spectrum with spatial distribution as shown by the left curve in the lower diagram. On the high potential side the electron beam generates waves propagating towards the anode, but damped out at 50 cm (500 MHz curve with spectrum extending from about 100 to 1000 MHz). However, the beam is essentially conserved up to the anode.

on the high potential side. These electrons supply the necessary negative charge upon reflection near the negative DL boundary.

Due to the loss cone there is always a net downward drift velocity on geomagnetic field lines. Electron absorption at 130 km altitude causes a net downward drift velocity equal to the thermal velocity of an initially isotropic distribution just below 300 km, i.e. where Albert and Lindstrom (1970) claimed to have demonstrated the existence of three ~100 V DL. Any upward parallel electric field will enhance the chances for DL formation in two ways: both drift velocity and the energetic trapped population due to backscattering will be enhanced. The latter effect will decrease the critical drift velocity (Kan and Lee, 1980). Hence, one DL could conceivably induce formation of an additional DL at higher potential, and so on. In other words, the available circuit voltage may well be shared by several layers in series (Block, 1972, cf. also the following section).

Lennartsson (1977) has considered this mechanism in rather great detail and proposed that the basic mechanism for production of parallel electric fields above the aurora is the magnetic mirror mechanism, but that when a current flows it is hardly possible to meet the quasi-neutrality condition everywhere. Hence, at least one DL must be formed somewhere. The mere existence of a magnetic mirror induced parallel field induces formation of one or more DL.

Particle velocity distributions

In the laboratory experiments made so far, no particular efforts have been made to realistically simulate the ionospheric and magnetospheric velocity distributions. In particular, no one has tried to inject a hot "plasma-sheet-like" plasma on the low potential side and a cool "ionospheric-like" plasma and energetic backscattered electrons on the high potential side. Even so, similarities have arisen, more or less by chance, as is discussed below.

The experiment by Baker et al. (1981) happens to provide an interesting example of another aspect, however. Fig. 2 shows the axial potential distribution in one of their layers along with the electron flux spectra before and after acceleration in the DL. The pronounced beam before acceleration is due to the sheath near the negative grid through which electrons (and ions) are injected. This is of course not realistic if we wish to simulate the normal plasma sheet. However, it is somewhat more similar to an already accelerated plasma sheet population. It has, in fact, been accelerated at the negative grid sheath, which from this point of view may be considered as another DL or any kind of parallel potential drop. Still, there are differences in the velocity distributions. As explained by Baker et al., the waterbag population at energies below that of the beam is due to ionization by the beam electrons between the sheath and the DL. Ionization is negligible thousands of km above the aurora. However, as will be explained in the section on potential distributions below, plasma may be injected below the highest, initially formed DL (corresponding to the sheath), and that plasma may act in the same way as the ionization in the laboratory, to create a potential plateau as a "platform" for the next DL. If that is correct, the experiment by Baker et al. supports the idea that there may be several layers in series on auroral field lines.

The plasmas injected on the low potential side in the experiments by Coakley et al. (1978) and Torvén and Lindberg (1980) were essentially Maxwellian, and in that respect they certainly simulate the plasma sheet rather well.

On the high potential side double-peaked spectra similar to auroral precipitation spectra have been observed in some experiments (Anderson et al., 1969; Coakley et al., 1978; Baker et al., 1981). The low energy peak, which is produced by local ionization and (in the case of Coakley et al. and Baker et al.) by a plasma source, may nevertheless contribute to the space charge in a way similar to the backscattered electrons seen in the auroral precipitation.

More experiments with different velocity distributions, separately controlled for injected electrons and ions on both sides of the DL are very desirable. Methods for backscattering or reflection of particles should also be used, e. g. magnetic mirrors.

Wave-emissions

As pointed out in an earlier section wave-particle interaction does not play a major role in the momentum balance, which determines the strength of the electric field. Whether that is true also in the auroral DL is a matter of controversy. No doubt, the beams produced by the DL must generate waves which are observed both in the laboratory and in space.

The experiment of Torvén and Lindberg (1980) provides a good example for comparisons. As shown in Fig. 3, the electron beam generates waves around the plasma frequency of 500 MHz in a spatially small region on the high potential side. The corresponding frequency spectrum covers about a decade in frequency. Phase velocity measurements showed that the power flow is from the electron beam to the waves. Nevertheless, the waves are damped out long before the beam, i.e. a plateau in the electron velocity distribution below the beam velocity is probably formed, such that wave generation ceases.

The VLF-emissions observed on auroral field lines may be of a similar nature provided the electron density at auroral DL is very low ($< 10\ cm^{-3}$) as is indicated by the S3-3 results (Mozer et al., 1979). However, they are not damped out as quickly. The reason for the damping and associated strong spatial confinement in the laboratory may be the limited radial extent of the plasma.

The ion cyclotron waves observed on S3-3 have no counterpart in the experiments since the ions are not magnetized. The 50 kHz fluctuations seen by Torvén and Lindberg in the DL region but not outside may be associated with trapped ionosonic waves that cause fluctuations in the DL potential and position.

Wave measurements in large plasma chambers, even without magnetized ions, are urgently needed, to see how the damping of different kinds of waves depends on radial density gradients.

Finally we note that the laboratory experiments do not indicate any necessity for waves to maintain a DL, but that waves are inevitably produced as a secondary effect. Neither does the theory of DL or of any BGK (Bernstein, Greene, Kruskal, 1959) equilibrium potential structures require that, although the stability is not yet understood. The laboratory DL experiments indicate that, at least under many circumstances, a constant non-fluctuating supply of charges is sufficient for stability. However, in space a great variability in charged particle supply must be expected, and hence it would indeed be surprising to find non-fluctuating DL there.

Magnetic field

From the above discussion we have seen that the too weak magnetic field is perhaps the most important deficiency of the experiments. Without a magnetic field
(1) the particle beams are not well confined,
(2) cyclotron waves cannot be excited,
(3) the three-dimensional potential distribution must differ from that around auroral DL (cf next section),
(4) magnetic mirror effects, important for the backscattering and trapping of energetic electrons that can contribute to the DL space charge balance, cannot be reproduced at all or must be replaced by other types of mirrors, e. g. electrostatic, which may influence the DL in other undesirable ways.

To produce strong magnetic fields in large plasma chambers is expensive. We may therfore have to wait a long time before DL with both magnetized ions and large radial extent can be studied.

Potential distribution

As has already been discussed, great progress has been made in the laboratory experiments to measure the 3-D potential distribution. In general, the radial field in laboratory discharges depends on the particle budget. The ion and electron losses in the radial direction differ by orders of magnitude, unless a radial electric field suppresses the dominant one and enhances the other. In the no magnetic field case the electron losses dominate, giving outward directed field as observed e.g. by Baker et al. (1981). With magnetized electrons, regardless of ion magnetization or not, the ion losses dominate and an inward electric field is necessary. Of course, the axial supply and losses have to be accounted for in order to determine the exact "turning point" (cf. discussion in Baker et al., 1981).

This view is confirmed both by Baker et al. (1981) as is seen in Fig. 1, and by the observed fields in space. In the ionosphere radial particle losses are of course of no importance since the plasma is essentially infinitely extended in the radial (horizontal) direction. However, inward directed electric fields usually prevail anyhow, because of current continuity with the upward directed Birkeland current.

Not only the losses but also the supply of particles depends on the potential distribution. This aspect deserves some further discussion, since some erroneous arguments against parallel field acceleration have been put forward, namely
(1) all particles in a precipitation spectrum must have been accelerated through the entire potential drop,
(2) no net acceleration can occur since the par-

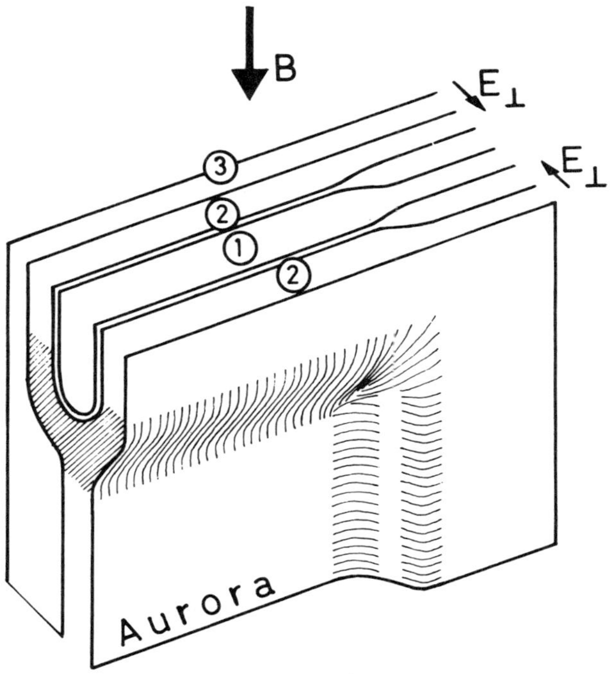

Fig. 4 Suggested equipotential surfaces above an auroral arc at a perpendicular electric field reversal. Electrons injected in region (1) will fall through the entire DL potential drop. Electrons drifting along surfaces in regions (2) at shaded altitudes will fall through part of the drop. Electrons on surface (3) will not see any parallel E-field.

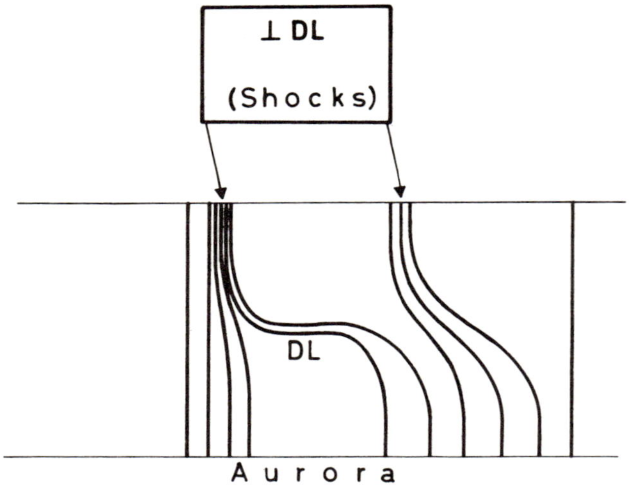

Fig. 5 Equipotential surfaces above an aurora with no perpendicular electric field reversal. Also in this case can electrons be injected sideways within the DL corresponding to the electrons formed through ionization in the experiment of Baker et al. (1981). Cf. Figure 2.

ticles cannot reach the high potential level without climbing the hill along the perpendicular electric field at high altitude.

The first argument presupposes no particle supply within the acceleration region. The second argument forgets that the particle supply is mainly along equipotentials, all of which can reach out to the solar wind. Besides, all generators lift up particles from low to high potential energy. There must be a generator in the circuit.

Figure 4 shows how it works. $\overline{E} \times \overline{B}$-drifting electrons in region 1 will fall through the entire parallel potential drop when they arrive above the arc. Electrons on equipotential surfaces in region 2, at altitudes corresponding to approximately the shaded area, will precipitate through part of the potential drop. The key is the indentations on the surfaces. Note that this also leads to a weaker ionospheric electric field within the arc than outside.

There must not necessarily be a perpendicular electric field reversal at the parallel field in the middle of the precipitation region. Displacement of the equipotential sideways as shown in Fig. 5 also gives a similar effect with one or possibly two shocks with the same, rather than reversed, perpendicular field, as sometimes observed on S3-3. Also in this case, a weakened southward or northward electric field will result in the arc.

A realistic model must also include an electric field component along the arc, since supply of particles at one end only would not suffice. It is easy to modify the potential distribution of Figures 4 and 5 accordingly, but difficult to make a good drawing of it.

Summary

Considerable progress in the understanding of DL has been achieved through recent laboratory studies. Many properties expected on auroral field lines have been reproduced:

(1) Two plasma sources at different potentials (ionosphere and magnetosphere) usually leads to formation of DL.

(2) A critical electron drift velocity is required on the low potential side. The loss cone will automatically lead to some finite drift velocity which can be enhanced by magnetic mirror induced fields. The critical drift requirement can be partly relaxed by backscattered or mirroring energetic electrons.

(3) The particle velocity distributions in the laboratory differ in some respects from those observed in space, but the electron distributions after acceleration are rather similar to auroral precipitation spectra. Ionization processes play a role in the laboratory, but that may have similar effects on the charge distribution and precipitation spectra as the injection of plasma within the acceleration region, which must occur in space.

(4) Waves do not play an essential role (or no role at all) in maintaining a layer, as shown by a momentum balance analysis of electron flow spectra in the laboratory. Waves are, however, generated by the electron beam on the high potential side, but they do not destroy the beam. A broad wave spectrum, centered at the plasma frequency, is observed in the laboratory and may correspond to VLF emissions from the acceleration region above auroras.

(5) Experimental DL with magnetized electrons but not ions have been studied in the laboratory. They are formed under the same conditions as DL with zero magnetic field. Since the ions are unmagnetized, no ion cyclotron waves are found, in contrast to what is seen on the S3-3 satellite. The magnetized DL results lend themselves more easily to demonstration of true DL properties than those of non-magnetized DL in the laboratory.

(6) The 3-D potential distribution with U-shaped equipotentials, oriented as is observed above auroras, can only be produced in the laboratory with a magnetic field. The reason is understood to be associated with the particle budget. To balance perpendicular losses of electrons and ions, the electric field must be directed towards the center with, but away without, a magnetic field.

The potentialities for further improvements in laboratory simulation of auroral DL are great. Good but incomplete wave measurements have been made in one experiment. Other experimenters have measured some particle spectra, but again with incomplete coverage. Stronger magnetic fields to magnetize the ions may be made within the near future but perhaps not in very large plasma chambers. More efforts should be made to separately control the velocity distributions of the different species in order to better simulate plasma sheet and ionospheric plasmas, and to improve our understanding of the dependence of DL properties on the velocity distributions.

References

Albert, R.D. , and P.J. Lindstrom, Evidence for Auroral Particle Acceleration and Trapping as a Result of Electrostatic Double Layers in the Ionosphere, Science, 170, 1398, 1970.

Andersson, D., Measurements of Electron Energy Distributions in Front of and Behind a Stationary Plasma Sheath, J. Phys.D: Appl. Phys., 10, 1549, 1977.

Andersson, D., M. Babić, S. Sandahl, and S. Torvén, On the Maximum Current Carrying Capacity of a Low Pressure Discharge, Proc. 9th Int. Conf. on Phenomena in Ionized Gases, Bucharest, Rumania, p. 142, 1969.

Baker, K.D., L.P. Block, R. Kist, W. Kampa, N. Singh, and H. Thiemann, Studies of Strong Laboratory Double Layers and Comparison with Computer Simulation, J. Plasma Physics (in print) 1981.

Bernstein, I.B., J.M. Greene, and M.D. Kruskal, Exact Nonlinear Plasma Oscillations, Phys. Rev., 108, 546, 1957.

Block, L.P., Potential Double Layers in the Ionosphere, Cosmic Electrodynamics, 3, 349, 1972.

Block, L.P., Interpretation of Laboratory Experiments of Interest to Space Physics, Physics of Solar Planetary Environments I, Proc. Int. Symp. on Solar-Terrestrial Phys., June 1976, Boulder, Col. (Ed. D.J. Williams). p.255. Publ. by AGU, 1976.

Block, L.P., A Double Layer Review, Astrophys. Space Sci., 55, 59, 1978.

Block, L.P., and C.-G. Fälthammar, Mechanisms that May Support Magnetic-Field-Aligned Electric Fields in the Magnetosphere, Ann. Geophys., 32, 161, 1976.

Bohm, D., Minimum Ionic Kinetic Energy for a Stable Sheath, The Characteristics of Electrical Discharges in Magnetic Fields (Ed. Guthrie and Wakerling) McGraw-Hill, N.Y., p. 77, 1949.

Carlqvist, P., On the Formation of Double Layers in Plasmas, Cosmic Electrodynamics, 3, 377 1972.

Coakley, P., N. Hershkowitz, R. Hubbard, and G. Joyce, Experimental Observations of Strong Double Layers, Phys. Rev. Letters, 40,230,1978.

Coakley, P., L. Johnsson, and N. Hershkowitz, Strong Laboratory Double Layers in the Presence of a Magnetic Field, Phys. Lett. 70A, 425,1979.

Crawford, F.W., and I.L. Freeston, Proc. VIth Int. Conf. on Phenomena in Ionized Gases, Paris, I, p. 461, 1963.

Goertz, C.K., and G. Joyce, Numerical Simulation of the Plasma Double Layer, Astrophys. Space Sci., 32, 165, 1975.

Kan, J.R., L.C. Lee, and S.-I. Akasofu, Two-Dimensional Potential Double Layers and Discrete Auroras, J. Geophys. Res., 84, 4305, 1979.

Kan, J.R., and L.C. Lee, On the Auroral Double Layer Criterion, J. Geophys. Res., 85,788,1980

Lennartsson, W., On the Role of Magnetic Mirroring in the Auroral Phenomena, Astrophys. Space Sci., 51, 461, 1977.

Lutsenko, E.I., N.D. Sereda, and L.M. Kontsevoi, Zh. Tekh.Fiz., 45, 789-96 (Engl. translation Sov. Phys. - Tech. Phys. 20, 498-502) 1975.

Mozer, F.S., C.A. Cattell, M. Temerin, R.B. Torbert, S. von Glinski, M. Woldorff, and J. Wygant, The DC and AC Electric Field, Plasma Density, Plasma Temperature, and Field-Aligned Current Experiments on the S3-3 Satellite, J. Geophys. Res., 84, 5875, 1979.

Quon, B.H., and A.Y. Wong, Formation of Potential Double Layers in Plasmas, Phys. Rev. Lett., 37, 1393, 1976.

Raadu, M.A., and P. Carlqvist, Electrostatic Double Layers and a Plasma Evacuation Process, Tech. Rep. TRITA-EPP-79-21, Dept of Plasma Physics, Royal Inst. of Technology, Stockholm, Sweden, 1979.

Schönhuber, M.-J., Über das Auftreten und Verhalten von Raumladungs-Doppelschichten in Hg-Niederdruckgasentladungen, Zeit. f. angew. Phys., 15, 454. 1963.

Singh, N., Computer Experiments on the Formation and Dynamics of Electric Doyble Layers, Plasma Physics (in press) 1980.

Swift, D.W., An Equipotential Model for Auroral Arcs: 2, Numerical Solutions, J. Geophys. Res., 81, 3935, 1976.

Torvén, S., Current Limitation and Electrical Clean-up in a Low Pressure Mercury Discharge, Arkiv för Fysik, 35, 513, 1968.

Torvén, S., and M. Babić, Current Chopping Space Charge Layers in a Low Pressure Arc Plasma, Proc. XIIth Int. Conf. on Phenomena in Ionized Gases, Eindhoven, the Netherlands, I, 124,1975.

Torvén, S., and D. Andersson, Observations of Electric Double Layers in a Magnetized Plasma Column, J. Phys. D.: Appl. Phys., 12, 717,1979.

Torvén, S., and L. Lindberg, Properties of a Fluctuating Double Layer in a Magnetized Plasma Column, Tech. Rep. TRITA-EPP-80-02, Dept of Plasma Physics, Royal Inst. of Technology, Stockholm, Sweden, 1980.

Wagner, J.S., T. Tajima, J.R. Kan, J.N. Leboeuf, S.-I. Akasofu, and J.M. Dawson, V-Potential Double Layers, Tech. Rep. PPG-457, Geophys. Inst., University of Alaska, Fairbanks, Alaska, and Phys. Dept, University of Calif., L.A., CA, 1980.

POTENTIAL DOUBLE LAYERS IN STRONGLY MAGNETIZED PLASMAS

R. L. Stenzel,[a] M. Ooyama, and Y. Nakamura

Institute of Space and Aeronautical Science, University of Tokyo, Komaba, Meguro-ku, Tokyo, Japan
[a] Department of Physics, University of California, Los Angeles, California 90024

Abstract. Experimental observations on a new potential double layer in a collisionless magnetoplasma are presented. The double layer is formed when an ion beam is injected along converging field lines and reflected at a boundary drawing electron saturation current. The double layer is detached from the boundaries, V-shaped for magnetized ions, establishes its potential self-consistently to the beam potential ($\phi_d \lesssim V_b$), is stationary over a wide potential range ($1 \lesssim e\phi_d/kT_e \lesssim 25$), and is produced free from ionization effects. Distribution functions measured on the high potential side show drifting Maxwellians subject to the Buneman instability ($v_d/v_e > 1$) and self-consistently produced trapped particles.

Introduction

Potential double layers are of current interest in basic plasma physics presenting a highly nonlinear state and in space plasma physics providing an explanation for generating fast particles. Previous research on double layers has included theories (Block, 1972; Carlqvist, 1972; Knorr and Goertz, 1974; Kan et al., 1979), computer simulations (Hubbard and Joyce, 1979; Wagner et al., 1980; Singh, 1980), observations in space (Wescott et al., 1976; Mozer et al., 1977) and laboratory experiments (Torven, 1979; Coakley and Hershkowitz, 1979; Crawford et al., 1979; Iizuka et al., 1979; Leung et al., 1980; Baker et al., 1980). While the latter have clearly established the existence of double layers they have so far been performed under conditions quite different from those in space. Double layers have been observed in connection with ionization phenomena in discharge columns with and without constrictions, upon injection of drifting electron distributions into unmagnetized or weakly magnetized plasmas in uniform magnetic fields. No double layer experiment with magnetized ions has been reported, to our knowledge, except for a Letter on the present work (Stenzel et al., 1980).

Experimental Setup

The experiment is performed in a double plasma device with surface magnetic field confinement as sketched in Fig. 1. Typical plasma parameters are for the density $n_e \simeq 10^8$ cm^{-3}, temperature $kT_e \simeq 2$ eV, $kT_i \simeq 0.2$ eV, gases H_2, Ar and Xe at pressures $2 \times 10^{-5} < p < 2 \times 10^{-4}$ Torr with base pressure of 10^{-6} Torr. By biasing the source plasma anode positively ($V_s > 0$), an ion beam of energy eV_b ($\simeq 0.85\ eV_s$) is injected into the target plasma whose anode is grounded. In the center of the target plasma (45 cm diam) a permanent magnet (6 cm diam, 2 cm length, $B \simeq$ 500 G at the pole face) is suspended with a dipole moment parallel to the chamber axis. A weak axial magnetic field ($B_o \simeq 20$ G) generated by external Helmholtz coils is applied in the direction opposite to the dipole moment. Thus, within the separatrix indicated by the dashed line there are closed dipole magnetic field lines, whereas, outside of it there are open field lines establishing some similarity with field configurations in the magnetosphere. The permanent magnet is made of several small samarium-cobalt magnets sandwiched between two soft-iron pole faces. These are insulated from each other and from ground to form two electrodes biased at V_m.

Plasma density and temperature are determined from plane Langmuir probes in regions of weak magnetic field. Electron temperatures are independently confirmed from sound speed measurements. The plasma potential ϕ is measured with an emissive probe (0.1 mm diam wire, 2 mm long) heated by 50 Hz ac current and its floating potential is measured with a vacuum tube voltmeter (75 MΩ input resistance). The emissive probe characteristic is also traced out so as to confirm that the floating potential closely matches the plasma potential ($\Delta\phi \lesssim 0.5$ V). Sampling the emissive probe characteristic with a boxcar integrator yields time resolved potential measurements. Ion and electron distributions are

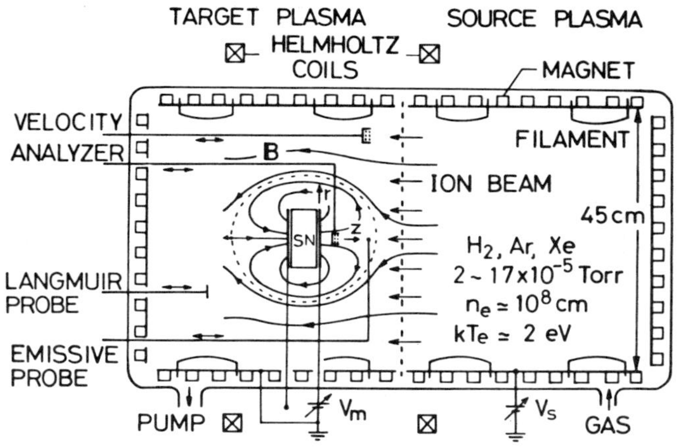

Fig. 1. Schematic view of the experimental setup. A plane ion beam of energy eV_b is injected into the target plasma against a biased magnet ($V_m > V_b > kT_e/e$). A uniform axial magnetic field is applied opposing the magnetic dipole.

measured with retarding grid velocity analyzers. A miniature analyzer (3 mm diam, 4 mm length) is used to determine the distributions near the front (N) pole face for velocities toward the magnet ($v_z < 0$).

Measurement Results

Sheath vs Double Layer

The axial potential profile $\phi_{(z)}$ near the front magnet surface has been measured in the absence and presence of a reflected ion beam. When $V_b = 0$ (no beam), we find an electron-rich sheath ($\rho = -\varepsilon_o \nabla^2 \phi < 0$) for a magnet voltage V_m positive with respect to the plasma potential. The sheath thickness at $\phi = V_m/2$ is approximately $4\ \lambda_D$. However, when an ion beam of energy eV_b is injected toward the magnet and reflected by the magnet potential ($V_m > V_b > kT_e/e$), we observe the formation of a stable double layer of a potential jump $\phi_d > kT_e/e$ and width $L = \mathcal{O}(10\ \lambda_D)$ well separated from the electrode sheath of residual potential drop $V_m - \phi_d$.

Some characteristic properties of the double layer are shown in Fig. 2. The measured potential profile is differentiated once to yield the electric field ($\vec{E} = -\nabla\phi$) and a second time to obtain the charge density ($\rho = -\varepsilon_o \nabla \cdot \vec{E} = q\delta n$). The potential ϕ_d of the double layer refers to the difference between the two potential plateaus connected by the double layer, and it is normalized to the electron temperature of the low potential target plasma ($\phi_d \simeq 22$ V, $e\phi_d/kT_e \simeq 9$). The double layer position z_{dl} refers to the position at which $\phi = \phi_d/2$. The double layer width is defined by $L = \phi_d/E_{max}$ and normalized to the Debye length in the target plasma ($L = 12.5$ mm, $L/\lambda_D \simeq 15$). The excess charge density δn_i of ions on the high potential side has a peak value $\delta n_{imax} \simeq 1.5 \times 10^7\ cm^{-3}$ and results in a surface charge of $\sigma_i = \int \delta n_i dz = 0.98 \times 10^7\ cm^{-2}$ which can be thought of as being distributed in a slab of thickness $\ell = \sigma_i/\delta n_{imax} \simeq 0.64$ cm. The electron surface charge is identical to that of the ions although we often find $\delta n_{emax} \neq \delta n_{imax}$. Normalized to the target plasma density ($n_{eo} \simeq 1.4 \times 10^8\ cm^{-3}$) the excess charge density is $\delta n_{emax}/n_{eo} \simeq 12\%$, i.e., considerably larger than $(m_e/m_i)^{1/2} \simeq 0.4\%$.

The stability of the double layer is determined from the potential fluctuations at different points along the z-axis. In the flat potential regions ($z \simeq 1$ cm, 5 cm in Fig. 2) the fluctuations $\delta\phi_{(t)}$ are small (0.1 V_{rms}) compared with those in the middle of the double layer ($z = z_{dl} \simeq 2.8$ cm, $\delta\phi \simeq 1\ V_{rms}$). From the level $\delta\phi_{max} \simeq 1\ V_{rms}$ and the potential gradient $(d\phi/dz)_{max} \simeq 17$ V/cm we obtain the spatial fluc-

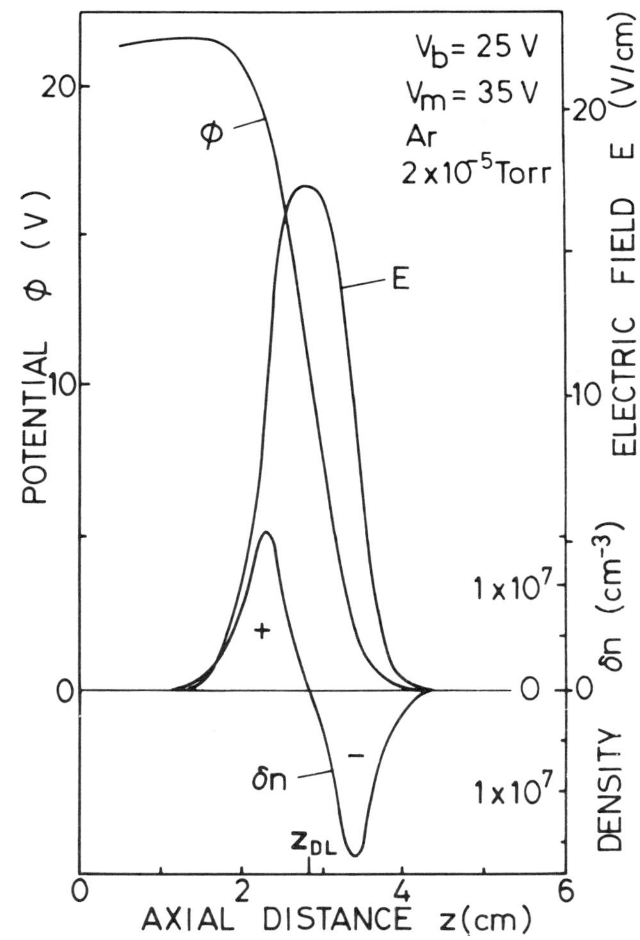

Fig. 2. Measured potential $\phi_{(z)}$ relative to target plasma potential, and deduced axial electric field $E(z)$ and space charge density $\delta n(z)$ (+ ions, - electrons).

tuation $\delta z \simeq 0.6$ mm. Thus, in contrast to many other observations our double layer is extremely stable ($\delta z < \lambda_D \simeq 0.8$ mm).

Potential, width, and position of the double layer depend on many parameters which will be discussed in the next subsection.

Parameter Dependence

In order to form the double layer two conditions have to be satisfied: (i) The magnet voltage has to exceed the plasma potential by $V_m - \phi_p > kT_e/e$ so that it is possible for the electron drift velocity to exceed the thermal velocity in some regions of the plasma. In fact, the transition from $v_d < v_e$ to $v_d > v_e$ marks the position for a possible double layer. (ii) The injected ion beam has to be reflected, i.e., $V_m > V_b$ so that an ion drift opposite to the electron drift is established. The second condition is demonstrated in Fig. 3 which shows axial potential profiles at a fixed beam voltage (V_b = 39 V) for different magnet voltages. For $V_m < V_b$ a sheath is observed, for $V_m > V_b$ a double layer is formed. The transition is gradual ($35 \lesssim V_m \lesssim 40$ V) since there is a spread toward lower parallel velocity components caused by mirror effects along converging magnetic field lines. With increasing V_m the double layer potential approaches a limiting value $\phi_d \lesssim V_b$ and the remaining potential difference $V_m - \phi_d$ drops off in the sheath. The double layer position z_{dl} increases with V_m. Associated with this is an increase in cross sectional area of the double layer ($S \propto B^{-1}$) and a corresponding increase in total current to the magnet, I_m.

When the beam voltage is varied at a fixed magnet voltage ($V_m > V_b$), the potential profiles are as shown in Fig. 4(a). The double layer potential increases with V_b as well as the position at which ϕ exceeds the target plasma potential by kT_e/e, the latter again causing an increase in I_m due to enhancement of the electron collecting area. The systematic increase in double layer potential ϕ_d with V_b is observed over a wide range of beam voltages as summarized in Fig. 4(b). For both weak and strong double layers ($1 \lesssim e\phi_d/kT_e \lesssim 26$), we find $\phi_d \lesssim V_b$. The beam voltage represents a natural upper limit for ϕ_d since for $\phi_d > V_b$ no ions can pass the double layer and neutralize the accelerated electrons. We also note that with increasing V_b the potential ϕ_d can be substantially lower than V_b which will be explained later.

Double layer potential and position also depend

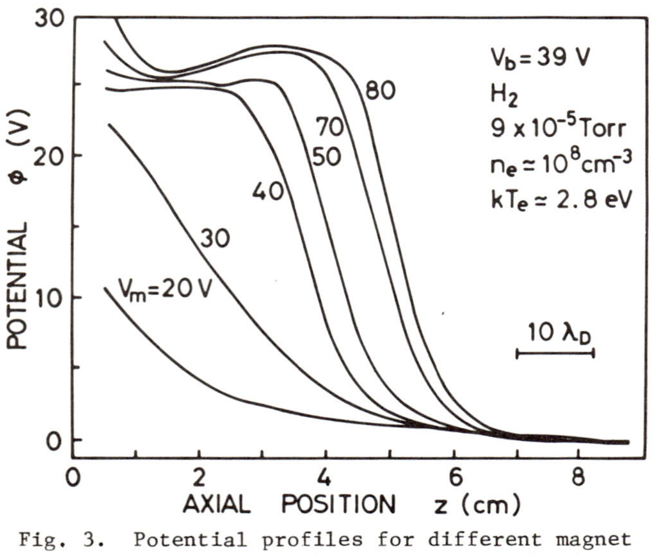

Fig. 3. Potential profiles for different magnet voltages V_m at a fixed beam voltage V_b showing double layer formation for $V_m \gtrsim V_b$, sheath formation for $V_m \lesssim V_b$.

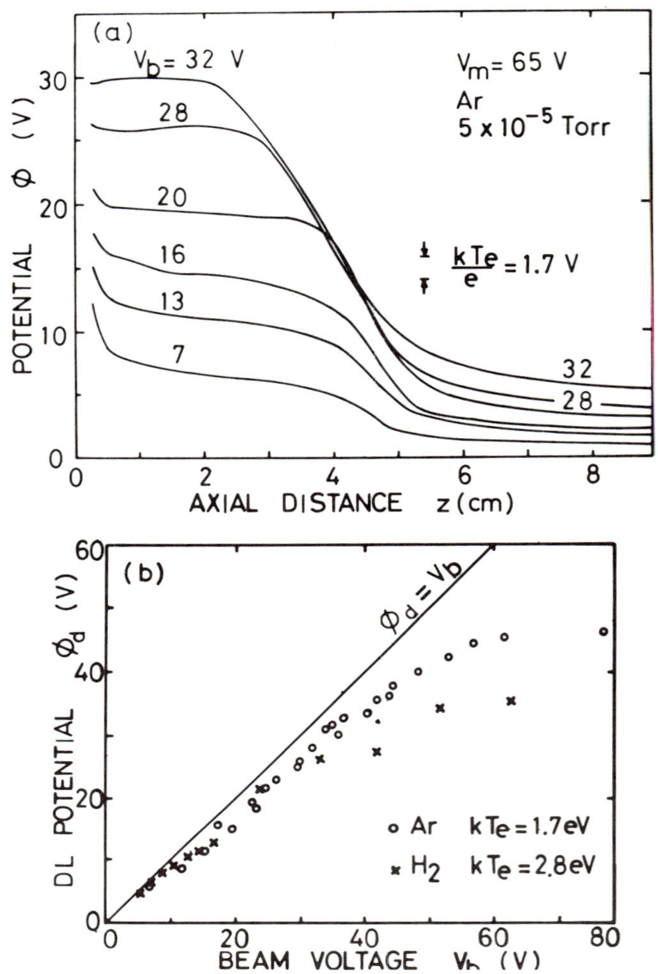

Fig. 4. Double layer scaling properties with ion beam energy. (a) Potential profiles for different beam voltages V_b at a fixed magnet voltage V_m showing increase of ϕ_d with V_b.
(b) Double layer potential ϕ_d vs beam voltage V_b.

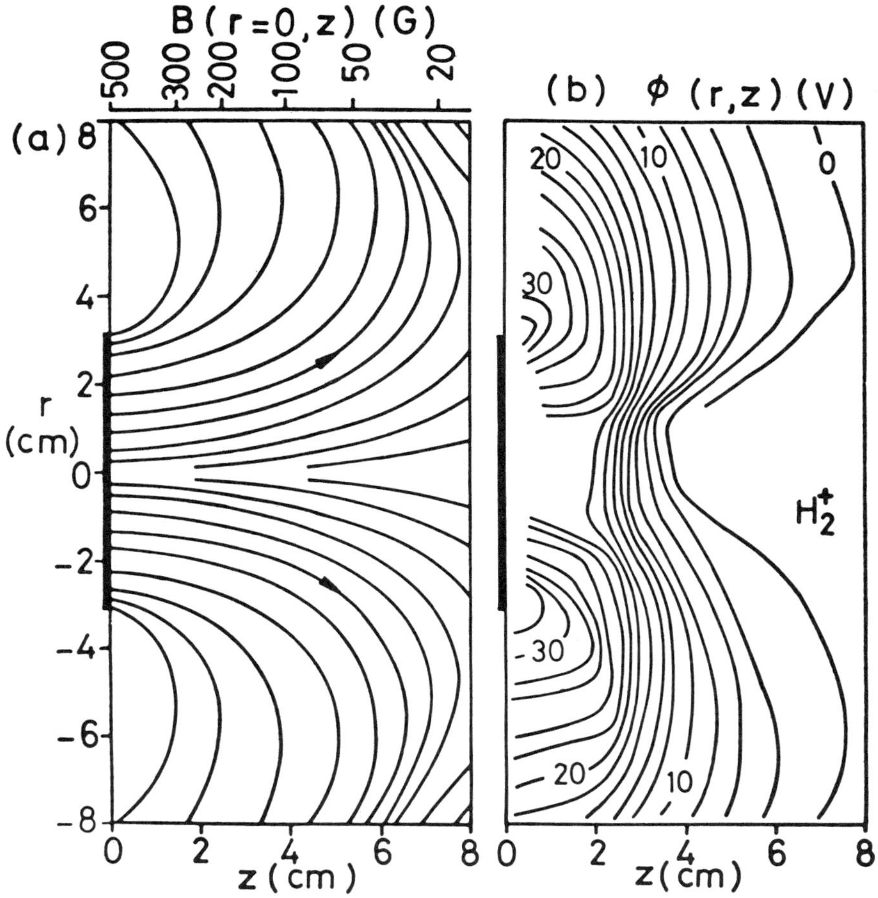

Fig. 5. Two-dimensional display of magnetic field lines and equipotential contours. (a) Magnetic field lines (arbitrarily spaced) in front of N-pole face (indicated by heavy line). Top scale gives field strength on axis (null point at r = 0, z = 10.5 cm; uniform field B_o = 21 G). (b) Contours of constant plasma potential ($\Delta\phi$ = 2 V between contours) showing a V-shaped double layer in H_2^+ (V_b = 30 V, V_m = 39 V, 9 x 10^{-5} Torr).

on ion beam and electron densities. With increasing ion beam density the double layer position is found to increase. Likewise, with increasing external magnetic field B the electron supply is restricted and the double layer position is observed to increase so that a larger cross-sectional area provides enough electrons for neutralization of the beam ions.

Two-Dimensional Potential Profiles

Since the dipolar magnetic field is highly nonuniform, it is important to investigate the potential distribution in more than one dimension. Because of the cylindrical geometry, it is sufficient to map the potential in the radial (r) and axial (z) directions. Figure 5 shows a comparison between the magnetic field topology and the potential distribution in a half-plane in front of the N-pole face indicated by the heavy solid line. The magnetic field lines [Fig. 5(a)] are arbitrarily spaced since they were obtained from an iron powder pattern but the variation of the magnetic field strength along the axis is indicated at the top of Fig. 5(a). A magnetic null point is at z ≃ 10.5 cm. The contours of constant plasma potential displayed in Fig. 5(b) clearly show the double layer at z ≃ 3 cm in front of the magnet surface. Only in the central region of about 3 cm diam the potential contours are normal to the magnetic field ($\vec{E}\|\vec{B}$) but in the adjacent regions they become field-aligned ($\vec{E}\perp\vec{B}$). These are the characteristics of V-shaped potential structures believed to exist in the auroral magnetospheric regions. The potential contours which point toward the magnet in Fig. 5(b) turn parallel to it in the sheath.

The results of Fig. 5(b) were obtained in a hydrogen discharge where the Larmor radius of the dominant H_2^+ ions is r_{ci} ≃ 0.61 cm (kT_i ≃ 0.2 eV, $\langle B \rangle$ ≃ 150 G). Since this is an order of magnitude smaller than the magnetic field gradient scale length the background ion motion is adiabatic. The beam ions, however, have various degrees of adiabaticity depending on their angle of incidence with respect to the magnetic field. Those injected along the axis parallel to $\vec{B}$ have the same Larmor radius as the background ions, but off-axis the radius can be as large as $r_{ci} = (2\ eV_b/m_i)^{1/2}/\omega_{ci} \approx 6$ cm (at an average $V_b = 20$ V) so that the motion becomes non-adiabatic. For the heavy rare gases investigated all ions are nearly unmagnetized [$r_{ci}(Xe^+) \simeq 4.9$ cm, $r_{ci}(Ar^+)$ 2.7 cm at $kT_i = 0.2$ eV, B = 150 G]. For comparison we have also mapped the equipotential contours in the case of unmagnetized ions to be discussed next.

Figure 6 shows the potential contours near both the front and back electrodes of the magnet biased to the same potential $V_m = 25$ V. The argon ion beam is incident from the right and reflected by the magnet ($V_m > V_b$) so that again a double layer is formed at $z \simeq 4$ cm in front of the N-pole face. Comparison with Fig. 5(b) shows, however, that the double layer is more planar for the unmagnetized Ar^+ ions than for $H_2{}^+$. Mirror effects lead to a decrease in parallel velocity or energy with increasing radius, hence ion reflection at larger distances z from the magnet which contributes to the V-shaped potential contour. The unmagnetized Ar^+ beam retains its parallel properties more easily than the $H_2{}^+$ beam. Although the double layer in Ar^+ appears geometrically plane, the curvature of $\vec{B}$ implies that off-axis (e.g., at r = ± 5 cm) the electric field is normal to the magnetic field. Thus, in both Ar and H_2 we have parallel fields and electron acceleration only in a central channel which is directly confirmed by current profile measure-

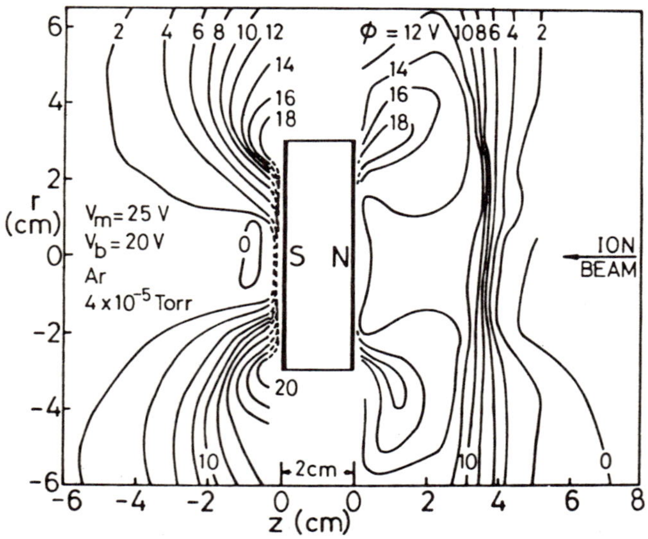

Fig. 6. Plasma potential contours in Ar^+ near both the front and back electrodes of the magnet biased to the same voltage ($V_m = 25$ V). The double layer is only formed on the side of ion beam injection.

ments. The potential contours in the wake of the ion beam (S-pole) are essentially field-aligned except in the sheath region. A small potential minimum is observed in front of the sheath, but the associated electric field is insignificant compared with the double layer field. Such structures have also been observed in confinement experiments (Hershkowitz et al., 1980). In the absence of the injected ion beam, the potential contours for z > 0 are similar to those shown for z < 0.

The ion beam injection creates a strong asymmetry in the potential distribution, a consequence of which is a large difference in the collected currents I_m for equal voltages V_m. On the N-pole electrode the electron current ($I_m \simeq 10$ mA) is approximately five times larger than on the opposite electrode without double layer. This is due to the fact that the double layer replaces the sheath as an effective electron collector and, with increasing distance z its cross section expands ($S \propto B^{-1}$). Since the S-pole electrode collects a relatively small current its bias has little influence on the shape of the double layer. In particular, the axial potential profile $\phi_{(z)}$ (z > 0) shows no difference whether the S-pole electrode is floating or biased to V_m.

Time-Space Evolution

Growth and decay of the double layer have been investigated by pulsing beam or magnet voltages between values for double layer presence and absence. When the beam voltage is pulsed the anode sheath begins to broaden after about one beam transit time ($t \simeq 15$ μsec) and subsequently the profile changes curvature ($t \simeq 30$ μsec), i.e., a weak double layer is formed near the sheath ($z \simeq 1$ cm). The double layer propagates away from the sheath while growing in amplitude and decreasing in width. Although clearly developed at $t \simeq 50$ μsec the double layer reaches its steady-state conditions at $t \simeq 400$ μsec. Some of the late evolution may be due to the adjustment of densities in the source and target plasmas.

When for a steady-state beam the magnet voltage is pulsed from $V_m = 0$ to $V_m > V_b$, we again see the double layer emerging from an expanding sheath. It propagates into the plasma with decreasing velocity ($v < c_s$) and reaches steady-state conditions after a few 100 μsec.

Distribution Functions

Besides the potential profiles the velocity distributions of particles are of primary importance in the problem of double layers. The knowledge of f(v) is essential for determining densities, instabilities, energy, and momentum transport. Perturbation-free measurements of f(v) are difficult to perform in a magnetized plasma with particle flows. The optimum location for measuring the electron distribution with a small Faraday cup is found to be near the magnet surface ($z \simeq 5$ mm, $r = 0$). With the outer grid and metallic shield of the velocity analyzer biased to the same potential as the magnet (V_m), the analyzer does not change the current flow I_m or potential profile which has been verified with the emissive probe. A small fraction of electrons passing through the analyzer grids is collected and its distribution is obtained from the collector I-V characteristics.

From the observed collector I-V curve we obtain by graphic derivation $dI/dV \propto f_e(V)$. The voltage is referenced with respect to the plasma potential ϕ measured in front of the analyzer with the emissive probe. The derivative is then displayed in Fig. 7 vs velocity rather than energy, $v_z = [2e(V-\phi)/m_e]^{1/2}$, with the direction pointing toward the magnet ($v_z < 0$). The observed distributions clearly show the acceleration of electrons by the double layer potential. Since there is no plasma source between the double layer and the magnet, we find for large electron accelerations almost no stationary electrons. However, the shifted distribution is not symmetric, but there are more particles below the drift velocity ($v = v_d$ at $f_e = 1$) than above it. With decreasing double layer potential the number of slow particles increases dramatically. This inverse velocity dependence indicates that the slow electrons are not produced by ionization phenomena but by a deceleration process which increases toward low drift speeds. The dominant mechanism is the magnetic mirror effect. For large parallel acceleration by the double layer most electrons are in the loss cone and can escape to the magnet, but with decreasing parallel velocity an increasing number of electrons will be slowed down and reflected. For a collisionless adiabatic motion a mirroring electron returns to the double layer with the same speed as it emerged from, hence it will also return through the double layer back into the low potential region. However, if the electron is scattered by a wave-particle interaction it can lose some parallel momentum and become trapped between the mirror point and the double layer. Particle-particle collisions are insufficient to build up a significant trapped electron distribution (electron-neutral mean free path $\ell_{en} \gtrsim 10$ m). But, the drifting distribution with $v_d/v_e > 1$ is subject to the strong hydrodynamic Buneman instability (Buneman, 1959) with large growth rate and short wavelength of order λ_D. Although a detailed instability analysis is still in progress, initial observations show a broad turbulence spectrum covering the frequency range of the Buneman instability $\omega \approx (m_i/m_e)^{1/6} \omega_{pi} \approx 2\pi \times 1.7$ MHz. This evidence leads us to believe that wave turbulence produces the population of trapped electrons. These are genuinely trapped particles in a potential well set by electrostatic and magnetic forces, which is in contrast to the usual (incorrectly called) trapped particles of one-dimensional double layer models which are only once reflected by the electrostatic potential.

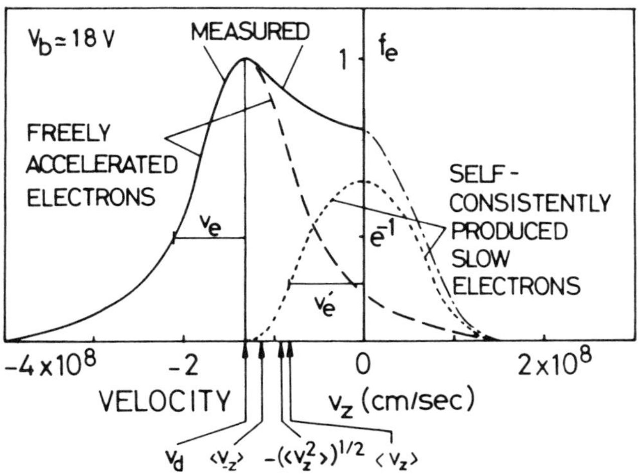

Fig. 7. Measured electron velocity distribution $f_e(v_z)$ (solid line), separated for $v_z < 0$ into stationary trapped electrons (dotted distribution) and accelerated free electrons symmetrically distributed about the drift velocity v_d (dashed/solid line). Assuming symmetry for both distributions the total function $f_e(v_z)$ is completed for $v_z > 0$ (dot-dashed line).

While scattering of free electrons can form trapped particles, it can also lead to their loss. Thus, a steady-state will be reached with a finite number of trapped particles. An exact measurement of its density is difficult but an analysis of the distribution function as shown in Fig. 7 should provide a reasonable estimate. The measured distribution f_e (v_z) ($v_z < 0$) for $\phi_d = 14$ V ($V_b \simeq 17.5$ V) is separated into a symmetric distribution of free electrons with drift velocity v_d and thermal velocity v_e, and a stationary population of trapped particles (dotted curve). Assuming symmetric distributions for the trapped electrons (which requires many bounces) and for the free electrons (which is the shifted distribution of electrons in the target plasma), we can add both contributions and complete the total distribution $f_e(v_z)$ for $v_z > 0$ (dash-dotted line). This analysis yields for the ratio of trapped to free electrons $n_t/n_f \simeq 0.48$.

From the measured distribution we can also determine the electron density correctly. The current density is given by $j_e = -ne\langle v_{-z}\rangle$ where

$$\langle v_{-z}\rangle = -\int_0^{-\infty} v_z\, f(v_z)\, d\, v_z \qquad (1)$$

is the average drift velocity of electrons moving toward the magnet ($\langle v_{-z}\rangle \simeq -1.15 \times 10^8$ cm/sec, $v_d \simeq -1.3 \times 10^8$ cm/sec). At the magnet the current density measured with a small cylindrical Langmuir probe has a bell shaped radial profile $f(r)$. From the total current to the magnet, $I_m \simeq 16$ mA, and its effective cross section,

$$S_m = 2\pi \int_0^{3cm} r\, f(r)\, dr \simeq 6.9\ cm^2 \qquad (2)$$

we determine the current density at r = o, z ≃ 0 to be $j_e(0,0) = 2.3\ mA/cm^2$. Thus, the axial electron density just in front of the magnet sheath is $n_e \simeq 1.25 \times 10^8\ cm^{-3}$ which is comparable to the density of the target plasma well outside of the double layer ($n_{eo} \simeq 1.4 \times 10^8\ cm^{-3}$). In contrast to one-dimensional geometries the density is not determined by flux conservation which would imply a large density decrease associated with the velocity increase $[n_e = n_{eo}\ (1 + 4\pi e\phi_d/kT_e)^{-1/2}]$. In the present two-dimensional magnetic field geometry the total current I_m has to be conserved implying a flux increase toward the magnet ($nv \propto B$). Drifting from the double layer toward the magnet the electrons lose parallel energy or velocity by mirror effects, hence the density has to rise even more rapidly than the magnetic field. Knowing the double layer potential energy ($e\phi_d \simeq$ 14 eV) and the observed kinetic energy near the magnet $[(1/2)m\ v_d^2 \simeq 4.8\ eV]$ as well as the magnetic field profile [Fig. 5(a)], we may extrapolate back to the double layer and obtain the density on the high potential edge, $n_{dl} \simeq 1.25 \times 10^8 \times (120\ G/400\ G) \times (4.8\ eV/14\ eV)^{1/2} \simeq 2.2 \times 10^7\ cm^{-3}$. Although this may be a low estimate since it does not take into account trapped electrons which carry no current, it indicates that this double layer separates plasmas of very different densities.

Finally, we have investigated the ion distribution function. In the absence of the double layer both the stationary background ions and the low density cold beam ions are observed near the magnet. However, in the presence of the double layer only one ion distribution is seen with drift velocity v_{di} and large velocity spread v_i. These ions are the beam ions decelerated by the double layer potential since all background ions are reflected. The large parallel velocity spread and enhanced ion beam flux are a result of the two-dimensional magnetic and electric field distributions. A V-shaped double layer has a focusing effect on the ion flow. An increased ion beam density is also required for space charge neutralization of the relatively large electron density near the magnet. Since the drifting ions encounter a sheath potential with $V_m > V_b$, they are all reflected and the distribution for $v_z > 0$ becomes a mirror image of that for $v_z < 0$. However, with increasing distance z scattering of ions by wave turbulence and geometric divergence will lead to a smaller reflected ion beam density compared with the incident one.

Physical Model

From the various experimental observations we may now propose a qualitative picture for the double layer formation. A more quantitative analysis is not yet possible since to our knowledge, theories on double layers in nonuniform magnetic fields have only recently been developed and require considerable numerical analysis (Kan et al., 1979).

The temporal evolution shows that the double layer originates when the stagnation of the injected ion beam creates a positive space charge layer in an electron-rich sheath. The potential profile is modified so that the location for ion beam deceleration, background ion reflection, and electron acceleration shifts away from the original sheath. The charge double layer is now maintained as sketched schematically in Fig. 8. In the target plasma the slowly drifting background electrons of density n_{eo} are space charge neutralized by stationary background ions n_{io} and beam ions n_b ($n_{eo} = n_{io} + n_b$). At the double layer of potential $\phi_d \gg kT_e/e$ the cold ion density rapidly vanishes, the beam ions density increases due to deceleration $[n_i = n_b\ (1 - \phi_d/V_b)^{-1/2}]$ and the electron density drops due to acceleration $[n_e = n_{eo}\ (1 + 4\pi e\phi_d/kT_e)^{-1/2}]$. At the low potential side of the double layer the negative excess space charge is due to the rapid drop in cold ion density, at the high potential side the excess ion density comes partly from reflection of trapped electrons n_{et} and partly from different axial density gradients for electrons and ions. In order to maintain the total current at decreasing cross section and velocity the free electron density n_{ef} must increase rapidly toward the magnet. The ion density also increases toward z = 0 but less rapidly since focused mainly by electric fields. A trapped electron population exists between double layer and magnet but little is known about its profile. At the magnet an electron-rich sheath is formed since the ion beam is completely reflected.

The double layer reaches a stable configuration when momentum balance is established. The momentum increase due to accelerating electrons gives rise to a pressure in +z direction, while the de-

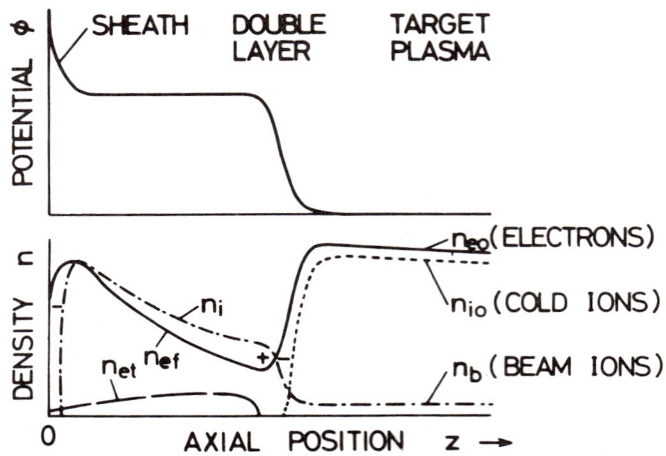

Fig. 8. Schematic picture of potential and particle densities in the double layer model.

celeration of the incoming ions and their subsequent reverse acceleration provides for the balancing pressure in -z direction. Additional but less important pressure terms arise from the reflected background ions, trapped electrons, and the magnetic pressure. In a collisionless plasma the double layer transforms potential energy from an external generator to kinetic electron energy which, upon impact at the magnet, is converted into thermal energy (heat). The ion beam changes momentum but not energy. However, in the presence of wave turbulence some of the directed kinetic particle energy is converted into random kinetic particle energy, i.e., electron and ion heating. This shifts the dissipation from the boundary into the plasma volume. In comparison, auroral electrons thought to be generated by a double layer also lose part of their energy through wave-particle interactions but deposit most of it at the ionospheric boundary.

Conclusion

Experimental observations on a new double layer have been presented. It differs from previous work in the following important points:

(i) A strong nonuniform magnetic field is applied. This gives rise to V-shaped double layers, causes mirror reflection and trapping of particles, axial, and radial density gradients, and modifies basic concepts from one-dimensional models such as axial particle flux conservation.

(ii) The double layer is produced by means of a reflected ion beam to provide positive charges on the high potential side. On the low potential side we do not have a drifting distribution with $v_d/v_e > 1$ which is often assumed to be a necessary condition for double layer formation.

(iii) Drifting nearly Maxwellian distributions with $v_d/v_e > 1$ are observed on the high potential side of the double layer as a result of free electron acceleration. The conditions required for the Buneman instability are satisfied and strong evidence for it is found in the fluctuation spectrum and the appearance of trapped particles. The absence of a plasma source on the high potential side implies that the observed trapped electrons are formed self-consistently. Genuine trapping between mirror points and the double layer is discussed.

Acknowledgments. The authors are grateful for the expert technical assistance from Mr. Y. Nomura. One of us (RLS) acknowledges support from the University of Tokyo during a sabbatical quarter at the Institute of Space and Aeronautical Sciences.

References

Baker, K.D., L. P. Block, R. Kist, and W. Kampa, Strong laboratory double layers and their relevance to auroral plasma physics, in Proceedings of International Conference on Plasma Physics, Vol. 1, 417, Nagoya, Japan, April 7-11, 1980.

Block, L. P., Potential double layers in the ionosphere, Cosmic Electrodynamics, 3, 349-376, 1972.

Buneman, O., Dissipation of Currents in ionized media, Phys. Rev., 115, 503-517, 1959.

Carlquist, P., On the formation of double layers in plasmas, Cosmic Electrodynamics, 3, 377-388, 1972.

Coakley, P. and N. Hershkowitz, Laboratory double layers, Phys. Fluids, 22, 1171-1181, 1979.

Crawford, F. W., J. S. Levine, and D. B. Ilic, Laboratory simulation of ionospheric double layers, in Wave Instabilities in Space Plasmas, P. J. Palmadesso and K. Papadopoulos, eds., 129-134, D. Reidel Publ. Co., 1979.

Hershkowitz, N., J. R. DeKock, and C. Chan, The effect of picket-fence surface magnetic fields at the end of a magnetized plasma, Nuclear Fusion, 20, 695-701, 1980.

Hubbard, R. F., and G. Joyce, Simulation of auroral double layers, J. Geophys. Res., 84, 4297-4304, 1979.

Iizuka, S., K. Saeki, N. Sato, and Y. Hatta, Buneman instability, Pierce instability, and double layer formation in a collisionless plasma, Phys. Rev. Lett., 43, 1404-1407, 1979.

Kan, J. Energization of auroral electrons by electrostatic shock waves, J. Geophys. Res., 80, 2089-2095, 1975.

Kan, J. R., L. C. Lee and S. -I. Akasofu, Two-dimensional potential double layers and discrete auroras, J. Geophys. Res., 84, 4305-4315, 1979.

Knorr, G., and C. K. Goertz, Existence and stability of strong potential double layers, Astrophys. Space Sci., 31, 209-223, 1974.

Leung, P., A. Y. Wong, and B. H. Quon, Formation of double layers, Phys. Fluids, 23, 992-1004, 1980.

Mozer, F. S., C. W. Carlson, M. K. Hudson, R. B. Torbert, B. Parady, and T. Yatteau, Observations of paired electrostatic shocks in the polar magnetosphere, Phys. Rev. Lett., 38, 292-295, 1977.

Singh, N. Computer experiments on the formation and dynamics of electric double layers, Plasma Physics, 22, 1, 1980.

Stenzel, R. L., M. Ooyama, and Y. Nakamura, V-shaped double layers formed by ion beam reflection, Phys. Rev. Lett., 45, 1498-1501, 1980 (also Phys. Fluids, April, 1981).

Torvén, S., Formation of double layers in laboratory plasmas, in Wave Instabilities in Space Plasmas, P. J. Palmadesso and K. Papadopoulos, eds., 109-128, D. Reidel Publ. Co., 1979.

Wagner, J. S., T. Tajima, J. R. Kan, J. N. Leboeuf, S. -I. Akasofu, and J. M. Dawson, Phys. Rev. Lett., 45, 803, 1980.

Wescott, E. M., H. C. Stenbaek-Nielsen, T. J. Hallinan, T. N. Davis, and H. M. Peek, The Skylab barium plasma injection experiments. 2. Evidence for a double layer, J. Geophys. Res., 81, 4495-4502, 1976.

EFFECTS OF AURORAL-PARTICLE ANISOTROPIES AND MIRROR FORCES ON HIGH-LATITUDE ELECTRIC FIELDS

Y. T. Chiu, J. M. Cornwall*, and Michael Schulz

Space Sciences Laboratory, The Aerospace Corporation, P. O. Box 92957
Los Angeles, CA 90009

Abstract. Auroral arcs result from the acceleration and precipitation of ring current and/or plasma-sheet plasma in narrow regions characterized by strong electric fields both perpendicular and parallel to the earth's magnetic field. The various mechanisms that have been proposed for the origin of such strong electric fields are not mutually exclusive. However, for most proposed mechanisms (with the possible exception of double layers of Debye length parallel scale), the effects of auroral particle anisotropy, of mirror forces due to the inhomogeneous geomagnetic field, of auroral electron backscatter by the atmosphere, and of electron trapping by the combination of magnetic mirroring and electrostatic forces must be taken into account in simulations of auroral electric fields. In addition, the effects of the very strong perpendicular electric field must also be taken into account in a kinetic description of the Poisson equation in order to achieve a unified theory of the auroral electrostatic structure. In this paper, progress in these areas in the past few years will be reviewed. It is shown that particle anisotropies and mirror forces can account for some basic electrostatic features of the quiet arc, while additional effects may be taking place in strong events in which the parallel potential drop exceeds ~ 10 kV.

Introduction

The latter half of the nineteen seventies has witnessed a spectacular increase in the understanding of the electrodynamics of the auroral arc. This happy circumstance is supported on the one hand by the advent of simultaneous high-resolution particle and field observations at altitudes of $\sim 1R_E$, such as those obtained by instruments on board the S3-3 satellite, and on the other hand by earlier rocket and radar backscatter observations at ionospheric altitudes. Although the optical aurora has stirred human imagination since time immemorial, it is perhaps not an exaggeration to say that only now are human beings able to grasp an outline of the mechanisms for the formation of the auroral arcs themselves. As we may reasonably expect, the observations indicate a complex but basically electrodynamical interaction between plasma particles and fields (electric as well as magnetic) in the auroral magnetosphere.

In this paper, we shall undertake to review certain unavoidable, indeed inevitable, aspects of the electrodynamical interaction between particles and fields in the auroral region of the magnetosphere. These effects, which conspire to support an electric potential drop along auroral field lines to accelerate magnetospheric electrons downward and ionospheric ions upward, are inevitable in the sense that they are imposed by the following unavoidable external circumstances: the mirroring motion of particles in the earth's dipole-like magnetic field, the differential anisotropy of injected plasma-sheet electrons and ions including the existence of different loss cones, the existence of the ionosphere with the inevitable implication of electron backscatter and current closure, and, last but not least, the existence of large electric potential drops perpendicular to the magnetic field. We shall show that an integrated model of these factors can account for some basic features of recent satellite observations of particles and fields in the quiet auroral arc (observed on the day-to-day basis). It is not our intention in this review to treat the electrodynamics of complex spectacular auroral events in which electron energies are greater than a few kilovolts.

Although the specific purpose of this brief review is to address the effects of the abovementioned factors in auroral dynamics, it would be unphysical to discuss these factors without considering their relationship to observations and to the theoretical interpretation of those observations. In particular, we need to clarify the relationship of these effects to present electrodynamical theories of the formation of

---

*Permanent Address: Department of Physics, University of California, Los Angeles, CA 90024

auroral arcs; thereby, we hope to formulate a unified model of auroral electrodynamics. For these reasons, brief reviews of observations and data interpretation will be included here.

In a brief review such as this one, it is inappropriate to give an exhaustive reference list; therefore, we shall attempt to give only a representative reference list, emphasizing the latest results, so that the reader can trace a more complete reference if he so chooses. Thus, works referenced here are not necessarily judged by us to be of special significance; similarly, works not referenced here are not judged to be comparatively any less significant.

Observations

Before the advent of high-resolution observations by auroral satellites, data on ionospheric and magnetospheric electric fields depended on balloon, rocket, and radar observations which were necessarily episodic; nevertheless, the basic physical properties of auroral ionospheric currents and electric fields (e.g., Cloutier, 1971; Mozer and Manka, 1971; Vondrak et al., 1971), together with their relationship to high-latitude convection electric fields (e.g., Cauffman and Gurnett, 1972; Heppner, 1972) measured by satellites, have been established.

Even as the large-scale auroral processes were being unraveled, certain microscopic features of auroral low-energy particle precipitations were being discovered. Frank and Ackerson (1971) noted that occasionally observations of low-energy (tens of keV) electron precipitation would show an inverted-V structure on an energy-time spectrum plot, i.e., the precipitating electron energy spectrum hardens and then softens as the Injun 5 satellite moves through the structure. Evans (1974; 1975) convincingly demonstrated that rocket measurements of auroral low-energy electrons indicated downward moving electron beams at keV energies, comparable to those of inverted-V structures. Further, by a careful study of electron backscatter from the atmosphere, Evans demonstrated that these auroral electron beams are indications of electric potential drops, along the magnetic field, existing between the equator and the ionosphere. At about the same time, observations of singly ionized energetic O^+ ions in the magnetosphere (Shelley et al., 1972; Sharp et al., 1974) also gave indication that microscopic processes in the aurora couple the ionosphere with the magnetosphere.

These observations of inverted-V structures, of electron beams, and of O^+ ions of probable ionospheric origin in the magnetosphere presage very interesting microscopic processes to be discovered in the auroral process in which the ionosphere plays an active rather than passive role. However, because of the episodic nature of rocket observations and because of the low resolution and low data rate of the early satellite observations, the scope of and inter-relationship between these phenomena were not understood until the launch of the polar-orbiting auroral satellite S3-3, which intercepts auroral field lines at altitudes up to ~ 8000 km , precisely in the region where ionospheric and magnetospheric plasmas are expected to interact. Included in the S3-3 payload are instruments to measure electric fields (Mozer et al., 1977), low energy electrons (Mizera et al., 1976), energetic ions (Shelley et al., 1976), and plasma waves (Kintner et al., 1978).

The S3-3 observations not only confirmed Evans' observations of downward-moving field-aligned electron beam at keV energies but also revealed the existence of upward-moving ion beams aligned with the magnetic field in inverted-V structures. This certainly indicates that the three phenomena are intimately related, but more importantly the S3-3 particle observations leave little doubt that an electric potential drop of several to tens of kilovolts, aligned with the magnetic field, exists between the ionosphere and the magnetospheric equator. Electrostatic field measurements also indicate paired regions of oppositely directed perpendicular electric fields, with latitudinal scale lengths of some 50 km, reflecting a negative space-charge region presumably associated with downward-streaming electrons. Figure 1, which is a composite of particle and electric field data, illustrates the above points. For further emphasis, Figure 2 shows an enlarged view of the perpendicular electric field data for the time period marked by the brace in the middle of Figure 1. A crucial, but seldom emphasized, feature which is brought out by the high sensitivity and high resolution of the S3-3 measurements is that the above correlated features are observed at the auroral zone pass after pass at all satellite altitudes. In other words, these are fundamental features of the aurora rather than episodic curiosities.

Theoretical Interpretations

The key theoretical issue concerning the interpretation of the S3-3 observations really involves the electrodynamics of the auroral arc itself. A key fact which must be recognized is that a magnetic-field-aligned electrostatic potential difference of kilovolt magnitude exists between the ionosphere and the equator. The crucial question is whether the mechanism for the buildup of such a field-aligned potential drop involves the ionosphere, for there is no doubt that the energy source of the aurora is derived from the interaction between the solar wind and the magnetosphere via plasmas which are injected by substorm processes onto the auroral field lines. Some aspects of the theories of auroral field-aligned potential drop have been reviewed (Shawhan et al., 1978; Hudson and Mozer, 1978), but our discussions will be primarily concerned with the particle anisotropy and particle mirroring aspects of these theories.

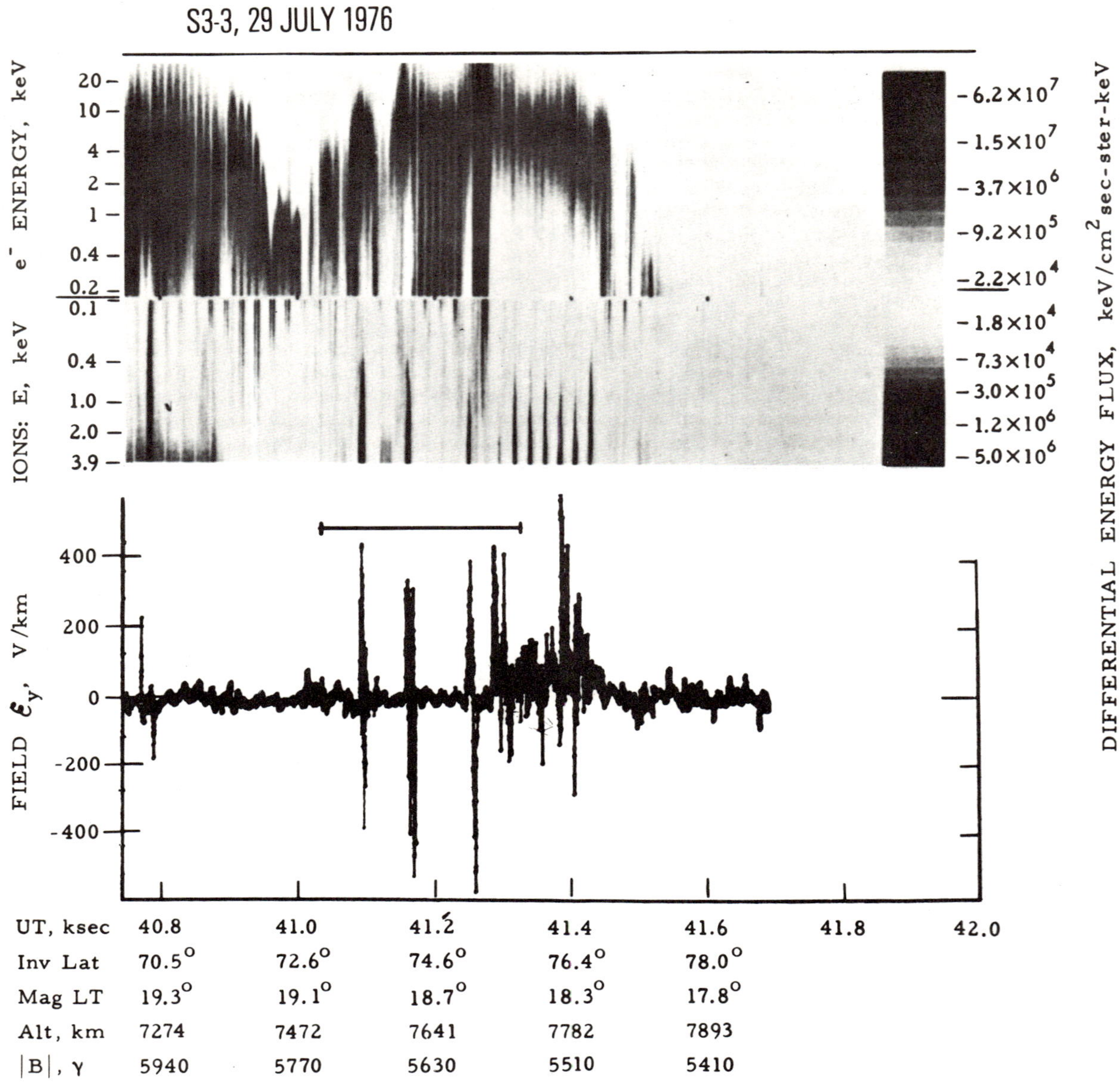

Fig. 1. Simultaneous observations of electron energy spectrum, ion energy spectrum and perpendicular electrostatic field structures on July 29, 1976, by the S3-3 satellite. (Courtesy F. S. Mozer, R. B. Torbert, P. F. Mizera and J. F. Fennell.)

Theories of auroral processes involving magnetic-field-aligned electrostatic potential differences can be roughly classified into four categories, although they are not mutually exclusive. These are: 1) Debye-length double layer, 2) oblique electrostatic shock, 3) anomalous resistivity, and 4) downward mapping of convection electric field discontinuities. These categories invoke theoretical arguments of varying degrees of sophistication and believability to show that kilovolt electrostatic potential drops may be produced in various assumed plasma distributions. Our brief discussion begins with the traditional view of these mechanisms in isolation; then we shall proceed to consider their relationship to the effects of particle dynamics mentioned in the Introduction.

The Debye-length double layer (Block, 1975; Shawhan et al., 1978) is a boundary layer between unmagnetized cold plasma on one side and hot plasma on the other. The potential drop across the layer is alleged to be $\sim kT_e/|e|$ and the

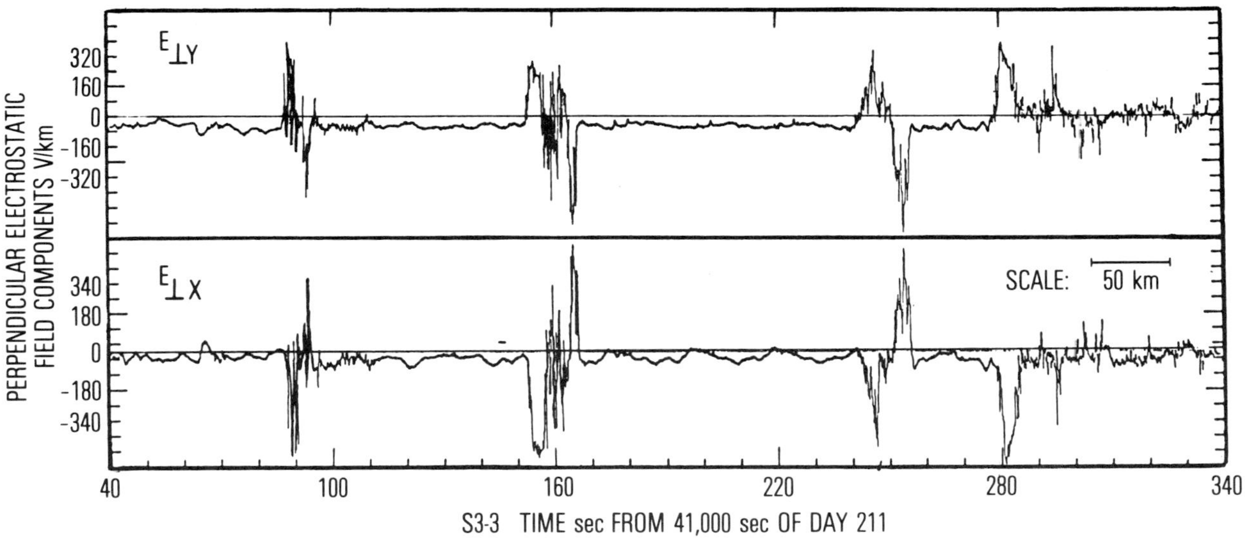

Fig. 2. Expanded view of the two perpendicular components of the electrostatic field for the period indicated by the brace in the middle of Fig. 1. (Courtesy F. S. Mozer and R. B. Torbert.)

layer thickness is of the order of several Debye lengths (~ 10 km); thus, the parallel electric field in double layers must be ~ (0.1 - 1) V/m. A current-driven instability is usually invoked as the formation mechanism of double layers, which requires a field-aligned current greater than a certain threshold value. If potential drops inferred by electron beam observations at S3-3 altitudes as high as ~ 8000 km are all due to double layers above the satellite, then evidently the ionosphere does not seem to be a factor in double layer formation because cold ionospheric plasma, particularly O^+, is not expected to be found at such high altitudes unless it is accelerated upward by other processes. Frequently, based on observations of both electron and ion beams on S3-3, one may infer that potential drops exist both above and below the satellite (Mizera and Fennell, 1977; Croley et al., 1978). Since it is highly improbable that the satellite just happened to pass through within the double layer thickness, such frequent occurrences seem to require more than one double layer to be formed on the same field line. Further, given the small scale length of the layer, it would be rather difficult for the mechanism to explain the coincident observation of downward electron beams and upward O^+ beams unless the double layer occurs at the topside ionosphere where O^+ is plentiful. Theories of double layer formation are mathematically difficult, even for very simple plasma distributions (Montgomery and Joyce, 1969), and a quantitative theory has yet to be developed for auroral plasmas in an inhomogeneous magnetic field. We shall not discuss the possible relationship between Debye-length double layers and the effects of particle anisotropies and mirror forces, although the latter are bound to enter into consideration of Debye-length double layers in magnetized plasmas.

Oblique electrostatic shocks (Swift, 1975; 1976; Kan, 1975) are similar to double layers except that they recognize the influence of the magnetic field and consider that the shock normal is at an arbitrary angle α to the magnetic field direction. For $\alpha \neq 0$ the shock thickness ℓ is measured in units of the ion gyroradius of a few km, a typical cross-field scale being some 15-20 gyroradii (~ 100 km). The field-aligned scale length is $\ell/\cos\alpha$ which can be quite extensive if the shock normal angle α approaches $\pi/2$. Swift has shown that self-consistent oblique shock solutions to Poisson's equation can be obtained with simple plasma distributions in an inhomogeneous magnetic field (Swift, 1979). The oblique shock geometry has certain advantages over the current-driven double layer in regard to the interpretation of S3-3 data, even though the theory was conceived prior to S3-3. This is because the field-aligned scale length $\ell/\cos\alpha$ can be chosen to be of the order of 1-2 R_e so that only one shock (or a pair of double reverse shocks) need be invoked to explain the existence of potential drops above and below the satellite. It is, of course, a disadvantage that the theory as developed by Swift does not predict α, or equivalently the cross-field scale length. As we discuss later, this scale length can be estimated by incorporating ionospheric physics. An oblique shock with parallel scale length ~ 1 R_e is almost certainly strongly coupled to the ionosphere, which at the very least supplies important boundary conditions for the shock. Presently, there seems to be some controversy concerning the usage of the terms "electrostatic shocks" and "double layers" of all parallel scales (Goertz, 1979; Kan, 1980). We feel that new terminology needs to be introduced to characterize the auroral potential structure but here we adopt the terminology of the initial author in order to maintain impartiality.

A third mechanism by which a magnetic-field-aligned electric potential drop can allegedly be generated is anomalous resistivity in the field-aligned direction (Hudson et al., 1978). Such anomalous resistivity may be due to a large number of possible modes of AC electric-field turbulence in the auroral plasma (e.g., Kindel and Kennel, 1971; Papadopoulos and Coffey, 1975). Hudson et al. (1978) estimated that turbulent electric fields in the electrostatic ion cyclotron mode with amplitudes ~ 50 mV/m may yield sufficient anomalous resistivity to generate parallel electrostatic (DC) fields of ~ 1 mV/m. It is not clear how the largely perpendicular AC fields can affect parallel electron currents (and their resistivity). One feature common to oblique-shock models and anomalous-resistivity models is that the potential smoothly varies over a scale of ~ 1 R_e extension in order to accommodate potential drops of ~ (1-10) kilovolts. It must be noted that the question of how such an extensive region of turbulence can be maintained at a high level (~ 50 mV/m AC), in the presence of non-linear stabilizing effects such as ion heating, must be addressed. However, to address this question, it seems to us that the effects of particle motion in an inhomogeneous magnetic field and the effects of ionospheric plasma cannot be ignored. Further, to study the generation of the electrostatic turbulence, the ultimate cause must be looked for in the characteristics of particle distribution functions. Again, we find ourselves returning to the question of particle distributions in an inhomogeneous magnetic field.

The fourth source of auroral electric fields that accelerate ions and electrons in opposite directions along the earth's magnetic field is the magnetospheric convection electric field. The convection electric field is perpendicular to the magnetic field at high altitudes, but its meridional (r,θ) component has a theoretical discontinuity at or near the boundary between closed and open magnetic field lines. Figure 3 shows the amplitudes of the diurnal variation of $\vec{E}$ at ionospheric altitudes (Chiu et al., 1979; Lyons, 1980). Ionospheric resistivity would partially connect electrostatic equipotentials across the discontinuity, but at too low an altitude to account properly for the observed component of $\vec{E}$ parallel to $\vec{B}$. However, the kinematical resistivity associated with magnetic-mirror forces on a hot plasma may increase the altitude at which the parallel (to $\vec{B}$) component of $\vec{E}$ would appear. The details of this latter effect, which (if it occurs) would produce the desired distribution of $\vec{E} \cdot \vec{B}$ with altitude,

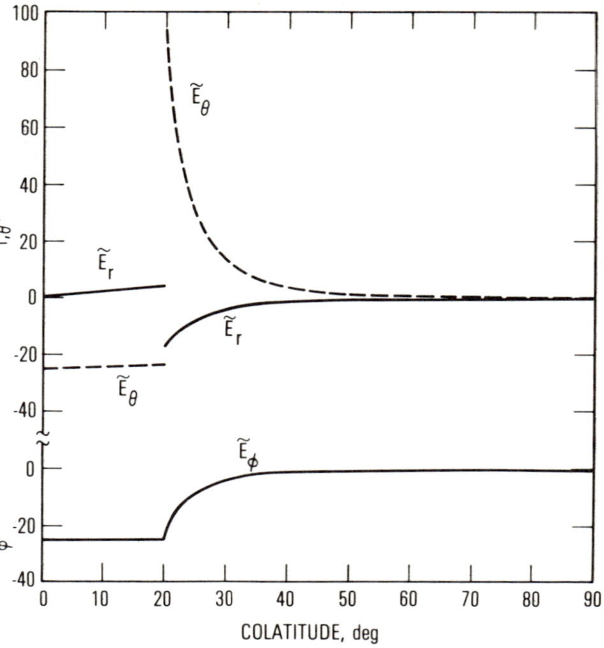

Fig. 3. Amplitudes of the diurnal variations of $\vec{E}_\perp$ (ionospheric altitudes).

remain to be worked out in detail. Lyons (1980) has considered the perpendicular distribution of the parallel potential drop in a model incorporating the ionosphere. The effect would be such as to produce an upward electric field in the PM sector (maximal at dusk) and a downward electric field in the AM sector (maximal at dawn) of the auroral oval. This expectation is in good agreement with the diurnal distribution of upgoing ion beams observed by Ghielmetti et al. (1978). However, to study the distribution of $\vec{E}$ parallel to $\vec{B}$, one must again face the question of particle distributions in an inhomogeneous magnetic field.

For the most part, these mechanisms have been considered in isolation of each other and of the ionosphere, not because physicists believe that it should be so, but because it is difficult to treat the couplings. In fact, a correct theoretical treatment of auroral phenomena will without doubt merge several of these mechanisms with each other and with ionospheric physics. It is unfortunate that much of the recent literature on auroral mechanisms pays so little attention to coupling with the ionosphere; some exceptions (with two of which the authors are connected) exist, though. As a general rule, the ionosphere couples neighboring field lines and allows for predictions of latitudinal structure and scale lengths. There is no such coupling in the individual mechanisms mentioned above (except that oblique shocks have an arbitrary structure which crosses field lines), so none can explain arc structure without going beyond the given mecha-

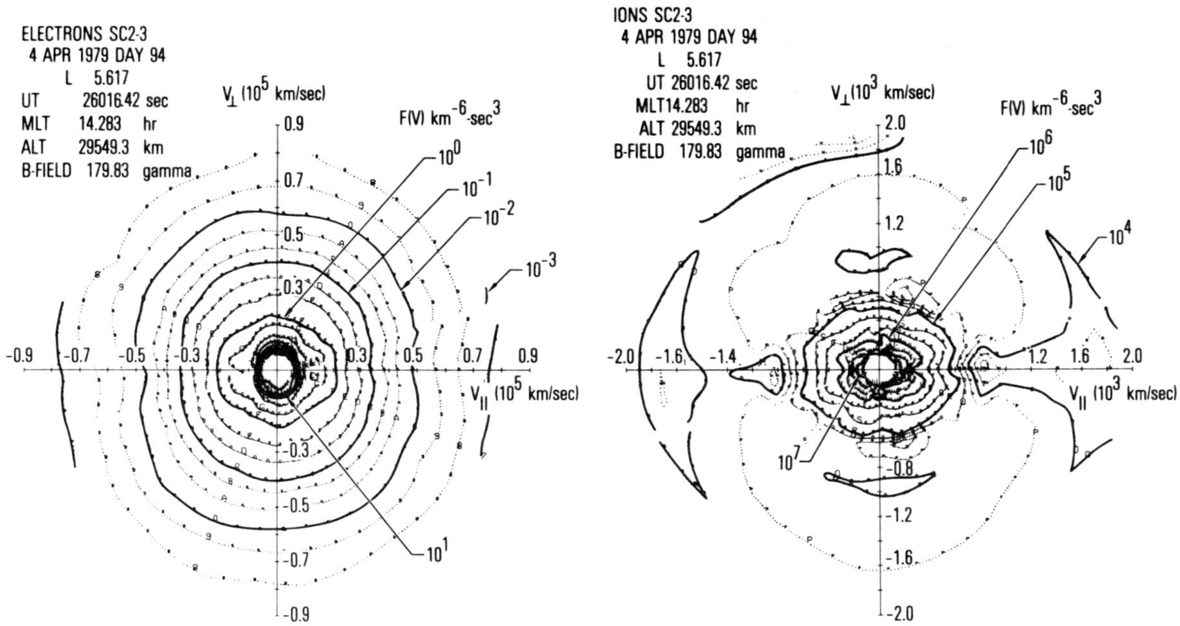

Fig. 4. Pitch angle distributions of ions and electrons near the equatorial regions of auroral field lines. (Courtesy J. F. Fennell and D. R. Croley, Jr.)

nism. The unifying thread running through the above consideration of the mechanisms is, of course, the necessity for establishing the proper self-consistency between particles and field distributions in an inhomogeneous mirroring magnetic field with proper ionospheric and magnetospheric boundary conditions.

It is with the consideration of this common thread in mind that we are led to the effects of auroral-particle anisotropies and mirroring forces on the distributions of electric fields in the auroral region.

Particle Anisotropy and Mirror Forces

The effects of particle anisotropies and mirror forces in an inhomogeneous magnetic field have traditionally been considered as a mechanism for generating a parallel potential drop in the same sense as the mechanisms considered above (see, for example, Lemaire and Scherer, 1974; Shawhan et al., 1978; Chiu et al., 1979). However, because these effects owe their existence to inevitable factors of magnetic field geometry, ionospheric and magnetospheric boundary conditions, they should properly be discussed apart from dynamical mechanisms. These effects do not act in isolation or in opposition to the mechanisms discussed above; rather, as we shall see, they may be the common thread which unifies the various mechanisms (except for the Debye-length double layer) into a single auroral mechanism. Figure 4 shows an example of the difference of pitch angle distributions between ions and electrons observed by instruments on board the SCATHA satellite near the equatorial regions of auroral field lines. So, differential anisotropy between ions and electrons is observed fact, not hypothesis.

As far as we can determine, Alfvén and Fälthammar (1963) were the first to point out that low density two-component monoenergetic plasma in an inhomogeneous mirroring magnetic field must support a parallel electric field unless the anisotropies of energy (i.e., ratio of parallel to perpendicular kinetic energy) are the same for both ions and electrons. Their results have been further developed by Persson (1966), yielding a potential drop $\Delta\phi$ between the ionosphere and the equator in the range of (~ 10-100) kilovolts. Ponyavin et al. (1977) pointed out that the potential drop for a two-component bi-Maxwellian plasma would be much less; $\Delta\phi$ would only be a few kilovolts. Similar results have also been discussed by Whipple (1977) and Lennartsson (1977).

While the geometrical and boundary-imposed properties of particle distributions are the basic causes of the parallel potential drops discussed above, i.e., ions and electrons of different pitch angle distributions mirror on the average at different locations of the field line thus creating a charge-separation electric field, the above results are arrived at by requiring a parallel potential distribution to maintain a balance of positive and negative charges on a field line (quasi-neutrality). For this reason, the effects of particle anisotropy and mirror forces are quite frequently intermingled with quasi-neutrality - to the extent that these effects have become inseparable from quasi-neutrality. In reality, we must be careful to distinguish quasi-neutrality, which is nothing more or less than an assumed method of deriving the parallel potential ϕ , from the more basic factors of particle anisotropy and mirror forces. Indeed, we shall discuss more general applications of these factors in the next section.

The application of quasi-neutrality to obtain solutions of kinetic equilibrium between particles and fields has had a long history. Lemaire and Scherer (1971) have pioneered a series of applications of these kinetic principles to a number of magnetospheric problems involving primarily isotropic Maxwellian distributions. Applications to auroral plasmas have become the dominant topic in recent years (Lemaire and Scherer, 1974; Knight, 1974; Lennartsson, 1977). Chiu and Schulz (1978) reexamined the problem in the light of the S3-3 observations and suggested the necessity of taking into account other inevitable factors such as electron backscatter from the ionosphere (eloquently noted by Evans (1974; 1975) a few years earlier) and electron trapping by a combination of electric and mirroring forces. A typical quasi-neutral solution for the equilibrium potential is shown in Figure 5. In general, the parallel electric field due to the factors discussed here is distributed over the entire auroral field line so that the parallel electric field strength is < 1 mV/m. Further, as has been noted in simple model calculations, the more complex models confirm that a potential drop of < 10 kV can be supported by these basic factors of particle kinetics; thus, particle anisotropy and mirror forces in quasi-neutral calculations cannot account for the more spectacular auroral events in which the potential drop can be as high as 30 kV.

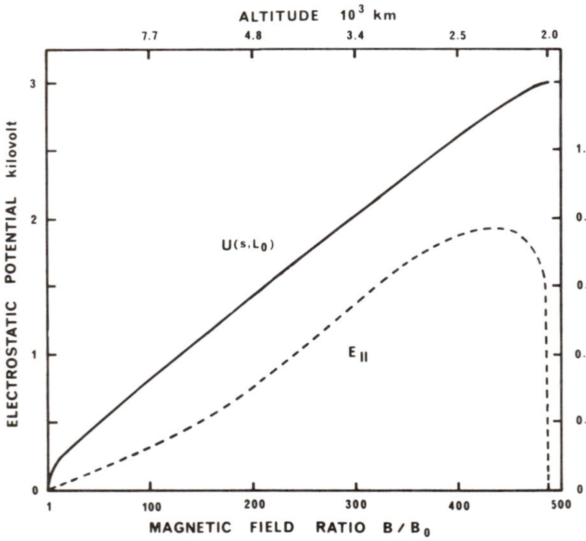

Fig. 5. A typical solution for the parallel potential and electric field for typical ionospheric distributions and assumed potential drops.

An important aspect of particle anisotropy and mirror forces in auroral dynamics is somewhat subtle and is not generally appreciated. By suggesting that the basic parallel electric field distribution is a direct kinetic consequence of externally imposed geometric and boundary factors, we are also suggesting that there is an inevitable kinetic resistivity to the auroral low-density plasma so that dynamical processes such as plasma waves do not play a major role in determining the current-field relationship ("Ohm's law") in the quiet arc. This viewpoint has been expressed as the "kinetic picture" by Lemaire and as the "current-carrying characteristics of the low density auroral flux tube" by Evans. A simple prediction of this kinetic picture is the relationship between field-aligned current $J_{\parallel}$ and the electric potential drop $\Delta\phi$

$$J_{\parallel} \simeq A\Delta\phi \qquad (1)$$

where A is derivable from the parameters of the particle distribution functions (Fridman and Lemaire, 1980; Chiu and Cornwall, 1980). Lyons et al., (1979) have essentially demonstrated the validity of (1) with observations.

Towards a Unified Model of the Quiet Arc

In Section III, we have noted that the kinematics of particle distributions in the inhomogeneous mirroring geomagnetic field is a common thread which links the various suggested mechanisms for the auroral parallel electric field. In Section IV, we have reviewed the development of this common thread - specifically, the effects of particle anisotropy and mirror forces in quasi-neutral models. Since these effects owe their existence to inevitable factors of magnetic field geometry and boundary conditions in the ionosphere and magnetosphere, the next logical step in the development of a model of quiet arc formation would be to consider the various mechanisms (except the Debye-length double layer) in the light of this unifying thread.

It is evident that these physical mechanisms do not exist entirely independently of one another. For example, if the restrictive assumption of strict charge neutrality is removed in the magnetic-mirror model, one has an oblique electrostatic shock in a mirroring field. To the extent that no dissipative mechanisms such as wave-particle turbulence are included in such a shock, the resulting electric-field structure is better described as a solution of Poisson's equation. From another point of view the oblique shock can be described as some sort of zero-frequency electrostatic ion-cyclotron (ESIC) mode. There surely is a great deal of ESIC turbulence connected with auroras, and the physical distinction between the oblique shocks of Swift and this turbulence is at best imprecise. Yet the merging of wave turbulence and shocks can lead to substantial parallel electric potential drops in the complete absence of anomalous resistivity. (The reader need not be reminded that turbulence is not synonymous with anomalous resistivity; in fact, it is quite difficult to make anomalous resistivity out of even the most turbulent waves.)

The effects of AC electric fields and electrostatic shock structures, together with the very important factor of ionospheric current transformation can all be folded into a model which includes the effects of particle anisotropy and mirroring forces. The formal development is based on Poisson's equation, of which quasi-neutrality is a truncated form, and current conservation in the ionosphere. Chiu and Cornwall (1980) have considered such a model in dipolar magnetic geometry, coupled with ionospheric physics. In such a model the parallel potential drop is intimately coupled to the perpendicular electrostatic field structure as indicated in Figure 1. Similar considerations have also been considered by Kan et al. (1979).

Poisson's equation with perpendicular electric field in a dipolar magnetic field

$$\nabla\cdot\vec{E}_\perp + B\frac{\partial}{\partial s}(B^{-1}E_\parallel) = 4\pi\sum_j n_j e_j \qquad (2)$$

becomes

$$\nabla\cdot(K\vec{E}_\perp) + B\frac{\partial}{\partial s}(B^{-1}E_\parallel) = 4\pi\sum_j \tilde{n}_j e_j \qquad (3)$$

where K is the plasma dielectric "constant"

$$K = 1 + 4\pi\sum_j \frac{\tilde{n}_j M_j c^2}{B^2} \qquad (4)$$

and the density $\tilde{n}_j$ is that which would be appropriate in the absence of the perpendicular electric field. Equation (3) is an approximate generalization of Swift's formulation (1975) to an inhomogeneous magnetic field and is in agreement with Swift's own generalization (1979). The quasi-neutrality equation of Chiu and Schulz (1978) is simply that the right-hand side of (3) vanish. In fact, taking $E_\perp$ from data and $\nabla\cdot\vec{E}_\perp \approx E_\perp/\ell_\perp$ where $\ell_\perp$ is the perpendicular scale, one finds that the left-hand side of (3) yields a fractional charge separation $\Delta n/n \approx 10^{-2}$–$10^{-3}$. Or, looked at the other way around, this sort of charge separation can drive perpendicular electric fields of $O(M^2/\ell_\perp e) \approx$ 100's of mV/m, as well as parallel fields of $O[(\ell_\perp/\ell_\parallel)E_\perp]$. It should be noted that the term $B\partial(E_\parallel/B)/\partial s$ in (3) is quite negligible; it is of order $\Lambda_D{}^2/\ell_\parallel{}^2 \approx 10^{-8}$ where Λ_D is the Debye length. This is why ordinary double layers need such small parallel scale lengths in order for there to be appreciable charge separation. In this generalized formulation, the existence of $E_\parallel$ (independent of whether the second term on the left side of (3) is included or not) is supported self-consistently not only by particle anisotropy and mirror forces implicit in $\tilde{n}_j$ but also by the existence of perpendicular potential drops in the magnetosphere implicit in $\nabla\cdot(K\vec{E}_\perp)$, since $K \gg 1$. Thus, a discontinuity in $\nabla\cdot\vec{E}_\perp$ in the magnetosphere (Lyons, 1980; Chiu et al., 1979) can also contribute to $E_\parallel$.

From (3), $\nabla\cdot\vec{E}_\perp$ introduces the perpendicular scale length $\ell_\perp$ of the auroral arc into the model. However, to self-consistently determine $\ell_\perp$, (3) is not enough because current and charge must also be conserved at the ionosphere. In other word, the scale $\ell_\perp$ must be such that ionospheric and magnetospheric conservation laws are both satisfied. Thus, the scale length of the perpendicular electrostatic field structure is related not only to the field-aligned current to the ionosphere but also to the ionospheric Pedersen conductivity. Thus, ionosphere-magnetosphere coupling is a crucial ingredient determining the geometric structure as well as the energetics of the quiet auroral arc in such a model.

To incorporate ionospheric current transformation in the (120-2000) km altitude region is a very difficult task. At present, only very rudimentary methods are used to prescribe the current flow in the ionosphere and the horizontal electric-field gradients there; they are coupled by the requirement that the ionosphere must be able to balance the current flow (as well as to balance sources and sinks of charge). Two simple equations express these facts: (e.g., Atkinson, 1970; Coroniti and Kennel, 1972) the equation of current continuity

$$J_\parallel = \frac{\partial}{\partial x}\left(\Sigma_p\frac{\partial\phi}{\partial x}\right) \qquad (5)$$

and of charge balance at the ionosphere

$$\frac{dN}{dt} = S - \alpha(N^2 - N_0{}^2) = 0 \qquad (6)$$

In these equations, x is a horizontal north-south coordinate, Σ_p is the height-integrated Pedersen conductivity, N is the height-integrated ionospheric electron density, α is a recombination coefficient, $\alpha N_0{}^2$ represents non-auroral sources, and S represents sources of charge from auroral electron precipitation, or from upward-going ions. A simple estimate of $\ell_\perp \sim 40$ km has been established using (1) and (5) (Chiu and Cornwall, 1980).

At present, no satisfactory solution of such a model has yet been obtained in the return-current region, although an approximate solution in the central electron beam region has been obtained (Chiu and Cornwall, 1980). A schematic illustration of this model is given in Figure 6. Figure 7 shows the latitudinal structure of the approximate solution which can be compared with the observed structures shown on Figure 2.

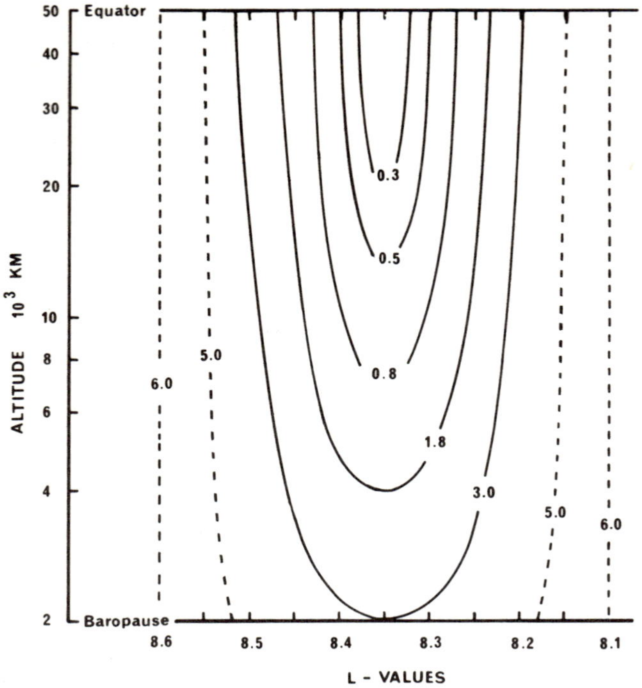

Fig. 6. Equipotential contour, in units of kilovolts, of an approximate solution of auroral electrostatic structure.

Conclusions and Prospects

We have undertaken a very brief review of the various mechanisms which have been invoked to account for the observed consequence of parallel potential drops in the auroral arc. Our purpose is, of course, not to address the details of these mechanisms separately but to emphasize that all the mechanisms (with perhaps the exception of Debye-length double layers) have a commonality underlying them: the unavoidable effects of particle anisotropy (including different loss cones for ions and electrons) and mirror forces, and the necessity to include ionospheric physics as well as proper magnetospheric boundary conditions. While specific dynamical mechanisms may be of varying importance in the wide range of auroral conditions, it seems to us that particle anisotropies and mirror forces must be taken into account for most electrostatic features of the quiet arc when the parallel potential drop does not exceed a few kilovolts. For more energetic events, additional dynamical factors such as electrostatic shock and perhaps anomalous resistivity may come into play.

To formulate a unified model of arc formation, we have pointed out that the basic geometrical factors of particle anisotropy and mirror forces must be merged into the considerations of the various mechanisms. In the light of these basic factors, consideration of the various mechanisms in isolation of each other has no a priori physical justification. Indeed, it is an unfortunate historical accident. Auroral physicists are now beginning to realize the necessity to unify the diverse mechanisms into a comprehensive model of auroral arc formation. The task is difficult because it must encompass both magnetospheric and ionospheric physics, but attempts in this direction are already being made (some features of these preliminary attempts are reviewed here).

The electrodynamical structure of auroral arcs is obviously much more complex than the very simple factors distilled into the models reviewed here. The immediate prospect for these models would be to consider the basic electrodynamical structure in relation to specific phenomena one at a time. For example, the particle distribution functions in self-consistent equilibrium with the electrostatic potential ϕ need not be stable with respect to various plasma wave modes; therefore, evaluation of wave generation under the basic unavoidable conditions of particle anisotropy and mirror forces in a strong DC electric field will be an important area of auroral theory in the near future. Further, although it

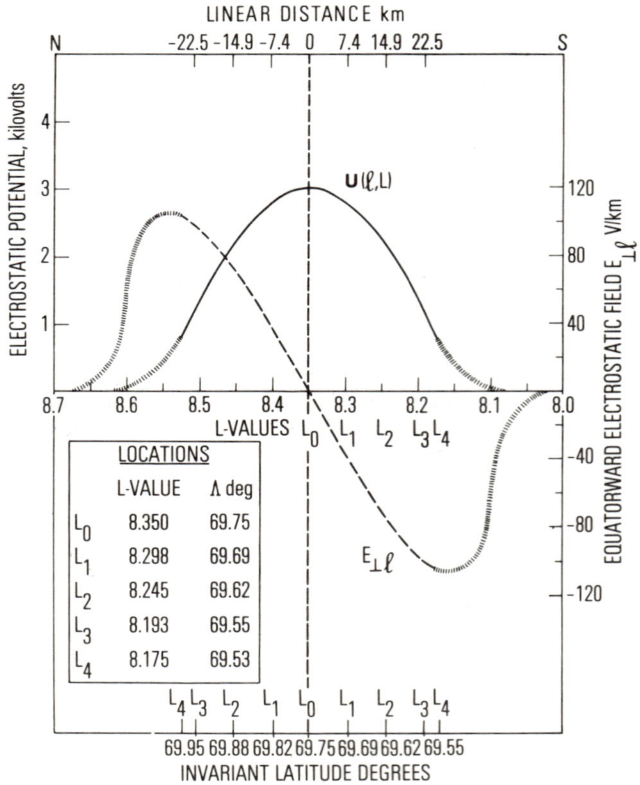

Fig. 7. Distributions of electrostatic potential U and the north-south perpendicular electrostatic field $E_\perp$ as functions of L at 2000 km altitude. The hashed portions of curves refer to regions where the model solution is uncertain.

is outside of the scope of the present review, the effects of the plasma wave AC electric fields are bound to feed back upon the DC electric field structure and the particle distribution. The next logical step of arc-model development will probably be in this direction, thus unifying some aspects of electrostatic turbulence theory into the basic factors considered here.

Acknowledgments. This work is supported by NASA Grant NASW-3327 and by the NASA Solar-Terrestrial Theory Program Grant NASW-3434.

References

Alfvén, H., and C.-G. Fälthammar (1963): Cosmical Electrodynamics, pp. 163-167, Clarendon Press, Oxford.

Atkinson, G. (1970): Auroral arcs: Result of interaction of a dynamic magnetosphere with the ionosphere, J. Geophys. Res., 75, 4796.

Block, L. P. (1975): Double layers, in Physics of the Hot Plasma in the Mag-netosphere, pp. 229-249, edited by B. Hultqvist and L. Stenflo, Plenum Press, N.Y.

Cauffman, D. P., and D. A. Gurnett (1972): Satellite measurements of high latitude convection electric fields, Space Sci. Rev., 13, 369.

Chiu, Y. T., and J. M. Cornwall (1980): Electrostatic model of a quiet auroral arc, J. Geophys. Res., 85, 543.

Chiu, Y. T., and M. Schulz (1978): Self-consistent particle and parallel electrostatic field distributions in the magnetospheric-ionospheric auroral region, J. Geophys. Res., 83, 629.

Chiu, Y. T., J. M. Cornwall, and M. Schulz (1979): Auroral magnetosphere-ionosphere coupling: A brief topical review, Solar-Terrestrial Pro-ceedings, ed. R. F. Donnelly, Vol. 2, p. 494, NOAA, Boulder, Colorado.

Cloutier, P. A. (1971): Ionospheric effects of Birkeland currents, Rev. Geophys. Space Phys., 9, 987.

Coroniti, F. V., and C. F. Kennel (1972): Polarization of the auroral electrojet, J. Geophys. Res., 77, 2835.

Croley, D. R., Jr., P. F. Mizera, and J. F. Fennell (1978): Signature of a parallel electric field in ion and electron distributions in velocity space, J. Geophys. Res., 83, 2701.

Evans, D. S. (1974): Precipitating electron fluxes formed by a magnetic field-aligned potential difference, J. Geophys. Res., 79, 2853.

Evans, D. S. (1975): Evidence for the low altitude acceleration of auroral particles, in Physics of the Hot Plasma in the Magnetosphere, pp. 319-340, edited by B. Hultqvist and L. Stenflo, Plenum Press, N.Y.

Frank, L. A., and K. L. Ackerson (1971): Observation of charged particle precipitation into the auroral zone, J. Geophys. Res., 76, 3612.

Fridman, M., and J. Lemaire (1980): Relationship between auroral electron fluxes and field-aligned electric potential difference, J. Geophys. Res., 85, 664.

Ghielmetti, A. G., R. G. Johnson, and E. G. Shelley (1978): The latitudinal, diurnal and altitudinal distributions of upward flowing energetic ions of ionospheric origin, Geophys. Res. Letters, 5, 59.

Goertz, C. K. (1979): Double layers and electrostatic shocks in space, Rev. Geophys. Space Phys., 17, 418.

Heppner, J. P. (1972): Electric field variations during substorms: Ogo 6 measurements, Planet. Space Sci., 20, 1475.

Hudson, M. K., and F. S. Mozer (1978): Electrostatic shocks, double layers, and anomalous resistivity in the magnetosphere, Geophys. Res. Letters, 5, 131.

Hudson, M. K., R. L. Lysak, and F. S. Mozer (1978): Magnetic field-aligned potential drops due to electrostatic ion cyclotron turbulence, Geophys. Res. Letters, 5, 143.

Kan, J. R. (1975): Energization of auroral electrons by electrostatic shock waves, J. Geophys. Res., 80, 2089.

Kan, J. R. (1980): Comment on "Double layers and electrostatic shocks in space" by C. K. Goertz, Rev. Geophys. Space Phys., 18, 338.

Kan, J. R., L. C. Lee and S.-I. Akasofu (1979): Two dimensional potential double layers and discrete auroras, J. Geophys. Res., 84, 4305.

Kindel, J. M., and C. F. Kennel (1971): Topside current instabilities, J. Geophys. Res., 76, 3055.

Kintner, P. M., M. C. Kelley, and F. S. Mozer (1978): Electrostatic hydrogen cyclotron waves near one earth radius altitude in the polar magnetosphere, Geophys. Res. Letters, 5, 139.

Knight, S. (1973): Parallel electric fields, Planet. Space Sci., 21, 741.

Lemaire, J., and M. Scherer (1971): Simple model for an ion-exosphere in an open magnetic field, Phys. Fluids, 14, 1683.

Lemaire, J., and M. Scherer (1974): Ionosphere-plasmasheet field-aligned currents and parallel electric fields, Planet. Space Sci., 22, 1485.

Lennartsson, W. (1977): On high-latitude convection field inhomogeneities, parallel electric fields and inverted-V precipitation events, Planet. Space Sci., 25, 89.

Lyons, L. R. (1980): Generation of large-scale regions of auroral currents, electric potentials, and precipitation by the divergence of the convection electric field, J. Geophys. Res., 85, 17.

Lyons, L. R., D. S. Evans, and R. Lundin (1979): An observed relation between magnetic field aligned electric fields and downward electron energy fluxes in the vicinity of the auroral forms, J. Geophys. Res., 84, 457.

Mizera, P. F., and J. F. Fennell (1977): Signatures of electric fields from high and low altitude particle distributions, Geophys. Res. Letters, 4, 311.

Mizera, P. F., D. R. Croley, Jr., and J. F.

Fennell (1976): Electron pitch-angle distributions in an inverted-'V' structure, Geophys. Res. Letters, 3, 149.

Montgomery, D. C., and G. Joyce (1969): Shock-like solutions of the electrostatic Vlasov equation, Plasma Phys., 3, 1.

Mozer, F. S., and R. H. Manka (1971): Magnetospheric electric field properties deduced from simultaneous balloon flights, J. Geophys Res., 76, 1697.

Mozer, F. S., C. W. Carlson, M. K. Hudson, R. B. Torbert, B. Paraday, J. Yatteau, and M. C. Kelley (1977): Observations of paired electrostatic shocks in the polar magnetosphere, Phys. Rev. Letters, 38, 292.

Papadopoulos, K. (1977): A review of anomalous resistivity for the ionosphere, Rev. Geophy. Space Phys., 15, 113.

Persson, H. (1966): Electric field parallel to the magnetic field in a low density plasma, Phys. Fluids, 9, 1090.

Ponyavin, D. I., M. I. Pudovkin, and S. S. Sazhin (1977): Self-consistent field-aligned electric field in the earth's magnetosphere, Geomagnet. Aeronomy, 17, 323.

Sharp, R. D., R. G. Johnson, E. G. Shelley, and K. K. Harris (1974): Energetic O^+ ions in the magnetosphere, J. Geophys. Res., 79, 1844.

Shawhan, S. D., C. G. Fälthammar, and L. P. Block (1978): On the nature of large auroral zone electric fields at 1-R_E altitude, J. Geophys Res., 83, 1049.

Shelley, E. G., R. D. Sharp, and R. G. Johnson (1976): Satellite observations of an ionospheric acceleration mechanism, Geophys. Res. Letters, 6, 54.

Shelley, E. G., R. G. Johnson, and R. D. Sharp (1972): Satellite observation of energetic heavy ions during a geomagnetic storm, J. Geophys. Res., 77, 6104.

Swift, D. W. (1975): On the formation of auroral arcs and acceleration of auroral electrons, J. Geophys. Res., 80, 2096.

Swift, D. W. (1976): An equipotential model for auroral arcs, 2, numerical solutions, J. Geophys. Res., 81, 3935.

Swift, D. W. (1979): An equipotential model for auroral arcs: The theory of two-dimensional laminar electrostatic shocks, J. Geophys. Res., 84, 6427.

Vondrak, R. R., H. R. Anderson, and R. J. Spiger (1971): Rocket-based measurements of particle fluxes and currents in an auroral arc, J. Geophys. Res., 76, 7701.

Whipple, E. C., Jr. (1977): The signature of parallel electric fields in a collisionless plasma, J. Geophys. Res., 82, 1525.

AURORAL DOUBLE LAYER

L. C. Lee and J. R. Kan

Geophysical Institute, University of Alaska, Fairbanks, Alaska 99701

Abstract. Theoretical models of double layers for the potential structures along auroral field lines are reviewed. The role of the backscattered and the trapped electrons in the formation of the auroral double layers is stressed. The importance of treating the auroral double layer as an integral part of the magnetosphere-ionosphere coupling system is also emphasized. Finally, the formation of discrete auroras due to an enhanced magnetospheric convection is demonstrated.

(I). Introduction

Alfvén [1958] proposed that auroral electrons are accelerated by a double layer electric field on auroral field lines. Double layers with potential drop localized in a few Debye lengths has been studied by a number of authors [e.g., Block, 1972; Knorr and Goertz, 1974; Goertz and Joyce, 1975; Kan, 1975; Torven and Babic, 1975; Quon and Wong, 1976; Coakley and Hershkowitz, 1979; Kan and Lee, 1980a]. However, particle observations [Mizera and Fennell, 1977; Sharp et al., 1979] indicate that parallel electric fields on auroral field lines are distributed over distances much greater than the Debye length. Extended potential structures distributed over several earth radii (R_E) can be supported by charged particles along the convergent auroral field lines [Lemaire and Scherer, 1974; Lennartson, 1977; Chiu and Schultz, 1978; Kan et al., 1979; Lee et al., 1980]. Such extended potential structures in the presence of field-aligned currents may be called extended auroral double layers, or simply, auroral double layers [Kan and Lee, 1981].

In a recent theoretical development, the auroral double layer has been treated as an integral part of the magnetosphere-ionosphere coupling system [Sonnerup, 1980; Kan and Lee, 1980c]. In the presence of parallel electric fields along auroral field lines, the magnetosphere-ionosphere coupling may be termed "imperfect" [Kan and Lee, 1980c].

The purpose of this paper is to review the general properties of auroral double layers in the context of the "imperfect" magnetosphere-ionosphere coupling.

(II). Auroral Double Layer Criterion

In this section, we discuss the criterion for the formation of a double layer along auroral field lines. Consider a one-dimensional double-layer which consists of five particle species as shown in Figure 1: (1) electrons of plasma sheet origin (of density n_{eo}); (2) ions of plasma sheet origin (n_{io}); (3) upstreaming ions of ionospheric origin (n_{i1}); (4) cold ionospheric electrons (n_{e1}); and (5) backscattered and trapped electrons (n_{et}). In Figure 1, ϕ is the electric potential, $n_\alpha(z)$ is the particle number density of the α-th species; z_o and z_1 are, respectively, the altitudes of the upper and lower boundaries of the acceleration region. The one-dimensional double layer model is a good approximation for the inverted V electron acceleration region in which the latitudinal width is much greater than the ion gyroradius.

In a model which consists of only the first four particle species listed above, the criterion for the existence of a strong double-layer can be written as [Bohm, 1949; Block, 1972]

$$m_e V_{eo}^2 \gtrsim \Gamma_S \theta_{eo} + \theta_{io} \qquad (1)$$

$$m_i V_{i1}^2 \gtrsim \Gamma_S \theta_{i1} + \theta_{e1} \qquad (2)$$

where V_{eo} is the preaccelerated streaming speed, θ is the thermal energy, Γ_S is the specific heat ratio, and m_e and m_i are, respectively, the electron and proton mass. The condition in (1) is to ensure that the space charge is negative in the region near the upper boundary z_o, where $\phi > 0$. With the required streaming speed, the electron number density would decrease more slowly than the ion number density, as the streaming electrons enter from the $\phi = \phi_o = 0$ region into the $\phi \gtrsim 0$ region. Similarly, the condition in (2) is to ensure a positive space

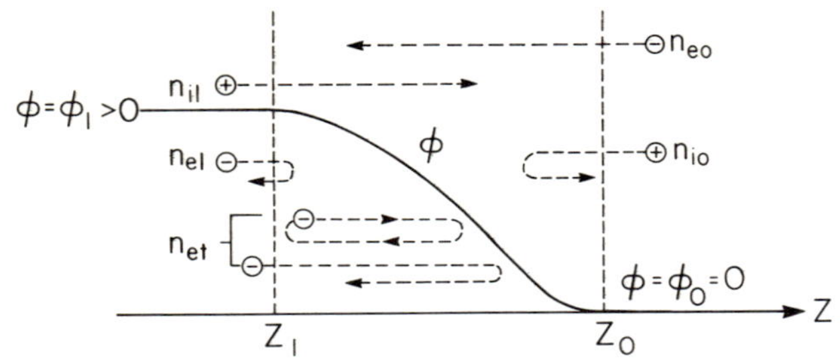

Figure 1. Schematic illustration of the electric potential profile $\phi(z)$ and the particle species along auroral field lines. The five particle species are: (1) precipitating magnetospheric electrons (n_{eo}); (2) magnetospheric ions (n_{io}); (3) ionospheric ions (n_{il}); (4) ionospheric electrons (n_{el}); and (5) backscattered and trapped electrons (n_{et}). Z_o and z_1 are, respectively, the upper and lower boundaries of the auroral acceleration region.

charge in the region near the lower boundary z_1, where $\phi \lesssim \phi_1$.

The necessary condition in (1) is in general not satisfied on auroral field lines at ~1 R_E altitude or higher, since at these altitudes the plasma is characterized by ions of a few keV and electrons of a few 100 eV. According to the condition in (1), the preaccelerated electron streaming energy is required to be greater than the thermal energy of the hot plasma sheet ions which is comparable to the observed peak energy of the auroral electrons (~ a few keV). If condition (1) were required, the acceleration of electrons by a double layer would be unnecessary for the formation of discrete auroras.

Kan and Lee [1980a] showed that the condition in (1) on the electron streaming energy can be relaxed in the presence of a sufficient number of nonthermal trapped and backscattered electrons with energies in the parallel motion ranging up to the double layer potential energy. They found that with the water-bag velocity distributions condition (1) becomes

$$m_e V_{eo}^2 \geq 2\,\theta_{eo} + \theta_{io}, \quad \text{for } N_{et} < N_c \qquad (3a)$$

$$V_{eo} > 0 \quad , \quad \text{for } N_{et} > N_c \qquad (3b)$$

where N_{et} is the number density of the trapped and backscattered electrons at z_1, $N_c = N_{eo}$ $[2e\phi_1/m_e(V_{eo} + a_{eo})]^{1/2}$ with $N_{eo} = n_{eo}(z_o)$ the number density and $a_{eo} = (2\theta_{eo}/m_e)^{1/2}$ the thermal energy of the streaming electrons of plasma-sheet origin. The results in (3a) and (3b) show that if $N_{et} \geq N_c$, the condition in (1) is relaxed; otherwise it remains except that Γ_S is replaced by 2 in the kinetic formulation. The trapped and backscattered electrons have been observed on auroral field lines and are of nonthermal character [e.g., Evans, 1974; Mizera and Fennel, 1977; Croley et al., 1978].

(III) Potential Drop Along Auroral Field Lines

An extended auroral double layer is characterized by: (1) the presence of a field-aligned current; (2) the presence of a converging magnetic field; and (3) a difference in the pitch-angle distribution of ions and electrons along field lines. A double layer must be a part of an electric circuit in which the field-aligned current connects the double layer to other parts of the electric circuit. The power consumed in the auroral double layer is supplied by the magnetospheric dynamo. The scale length of an extended auroral double layer is characterized by the scale length of the converging geomagnetic field lines. A difference in the pitch-angle anisotropy between the ions and electrons can contribute to the field-aligned potential drop along auroral field lines [Alfvén and Fälthammar, 1963; Whipple, 1977; Ponyavin et al., 1977]. Note that the field-aligned potential drop due to differential pitch-angle anisotropy is limited by the thermal energy of the less energetic particle species [Ponyavin et al., 1977].

The parallel electric field models studied by Chiu and Cornwall [1980] can be considered as extended double layers since the field-aligned currents are included in their models. They found that the maximum potential drop along auroral field lines is limited to

$$e\Psi_m \simeq 2\theta_{eo} \qquad (4)$$

where θ_{eo} is the thermal energy of the plasma-sheet electrons. It may be noted that the limitation on the potential drop in (4) is due to the conditions imposed on the trapped and backscattered electrons in their models.

The importance of the trapped and backscattered electrons in maintaining the charge-neutrality in the auroral double-layer with a potential drop much greater than the thermal energy θ_{eo} can be understood as follows. Electrons with a suitable energy and pitch-angle can be trapped between any two points along converging auroral field lines in the presence of an upward parallel electric field [Lee et al., 1980]. The point at a lower altitude is a magnetic mirror point and the point at a higher altitude is a turning point due to the electric force. In this manner, the trapped electrons can provide the necessary negative space charge to maintain quasi-neutrality everywhere in an extended double layer. The existence of double-layer solutions with a potential drop (Ψ) much larger than the thermal energy of the plasma-sheet electrons (i.e., $e\Psi >> \theta_{eo}$) can be demonstrated by an example discussed in Lee et al. [1980], in which $\Psi = 16\theta_{eo}$.

(IV). The Relationship Between Field-Aligned Current and Potential Drop

The magnetosphere and ionosphere are coupled through field-aligned currents flowing along the converging geomagnetic field lines. The downward field-aligned currents are carried primarily by the ionospheric thermal electrons while the upward field-aligned currents are carried dominantly by the precipitating electrons of magnetospheric origin [Knight, 1973]. In the absence of a parallel potential difference along auroral field lines, the upward field-aligned current density ($J_\parallel$) is limited to J_o, which is the current density carried by the thermal flux of the magnetospheric electrons in the loss-cone. For typical plasma parameters in the plasma sheet, J_o is $\sim 10^{-6}$ Amp/m^2.

However, following enhanced magnetospheric convection, the field-aligned current density must increase in order to speed up ionospheric convection [Kan and Lee, 1980c]. When the upward field-aligned current ($J_\parallel$) exceeds J_o, a parallel potential drop (Ψ) along auroral field lines would be needed to accelerate the magnetospheric electrons downward to carry the required upward current. The relationship between the upward field-aligned current density $J_\parallel$ and the parallel potential difference Ψ has been given by Knight [1973] and Fridman and Lemaire [1980]. From their formulation, it can be seen that the $J_\parallel - \Psi$ relationship depends on the altitude z_o of the upper boundary of the acceleration region. Assume that the plasma sheet electrons have a Maxwellian distribution with number density N_{es} and thermal energy θ_{eo} in the plasma sheet source region. The electric potential profile $\phi(z)$ is assumed to increase monotonically from z_o to z_1. Following Knight [1973], the field-aligned current density $J_\parallel$ at the ionospheric altitude (z_i) can be written as

$$J_\parallel = (B_i/B_o)\, N_{es}\, (2\theta_{eo}/\pi m_e)^{1/2} \cdot \{1 - (1-B_o/B_i)\exp[-\frac{e\Psi}{(B_i/B_o - 1)}]\} \quad (5)$$

where B_o and B_i are, respectively, the magnetic fields at z_o and z_i. Assuming that the geomagnetic field is dipole-like, we have

$$B_i/B_o \simeq (1 + z_o)^3 \quad (6)$$

where z_o is measured in units of R_E. Figure 2 shows the $J_\parallel - \Psi$ relationship for several values of z_o. The normalization constant J_o in Figure 2 is the current density carried by the plasma sheet electrons in the absence of a potential drop along field lines. By setting $\Psi = 0$ in (5) we have

$$J_o = N_{es}\,(2\theta_{eo}/m_e)^{1/2} \simeq 7.6 \times 10^{-7}\ \text{Amp/m}^2 \quad (7)$$

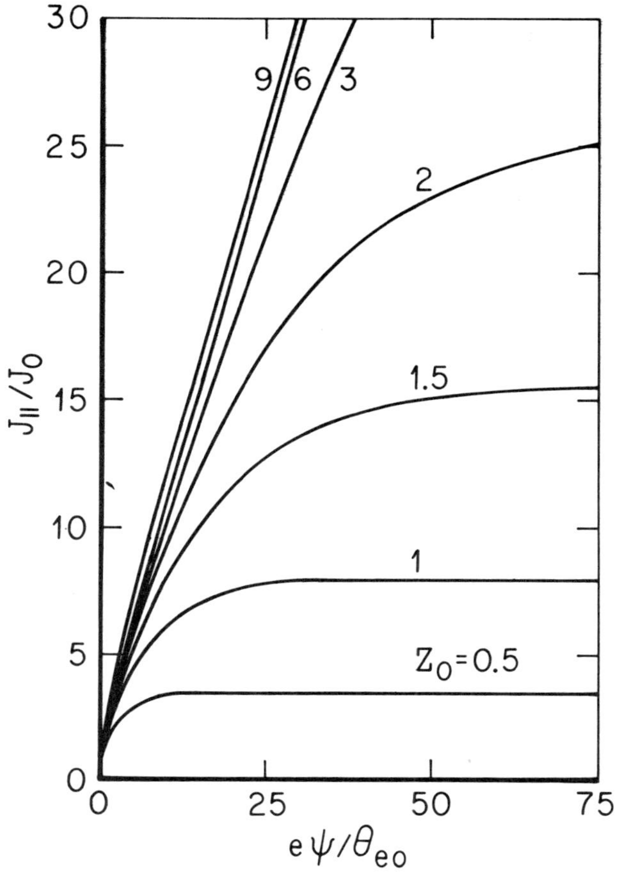

Figure 2. The relationship between the field-aligned current density $J_\parallel$ and the parallel potential drop Ψ is shown for z_o = 0.5, 1, 1.5, 2, 3, 6, and 9 R_E. Here z_o is the altitude of the upper boundary of the auroral acceleration region, and θ_{eo} is the thermal energy of the plasma-sheet electrons.

for $N_{es} \simeq 1$ cm^{-3} and $\theta_{eo} \simeq 200$ eV. As shown in Figure 2, for a given potential drop Ψ, the current density $J_\parallel$ increases with z_o. This can be understood physically in the following way. As the altitude of the acceleration region increases, more plasma sheet electrons will be accelerated downward to reach the ionosphere before mirrored upward by the converging geomagnetic field.

Figure 2 indicates that for a given upward field-aligned current density $J_\parallel$, the parallel potential difference Ψ cannot be determined unless z_o is known. Based on the extended double layer model discussed above, it can be shown that for a given $J_\parallel$, the parallel potential difference Ψ and the altitude z_o are controlled by the energetic component of the backscattered and trapped electrons. The detailed calculations will be reported

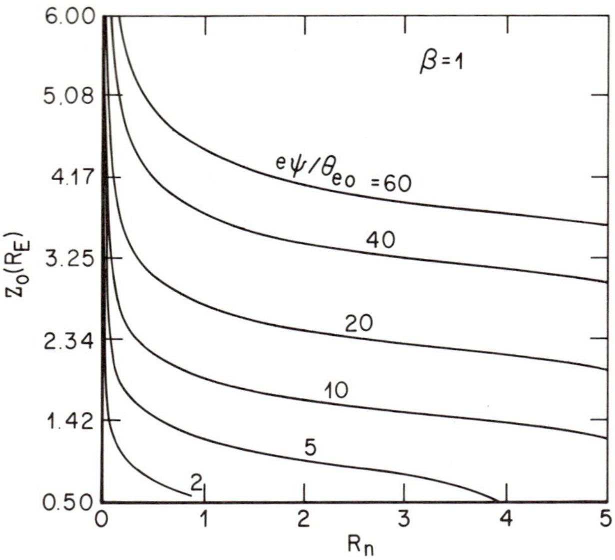

Figure 3. The altitude of the upper boundary of the auroral acceleration region, z_o, is shown as a function of the number density ratio R_n for various values of $e\Psi/\theta_{eo}$. β is defined in the text.

elsewhere; the main results are presented below.

Let $R_n = n_{et}(z_1)/n_{eo}(z_1)$. The observed value of R_n is typically 0.2. Figure 3 shows the altitude z_o as a function of R_n for $e\Psi/\theta_{eo} = 2$, 5, 10, 20, 40 and 60, based on the double layer criterion in (3b). In Figure 3, the lower boundary of the acceleration region is set to $z_1 = 0.5\ R_E$, and $\beta = \lim_{z \to z_o^-} (\phi - \phi_o)\ (B_1/B - 1)/[(\phi_1 - \phi_o)\ (1 - B_o/B_z)]$ is a parameter measuring the potential profile near z_o. For a given density ratio R_n, Figure 3 provides the relationship between Ψ and z_o. Combining Figures 2 and 3, we are able to obtain the $J_\parallel - \Psi - z_o$ relationship for various values of R_n. As can be seen in Figure 4, the $J_\parallel - \Psi$ relationship is not sensitive to the value of R_n. The $J_\parallel - \Psi$ relationship is also insensitive to the β value. As shown in Figure 4, the relationship between Ψ and $J_\parallel$ is approximately linear, which is consistent with the rocket observations as reported by Lyons et al. [1979]. It should be mentioned that the altitudes z_o shown in Figures 3 and 4 are different from those obtained by Kan and Lee [1980b], in which the double layer is assumed strong, i.e., $e\Psi/\theta_{eo} > (B_1/B_o - 1)$. Under auroral conditions, strong double layers with $e\Psi/\theta_{eo} > (B_1/B_o - 1)$ occur rarely.

It can be seen from Figure 4 that if the field-aligned current density $J_\parallel$ is enhanced by 10 times the quiet-time value, i.e., $J_\parallel \simeq 10\ J_o \simeq 10^{-5}$ Amp/m^2, a parallel potential difference $e\Psi \simeq 10\ \theta_{eo} \simeq 2$ keV is needed to accelerate the magnetospheric electrons downward along auroral field lines to carry the required current density.

(V). Imperfect Magnetosphere-Ionosphere Coupling and Two-Dimensional Auroral Double Layer

The magnetosphere-ionosphere coupling may be termed "perfect" if the geomagnetic field lines are equipotential. On the other hand, the coupling may be termed "imperfect" if part of the oval field lines are non-equipotential. As discussed in the last section, the development of a parallel potential difference Ψ along converging field lines is a consequence of the enhancement of the upward field-aligned current density $J_\parallel$ in the magnetosphere-ionosphere coupling. This parallel potential difference Ψ will accelerate precipitating electrons to several kilovolts, resulting in the formation of discrete auroras.

Since magnetosphere-ionosphere coupling leads to variations of the potential profile in the latitudinal direction (x direction), the auroral double layer must necessarily be two-dimensional in nature. Studies of the two-dimensional auroral potential structure as an isolated entity have been reported by Swift [1976, 1979], Kan et. al. [1979], Chiu and Cornwall [1980] and Wagner et al. [1980]. In these models, the loading effect on the magnetosphere due to the energy dissipation in the ionosphere and in the auroral double layer is neglected.

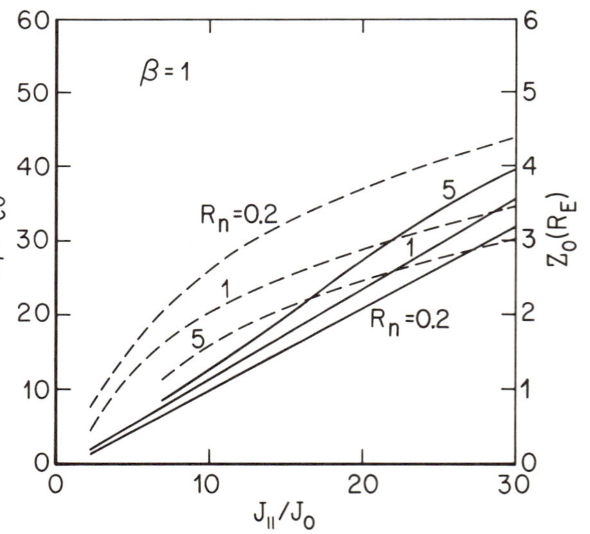

Figure 4. The normalized parallel potential drop, $e\Psi/\theta_{eo}$, (shown as solid lines) and the altitude of the upper boundary of the acceleration region z_o (dashed lines) are plotted as functions of the normalized field-aligned current density $(J_\parallel/J_o)$ for the density ratio R_n = 0.2, 1 and 5.

A consistent two-dimensional potential structure on auroral oval field lines with a latitudinal width $L_x >> \rho_i$ (ion gyroradius) can be constructed in the following manner. First, the latitudinal profiles of field-aligned potential difference, Ψ (x), and field-aligned current density, $J_\parallel$ (x), can be obtained from the imperfect coupling between magnetosphere and ionosphere [Kan and Lee, 1980c]. Second, a one-dimensional kinetic potential model as discussed earlier can then be constructed along each field line with the required values of Ψ (x) and $J_\parallel$ (x) from the coupling study. In the following, we will briefly discuss the theory of imperfect magnetosphere-ionosphere coupling based on the paper by Kan and Lee [1980c].

Consider a coordinate system in which the x axis is positive northward, the y axis is positive westward and the z axis is positive upward. To simplify the geometry of the converging geomagnetic field, x and y are assumed constant along a given field line and measure the actual distance only at the ionospheric altitude. For simplicity, all magnetospheric quantities are scaled to the ionospheric altitude. Following Kan and Lee [1980c], the basic equations describing the magnetosphere-ionosphere coupling system can be written as

$$-\gamma S E_m = I_m \tag{8}$$

$$dI_m/dx = J_\parallel = -dI_p/dx \tag{9}$$

$$J_\parallel = \begin{cases} \alpha_o \Psi + J_o \; , & J_\parallel > J_o \\ -d(\Sigma_p E_m)/dx \; , & J_\parallel \leq J_o \end{cases} \tag{10}$$

$$d\Psi/dx = E_m - E_i \tag{11}$$

$$I_p = \Sigma_p E_i \tag{12}$$

$$\Sigma_p (x) = \Sigma_o (1 + \beta J_\parallel \Psi)^{1/2} \tag{13}$$

where γS measures the slowing-down rate of plasma convection, I_m is the dynamo current, and E_m is the convection electric field in the magnetosphere; I_p is the Pedersen current, Σ_p is the Pedersen conductivity and E_i is the electric field in the ionosphere; α_o, Σ_o and β are constants. Equation (8) relates the dynamo current to the slowing-down rate of the magnetospheric convection, in which effects due to the pressure gradient and viscous force are neglected. Equation (9) relates the field-aligned current to the dynamo current and to the Pedersen current. Equation (10) describes the $J_\parallel - \Psi$ relationship as shown in Figure 4, from

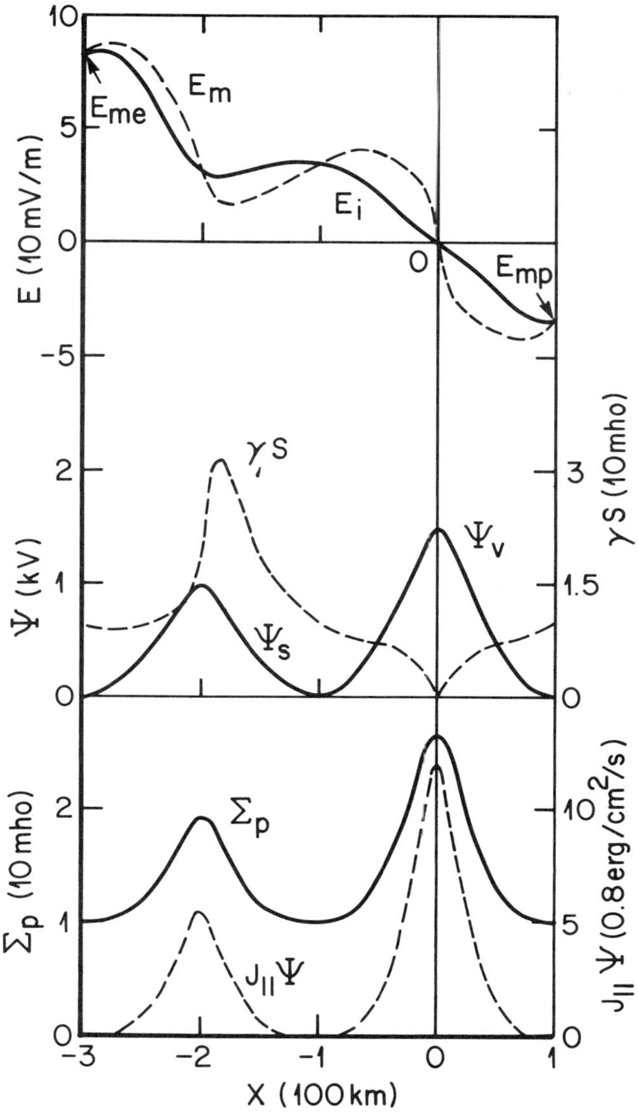

Figure 5. Configurations of various quantities in a solution of the imperfect magnetosphere-ionosphere coupling. The notations are given in the text.

which it is found $\alpha_o \simeq 5$ μA/m²/kV. Note that $\Psi \neq 0$ for $J_\parallel > J_o$; otherwise $\Psi = 0$. Equation (11) follows from $\nabla \times E = 0$; (12) is the Ohm's law in the ionosphere. Equation (13) gives the Pedersen conductivity in the presence of precipitating electron energy flux; β in (13) is a function of the ionization and the recombination coefficients.

Figure 5 shows an example of the solutions in which latitudinal profiles of E_m, E_i, γ_S, Ψ, Σ_p and $J_\parallel \Psi$ are given. The parameters used are: $\alpha_o = 5 \mu A/m^2/kV$, $\Sigma_o = 10$ mho, $J_o = 0.5 \mu A/m^2$, $\beta = 5 \times 10^{-5} (kV \cdot A/m^2)^{-1}$. These parameters are chosen to produce results comparable in magnitude to observations. The two peaks in the Ψ profile in Figure 5 will lead to two inverted V electron precipitation regions which produce discrete auroras.

Based on the $J_{\parallel}\Psi$ and Ψ profiles shown in Figure 5, one can calculate the distribution of potential drop along each field line with the required $J_{\parallel}(x)$ and $\Psi(x)$ [Chiu and Schulz, 1978; Lee et al., 1980] . The resulting two-dimensional equipotential contours are V-shaped near the convective reversal boundary (x = 0) and S-shaped near x = -200 km for the example shown in Figure 5.

(VI). Summary

In summary, we have demonstrated the possibility for the existence of double layers with a parallel potential drop (Ψ) much larger than the thermal energy of the plasma-sheet electrons (i.e., $e\Psi >> \theta_{eo}$). The parallel potential drop in the auroral double layers will accelerate the precipitating plasma-sheet electrons to several kilovolts to produce the observed discrete auroras.

On the other hand, auroral double layers are a consequence of the imperfect magnetosphere-ionosphere coupling following an enhancement of magnetospheric convection. The parallel electric fields in the auroral double layers are needed to accelerate the plasma-sheet electrons downward to carry the required upward field-aligned currents ($J_{\parallel}$) in the enhanced magnetosphere-ionosphere coupling. The relationship between the upward field-aligned current $J_{\parallel}$, the parallel potential drop Ψ, and the upper boundary z_o of the auroral double layer is shown to be controlled by the energetic component of the backscattered and trapped electrons. It is shown that latitudinal profiles of the parallel potential drop, $\Psi(x)$, and the field-aligned current density, $J_{\parallel}(x)$, can be obtained from the coupling study to construct a two-dimensional auroral double layer whose equipotential contours can be either V-shaped or S-shaped.

Acknowledgments: This work was supported in part by the Atmospheric Science Section of NSF under grant ATM79-16470 and by the U.S. Air Force under contract F19628-79C-0046 to the University of Alaska.

References

Alfvén, H., On the thery of magnetic storms and aurorae, Tellus, 10, 104, 1958.

Alfvén, H. and C.-G. Fälthammar, Cosmical Electrodynamics, Fundamental Principles, p. 162, Clarendon Press, Oxford, 1963.

Block, L.P. Potential double layers in the ionosphere, Cosm. Electrodyn., 3, 349, 1972.

Bohm, D., in The Characteristics of Electrical Discharges in Magnetic Fields, ed. by A. Guthrie and R.K. Wakerling, p. 77, McGraw-Hill, New York, 1949.

Chiu, Y.T. and J.M. Cornwall, Electrostatic model of a quiet auroral arc, J. Geophys., Res., 85, 543, 1980.

Chiu, Y.T. and M. Schulz, Self-consistent particle and parallel electrostatic field distributions in the magnetospheric-ionospheric auroral region, J. Geophys. Res., 83, 629, 1978.

Coakley, P. and N. Hershkowitz, Laboratory double layer, Phys. Fluids, 22, 1171, 1979.

Croley, D.R., Jr., P.F. Mizera, and J.F. Fennell, Signatures of a parallel electric field in ion and electron distributions in phase space, J. Geophys. Res., 83, 2701, 1978.

Evans, D.S. Precipitating electron fluxes formed by a magnetic field-aligned potential difference, J. Geophys. Res., 79, 2853, 1974.

Fridman, M. and J. Lemaire, Relationship between auroral electron fluxes and field-aligned electric potential difference, J. Geophys. Res., 85, 664, 1980.

Goertz, C.K. and G. Joyce, Numerical simulation of the plasma doauble layer, Astrophys. Space Sci., 32, 165, 1975.

Kan, J.R., Energization of auroral electrons by electrostatic shock waves, J. Geophys. Res., 80, 2089, 1975.

Kan, J.R., L.C. Lee, and S.-I. Akasofu, Two-dimensional potential double layers and discrete auroras, J. Geophys. Res., 84, 4305, 1979.

Kan, J.R. and L.C. Lee, On the auroral double layer criterion, J. Geophys. Res., 85, 788, 1980a.

Kan, J.R. and L.C. Lee, Double-layer criterion on the altitude of the auroral acceleration region, Geophys. Res. Lett., 7, 429, 1980b.

Kan, J.R. and L.C. Lee, Theory of imperfect magnetosphere-ionosphere coupling, Geophys. Res. Lett., 7, 633, 1980c.

Kan, J.R. and L.C. Lee, Formation of auroral arcs and inverted V precipitations: an overview, in Plasma Processes of Discrete Auroras, Geophys. Monograph Series, AGU, 1981.

Knight, S., Parallel electric fields, Planet. Space Sci., 31, 209, 1974.

Lee, L.C., J.R. Kan, and C.S. Wu, Generation of auroral kilometric radiation and the structure of auroral acceleration region, Planet. Space Sci., 28, 703, 1980.

Lemaire, J. and M. Scherer, Ionosphere-plasmasheet field-aligned currents and parallel electric fields, Planet. Space Sci., 72, 1485, 1974.

Lennartsson, W., On high-latitude convection field inhomogeneities, parallel electric fields and inverted-V precipitation events, Planet. Space Sci., 25, 89, 1977.

Lyons, L.R., D.S. Evans, and R. Lundin, An observed relation between magnetic field aligned electric fields and downward electron energy fluxes in the vicinity of auroral forms, J. Geophys. Res., 84, 457, 1979.

Mizera, P.F. and J.F. Fennell, Signature of electric fields from high and low latitude particle distributions, Geophys. Res. Lett., 4, 311, 1977.

Ponyavin, d.I., M.I. Pudovkin, and S.S. Sazhin, Self-consistent field-aligned electric field in the earth's magnetosphere, Geomag. and Aeron., 17, 323, 1977.

Quon, B.M. and A.Y. Wong, Formation of potential double layers in plasmas, Phys. Rev. Lett., 37, 1393, 1976.

Sharp, R.D., R.G. Johnson, and E.G. Shelley, Energetic particle measurements from within ionospheric structures responsible for auroral acceleration processes, J. Geophys. Res., 84, 480, 1979.

Sonnerup, B.U.Ö., Theory of the low-latitude boundary layer, J. Geophys. Res., 85, 2017, 1980.

Swift, D.W., An equipotential model of auroral arcs, 2. Numerical solutions, J. Geophys. Res., 81, 3935, 1976.

Swift, D.W., An equipotential model for auroral arcs: The theory of two-dimensional laminar electrostatic shocks, J. Geophys. Res., 84, 6427, 1979.

Torvén, S. and M. Babic, Current chopping space charge layers in a low pressure arc plasma, Proc. 12th Int. Conf. on Phenomena in Ionized Gases, Eindhoven Netherlands, Aug. 18-22, 124, North-Holland Publ. Co., 1975.

Wagner, J.S., T. Tajima, J.R. Kan, J.N. Leboeuf, S.-I. Akasofu, and J.M. Dawson, V-potential double layers and the formation of auroral arcs, Phys. Rev. Lett., 45, 803, 1980.

Whipple, E.C., The signature of parallel electric fields in a collisionless plasma, J. Geophys. Res., 82, 1525, 1977.

THE FIELD-ALIGNED CURRENT VERSUS ELECTRIC POTENTIAL RELATION AND AURORAL ELECTRODYNAMICS

L. R. Lyons

Space Environment Laboratory, NOAA, Boulder, Colorado 80303, USA

Abstract. Observations over aurora are inconsistent with the concept that instabilities leading to anomalous resistivity govern the relation between the field-aligned current density $j_{\parallel}$ and the field-aligned potential difference $V_{\parallel}$. The observations are also inconsistent with the analogy to laboratory double layers that significant $V_{\parallel}$'s result from $j_{\parallel}$ reaching a critical value that cannot be exceeded. However, the observed magnitudes of $j_{\parallel}$ and $V_{\parallel}$ over aurora, and the observed relation between them, are quantitatively explained by considering nothing more than the uninhibited, single-particle motion along field lines. It is shown that the $j_{\parallel}$ versus $V_{\parallel}$ relation from single-particle motion, together with the ionospheric current continuity equation, can account for the overall electrodynamics (i.e., particle acceleration, precipitating electron energy fluxes, currents, and electric potentials) of large scale "inverted-V" (∿200 km in latitudinal width) and discrete auroral (∿10's km in latitudinal width) precipitation regions. The inverted-V scale size is a natural result of the current versus electric potential relations along field lines and in the ionosphere, whereas the discrete auroral scale size results from specific structure in the electric potential distribution deep within the magnetosphere.

Introduction

Electrons responsible for discrete aurora are accelerated by a potential difference $V_{\parallel} \gtrsim 1$ kV along magnetic field lines, and these electrons carry a current along the field lines. To understand why there are aurora, it is necessary to nderstand the physics of the relation between the the field-aligned current density $j_{\parallel}$ and $V_{\parallel}$. In this paper, I first discuss this $j_{\parallel}$ versus $V_{\parallel}$ relation and I then show how this relation can be applied to understand the overall electrodynamics of the auroral processes along field lines and in the ionosphere. The diffuse aurora, which can simply be explained by wave driven diffusion of trapped electrons into the loss cone with $V_{\parallel} = 0$, is not considered.

Relation Between $j_{\parallel}$ and $V_{\parallel}$

Three basic approaches to understanding the relation between $j_{\parallel}$ and $V_{\parallel}$ have received considerable attention over the past several years: anomalous resistivity, double layers/electrostatic shocks, and single particle motion.

The "anomalous resistivity" approach (e.g., Kindel and Kennel, 1971; Papadopoulos and Coffey, 1974) is based on the concept that the Coulomb collision frequency becomes essentially zero along auroral magnetic field lines so that the field lines become "infinitely conducting". Such an infinite conductivity would short-out any nonzero $V_{\parallel}$ unless there were some resistivity in addition to that from Coulomb collisions. The anomalous resistivity is then postulated to result from particle interactions with wave turbulence.

However, for the resistivity derived for Coulomb collisions to be valid, the energy gained by the particles from an electric field between collisions must be negligible compared with the particles' thermal energy (Spitzer, 1962). This condition is clearly violated in the presence of the observed (e.g., Evans, 1974, 1976) 1-10 kV potential differences along auroral field lines, so that it is incorrect to conclude that auroral field lines are infinitely conducting without calculating the current from uninhibited, single-particle motion along field lines.

Two particle populations can contribute to a field-aligned current: the ionospheric plasma and the high altitude plasma in the plasmasheet or magnetosheath. The ionospheric plasma can supply only an upward field-aligned current of magnitude $\lesssim 10^{-7}$-10^{-6} A/m^2 (Lennartsson, 1980; Lyons, 1980a), while field-aligned currents associated with aurora are upwards and generally from 10^{-6} A/m^2 to a few times 10^{-5} A/m^2 (Kamide and Rostoker, 1977; Anderson, 1978; and references therein). Thus the upward current from the ionosphere can generally be neglected over aurora.

The field-aligned current density from the high-altitude electron population, assuming an

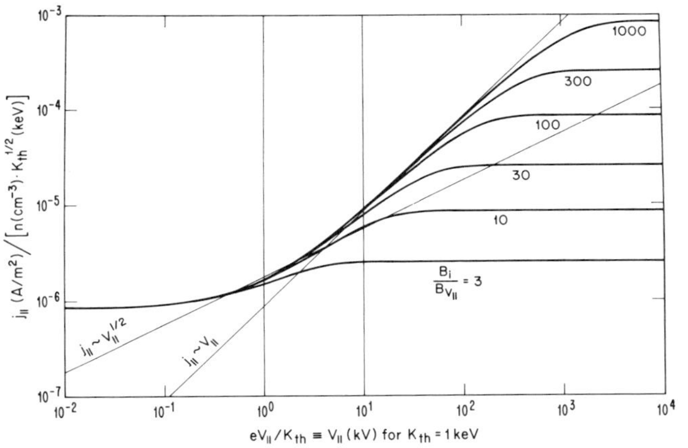

Fig. 1 The $j_\parallel$ versus $V_\parallel$ relation for uninhibited, single-particle motion along field lines for a high-altitude electron plasma with $n=1$ cm^{-3} and $K_{th} = 1$ keV. Results for other values of n and K_{th} can be obtained by multiplying the current densities on the vertical axis by $n(cm^{-3}) \cdot K_{th}^{1/2}$ (keV) and the potential differences on the horizontal axis by K_{th} (keV). Curves are shown for $B_i/B_{V_\parallel}$ from 3 to 1000. Lines for $j_\parallel \sim V_\parallel^{1/2}$ and $V_\parallel$ and for $eV_\parallel/K_{th} = 1$ and 10 are shown for reference.

isotropic Maxwellian distribution of density n and thermal energy K_{th}, is related to the total field-aligned potential difference $V_\parallel$ by (Knight, 1973; Lemaire and Scherer, 1974; Antonova and Tverskoy, 1975):

$$j_\parallel = en\left(\frac{K_{th}}{2\pi m_e}\right)^{1/2} \frac{B_i}{B_{V_\parallel}} \left[1 - \left(1 - \frac{B_{V_\parallel}}{B_i}\right) \exp\left\{- \frac{eV_\parallel}{K_{th}\left(\frac{B_i}{B_{V_\parallel}} - 1\right)}\right\}\right] \quad (1)$$

where e and m_e are the electron charge and mass, respectively. B_i is the ionospheric magnetic field strength, and $B_{V_\parallel}$ is the magnetic field strength at the top of the potential variation along field lines. This relation is independent of the distribution of the potential along field lines, other than the assumption that no particles incident upon the top of the parallel potential variation mirror before falling through the entire potential difference. However, this assumption is not a significant restriction, since $B_i/B_{V_\parallel}$ can be varied in (1) and $j_\parallel$ has little dependence upon $B_i/B_{V_\parallel}$ for $V_\parallel \lesssim 10$ kV as long as $B_i/B_{V_\parallel} > eV_\parallel/K_{th}$, and this inequality is generally satisfied along auroral field lines. Figure 1 shows $j_\parallel$ versus $V_\parallel$ as obtained from (1) for $n = 1$ cm^{-3} and $K_{th} = 1$ keV, which are reasonable values for the plasmasheet (Vasyliunas, 1968, Schield and Frank, 1970) and the high-latitude magnetosheath (Bahnsen, 1978). The results for other values of n and K_{th} can easily be obtained from Figure 1 by multiplying the current densities on the vertical axis by $n(cm^{-3}) \cdot K_{th}^{1/2}$(keV) and the potential differences on the horizontal axis by K_{th}(keV). Curves are shown for values of $B_i/B_{V_\parallel}$ ranging from 3, corresponding to a parallel potential variation entirely below ∿2000 km in altitude, to 1000, corresponding to a parallel potential variation extending to many earth radii in altitude. S3-3 satellite observations show that the potential variation generally occurs at altitudes between 5000 km and 10,000 km (Gorney et al. 1980), which gives $B_i/B_{V_\parallel} \gtrsim 30$ with 10 as a reasonable lower limit.

Figure 1 shows that the concept of "infinite conductivity" along auroral field lines is invalid. Uninhibited, single-particle motion in the presence of the observed potential differences of 1-10 kV does not give a very large $j_\parallel$. To the contrary, it gives $j_\parallel$'s of the observed magnitude of 10^{-6} - 10^{-5}A/m^2. Thus no "anomalous resistivity" is required along auroral field lines to account for the observed parallel potential differences. In fact, quite the opposite is true. The observed potential differences are required for the observed field-aligned current densities to exist (Knight, 1973).

For anomalous resistivity to be significant it must significantly reduce the current density for a given $V_\parallel$ from what it would be without the additional resistivity. If anomalous resistivity were to reduce the current by, say, a factor of 10, then unreasonably large $V_\parallel$'s of 10-100 kV would be required to account for the observed auroral $j_\parallel$'s. In addition, if anomalous resistivity were significant, then more energy (say a factor of ten more) would be dissipated by the resistance than is carried by the precipitating auroral electrons. I know of no evidence suggesting that energy dissipation along auroral field lines above the ionosphere is even comparable with that directly associated with the discrete aurora.

The above arguments show that the concept of anomalous resistivity cannot account for the upward currents and field-aligned potential differences associated with discrete aurora. Next we consider double layers.

Since a plasma has a finite particle density, the current that a plasma can carry is limited to a finite value. In analogy with laboratory double layers, when the current along auroral field lines reaches this critical value, no additional current enhancement is possible and double layers may form (Block and Fälthammar, 1975; Block, 1975). The double layers give a potential difference along field lines, but the current remains at its critical value. Figure 1 shows that the maximum current that can be carried along field lines with $V_\parallel = 0$ is ∿10^{-6} A/m^2; however $j_\parallel$ increases with increasing $V_\parallel$ and can reach as high as ∿10^{-3}A/m^2 assuming nothing limits the altitude to which the potential variation along field lines can extend. Thus the auroral field-aligned currents are much below the maximum current that can be carried by the plasma along the field lines, though a $V_\parallel > 0$ is required to obtain the observed current densities.

It is easy to use rocket observations to test whether currents are at a maximum value over aurora. If $j_\parallel$ were required to be at its maximum value to obtain a significant $V_\parallel > 0$, then $j_\parallel$ should be independent of $V_\parallel$ for $V_\parallel > 0$. On the other hand, from Figure 1, it is expected that $j_\parallel \sim V_\parallel^{0.5-1.0}$ if the maximum current is not reached. For $eV_\parallel / K_{th} \gtrsim 3$, most of the particles precipitating into the ionosphere will have an energy of $\sim eV_\parallel$, so that the total precipitating electron energy flux $\epsilon_\parallel$ will be $\sim j_\parallel V_\parallel$. Thus $\epsilon_\parallel \sim V_\parallel$ for all $V_\parallel > 0$ if a maximum current is reached. On the other hand, if a maximum current is not reached, then it is expected that $\epsilon_\parallel \sim V_\parallel^{1.5-2.0}$ for $eV_\parallel / K_{th} \gtrsim 3$. Lyons et al. (1979) found that $\epsilon_\parallel$ was approximated well by the functional form $KV_\parallel^2$ throughout three rocket flights, where K is an adjustable constant. The comparison between $\epsilon_\parallel$ and $KV_\parallel^2$ for one of the flights (POLAR 3, see Maynard et al., 1977 and Evans et al., 1977) is shown in Figure 2. Although no attempt was made to find the best value of the exponent of $V_\parallel$, it is evident that the relation $\epsilon_\parallel \sim V_\parallel$ would be in substantial disagreement with the observations.

This comparison between $\epsilon_\parallel$ and $V_\parallel$ implies that the parallel potential difference does not generally form because a maximum possible $j_\parallel$ has been reached (though this result has not yet been checked for a large number of aurora). However, the above discussion gives no information about the distribution of the potential along field lines. Calculations of this potential distribution using single particle motion show that both smoothly varying distributions (Lemaire and Scherer, 1973; Chiu and Schulz, 1978) and distributions with a large variation over a short distance as occur in double layers (Lemaire and Scherer, 1978; Lennartsson, 1980) are possible. A definitive conclusion as to how the potential is distributed along field lines has not yet been obtained. In addition, it is possible that the effects of particle scattering by plasma waves will have to be included to quantitatively evaluate this potential distribution, even though the scattering by plasma waves does not affect the relation between $j_\parallel$ and $V_\parallel$.

Nevertheless, the relation between $j_\parallel$ and $V_\parallel$ does not appear to be governed by $j_\parallel$ reaching a maximum value, in analogy with laboratory double layers, nor by anomalous resistivity. However the observed magnitudes of $V_\parallel$ and $j_\parallel$ and the observed relationship between them are easily a explained by considering nothing more than the single-particle motion along field lines. In the subsequent section, I show that the $j_\parallel$ versus $V_\parallel$ relation, (1), together with the ionospheric current continuity equation can quantitatively account for the precipitating electron energy fluxes and electric potentials associated with aurora.

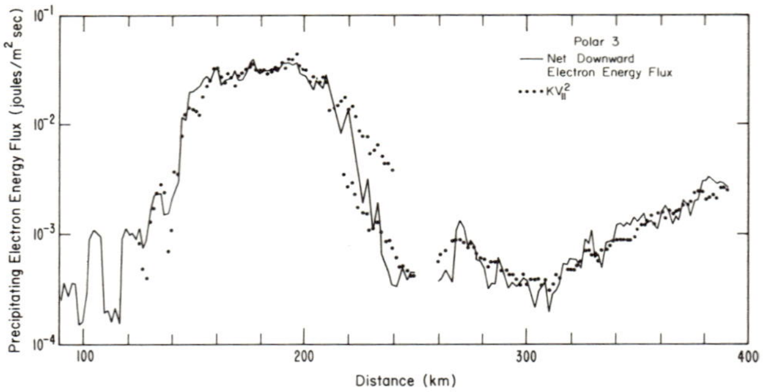

Fig. 2 Net downward electron energy flux throughout the POLAR 3 flight (Maynard et al., 1977; Evans et al., 1977) and the functional form $KV_\parallel^2$. Values for $KV_\parallel^2$ are shown from 120 km until the end of the flight. $V_\parallel$'s $\gtrsim$1 kV were inferred from the peak in the energy spectra of the precipitating electrons throughout this period. Two values of the constant K have been used, the higher value ($4.7\text{x}10^{-4}$ joules/m^2-s-kV^2) giving a good fit before 220 km and the lower value ($1.0\text{x}10^{-4}$ joules/m^2-s-kV^2) giving a good fit after 225 km. (From Lyons et al., 1979).

Large-scale "Inverted-V" Precipitation Regions (Lyons, 1980)

In this section, it is shown that discontinuities in the magnetospheric convection electric field $\underline{E}$ such that $\underline{\nabla} \cdot \underline{E} < 0$ can generate large regions of parallel currents with associated parallel potential differences and electron precipitation of the observed magnitudes and latitudinal extent (order of 200 km) of inverted V's. (Smaller scale (10's of km) structure associated with discrete aurora is addressed in the next section.) A commonly observed example of such an electric field divergence exists along the evening boundary between anti-sunward convection over the polar cap and sunward convection at lower latitudes, and inverted-V precipitation regions are often observed to be coincident with this boundary (Frank and Gurnett, 1971; Gurnett and Frank, 1973). If the convection electric field is driven by magnetosheath flow across polar cap magnetic field lines opening into the magnetosheath, then discontinuities in the magnetosheath flow can also cause electric field discontinuities with $\underline{\nabla} \cdot \underline{E} < 0$. Such discontinuities may result in regions of parallel currents, potential differences, and electron precipitation over the polar cap that are similar to those at the boundary between sunward and anti-sunward convection.

The initial assumption is that a discontinuity in the magnetospheric convection electric field, with $\underline{\nabla} \cdot \underline{E} < 0$, exists well above the ionosphere. This upper boundary condition is then applied to obtain the ionospheric potential by solving the ionospheric current continuity equation.

$$j_{\parallel} = \frac{d}{dX_i} \left(\Sigma_p \frac{d\phi_i}{dX_i}\right) \qquad (2)$$

Here $j_{\parallel}$ is the upward, field-aligned current density, ϕ_i is the electric potential in the ionosphere as a function of horizontal distance X_i, and Σ_p is the height-integrated Pederson conductivity. The problem has been reduced to one horizontal dimension by assuming $E_x = -d\phi_i/dX_i$ and Σ_p are homogeneous in the Y_i direction and neglecting the divergence of the Hall current from an electric field in the Y_i direction.

Equation (2) is a soluble, second-order differential equation in the ionosphere potential ϕ_i for a specified high-altitude potential distribution $\phi(X_i)$ provided $j_{\parallel}$ and Σ_p can be expressed as a function of ϕ_i and ϕ. The upper boundary potential $\phi(X_i)$ is at altitudes above any field-aligned potential difference, but is specified as a function of distance as mapped along field lines into the ionosphere so that it is unnecessary to specify a specific upper boundary. Equation (1) gives $j_{\parallel}$ as a function $V_{\parallel} = \phi_i - \phi$ for single-particle motion, and Σ_p is taken to vary with the precipitating electron energy flux $\varepsilon_{\parallel}$ as (Harel et al., 1977, 1980):

$$\Sigma_p = 0.5+160\ [\varepsilon_{\parallel}(\text{joules/m}^2\text{-s})]^{1/2}\text{mhos} \qquad (3)$$

Under the same conditions used in deriving (1), $\varepsilon_{\parallel}$ is related to $V_{\parallel}$ by (Lundin and Sandahl, 1978)

$$\varepsilon_{\parallel} = nK_{th}\left(\frac{K_{th}}{2\pi m_e}\right)^{1/2} \frac{B_i}{B_{V_{\parallel}}} \cdot$$

$$\left\{\left[2+\frac{eV_{\parallel}}{K_{th}}\right] - \left[\frac{eV_{\parallel}}{K_{th}} + 2\left(1-\frac{B_{V_{\parallel}}}{B_i}\right)\right] \exp\left[-\frac{eV_{\parallel}/K_{th}}{\frac{B_i}{B_{V_{\parallel}}}-1}\right]\right\} \qquad (4)$$

The solid line in Figure 3 gives the electric potential distribution used to represent an electric field change with $\underline{\nabla}\cdot\underline{E} < 0$ such as exists along the evening convection boundary. This assumed electric potential variation at high altitudes, above any significant field-aligned potential differences, is shown as a function of distance X_i across field lines as mapped into the ionosphere. Using this assumed upper boundary potential distribution, the current continuity equation in the ionosphere (2) was solved numerically to obtain the ionospheric potential distribution illustrated by the dashed line in Figure 3. The boundary condition $V_{\parallel} \to 0$ as $|X_i| \to \infty$ was applied, since the only other condition consistent with (2) is $V_{\parallel} \to \infty$. The difference between the solid and dashed curves in the figure gives the field-aligned potential difference. Note that the only way current continuity can be maintained in the ionosphere in the ionosphere in the presence of the discontinuity in the upper boundary electric field is by means of an ionospheric potential distribution giving a significant field-aligned potential difference as illustrated in Figure 2

When realistic parameters are used (i.e., an electric field charge of ~100 mV/m, and a thermal energy of ~1 keV and density of 1 cm^{-3} for the high altitude electron distribution), the solution to the current continuity equation gives auroral precipitation regions with $V_{\parallel} \gtrsim 1$ kV over latitudinal widths of about 200 km. This width is typical of the inverted-V precipitation regions (Frank and Ackerson, 1971). However it is larger than that generally associated with discrete auroral features. The solution to the current continuity equation gives maximum field-aligned potential differences of ~1-10 keV, maximum field-aligned current densities of ~1-10 $\mu A/m^2$, and maximum precipitating electron energy fluxes of ~3-100 ergs/cm^2-s. The values are all typical of auroral precipitation regions.

A comparison of the theoretical results with particle observations of a specific inverted-V region near 1800 LT (from Gurnett and Frank, 1973) is shown in Figure 4. Gurnett and Frank give, as a function of latitude, the simultaneously measured differential electron energy flux as a function of energy (from which an approximation for the parallel potential difference has been inferred from the peak in the energy spectrum), the total precipitating electron energy flux, and the horizontal electric field (see their Plate 1 and Fig. 6). The electric field was measured along a direction ~ 45° to the N-S direction and a change of ~ 8.5x10-2 V/m was measured across the precipitation region. Assuming the actual horizontal electric field was in the N-S direction, corresponding to the idealized model in Figure 2, the actual electric field change was ~.12 V/m. Thus .12 V/m was taken as the upper-boundary electric field change as mapped into the ionosphere.

A reasonable fit to the magnitudes and spatial extent of the observed parallel potential differences and total precipitating electron fluxes was obtained for a density n=1 cm^3 and thermal energy K_{th}= 500 eV. The model results are compared with the observations in Figure 4 where latitude has been converted to horizontal distance in the ionosphere. The measured precipitating energy fluxes, presented per steradian by Gurnett and Frank, have been multiplied by π to estimate the total precipitating electron energy flux. This presumes that the precipitating electrons were isotropic in pitch angle. Note in Figure 4 that the detailed structures within the inverted-V, which have magnitudes and scale sizes comparable with those of discrete auroral structures, are not reproduced by the model. However, the model does reproduce the overall magnitudes and spatial extent of the observed precipitating energy fluxes and parallel potential differences, showing that the overall precipitation region can result from the observed electric field change.

Remember that, although the electric field change was measured at low altitudes below any significant parallel potential difference, the inferred parallel potential difference approaches 0 away from the center of the precipitation region. Thus the measured electric field change can be mapped along magnetic field lines to any desired altitude. The assumption of a discontinuity in the upper boundary electric field is valid as long as the upper boundary electric field changes over a distance in X_i much smaller than does the ionospheric electric field. A more gradual electric field change would increase the size of the precipitation region and decrease the values of $V_{\parallel}$ and $j_{\parallel}$.

Smaller Scale Discrete Auroral Structure (Lyons, 1980b)

The above analysis shows that the electric field change associated with the evening convection boundary can account for large scale inverted-V precipitation regions, but not for the smaller scale precipitation structure associated with discrete aurora. Here I address the question of whether a modification of the structure of the

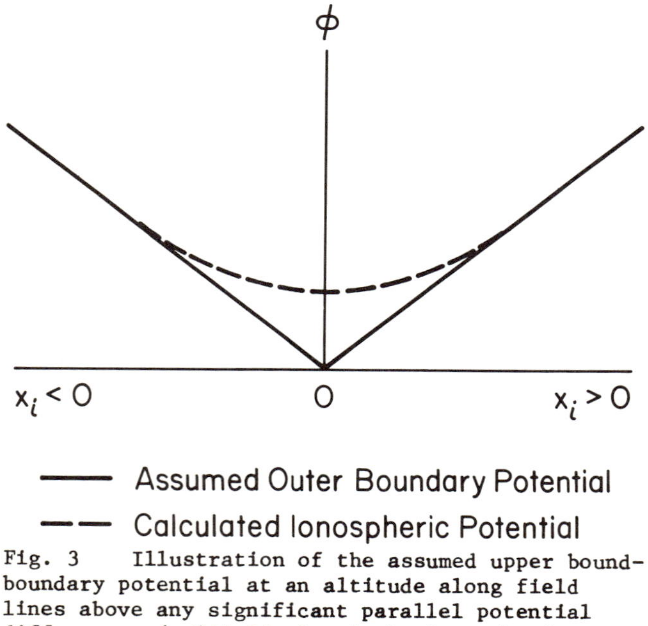

Fig. 3 Illustration of the assumed upper boundboundary potential at an altitude along field lines above any significant parallel potential differences (solid line) and the ionospheric potential calculated (dashed line) as a function of horizontal distance across the field lines in the ionosphere. (From Lyons, 1980)

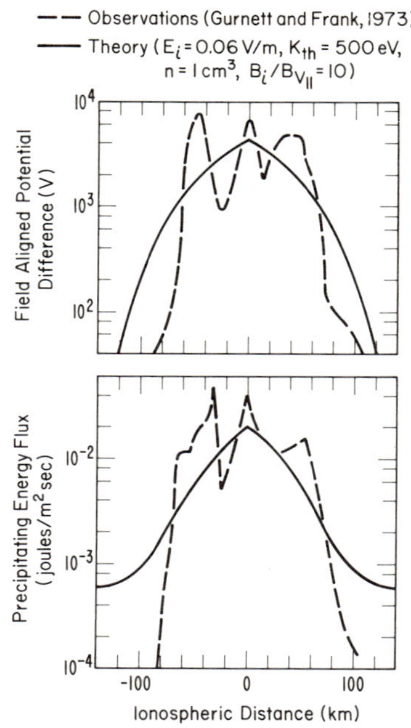

Fig. 4 Comparison of the theoretically calculated field-aligned potential differences and precipitating electron energy fluxes with observations within an inverted-V region at ∿1800 LT from Gurnett and Frank (1973). The calculations used the electric field change measured simultaneously with the precipitating electrons as described in the text, and the density n and thermal energy K_{th} of the electron population incident upon the parallel difference have been adjusted in an attempt to obtain reasonable agreement with the observations. The "observed" parallel potential differences have been inferred from the peak in the electron energy spectra, and the measured precipitating electron fluxes, presented per steradian by Gurnett and Frank, have been multiplied by π to estimate the total precipitating electron energy fluxes. (From Lyons, 1980)

upper boundary potential distribution in the vicinity of the electric-field change can account for the discrete aurora.

This analysis uses auroral observations of electric fields and precipitating electrons from the POLAR 3 rocket flight, the same flight used in Figure 2. The POLAR 3 rocket was launched towards the north over a large, discrete, auroral arc. The measured electric field was primarily normal to the arc, and the ionospheric electric potential obtained from integrating the measured normal electric field is shown versus horizontal distance along the rocket trajectory (which was approximately normal to the arc) in Figure 5. This electric potential is at ∿ 200 km altitude and is below any significant field-aligned potential differences. Given the total field-aligned potential difference, the high-altitude potential distribution of the auroral generator can be ob-

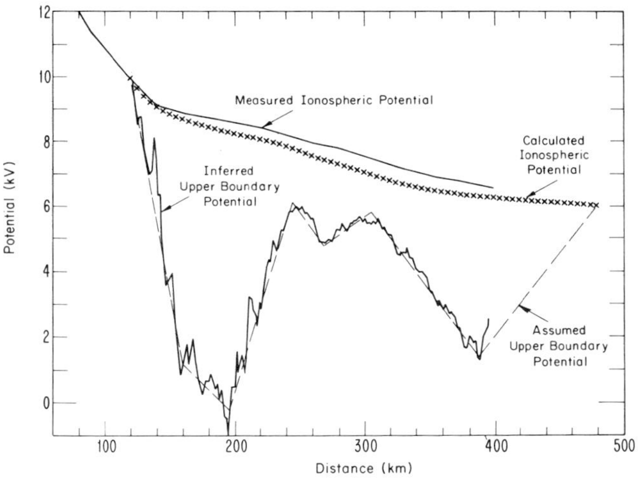

Fig. 5 Shown as a function of distance in the ionosphere, in a direction nearly normal to a large, discrete, auroral arc encountered between ∿150 km and ∿200 km, are the measured ionospheric potential from the POLAR 3 auroral rocket flight (Maynard et al., 1977; Evans et al., 1977), the inferred and assumed upper boundary potentials above all field-aligned potential differences, and the ionospheric potential obtained by solving the current continuity equation in the ionosphere. (From Lyons, 1981)

tained by subtracting the field-aligned potential difference from the measured ionospheric potential.

Assuming that the electrons over the aurora were accelerated by a field-aligned potential difference, the energy of the peak in the measured precipitating electron energy spectra that typically occurs at an energy between 1 and 10 keV gives an estimate of the magnitude of the field-aligned potential difference. The result of subtracting this inferred potential difference from the measured ionospheric potential gives the inferred upper boundary potential shown by the jagged, solid line in Figure 5. Field-aligned potential differences $\geq$ 1 kV were inferred to exist from 120 km into the flight until the payload re-entered the atmosphere after traversing ∿400 km in horizontal distance. The large auroral arc was encountered between ∿150 and ∿200 km, and a weaker, broader auroral form was entered at the end of the flight.

The inferred upper boundary potential in Figure 5 shows an overall electric field change, with $\underline{\nabla} \cdot \underline{E} < 0$, consistent with that used in the earlier analysis. However, the inferred upper boundary potential associated with the discrete aurora has considerable significant structure which was not included in the earlier analysis. This structure gives a total electric field change as mapped into the ionosphere of several hundred mV/m above the auroral arc, which is considerably greater than the electric field change of the convection boundary. In addition, at the southern (equatorward) edge of the region of field-aligned potential differences (120 km), the upper boundary electric field had $\underline{\nabla} \cdot \underline{E} > 0$. Thus the southern edge of the generator was positively charged, which can drive a downward field-aligned current not requiring a significant field field-aligned potential difference. In Lyons (1980b) it is shown that the southern edge was sufficiently positively charged to drive a downward current and require $V_{\parallel} \approx 0$.

The current continuity equation in the ionosphere was then solved numerically for the ionospheric potential using the upper boundary potential distribution inferred from the POLAR 3 observations. Rather than using every data point, the dashed line through the data in Figure 5 was input for simplicity. This upper boundary potential was extrapolated from 400 km to 480 km to give a poleward boundary condition with $V_{\parallel} = 0$; however the results at distances < 400 km have very little dependence on the assumed upper boundary potential beyond 400 km. It should be noted that, while the measured ionospheric potential is used when inferring the upper boundary potential, the measured ionospheric potential itself is not used when solving the current continuity equation.

From the observed relation between the precipitating electron energy flux and $V_{\parallel}$ (Lyons et al., 1979), the ratio $n/K_{th}^{1/2}$ can be inferred for the high altitude electron plasma (Lundin and Sandahl, 1978; Fridman and Lemaire, 1980; Lyons, 1981). This ratio is more important than the absolute values of the electron density n and thermal energy K_{th} in the solution of the current continuity equation, and the measured value has been taken. Specifically $n = .5\ cm^{-3}$, $K_{th} =$ 865

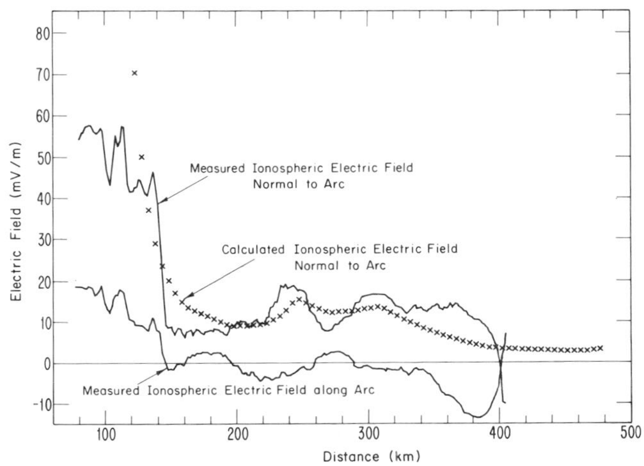

Fig. 6 Measured electric fields normal and parallel to the large, discrete auroral arc from the POLAR 3 rocket flight and the normal electric field calculated from the solution to the ionospheric current continuity equation. (From Lyons, 1981)

keV was used for distances <225 km, and n = 106 cm^{-3}, K_{th} = 865 keV was used for distances $\geq$ 225 km. (The value of $n/K_{th}^{1/2}$ was observed to decrease by a factor of 4.7 at 225 km).

The calculated ionospheric potential is given by the x's in Figure 5, and the results agree with the measured ionospheric potential to within ∿0.5 kV. The measured electric field normal to the arc and the electric field obtained from differentiating the calculated ionospheric potential are compared in Figure 6, and the fields agree to within a factor of two. Even the overall structure of the increases in the ionospheric field poleward of the main arc, which result from a marked decrease in the ionospheric Pederson conductivity in the calculations, is reproduced by the calculations. The measured and calculated precipitating electron energy fluxes are compared in Figure 7, and the agreement is good. However the agreement must be good whenever the field-aligned potential difference is in agreement with the observations and the measured value for $n/K_{th}^{1/2}$ is used.

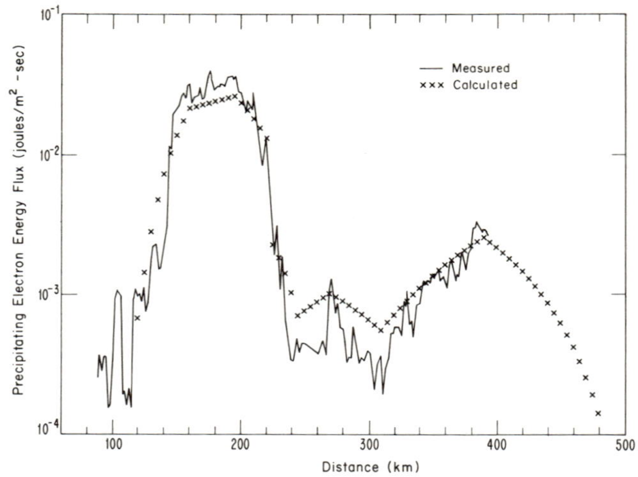

Fig. 7 Measured precipitating electron energy flux from the POLAR 3 flight and the calculated precipitating electron energy flux from the solution to the ionospheric current continuity equation. (From Lyons, 1981)

Conclusion

The relation between the field-aligned current density $j_\parallel$ and the field-aligned potential difference $V_\parallel$ is a crucial part of the electrodynamics of aurora. Observations over aurora are inconsistent with the concept that instabilities leading to anomalous resistivity govern the $j_\parallel$ versus $V_\parallel$ relation, and the observations are also inconsistent with the concept that a significant $V_\parallel$ results from $j_\parallel$ reaching a critical value that cannot be exceeded. However, the observed magnitudes of $j_\parallel$ and $V_\parallel$ over aurora and the observed relation between them are quantitatively explained by considering the simplest of all possibilities: the uninhibited single-particle motion along field lines.

Using the $j_\parallel$ versus $V_\parallel$ relation for single-particle motion and the ionospheric current continuity equation, a quantitative explanation has been presented for the currents and electric potentials along field lines and in the ionosphere associated with large-scale, inverted-V, precipitation regions and smaller scale discrete aurora.

The width of large-scale, inverted-V, precipitation regions appears to be a natural result of the current versus electric potential relations along auroral field lines and in the ionosphere. A negative divergence of the high altitude, magnetospheric electric field is required, however the inverted-V scale size need not be imposed by structure in the high-altitude electric field.

On the other hand, the width of smaller scale, discrete auroral precipitation structure appears not to be a natural result of these current versus electric potential relations. However, when the same current versus electric potential relations are used, discrete auroral structure can result from structure in the high-altitude magnetospheric electric field. The analysis shows discrete auroral forms to be approximately centered along field lines that connect to regions where the high-altitude electric field has $\nabla \cdot E < 0$, as is the case for the inverted-V precipitation regions. However the high-altitude electric field must have $\nabla \cdot E$ sufficiently greater than zero at the edges of discrete aurora to force the auroral width to be narrow. In addition, the magnitude of the total, high-altitude, electric field change giving ∇:E < 0 must be several hundred mV/m as mapped into the ionosphere, which is several times greater than that required for the inverted-V precipitation regions.

Although inverted-V and discrete auroral precipitation regions have been treated separately, the analysis is not intended to imply that the two sizes of precipitation regions are generally physically distinct from each other. It is reasonable that, at least at times, the discrete auroral precipitation is imbedded in a larger scale precipitation region. In this situation, the high altitude electric field structure causing the discrete aurora would be expected to be imbedded in a larger scale region with a total electric field change less than zero and of the magnitude required for the inverted-Vs. The inferred, high-altitude, potential distribution (Fig. 4) indicates that this may have been the case for the aurora traversed by POLAR 3.

A high-altitude electric potential distribution that can cause discrete aurora has been inferred in this analysis, but no attempt has been made here to explain this distribution. The results presented here suggest that studies leading to an understanding of how such an electric potential distribution may be generated within the magnetosphere will be crucial to fully understanding how discrete aurora are generated.

References

Anderson, H. R., Birkeland currents and auroral, J. Geomag. Geolec., 30, 381, 1978.

Antonova, YE. YE., and B. A. Tverskoy, Nature of the electron precipitation band of the "inverted-V" type and of the Harang discontinuity in the evening sector of the auroral ionosphere, Geomagnetism and Aeronomy, 15, 85, 1975.

Bahnsen, A., Recent techniques of observations and results from the magnetopause regions, J. Atmos. Terr. Phys., 40, 235, 1978.

Block, L. P., Double layers, Physic of the Hot Plasma in the Magnetosphere, ed. B. Hultqvist and L. Stenflo, Plenum Press, New York, 229, 1975.

Block, L. P., and C.-G. Fälthammar, Mechanisms that may support magnetic-field-aligned electric fields in the magnetosphere, Royal Inst. of Technology, Stockholm, report TRITA-EPP-75-22, 1975.

Chiu, Y. T., and M. Schulz, Self-consistent particle and parallel electrostatic field distributions in the magnetospheric-ionospheric auroral regions, J. Geophys. Res., 83, 629, 1978.

Evans, D. S., Precipitating electronf fluxes formed by a magnetic field aligned potential difference, J. Geophys. Res., 79, 2853, 1974.

Evans, D. S., The acceleration of charged particles at low altitudes, in Physics of Solar Planetary Environments, ed. D. J. Williams, Amer. Geophys. Union, 730, 1976.

Evans, D. S., N. C. Maynard, J. Troim, T. Jacobsen, and A. Egeland, Auroral vector electric field and particle comparisons, 2. Electrodynamics of an arc, J. Geophys. Res., 83, 2235, 1977.

Frank, L. A., and K. L. Ackerson, Observations of charged particle precipitation into the auroral zone, J. Geophys. Res., 76, 3612, 1971.

Frank, L. A., and D. A. Gurnett, Distributions of plasmas and electric fields over the auroral zones and polar caps, J. Geophys. Res., 76, 6829, 1971.

Fridman, M., and J. Lemaire, Relationship between auroral electron fluxes and field-aligned electric potential difference, J Geophys. Res., 85, 664, 1980.

Gorney, D. J., A. Clarke, D. Croley, J. Fennell, J. Luhmann, and P. Mizera, The distribution of ion beams and conics below 8000 km, J. Geophys. Res., 86, 83, 1981.

Gurnett, D. A., and L. A. Frank, Observed relationships between electric fields and auroral particle precipitation, J. Geophys. Res., 78, 145, 1973.

Harel, M., R. A. Wolf, P. H. Reiff, and H. K. Hillis, Study of plasma flow near the earth's plasmaspause, U. S. Air Force Geophysics Lab. report, AFGL-TR-77-0286, 1977.

Harel, M., R. A. Wolf, P. H. Reiff, and R. W. Spiro, Quantitative simulation of a magnetospheric substorm, 1. Model logic and overview, submitted to J. Geophys. Res., 1980.

Kamide, Y., and G. Rostaker, The spatial relationships of field-aligned currents and auroral electrojets to the distribution of nightside auroras, J. Geophys. Res., 82, 5589, 1977.

Kindel, J. M., and C. F. Kennel, Topside current instabilities, J. Geophys. Res., 76, 3055, 1971.

Knight, L., Parallel electric fields, Planet. Space Sci., 21, 741, 1973.

Lemaire, J., and M. Scherer, Plasma sheet particle precipitation: a kinetic model, Planet. Space Sci., 21, 281, 1973.

Lemaire, J., and M. Scherer, Ionosphere-plasma-sheet field-aligned currents and parallel electric fields, Planet Space Sci., 22, 1485, 1974.

Lemaire, J., and M. Scherer, Field aligned distribution of plasma mantle and ionospheric plasma, J. Atmos. Terr. Phys., 40, 337, 1978.

Lennartsson, W., On the consequences of the interaction between the auroral plasma and the geomagnetic field, Planet. Space Sci., 28, 135, 1980.

Lundin, R., and I. Sandahl, Some characteristics of the parallel electric field acceleration of electrons over discrete auroral arcs as observed from two rocket flights, Symposium on European Rocket Research, Ajaccio, Corsica, 1978, ESA SP-135, 125, 1978.

Lyons, L. R., Generation of large-scale regions of auroral currents, electric, potentials, and precipitation by the divergence of the convection electric field, J. Geophys. Res., 85, 17, 1980.

Lyons, L. R., Discrete aurora as the direct result of an inferred, high-altitude generating potential distribution, J. Geophys. Res., 86, 1, 1981.

Lyons, L. R., D. S. Evans, and R. Lundin, An observed relation between magnetic field aligned electric fields and downward electron energy fluxes in the vicinity of auroral forms, J. Geophys. Res., 84, 457, 1979.

Maynard, N. C., D. S. Evans, B. Maehlum, and A. Egeland, Auroral vector electric field and particle comparisons, 1. Premidnight convection topology, J. Geophys. Res., 82, 2227, 1977.

Papadopoulos, K., and T. Coffey, Nonthermal features of the auroral plasma due to precipitating electrons, J. Geophys. Res., 79, 674, 1974.

Schield, M. A., and L. A. Frank, Electron observations between the inner edge of the plasma sheet and the plasmapause, J. Geophys. Res., 75, 5401, 1970.

Spitzer, L., Jr., Physics of fully ionized gases, Second Revised Edition, Interscience, New York, p. 139, 1962.

Vasyliunas, V. M., A survey of low energy electrons in the evening sector of the magnetosphere with OGO-1 and OGO-3, J. Geophys. Res. 73, 2839, 1968.

Electrostatic Shocks in the Auroral Magnetosphere

Mary K. Hudson and Douglas W. Potter

Space Sciences Laboratory
University of California
Berkeley, California 94720

Abstract. We present a 1d unmagnetized double layer simulation along with analytic considerations of the Langmuir-Bohm criterion for double layer formation. This requirement of large electron drift is not reduced by trapped thermal electron and ion populations; the assumption of a nonthermal population which requires the pre-existence of the double layer also cannot reduce the required electron drift. The double layer produced in the simulation by an electron drift equal to the electron thermal speed has $e\phi/T_e \sim 2$ across a distance of $50\lambda_D$ in a system $1024\lambda_D$ long, where λ_D is the Debye length. Negative charge density spikes and holes in ion phase space accompany the double layer. A magnetized BGK mode solution to Poisson's equation for the potential profile perpendicular to **B** for cold counterstreaming electrons and ions demonstrates how the spatial structure changes in a magnetized plasma.

This paper presents a 1d unmagnetized simulation and discusses analytic steady state models of double layers. We also discuss the relevance of these results to observations of large, spatially confined electric fields observed at $1R_E$ on the S3-3 satellite and to auroral particle acceleration and the production of discrete auroral arcs.

The first section presents time stationary analytic double layer models. In the second section, a simulation of an unmagnetized double layer provides information about temporal evolution. The third section is a discussion of magnetized potential structures observed in space and the laboratory in the context of *Swift's* [1975, 1976, 1979] time stationary model for such structures.

Time Stationary Analytic Double Layers

Bernstein et al. [1957] show that the steady state double layer problem can be treated analytically as a BGK mode. A potential $\phi(x)$ with continuous $\phi''(x)$ is a steady state solution to Poisson's equation if there is a reference frame in which the potential and particle populations required to support it are time stationary. In general, four particle populations are required: free electrons and ions with total energy positive everywhere, trapped ions in negative potential wells, and trapped electrons in positive potential wells. The trapped particles are those excluded from some region of space, and therefore include reflected particles. With this definition, the steady state trapped particle distribution functions $f(v)$ must be symmetric in $\pm v$; as many particles must slosh right as left. There is no such requirement on untrapped particles. Therefore, in principle one could arbitrarily adjust the free particle streaming velocities and mass flow with respect to the potential $\phi(x)$, or equivalently, the potential sheath propagation velocity with respect to the plasma. However, the boundary condition that the electric field vanish outside the double layer leads to a constraint on the streaming velocities of electrons and ions into the sheath. In the simple case of cold counterstreaming electrons and ions only, the Langmuir condition results [*Block,* 1972]. Integrating Poisson's equation once using $d\phi/dx=0$ at $\phi=\phi_0$ on the high potential side and $\phi=0$ on the low potential side of the double layer (Figure 1) yields

$$\frac{1}{2}\left(\frac{d\phi}{dx}\right)^2 = 4\pi e\int_0^{\phi_0} (n_e(\phi) - n_i(\phi))\, d\phi \tag{1}$$

The cold streaming electron and ion densities as functions of ϕ are

$$n_e(\phi) = \frac{n_{e1}u_{e1}}{u_e(\phi)} = n_{e1}\left(1 + \frac{2e\phi}{mu_{e1}^2}\right)^{-1/2} \tag{2a}$$

where n_{e1} and u_{e1} are the electron density and streaming velocity before entering the double layer from the low potential (magnetosphere) side, and

$$n_i(\phi) = \frac{n_{i0}u_{i0}}{u_i(\phi)} = n_{i0}\left(1 + 2e\frac{(\phi_0 - \phi)}{Mu_{i0}^2}\right)^{-1/2} \tag{2b}$$

where n_{i0} and u_{i0} are the ion density and streaming velocity before entering the double layer from the high potential (ionosphere) side. Integration of (2a) and (2b) substituted into (1) subject to the requirement that $d\phi/dx=0$ at $\phi=\phi_1=0$ as well as at $\phi=\phi_0$ yields the Langmuir condition for strong sheaths $e\phi_0 >> \frac{1}{2}mu_{e1}^2, \frac{1}{2}Mu_{i0}^2$

$$n_{e1}u_{e1} = n_{i0}u_{i0}\left(\frac{M}{m}\right)^{1/2} \tag{3}$$

For equal electron and ion densities outside the sheath $n_{e1}=n_{i0}$, this requires that the electron to ion streaming velocities scale $\sim(M/m)^{1/2}$.

If there are thermal electrons on the high potential side as indicated in Figure 1 ions must enter the sheath at $u_{i0} > c_s = (T_e/M)^{1/2}$. This is the *Bohm* [1949] criterion which can be derived by adding a Boltzman electron population to (1), integrating subject to the same boundary conditions, and expanding the result in the limit $e\phi(x)/T_e << 1$. Since the absolute value of the potential is arbitrary and is often taken to be zero on the high potential side [*Chen,* 1974], this expansion is appropri-

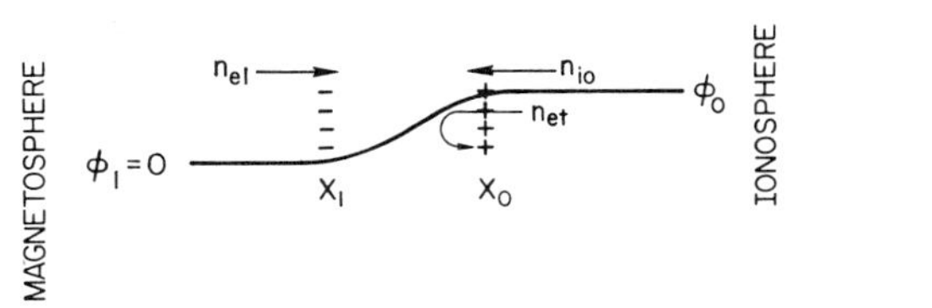

Fig. 1. Schematic double layer potential profile with three particle populations: cold streaming electrons n_{e1}, cold streaming ions n_{i0}, and trapped electrons n_{et}. The subscripts 0 and 1 refer to the right side (ionosphere) and left side (magnetosphere) respectively.

ate at the entry point of ions into the sheath. Physically, ion density must decrease slower than electron density on the high potential side to maintain the positive charge sheet which is self-consistent with the sheath. Combining this result with the Langmuir condition requires $u_{e1} > v_{th,e}$, the electron streaming velocity exceeds the electron thermal speed, which is the Langmuir-Bohm criterion for double layer formation. Drifts exceeding the electron thermal speed correspond to the largest currents which have been measured in the auroral magnetosphere [*Cattell*, 1981].

Kan and Lee [1980] show that the addition of a non-Maxwellian trapped electron population on the high potential side of the sheath reduces the requirement on u_{e1}. If instead, one includes stationary thermal electrons and ions, as well as cold counter-streaming electrons, and ions in (1), the Langmuir-Bohm criterion is not reduced. Substituting a thermal electron population

$$n_{et} = n_{0t} e^{e(\phi - \phi_0)/T_t} \rightarrow n_{0t}, \quad \phi = \phi_0 \tag{4a}$$

and a thermal ion population

$$n_{it} = n_{1t} e^{e\phi/T_t} \rightarrow n_{1t}, \quad \phi = \phi_1 = 0 \tag{4b}$$

into (1) along with the cold electron and ion populations, (2a) and (2b), and integrating subject to the preceding boundary conditions yields

$$n_{e1} m u_{e1}^2 \left[\left(1 + \frac{2e\phi_0}{m u_{e1}^2} \right)^{1/2} - 1 \right] - n_{i0} M u_{i0}^2 \left[\left(1 + \frac{2e\phi_0}{M u_{i0}^2} \right)^{1/2} - 1 \right]$$
$$+ n_{1t} T_t (e^{e\phi_0/T_t} - 1) - n_{0t} T_t (1 - e^{-e\phi_0/T_t}) = 0 \tag{5}$$

We have assumed, consistent with S3-3 observations [*Mozer et al.*, 1980], that the thermal electrons and ions have the same temperature T_t. Expanding in the strong sheath limit $e\phi >> \frac{1}{2} m u_{e1}^2, \frac{1}{2} M u_{i0}^2$ yields

$$(n_{e1} m^{1/2} u_{e1} - n_{i0} M^{1/2} u_{i0})(2e\phi_0)^{1/2}$$
$$+ n_{1t} T_t \frac{e\phi_0}{T_t} - n_{0t} T_t \frac{e\phi_0}{T_t} = 0 \tag{6}$$

For the reasonable assumption of equal densities of thermal electrons and ions outside the sheath, this reduces, independently of the assumption of equal thermal electron and ion temperatures, to the Langmuir condition (3).

By assuming a non-Maxwellian trapped electron population, *Kan and Lee* [1980] obtained a reduction in the required cold electron streaming velocity. Physically, trapped electrons with $T_t \sim e\phi_0$ can penetrate the potential barrier and contribute to the negative charge sheet (Figure 1). Otherwise, analogous to the Bohm criterion for ions $u_{i0} > c_s$ on the high potential side, electrons must stream into the sheath with sufficient $u_{e1} > v_{th,e}$ on the low potential side of the sheath such that electron density falls off slower than ion density. However, the trapped electrons must have $T_t \sim e\phi_0$ to effectively reduce the Langmuir-Bohm criterion. *Kan and Lee* [1980] assume that degraded primary and secondary electrons reflected or backscattered from the ionosphere form such a population, as indeed they do, but only after the double layer has formed and accelerated the auroral primaries. Since the double layer creates the trapped electron and ion populations, one cannot appeal to these populations to reduce the Langmuir-Bohm criterion for double layer formation.

In summary, the Langmuir criterion $n_{e1} u_{e1} > n_{i0} u_{i0} (M/m)^{1/2}$ is a condition on the ratio of electron to ion flux into a steady state sheath. Derived for cold counterstreaming electrons and ions, it is not modified by the presence of thermal electrons and ions of equal density. What sets the velocity scale is the Bohm criterion $u_{i0} > c_s$, a requirement on the cold ion streaming velocity into the sheath in the presence of thermal electrons on the high potential side. If one assumes equal densities of cold counterstreaming electrons and ions and equal thermal densities, the requirement $u_{e1} > v_{th,e}$ results. There is, however, enough freedom in the four density parameters of (6) to set up zero current solutions and still satisfy quasi-neutrality outside the double layer [*Montgomery and Joyce*, 1969]. We are using plasma simulations to further study the initial criteria, evolution, and final steady state of double layers.

Double Layer Simulations

The above time stationary solutions to Poisson's equation say nothing about the temporal evolution of the potential or particle populations. Plasma particle simulations are a tool for examining this temporal evolution. We have done a 1d unmagnetized simulation with periodic boundary conditions using the particle code ES1 [*Birdsall and Langdon*, 1978]. The electrons are a drifting Maxwellian with a drift equal to their thermal speed, and the ions are a stationary Maxwellian with $T_e/T_i = 20$. We observe the evolution of a double layer (Figure 2) for approximately the same simulation parameters as *Sato and Okuda* [1980a, b]. The amplitude is $e\phi/T_e \sim 2$ and the potential drop occurs over a scale length of $50\lambda_D$ out of $1024\lambda_D$.

Since a double layer is the collective effect of a Debye sheath, the number of particles in a Debye length must be large. To minimize the effects of periodic boundary conditions, the system must be many Debye lengths long. Finally, double layers evolve on a time scale of $1000\omega_{pe}^{-1}$ as discussed below. We have chosen

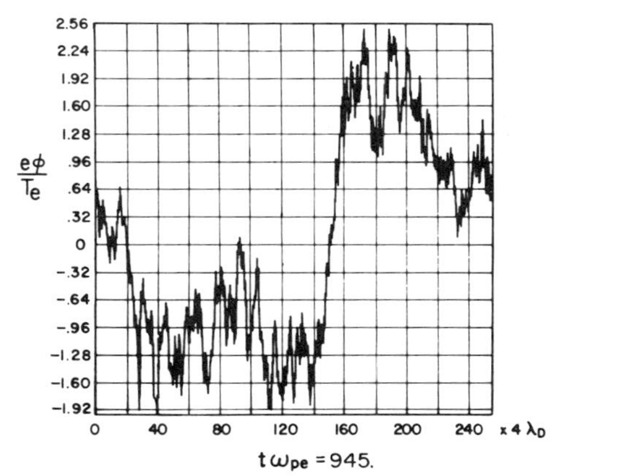

Fig. 2. Plasma simulation double layer potential profile for electron drift $u_{e1} = v_{th,e}$ and $T_e/T_i = 20$ in a system $1024\lambda_D$ long.

the plasma parameter $n\lambda_D=64$, the system length $L=1024\lambda_D$, and the time $T=1000\omega_{pe}^{-1}$. A double layer simulation is therefore a substantial computing effort in terms of the number of particles to be followed over many time steps.

Since the electron drift equals the electron thermal speed and $T_e/T_i=20$, the evolution of a double layer in our system is preceded by the growth of the ion acoustic instability. The most persistent features in the evolution are negative charge density spikes (Figure 3a) and holes in ion phase space (Figure 3b). These features persist and are stationary on time scales of hundreds of ω_{pe}^{-1}. Presumably, they are due to the trapping of electrons and ions in the potential of an ion acoustic wave or soliton as discussed below.

Hasegawa and Sato [1980] suggest that the negative charge spikes lead to double layer formation: current carrying electrons drifting from the left are reflected by a negative charge spike, enhance the negative charge density at that point, and leave a positive charge excess on the downstream side which yields the double layer potential structure. This explains why a long system

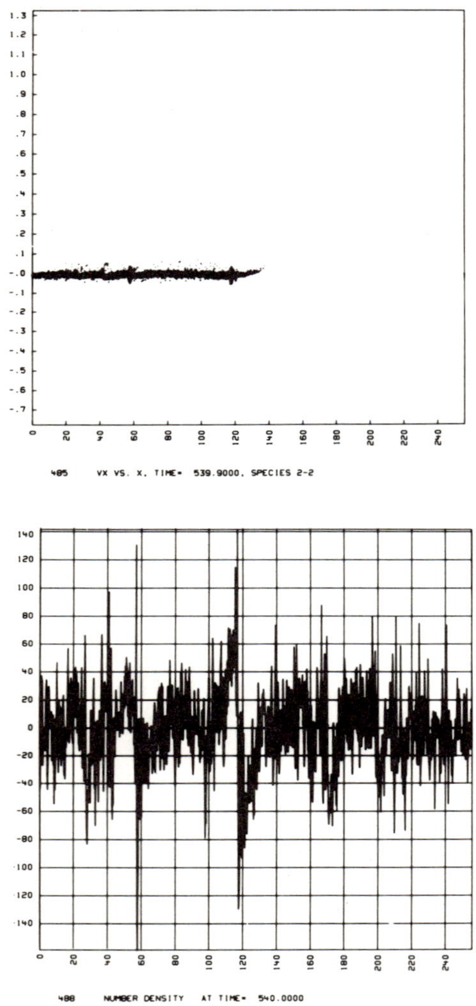

Fig. 3. a, b) Negative charge density spikes aligned with holes in half of ion phase space which persist for hundreds of ω_{pe}^{-1}.

is required. The hole in electron phase space on the downstream side created by reflection of electrons from the left must not be filled by reflected electrons from the next double layer over on the right in either a periodic system such as ours or a longer system such as *Sato and Okuda's* [1980b].

Sato and Okuda [1980b] have performed longer simulations in longer systems and confirm that $\sim1000\lambda_D$ is a characteristic scale length between double layers, and $\sim1000\omega_{pe}^{-1}$ is a characteristic time scale of growth and decay. They suggest that the temporal repetition rate is due to the propagation time of ion acoustic solitons across the $50\lambda_D$ width of a double layer which is consistent with the ion acoustic speed: $1000\omega_{pe}^{-1}=50\lambda_D/c_s$. They suggest that double layers decay by emitting ion acoustic solitons which propagate at c_s relative to the double layer. The relationship between ion acoustic solitons and double layers requires further investigation, both analytically [*Lotko and Kennell,* 1980] and in simulations.

It is significant that a series of small amplitude double layers with $e\phi/T_e\sim1$ is observed in longer simulations *vs.* a single large amplitude potential drop [*Sato and Okuda,* 1980b]. This suggests that the large amplitude potential drop necessary to accelerate auroral electrons to keV energies might be a series of small amplitude double layers. To accelerate 0.1 keV plasma sheet electrons to auroral electron energies requires 100 double layers with $e\phi/T_e\sim1$. If these are spaced $1000\lambda_D$ = 1.5 km apart based on $n=25\,\mathrm{cm}^{-3}$ and a thermal plasma $T_e=1$ eV, the acceleration region must be 150 km long. S3-3 particle data suggests that acceleration regions may be an order of magnitude longer [*Cladis and Sharp,* 1979]. However, the $1000\lambda_D$ spacing found in the unmagnetized simulations may not be appropriate for magnetized geometry.

Magnetized Structures

The foregoing conclusions are based upon unmagnetized simulations. A magnetic field probably alters the double layer spatial and temporal structure. The large, spatially confined electric fields observed on the S3-3 satellite are quasi-perpendicular to the magnetic field. Such a signature would be expected both from an oblique structure (Figure 4a) and from the equipotential mapping of a 1d double layer (Figure 4b) along most of the geomagnetic field line. Comparison of the electric field signatures with the S3-3 particle data sometimes, however, shows a switch from downward accelerated electron beams to upward accelerated ion beams and vice versa as the satellite passes through a region of intense quasi-perpendicular electric field. This may be due to a passage of the satellite through an oblique electric field structure that switches from an upward parallel electric field component above the satellite accelerating electrons down, to an upward parallel electric field component below the satellite accelerating ions up and vice versa. Often the other half of the V appears and the scale size of the entire structure is that of inverted V's [*Temerin,* 1980a]; the quasi-perpendicular electric field is large over about an energetic ion gyro-radius [*Mozer et al.,* 1977]. Many more complicated structures are seen in the S3-3 data [*Temerin,* 1980b], but the one dimensional double layer of Figure 4b is not a good explanation of the above feature at 1 R_E.

In a laboratory experiment with magnetized ions, *Stenzel et al.* [1980] have produced a V-shaped equipotential structure with an ion beam as the source of free energy. These structures appear to be much more stable than the unmagnetized double layers produced by electron current in the laboratory, [*Quon and Wong,* 1976; *Coakley et al.,* 1978; *Leung et al.,* 1980]. The transient

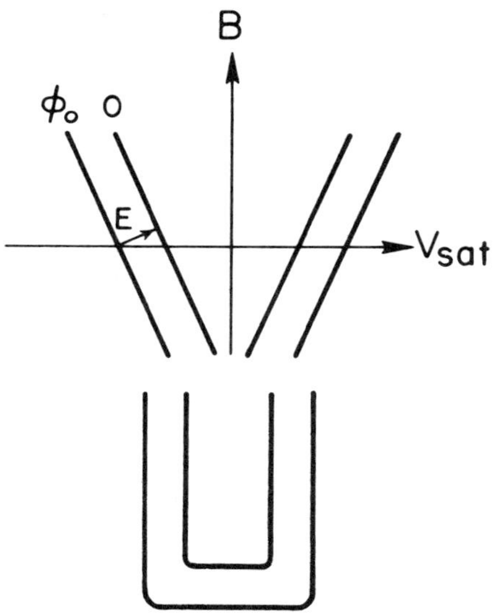

Fig. 4. a, b) Oblique equipotential structures inferred from S3-3 satellite data contrasted with the perpendicular mapping of a parallel potential double layer.

nature of the latter structures agree with the 1d unmagnetized simulations, although the details depend upon the specific plasma properties in a given experiment, e.g., the ionization potential of the background neutrals determines the maximum $e\phi/T_e$ of the double layer, whereas neutrals are neither included in the simulations nor important at 1 R_E.

Swift [1975, 1976, 1979] obtains a magnetized time stationary solution to Poisson's equation for cold counterstreaming electrons and ions which support an equipotential structure oblique to the magnetic field. Poisson's equation for planar equipotentials at an oblique angle $\pi/2+\theta$ with respect to the magnetic field in Figure 5 is [*Swift*, 1975]

$$\frac{d}{d\phi}[\phi'^2(1+\alpha\kappa)] = -8\pi e(n_i(\phi) - n_e(\phi)) \qquad (7)$$

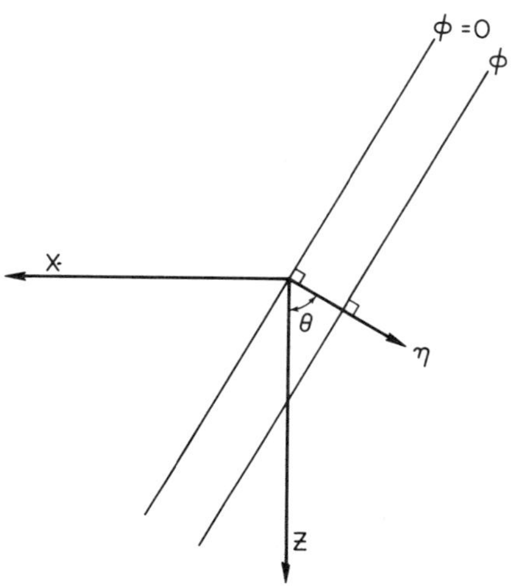

Fig. 5. Oblique planar equipotential structure; half of Figure 4a expanded to show the geometry.

where the prime denotes a derivative with respect to $\eta = z - \alpha x$, the direction of the oblique electric field, and α is the tangent of the angle the electric field makes with **B**. Integrating once yields the magnetized version of (1) independent of the dependence of the low frequency dielectric function $\kappa = 1 + c^2/v_A^2$ [*Chen*, 1974] on ion density and therefore potential through the Alfvén speed $v_A^2 = B^2/4\pi n_i M$.

To proceed one must specify $n_e(\phi)$ which is the same as the unmagnetized case (2a) for strongly magnetized electrons, and $n_i(\phi)$ which must be modified to include the **E** × **B** and polarization drift of ions in the perpendicular electric field component which changes in time for ions drifting along **B**. Thus [*Swift*, 1979]

$$n_i(\phi,\phi',\phi'') = \frac{n_{i0}u_{i0}}{u_i(\phi,\phi',\phi'')}$$

$$= n_{i0}\frac{1 + e\phi''/M\Omega^2}{\left(1 + \dfrac{2e(\phi_0 - \phi)}{Mu_{i0}^2} - \dfrac{c^2\phi'^2}{u_{i0}^2 B^2}\right)^{1/2}} \qquad (8)$$

replaces (2b). The **E** × **B** drift term appears in the denominator, the polarization drift term appears in the numerator, and Ω is the ion cyclotron frequency. The prime now denotes derivatives with respect to x which we assume to be much larger than gradients parallel to **B**. This simplifying assumption is consistent with the S3-3 electric field measurements [*Mozer et al.*, 1980], but rules out 1d magnetic field aligned double layers as a possible solution.

Swift [1979] has integrated (7) for the preceding $n_e(\phi)$ and $n_i(\phi,\phi',\phi'')$ for the case of cold stationary ions uniformly distributed in x at z_0 in Figure 6a. Swift's attempts to numerically integrate (7) for ions streaming into the equipotential structure fail to satisfy the condition on vanishing electric field at the boundary $\phi'(a)=0$, where a is the halfwidth of the equipotential structure which varies with height. We have Taylor series expanded (7) about $\phi(0)=0$ at z_1 and integrated it subject to the same boundary conditions for the case of downstreaming electron energy $W_e = 100\,\text{eV}$ and upstreaming ion energy $W_i = 10\,\text{eV}$. The result in Figure 6b is a plot of $\phi(x)$ *vs.* x within the range of validity of the expansion for different values of $\phi(0)$ which correspond to different altitudes $z_0 < z < z_1$. This plot can be converted to equipotential contours in X-Z space similar to Figure 6a [*Swift*, 1979], but the z scale is arbitrary since we have dropped the z dependence in (7). The significant feature is that, in agreement with the S3-3 observations [*Mozer et al.*, 1977; *Mozer et al.*, 1980], the width of the potential structure in x is comparable to the gyroradius of an ion whose perpendicular energy equals the potential drop across the structure.

Discussion

We have focused on equipotential structures at $\sim 1\ R_E$ where S3-3 data indicate a substantial fraction of auroral particle acceleration occurs [*Mozer et al.*, 1980]. We have assumed that a current is imposed on this region of space and not addressed the question of magnetosphere-ionosphere coupling, which is important to the overall problem of auroral arc formation. Kinetic Alfvén waves have been suggested as a source of this field-aligned current [*Hasegawa*, 1976; *Mallinckrodt and Carlson*, 1978; *Goertz and Boswell*, 1979; *Lysak and Carlson*, 1981]. These waves carry to the ionosphere the magnetic field response to changes in the magnetotail due, for example, to changes in the solar wind pressure at the magnetopause or reconnection. *Lysak and Carlson* [1981] have shown that the field-aligned current associated

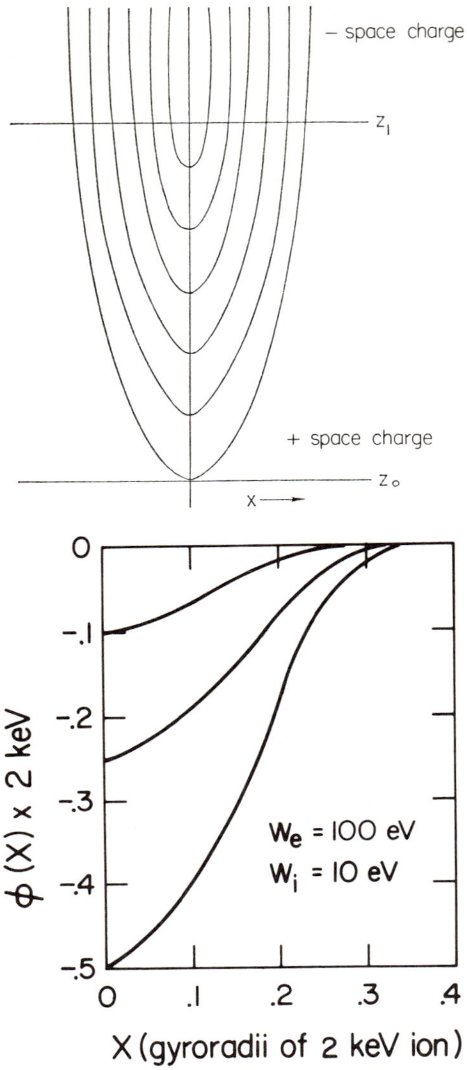

Fig. 6. a) Schematic 2d equipotential contours modeling Figure 4a. b) Solution to Poisson's equation (7) Taylor series expanded about $\phi(0)$ neglecting z-derivatives. Plotted is $\phi(x)$ *vs.* x for different $\phi(0)$ corresponding to different z. Conversion of $\phi(x)$ *vs.* x into equipotential contours in $x-z$ space requires knowledge of $\phi(0)$ *vs.* z lost in dropping the z dependence in (7).

with these waves may exceed the electron thermal speed, and thus the Langmuir-Bohm criterion, as well as electrostatic instability thresholds. They show that generation of the electrostatic ion cyclotron (EIC) instability can damp out the Alfvén wave before it reaches the ionosphere. A perpendicular scale size for the auroral arc is determined by wave conductivity rather than ionospheric conductivity. The result, ~ 1.5 km at 1 R_E, is in good agreement with the small scale dc electric field structures observed on S3-3 [*Mozer et al.,* 1977] and maps to ionospheric scale sizes that correspond to visual discrete arcs [*Maggs and Davis,* 1968; *Davis,* 1978]. The perpendicular scale sizes obtained by *Chiu and Cornwall* [1980], coupling ionospheric conductivity to the same type of current-voltage relation [*Fridman and Lemaire,* 1980], yield a perpendicular scale size of 100 - 200 km at 1 R_E in agreement with inverted-V scale size electric fields.

In the model of *Lysak and Carlson* [1981], ionospheric coupling results from smaller amplitude or larger perpendicular wavelength kinetic Alfvén waves which carry less field-aligned current than is required to produce wave resistivity at 1 R_E. They have not included the effect of double layer or electrostatic shock generation on kinetic Alfvén wave propagation. It should have a dissipative effect and also decouple the ionosphere. Finally, the small scale dc electric field structures observed on S3-3 are imbedded in regions of EIC waves, so both must be included in the complete calculation of field-aligned current generation of an auroral particle acceleration region.

Conclusions

We have briefly reviewed current supported electrostatic potential drops as a means of auroral particle acceleration at ~ 1 R_E. We have intentionally focused on BGK mode solutions to Poisson's equation in unmagnetized and uniformly magnetized plasmas, which may provide the spatially confined electric fields necessary to explain S3-3 electrostatic shock observations. Other parallel electric field mechanisms such as the mirror effect [*Chiu and Cornwall,* 1980], and anomalous resistivity [*Papadopoulos,* 1977; *Hudson et al.,* 1978] contribute more extended parallel potential drops and weaker electric fields. The correlation between the electrostatic shocks observed on S3-3 and discrete auroral arcs has been discussed by *Torbert and Mozer* [1978].

We have examined the analytic criterion for unmagnetized double layer formation and found that the Langmuir-Bohm criterion is not modified by equal density populations of thermal electrons and ions. We then questioned the assumption of nonthermal populations of trapped particles which reduce the double layer criterion but require the pre-existence of the double layer to produce such nonthermal populations.

Next, we examined the temporal evolution of a double layer in a 1d unmagnetized simulation. We found that the most persistent features in the evolution are negative charge density spikes and holes in ion phase space, presumably due to the trapping of electrons and ions in the potential of ion acoustic waves or solitons. These features were related to models discussed by *Hasegawa and Sato* [1980] and *Sato and Okuda* [1980b] for double layer evolution. However, our understanding of the temporal evolution of even unmagnetized double layers is still at the qualitative stage.

The S3-3 satellite observations demonstrate that the equipotential structures associated with auroral particle acceleration are greatly modified by the geomagnetic field. The electric fields measured are predominantly perpendicular to the geomagnetic field, and a structure such as shown in Figure 4a (not 4b) is required to explain some of the correlated particle and electric field data obtained on S3-3. Both halves of a V-shaped equipotential structure, however, are not always seen on a satellite pass; it is not clear whether this is due to temporal variations or the existence of a single S-shaped structure.

We thus far have examined only one half of such a structure analytically. The V-shaped solutions are just reflections through x=0, while S-shaped solutions are reflections through the origin which place additional constraints on (7). Extending Swift's numerical integration of (7) to the case of cold streaming ions as well as electrons yields Figure 6b, a plot of $\phi(x)$ *vs.* x for different $\phi(0)$ corresponding to different z. The conversion of Figure 6b into a plot of equipotentials in $x-z$ space like Figure 6a requires some assumption about $\phi(0)$ *vs.* z since the z dependence was dropped from (7) by the assumption, consistent with S3-3 measurements, that perpendicular exceed parallel gradients. However, we have lost all information on the altitude extent of the acceleration region and excluded one-dimensional double layers of the type discussed in the first two sections as possible solutions by dropping parallel gradients in (7). Nevertheless, the perpendicular scale size is consistent with S3-3 observations.

Finally, we comment on the distinction between electrostatic shocks, which appear in the title of this paper, and double layers, magnetized and unmagnetized, discussed throughout this paper. A useful distinction between the two in the laboratory plasma physics literature has been noted by *Hershkowitz* [1980]. Namely, the free streaming (not reflected) ions are accelerated in laboratory double layers and decelerated in electrostatic shocks produced in laboratory plasmas. If one applies this definition to the magnetosphere, one must call those potential structures which accelerate ions upward from the ionosphere and have potential drops much greater than plasma sheet ion energies double layers. Historically, this was not done to distinguish the magnetized structures observed on S3-3 from conventional unmagnetized double layers, and to relate those observations to the work of *Swift* [1975, 1976] on magnetized BGK mode solutions to Poisson's equation which preceded the S3-3 observations. The latter has turned out to be a difficult analytic problem. Hopefully the magnetized simulations will shed further light on the temporal evolution and stability of the oblique equipotential structures observed on S3-3 which appear to play a key role in auroral particle acceleration and the formation of discrete auroral arcs [*Mozer et al.*, 1980].

Acknowledgments. We thank A. Hasegawa, F. Mozer, T. Sato, and M. Temerin for useful discussions and L. Block and C. Goertz for their comments after the talk. We especially thank Charles Birdsall and Alex Friedman for providing much help and guidance in the care and feeding of the computer code. The original version of the code ES1 was written by A. Bruce Langdon. NASA grants NSG-7401 and NAGW-75 support this research.

References

Bernstein, I. B., J. M. Greene, and M. D. Kruskal, "Exact non-linear plasma oscillations," *Phys. Rev. Lett.*, **108**, 546, 1957.

Birdsall, C. K., and A. B. Langdon, "Plasma physics via computer simulation," v. 1-4, EECS Dept., U. C. Berkeley, 1978.

Block, L. P., "Potential double layers in the ionosphere," *Cosmic Electrodynamics*, **3**, 349, 1972.

Bohm, D., in *The Characteristics of Electrical Discharges in Magnetic Fields*, edited by A. Gutherie and R. K. Walkerling, McGraw-Hill, New York, p. 77, 1949.

Cattell, C. A., "The relationship of field-aligned currents to electrostatic ion cyclotron waves," *J. Geophys. Res.*, in press 1981.

Chen, F. F., *Introduction to Plasma Physics*, Plenum Press, 1974.

Chiu, Y. T., and J. M. Cornwall, "Electrostatic model of a quiet auroral arc," *J. Geophys. Res.*, **85**, 543, 1980.

Cladis, J. B., and R. D. Sharp, "Scale of electric field along magnetic field in an inverted-V event," *J. Geophys. Res.*, **84**, 6564, 1979.

Coakley, P., N. Hershkowitz, R. Hubbard, and G. Joyce, "Experimental observations of strong double layers," *Phys. Rev. Lett.*, **40**, 230, 1978.

Davis, T. N., "Observed characteristics of auroral forms," *Space Sci. Rev.*, **22**, 77, 1978.

Fridman, M., and J. Lemaire, "Relationship between auroral electron fluxes and field-aligned electric potential difference," *J. Geophys. Res.*, **85**, 664, 1980.

Goertz, D. K., and R. W. Boswell, "Magnetosphere-ionosphere coupling," *J. Geophys. Res.*, **84**, 7239, 1979.

Hasegawa, A., "Particle acceleration by MHD surface wave and formation of aurora," *J. Geophys. Res.*, **81**, 5083, 1976.

Hasegawa, A. and T. Sato, "Existence of negative potential solitary-wave structure and formation of double layer in a non-equilibrium plasma," submitted to *Phys. Rev. Lett.*, 1980.

Hershkowitz, N., "Double layers and electrostatic shocks," submitted to *Geophys. Res. Lett.*, 1980.

Hudson, M. K., R. L. Lysak, and F. S. Mozer, "Magnetic field-aligned potential drops due to electrostatic ion cyclotron turbulence, *Geophys. Res. Lett.*, **5**, 143, 1978.

Kan, J. R., and L. C. Lee, "On the auroral double layer criterion," *J. Geophys. Res.*, **85**, 788, 1980.

Leung, P., A. Y. Wong, and B. H. Quon, "Formation of double layers," *Phys. Fluids*, **23**, 992, 1980.

Lotko, W., and C. F. Kennel, "Ion-acoustic like solitons in auroral plasma," Proceedings of the AGU Chapman Conference on the Formation of Auroral Arcs, Fairbanks, 1980.

Lysak, R. L., and C. W. Carlson, "The effect of microscopic turbulence on magnetosphere-ionosphere coupling," *Geophys. Res. Lett.*, in press 1981.

Maggs, J. E., and T. N. Davis, "Measurements of the thickness of auroral structures," *Planet. Space Sci.*, **16**, 205, 1968.

Mallinckrodt, A. J., and C. W. Carlson, "Relations between transverse electric fields and field-aligned currents," *J. Geophys. Res.*, **83**, 1426, 1978.

Montgomery, D., and G. Joyce, "Shocklike solutions of the electrostatic Vlasov equation," *J. Plasma Phys.*, **3**, 1, 1969.

Mozer, F. S., C. W. Carlson, M. K. Hudson, R. B. Torbert, B. Parady, J. Yatteau, and M. C. Kelley, "Observations of paired electrostatic shocks in the polar magnetosphere," *Phys. Rev. Lett.*, **38**, 292, 1977.

Mozer, F. S., C. A. Cattell, M. K. Hudson, R. L. Lysak, M. Temerin, and R. B. Torbert, "Satellite measurements and theories of low altitude auroral particle acceleration," *Space Sci. Rev.*, **27**, 155, 1980.

Papadopoulos, K, "A review of anomalous resistivity for the ionosphere," *Reviews of Geophys. and Space Phys.*, **15**, 113, 1977.

Quon, B. H., and A. Y. Wong, "Formation of potential double layers in plasma," *Phys. Rev. Lett.*, **37**, 1393, 1976.

Sato, T., and H. Okuda, "Ion acoustic double layers," *Phys. Rev. Lett.*, **44**, 740, 1980a.

Sato, T., and H. Okuda, "Numerical simulations on ion acoustic double layers," submitted to *J. Geophys. Res.*, 1980b.

Stenzel, R. L., M. Oyama, and Y. Nakamura, "V-shaped double layers formed by ion beam reflection," *Phys. Rev. Lett.*, **45**, 1498, 1980.

Swift, D. W., "On the formation of auroral arcs and acceleration of auroral electrons," *J. Geophys. Res.*, **80**, 2096, 1975.

Swift, D. W., "An equipotential model for auroral arcs," *J. Geophys. Res.*, **81**, 3935, 1976.

Swift, D. W., "An equipotential model for auroral arcs: the theory of two-dimensional laminar electrostatic shocks," *J. Geophys. Res.*, **84**, 6427, 1979.

Temerin, M., "Plasma waves on auroral field lines," Proceedings of the AGU Chapman Conference on the Formation of Auroral Arcs, Fairbanks, 1980a.

Temerin, M., "The small scale structure of electrostatic shocks," submitted to *J. Geophys. Res.*, 1980b.

Torbert, R. B., and F. S. Mozer, "Electrostatic shocks as the source of discrete auroral arcs, *Geophys. Res. Lett.*, **5**, 135, 1978.

FORMATION OF AURORAL ARCS BY PLASMA SHEET PROCESSES

W.J. Heikkila

University of Texas at Dallas, Richardson, Texas 75080

Abstract. There is a cross-tail electric field, and a neutral sheet current in the plasma sheet, both in the dawn-dusk direction; thus E.J is positive, and there is particle energization. This corresponds to a power dissipation of about 10^{12} watts out to lunar orbit under average conditions. If the first adiabatic invariant is conserved, the mirror points of all particles will be lowered as they are energized, by any mechanism. Particles which already mirror at low altitudes will be precipitated. In the distant plasma sheet, it is likely that curvature drift is the most important source of drift parallel to the electric field, leading to what is commonly called Fermi acceleration of the particles. This energization mechanism is proportional to the neutral sheet current density. It is a form of field-aligned acceleration, with rapid lowering of mirror points, due to the transverse electric field in the plasma sheet. The process will work for both negative and positive particles. We now postulate a filamentation of the neutral sheet current. The maximum energization by curvature drift, and the accompanying intense precipitation, will form an auroral band or arc along the sheet of magnetic field lines that maps out to the local enhancement of the cross-tail current, explaining inverted V events. It seems possible that acceleration by a parallel component of the electric field at low altitudes may be a minor effect under normal conditions, but may take on increased importance with intense arcs, or during substorms.

Introduction

There now seems little doubt that a component of the electric field parallel to the magnetic field lines occurs above discrete auroral arcs. In a plasma, this $E_\parallel$ is dependent upon a strong upward current, probably due to precipitating electrons in the keV energy range. One theory of the formation of auroral arcs is that a hot magnetospheric plasma above the arc, accelerated by this $E_\parallel$, will produce the arc; accordingly, it does not matter whether the magnetic field lines are open or closed.

However, a number of difficulties exist with this rather simplistic view. We list a few as follows:

1. What causes the field aligned current in the first place, that seems to be a necessary condition for the creation of $E_\parallel$?
2. How can electron and positive ion precipitation sometimes occur in the same location at the same time, with both showing signs of field aligned acceleration? (Hultqvist et al., 1971; Bryant et al., 1977).
3. How can field alignment occur over a very large energy range, sometimes with no enhancement near the characteristic peak in the energy spectrum (Whalen and Daly, 1979; Bryant, 1980 at this meeting)?
4. Why do auroral electrons show two loss cones in their pitch angle distribution (Sharber and Heikkila, 1972; Venkataragan et al., 1975)?
5. Why do auroral electrons show field alignment predominantly at low energies, at both edges of the auroral structures (Whalen and Daly, 1979; Hoffman, 1980 at this meeting)?
6. How can there be a decrease in the flux of precipitating electrons over a wide energy range (0.1 to 50 keV), just before auroral breakup, to cause auroral fading and other concurrent phenomena (Pellinen and Heikkila, 1978)?

We suggest that plasma sheet processes play an essential role in the formation of auroral arcs. In particular, there is normally energization of plasma sheet particles amounting to 10^{12} watts out to lunar orbit under average conditions (Heikkila et al., 1979). Energization by curvature drifting particles in the distant plasma sheet on closed magnetic field lines must be accompanied by intense precipitation, as long as the first invariant of the particles is conserved. Because of the different mobilities of the charge carriers, this will lead to field aligned current. It will be shown that filamentation of the cross-tail current (a current sheet pinch, see Figure 1) will result in inverted V structures. The auroral forms

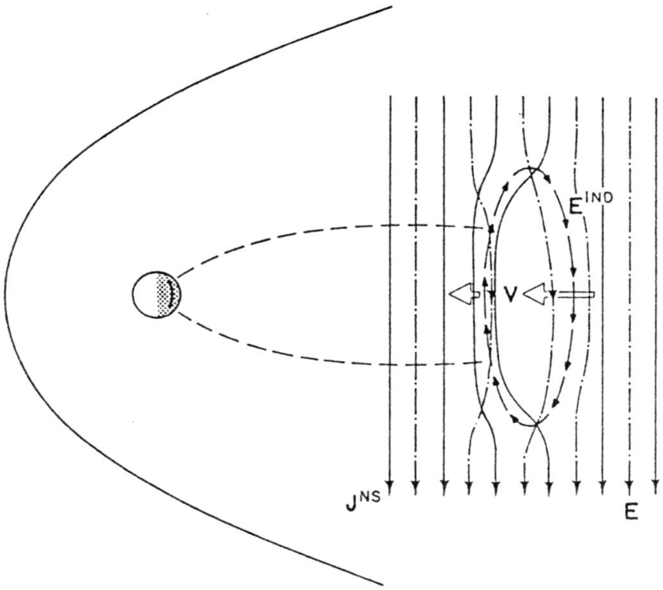

Fig. 1. Energization by curvature drift is proportional to cross-tail current density times the total electric field. A filamentation of the current can explain inverted V events, and several other features of auroral arcs. With rapidly changing current there will be an induction electric field as well, but for quiescent arcs this can be ignored.

will be even more concentrated in discrete arcs by the action of $E_{\parallel}$ at lower altitudes. These two processes must be treated together, rather than separately.

Energization of Plasma Sheet Particles

Charged particles will experience magnetic gradient and curvature drifts that will in general have components parallel (or anti-parallel) to an imposed transverse electric field in the magnetotail. The rate of gain in energy by a particle with energy W and pitch angle α is given by the dot product of the electric force $q\underline{E}$ and the drift velocity $\underline{V}_d$, which is due to a combination of factors:

$$\underline{V}_d = \underline{V}_e + \underline{V}_g + \underline{V}_c \tag{1}$$

$$= \frac{\underline{E} \times \underline{B}}{B^2} + \frac{W\sin^2\alpha(\underline{B} \times \nabla B)}{qB^3} + \frac{2W\cos^2\alpha(\underline{B} \times \underline{R})}{qB^2R^2} \tag{2}$$

$$q\underline{E} \cdot \underline{V}_d = 0 + q\underline{V}_g \cdot \underline{E} + q\underline{V}_c \cdot \underline{E} \tag{3}$$

Here, we have assumed that the particle motions are adiabatic. The first term on the right is the electric drift; it provides no energization, since the triple vector product $\underline{E} \cdot \underline{E} \times \underline{B}$ is identically zero. The second and third terms are the magnetic gradient (∇B) and the curvature drifts ($\underline{R}$ is from the line to the center of curvature), and their dot product with $\underline{E}$ need not vanish. The rate of energization depends in a complicated way on the pitch angle α and on the structure of the magnetic field.

When the first invariant is conserved, the mirror point of a particle which gains energy will be lowered (in terms of geocentric distance), as can be shown very easily (this invariant is the transverse energy divided by the magnetic field strength). Assume that the particle at two different times has energies W^1 and W^2, with

$$t^2 \gg t^1 \qquad W^2 > W^1 \tag{4}$$

At the mirror points this energy is entirely in the transverse component, by definition, so that

$$W_{\perp}^2 > W_{\perp}^1 \quad \text{at mirror points} \tag{5}$$

Then conservation of the invariant implies that

$$W_{\perp}^2/B_m^2 = W_{\perp}^1/B_m^1 \tag{6}$$

Thus the magnetic field B_m at the mirror point must increase in accordance with the relation

$$B_m^2 = \frac{W^2}{W^1} B_m^1 \tag{7}$$

The geomagnetic field strength decreases with decreasing latitude, at constant geocentric distance, so that the mirror point will be lowered quite rapidly with plasma convection toward lower L-shells. This statement is true for any energization process whatever, as long as the first invariant is conserved in that process.

In the past, energization of plasma sheet particles has generally been taken to mean adiabatic compression, using the concept of moving magnetic field lines. Particles with 90° pitch angles in the equatorial plane are thought to be compressed as they move inward; this is commonly called betatron acceleration. Particles which have a large component of velocity along magnetic field lines undergo longitudinal compression, thought to be associated with conservation of the second invariant; this is the so-called Fermi acceleration process.

An alternative way (and the more fundamental) is to do orbit calculations on individual particles. Hines (1963) has considered the equivalence of the two methods. He finds good consistency with regards to betatron acceleration, although the apparent physical interpretation is quite different in the two approaches. As one example, with MHD only $\underline{V}_e$ is important, whereas in the particle approach, $\underline{V}_g$ is the relevant quantity. On Fermi acceleration, he has concluded that although the second invariant may be preserved in longitudinal compression on a differential scale, this might not be true on an integral scale. A similar situation exists with infinitesimal and finite rotations in mechanics. In any case, as he recalls, MHD theory is only approximate, and it must be based on particle orbit theory.

Consider a particle traversing the equatorial plane far out in the magnetotail, with a mirror point at low altitudes. Out there, we can make the assumption that $(\nabla B)_x$ (the component of the gradient in the magnetic field toward the earth) is quite small in the equatorial plane; the gradient is mostly along the field lines $\nabla B = |\nabla B|\ \underline{z}$. Consequently, there is a large curvature to the field lines: $\underline{R}/R^2 = (\underline{B}\cdot\nabla)\ \underline{B}/B^2$. Curvature drift (the last term in eq. 2) carries particles across the tail, so that $q\underline{E}\cdot\underline{V}_c > 0$: the particles gain energy. What fraction of the energy appears in the transverse component, and what fraction in the parallel component? Since $(\nabla B)_x$ is small, the transverse energy cannot increase much without violating conservation of the first invariant. Consequently, we conclude that most of the energy gain may be in the parallel component; this would lead to particularly rapid lowering of mirror points. We have

$$\frac{\underline{R}}{R^2} = \frac{\underline{B}}{B}\cdot\nabla\frac{\underline{B}}{B} = -\frac{1}{B^2}\,\underline{B}\times(\nabla\times B) + \frac{1}{B}(\nabla B)_\perp \qquad (8)$$

We can write $\nabla\times\underline{B} = \mu_0\,\underline{J}$, and assume that $(\nabla B)_\perp = 0$. Hence, the curvature drift is

$$\underline{V}_c = \frac{2W_\parallel}{qB^2}\,\frac{\underline{B}}{B}\times(\mu_0\underline{J}\times\frac{\underline{B}}{B}) = \frac{2\mu_0 W_\parallel}{qB^2}\,\underline{J}_\perp \qquad (9)$$

where $\underline{J}_\perp$ is the current transverse to $\underline{B}$, i.e. the cross-tail current. Eq. 9 shows that the curvature drift speed is directly proportional to the local cross-tail current density. The rate of increase of the particle's energy W (with $W \sim W_\parallel$) is

$$\frac{dW_\parallel}{W_\parallel} = q\underline{E}.\underline{V}_c = \frac{2\mu_0\underline{E}\cdot\underline{J}_\perp}{B^2}dt \qquad (10)$$

Hence, the energy growth is exponential in time if $(\underline{E}.\underline{J}_\perp)/B^2$ stays constant. In practice this term responds to the changes in $\underline{J}_\perp$, by induction electric fields, polarization currents, and changes in $\underline{B}$.

One likely possibility is that of filamentation of the cross-tail current, in other words, a current sheet pinch. Resorting to physical intuition, it seems highly unlikely the cross-tail current is prefectly uniform (except for a slow decrease with distance in the tailward direction). Certainly, in the case of water, there are always meanders in the flow, whether it be rivers, lakes, or oceans. Such a slow meander in the case of the cross-tail current is shown in Figure 1. Equation 10 shows that energization by curvature drift is enhanced where the current density is locally increased, provided other factors remain the same. Furthermore, the previous discussion indicates that the energization is primarily in the longitudinal component, with enhanced precipitation. The latitudinal profile of particle energy in an inverted V will reflect the profile of the cross-tail current in the equatorial plane.

Because of the high mobility of electrons, an upward current will result. This in turn may be associated with an electric field along the magnetic field lines, which will depend on the strength of the field aligned current. Thus the auroral arc formation, especially its thinness, may depend on ionospheric parameters as well.

DISCUSSION

The first question listed in the introduction, the need for an upward field-aligned current, demands consideration of magnetospheric processes before we can truly understand the formation of auroral arcs. Three significant factors of these processes are: (1) the continual energization of plasma sheet particles of both signs, since $\underline{E}\cdot\underline{J} > 0$, (2) field-aligned acceleration in the plasma sheet, due to curvature drifting particles, in the presence of a dawn-dusk electric field, and (3) lowering of mirror points accompanying this energization if the first adiabatic invariant is conserved; this can lead to intense precipitation of electrons, and an upward current.

It is interesting to note that a purely transverse electric field can cause longitudinal acceleration, very similar to magnetic field-aligned electric fields $\underline{E}_\parallel$. More mundane examples can be found to illustrate this process, e.g. gradual shortening of a swinging pendulum. The main difference between energization by $\underline{E}_\parallel$ and energization by curvature drift in the presence of $\underline{E}_\perp$ is that the former produces clear electron and proton differences, while the latter can produce proton and electron beams along field lines in the same direction at the same location.

The other questions in our list now have at least a conceptual answer. Acceleration by curvature drift works for both electrons and protons, over a large energy range. Repeated bounces between hemispheres are required, with particles being forced into the loss cones. The action of $E_\parallel$ is evident primarily at low electron energies.

An inductive electric field (as shown in Fig. 1), vectorially added to the electrostatic cross-tail field, will result in a local weakening of the total electric field where the cross-tail current density is increasing (by Lenz's law). This weakening would cause reduced precipitation of particles, both negative and positive, and of all energies, as long as their first invariant is conserved. This can explain auroral fading, decreased riometer absorption, reduction of the auroral electroject, and darkening of the poleward sky (Snyder and Akasofu, 1972) before substorm onset (see the paper by Opgenoorth et al., 1980, for examples).

Still outstanding questions are: What is the primary role of $E_{\parallel}$ at low altitudes, and how large a voltage drop is involved? Three papers at this conference may go a long way to answering these questions. Larry Lyons (1980) claimed that a parallel electric field is requiredfor proper current continuity. This claim was reinforced by laboratory experiments reported by Block (1980) and by Stenzel (1980); an $E_{\parallel}$ was developed in order to suppress the secondaries, which otherwise would have carried a reversed current. A downward current would quickly lead to cancellation of the dawn-dusk electrostatic field, at least locally, preventing further energization and precipitation by curvature drift, and causing disappearance of the field-aligned current.

The division of auroral forms into two categories, namely discrete poleward arcs, (BPS precipitation, in the terminology of Winningham et al., 1975), and the diffuse equatorward forms (CPS precipitation) may now be explained on the basis of how the particles got their energy (see Heikkila, 1974, p. 2499). Gradient drift is more important in the CPS region, while curvature drift is more important in the BPS region. Although $\underline{E}_{\parallel}$ may also be involved at lower altitudes, we believe that $\underline{E}_{\perp}$ in the magnetotail plays the primary role by causing the field-aligned current in the first place.

Acknowledgements. I wish to acknowledge helpful comments by Dr. Chris Goertz. This was supported by NASA Grant NAGW-53, and NSF Grant ATM78-03603.

References

Block, L.P., Double Layers in the Laboratory and Above the Aurora, A.G.U. Chapman Conference, The Formation of Auroral Arcs, Fairbanks, Alaska, July 1980.

Bryant, D.A., D.S. Hall, D.R. Lepine and R.W.N. Mason, Electrons and Positive Ions in an Auroral Arc., Nature, 266, 148-149, 1977.

Bryant, D.A., Auroral Particles and Visual Auroral Arcs, A.G.U. Chapman Conference, The Formation of Auroral Arcs, Fairbanks, Alaska, July 1980.

Heikkila, W.J., Outline of a Magnetospheric Theory, J. Geophys. Res., 79, 2496, 1974.

Heikkila, W.J., R.J. Pellinen, C.-G. Fälthammar, and L.P. Block, Potential and Inductive Electric Fields in the Magnetosphere during Auroras, Planet. Space Sci., 27, 1383, 1979.

Hines, C.O., The Energization of Plasma in the Magnetosphere: Hydromagnetic and Particle-Drift Approaches, Planet. Space Sci., 10, 239, 1963.

Hoffman, R.A. and C.S. Lin, Study of Inverted-V Auroral Precipitation Events, A.G.U. Chapman Conference, The Formation of Auroral Arcs, Fairbanks, Alaska, July 1980.

Hultqvist, B., H. Borg, W. Riedler, and P. Christophersen, Planet Space Sci., 19, 279, 1971.

Lyons, L.R., The Field-Aligned Current Versus Electric Potential Relation and Auroral Electrodynamics, A.G.U. Chapman Conference, The Formation of Auroral Arcs, Fairbanks, Alaska, July 1980.

Opgenoorth, H.J., R.J. Pellinen, H. Maurer, F. Küppers, W.J. Heikkila, K.U. Kaila, and P. Tanskanen, Ground-Based Observations of an Onset of Localized Field-Aligned Currents During Auroral Breakup Around Magnetic Mid-Night, J. Geophys., 48, 101-115, 1980.

Pellinen, R.J., and W.J. Heikkila, Observations of Auroral Fading Before Breakup, J. Geophys. Res., 83, 4207, 1978.

Sharber, J.R., and W.J. Heikkila, Fermi Acceleration of Auroral Particles, J. Geophys. Res., 77, 3397, 1972.

Snyder, A.L., and S.-I. Akasofu, Observations of the Auroral Oval by the Alaskan Meridian Chain of Stations, J. Geophys. Res., 77, 3419, 1972.

Stenzel, R.L., Reporting on Joint Experiment with N. Nakamura in Japan, at the A.G.U. Chapman Conference, Fairbanks, Alaska, July 1980.

Whalen, B.A., and P.W Daly, Field-Aligned Auroral Particle Distributions Imply Acceleration by Quasi-Static Parallel Electric Fields? J. Geophys. Res., 84, 4175, 1979.

Winningham, J.D., F. Yasuhara, S.-I. Akasofu and W.J. Heikkila, The Latitudinal Morphology of 10-eV to 10-keV Electron Fluxes During Magnetically Quiet and Disturbed Times in the 2100-0300 MLT Sector, J. Geophys. Res., 80, 3148, 1975.

Venkatarangan, P., J.R. Burrows, and I.B. McDiarmid, On the Angular Distributions of Electrons in 'Inverted V' Substructures, J. Geophys. Res., 80, 66, 1975.

SIMULATION OF SPACE PLASMA PHENOMENA

John M. Dawson

Department of Physics, University of California, Los Angeles, CA 90024

Introduction

The traditional means for investigating physical phenomena are laboratory experiments and through the analytic application of well-established physical laws. In the case of large-scale natural phenomena one must often substitute observations of what is taking place for controlled experiments. The traditional methods have their limitations; often the complexity of the phenomenon and the simultaneous interaction of many effects makes anything like complete analysis impossible. On the experimental side, one is limited to measurements of only a small fraction of the quantities of interest in a process and even these may be only sampled at a few times and positions and with a limited degree of accuracy. This is particularly true for observations of natural phenomena such as are encountered with space plasmas. Thus, one is then faced with the task of interpreting limited observations with theories which are incomplete, and often there is more than one theory which fits the known facts.

Recently, a new powerful tool has been added to the traditional method; it is that of computer modeling of the physical system. Modern computers are able to follow the time evolution of systems containing many millions of degrees of freedom, all of which are simultaneously interacting with each other. One constructs numerical models of the system of interest and then carries out numerical experiments on these. One starts the system out in some configuration of interest and observes its behavior. One can obtain as complete details about the motion as one desires. One can often test theoretical predictions and the assumptions and approximations which go into them in ways not possible with real experiments. For example, it is possible to turn effects on and off in such models and see how the results change. Needless to say, this is something which is often impossible in the real world. Examples of this will be shown later in the paper.

This approach complements the other two. The models are simpler and more idealized than the actual physical system; however, they are far more complete and realistic than we can hope to handle analytically. They reproduce both linear and nonlinear behavior. One can compare the results of such calculations with the behavior of real physical systems and to theory; the results can be used to predict the behavior, test theoretical predictions, and to gain an understanding of the phenomenon involved.

The real power of numerical simulation does not lie in reproducing complex physical phenomena. This would simply replace a set of complex experimental data with equally complex sets of computer output. Piles of such output are of little value unless we are able to construct a relatively simple theoretical framework which condenses their essence into easily understood models. The results of such calculations often show us which are the important effects among the many possibilities and thus let us construct useful simple physical models which we would be hard pressed to justify a priori. Such calculations do not reduce the amount of physics we are called on to do, but rather increase the time spent on physics and put a premium on physical intuition.

Models

As I have mentioned, there are a large variety of models for simulating plasmas. Basically, there are two types: particle models and fluid models. In a particle model, we attempt to emulate nature by following the motion of a large number of charged particles in their self consistent electric and magnetic fields. Such models follow the motion of the plasma on the finest space scale and on the most rapid time scale at which things happen. They, therefore, are limited to looking at the phenomena in a relatively small sample of plasma and over relatively short periods of time. However, the phenomena at this scale influence the macroscopic behavior of the plasmas and are fundamental to understanding plasmas.

The second type of model is a fluid model. Here one adopts a set of fluid equations to describe the plasma. These generally can describe the plasma behavior on large space and time scales. However, one often must insert into such models ad-hoc transport coefficients and other assumptions related to microscopic processes and the validity of these may be questionable. Often, they are simply chosen to fit observed data.

Here, I will discuss particle models and some recent results we have obtained relevant to the diffuse Aurora. Figure 1 shows some examples of electrostatic particle models. They come in one, two and three dimensional versions. The one dimensional version may be thought of as a large number of charge sheets perpendicular to the x axis. The two dimensional model is a series of charged rods parallel to say the z axis. A three dimensional model, of course, consists of particles in three dimensions. The one dimensional model can be generalized to what we call a one and a half dimensional model by allowing the charge sheets to also move in the y direction. There is then motion and current in the y direction, but none of the quantities such as charge density, potential, electric field, magnetic field, etc., vary in that direction. We would say x is a full dimension, but that y is a half dimension. We can further generalize the model by allowing the sheets to also move in the z direction. Then both the y and z directions would be considered half dimensions and the model is one and two halves dimensional. Similarly, the two dimensional model can be generalized by allowing the rods to move in the z direction. Again, there would be z directed currents but no quantities would vary in the z direction. The x and y directions would be considered full dimensions and the z direction a half dimension and the model would be considered a two and one half dimensional model. The 1-1/2, 1-2/2, and 2-1/2 dimensional models are particularly important for treating electromagnetic effects which involve currents perpendicular to k. Figure 1 illustrates the electrostatic versions of these models. The electric field is determined from Poisson's equation:

$$\underset{\sim}{\nabla} \cdot \underset{\sim}{E} = 4\pi \sum_i q_i \delta(\underset{\sim}{r} - \underset{\sim}{r}_i), \tag{1}$$

and is given by

$$\underset{\sim}{E}(\underset{\sim}{r}) = 2\pi \frac{(1-n)!}{2} \Gamma\left(\frac{n}{2}\right) \frac{q\,(\underset{\sim}{r} - \underset{\sim}{r}_i)}{|\underset{\sim}{r} - \underset{\sim}{r}_i|^{n}}, \tag{2}$$

where n is the dimensionality and $\Gamma\left(\frac{n}{2}\right)$ is the Gamma function. I will restrict this discussion to electrostatic models (Birdsall et al., 1970) although very similar techniques can be applied when the full set of Maxwell's equations are employed (Lin et al., 1974) or to magnetostatic or Darwin models (Busnardo-Neto et al., 1977).

Basically such particle models are simple; for the electrostatic case we wish to solve the equations

$$\ddot{\underset{\sim}{r}}_i = \frac{q_i}{M_i} \sum_j \underset{\sim}{E}_j(\underset{\sim}{r}_i) = \frac{q_i}{M_i} \sum_j \frac{q_j(\underset{\sim}{r}_i - \underset{\sim}{r}_j)}{|\underset{\sim}{r}_i - \underset{\sim}{r}_j|^n} \tag{3}$$

for a large number of particles. Here i refers to the i-th particle and q_i and M_i are its charge and mass; n is still the dimensionality. Typically we desire to solve these equations for $10^4 \sim 10^6$ particles; a simple straightforward estimate shows that it is not possible even on forseeable computers. Suppose we have N particles, then for each particle the sum over j contains N terms and we must evaluate it for all N particles so that the number of operations is proportional to N^2. The evaluation of each term in the sum involves a number of arithmetical operations. For the purpose of evaluation let us assume that $10N^2$ operations are involved per time step Δt, so that

$$N_{op} \sim 10N^2/\Delta t. \tag{4}$$

If we used 5×10^4 particles and on the average the machine carried out one arithmetic operation in 10^{-7}sec, just evaluating the forces would require 2.5×10^3 sec per time step. Simulations involving 10^3 to 10^4 time steps would require from one month to one year to do. Clearly such calculations are not feasible unless the results are extremely valuable. We must find a faster, more efficient method for evaluating the forces; fortunately, these exist (Kamimura et al., 1975).

A second important consideration is that of particle collisions. In fact, this consideration sets the requirement on the number of particles we must use. For a particle model there are collisions between particles just as there are in a real plasma. The kinetic theory of one, two and three dimensional plasmas can be carried out (Birdsall et al., 1970). The collision times, τ_{coll}, obtained are roughly given by the following expression

$$\omega_p \tau_{coll} \simeq \begin{cases} 2n\lambda_D & 1D, \\ 2\pi^2 n\lambda_D & 2D, \\ 2\pi n\lambda_D^3/\ln\Lambda & 3D, \end{cases} \tag{5}$$

where ω_p is the plasma frequency, τ_{coll} is the collision time, n is the charge density, λ_D is the Debye length and $\ln\Lambda$ is the Coulomb logarithm.

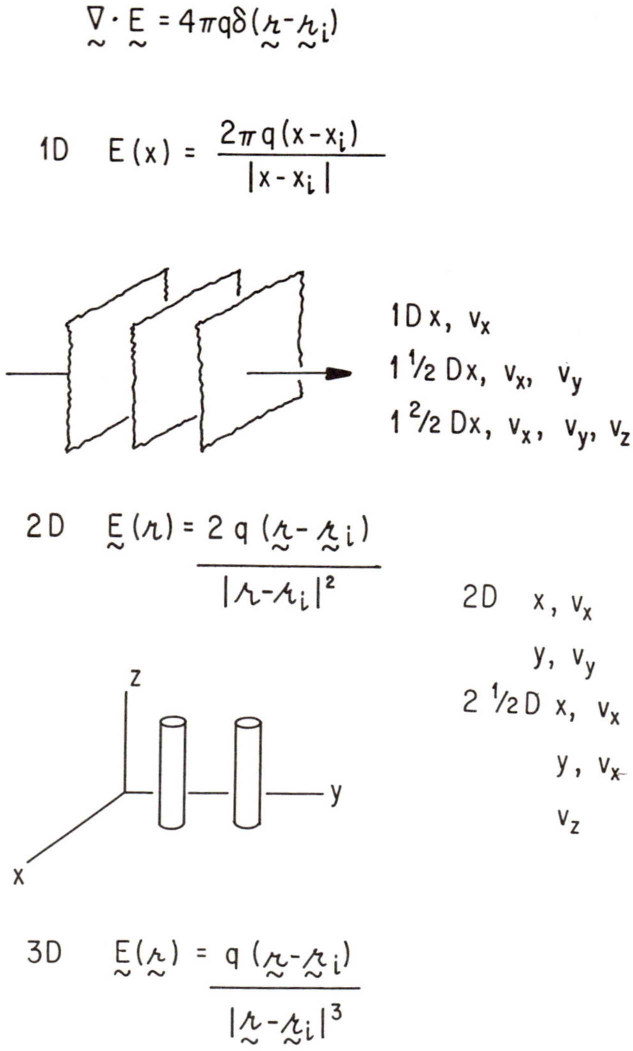

Fig. 1. One, two and three dimensional electrostatic models.

These quantities are defined as

$$\omega_p^2 = \frac{4\pi ne^2}{m_e}\,,$$

$$\lambda_D^2 = \frac{\kappa T}{4\pi ne^2}\,, \qquad (6)$$

$$\ln\Lambda = \ln(n\lambda_D^3)\,,$$

where T is the temperature, e and m_e are the electronic charge and mass (those for the lighter particles). We see from these formulae that the collision times are proportional to the number of particles in a Debye sphere in 1, 2 and 3D.

Now for a plasma, the long range nature of the Coulomb force gives rise to collective modes - modes involving coherent motions of many particles. These are the modes of interest. The collective modes are modes with wavelengths greater than the Debye length. Also for plasmas of interest $\omega_p\tau_{coll}$ should be a lot larger than one; collisional effects should be small on the time scale of plasma oscillations. This requires a large number of particles in a Debye sphere; thus we want

$$N_D \gg 1\,,$$

$$N_D = n\lambda_D \quad 1D\,,$$

$$n\lambda_D^2 \quad 2D\,, \qquad (7)$$

$$n\lambda_D^3 \quad 3D\,.$$

Further, for a significant simulation there should be many collective modes in the system. For the two dimensional case a significant simulation might involve a system 100 Debye lengths

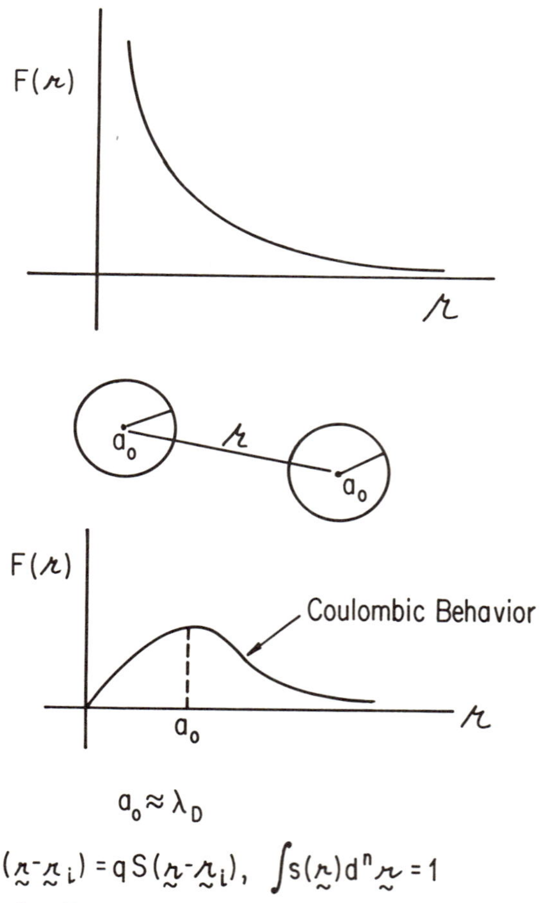

Fig. 2. Force variation between point particles and finite sized particles.

on a side. Thus the total number of particles must be large compared to 10^4 (hence, the number of particles used in the above estimate).

Now the force between two charged particles has the general form shown at the top of Fig. 2 (except for one dimension where it is $x/|x|$). Collisions come about due to the rapid variation of the force as particles pass close to each other. If we could eliminate the strong variation of the force for close encounters we could greatly reduce the collision rate. This can be done as shown in Figure 2. Instead of point charges we consider particles made of finite size charge clouds of radius a_o. For separations r which are large compared to a_o the force between the particles is Coulombic. However, when the particles overlap (we allow them to pass freely through each other) the force falls to zero as r goes to zero as illustrated in the lower drawing of Fig. 2.

Since we desire to accurately model modes with scalelengths greater than the Debye length, we generally take $a_o \lesssim \lambda_D$. We can represent the charge density ρ of such a particle by

$$\rho(\mathbf{r} - \mathbf{r}_i) = qS(\mathbf{r} - \mathbf{r}_i), \tag{8}$$

where $\mathbf{r}_i$ is the center of the particle, q is its total charge and S is a shape factor

$$\int S(\mathbf{r})d\mathbf{r}^n = 1. \tag{9}$$

The shape and size of the charge cloud is at the choice of persons making the calculations. Two possible shapes, a uniformly charged sphere and Gaussian charge distribution are shown in Fig. 3 along with a sketch of the ratio of the colli-

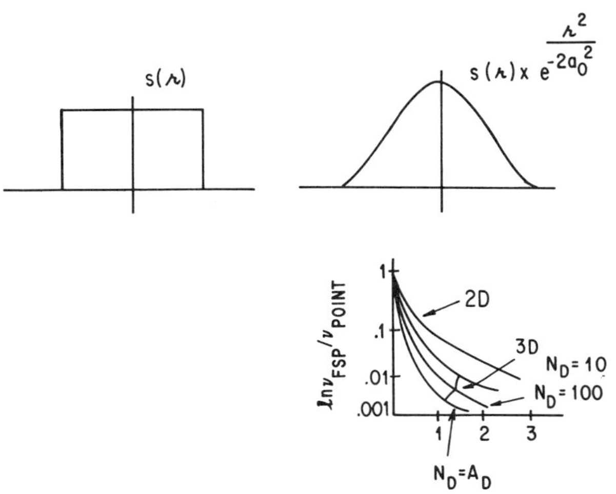

Fig. 3. Some typical charge distributions used for finite sized particles. The effect of finite size on collision frequency.

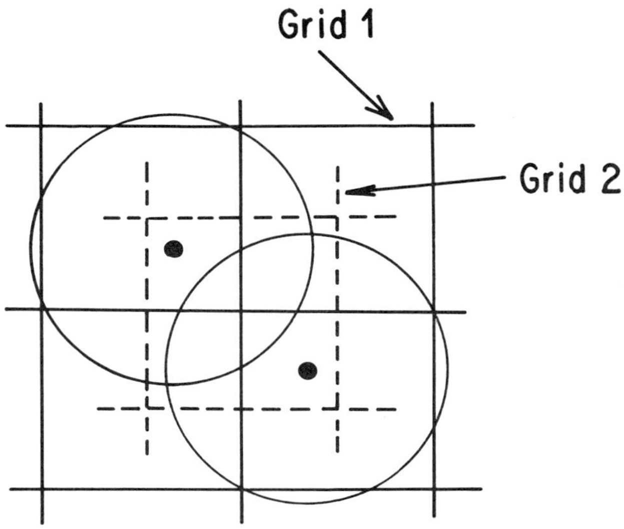

Fig. 4. Grids 1 and 2 used for computing fields for finite sized particles.

sion rates for finite size particles (FSP) and those for point particles for two and three dimensional models. As already stated, we generally choose the particle size to be a Debye length. If we choose it smaller than this, the reduction in collision rate is not so large; on the other hand, if we choose it larger than this, we cease to accurately model modes with wavelengths a few Debye lengths long.

The use of finite size particles brings with it another advantage, a method for rapidly calculating the force on a particle. This comes about as follows:

We can also calculate the force on a particle by use of the electric field. For point particles this is given by

$$\mathbf{F}_i = q_i\mathbf{E}(\mathbf{r}_i). \tag{10}$$

For our finite sized particles this must be replaced by

$$\mathbf{F}_i = q_i \int S(\mathbf{r} - \mathbf{r}_i)\mathbf{E}(\mathbf{r})d\mathbf{r}^n . \tag{11}$$

The electric field can be determined from Parsson's equation

$$\nabla \cdot \mathbf{E} = -\nabla^2\phi = 4\pi\rho(\mathbf{r}) . \tag{12}$$

If the charge density, $\rho(r)$, is that given by point particles, then calculating the field is equivalent to computing the sum in Eq. (3) and we have gained nothing. However, if the particles have finite size we cannot resolve charge density variations smaller than the size of the particle. Therefore, we can divide the space by a uniform grid with grid spacing about equal to a particle radius as shown in Fig. 4. A second grid (grid 2) connecting the centers of the squares formed by the first grid (grid 1) is also introduced as shown by the dashed grid. Around each grid point of grid 1 is a square of grid 2. We may associate the charge of particles whose centers lie in this square with its associated grid 1 point; i.e., we can place all particles within a cell of grid 2 at its center. If the particle size is a grid spacing or greater, this is a reasonably accurate approximation and is called the nearest grid point approximation (NGP) (Hockney, 1966; Birdsall and Fuss, 1969). We can improve upon this approximation by also putting a dipole at the grid point which is equal to the dipole moment of the particle with respect to the grid 1 point (Kamimura et al., 1975; Birdsall and Fuss, 1969). We then get the following charge distribution:

$$\rho(r) = \sum_g [Q(\underset{\sim}{r}_g)S(\underset{\sim}{r} - \underset{\sim}{r}_g) + \underset{\sim}{D}(\underset{\sim}{r}_g) \cdot \underset{\sim}{\nabla}_g S(\underset{\sim}{r} - \underset{\sim}{r}_g)] , \quad (13)$$

$$Q(\underset{\sim}{r}_g) = \sum_{i \varepsilon g} q_i, \quad \underset{\sim}{D}(\underset{\sim}{r}_g) = \sum_{i \varepsilon g} q_i(\underset{\sim}{r}_i - \underset{\sim}{r}_g). \quad (14)$$

A modification is to put fractional charges on the surrounding grid points so that the configuration has the same charge and dipole moment (charge sharing) (Kamimura et al., 1975; Birdsall and Fuss, 1969; Kruer et al., 1973). For either approximation, the result is a set of finite sized charged particles located on a regularly spaced grid system.

There exists fast methods for solving for the fields on such a regularly spaced grid. These are fast Poisson solvers (Hockney, 1970) and fast Fourier Transforms (Kamimura et al., 1975). Once we obtain the fields we compute the forces on the particles; if the nearest grid point method is used this is done by evaluating the field at the nearest grid point; if the dipole approximation is used, we either interpolate the fields at the position of the particles (Kamimura et al., 1975; Kruer et al., 1973) or in the charge sharing scheme we take an appropriate weighting (Birdsall and Fuss, 1969) of the forces on the nearby grid points.

The speed of this method of computing forces is far faster than the direct calculation method used in Eq. (3). The time to calculate the fields from the charges is proportional to MlnM where M is the number of grid points and the time required to put the charges on the grids, compute the forces on the particles from the fields and to advance the particles according to the equations of motion (Kamimura et al., 1975; Kruer et al., 1973) is proportional to N so that the total time is given by

$$\tau = A\, M \ln M + BN, \quad (15)$$

as compared to N^2. The MlnM term is much smaller than N because it is not quadratic and M is much smaller than N^2 since there are many particles per cell; the cells must simply give an adequate representation of the field while the particles must represent both the spatial charge variation and the velocity distribution function. We have achieved computation times of

$$\tau = \frac{15 \text{ sec}}{\text{particle x } \Delta t} , \quad (16)$$

for a two and one half dimensional electrostatic particle model with fixed magnetic field using up to 10^6 particles on the UCLA CHI machine (Dawson et al., 1978). Comparable speeds have been achieved on the CDC-7600 and several times this speed have been achieved on the Cray 1. With such speeds it is possible to do calculations involving 10^5 to 10^6 particles and 10^4 time steps. For a system of $100\lambda_D$ by $100\ \lambda_D$, the collision rates are as low as

$$\omega_{pe}\tau_{coll} \sim 6 \times 10^{-4} \text{ to } 6 \times 10^{-5}. \quad (17)$$

Simulation of Instabilities Associated with the Diffuse Aurora

Observations of electric field fluctuations in the magnetosphere during periods of the diffuse Aurora show oscillations at many frequencies, but in particular there are oscillations at roughly 3/2, 5/2, 7/2 ω_c. This is illustrated in Fig. 5 which is a frequency time spectrogram taken by D. A. Gurnett on ISEE 1 (Gurnett et al., 1978). Here frequency is plotted vs. time; the satellite is approaching the earth so the magnetic field and associated cyclotron frequency are increasing with time. The intensity of the fluctuations is denoted by the color, from red (very intense) to blue (weak). We can see just outside the plasmapause a very intense upper hybrid wave near 2120 UT. Just below the upper hybrid, we observe a series of odd-half harmonic waves (ω/ω_c = 3/2, 5/2, 7/2). Their amplitudes are weak (1-10μV/m) and intensify around 1900 UT (.1 mV/m) (Gurnett, et al., 1978). This intensification corresponds to an increase in the flux of a few keV electrons (Sentman et al., 1979). Such waves are believed to be due to "loss cone" velocity distribution functions which are more easily destabilized by the presence of a cold component (Ashour-Abdalla and Kennel, 1978).

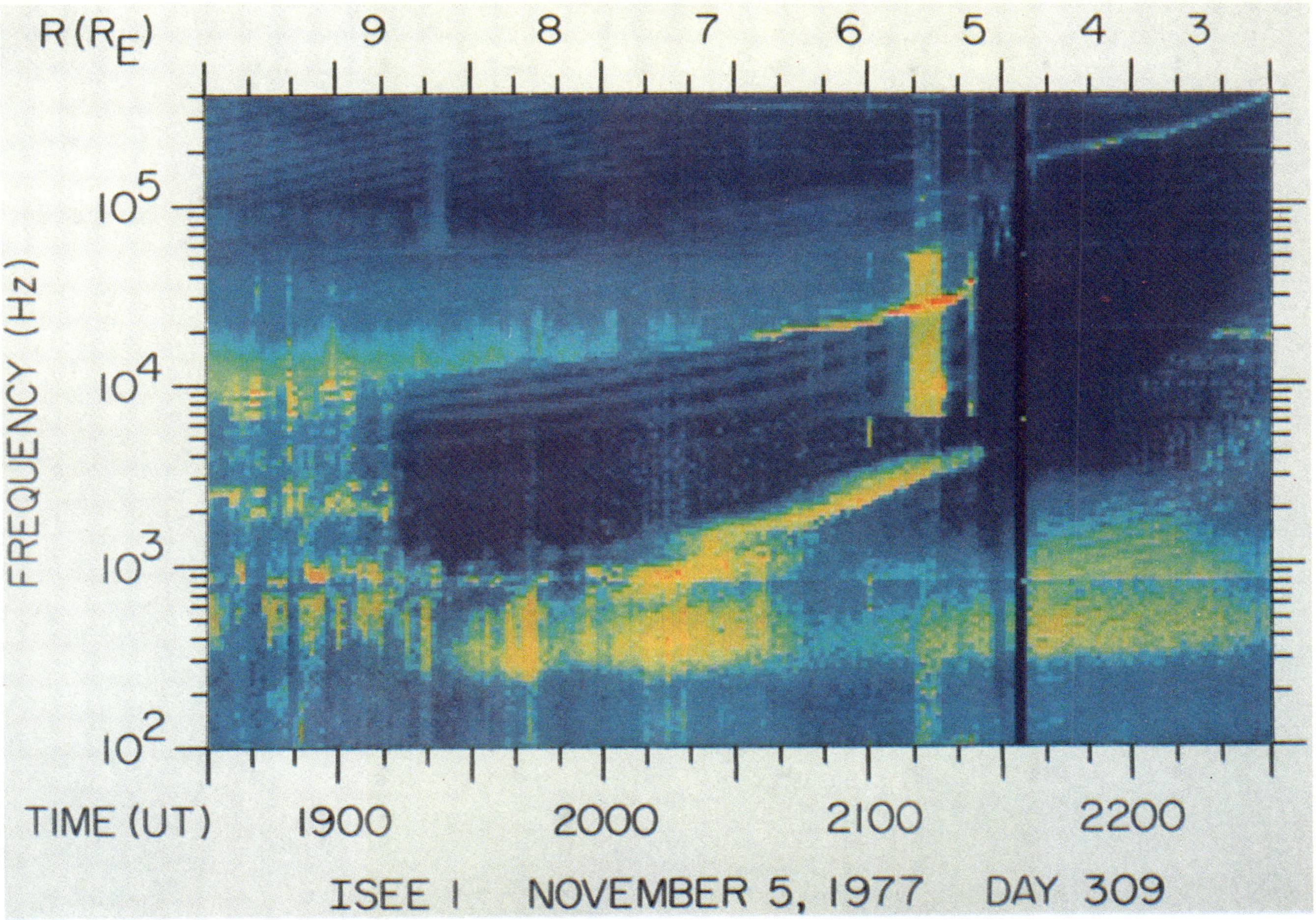

Fig. 5. Frequency time spectrogram from ISEE-1.

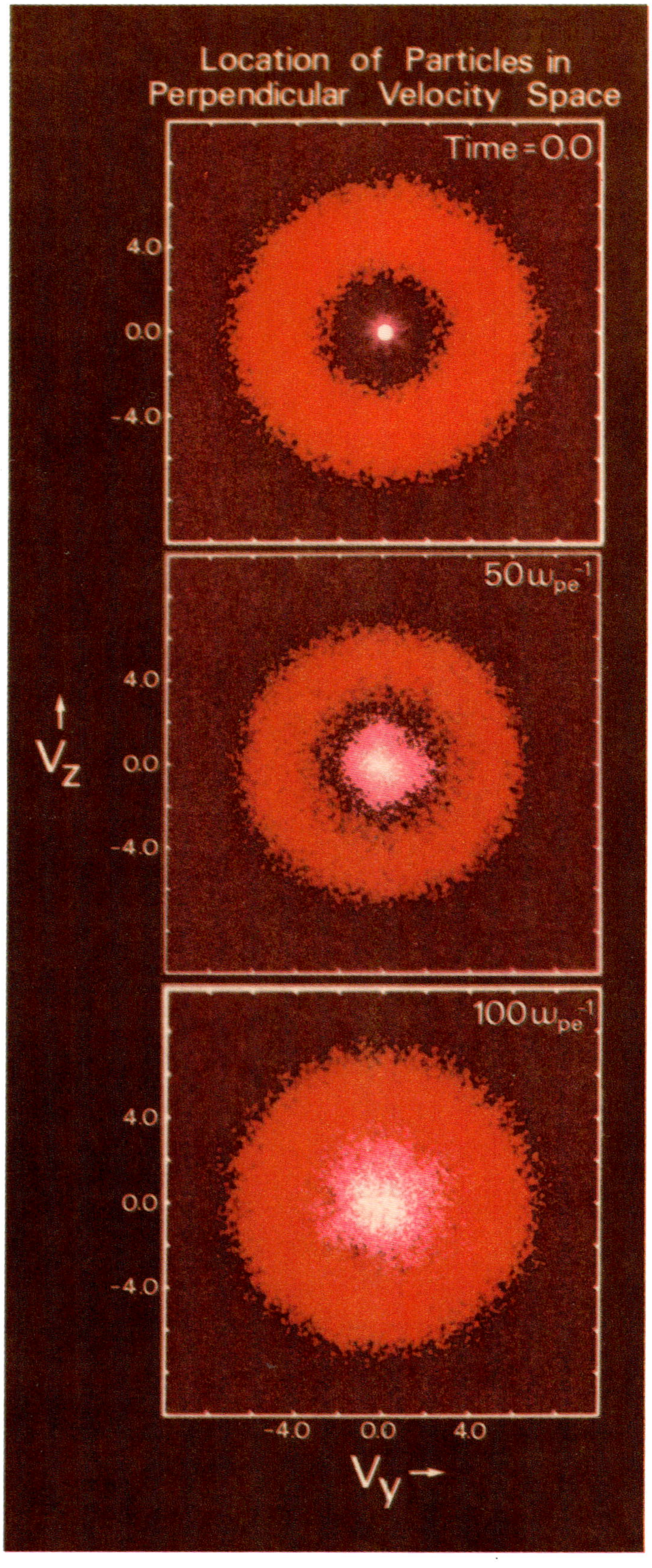

Fig. 15. Plots of $v_\perp$ (v_y, v_z) at a number of times.

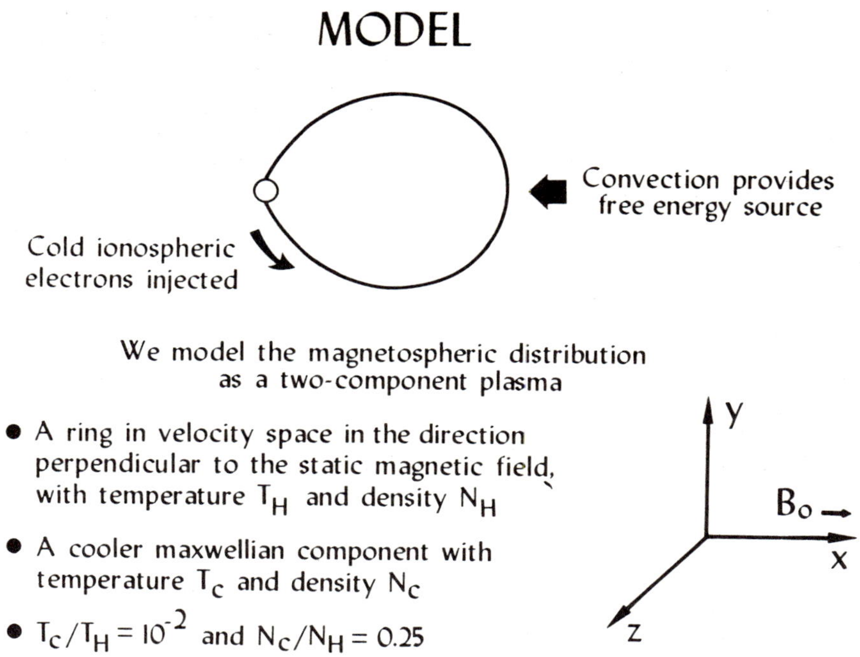

Fig. 6. Model used to simulate the diffuse aurora.

One possible scenario for the formation of such distributions is shown in Fig. 6. We model the distribution as consisting of a cold component and a ring distribution with 25% of the electrons in the cold component and the temperature ratio of the ring and cold electrons being 10^2. The mean velocity of the ring was varied from two to five times its thermal velocity. For most of the simulations, the cyclotron frequency was taken as 0.3 of the plasma frequency although some other ratios were also tried. It has been shown that such a situation is unstable to oscillations at $3/2\omega_c$; (Ashour-Abdalla and Kennel, 1978) for higher ratios of ω_{pe} to ω_c oscillations at 5/2, $7/2\omega_c$, etc. are also unstable. The observed distributions, of course, are what exists in an almost steady state; since strongly unstable distributions cannot exist for appreciable lengths of time only marginally unstable distributions will be observed.

The computer model used to simulate this instability is the two and one half dimensional electrostatic model with fixed B field shown in Fig. 7. The B field was taken parallel to the x axis. Systems of 64 x 64 Debye lengths were used and the number of particles varied from 3 x 10^4 to 5 x 10^5.

The time development of the total electrostatic field energy, the kinetic energy of the energetic electrons and of the cold electrons are shown in Fig. 8 for the case in which the initial ring velocity is five times its thermal velocity. Fig. 9 shows the electrostatic energy vs. time for two modes, one with one wavelength in the x-direction (|| to B) and two in the y-direction ($\perp$ to B), the other with one wavelength in the x-direction and three in the y-direction. These were the two dominant modes for this case. Fig. 10 shows the frequency spectrum for these two modes. The assymmetry is due to small fluctuations associated with the initial distribution. There are clear peaks at frequencies $\omega = \pm .45\omega_{pe}$ which is $\pm 3/2\omega_c$ ($\omega_c = 0.3\omega_{pe}$). For other choices of the ratio of plasma frequency to cyclotron frequency, we also have seen 5/2 and $7/2\omega_c$.

From Fig. 8 we see that the electrostatic field energy ceases to grow at about $\omega_{pe}t = 50$. If one looks at the velocity distribution function at that time, one finds that linear theory predicts that the system should still be unstable. One also sees from this figure that the cold electrons are being rapidly heated at this time. The electrons in the energetic velocity ring are losing energy at this time as they must to conserve energy.

There are two important questions which we should like the answer to (1) What is the saturation mechanism? (2) By what mechanism are the cold electrons heated? A number of saturation mechanisms have been previously proposed but none of them fit the facts observed in the simulations. Along the mechanisms proposed are the following: (1) collisional damping of the waves is large enough to stabilize the waves when the growth rate becomes small enough, (2) electrostatic trapping of electrons upsets the driving mechanism, (3) quasilinear diffusion stabilizes the distribution, (4) convective cells or vortex motion scrambles the phases of the particles

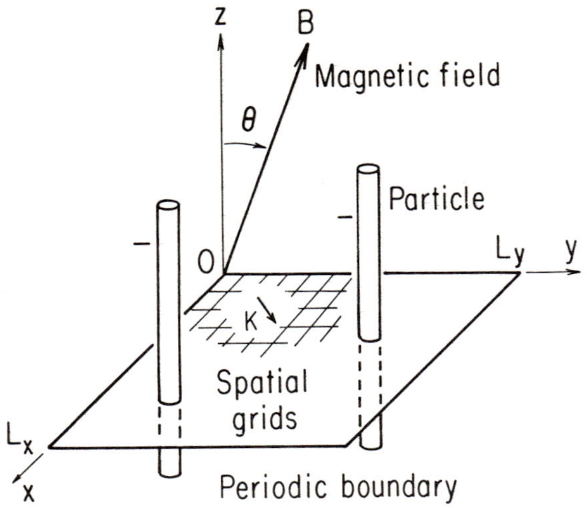

Fig. 7. Two and one half dimensional model used for the simulation.

destroying the coherent wave motion and giving rise to an effective damping rate which is equal to the linear growth rate (Okuda et al., 1974) and (5) mode coupling scatters energy out of the unstable modes into stable or damped modes producing a damping equal to the linear growth rate. As far as these mechanisms go we can list the following objections. For mechanism (1) the collision rate for the case shown in Figs. 8-10 was $\omega_{pe}\tau_{coll} > 1500$ as compared to a saturation time of $\omega_{pe}t = 50$. We have even run situations where $\omega_{pe}\tau_{coll}$ was as large as 20,000 with no observed change in behavior. As far as electrostatic trapping of particles in the waves is concerned the phase velocity of the waves along the field is 4.5 times the thermal velocity of the hot particles and no particles are observed at this high velocity. With regards to mechanism (3) quasilinear theory would still predict growth of the waves as linear theory does; a small amount of spreading of the ring in velocity space at the time of saturation is observed but not enough to give stability. For mechanism (4) to be effective, convective motion must exist and it must be such as to scramble the coherent wave motion (Okuda et al., 1974). Convective cells have zero frequency. From Fig. 10 for the spectrum we see no strong zero frequency component for these two wave numbers. This is true for other wave numbers as well. There is another point to be made here. The convective motion is associated with the E x B drifts. B is in the x direction and E has only x and y components; E x B is in the z direction or the ignorable direction. Such motion does not destroy coherence in the x-y plane. We shall see shortly that the instability does give rise to enhanced diffusion (displacement of a point on the rods) in the z direction due to another interesting mechanism but this diffusion does not produce any damping. This is an example of how an effect can be turned off in the model so that its effect on the process can be evaluated. It is possible to modify the model so that this mechanism does play a role; this is done simply by tilting the magnetic field in the x-z direction so that the E x B drift has x and y components. We are

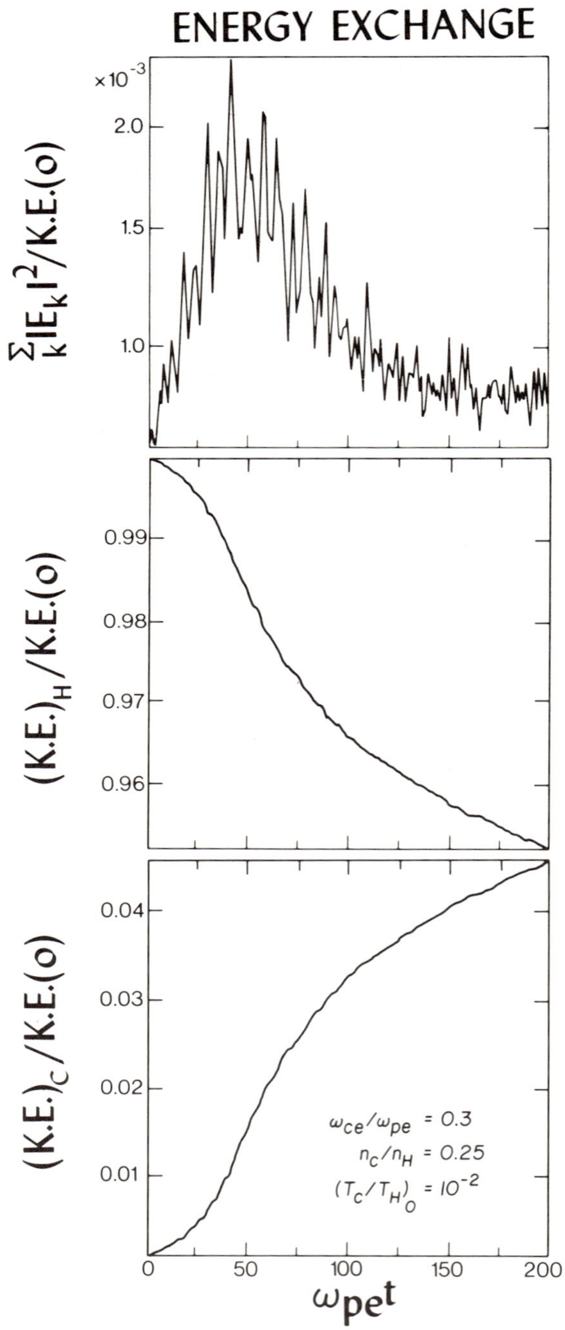

Fig. 8. Electric field energy, hot ring energy and cold particle energy.

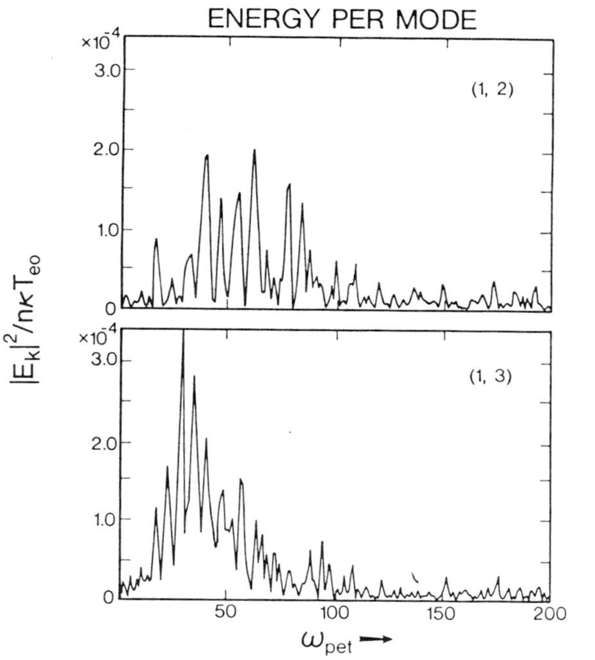

Fig. 9. Electric field energy in the two strongly unstable modes.

presently making such extensions and it appears that enhanced x-y diffusion may play some role although results to date do not indicate that it has a major influence. With regard to mechanism (5) the scattered waves should have frequencies $\omega = \omega_1 \pm \omega_2$, that is frequencies of zero or $3\omega_c$, and wave numbers with $k = k_1 + k_2$, where 1 and 2 refer to two of the strong waves observed. We have not observed any effects which can be identified as mode coupling; for example if scattering to undamped waves were involved the electric field energy should have continued to increase linearly with time. If the coupling involved two heavily damped modes it might not be observed or it might be possible to detect it only with sophisticated correlation measurements which have not been carried out. If this were the case however, it would have to primarily dump energy in the cold electrons rather than diffuse the hot electrons in velocity as is observed.

The Saturation and Heating Mechanism

The simulations give clear evidence that a type of nonlinear cyclotron resonance process for the cold particles is responsible for the saturation observed in these calculations. It is also responsible for the cold electron heating and it is observed that at saturation the rate of heating of the cold electrons essentially balances the rate at which energy is fed into unstable waves.

The mechanism is illustrated in Fig. 11. In this example, the unstable waves propagate at shallow angles to the magnetic field between 16 and 30 degrees. Make a transformation to a frame of reference moving with the phase velocity of one of the waves along the magnetic field as shown in Fig. 11. In this frame the wave is static or more precisely it is growing but at a relatively slow rate and it has no real frequency. The cold particles are streaming over this wave and see a frequency of $3/2\ \omega_c$. The wave sets the electrons to oscillating in the y direction and imparts a y velocity to them which is roughly

$$\delta v_y \simeq \frac{-eE_y k_x v_x'}{m(k_x^2 v_x'^2 - \omega_c^2)}\ , \tag{18}$$

where v_x' is the velocity in the wave frame. Now energy must be conserved in this frame of reference since ω is zero and so the y motion can be produced only through a reduction in the x velocity relative to the wave. Conservation of energy in the wave frame requires that for small changes in v_x', δv_x is given by

$$\delta v_x = \frac{\delta v_y^2}{2\omega/k_x}\ . \tag{19}$$

Here we have assumed that the velocity of the cold electrons in the wave frame is ω/k_x. As the electrons slow down in the wave frame the Doppler shifted frequency moves closer to cyclotron resonance. When the wave amplitude becomes large enough the cold particles can be slowed to such an extent that they are brought into resonance with the wave at which point they are strongly accelerated and damp the wave. Their x velocity also changes strongly at this time. To become resonant with the wave requires that

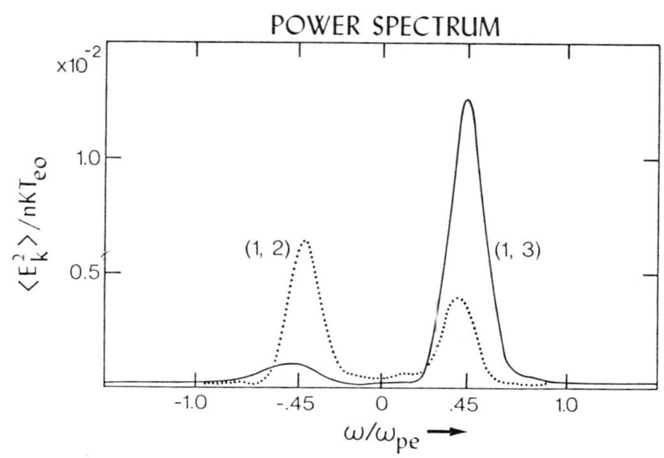

Fig. 10. Frequency spectrum of the two strongly unstable modes.

$\omega = \omega_c + k_x v_x$ which means that in the laboratory frame of reference the electrons must be accelerated to 1.5 v_{th} (v_{th} is the thermal velocity of the hots). To test this mechanism, the number of electrons with x velocities equal to or greater than 1.5 v_{th} were counted. The temperature of the cold component of this sample was also measured; the reason a small region was sampled was to avoid the appearance of random motion due to averaging over different parts of the wave which have different phases. The results of this measurement are shown in Fig. 12. The top part of the figure shows the fraction of the particles which have acquired velocities of 1.5v_{th} or above; the lower part of the figure shows the ratio of the temperature (determined from $[v - \bar{v}]^2$) of the cold electrons. We see that the number of cold electrons accelerated to 1.5v_{th} increases rapidly at $\omega_{pe}t = 50$ and that shortly thereafter the temperature of the cold electrons also increases rapidly; the time lag is associated with the finite ratio of heating of the resonant particles. The rate of increase of the resonant electrons occurs at roughly $\omega_{pe}t = 75$ which is also the time at which the wave energy starts to decrease as seen from Fig. 8.

The correctness of this interpretation is further shown by the results presented in Figs. 13 and 14. Figure 13 shows the mean velocity of the cold electrons parallel to $B(v_{\parallel})$ at the top and the electric field energy at the bottom; both are plotted vs. $\omega_{pe}t$. This simulation used 5×10^5 particles but otherwise was the same as that shown in the earlier figures. We see that $v_{\parallel}$ rises rapidly at the time the waves are strong and then levels out after the waves die out. If one plots the mean velocity of the hot ring electrons, it is the negative of that for the cold electrons as is required by conservation of momentum. The explanation of this effect is as follows. We see from the spectrum shown in Fig. 8 that positive and negative frequencies are not equally excited. This is a statistical effect resulting from the choice of initial conditions; one wave is initially more strongly excited. Opposite frequency waves propagate in opposite directions. Now the waves carry momentum as well as energy. When the waves are excited, they absorb energy and momentum from the hot particles. When the waves are absorbed their energy and momentum is taken up by the absorbing particles; in this case the initially cold electrons. These results indicate the potential for instabilities to generate currents. Here, of course, no current is generated because the hot and cold electrons have equal and opposite drifts. However, the hot electrons would tend to be expelled from an unstable region in one direction and the cold electrons in the other direction. Also, if we were dealing with an instability involving ions then such an effect would produce a current directly.

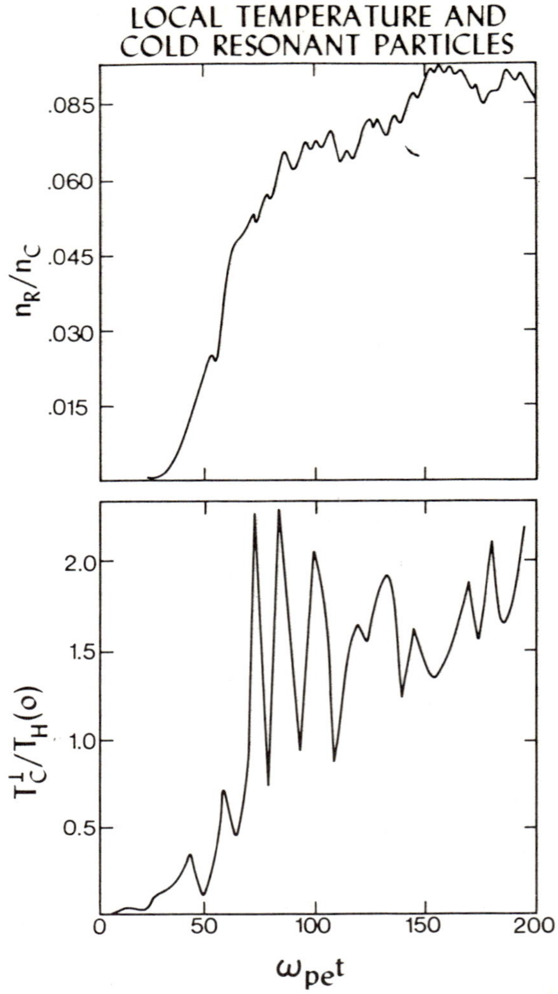

Fig. 12. Top, fraction of cold particles moving fast enough to resonate with the wave vs. time. Bottom, ratio of the cold electron temperature to the initial hot electron temperature vs. time.

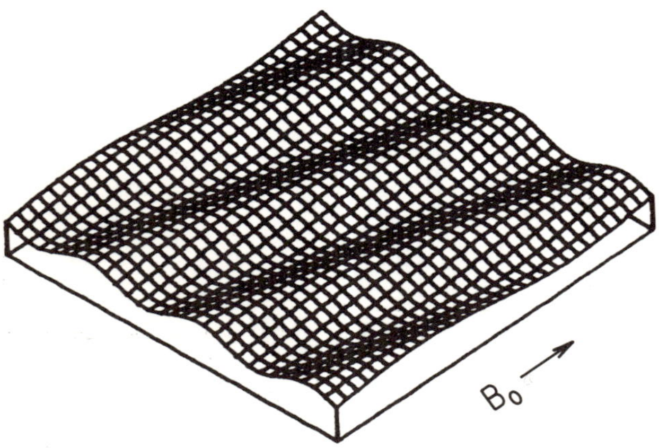

Fig. 11. Wave particle interaction situation.

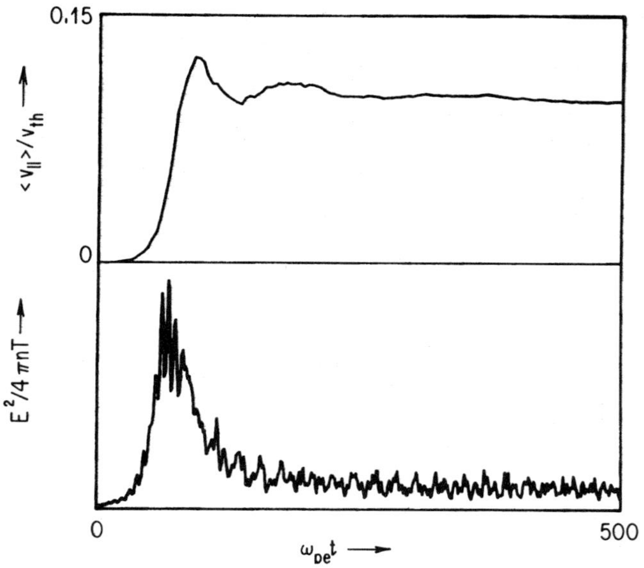

Fig. 13. Velocity of cold electrons parallel to B and total electric field energy vs. time.

Figure 14 shows plots of the temperature of the cold electrons perpendicular and parallel to the magnetic field at the top, and a plot of the electric field energy at the bottom. We see that both the perpendicular and parallel temperature rise rapidly at the time of major wave activity and then rise much more slowly after the waves die down. The ratios of $T_{\|}$ to $T_{\perp}$ can be predicted from Eqs. (18) and (19) and these results are in good agreement with those predictions.

The strong heating of the cold electrons is also shown in Fig. 15. This shows a plot of the particle velocities perpendicular to the magnetic field. The red particles are the energetic particles, the blue ones are the cold electrons. Each point is the velocity of a particle; the velocities are shown at three different times. Initially the hot ring had a velocity of $5v_{th}$. As the instability gets going the perturbation of the cold particles is clearly visible. Their heating and spreading in $v_{\perp}$ is very evident for the late time frame. Little distortion of the hot ring is evident.

Another interesting effect which is observed is enhanced diffusion of guiding centers in the z direction. While none of the quantities, E, J, n, etc., depend on z and the particles are rods parallel to z, we can keep track of the displacement of a point on a rod by integrating v_z with respect to time.

$$\Delta z = \int_0^t v_z(t')dt'. \qquad (20)$$

This would correspond to the actual displacement of a particle perpendicular to the plane of B and k. Fig. 16 shows a plot of the mean square displacement of the z position of the guiding center for a set of test particles. There are two curves, one for the hot energetic ring and the other for the cold background electrons. We see that rapid diffusion sets in when the instability gets going. It can be shown that the guiding center diffusion is produced by $\langle E^2_{sbp}(\omega = 0)\rangle$ where sbp stands for seen by the particles, i.e., the zero frequency component of the E field which the particle sees gives the diffusion through an E x B drift. However, examination of frequency spectra like that shown in Fig. 10 show that there is virtually no noise at $\omega = 0$. The particles, however, can see a zero frequency component due to a combination of Doppler shift and finite orbit effects ($k_{\perp}\rho_{\ell} \simeq 1$); the particle sees a frequency component with frequency $\omega - k_{\|} v_{\|} \pm \omega_c$. It is interesting to note from Fig. 16 that the energetic particles initially have the larger diffusion but end up with less diffusion than the cold particles. The cold particles initially have small Larmor radii and small values of $k_{\|}v_{\|}$ and cannot beat with the wave to produce a zero frequency component; however, the particles in the hot ring can. Once the cold particles are brought into resonance by the saturation mechanism they can and do start to diffuse rapidly. At late times the cold particles diffuse more rapidly than the hot particles because for them $k_{\perp}\rho_{\ell} \simeq 1$ while for the hot particles $k_{\perp}\rho_{\ell} > 1$; the large orbits of the hot particles result in their averaging out the turbulent fields to a certain extent and thus not being diffused by them (Naitou et al., 1978).

This type of diffusion could play a role in plasmas heated by electron cyclotron waves;

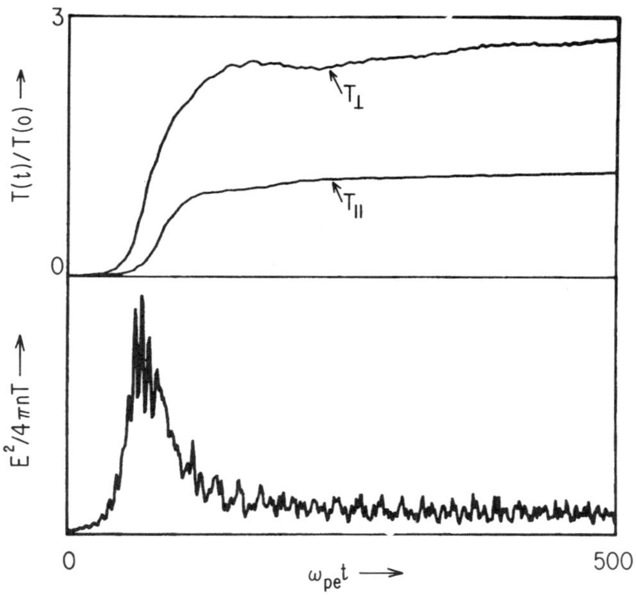

Fig. 14. The perpendicular temperature, parallel temperature and electric field vs. time.

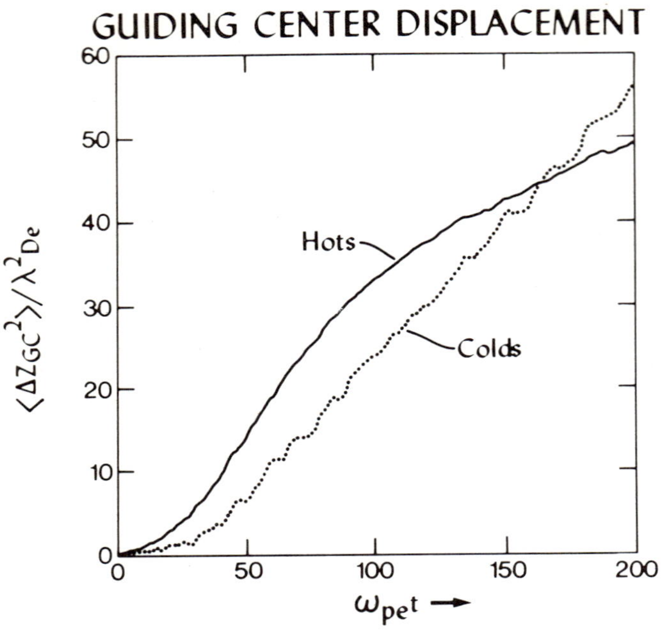

Fig. 16. Diffusion of guiding centers of hot and cold electrons.

similar processes might be important for ion cyclotron heated plasmas. Much effort is being put into such heating of fusion plasmas and its effects should be examined and looked for.

Acknowledgements. The author wishes to acknowledge that most of the work reported here was carried out by J. N. Leboeuf, M. Ashour-Abdalla and R. W. Huff. He is also indebted to Dr. D. A. Gurnett for allowing him to use his satellite data. Work supported by NSF contract ATM79-26492, DOE contract DE-AM03-76SF000010 PA 26, Task III, and NAGW-78.

References

Ashour-Abdalla, M., and C.F. Kennel, Nonconvective and convective electron cyclotron instabilities, J. Geophys. Res., 83, 1531, 1978.

Birdsall, C.K. and D. Fuss, Clouds-in-clouds, clouds-in-cell physics for many-body plasma simulation, J. Comp. Phys., 3, 494, 1969.

Birdsall, C.K., A.B. Langdon, and H. Okuda, Finite size particle physics applied to plasma simulation, in Methods in Computational Physics, ed. by B. Alder et al., Academic Press, New York, Vol. 9, 241, 1970.

Busnardo-Neto, J., P. Pritchett, A.T. Lin, and J.M. Dawson, A self-consistent magnetostatic code for numerical simulations of plasmas, J. Comp. Phys., 23, 300, 1977.

Dawson, J.M., R.W. Huff, and C.C. Wu, Plasma simulation on the UCLA CHI computer system, AFIPS Conf. Proc., National Computer Conference, 47, 395, 1978.

Gurnett, D.A., F.L. Scarf, R.W. Fredricks, and E.J. Smith, The ISEE-1 and ISEE-2 plasma investigation, Geoscience Electronics, GE-16, 225, 1978.

Hockney, R.W., Computer experiments of anomalous diffusion, Phys. Fluids 9, 1826, 1966.

Hockney, R.W., The potential calculation and some applications, in Methods in Computational Physics, ed. by B. Alder et al., Academic Press, New York, Vol. 9, 135, 1970.

Kamimura, T., J.M. Dawson, B. Rosen, G.J. Culler, R.D. Levee, and G. Ball, Plasma simulation on the CHI microprocessor system, U. of California Res. Report PPG-248, University of California, Los Angeles, 1975.

Kruer, W.L., J.M. Dawson, and B. Rosen, The dipole expansion method for plasma simulation, J. Comp. Phys., 13, 114, 1973.

Lin, A.T., J.M. Dawson, and H. Okuda, Application of electromagnetic particle simulation to the generation of electromagnetic radiation, Phys. Fluids 17, 1995, 1974.

Naitou, H., T. Kamimura, and J.M. Dawson, Kinetic effects on the convective plasma diffusion and the heat transport, J. Phys. Soc., Japan, 46, 258, 1979.

Okuda, H., C. Chu, and J.M. Dawson, Turbulent damping of the convective cells and the lower hybrid waves, Phys. Fluids, 18, 243, 1975.

Sentman, D.D., L.A. Frank, C.F. Kennel, D.A. Gurnett, and W.S. Kurth, Electron distribution functions associated with electrostatic emissions in the dayside magnetosphere, Geophys. Res. Lett., 6, 781, 1979.

ANOMALOUS DIFFUSION AND ION HEATING IN THE PRESENCE OF ELECTROSTATIC HYDROGEN CYCLOTRON INSTABILITIES

H. Okuda, C. Z. Cheng, and W. W. Lee

Plasma Physics Laboratory, Princeton University, Princeton, NJ 08544

Abstract. One- and two-dimensional particle simulations have been carried out to study electrostatic ion cyclotron instabilities for a hydrogen plasma in a strong magnetic field. It is found that strong ion heating and anomalous cross-field diffusion comparable to Bohm diffusion take place associated with the instability. Implications of the instability to the recent observations in fusion devices and space plasmas are discussed.

It is well known that the electrostatic ion cyclotron waves may be destabilized by drifting electrons along magnetic fields in an isothermal plasma ($T_e \simeq T_i$) where the ion acoustic waves are damped (Drummond and Rosenbluth, 1962). Such waves have been observed recently in fusion devices (TFR Group, 1978) as well as in space plasmas along auroral field lines (Mozer et al., 1977). It is important, therefore, to study the nonlinear consequences of such instabilities and, in particular, anomalous cross-field diffusion, anomalous resistivity and heating in the presence of cyclotron instabilities may be much greater than what are expected from classical theory. A number of laboratory experiments have been reported and enhanced plasma transport as well as plasma heating have been reported (Yamada and Hendel, 1978; Bohmer and Fornaca, 1979).

Here we would like to present one- and two-dimensional particle simulation studies of the hydrogen cyclotron waves for a collisionless plasma in a low β limit. The model is electrostatic and drift kinetic approximation is used for the electrons while full dynamics of ions are followed in time (Lee and Okuda, 1978).

Figure 1 shows the time development of the potential, $e\phi_k(t)/T_e$, for a long wavelength, $k_\perp\rho_i = 0.16$ (a), and a short wavelength, $k_\perp\rho_i = 1.2$ (b), and the frequency spectrum obtained from one of the one-dimensional simulations. Here, $m_i/m_e = 1837$, $T_e/T_i = 1$, $\Omega_e/\omega_{pe} = 5$ (electron gyrofrequency/plasma frequency), $k_\parallel/k_\perp = 0.1$, $v_d/v_{te} = 1.4$, and $L = 1024\Delta$ are used where v_d, v_{te} are initial electron drift speed along the magnetic field, electron thermal speed, and Δ is the computational mesh size. Note the system length L is long compared with the ion gyroradius $\rho_i = 8.6\ \Delta$, and there are more than 100 unstable Fourier modes in this simulation. Both the frequency and the growth rate observed in Fig. 1 agree well with the linear theory. For $k_\perp\rho_i = 0.16$, only the fundamental cyclotron frequency is unstable, while for $k_\perp\rho_i = 1.2$ several higher harmonics grow in time giving rise to modulation in amplitude due to the beating among different frequencies. This is quite similar to the satellite observations along the earth magnetic field (Mozer et al., 1977).

Saturation amplitudes for these modes are relatively low and are typically $|e\phi_k/T_e| \lesssim 0.2$. Nonlinear saturation is a result of electron velocity space diffusion along the magnetic field shown in Fig. 2. Clearly the diffusion flattens the positive slope of the distribution generating a plateau, and at the same time the diffusion process reverses the sign of $\partial f/\partial v_\parallel$ for a region of $v_\parallel > v_d$ from negative to positive, thereby destabilizing the modes which were initially stable. From many other one-dimensional simulations and also from two-dimensional simulations which include modes propagating at different angles, we find nonlinear saturation is a result of electron velocity space diffusion creating a plateau which gives the lowest level of saturation. Two-dimensional effects, such as mode-coupling among Fourier modes at different angles of propagation or scattering of ion orbits by $c\underset{\sim}{E} \times \underset{\sim}{B}/B^2$ motion (Dum and Dupree, 1970), appear small for the level of saturation given by the electron velocity space diffusion. This is partly because the ion cyclotron waves are nondecay type, $\omega_{k1+k2} \neq \omega_{k1} + \omega_{k2}$, and, therefore, a large amplitude is required for mode-coupling. Also, $e\phi/T_e \sim 1$ is necessary for the scattering of ions by the fluctuating electric field (Dum and Dupree, 1970), which is well above the saturation level given by the electron diffusion. Associated with the instability is

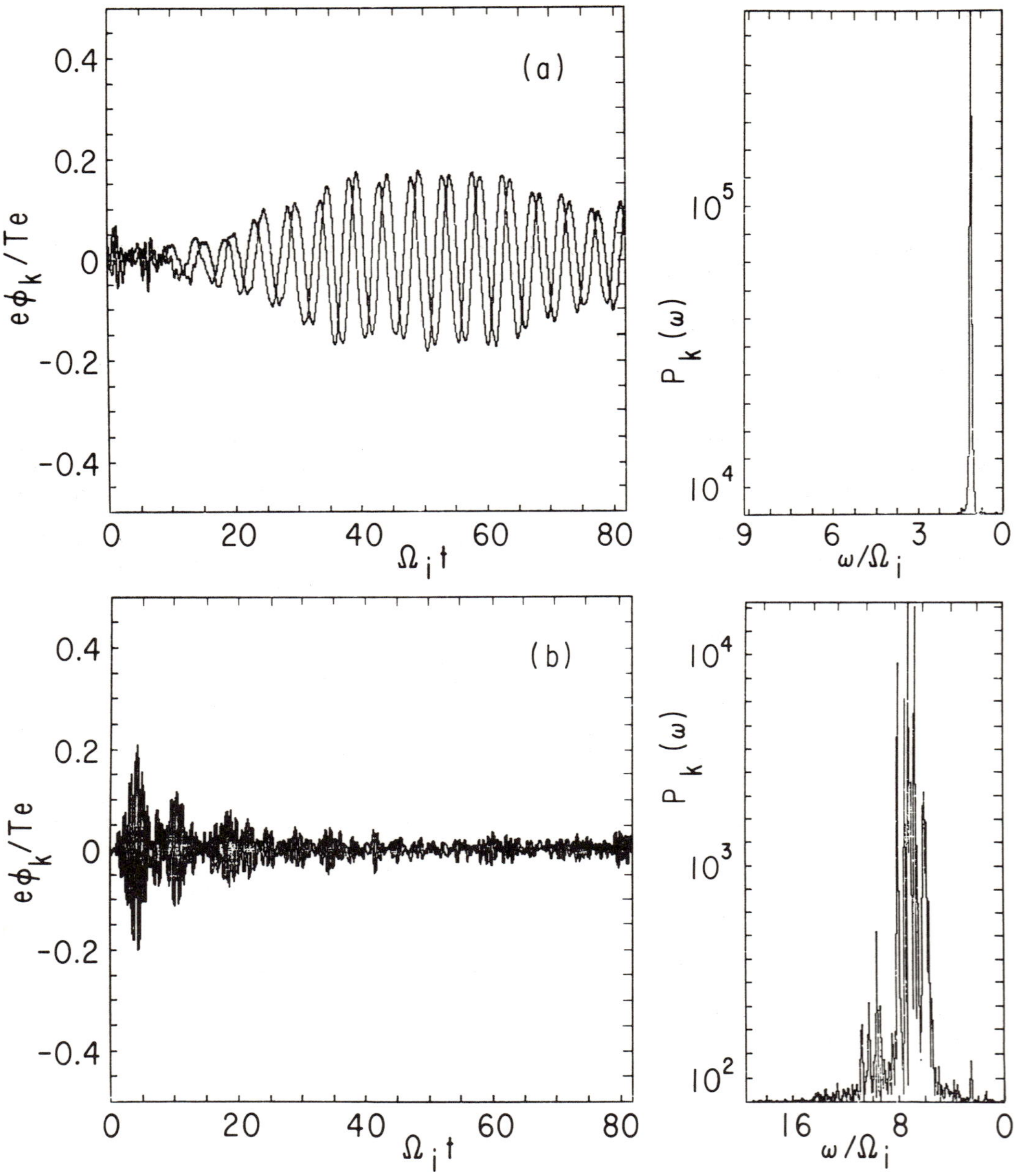

Fig. 1 Temporal growth of the cyclotron waves for (a) $k_\perp \rho_i = 0.16$ and (b) $k_\perp \rho_i = 1.2$ (c) and the frequency spectrum. Both real and imaginary parts of $e\phi_k(t)/T_e$ are shown. One-dimensional simulation.

the anomalous resistivity η which is found $\eta/\eta_o \approx 2 \times 10^{-4}$ where $\eta_o = 4\pi/\omega_{pe}$ and $\eta = -(m_e/ne^2)(dv_d/dt)/v_d$. This value is much larger than the classical value and is comparable to the anomalous resistivity associated with ion acoustic instabilities (Sato and Okuda, 1980). Since the saturation takes place via quasilinear diffusion, it is straightforward to estimate the decay rate of the electron drift along the magnetic field.

From a quasilinear equation (Kadomtsev, 1965), one finds

$$\frac{d}{dt} \int f v_\parallel \, dv_\parallel = \frac{\pi e^2}{m_e^2} \sum_k |k_\parallel| |\phi_k|^2 \left. \frac{\partial f}{\partial v_\parallel} \right|_{v_\parallel = \omega_k / k_\parallel} \tag{1}$$

Using the observed level of fluctuations $|e\phi_k/T_e| \lesssim 0.2$ and summing over the unstable modes, we find from Eq. (1) the decay of the electron drift speed is

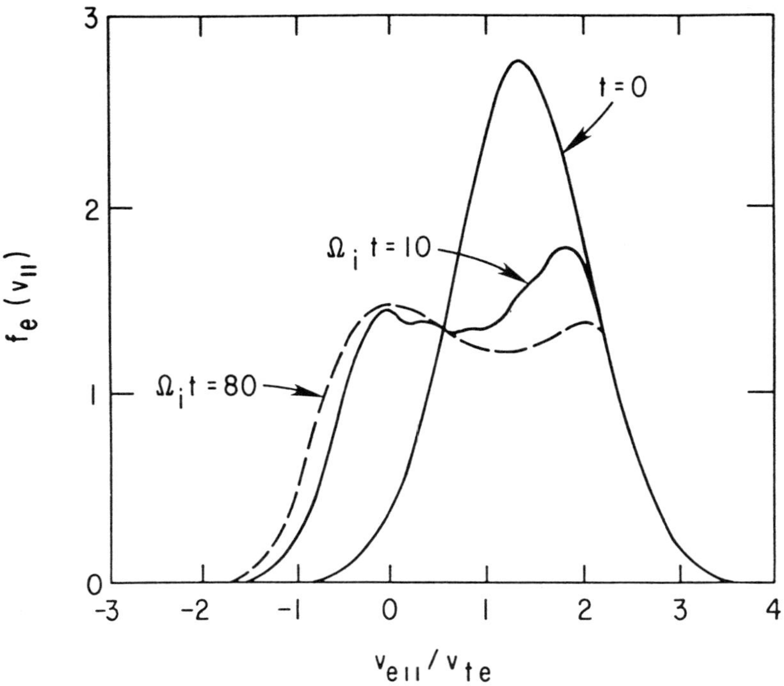

Fig. 2 Velocity space diffusion of electrons along magnetic field. One-dimensional simulation.

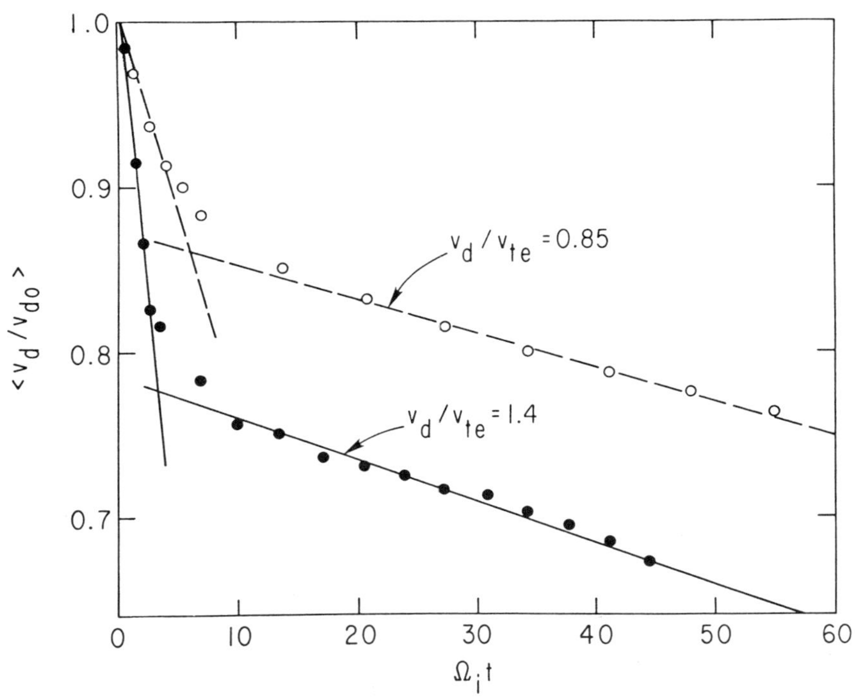

Fig. 3 Decay of the average electron drift for two different initial drifts, v_d/v_{te} = 1.4 and 0.85. One-dimensional simulation.

$d(v_d/v_{te})/d(\omega_{pe}t) \approx 10^{-4}$ consistent with the anomalous resistivity observed in the simulations.

Figure 3 shows the decay of the average electron drift along the magnetic field for two one-dimensional simulations using $v_d/v_{te} = 1.4$ and 0.85. All the other parameters remain the same. We clearly observe that there are two distinct stages of resistivity, initial rapid decay followed by a much slower but finite decay. The initial rapid decay is clearly due to the quasilinear effects associated with the unstable ion cyclotron waves. We believe the latter part is caused by the long wavelength modes ($\omega/k_\parallel > v_d$) which are initially stable and become unstable as a result of velocity space diffusion.

In addition to anomalous resistivity, ion heating perpendicular to the magnetic field takes place as a result of cyclotron damping in which the Doppler-shifted frequency of an ion with the velocity $v_\parallel^i$ along the magnetic field, $\omega - k_\parallel v_\parallel^i$, coincides with the gyro-frequency, Ω_i. It is found that the heating of the ions can easily be (20 ~ 50)% of the initial temperature. The energy fed into the ions is roughly equal to the energy lost from the electron motion along the magnetic field and this amount is also in reasonable agreement with quasilinear theory (Yamada et al., 1975).

It is natural to expect enhanced cross-field particle diffusion associated with the cyclotron instability. Since the resonant electrons moving at the wave phase velocity along magnetic field experience almost the constant electric field, they suffer large diffusion by $c\underset{\sim}{E} \times \underset{\sim}{B}/B^2$ motion. It is estimated that the electron diffusion may be comparable to Bohm diffusion (Drummond and Rosenbluth, 1962). Ion diffusion is somewhat more complicated since the wave frequency is close to ion gyro-frequency and the wavelength is comparable to ion gyro-radius. Nonetheless it is natural to expect large ion diffusion since the resonant ions which satisfy cyclotron resonance, $\omega - k_\parallel v_\parallel^i = \Omega_i$, experience the constant electric field and thereby drift across magnetic field by the $c\underset{\sim}{E} \times \underset{\sim}{B}/B^2$ motion. Since the cyclotron resonance is responsible for ion heating, it is clear anomalous diffusion and heating are always accompanied with each other.

Using a two-dimensional model in which the variations are in (x,y) plane with the magnetic field in the (y,z) plane with $B_z \gg B_y$ (Lee and Okuda, 1978), one can indeed measure diffusion coefficients of both ions and electrons across magnetic field associated with the cyclotron instability.

Figure 4 plots the position of test particles, both ions and electrons, at $t = 8.2\,\Omega_i^{-1}$ which were located at x = 0 initially. Here m_i/m_e = 1837, $v_d/v_{te} = 1.4$, $\Omega_e/\omega_{pe} = 5$, and $T_e/T_i = 1$ are used. It is clearly seen that the spread of the

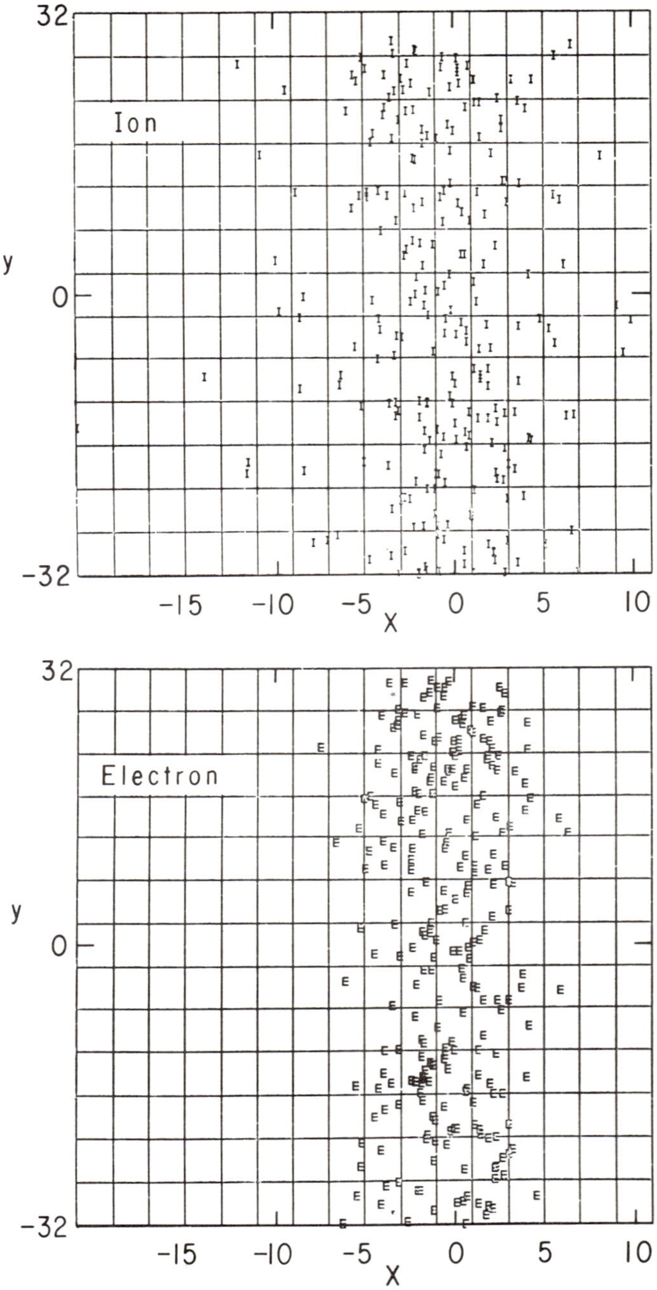

Fig. 4 Plot of the test particle position at $t = 8.2\,\Omega_i^{-1}$ initially located at x = 0. Two-dimensional simulation.

ions is greater than the spread of electrons. Diffusion coefficient defined by $D^{i,e} = \langle[x(t) - x(0)]^2\rangle/t$ is found $D_x^e = 4.5 \times 10^{-3}\Delta^2\omega_{pe}$ and $D_x^i = 6.0 \times 10^{-3}\Delta^2\omega_{pe}$ which are about one-half of Bohm diffusion $D_B = (1/16)(cT_e/eB) = 1.25 \times 10^{-2}\,\Delta^2\omega_{pe}$. This estimate is also in reasonable agreement with earlier work (Drummond and Rosenbluth, 1962). Since the hydrogen cyclotron waves have been observed in tokamaks (TFR Group, 1977) and are expected from high powered neutral injection experiments, hydrogen cyclotron waves, if excited, may cause not only rapid ion heating but also anomalous plasma diffusion in addition to low frequency drift type fluctuations (Seiler and Yamada, 1979).

Finally we would like to emphasize that double layers have never been observed in one-dimensional simulations using a very long system. This is in sharp contrast to the ion acoustic instability simulations where double layers were almost always found (Sato and Okuda, 1980). It appears hydrogen cyclotron waves observed along auroral field lines may be an effect of double layers rather than a cause for generating them (Mozer et al., 1977).

It is a pleasure to acknowledge Dr. M. Ashour-Abdalla, Dr. J. M. Dawson, Dr. C. T. Dum, Dr. F. W. Perkins, Dr. T. Sato and, in particular, Dr. M. Yamada for their interest and encouragement.

Acknowledgment. This work was supported by the United States Department of Energy Contract No. DE-AC02-76-CH03073.

References

Bohmer, H. and S. Fornaca, Experiments on Nonlinear Effects of Strong Ion-Cyclotron Wave Turbulence, J. Geophys. Res. 84, 5234 (1979).

Drummond, W. E. and M. N. Rosenbluth, Anomalous Diffusion Arising From Microinstabilities in a Plasma, Phys. Fluids 5, 1507 (1962).

Dum, T. and T. H. Dupree, Nonlinear Stabilization of High-Frequency Microinstabilities in a Magnetic Field, Phys. Fluids 13, 2064 (1970).

Kadomtsev, B. B., Plasma Turbulence, Academic Press, New York, 1965, p. 84.

Lee, W. W. and H. Okuda, A Simulation Model for Studying Low Frequency Microinstabilities in a Magnetic Field, J. Comput. Phys. 26, 139 (1978).

Mozer, F. S., C. W. Carlson, M. K. Hudson, R. B. Torbet, B. Paraday, J. Yatteau, and M. C. Kelley, Observations of Paired Electrostatic Shocks in the Polar Magnetosphere, Phys. Rev. Lett. 38, 292 (1977).

Sato, T. and H. Okuda, Ion Acoustic Double Layers, Phys. Rev. Lett. 44, 740 (1980).

Seiler, S. and M. Yamada, Ion-Beam-Driven Lower Hybrid Instability and Resultant Anomalous Beam Slowing, Nucl. Fusion 19, 469 (1979).

TFR Group, Ion-Cyclotron Instability in a TFR Tokamak, Phys. Rev. Lett., 41, 113 (1978).

Yamada, M. and H. W. Hendel, Current-Driven Instabilities and Resultant Anomalous Effects in Isothermal, Inhomogeneous Plasmas, Phys. Fluids 21, 1555 (1978).

Yamada, M. H. W. Hendel, S. Seiler, and S. Ichimaru, Anomalous dc Resistivity and Turbulent Ion Heating in Isothermal Plasmas, Phys. Rev. Lett., 34, 650 (1975).

THE GENERATION OF ELECTRIC POTENTIALS RESPONSIBLE FOR THE ACCELERATION OF AURORAL ELECTRONS

Daniel W. Swift

Geophysical Institute, University of Alaska, Fairbanks, Alaska 99701

Abstract. The acceleration of electrons responsible for the discrete and inverted-V aurora requires some processes for the creation of potential differences perpendicular to magnetic field lines in the magnetosphere. It is shown that the two-dimensional, E x B, turbulent cascade can be an effective mechanism for the generation of potential variations over a wide range of size scales perpendicular to magnetic field lines. It is also shown by means of two-dimensional particle simulations that plasma microinstabilities can inject energy into the E x B turbulent cascade and that microinstabilites generated by ion beams in the magnetosphere can generate potential variations of a magnitude sufficient to account for the energy of auroral electrons. These processes have a strong tendency toward directional isotropy, which is inconsistent with the preferred east-west alignment. It is suggested that the modified two-stream instability, excited by ions streaming across magnetic field lines through strongly magnetized electrons, may be the instability which generates auroral potentials. A compelling feature of this instability is that it should be excited in a region of magnetic merging, which would allow us to directly identify the presence of auroral arcs with magnetic merging taking place elsewhere along the field lines.

1. Introduction

Electrons responsible for the discrete and inverted-V aurora are believed to be accelerated by potential difference parallel to the magnetic field lines (Swift, 1981a). The parallel potential differences can be viewed as a consequence of a process that creates potential differences perpendicular to field lines in the magnetosphere. The parallel potential differences occur because the ionosphere, which conducts perpendicular to field lines, gives rise to parallel currents. The parallel electric fields occur because of processes, such as double layers, anomalous resistivity, magnetic mirror forces, and particle inertia, which inhibit the free flow of electrons and ions parallel to the magnetic field.

In this paper we address the problem of creating potential differences perpendicular to the magnetic field. In particular we examine the possible role of electrostatic turbulence in generating potential variations. Recent observations of magnetospheric electric fields have shown broad-band electrostatic noise to be an important feature on field lines passing through the auroral zone. Gurnett and Frank [1977] have observed intense electrostatic turbulence on auroral field lines extending from just above the ionosphere to 40 Re in the magnetotail, with electric fields perpendicular to the magnetic field and extending in frequency from less than the local ion gyrofrequency up to frequencies the order of the electron gyrofrequency. The spectra show no special features at any of the plasma resonance fequencies, i.e. the gyrofrequencies, plasma frequencies or hybrid resonance. Aggson and Heppner [1977] have observed large transient, essentially D C, electric fields in the same region. Moreover, Mozer et al. [1977] have systematically observed electrostatic turbulence and large transient, shock-like electric field structures on auroral field lines at 1 Re altitude, and Temerin [1978] has identified some of the turbulence as zero frequency. This turbulence may be an important clue to processes responsible for the generation of the accelerating potentials.

The purpose of this paper is to examine properties of turbulence as it relates to generation of potential differences across field lines in the magnetosphere, and to identify mechanisms for the generation of the turbulence. Of interest is not only the magnitude of potential variation, but also the spatial distribution since that determines the structure of auroral forms.

The next section of this paper will describe the properties of two-dimensional, guiding center, or E x B, turbulence. This will relate to the generation of large scale, yet smaller than the scale size of the convection electric field, inverted-V precipitation regions. The following section describes a kinetic approach to the generation of turbulence. Turbulence generated on this much smaller scale can serve as an energy source for the E x B cascade and also generate potential differences over size scales consistent with the widths of discrete auroral arcs. The final section will assess our current understanding of auroral voltage generation mechanisms and suggest directions for future research.

2. Two-dimensional E x B turbulence

In the magnetosphere, where the conductivity parallel to the magnetic field is large, potential variations will be primarily perpendicular to the field. At time and distance scales greater than an ion gyroperiod and gyroradius respectively, the plasma moves according to the $\vec{E} \times \vec{B}$ drift. The charge density source of the electric potential can be shown [Swift, 1979a] to be governed by the equation

$$\frac{\partial \rho}{\partial t} + \vec{u}\cdot\nabla_{\perp}\rho = q - \frac{\partial j_z}{\partial z} \qquad (1$$

where $\vec{u}$ is the $\vec{E} \times \vec{B}$ drift given by

$$\vec{u} = -\frac{c}{B}\left(\vec{1}_x \frac{\partial \phi}{\partial y} - \vec{1}_y \frac{\partial \phi}{\partial x}\right) \qquad (2$$

where ϕ is the electric potential, B is the magnetic field intensity, q represents a microscopic source and j_z is a field-aligned current. The coordinates x and y are assumed perpendicular to the magnetic field--while z is parallel to the field. In the absence of sources or parallel currents, the charge density is proportional to the two-dimensional vorticity, ω given by

$$\omega = \frac{\partial}{\partial x} u_y - \frac{\partial}{\partial y} u_x \qquad (3$$

This can be seen by substituting (2) into (3) and making use of the Poisson equation for the potential, with the correspondence

$$\omega = -\frac{4\pi c}{B\varepsilon}\rho \qquad (4$$

when ε is the plasma dielectric constant given by

$$\varepsilon = 1 + \frac{\Pi_i^2}{\Omega^2} \qquad (4'$$

where Π_i and Ω are the ion plasma and gyro frequencies respectively.

The point is that a strongly magnetized plasma behaves as a perfect two-dimensional fluid, and the theory of two-dimensional turbulence [Kraichnan, 1975, 1976] may be applied.

The most convenient way to represent turbulent flow is in terms of the Fourier coefficients of the vorticity, $\omega_{mn}(t)$, which are related to the spatial representation of ω by

$$\omega(x,y,t) = \sum_{m,n} \omega_{mn}(t) \exp \frac{2\pi i}{D}(mx + ny) \qquad (5$$

where D is the size of a spatial domain. Since, in well-developed turbulence, ω and the fluid velocities are random variables, the turbulence may be characterized by an energy spectrum

$$E(k) = \frac{1}{2}\frac{\omega_{mn}\,\omega_{mn}^*}{k^2} \qquad (6$$

where $k^2 = (2\pi/D)^2 (m^2 + n^2)$ is the wave number and D is the size of a square box.

In an infinite domain, k becomes a continuous variable. In this case if there is an energy input to the turbulent cascade near wave number k_i, and an energy sink at large k, Kraichnan [1967] has shown that an equilibrium spectrum exists and that for $k > k_i$, the spectrum is given by

$$E(k) \sim k^{-8/3} \qquad (7$$

The spectrum is a result of the flow of energy toward ever larger wavelengths. This spectral distribution was subsequently verified by Fyfe et al. [1977] by use of numerical simulations in which they integrated in time the equations for ω_{mn} and found that the average distribution for $\omega_{mn}\omega_{mn}^*$ agreed rather well with the predictions.

In the case there is no dissipation, and there are upper and lower bounds in k, Kraichnan [1975] has shown that the requirement for the conservation of both energy and mean square vorticity, also referred to as enstrophy, leads to a spectrum of the type

$$E(k) = \frac{1}{\alpha + \beta k^2} \qquad (8$$

where the parameters α and β are determined from the initial spectral distribution. This spectral distribution was also confirmed by numerical simulations [Fyfe et al., 1977].

Figure 1 shows examples of equilibrium turbulent modal energy spectra for various input wave energy in which the allowed Fourier modes are in a 10 x 10 square box in Fourier space. The significant feature of the curves and the spectrum in (8) is that, except for energy injected at the shortest wavelengths, the energy per mode increases toward long wavelengths. This means that there are larger electric fields over longer wavelengths, which mean larger potential differences. Hence, through the turbulent cascade, turbulent energy injected as small electric field fluctuations over short distance can generate large potential differences.

This effect is illustrated in Figures 2a,b which show the results of a numerical simulation. Electric field fluctuations were injected at modes near a wave number of 6.5 in a 10 x 10 square box in Fourier space. The equilibrium spectrum would correspond to the $k_i = 6.5$ curve shown in Figure 1. Figure 2a shows potential contours corresponding to the $k_i = 6.5$ scale sizes. In this simulation energy is injected near this wavelength at a steady rate.

As the level of energy in the system increases, the rate of energy transfer among modes increases to the point where the modal spectrum assumes a distribution approximating an equilibrium spectrum. Figure (2b) shows a snapshot of equipotential contours taken at this time. Observe that even though energy is still being fed into the system near $k_i = 6.5$, the scale sizes of the potential contours are much larger. This illustrates a property of E x B turbulence, namely that the observed spectrum need not have any obvious relation to the wavelengths at which it is excited.

Given an energy source for excitation, the turbulent cascade serves as a voltage amplification mechanism. However, the turbulent cascade by itself is incapable of accounting for the structure of auroral forms, in particular the auroral arc structure in which one dimension is much longer than another. This is because the turbulent cascade has strong tendency toward directional isotropy. This can be seen from the mode coupling equations, which can be derived by substituting (5) with (1) with the terms on the right-hand-side deleted making use of (2), (3), and Poisson's equation. The Equation for the time evolution of each Fourier amplitude can be written

$$\frac{d}{dt}\omega_{\vec{k}} + \frac{2\pi c}{B\varepsilon}\sum_{\vec{k}=\vec{p}+\vec{q}} M(\vec{p},\vec{q})\,\omega_{\vec{p}}\,\omega_{\vec{q}} = 0 \qquad (9$$

where the coupling coefficient, $M(\vec{p},\vec{q})$ is given by

$$M(\vec{p},\vec{q}) = \frac{1}{2}(\vec{p} \times \vec{q})\left(\frac{1}{p^2} - \frac{1}{q^2}\right) \qquad (10$$

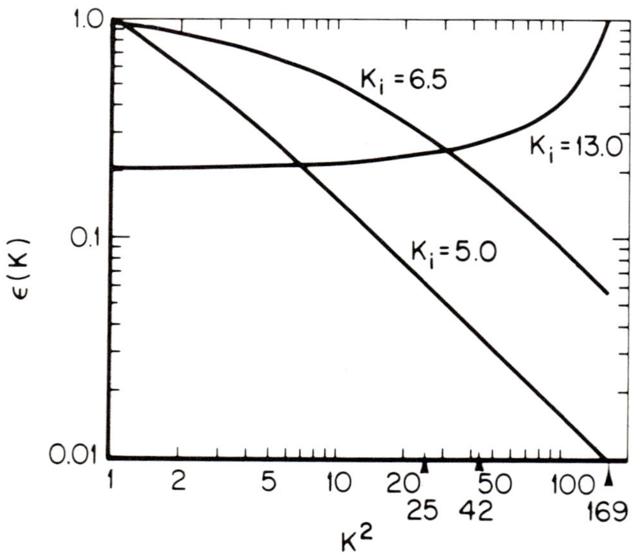

Figure 1. Plots of the expectation modal energy distribution for three input wave numbers. These curves pertain to the distribution of turbulent energy within a 10 x 10 box in Fourier space.

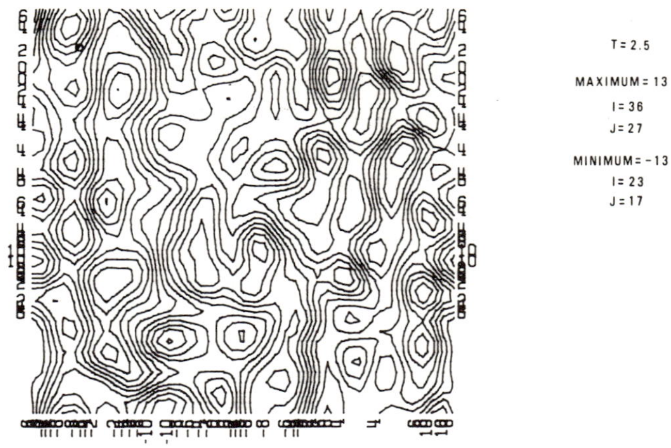

Figure 2a. Plot of equipotential contours computed with input wave numbers near $k_i = 6.5$ in a 10 x 10 Fourier space. Each contour reprints a potential difference of two units. The spatial size scale of the contours corresponds to the input wavelength.

Because of the cross product, there is no coupling among modes with parallel or anti-parallel wave numbers, but there is strong coupling among mutually perpendicular wave numbers.

Swift [1979a] has suggested that the arc-like structure may be a consequence of the fact that the V-shocks where electrons are accelerated which are likely responsible for the discrete auroral arc [Swift et al., 1976] serve to limit field-aligned currents. Since the width of the shocks are confined to be slightly less than the

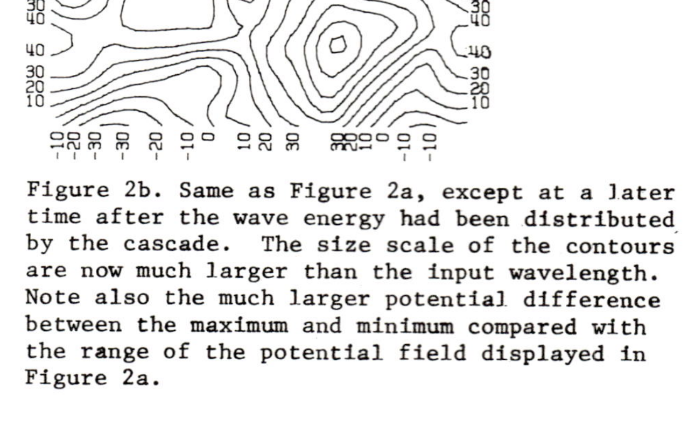

Figure 2b. Same as Figure 2a, except at a later time after the wave energy had been distributed by the cascade. The size scale of the contours are now much larger than the input wavelength. Note also the much larger potential difference between the maximum and minimum compared with the range of the potential field displayed in Figure 2a.

gyrodiameter of an ion whose energy is equal to the potential change across the shock [Swift, 1979], very severe limits are placed on the shock thickness. Hence the discrete auroral arc structure may be more a consequence of the dynamics of the interaction between the ionosphere and the magnetosphere than the result of a process taking place largely within the magnetosphere, because certain modes with scale sizes which match the scale sizes of the V-shock are less attenuated by ionospheric dissipation. Inverted-V electron precipitation which appears to have a much larger scale size than the discrete arc would likely be a consequence of the larger size potential variations.

3. Generation of Turbulence on a Kinetic Scale

The E x B turbulent cascade, as applied to the magnetosphere, is incomplete in that no energy source has been specified. A possible source of energy may be in the free energy in certain types of ion velocity distribution functions. Wave energy could be generated as a result of a plasma instability at wavelengths the order of an ion gyrodiameter and smaller. This is considerably smaller than the range of validity for the E x B cascade. In this section we wish to examine processes for the generations of turbulent wave energy from plasma microinstabilities, in the form of electrostatic waves with wave normals nearly perpendicular to the magnetic field.

There are two ion instabilities involving ions that could lead to perpendicularly propagating waves. One is the flute-mode instability driven by a ring-type velocity distribution. However, there is no obvious mechanism for creating such a distribution in the magnetosphere.

The other is the two-stream instability excited by ions streaming perpendicular to the magnetic field. A means whereby ions can be caused to stream across field lines in the magnetosphere is illustrated in Figure 3. This shows magnetic field lines in the region of the neutral sheet in the noon-midnight meridian plane. A dawn-to-dusk convection electric field, E_y, directed perpendicular to the noon-midnight meridian plane can be eliminated by transforming to an inertial frame moving earthward at $u = c\, E_y/B_z$, where B_z is the z component of the magnetic field. In this inertial frame, a low-energy ion will be seen streaming tailward along the magnetic field lines at approximately this velocity. The trajectory of an ion in this inertial frame is also illustrated in the Figure 3. It can be seen that if the field line curvature is large enough, an ion will not follow the field lines but will instead stream across them. Electrons on the other hand, because of their very low mass will follow the field lines, resulting in a situation that can give rise to the modified two-stream instability [Krall and Liewer, 1971].

Ions streaming perpendicular to the magnetic field through another population of ions can also give rise to the simple two-stream instability. Colliding ion beams would occur if there were ions incident from below the neutral sheet in a region where the two beams met. In a portion of the interaction region ions that have passed through the neutral sheet would stream with a large component perpendicular to the field line through ions that had not yet encountered the neutral sheet.

In either case, the criterion for the excitation of the two instabilities is that the streaming speed exceed the ion themal velocity, a condition easily met in the magnetotail. For a cross-tail electric field, E_y, of 1.5 mv/m and a B_z of 3 γ, u = 500 km/sec, corresponding to an energy of 1.3 kev for protons. Thermal energies would be the order of 100 eV for ions of magnetosheath origin or a few eV for ions of ionospheric origin.

The modified two-stream instability involves electron motion parallel to the field lines. The effects of colliding ion beams can be treated within the context of the two-dimensional model used to simulate the instability generated by the ring-type velocity distribution. Such a model would, of course, ignore the fact that over part of the region where the ion beams interact, there will be a significant component of the relative motion of two beams parallel to the magnetic field, which would excite waves with a component of the wave normal parallel to the magnetic field. The model also ignores the fact that in certain regions the plasma energy densities are likely to be larger than the energy density of the containing magnetic field, which suggests that important electromagnetic effects are being ignored. However, appeal is made to the fact that the waves observed by Aggson and Heppner [1977] and Gurnett and Frank [1977] are primarily electrostatic and the fact

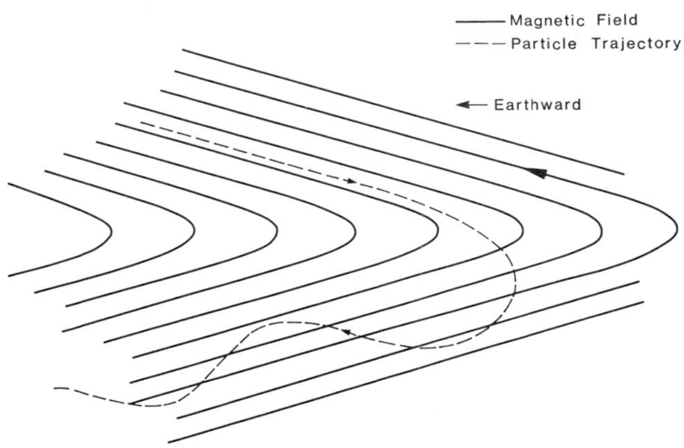

Figure 3. An illustration showing the motion of an ion crossing field lines in passing through the neutral sheet region of the magnetotail. This is shown in the inertial frame where the dawn-to-dusk electric field vanishes.

that we are seeking a source of electrostatic energy to drive the turbulent cascade.

Swift [1981b] used a two-dimensional numerical model to simulate the effect of colliding ion beams in a plane perpendicular to the magnetic field. The simulations were run on a 128 x 128 point grid, with 102400 ions. The electrons were assumed to form a neutralizing background. Electron dynamics could be ignored in this model because, due to their low mass, electrons can be assumed to move only according to the E x B drift. Since this drift is divergenceless, an initially uniform electron density will always remain uniform.

In this ion beam model, one-half the population was assumed to be accelerated perpendicular to the magnetic field by a centrifugal force, F_g, that varies with time in a way that would be seen by an observer moving at constant speed along a magnetic field line with a neutral sheet configuration like that shown in Figure 3. In this model field, the x component varies as a hyperbolic tangent with a thickness of 0.2 Re with an asymptotic value of 25 γ. The z component is assumed constant at 3 γ. The speed of an observer is 300 km/sec, which is the speed the magnetic field would translate in a 1 mv/m cross-tail, convection, electric field, E_y. The intensity of the magnetic field varies in a way consistent with this model. This means that the centrifugal force goes through a sharp maximum at the time the magnetic field is a minimum, so that the ions become partially demagnetized during this acceleration process. The changing magnetic field induces a solenoidal electric field, the E x B drift of which causes an expansion and contraction of the simulation domain. The details of the model and method of calculation are described in Swift [1981b].

The results of this model simulation are summarized by the modal electric field energy spectra shown in Figure 4. The simulation parameters at times corresponding to the spectral plots are listed in Table 1. The spacing between the grid points along the x axis is variable in time because of the solenoidal electric field, but the grid-point spacing in the y direction is constant. The wave number scale in Figure 4 is referenced to the grid point spacing in the y direction, but a wave number assigned to a wave refers to the same physical wavelength irrespective of the direction of the wave normal. The simulation domain and the grid point lattice are perfect squares at $\Omega = 1$; when $\Omega < 1$ the domain is a rectangle elongated in the y direction; and when $\Omega > 1$, it is for-shortened in this direction.

The peak in the acceleration pulse, and the minimum in Ω, occur at t = 5.7. The particle acceleration process is essentially complete at t = 9.6, and subsequent kinetic energy increase is due to the increase in magnetic field intensity and contraction of the domain.

The wavelengths where the instability is

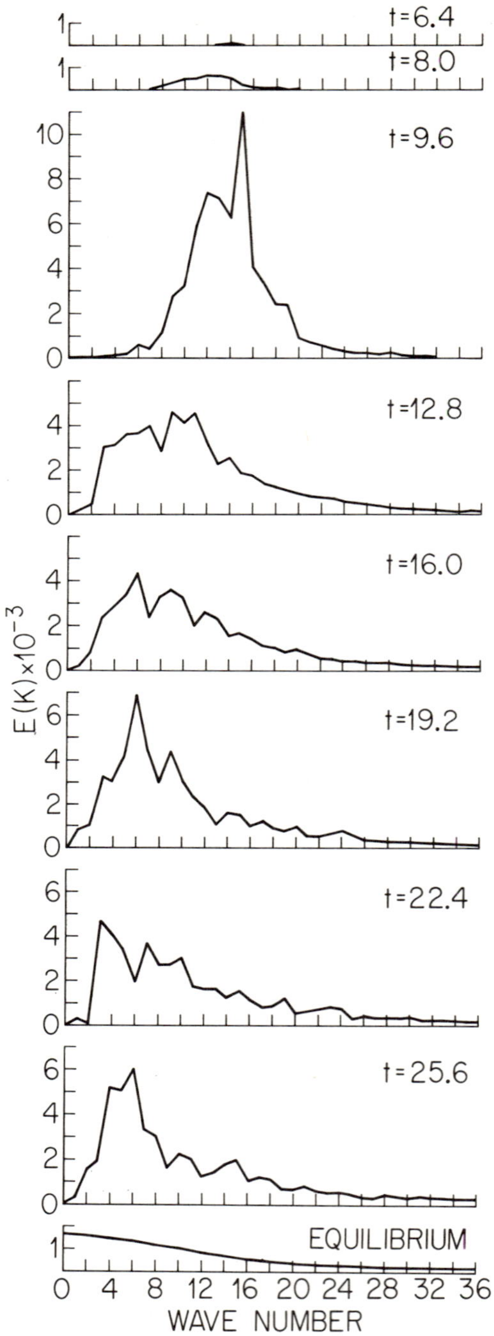

Figure 4. The modal energy spectrum excited by partially magnetized ion beams. The wave number scale is such that a wave tumber of L corresponds to a wavelength of 128 grid points. The relationship between wave numbers and particle gyrodiameter is listed in Table 1. Shown in the bottom panel is the asymptotic equilibrium spectrum of a plasma with an average kinetic energy equal to the average particle energy at t = 25.6.

Table 1.

Time	Ω	Π_i	$\bar{v}$	2ρ	n_d
6.4	0.44	3.32	5.5	25.0	5.0
8.0	0.72	4.24	8.6	24.1	5.3
9.6	0.93	4.82	10.3	27.1	5.8
12.8	1.25	5.59	11.6	18.5	6.9
16.0	1.49	6.10	12.6	16.9	7.6
19.2	1.69	6.50	13.5	16.0	8.0
22.4	1.84	6.78	14.1	15.3	8.4
25.6	1.98	7.04	14.5	14.7	8.7
∞	--	7.04	14.5	--	--

Parameters at times corresponding to spectra plotted in Figure 4. Ω is the gyrofrequency and Π_i is the ion plasma frequency in simulation time units. $\bar{v}$ is the average particle velocity, 2ρ is the particle gyrodiameter, and n_d is the wave number corresponding to a gyrodiameter.

excited are considerably shorter than an ion gyrodiameter. The maximum field energy occurs near t = 9.6. By this time the beam and background plasma have merged, and there is little free energy left in the velocity distribution. Following this time there is a rapid decay toward equilibrium levels at wavelengths shorter than a gyrodiameter, and nonlinear transfer of energy to long wavelengths. By t = 25.6, the part of the spectrum above equilibrium levels is almost entirely at wavelengths longer than an ion gyrodiameter. This seems to provide a clear demonstration that a beam-driven microinstabilty can feed energy into the E x B turbulent cascade.

Although the electric field energy at the long wavelengths is not large in comparison to the total electric field energy at short wavelengths, it should be remembered that the plasma has a finite dielectric constant, which at long wavelengths and low frequencies is given by (4'). Therefore the total wave energy, which is the sum of the vacuum electric field energy and the kinetic energy in E x B drift is a factor of ε larger than the electric field energy. At t = 25.6, this factor is $\varepsilon = 13.5$.

Another significant feature of the simulations is that the turbulence, at both small and large wavelengths, appears to be directionally isotropic, except in the early stages at t = 6.4 and 8.0. Early in the simulation, the wave normals show a preference for the x direction, consistent with the beam direction. However, by t = 9.6 the contour plots show the spectrum to be isotropic, in spite of the anisotropic excitation and the anisotropic contraction of the simulation domain. Another significant feature is that the maximum voltage differences generated occurred at short wavelengths, at t = 9.6, where the voltage differences were comparable to the kinetic energy of the beam particles.

4. Summary and Conclusions

Hydrodynamical and kinetic models have been used to investigate processes for the generation of turbulence and the transfer of turbulent energy from shorter to longer wavelengths. The motivation for this study has been to understand the processes for the generation of potential differences needed to accelerate auroral electrons. This section presents an assessment of our understanding of these processes in terms of the observed characteristics of the aurora.

The simulations show that electric field fluctuations at all size scales can be excited by microinstabilities. The simulations also show a strong tendency toward directional isotropy, which is inconsistent with the observed structure of auroral forms. The arc structure may be a consequence of the interaction between the ionosphere and magnetosphere which results in the formation of V-shock structures with very definite size scale limitations, but this would not explain the preferential east-west alignment of auroral forms. The east-west alignment is consistent with the excitation of some large amplitude electrostatic wave with wave normals perpendicular to surfaces of L-shells. This is the direction in which the ion beam model would excite waves. The ion beam model can also generate potential variations of the magnitude required for the acceleration of auroral electrons. However, from Figure 4 and Table 1, the wavelengths excited are about a factor of two smaller than the ion gyrodiameter, so it is probable that the potentials would map down into ionospheric levels at size scales which may be too small to be consistent with the thickness even of the discrete auroral arcs.

There are also auroral forms which are the order of tens of kilometers across. These have been identified with inverted-V precipitation events. Such scale sizes can be generated by the turbulent cascade, but the preferred east-west alignment of these forms is even more troubling; because the cascade process leads so strongly to directional isotropy.

It therefore does not appear that two-dimensional models are by themselves adequate to account for auroral structure even though they seem capable of generating the required acceleration voltages and size scales. The most likely instability for the generation of wave energy is not the colliding ion beam, but the modified two-stream instability, involving motion of electrons parallel to the field lines. The next obvious step toward understanding the aurora is to examine the nonlinear effects of this instability in a two-dimensional simulation in a plane containing a magnetic field of the type shown in Figure 3. A simulation of this type would allow us to model a continuous process, rather than a one-time event. Ions and electrons could continuously stream into the simulation from one or both sides. The instabilities would be initiated where the field line curvature is large, and the propagation to other parts of the domain where the curvature is small could be observed.

Such a model would naturally give rise to potential variations that would be consistent with the east-west alignment of multiple auroral arcs. Because of the close similarities between the dispersion relations for the ion beam and modified two-stream instabilities [Swift, 1981b], the excited wavelengths perpendicular to the field line would be similar to the wavelengths excited in the ion beam instability shown in Figure 4 in the t = 9.6 and earlier time frames. The parallel wavelength would be the order of the square root of the ion-to-electron mass ratio times the perpendicular wavelength [Krall and Liewer, 1971]. The perpendicular wavelengths projected to ionospheric heights would again likely be smaller than observed widths of auroral arcs, and would certainly be smaller than the widths of inverted-V aurora.

Clearly, a process for cascading of wave energy to longer wavelengths is necessary. In the simulations involving ion motion perpendicular to field lines, the presence of two degrees of freedom perpendicular to the field lines is necessary for the cascade of energy to long wavelengths. This is evident in the mode-coupling matrix of (10). The one-dimensional particle simulation of Byers and Grewal [1970] of instabilities generated by ring-type distribution and the ion beam simulation of Papadopoulos et al. [1971] indicate no transfer of energy to long wavelengths. These results suggest that in the case of the modified two-stream instability that transfer of energy to long wavelengths might also require two degrees of freedom perpendicular to the magnetic field. This would be necessary to break the invariance of the y component of cannonical momentum. This suggests that a definitive investigation of the effects of the modified two-stream instability would require a three-dimensional simulation. Although energy transfer to longer wavelengths would be a desirable outcome of such simulations, the rapid trend toward directional isotropy in direction perpendicular to the magnetic field, as exhibited by the two-dimensional ion simulation, would certainly not be a desirable outcome.

The two-dimensional ion simulation has demonstrated a process for generating auroral acceleration voltages and of spreading wave energy to long wavelengths, but the extreme directional isotropy exhibited by the ion simulations is incompatible with observed auroral structure. A promising mechanism for generating the required auroral accelerating potentials is the modified two-stream instability in which partially demagnetized ions stream across strongly magnetized electrons. However, much work remains to be done in demonstrating its role in generating auroral potentials. A compelling feature of this instability is that it is a mechanism for generating auroral potentials wherever magnetic field line curvature is large and there is a convection electric field perpendicular to the plane containing the magnetic field such that the E x B drift is toward the neutral plane. This would occur in the neighborhood of any region in which there is magnetic merging.

Acknowledgements. This work was supported by the Atmospheric Science Section of the National Science Foundation under Grant ATM - 7923614.

References

Aggson, T. L. and J. P. Heppner, Observations of large transient magnetospheric electric fields, J. Geophys. Res., 82, 5155, 1977.

Byers, J. A. and M. Grewal, Perpendicularly propagating plasma cyclotron instabilities simulated with a one-dimensional computer model, Phys. Fluids, 13, 1819, 1970.

Fyfe, D., D. Montgomery, and G. Joyce, Dissipative and forced turbulence in two-dimensional magnetohydrodynamics, J. Plasma Phys., 17, 369, 1977.

Gurnett, D. A. and L. A. Frank, A region of intense plasma wave turbulence on auroral field lines, J. Geophy. Res., 82, 1301, 1977.

Kraichnan, R. H., Inertial ranges in two-dimensional turbulence, Phys. Fluids, 10, 1417, 1967.

Kraichnan, R. H., Statistical dynamics of two-dimensional flow, J. Fluid Mech., 67, 155, 1975.

Krall, N. A., and C. D. Liewer, Low-frequency instabilities in magnetic pulses, Phys. Rev., A4, 2049, 1971.

Mozer, F. S., C. W. Carlson, M. K. Hudson, R. B. Torbert, B. Parady, J. Yatteau, and M. C. Kelley, Observations of paired electrostatic shocks in the polar magnetosphere, Phys. Rev. Lett., 38, 292, 1977.

Papadopoulos, K., R. C. Davidson, J. M. Dawson, F. Haber, D. A. Hammer, N. A. Krall, and R. Shanny, Heating of counterstreaming ion beams in an external magnetic field, Phys. Fluids, 14, 849, 1971.

Seyler, C. E., Jr., Y. Salu, D. Montgomery, and G. Knorr, Two-dimensional turbulence in inviscid fluids or guiding center plasmas, Phys. Fluids 18, 803, 1975.

Swift, D. W., An equipotential model for auroral arcs 2. Numerical solutions, J. Geophys. Res., 80, 2096, 1976.

Swift, D. W., On the structure of auroral arcs: The results of numerical simulations, J. Geophys. Res., 84, 469, 1979a.

Swift, D. W., An equipotential model for auroral arcs: The theory of two-dimensional laminar electrostatic shocks, J. Geophys. Res., 84, 6427-6434, 1979b.

Swift, D. W., Mechanism for auroral precipitation--A review, Rev. Geophys. Space Phys., [to be published] 1981a.
Swift, D. W., Numerical simulation of the generation of electrostatic turbulence in the magnetotail, J. Geophys. Res., 86 [to be published] 1981b.
Swift, D. W., H. C. Stenbaek-Nielsen, and T. J. Hallinan, An equipotential model for auroral arcs, J. Geophys. Res., 81, 3931, 1976.
Temerin, M., The polarization, frequency and wavelengths of high-latitude turbulence, J. Geophys. Res., 83, 2609, 1978.

ANOMALOUS DC RESISTIVITY AND DOUBLE LAYERS IN THE AURORAL IONOSPHERE

J. M. Kindel, Chris Barnes, and D. W. Forslund

Los Alamos National Laboratory

Abstract. There are at least four candidate instabilities which might account for anomalous dc resistivity in the auroral ionosphere. These are: the ion-acoustic instability, the Buneman instability, the ion-cyclotron instability and double layers. We report here results of computer simulations of these four instabilities which suggest that double layers are most likely to be responsible for resistivity in the auroral zone.

Introduction

The polar magnetosphere-ionosphere interface is characterized by field-aligned currents connecting regions of plasma with vastly different particle density and energy. Observational and theoretical studies indicate that parallel and perpendicular electric fields exist in the region.[1] It would not be surprising to find micro-instabilities there which grow to significant amplitude and lead to resistivity and particle acceleration. Although the "boundary" conditions in this region are not well known, particularly particle distribution functions, it would seem useful to study these micro-instabilities in a simplified, but more controllable form. If their behavior and effect under these conditions are known, then additional complexity and reality can be built into the model with some hope of producing results which can be compared with experimental observations.

The contents of this paper are therefore a summary of our work in simulating several instabilities in simple configurations. From these results, one can then say something about what physics might still be missing from a simulation model which explains physical observations. In a sense, we are presenting only a progress report in a larger ongoing program aimed at matching simulation with the physical world.

We are concerned with electrostatic instabilities and waves which appear in a current carrying plasma with a possible field aligned electric field in which both particle species initially have Maxwellian distribution functions. When $v_d < v_e$ and $T_i \ll T_e$, we expect ion acoustic waves to grow and produce d-c resistivity. As v_d exceeds the electron thermal speed, the Buneman instability will arise for all ion temperatures.[2] In the presence of a strong magnetic field, $\omega_{ce} \simeq \omega_{pe}$, ion-cyclotron waves will grow.[3] There is also the possibility of double layers, either with or without a magnetic field present.[4] We find that significant resistivity occurs as a result of ion-acoustic and Buneman instability and of double layers, but not ion-cyclotron waves directly. The former effects give rise to large changes in electron and ion distribution functions. Figure 1 shows the regimes of these instabilities as determined by the results of computer experiments in our simplified configurations (uniform, periodic).

The Model

A standard electrostatic particle code was used throughout. An arbitrary strength, uniform magnetic field could be applied in any direction relative to the coordinate axes. The bulk of the calculations were done with periodic boundary conditions on the fields and particles. However, some runs were done in a convective system where the potential was held fixed at two boundaries and particles striking these walls were collected. New particles could then be injected at the boundaries with any desired distribution function. In this configuration, heat could be continually drained from the system by particle convection. A fixed or time-varying potential could be placed across the boundaries of this system to accelerate electrons or maintain a constant current. In the periodic system, a uniform electric field could also be applied.

Ion-Acoustic Turbulence

It is well known that electrons drifting through an ion background will excite ion-acoustic waves which, if the ions are cold enough, will grow to large amplitude. The threshold drift velocity for growth is predicted from theory to be only several times the ion sound speed $v_e/\sqrt{M_i/m_e}$. It has been found from simulations that drift velocities must exceed roughly $v_e/3$ to produce a significant level of

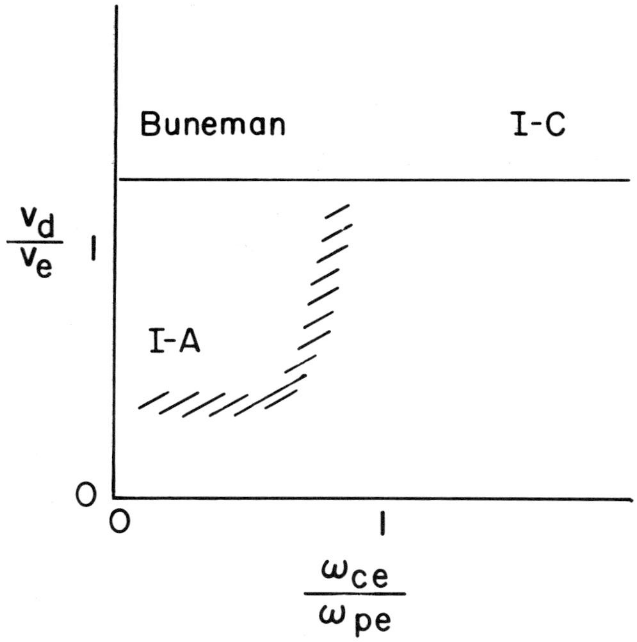

Fig. 1. Regions at which various instabilities will be found as a function of v_d/v_e, the electron drift and ω_{ce}/ω_{pe}, the magnetic field strength.

ion wave fluctuations. When $v_d > v_e/3$, modification of the electron distribution function is observed as is an enhanced level of dc resistivity. The ion-acoustic waves produced in such a system are characterized by a large angular spectrum of modes. Although each mode saturates at a fairly low level, taken as whole they can produce ion fluctuation levels of ≈30-40%, particularly for drift velocities exceeding $v_e/2$. The drifting electrons then scatter off these ion-fluctuations, leading to an anomalously high effective collision frequency in the several x 10^{-3} ω_{pe} range. The effect on the electron distribution function is to first flatten it, then skew it towards low velocities so that the peak appears very near v = 0. The high velocity tail then runs away (in the presence of an externally applied field) at nearly the free acceleration rate since fast electrons do not match well the parallel components of the ion wave phase velocity. Figure 2` is a plot of a typical electron distribution function showing the skewing of the peak and the runaway tail due to the presence of ion-acoustic turbulence.

The wide angular spread of the ion modes suggests that the level of ion-acoustic turbulence in a one-dimensional system would be much smaller since only one k-directional is allowed , producing a low level of resistivity. One-dimensional simulations show this to be true, giving one to two orders of magnitude lower resistivity than is found in two-dimensional systems.

In a magnetized system with a uniform magnetic field parallel to the direction of drift, the level of ion waves excited is reduced, more so as the field increases. For instance when $\omega_{ce} = \omega_{pe}$, the wave amplitude is less than half of the equivalent unmagnetized value. The anomalous resistivity resulting from the ion turbulence decreases rapidly as magnetic field strength increases. Figure 3 shows the variation of measured resistivity in four 2-d runs with $v_d = v_e$ in which the magnetic field increases from $\omega_{ce} = 0$ to $\omega_{ce} = 2\omega_{pe}$. The resistivity varies over almost an order of magnitude in these four runs.

Buneman Instability

When the electron drift exceeds roughly 1.3 v_e, the Buneman instability takes over. This instability is a hydrodynamic type instability which is almost insensitive to ion temperature and magnetic field strength. It leads to high resistivity and rapid electron heating. Because the unstable waves are bunched around the drift direction, a one-dimensional or magnetized system will exhibit this instability as strongly as a multi-dimensional or unmagnetized system.

This instability effectively places an upper limit on the relative electron-ion drift in a Maxwellian plasma. Because resistivity rapidly increases with drift, v_d/v_e is effectively limited to the order of 1.5.

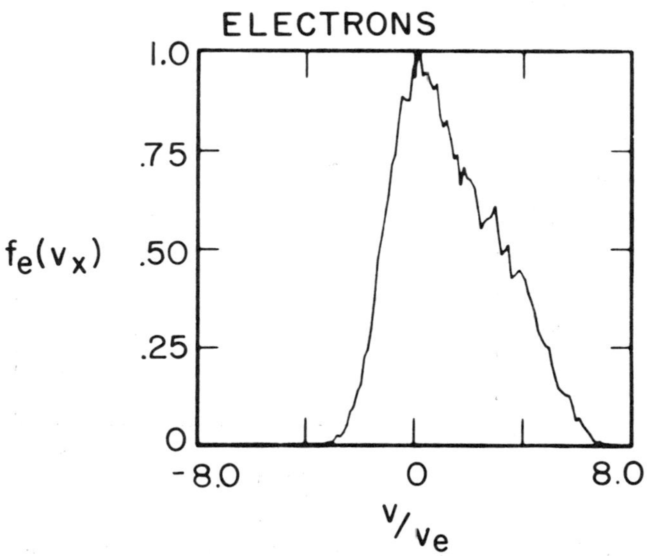

Fig. 2. The effect of ion-acoustic resistivity on the electron distribution function. Initially a drifting Maxwellian, the peak has skewed towards zero velocity while the tail has extended, indicating runaway electrons.

Ion-Cyclotron Waves

Because of the presence of a magnetic field and the likelihood that the ion temperature equals or exceeds the electron temperature in the magnetosphere, ion-acoustic waves may not provide the principal resistivity mechanism there. It has been shown that ion cyclotron waves are excited by relative electron-ion drift in a magnetized plasma. The question is whether these waves will lead to large enough ion density variations to produce resistivity.

We have found that in order to excite large amplitude ion-cyclotron waves in a uniform plasma, electron drift must exceed the Buneman limit. Not surprisingly, Buneman resistivity overshadows any due to ion-cyclotron waves. In a run in which $v_d = 2v_e$ initially and in which subsequently the current is kept constant by the application of an external electric field, the Buneman instability grows rapidly to saturation producing a peak in resistivity. Figure 4a shows a time history of electrostatic field energy in this run. The rapid growth of the Buneman instability can be seen. There is rapid electron heating so that v_d/v_e declines, quenching the instability. Figure 4b shows the parallel electron temperature history for the same run. Meanwhile, ion-cyclotron waves are growing to large amplitude. They persist throughout the remainder of the run, although they seem to be weakly driven. This is apparently the case since v_d/v_e is now only of order unity. The modes are nearly perpendicular to the magnetic field and occur as the first three or four harmonics of the ion cyclotron frequency. The resistivity declines late in the run an order of magnitude from its peak value. Even this value is possibly attributable to residual Buneman induced resistivity. Figure 5 shows a plot of a potential surface taken during the Buneman instability phase of this run. Electron drift and magnetic field direction is nearly perpendicular to the horizontal axis. Figure 6 shows a similar plot later in the run when the Buneman waves have died down and ion-cyclotron waves are fully developed. The model was arranged to favor an axis aligned ion-cyclotron mode.

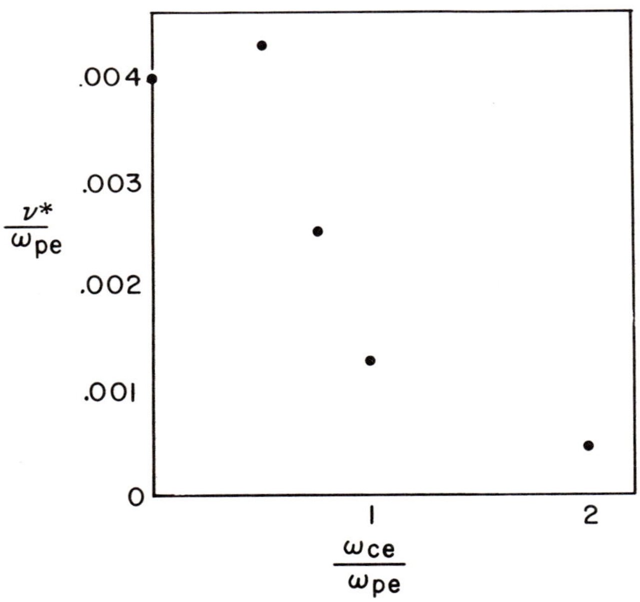

Fig. 3. The effect of magnetic field strength on ion-acoustic resistivity. Shown is a plot of effective collision frequency normalized to ω_{pe} as a function of ω_{ce}/ω_{pe}. In all runs, the electron drift was maintained at $v_d = v_e$. The mass ratio was 100 and T_i/T_e was 0.05.

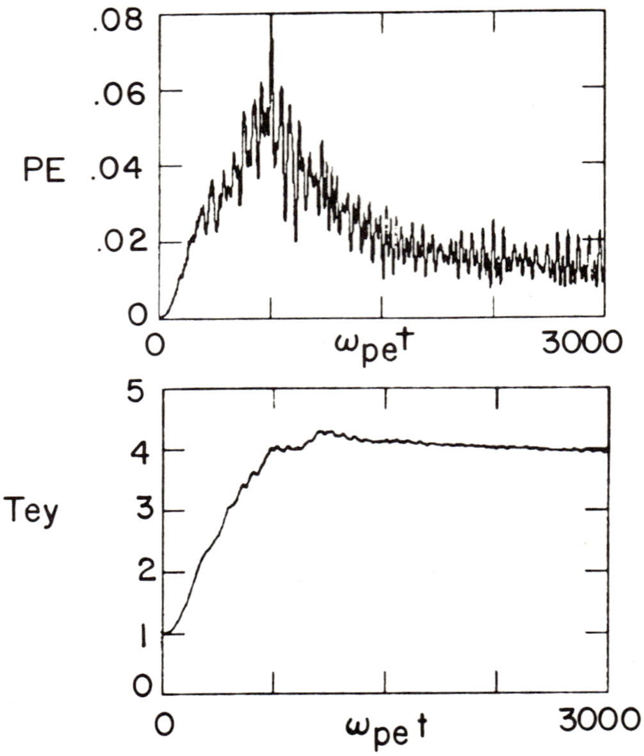

Fig. 4. A two-dimensional simulation of ion-cyclotron wave excitation in which $v_d = 2v_e$, $\omega_{ce} = 2\omega_{pe}$ and $T_i = T_e$. The mass ratio was 1000. (a) The field energy (normalized to electron thermal energy) as a function of time with $\omega_p dt = 0.3$. The Buneman instability rapidly saturates and heats electrons. The heated electrons reduce v_d/v_e which largely kills the instability. Later, ion-cyclotron waves grow to large amplitude. (b) A time history plot of parallel electron temperature. One can readily see the heating effect of the Buneman instability. The perpendicular electron temperature slowly increases by 40% during the run as parallel thermal energy is isotropized.

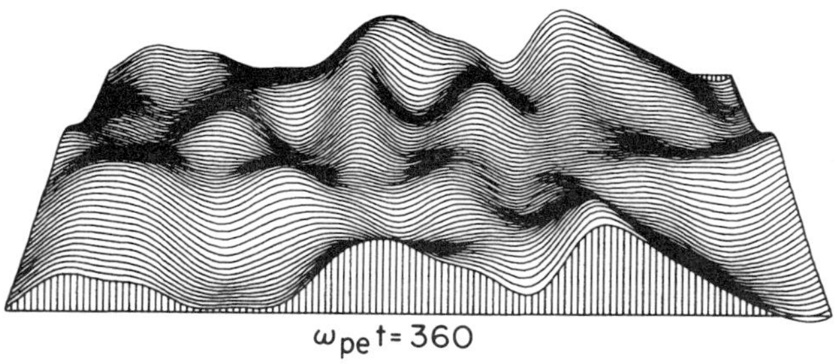

Fig. 5. A potential surface plot of the run of Fig. 4 during the early time when Buneman waves are strongest. Electron drift and magnetic field are away from the viewer but at a slight angle (5°).

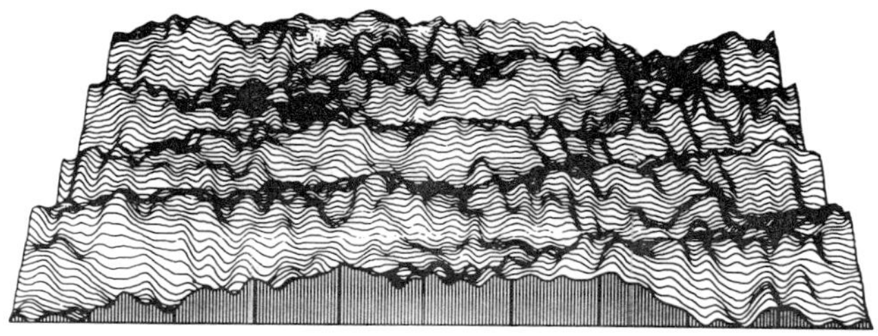

Fig. 7. A potential surface plot for a run in which an angular spectrum of ion-cyclotron waves is allowed. Several modes are present but nearly perpendicular to the magnetic field which is along the horizontal axis here.

Figure 7 shows a potential surface plot for a similar run in which the drift and magnetic field were aligned with the horizental axis. Ion-cyclotron waves can be seen propagating obliquely to the horizontal axis. The limited angular spectrum can be inferred from this plot.

In order to assess the level of resistivity due to ion-cyclotron waves alone, we performed a series of one-dimensional calculations in which the mode vector was at a large angle (85°) from the magnetic field and drift direction. This prevents the Buneman instability from growing while allowing ion-cyclotron modes to grow freely. Drift velocities between 1.5 v_e and 3 v_e were used. Anomalous collision frequencies bebetween 10^{-5} and 10^{-4} ω_{pe} were measured. Even this must be taken as an upper limit since there was some evidence of off angle Buneman waves. Ion density variations of order 10% were seen. Figure 8 shows a plot of anomalous collision frequency as a function of drift velocity measured in these one-dimensional runs.

The feeble resistivity observed must at least in part be due to the restricted mode spectrum characteristic of ion-cyclotron waves in this system. Unlike ion-acoustic waves, ion-cyclotron waves do not have a large angular spectrum. The waves do not steepen in saturation as ion-acoustic waves do, producing a smaller density variation.

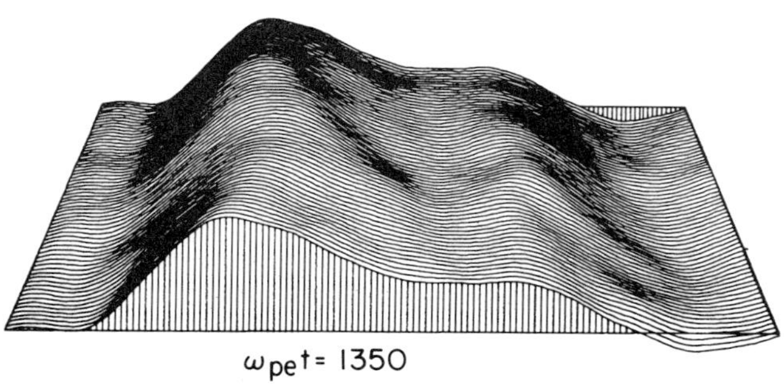

Fig. 6. A similar potential surface later in the run when ion-cyclotron waves have reached large amplitude.

Double Layers

Double layers are non-neutral regions in a current carrying plasma across which there can be a large potential jump. Electrons are accelerated across this jump and downstream are thermalized by electron-electron two stream turbulence, providing a mechanism for converting $\underline{E} \cdot \underline{j}$ work into thermal energy. Double layers are formed and driven by the drifting electron component at small negative charge perturbations. As shown by Hasegawa and Sato,[5] there is a self consistent solution to Poisson's equation and the Vlasov equation in a non-drifting plasma in which a negative potential well exists of depth less than the electron thermal potential. If now the electrons are given a drift across such a local structure, reflected electrons upstream will have greater density than trapped electrons downstream causing a net potential jump across the structure. This double layer is driven by

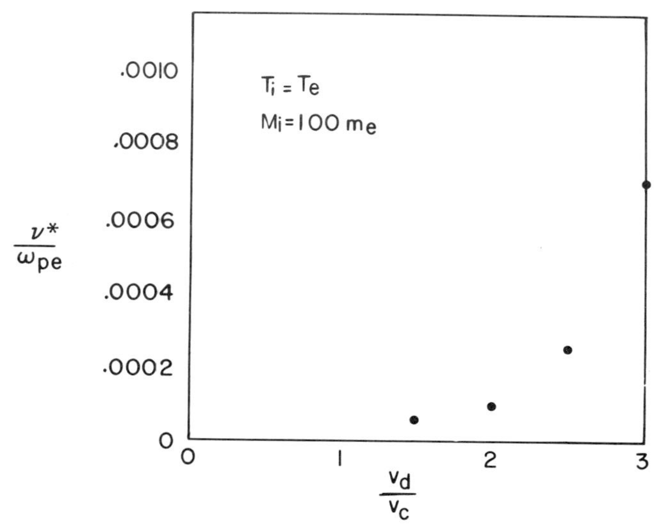

Fig. 8. Anomalous collison frequency in several 1-d runs due to ion-cyclotron waves.

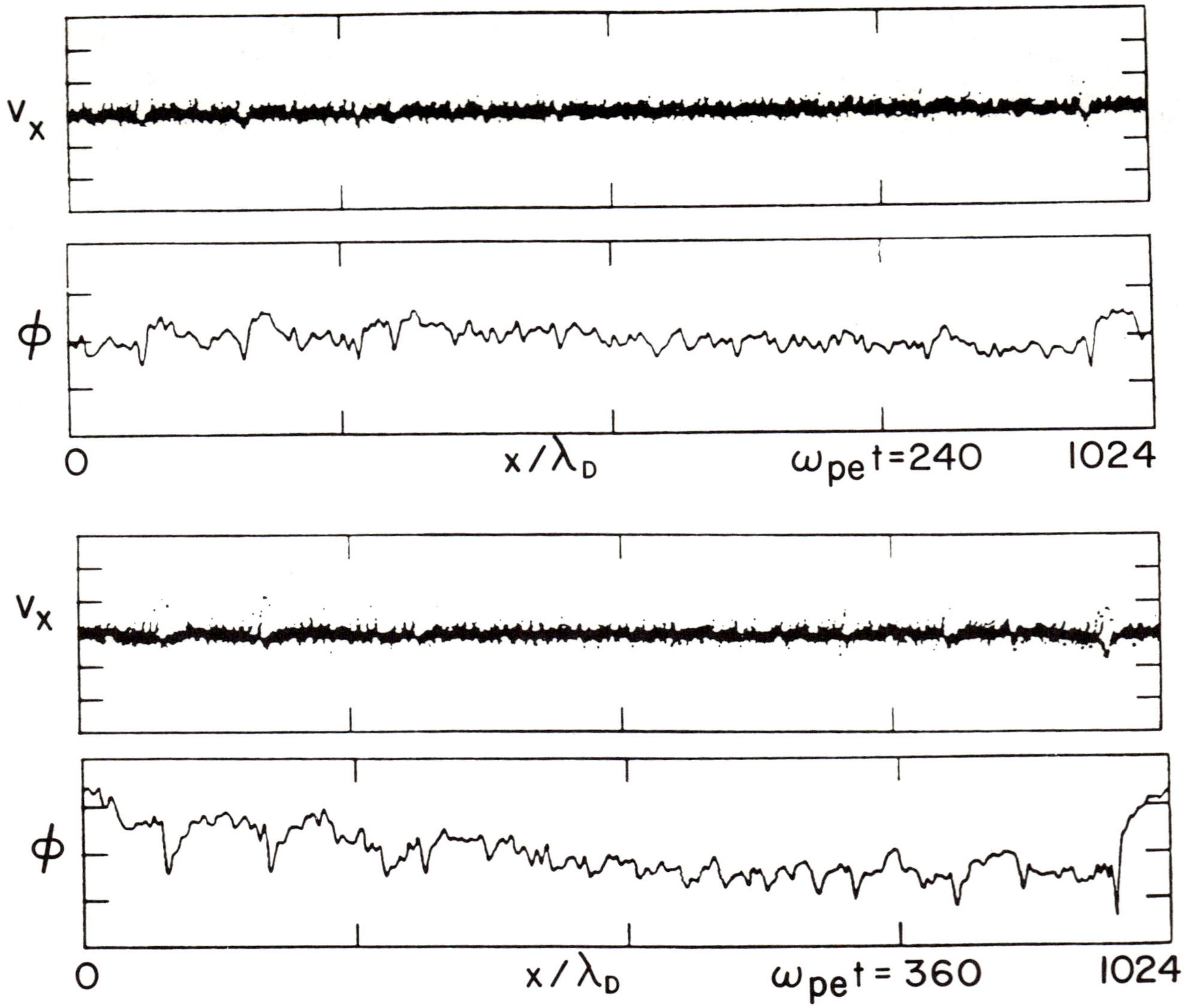

Fig. 9. Results of 1-d spatially periodic simulations of current driven double layers. Electron drift is to the right. (a) Ion phase space and electrostatic potential at time $\omega_{pe}t = 240$, (b) Similar plots at $\omega_{pe}t = 360$. One can easily see a potential dip followed by a jump near the right hand side.

electron drift energy. In one-dimensional simulations it is seen that a small negative potential disturbance will form moving in the direction of electron drift at nearly the sound speed. As it grows in amplitude, it decelerates. It becomes stationary or nearly stationary as it reaches maximum amplitude at which time there is a fully developed hole in ion phase space. As it decays into an ion-acoustic solitary wave, it again picks up speed, usually in the direction opposite to the electron drift. It then disappears as it reaches the ion sound speed. Its lifetime as a nearly stationary double layer can vary considerably, up to many hundreds of plasma periods.

The above described double layer is a current driven instability in which the potential jump cannot greatly exceed the electron thermal energy, assuming that the drift of incoming electrons will be limited to roughly their thermal speed by ion-acoustic or Buneman instabilities. The ion phase space has a hole in it caused by the potential dip and ions are trapped because downstream ions accelerated upstream across the jump do not gain sufficient energy to free stream upstream once they have passed the potential dip. Figure 9 shows an example of double layers formed in one-dimensional computer simulations. Shown are plots of ion phase space and time averaged electrostatic potential for

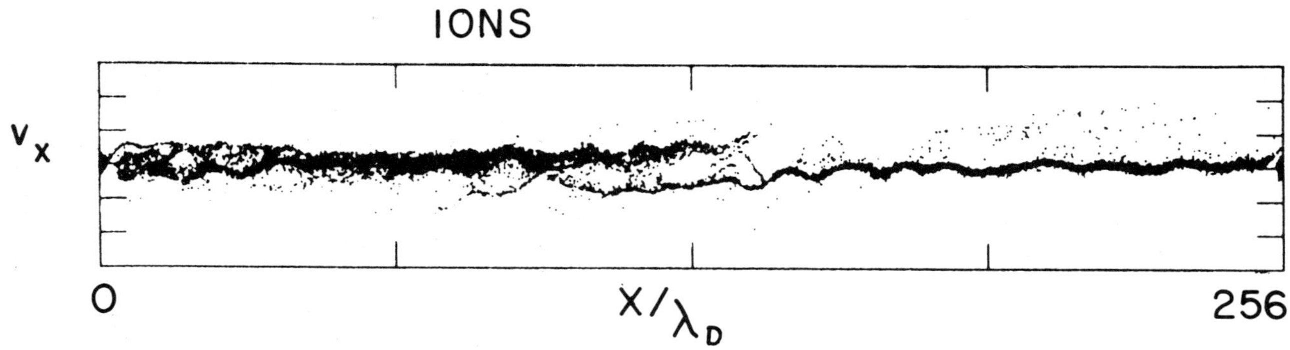

Fig. 10. Ion phase space for a field driven double layer. Notice how the ions are not trapped but stream freely to the left (upstream).

several instants of time. One can see the negative potential dip in front of the potential jump.

Suppose now that an external potential drop is applied across the plasma which contains the double layer.[6] The neutral plasma will exclude the applied field and force it to appear solely in the non-neutral double layer regions. If this potential exceeds several kT_e/e, the ions accelerated upstream by the jump will become untrapped and stream upstream freely past the original potential dip. The dip itself will disappear because the fractional change in ion density through the dip will be small due to the high ion entrance velocity. There will be less excess negative charge there, and therefore a smaller dip. Since the double layer is now field driven, the dip which allowed a current driven double layer to form is not necessary and in fact disappears when the potential jump becomes large enough. Now the ion and electron phase space plots are near mirror images of each other, with ions beaming upstream of the double layer. Figure 10 shows a plot of ion phase space for a one-dimensional simulation of a field driven double layer across which there is a potential drop of 10 kT_e/e. Notice that ions are not trapped and freely stream to the left against the electron drift direction.

In one-dimensional simulations, double layers are seen to form spontaneously with drift velocities as low as 0.6 v_e, particularly in very long systems. There is a bootstrapping mechanism in one dimension due to the fact that fields do not decay with distance as in two and three dimensions. If the boundary conditions are fixed potential or periodic, a potential jump will create a decelerating field elsewhere which will extract energy from global electron drift and feed it into the double layer.

In multidimensional simulations, double layers do not as readily form. This is because they form locally and only become one-dimensional by spreading sideways as they grow. The local potential dip does not tend towards a potential jump because the excess negative charge upstream can be deflected sideways and dissipate. Also, the retarding field generated by a jump only extracts drift energy from electrons locally since fields fall off with distance.

If a planar density perturbation is applied as an initial condition, upstream excess electrons cannot be deflected sideways (there is no transverse E) and fields again do not fall of with distance. It is observed that double layers under such conditions form more readily in two-dimensional simulations.

If the plasma is magnetized, a double layer growing from a local perturbation will not spread sideways and become one-dimensional. Instead, it remains local with a transverse width roughly twice its length. These local double layers form more easily in a magnetized plasma because ion-acoustic turbulence is suppressed with far less electron heating as a result.

In convective systems, where particles leave

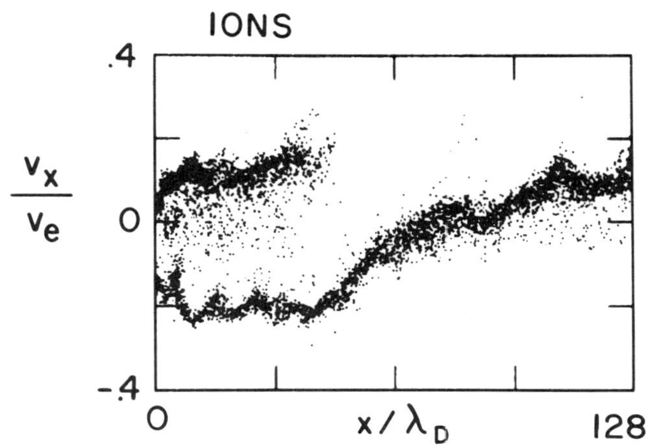

Fig. 11. A 2-d simulation of a field driven double layer. The double layer is one-dimensional (planar) because it was launched from the left boundary which itself is flat.

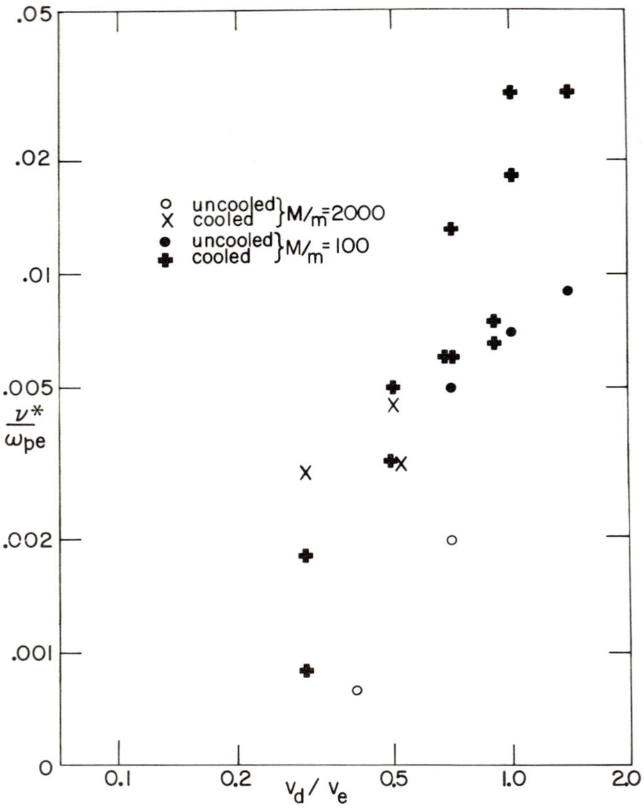

Fig. 12. A plot of anomalous collision frequency as a function of drift for a number of simulations in which ion-acoustic turbulence was seen. For runs at the larger drifts, double layers formed. One can see the rapid increase as electron drift exceeds the thermal speed. Shown are runs with two different mass ratios (100 and 2000). The electrons were cooled artificially in some runs to better approximate steady state conditions in a convective system.

and new ones are injected at the boundaries, field driven double layers form easily at the entrance boundary where a sheath acts as an initial perturbation. The double layer moves downstream and across the system. As it disappears, a new one forms at the entrance plane. Figure 11 shows a 2-d simulations of a field driven double layer in which a potential of 10 kT_e/e is applied across a system 256 debye lengths long. In the presence of a magnetic field, large amplitude ion-cyclotron waves are excited just downstream of the double layer where the electrons are beaming with high velocity. The ion-cyclotron waves then propagate downstream but nearly obliquely to the magnetic field. Although they grow to large amplitude, they apparently do not affect the double layer significantly.

Resistivity

As suggested in preceeding paragraphs, ion-acoustic turbulence, double layers and the Buneman ion-electron instability all can produce large resistivity in current carrying plasmas. Of the four instabilities discussed, the ion-cyclotron instability produces by far the lowest resistivity. Figure 12 shows the anomalous collision frequency measured in a number of two-dimensional runs as a function of electron drift velocity. In all cases, the plasma was unmagnetized with $T_i \ll T_e$. Double layers formed in some of the runs at the higher drift velocities.

Conditions in the auroral zone may not lead to excitation of all these modes since each has drift, temperature, or magnetic field requirements for its existence which may not be met there. Ion-acoustic waves require low temperature ions and low magnetic fields. The ion-cyclotron and Buneman instabilities require electron-ion drift velocities exceeding the electron thermal speed. Double layers are relatively insensitive to magnetic field strength and ion-electron temperature ratio although they apparently do need a perburbation from which to grow.

It is thought that ion temperature in the auroral zone is comparable to electron temperature. That and the presence of a magnetic field suggests that ion-acoustic waves will not be driven to large amplitude and will produce little resistivity. The two best candidates for the production of auroral zone resistivity are double layers and the Buneman instability. Both occur when electron drift is comparable to the electron thermal speed. Because both can account for large resistivity, electron drift driven by external fields will not much exceed the electron thermal velocity.

In situations where the instability is driven by large external potential drops (such as are found in the auroral zone), simulations have shown that double layers will dominate over the other three candidates. Once formed, they absorb most of the potential drop so that electron drift remains at or below the thermal speed except near the double layer. Downstream, after the beaming electrons become thermalized by electron-electron two stream effects, their temperature may be quite high.

It should be mentioned that the form of double layers in the auroral ionosphere may be quite different from that suggested by simulations reported here. The particle distribution function applicable as boundary conditions are most likely quite different there and could cause much more extended double layer like structures almost unrecognizable as being related to the more classical form reported here. Whatever their form, double layers can very efficiently convert beaming energy into thermal energy, leading to high levels of resistivity.

Acknowledgements

This work performed under the auspices of the U. S. Department of Energy. This work is also partially supported by the NASA STTP program under grant #10-23726.

References

J. R. Kan and L. C. Lee, "Formation of Auroral Arcs and Inverted V Precipitations: An Overview," in Physics of Auroral Arc Formation, AGU Publication, 1981.

O. Buneman, Phys. Rev., 115, 503 (1959).

J. M. Kindel and C. F. Kennel, "Topside Current Instabilities," J. Geophys. Res. 3055 (1971).

J. S. DeGroot, C. Barnes, and A. E. Walstead, "Localized Structures and Anomalous dc Resistivity," Phys. Rev. Lett., 38, 1283 (1977).

A. Hasegawa and T. Sato, "Existence of Negative Potential Solitary-Wave Structure and Formation of Double Layer in a Non-equilibrium Plasma," submitted for publication.

M. K. Hubbard and G. Joyce, "Simulation of Auroral Double Layers," J. Geophys. Res., 84, 4297 (1979).

A SIMULATION STUDY OF V-POTENTIAL DOUBLE LAYERS AND AURORAL ARC DEFORMATIONS

J. S. Wagner, J. R. Kan, S.-I Akasofu

Geophysical Institute, University of Alaska, Fairbanks, Alaska 99701

T. Tajima

Institute for Fusion Studies, University of Texas, Austin, Texas 78712

J. N. Leboeuf and J. M. Dawson

Department of Physics, University of California, Los Angeles, Los Angeles, California 90024

Abstract. We present a simulation study of the interaction of a field-aligned electron current sheet shielded by ambient ions with a conducting ionosphere. Provided that a suitable population of electrons is trapped between the dipole mirror and the developing field-aligned potential drop, a stable V-shaped potential double layer develops and accelerates the current carrying electrons into the ionosphere at near and above thermal energies. The double layer potential depends critically on the electron drift speed and ion temperatures at the plasma sheet and ionospheric boundaries. The field-aligned potential is limited by the plasma sheet ion temperature when the electron drift speed is near zero (a half-Maxwellian source), but when the drift speed becomes greater than about 0.5 times the thermal, anomalous resistivity through wave trapping of the current carrying electrons sets in, and the potential drop can exceed the plasma sheet ion thermal energy. In addition to this study, we examine the thin (1-10 Debye lengths) electron current sheet in the plane parallel to and elevated above the ionosphere. In this plane the current sheet is deformed by electrostatic guiding center velocity shear driven by unstable Kelvin-Helmholtz waves and forms either auroral curls or auroral folds, depending on whether the shielding ion gyroradius is much greater than, or approximately equal to, the electron current sheet thickness, respectively.

Introduction

We have begun studying the interaction of a magnetized, field-aligned, current carrying, ion shielded electron sheet with a conducting ionosphere boundary using plasma simulation techniques. Our motivation for this study is that the acceleration mechanism which produces auroral electrons is one of the most challenging, outstanding problems in cosmic electrodynamics. Recent satellite observations have revealed two intriguing features associated with the thin curtain-like auroral forms known as auroral arcs. The first is that an auroral arc is produced by a thin sheet of electrons streaming down along geomagnetic field lines and these electrons carry a significant amount of field-aligned (upward-directed) currents [Cloutier, et al., 1973; Arnoldy et al., 1974]. The second is the indication that this current sheet is associated with a V-shaped potential structure in which a potential difference on the order of 5 kilovolts is dropped along the geomagnetic field lines above the auroral ionosphere [Gurnett, 1972; Mozer, 1977]. Several mechanisms, including double layers, electrostatic shocks, anomalous resistivity, thermoelectric effects, pitch angle anisotropy and Alfvén waves have been proposed for generating and sustaining such a field -aligned potential difference in the collisionless auroral plasma; see the general reviews by Fälthammer [1977] and Shawhan [1978]. Laboratory double layers have been studied recently by Baker et al. [1980], Coakley et al. [1979], Quan et al. [1976], and Stenzel et al. [1980], while double layer theory has been studied by Block [1972] and Kan et al. [1979, 1980] and others. Double layer simulation studies include Degroot et al. [1977], Joyce et al. [1978], Hubbard et al. [1979], Singh et al. [1980], Sato et al. [1980] and Wagner et al. [1980]. Reviews of the effects of anomalous resistivity can be found in Papadopoulos et al. [1975] and Hudson et al. [1978]. Recent work on pitch angle anisotropy can be found in Ponyavin et al. [1977]. The thermoelectric effect has been studied by Hultqvist [1972]. Recent studies of the effects

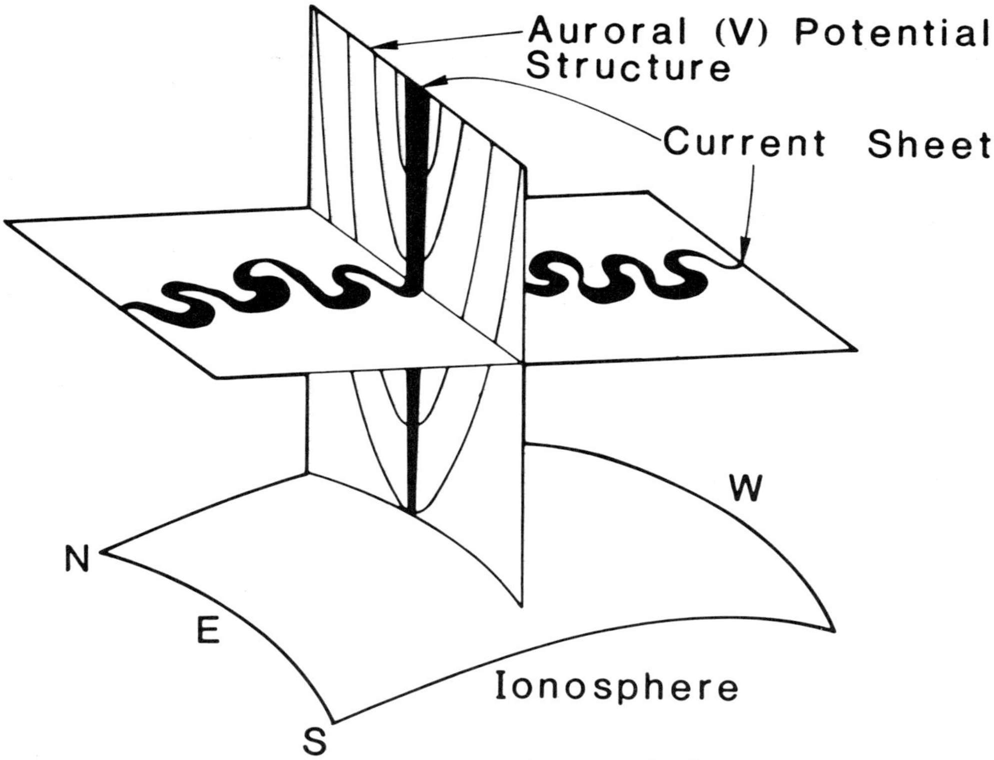

Figure 1. A schematic view of the interaction of the auroral electron current sheet with the polar ionosphere. Two perpendicular planes intersecting the current sheet are shown, and are the topic of this study.

of oblique Alfven waves include Goertz and Boswell [1980]. In order to discover the roles these mechanisms might play we have modeled the auroral environment using realistic initial and boundary conditions, and then applied them using self-consistent, time-dependent electrostatic particle simulation techniques. Computer capabilities restrict us to making only two-dimensional cuts through the auroral current sheet as shown in Figure 1. The two most interesting planes are shown, one containing the V-potential structure and the other containing the deformations produced by the guiding center shear (electrostatic Kelvin-Helmholtz instability). These small scale deformations are observationally known as curls and folds. In this paper we describe our results from simulations in both of these planes, and we begin in the next section with the plane containing the V-potential structure.

V-Potential Structure

In order to simulate the formation of the V-potential structure, we must specify the initial and boundary conditions, modeling the auroral environment as closely as possible under the constraint of limited computer time and core size. In this section we describe our model of the initial and boundary conditions and describe the resulting physical picture to explain how a V-shaped potential structure can form above the auroral ionosphere and accelerate plasma sheet electrons into the ionosphere in a self-consistent, time-dependent manner. The boundary conditions we have used to model the auroral environment in this study are as follows:

1) A lower conducting boundary to model the ionospheric effect. The Pederson conductivity is much greater in the ionosphere than in the accelerating region above it, so that we can model the ionosphere as an equipotential surface at the lower edge of our simulation. We assume $\phi = 0.0$ at $X = 0.0$.
2) An upgoing flux of electrons from the ionospheric boundary, representing three different electron sources. They are: mirroring primary electrons reflected at their incident energy, backscattered secondary electrons which are energy degraded by collisions in the ionosphere (by 6% in the simulations shown), and cold ionospheric electrons. The flux of the upgoing electrons is varied to maintain overall charge neutrality in the simulation. (Other combinations of handling the particles at the boundaries were tried, but resulted in unstable accelerating potentials).

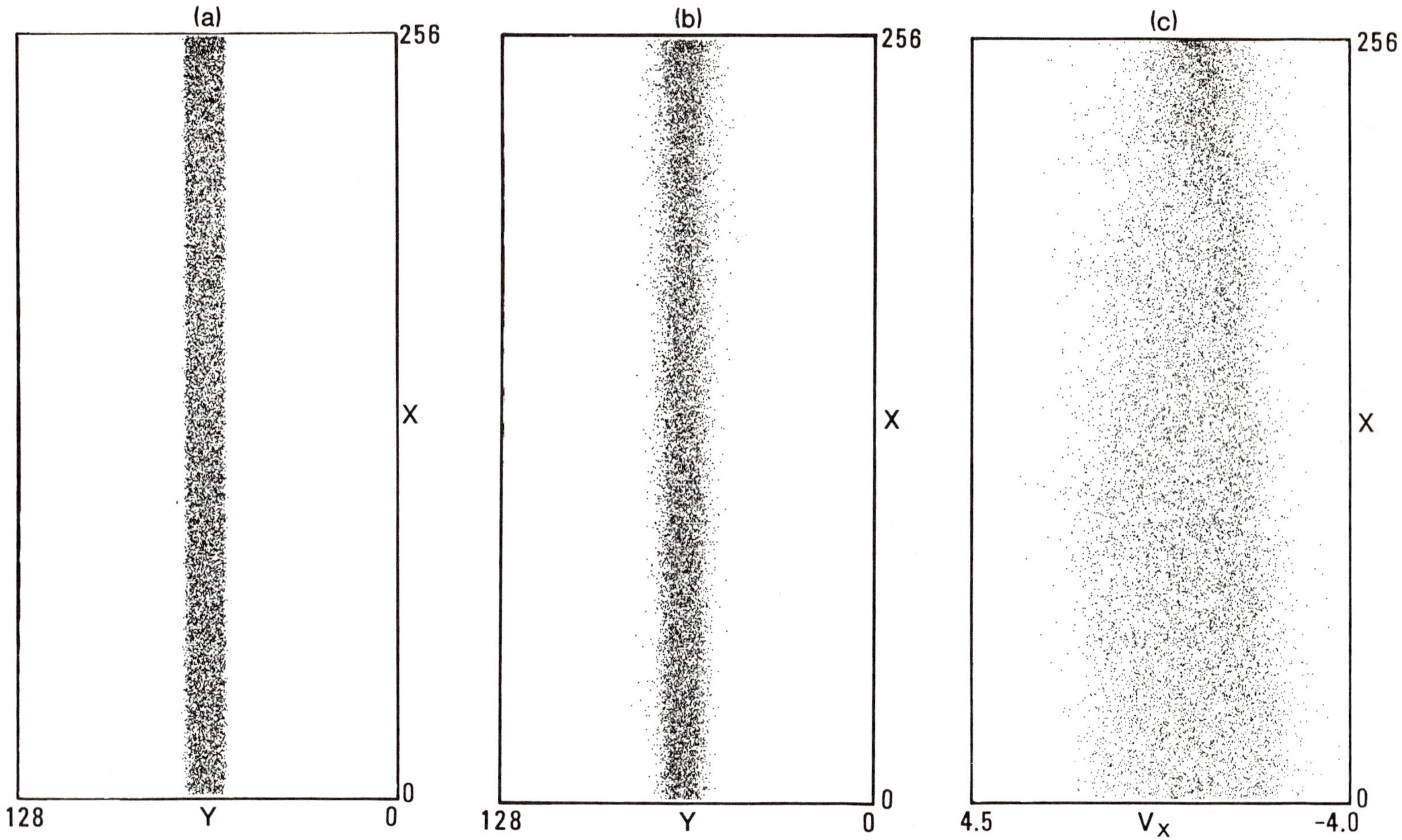

Figure 2. Results taken from a V-shaped simulation with an initial electron drift equal to zero at a later double layer time $\tau = 400\ \omega_{pe}^{-1}$; (a) the electron positions with the ionospheric boundary at the bottom and the plasma sheet boundary at the top, (b) the ion positions, (c) the electron phase space x vs. V(x), (d) the ion phase space (all velocities are in units of plasma sheet electron thermal speeds), (e) the V-shaped electric potential contours, and (f) the potential as a function of altitude down the center of the current sheet.

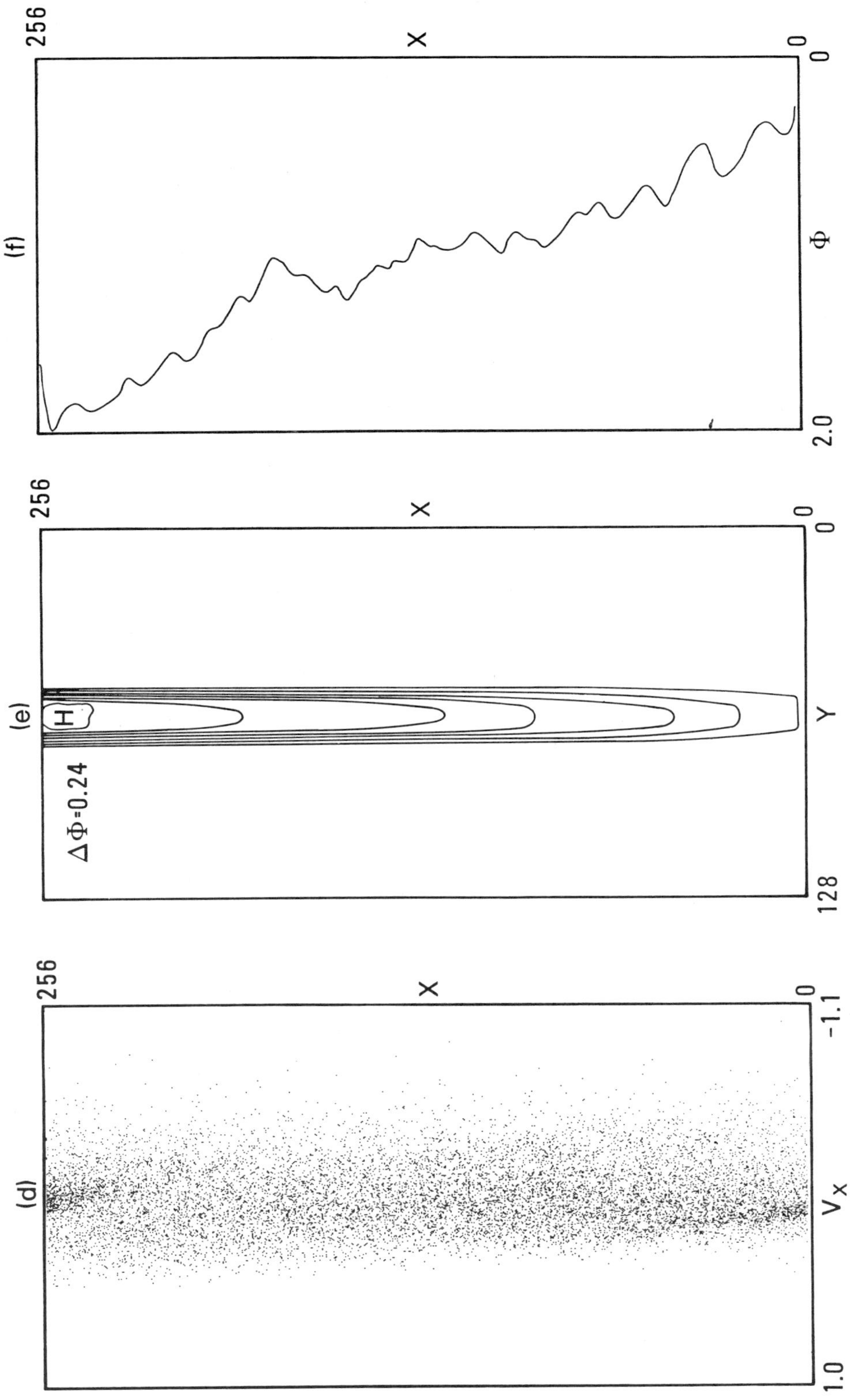
(d)
(e)
(f)
256
X
0
-1.1
1.0
V_x
128
Y
$\Delta\Phi$=0.24
H
2.0
Φ

3) An upgoing constant flux of ionospheric ions having a temperature ratio $T_e/T_i = 1$ with the ionospheric electrons, from a half-Maxwellian reservoir.
4) A constant current electron sheet is introduced at the upper "plasma sheet" boundary. The sheet is assumed to be a drifting Maxwellian. In the limiting case of zero drift we use a half-Maxwellian source.
5) A constant flux of hot plasma sheet ions having a temperature ratio $T_e/T_i < 1$ with respect to the plasma sheet electrons introduced from a half-Maxwellian source from the upper "plasma sheet" boundary.
6) A converging dipole magnetic field. This magnetic field magnetizes the electron sheet and provides for the mirroring of both ions and electrons of plasma sheet origin. It also provides for a population of electrons trapped between the mirror and the accelerating potential after it develops.
7) A cold background plasma uniformly distributed throughout the simulation grid.

As initial conditions the electrons are uniformly loaded with a thermal and a small drift velocity to form a two-dimensional current sheet. A neutralizing background of thermal ions is loaded with their guiding centers on top of the electrons, but due to their larger gyroradius they occupy a larger volume of space and set up an electric field perpendicular to the sheet. The details of the initial particle loading are not observed to be important to the long term formation of the accelerating potential, which depends primarily on the boundary condition at the ionospheric and plasma sheet boundaries.

The potential drop along the field lines is not applied, but is allowed to evolve self-consistently as a result of the interaction. The total current is not fixed by the particle sources at the boundaries because the injection of upgoing electrons fluctuates to maintain overall charge neutrality and particles are free to leave the system through any boundary. However, when the double layer potential becomes large enough to trap upstreaming electrons and downstreaming ions between the double layer and the boundaries, then the current is controlled by the electron current source and the upstreaming ions. This is different from previous double layer simulations and laboratory experiments where the potential drop and/or the total current was externally given so that the origin of the electron acceleration process remains unexplained.

On the basis of the present simulation study the following physical picture has emerged in an attempt to explain the above observations. A magnetized sheet of current-carrying electrons and hot shielding ions (of plasma sheet origin) is injected through a cool background plasma towards a conducting (ionospheric) boundary. As the hot ions try to shield the current sheet, a potential drop forms perpendicular to the magnetic field as a result of their larger Larmor radius (the potential contours being parallel to the magnetic field). The partially shielded electron beam approaches the conducting boundary and is accelerated towards it, because the electrons near the boundary "see" their oppositely charged images on the other side of the boundary. As the electron sheet beam reaches the conductive layer, the potential contours close, forming a V-shaped potential structure. The resulting potential drop is limited by the plasma sheet ion kinetic energy T_i, which is in agreement with a two-dimensional, double layer solution to Poisson's equation for a steady-state condition [Kan et al., 1979].

However, once the electrons become accelerated, or if their initial drift velocity is high enough, they excite electrostatic waves through the electron-ion two-stream instability, and they become trapped in the waves. When this occurs, the total potential drop results from the interaction potential (limited by T_i), and anomalous resistivity (unlimited by T_i). Thus, the double layer develops in a two-step process involving both the conducting boundary and the effects of anomalous resistivity. The back-scattered and mirroring electrons play a crucial role because without them the double layer is not stable and the collisionless plasma is unable to sustain a steady-state voltage drop. Our simulation shows that not only does a double layer solution exist, but it forms spontaneously and remains stable provided that the proper initial and boundary conditions are met. This is the first two-dimensional simulation which shows that a double layer can form under auroral conditions, and demonstrates that it is capable of accelerating the current-carrying electrons to the observed energies.

Simulation Results

We show here some of our simulation results, one computer run with zero drift velocity for the electron current sheet (the current is from a half-Maxwellian source) and another computer run with a drift velocity of $0.75\ v_{th}$, where v_{th} is the electron thermal velocity at the plasma sheet boundary. The simulation is carried out according to a bounded $2\frac{1}{2}$-dimensional finite size particle electrostatic algorithm [Decyk and Dawson, 1979; Birdsall et al., 1970]. The boundary conditions used are periodic in the y direction and bounded in the x direction with the potential $\phi = 0$ specified at $x = 0$ (the conducting boundary) and $E_x = 0$ at $x = N_x\ \Delta$ ($\Delta = 2\ \lambda_{De}$ is the grid distance). Typical parameters employed are grid size $N_x \times N_y = 256 \times 128\Delta$, L_y (electron sheet thickness) = 6Δ, the particle number 65,536, the electron cyclotron and plasma frequencies $\Omega_e = 4.4\ \omega_{pe}$ at the ionospheric boundary and $\Omega_e = 4.0\ \omega_{pe}$ at the plasma sheet boundary (this

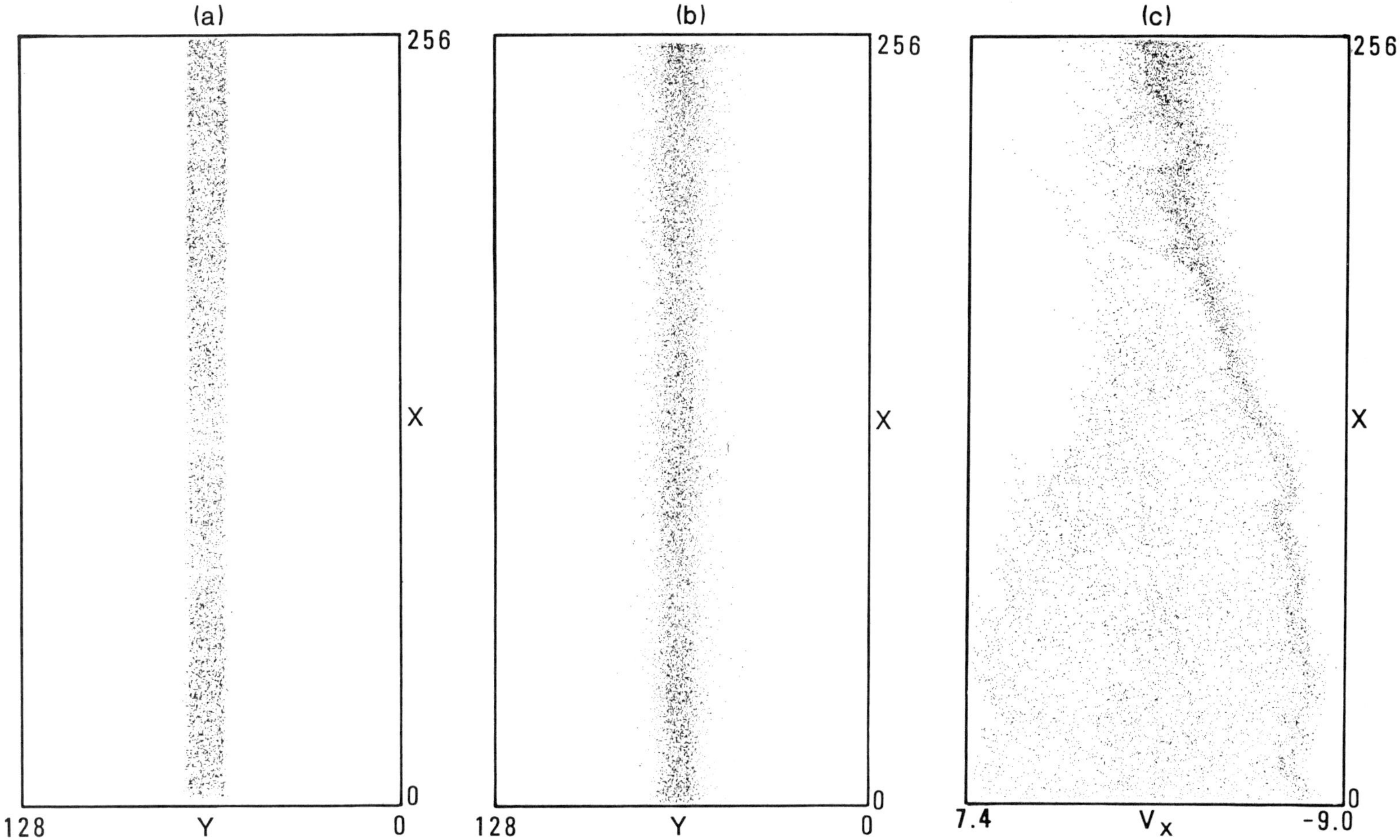

Figure 3. Results taken from a V-potential simulation with initial electron drift speed equal to 0.75 V_{th} at a later time $\tau = 400\ \omega_{pe}^{-1}$. (a) the electron position, (b) the ion position, (c) the electron phase space, the acceleration of the beam to the right as it comes down is shown, along with the population of trapped backscattered electrons, (d) ion phase space with the plasma sheet ions being clearly relfected near the top, and the ionospheric ions being accelerated through the potential, (e) the distribution of electron velocities in the x-direction, (f) the electron velocity distribution in the y-direction, and (g) the electric potential as a function of altitude.

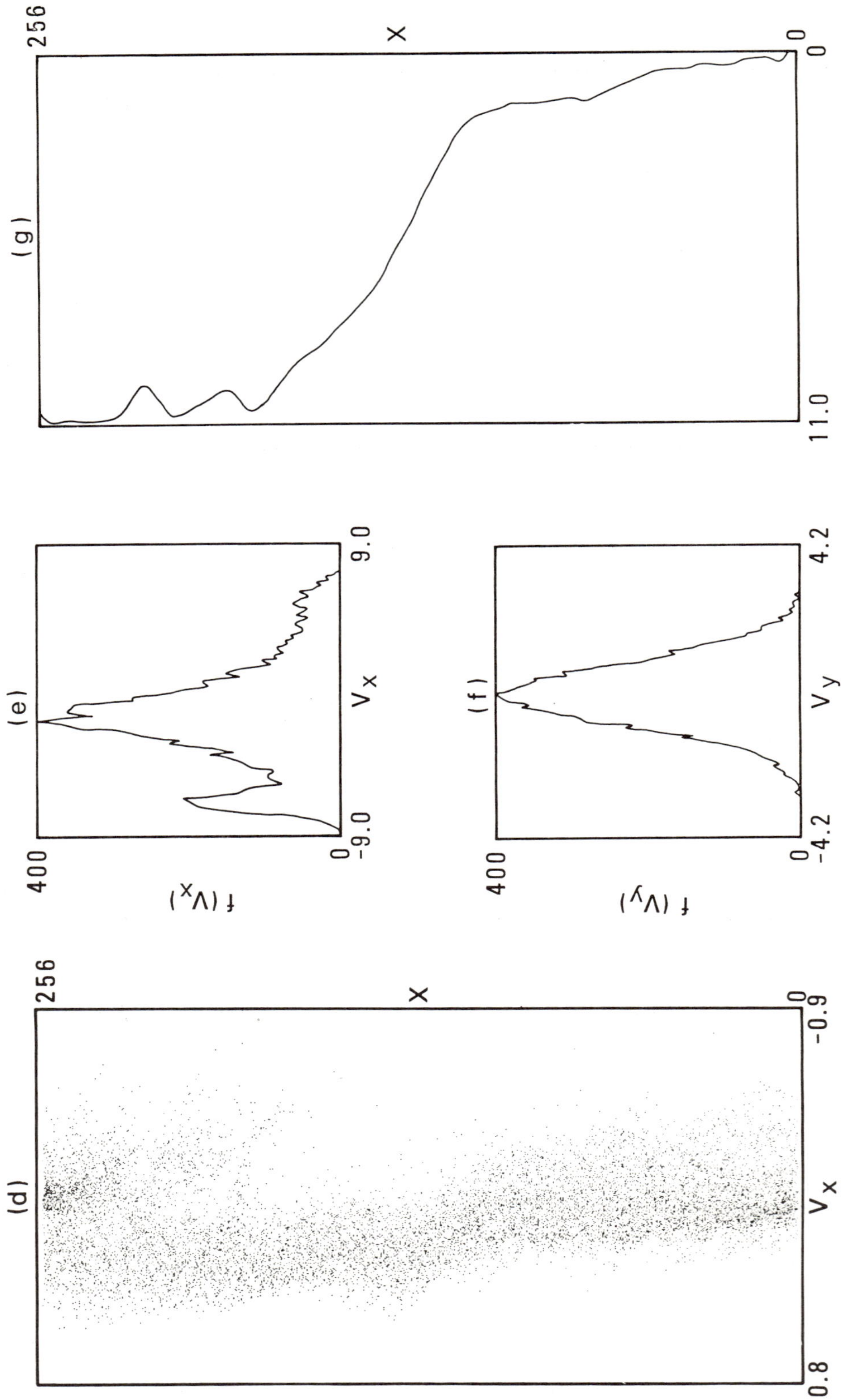

(d)
(e)
(f)
(g)
X
f (Vx)
f (Vy)
Vx
Vy
256
400
0
0.8
-0.9
-9.0
9.0
-4.2
4.2
11.0

is a magnetic field convergence of 10% over 512 Debye lengths). These parameters model the auroral environment as accurately as our computer resources will allow. Results from the $v_D = 0$ case are shown in Figure 2, taken at time $\tau = 400\ \omega_{pe}^{-1}$. The electron positions are shown in Figure 2(a), the ion positons in 2(b). The slight field line convergence is evident. The hotter ions near the plasma have a higher temperature, and hence a larger gyroradii. This causes a larger potential drop perpendicular to the sheet (near the upper boundary) as seen in the contour plot 2(e). The electron phase space is shown in 2(c) and the ion phase space in 2(d).

The most important feature of this simulation is that a quiet, stable double layer has formed and the resulting potential drops linearly [Figures 2(e) and 2(f)] between the upper and lower boundaries. The net potential drop is slightly less than the ion kinetic energy at the plasma sheet, 0.8 T_i. It is important to notice in the electron phase space 2(c) that the electrons from the plasma sheet are accelerated into the ionosphere with a beam velocity greater than the electron thermal velocity in the plasma sheet. A very large population of electrons trapped between the mirror and the potential drop, and a large population of backscattered degraded primaries are also evident (the degraded primary is almost, but not quite, able to reach the upper plasma sheet boundary). In the ion phase space 2(d) it is clear that some, but very few, of the plasma sheet ions have sufficient energy to reach the ionospheric boundary. The cold ionospheric ions can be seen to be heated and accelerated as they pass through the double layer, which has a thickness of at least 512 λ_D. This is in contrast to our previous results in a uniform magnetic field, where the double layer under similar conditions was confined to a region ~10 λ_{De} near the conducting boundary. In addition, the fact that the electron current source is half-Maxwellian, the drift velocity is too low to satisfy the Böhm-Block criterion. The presence of the trapped electrons, not included in the Böhm-Block model, still allows the double layer to form, in agreement with the recent theoretical results of Kan et al. [1979].

The second simulation example with ($v_D = .75 v_{th}$) is shown in Figure 3. The figures are taken at time $\tau = 400\ \omega_{pe}^{-1}$. Figures 3(a) and 3(b) show the electron and ion positions respectively. A noticeable density depression occurs in the region of maximum potential drop, from x = 80 Δ to x = 200 Δ. The electron phase space 3(c) shows the plasma sheet electrons being accelerated through a double layer into the ionosphere at 6-8 times their initial thermal velocity. A large population of trapped, degraded primary and secondary electrons are also evident in phase space, although none are able to pass through the double layer to reach the upper boundary. Significant electrostatic waves due to the Buneman electron-ion two stream instability are excited and are capable of trapping some of the current-carrying electrons. The most unstable wave number for Buneman waves is $k \sim \omega_{pe}/v_d$, which changes as the electron beam is accelerated, so that the electron phase space modulation wavelength increases as the beam propagates toward the conducting boundary. The wavelength of this vortex structure can be scaled by $\lambda(x) = 2\pi v_d\, \omega_{pe}^{-1}$. The trapping of electrons by these waves can lead to an increase of the effective resistivity (anomalous electron dragging) and thus contributes to the additional voltage drop distributed over the length where the instability is active. The spatially averaged electron velocity distribution functions for v_x and v_y are shown in Figures 3(e) and 3(f), respectively (notice the velocity scales in the two figures are not the same). Figure 3(e) shows the various populations of electrons quite well, the accelerated beam with negative velocities 4-6 times the thermal velocity, the trapped electrons near the thermal velocity, and the energy degraded primaries streaming upwards with positive velocities. Note that there is no significant heating of the electrons in $f(v_y)$, since 3(f) shows an undisturbed Maxwellian distribution. Figure 3(g) shows the potential drop a function of x at y = 64 Δ (the center of the sheet). The magnitude of the potential drop is 16 T_e, or 6 times the ion thermal energy at the plasma sheet. We observe these quasi-stable strong double layers to occur whenever the accelerated drift velocity of the beam exceeds the Buneman threshold over a significant region of the simulation grid.

Auroral Arc Deformations

Ground based observations of auroral curtain deformations, in particular the small scale distortions known as auroral curls and folds, have been going on for some time. Stormer [1955], from his visual observations, suggested that rayed arcs might actually be thin curtains with many tiny folds viewed from the side. Akasofu and Kimball [1967] published photographs of folded arcs seen in the magnetic zenith which convincingly demonstrated that rayed arcs viewed from the side are actually folded curtains viewed from below.

Problems with length exposure times compared to deformation times and the resulting photographic averaging limited further study until Hallinan [1970] and Hallinan and Davis [1970] carefully analyzed their results from TV images and categorized the small scale distortions into classes according to shape, rotational sense, and statistical occurrence.

Parallel to the study of small scale auroral forms the experimental results of Kyhl and Webster [1956], Pierce [1956], Cutler [1956] and

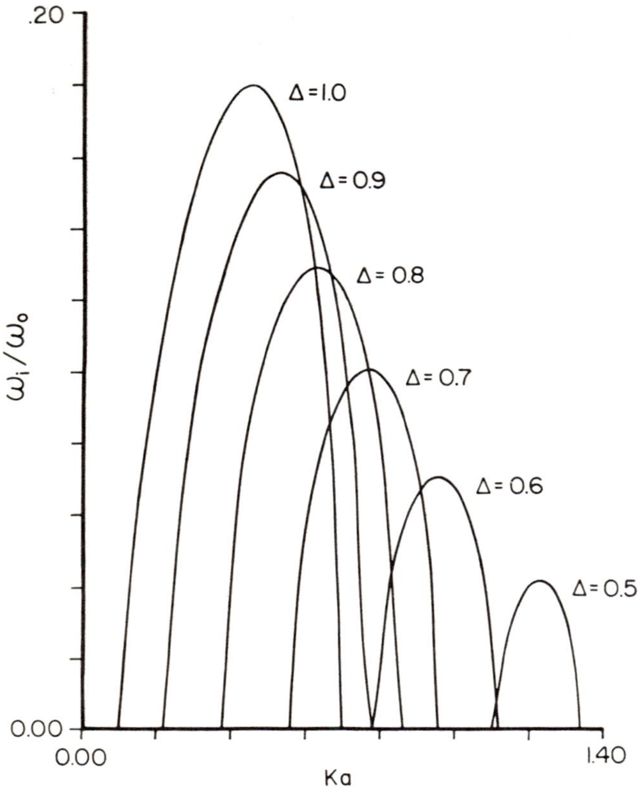

Figure 4. The imaginary part (growth rate) of the frequency of electrostatic Kelvin-Helmholtz waves for various values of charge separation.

Webster and Hallinan [1973], concerning the breakup of hollow and sheet electron beams in vacuum tubes, showed a structure remarkably similar to those seen in the aurora, and suggested that the mechanism producing the distortions is closely related. Levy and Hockney [1968] were the first to apply the techniques of plasma simulation to study the development of electrostatic Kelvin-Helmholtz waves in a magnetized electron sheet, and observed that the nonlinear development of the instability was consistent with the experimental vacuum tube observations.

Unfortunately, the theoretical works [Buneman et al., 1966] and early simulation results were primarily concerned with the deformation of an electron charge sheet in vacuum tubes and it was not obvious that they had any direct applicability to the aurora. The primary difference between the auroral environment and the vacuum tube studies is that quasi-neutrality considerations in the aurora demand that the electron current sheet producing the aurora be imbedded in a neutralizing background of ions [Miura et al., 1978 and Wagner et al., 1980]. The ion temperature of this background determines the magnitude of the electric field perpendicular to the shield, and therefore has considerable influence on the resulting deformations. In this paper we demonstrate, through the use of plasma simulation, that in the early stages of development the most unstable wavelength forming the deformations of the electron sheet can be computed under the assumption that the electric field profile is established by the ions and the electron sheet responds to the profile. In addition, the growth rate is substantially reduced due to the decreased shielding electric field. In later stages, when the vortices formation reaches the nonlinear stage of development, the ions can in some cases modify the deformation by limiting the twisting of the vortices, which causes the electron clumps to elongate. In this paper we conform to the definitions of curl, fold and spiral auroral forms given by Hallinan [1970] and Hallinan and Davis, [1970].

The Model

We model the auroral formations by assuming that at t = 0 there is a uniform homogeneous electron charge sheet having a thickness 2a. Shielding the charge sheet is a background of ions, with temperatuare ratio T_e/T_i (<1) with their guiding centers located on top of electron guiding centers. Because of the finite size of the ion gyroradii if they are magnetized, and because of their thermal motion if they are not magnetized, an electric field is set up perpendicular to the charge sheet. When the ions are unmagnetized, the resulting electric potential turns out to be proportional to the ion kinetic energy, $e\phi \approx kT_i$. We further assume that the electron sheet is magnetized by a uniform magnetic field, B_z, parallel to the sheet. We are interested in electrostatic deformations perpendicular to the magnetic field, so we examine a cross section (the x-y plane) of the sheet and ignore variations along the z-axis.

The zeroth order electric field set up by the ions points toward the center of sheet from both sides causing the electron guiding centers to E x B drift in opposite directions on both sides of the sheet. The velocity shear is capable of exciting electrostatic Kelvin-Helmholtz waves. The previously published Kelvin-Helmholtz theories have only been applied to the case of an unshielded electron sheet, and its application to the ion shielded case is not obvious. As a first approximation it is reasonable to extend the theory by assuming the ions are responsible only for the electric potential profile, while the electrons respond to it and excite the Kelvin-Helmholtz deformation.

Since the ion guiding centers lie on top of the electron guiding centers at $\tau = 0$, the electron and ion charge distributions at early times in the simulation form profiles given approximately by (in this case the ions are magnetized)

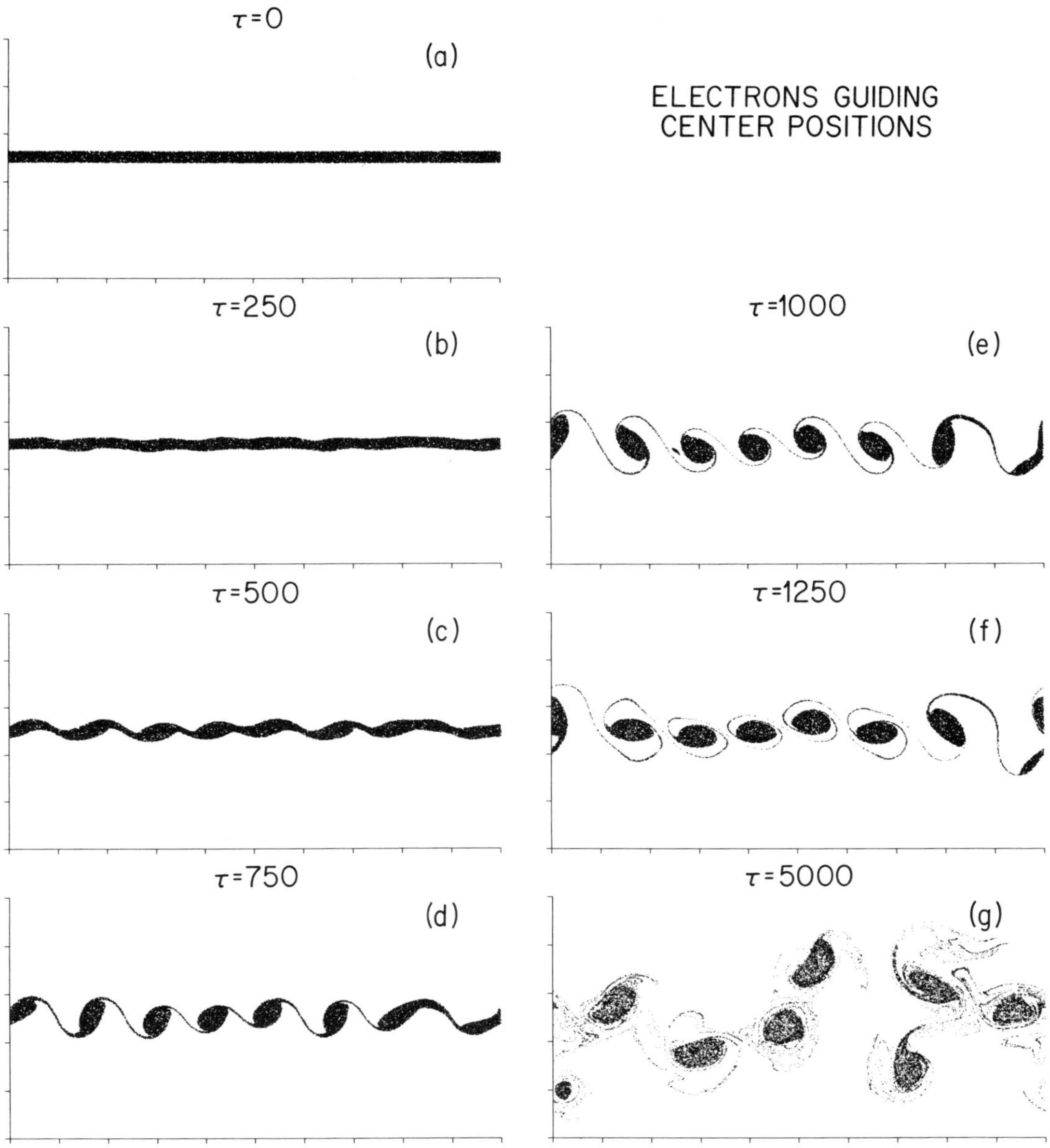

Figure 5. Electron guiding center positions at various times from a simulation of auroral curl formation.

$$n_i(y) = \begin{cases} n_{io}, & |y| < a + \rho_i \\ o, & |y| > a + \rho_i \end{cases} \quad (1)$$

and

$$n_e(y) = \begin{cases} n_{eo}, & |y| < a \\ o, & |y| > a \end{cases} \quad (2)$$

where $n_{io} = a/(a + \rho_i)n_{eo}$, and $\rho_i = m_i V_{th}/q B_z$ is the mean ion gyroradius with m_i the ion mass, q the electron charge, B_z the magnetic field, V_{th} the ion thermal velocity, and n_{eo} and n_{io} are constant electron and ion number densities, respectively. This allows us to estimate the charge separation Δ, defined to be one minus the ratio n_{io}/n_{eo}. In this case $\Delta = 1 - a / (a + \rho_i)$. Notice Δ = 1 reduces to the case with no ion shielding while Δ = 0 indicates perfect shielding and hence no electric field. This density profile is not an equilibrium profile, [Albano et al., 1979] but it is close enough for our purposes. Lack of equilibrium causes small amplitude ion plasma oscillations perpendicular to the charge sheet. These oscillations are

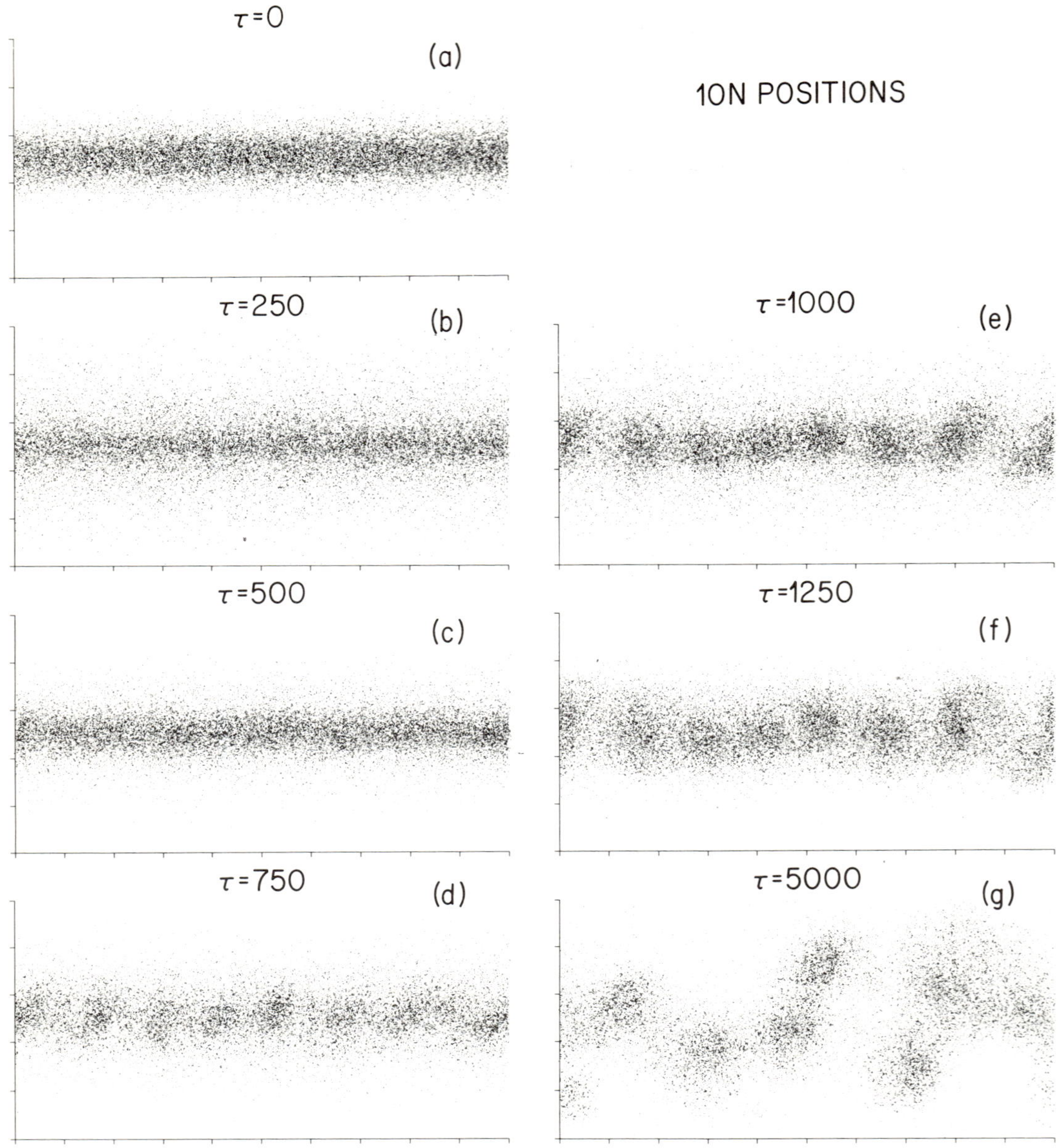

Figure 6. Ion positions at various times, taken from same simulation shown in Figure 5.

observed to be slowly damped in the simulation and do not have any effect on the deformations. Furthermore, there is no reason to believe that the auroral charge sheet should be in equilibrium, and in fact, observations of electrostatic field energy perpendicular to B at a frequency of ω_{pi} should be expected.

It is possible to compute the growth rate of the standing velocity shear waves in the electron sheet from the dispersion relationship [Hasegawa, 1967 and Sydora et al., 1980],

$$\left(\frac{\omega}{\omega_o}\right)^2 - \frac{1}{4}(1 - 4ka\Delta + 4k^2a^2\Delta^2) + \frac{1}{4}e^{-4ka} = 0 \text{ , (3)}$$

where $\omega_o = \omega^2_{pe}/\omega_{ce}$.

This gives the classical result for $\Delta = 1$ which is complete charge separation, or the unshielded case. The imaginary part of ω, the wave frequency, has been plotted in Figure 4 for the various values of Δ. The important features of this dispersion relation are that: 1) the highest growth rates occur for largest values of

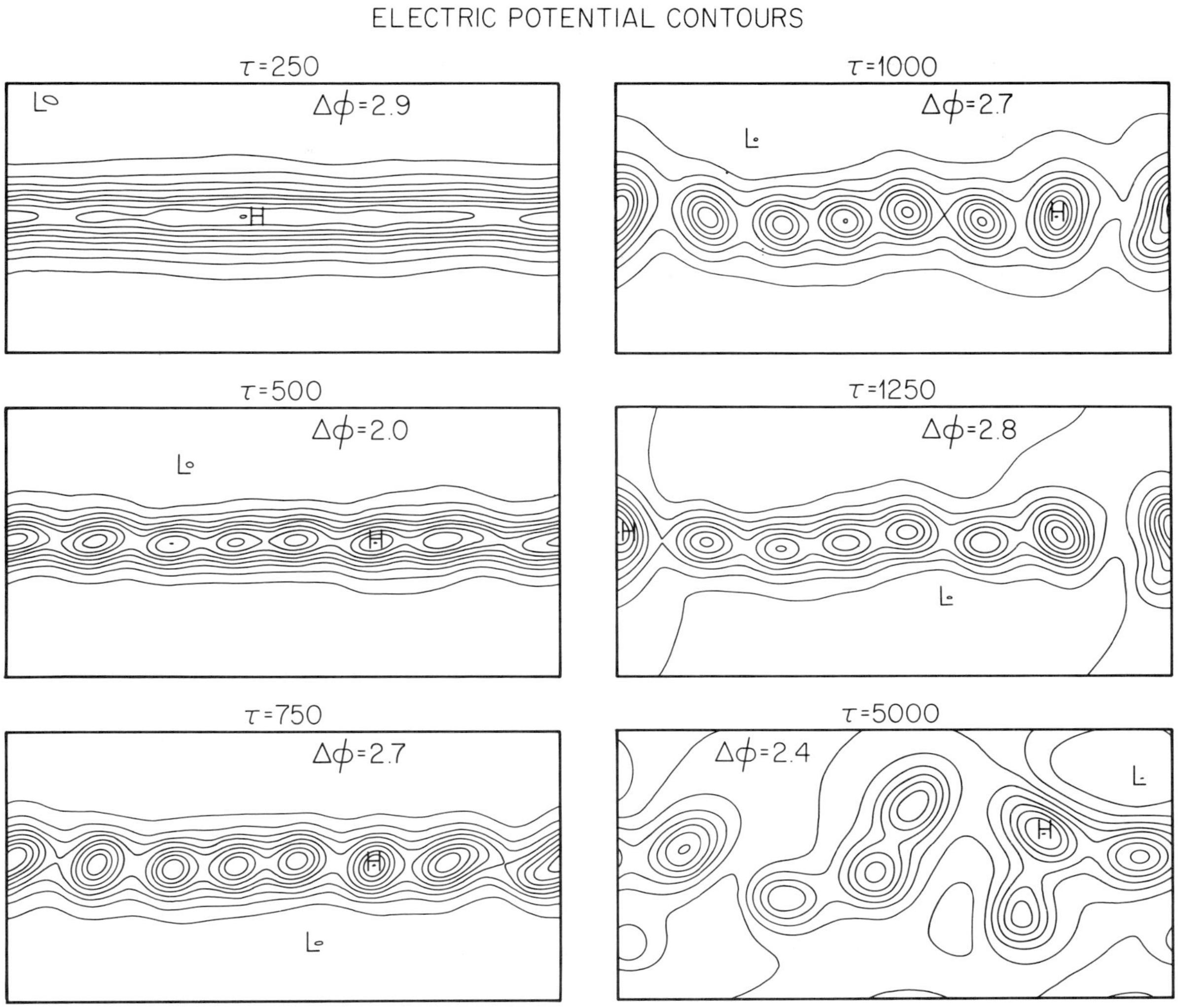

Figure 7. Electric potential contours at various times, taken from the same simulation as in Figure 5.

charge separation, and 2) the most unstable wave number k_{max}, which is given by the peak of each curve, is an increasing function of the charge separation. This means a tendency towards shorter wavelengths and smaller vortices for increased ion shielding.

The Simulation

With this model in mind, we have used the two-dimensional electrostatic particle algorithm to simulate the time development of the resulting charge sheet deformations. The electrons are pushed using the guiding center approximation, while the full dynamics of the ions are followed [Lee and Okuda, 1976]. This is different from most previous simulations, but is an important approximation because it allows us to use realistic plasma parameters, particularly the realistic mass ratio (m_e/m_i) and magnetic field (B_z).

We use a two-dimensional computational mesh size $N_x \times N_y = 64 \times 128$. The grid spacing Δ used was $\Delta = 1/3\ \lambda_{De}$ where λ_{De} is the electron Debye length. The mass ratio $m_e/m_i = 1/1837$ was used. The electrons and ions are assumed to be finite sized charge clouds having a radius $a_x = a_y = \Delta$. We assume an initial sheet half-thickness $a = 3\ \lambda_{De}$, a temperature ratio $T_e/T_i = .6$, and a magnetic field which gives an electron plasma to electron cyclotron frequency of $\omega_{pe}/\omega_{ce} = 1/20$. The simulation was run using a time step of $1\ \omega_{pe}^{-1}$ for 10,000 steps, or until $10{,}000\ \omega_{pe}^{-1}$. This was sufficient time for the deformations to form, saturate, and diffuse to the boundaries of the simulation grid. Periodic boundary conditions were assumed.

This choice of parameters corresponds to con-

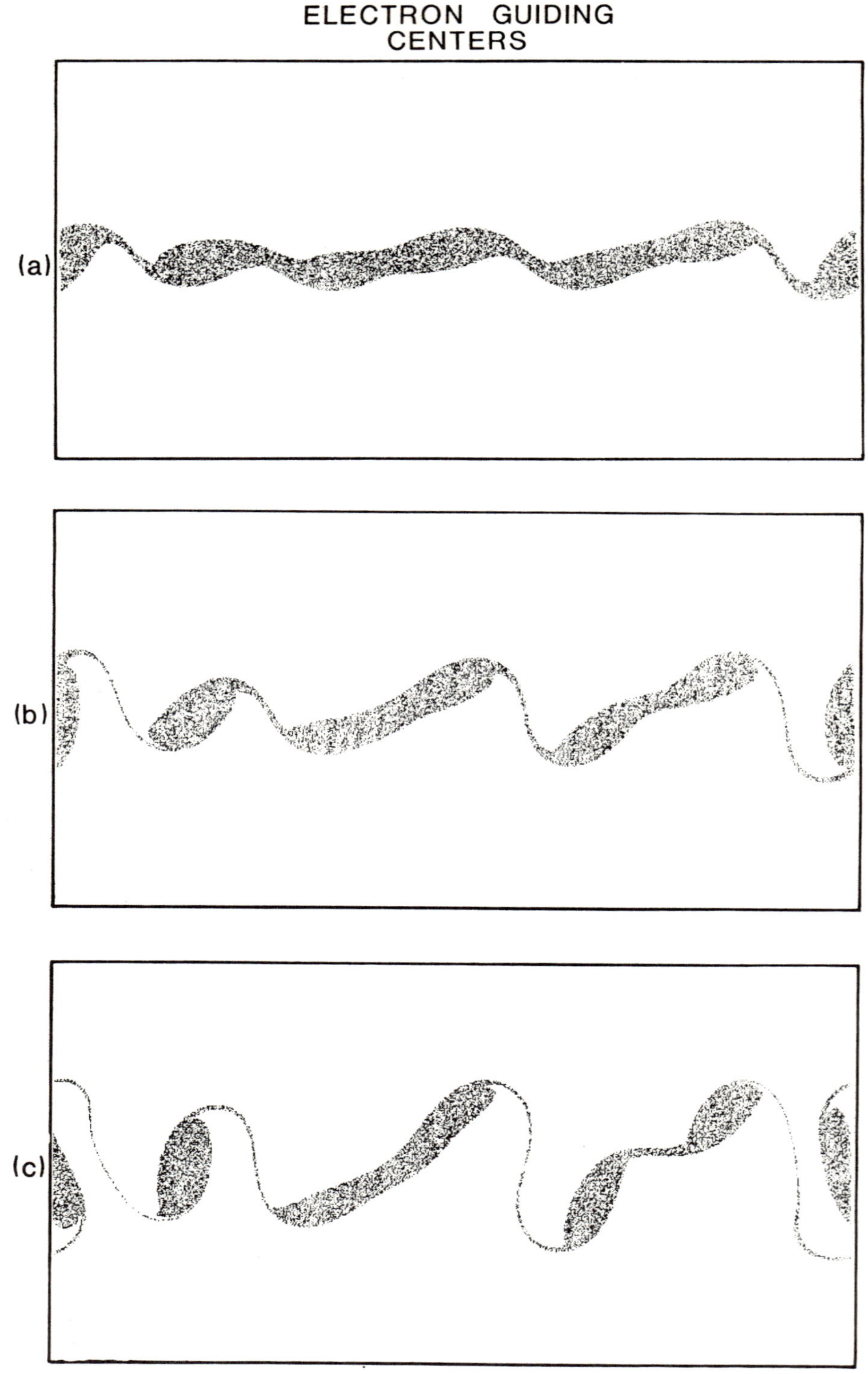

Figure 8. An example of auroral folds unstable to further "curl-type" instabilities. In this simulation $N_x \times N_y = 256 \times 128$, $T_e/T_i = .6$, $\omega_{ce}/\omega_{pe} = 75$, sheet width = 6λ, and Δ (grid size) = 1/3 λD_e. The electron guiding center positions (a), (b), (c), ion positions (d), (e), (f), and potential contours (g), (h), (i) are shown at three successive times, 1500 ω_{pe}^{-1}, 2000 ω_{pe}^{-1}, and 2500 ω_{pe}^{-1}.

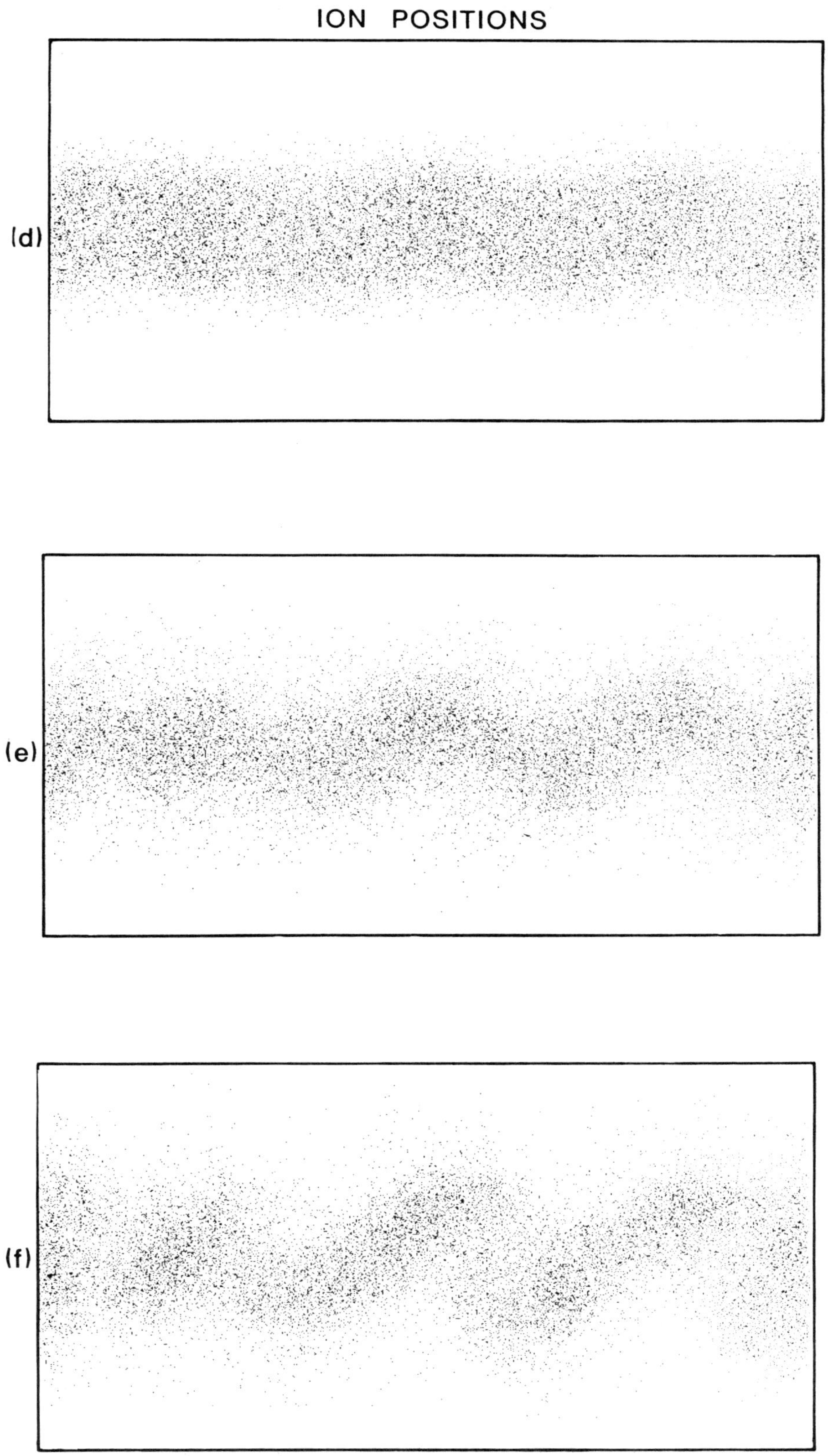
ION POSITIONS
(d)
(e)
(f)

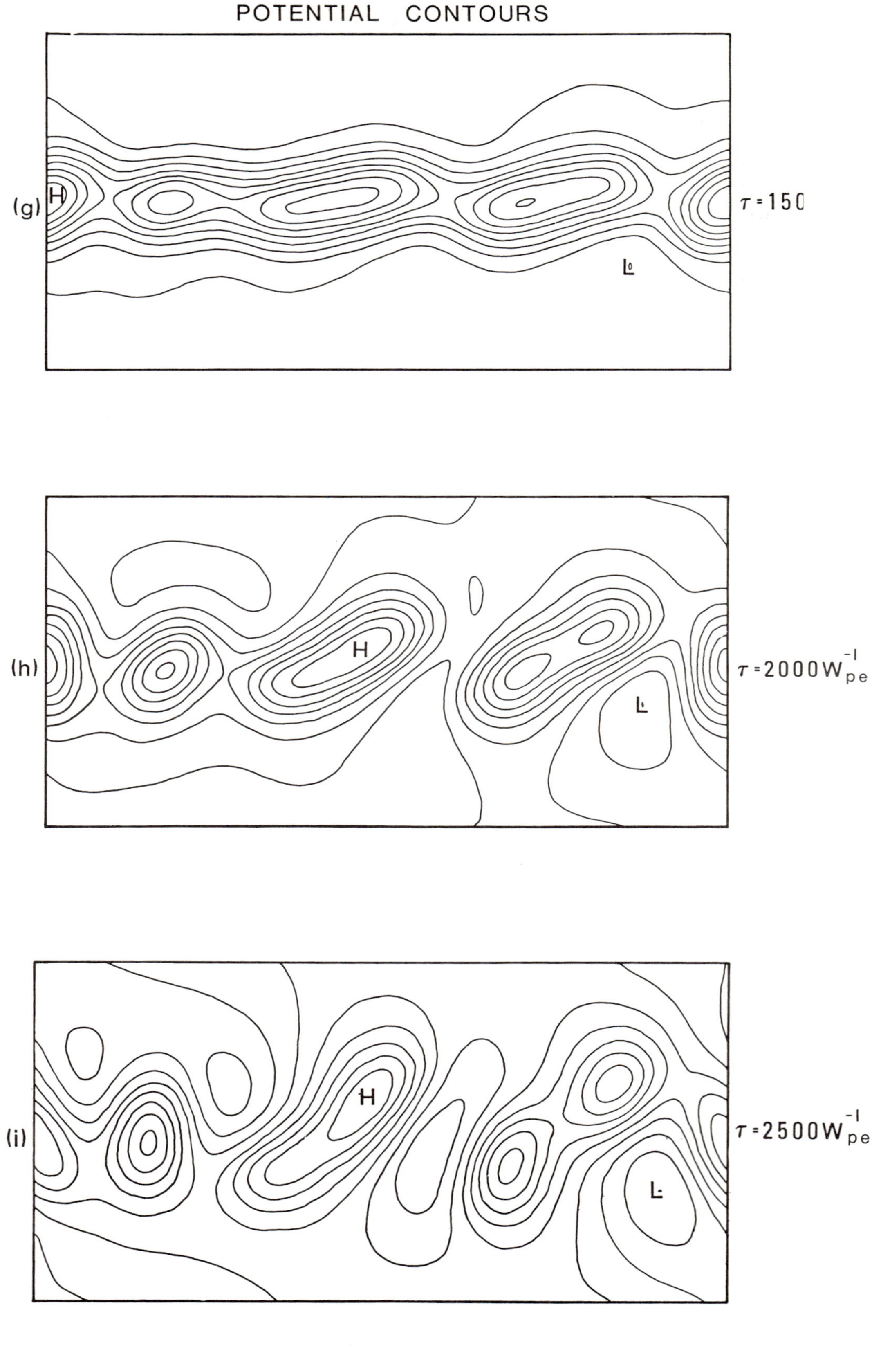
POTENTIAL CONTOURS
(g)
H
L
τ=150
(h)
H
L
τ=2000W$_{pe}^{-1}$
(i)
H
L
τ=2500W$_{pe}^{-1}$

ditions at roughly a (2-3) R_E altitude of the earth ionosphere. Note that at lower altitudes we might expect cooler ions (T_e/T_i = 1 in the ionosphere) and higher magnetic fields, while at higher altitudes we expect hot ions ($T_e/T_i \simeq .2$ in the plasma sheet) and a lower magnetic field. By finding the plasma parameters which closely match the visual observations it is possible to discover the environment and altitude in which the instability occurred.

Simulation Analysis

Figure 5(a) shows the initial positions of the electron guiding centers. Figure 6(a) shows the initial positions of the ions. The ion guiding centers lie on top of the electrons, so that the ions occupy a larger area due to larger gyroradii.

The linearly unstable phase of the distortions is evident in Figures 5(b) and 6(b) at $\tau = 250\ \omega_{pe}^{-1}$. The electric potential contours at this time (see Figure 7) show a potential drop of $\Delta\phi = 2.9\ (T_e \Delta\lambda_{De}^{-1})$ between contour lines. The total potential drop on each side of the sheet is about 1.6 T_e. This corresponds well to the ion kinetic energy which is about 1.5 T_e. The simulation shows that the most unstable mode is m = 8. Since $\rho_i \approx 11\ \lambda_{De}$ and $a = 3\ \lambda_{De}$ we have a charge separation of 0.8, and a predicted unstable mode number ka ≈ .6, obtained from the theoretical result in Fig. 4, or $\frac{2\pi ma}{L_x} = .6$, which corresponds to mode 8 in agreement with the simulation (periodic boundary conditions in the simulation require integral mode numbers).

In general, our results illustrate that the assumption that the ions produce the electric potential profile and the electrons respond to it, is good as long as $\rho_i \gg a$, and the ions appear almost unmagnetized. If $\rho_i \cong a$ our results show that curl formation is limited and that a different deformation, one known as an auroral fold occurs. When $\rho_i \cong a$, folds (not curls) occur because the ions tend to shield the electrons much more closely, but lag behind them spatially as the deformation proceeds in time. This causes the deformations to grow with an additional perturbing electric field which inhibits vortex twisting. As a result the vortices elongate into thin bars or folds. These thin folds, which thin as they elongate, maintain a constant density and eventually become long enough (one or more wavelengths) to be once again unstable against Kelvin-Helmholtz waves (see Figure 8). This process continues until the sheet is thin enough for $\rho_i \gg a$, when quasi-stable curls form. This is in agreement with the observational fact that curls are observed more frequently than folds, and folds are often seen to evolve into curls, not vice-versa. Hence our results show that curls and folds are caused by the same process, the difference being the degree to which the ions interact with the deforming electron sheet, a process which is dependent on the ion gyroradius to electron sheet thickness ratio ρ_i/a.

Acknowledgments. We would like to thank Dr. V. Decyk, Dr. T. J. Hallinan, and Dr. L. C. Lee, for their helpful discussions. We would like to thank the National Center for Atmospheric Research and the Geophysical Institute for their support of the computations. This work was supported in part by the Department of Energy under contract EY-76-S-06-2229 #5 and the Atmospheric Sciences Section of the National Science Foundation under grant ATM77-26522 to the University of Alaska and ATM78-19253 to the University of California, Los Angeles.

References

Akasofu, S.-I. and D. S. Kimball, The dynamics of the aurora - I: Instabilities of the aurora, J. Atmos. Terr. Phys., 26, 205, 1967.

Albano, R. K., L. C. Lee, and J. R. Kan, Non-neutral field-aligned current sheet and the auroral electic field, J. Atmos. Terr. Phys., 42, 317, 1979.

Arnoldy, R. L., P. B. Lewis, and P. O. Issacson, Field-aligned auroral electron fluxes, J. Geophys. Res., 79, 4208, 1974.

Baker, K. D., L. P. Block, R. Kist, and W. Kampa, Strong laboratory double layers and their relevance to auroral plasma physics, Proc. ICPP, 417, 1980.

Birdsall, C. K., A. B. Langdon, and H. Okuda, Methods in Computational Physics, ed. by Ahler et al., (Academic, New York, 1970), Vol. 9, p. 241.

Block, L. P., Potential double layers in the ionosphere, Cosmic Electrodynamics, 3, 349, 1972.

Buneman, O., R. H. Levy, and L. Linson, The stability of crossed-field electron beams, J. Appl. Phys., 37, 3203, 1966.

Cloutier, P. A., B. R. Sandel, H. R. Anderson, P. M. Pazich, and R. J. Spiger, Measurement of auroral Birkeland currents and energetic particle fluxes, J. Geophys. Res., 78, 640, 1973.

Coakley, P. and N. Hershkowitz, Laboratory double layers, Phys. Fluids, 22, 1171, 1979.

Cutler, C. C., Instability in hollow and strip electron beams, J. Appl. Phys., 27, 1028, 1956.

Decyk, V. K., and J. M. Dawson, A computer model for a bounded plasma, J. Comp. Phys., 30, 407, 1979.

Degroot, J. S., C. Barnes, A. E. Walstead, and O. Buneman, Localized structures and anomalous dc resistivity, Phys. Rev. Lett., 38, 1283, 1977.

Fälthammar, C.-G., Problems related to macroscopic electric fields in the magnetosphere, Rev. Geophys. Space Phys., 15, 457, 1977.

Goertz, C. K., and R. W. Boswell,

Magnetospheric-ionospheric coupling, J. Geophys. Res., 84, 7239, 1979.

Gurnett, D. A., Critical Problems of Magnetospheric Physics, ed. by E.R. Dyer, (Nat. Acad. Sci., Washington, D.C., 1972) 123.

Hallinan, T. J., and T. N. Davis, Small-scale auroral arc distortions, Planet. Space Sci., 18, 1735, 1970.

Hallinan, T. J., Small-scale auroral arc distortions, M.S. Thesis, Univ. of Alaska, Fairbanks, 1970.

Hasegawa, A., Plasma Instabilities and Nonlinear Effects, Springer-Verlag, Berlin, Heidelberg, 129, 1957.

Hubbard, R. F. and G. Joyce, Simulation of auroral double layers, J. Geophys. Res., 84, 4297, 1979.

Hudson, M. K., and F. S. Mozer, Electrostatic shocks, double layers, and anomalous resistivity in the magnetosphere, Geophys. Res. Lett., 5, 131, 1978.

Hultqvist, B., On the interaction between the magnetosphere and the ionosphere, in Solar-Terrestrial Physics Proceedings of the International Symposium, Leningrad, U.S.S.R., 1970, ed. by E. Dyer, p. 176, D. Reidel, Dordrecht-Holland, 1972.

Joyce, G., and R. F. Hubbard, Numerical simulation of plasma double layers, J. Plasma Phys., 20, 391, 1978.

Kan, J. R., L. C. Lee, and S.-I. Akasofu, Two dimensional potential double layers and discrete auroras, J. Geophys. Res., 84, 4305, 1979.

Kan, J. R., and L. C. Lee, On the auroral double-layer criterion, J. Geophys. Res., 85, 788, 1980.

Kyhl, R. L., and H. F. Webster, Breakup of hollow cylindrical electron beams, IRE Trans. Electron Devices, 3, 172, 1956.

Lee, W. W., and H. Okuda, A simulation model for studying low frequency micro-instabilities, Princeton Plasma Physics Lab Report, MATT 1231, 1976.

Levy, R. H., and R. W. Hockney, Computer experiments on low density crossed-field electron beams, Phys. Fluids, 11, 766, 1968.

Miura, A., and T. Sato, Shear instability: auroral arc deformation and anomalous momentum transport, J. Geophys. Res., 33, 2109, 1978.

Mozer, F. S., C. W. Carlson, M. K. Hudson, R. B. Torbert, B. Parody, Y. Yatteau, and M. C. Kelley, Observations of paired electrostatic shocks in the polar magnetosphere, Phys. Rev. Lett., 38, 292, 1977.

Papadopoulos, K., A review of anomalous resistivity for the ionosphere, Rev. Geophys. Space Phys., 15, 113, 1977.

Pierce, J. R., Instability of hollow beams, IRE Trans. Electron Devices, 3, 183, 1956.

Ponyavin, D. I., M. I. Pudovlan, and S. S. Shazhin, Self-consistent field-aligned electric field in the earth's magnetosphere, Geomag. and Aeron., 17, 323, 1977.

Quon, B. M., and A. Y. Wong, Formation of potential double layers in plasmas, Phys. Rev. Lett., 37, 1393, 1976.

Sato, T. and H. Okuda, Ion acoustic double layers, Phys. Res. Lett., 44, 740, 1980.

Shawhan, S. D., C.-G. Fälthammar, and L. P. Block, On the nature of large auroral zone electric fields at 1 R_E altitude, J. Geophys. Res., 83, 1049, 1978.

Singh, N., Reflection of probing beams by an electric double layer, Cospar Space Research, 20, Pergamon Press, 1980.

Stenzel, R. L., M. Ooyama, and Y. Nukamura, V-shaped potential double layers in a strongly magnetized laboratory plasma, EOS, 61, 1083, 1980.

Störmer, C., The Polar Aurora, Oxford Univ. Press, Oxford, 1955.

Sydora, R. D., J. S. Wagner, L. C. Lee, E. M. Wescott, and T. Tajima, Electro-static Kelvin-Helmholtz instability in a radially injected plasma cloud, Phys. Fluids, (submitted) 1980.

Wagner, J. S., R. D. Sydora, L. C. Lee, T. J. Hallinan, S.-I. Akasofu, and T. Tajima, The formation of auroral folds, J. Geophys. Res., (submitted) 1980.

Wagner, J. S., T. Tajima, J. R. Kan, J. N. Leboeuf, S.-I. Akasofu, and J. M. Dawson, V-potential double layers and the formation of auroral arcs, Phys. Rev. Lett., 45, 803, 1980.

Webster, H. F., and T. J. Hallinan, Instabilities in charge-sheets and current sheets and their possible occurrences in the aurora, Radio Sci., 8, 475, 482, 1973.

GLOBAL SIMULATION OF AURORAL ARCS

Akira Miura[1*] and Tetsuya Sato[2]

[1]Institute of Geophysics and Planetary Physics, University of California, Los Angeles, CA 90024
*Permanent address: Geophysics Research Laboratory, University of Tokyo, Tokyo 113, Japan
[2]Institute for Fusion Theory, Hiroshima University, Hiroshima 730, Japan

Abstract. Numerical simulation of global formation of auroral arcs is performed for a three-dimensional coupled ionosphere-magnetosphere system. Under the presence of the stationary convection electric field, the field-aligned current driven by the polarization and drift of the ionospheric conductivity enhancement is shown to be responsible for the feedback instability (Sato, 1978) of the shear Alfvén wave. The simulation result demonstrates the dynamic process of auroral arc formation, in which an ionospheric perturbation nearly resonant with the magnetospheric Alfvén wave (toroidal mode) can grow eventually into auroral arcs. Many of the important characteristics of quiet auroral arcs are demonstrated, and the overall distribution of auroras is shown to depend critically on the stationary state of the auroral oval characterized by the large-scale Birkeland current, the electric field, and the electron density distribution.

Introduction

The coupling between the ionosphere and the magnetosphere is the key to understanding the formation of auroral arcs. A self-consistent dynamic treatment is necessary to clarify the chicken-egg cause and effect relationship among the field-aligned current, electrojets, conductivity enhancement, and electric field change, etc., associated with auroral arcs. The principal purpose of this paper is to clarify the basic mechanism underlying the feedback instability (Sato, 1978) and simulate a global formation of auroral arcs by the feedback instability in a three-dimensional ionosphere-magnetosphere coupling system.

A scenario of the arc formation in the present study might be summarized as follows: the stationary convection electric field, which is a permanent feature at auroral latitudes, is responsible for destabilizing the shear Alfvén wave in the magnetosphere (feedback instability). When the field-aligned current associated with this localized standing Alfvén mode reaches some critical value, a current-driven instability is excited at 1 R_E altitude (Kindel and Kennel, 1971). At this stage auroral arcs are formed in the polar ionosphere by precipitation of auroral electrons, which are accelerated by parallel electric fields maintained by the anomalous resistivity and/or double layer resulting from the current-driven instability. Although parallel electric fields are important in auroral acceleration, the present simulation assumes the parallel electric field is zero.

In section 2 we clarify the feedback mechanism in terms of a simple physical picture. Governing equations and a model of the auroral oval are given in sections 3 and 4, respectively. Simulation results are given and discussed in sections 5-8.

Physical Picture of the Feedback Instability

Let us consider the physical mechanism of the feedback instability, which arises under the action of the stationary electric field (current) in the ionosphere. We should first point out that the feedback instability derives its energy from the relative motion or slip between the magnetic field lines and the ionospheric perturbation driven by the stationary electric field. Figure 1 shows how this relative motion occurs in the coupling system. For comparison with the numerical results given in section 5, let us assume that the left and right sides of each panel (a-c) correspond to the north and south, respectively, and the magnetic field is downward. Given a northward component of the large-scale horizontal current J_{10} which may be the Pedersen current closing the large-scale Birkeland (field-aligned) current, a small density perturbation δn in the ionosphere, however small and by whatever mechanism it may be produced, is naturally accompanied by polarization charges represented by positive and negative signs in panel a: posi-

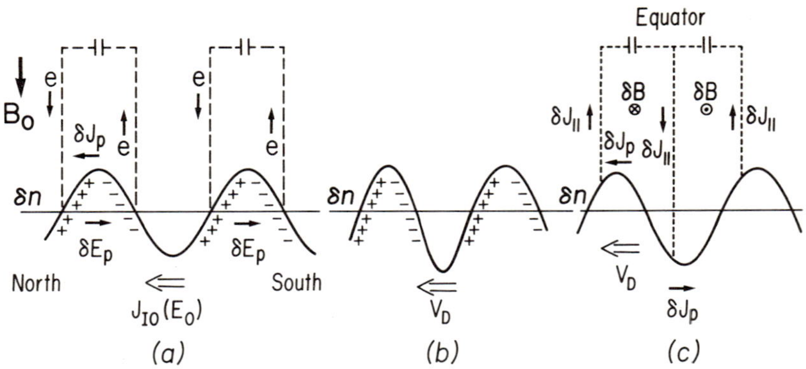

Fig. 1. Physical picture of the ionosphere-magnetosphere feedback instability. The field-aligned current is induced in the coupling system by the drift of the ionospheric density perturbation relative to the magnetic field line (panel c).

tive charges accumulate in the northern part of the density peak, and negative charges accumulate in its southern part. Owing to these polarization charges, the polarization electric field denoted by δE_p is set up in the ionosphere. In the actual coupled system, however, electrons can freely move along the field lines to neutralize the ionospheric space charges. More specifically, electrons come down along the field lines towards the positive part of the potential perturbation and go up from the negative part in order to neutralize space charges. As a result of this field-aligned motion of electrons and the horizontal motion of ions, a closed current loop, composed of perfectly conducting magnetic field lines, and the polarization current across the magnetic field is set up, as indicated by dashed lines in panel a. Hence the plasma density in the positive part of the potential perturbation increases (convergence of the plasma) but decreases in the negative part (divergence of the plasma). This means a net drift (denoted by V_D hereafter) of the original density perturbation in the direction of J_{10} relative to the magnetic field lines (panel a → panel b) [see Appendix A].

An important effect arises from this relative drift or slip under the action of the stationary electric field; namely, the ionospheric perturbation can now be considered as an ac perturbation having a Doppler-shifted frequency $\omega = k_\perp V_D$ seen from the reference system fixed to the zeroth-order magnetic field line motion, where $k_\perp$ is the wave number of the ionospheric perturbation. Therefore it is necessary to consider a temporal or MHD response of the magnetosphere to the ionospheric perturbation. For simplicity, let us assume that the stationary electric field $\vec{E}_o$ has only a northward component. The north-south ionospheric perturbation current can then be written as

$$\delta J_{NS} = \Sigma_p \delta E + \delta\Sigma_p E_o = \mu_o n_o e \delta E + \mu_o e \delta n E_o \qquad (1)$$

where μ_p is the ion Pedersen mobility and δE is the perturbed electric field. Note that the second term, which is hereafter denoted by δJ_p, is the perturbed current responsible for the polarization of the density perturbation. This perturbed current is caused by the motion of δn relative to the field lines. Since the first term, proportional to δE, contributes only to the damping of the perturbation (Joule dissipation), we consider only the response of the magnetosphere to δJ_p, which plays an active role in the feedback instability. Here, we should note that δJ_p accompanies naturally the toroidal magnetic field perturbation (shear Alfvén wave) in the magnetosphere. Therefore, if a fictitious current loop (dotted lines in panel c), composed of perfectly conducting magnetic field lines and the polarization current in the equator, is assumed fixed to the magnetic field lines, the toroidal magnetic flux inside it is time varying because of the ac variation of δJ_p seen from the frame fixed to the field lines. This temporal change of the magnetic flux brings an important physical effect in the time development of the perturbation.

Let us see in detail what happens in panel c. In panel c the magnetic flux in the left current loop due to δJ_p is into the page and is decreasing, since the northward current in the bottom of this loop is decreasing owing to the motion of δn. Therefore, Faraday's law says that the emf is induced in the current loop and the induction current flows in such a direction as indicated by solid arrows to keep the magnetic flux inside the loop from changing with time. In the right loop, however, the current is induced in the opposite sense, because the magnetic flux out of the page is decreasing. Consequently, upward field-aligned current appears wherever the density is enhanced, but downward current appears wherever the density is depleted. If we assume that the field-aligned current is carried by local acceleration of ambient electrons, the electrons come down to the position of

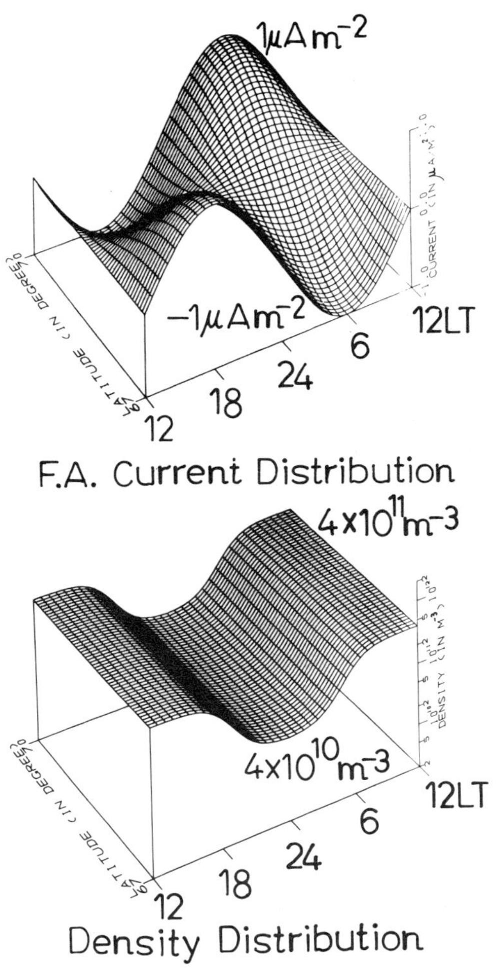

Fig. 2. Perspective view of the stationary models of the field-aligned current (upper panel) and the ionospheric plasma density (lower panel). Downward current is defined positive.

the density enhancement but go up from the density depletion; therefore, the density perturbation grows with time in panel c. In summary, the induction current associated with δJ_p flows so that it increases an original density perturbation due to the proper phase lag caused by motion of the ionospheric perturbation relative to the field lines. On the other hand, when the current is closed substantially by the polarization current associated with the Aflvén wave, the ionospheric perturbation is damped, since the phase relationship is shown to be just the opposite of that of the induction current (next section). Thus the phase of the field-aligned current perturbation plays a crucial role in discussing the feedback instability. Note that the frequency or the wavenumber $k_\perp$ of the perturbation is simply determined by the resonance condition

$$k_\perp V_D \simeq V_A/\ell \tag{2}$$

where ℓ is the field line length and V_A is the average Alfvén speed along the field line. Equation (2) says that the Doppler-shifted frequency of the ionospheric perturbation is nearly equal to the resonance frequency of the field line oscillation (standing Alfvén mode).

Finally, we should briefly mention the energetics of the present instability. The energy dissipation in the ionosphere can be calculated by multiplying (1) by δE^* and then taking the time-average during one wave period. It is obvious that the first term on the R.H.S. gives a usual Ohmic loss; however, the second term gives positive or negative dissipation depending on the phase difference between δn and δE. If δE is determined solely by the polarization of the density enhancement as shown in panel a of Figure 1, the second term gives a negative Ohmic dissipation, since $<\delta E_p \delta n> < 0$; this indicates that a combined effect of the density perturbation and the associated polarization electric field can feed energy into the standing Alfvén mode in the magnetosphere; in other words, the time average of the Poynting flux associated with the standing Alfvén mode becomes upward out of the ionosphere (Miura et al., 1980).

Governing Equations

In order to simulate the physical process described in the previous section we must self-consistently follow the time evolution of three unknowns $(n, j_{\shortparallel}, \phi)$ at the ionospheric level (110 km). To accomplish this generally for a realistic coupled ionosphere-magnetosphere system is a very difficult task; however, coupling by the shear Alfvén wave, as described above, is described by the following three governing equations (see Miura and Sato, 1980, for details).

<u>Continuity Equation of Ionospheric Plasma Density</u>

$$\frac{\partial n}{\partial t} + \frac{\vec{E} \times \vec{B}_o}{B_o^2} \cdot \vec{\nabla}_\perp n = \frac{-j_{\shortparallel}}{eh} - \gamma(j_{\shortparallel}) \frac{j_h}{eh} - \alpha(n^2 - n_o^2) \tag{3}$$

<u>Current Continuity Equation in the Ionosphere</u>

$$\vec{\nabla}_\perp \cdot \vec{j}_\perp = \frac{j_{\shortparallel}}{h} \tag{4}$$

where on neglecting the contribution of the neutral wind to the ionospheric current

$$\vec{j}_\perp = -neM_H \frac{\vec{E} \times \vec{B}_o}{B_o} + neM_p \vec{E} \tag{5}$$

$$\vec{E} = -\vec{\nabla}_\perp \phi \tag{6}$$

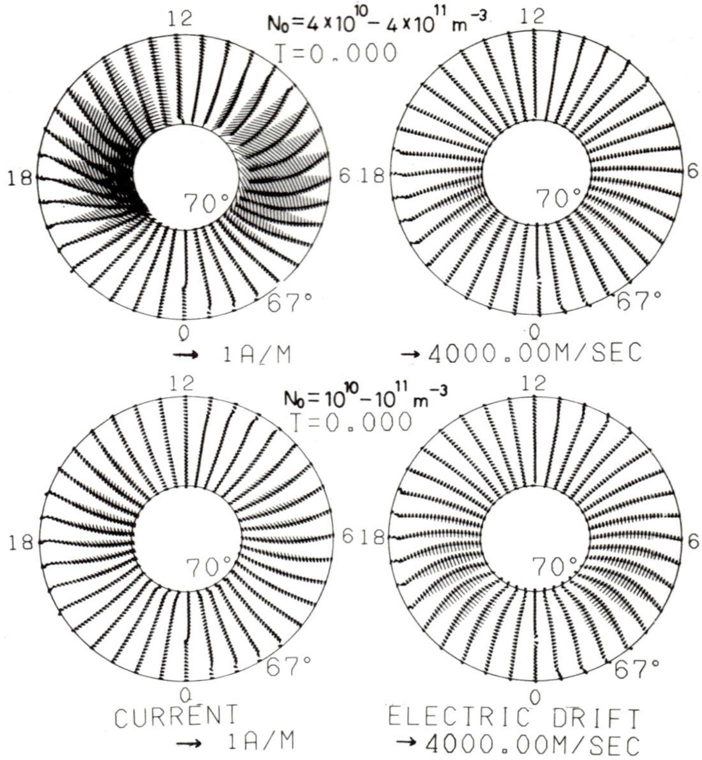

Fig. 3. Stationary current (left panels) and $\vec{E} \times \vec{B}$ drift (right panels) patterns for two ionospheric density models

Model Equation for Magnetospheric Response to the Ionospheric Perturbation

$$\frac{2\mu_o \ell}{\pi^2} \frac{d\tilde{j}_{\shortparallel}}{dt} = \left[\left(\frac{\ell}{\pi V_A} \right)^2 \frac{d^2}{dt^2} + 1 \right] \cdot \nabla_{\perp}^2 \tilde{\phi} \qquad (7)$$

where

$$\frac{d}{dt} = \frac{\partial}{\partial t} + \frac{\vec{E} \times \vec{B}_o}{B_o^2} \cdot \vec{\nabla}_{\perp}, \qquad (8)$$

and the tilde denotes the time-dependent term. In the above we have defined $j_{\shortparallel} > 0$ for downward current; h is the effective height range of the E layer, j_h is the current density carried by auroral electrons, $\gamma(j_{\shortparallel})$ is the ion-pair production rate depending on $j_{\shortparallel}$, α is the recombination coefficient, $M_H \simeq \mu_{\bar{H}} \simeq B_o^{-1}$ is the electron Hall mobility, $M_p \simeq \mu_p$ is the ion Pedersen mobility, ℓ is the field line length from the ionosphere to the equator, and V_A is the average Alfvén speed along the field line. The typical values for the above constants are used in simulation runs.

Equation (3) governs the time evolution of the ionospheric plasma density n. The potential ϕ in the ionosphere is determined by (4), which, after substituting (5) and (6) into (4), becomes an elliptic partial differential equation on ϕ. Thus the three governing equations (3), (4) and (7) for the three unknowns are sufficient to describe the dynamic process of quiet auroral arc formation by the ionosphere-magnetosphere coupling in the self-consistent manner. The only process which is beyond the description of the present fluid simulation is the precipitation of auroral electrons, which requires detailed knowledge of the acceleration process in the topside ionosphere. Therefore, we include this effect in (3) by assuming a dependence of γ on $j_{\shortparallel}$, such that $\gamma(j_{\shortparallel}) = 0$ for $|j_{\shortparallel}| < |j_{cr}|$, but $\gamma = 10$ for $j_{\shortparallel} \leq j_{cr} < 0$. This assumption is based on the previous scenario that electrons are accelerated by a parallel electric field maintained by an anomalous resistivity and/or a double layer, which occur when the upward current density exceeds some critical current density for the current-driven instability.

The response equation (7) can be derived from the ideal MHD equations ($E_{\shortparallel} = 0$) for a low frequency standing Alfvén mode, which satisfies $\delta j_{\shortparallel} = 0$ at the equator (see Miura and Sato, 1980, for details). Great care has been paid in solving (7) numerically (see Appendix B). Since the magnetospheric response to the ionospheric per-

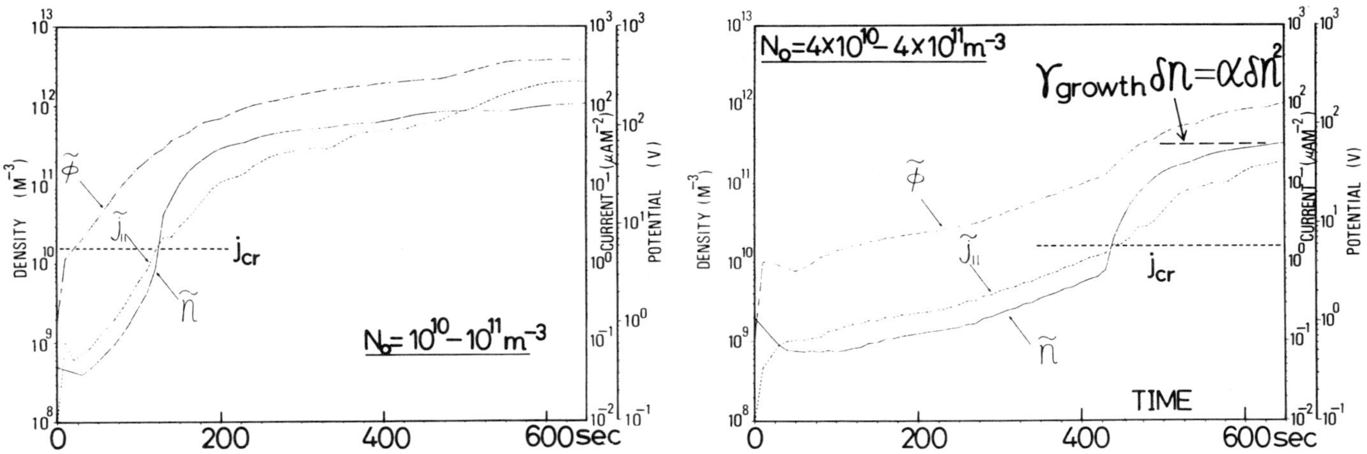

Fig. 4. Temporal evolution of peak amplitudes of the density (solid line), field-aligned current (dotted line) and potential (dot-dash) disturbances for two ionospheric density models. The horizontal dotted lines represent the assumed critical current density for the precipitation of auroral electrons. The horizontal dashed line in the lower panel shows the calculated saturation amplitude by nonlinear recombination damping.

turbation via Alfven wave is the key to understanding the ionosphere-magnetosphere coupling, let us consider the physical property of (7) in detail. We consider this in a reference frame fixed to the magnetic field lines; therefore, we rewrite (7) as

$$\left(\frac{2\mu_o \ell}{\pi^2}\right) \frac{\partial \tilde{j}_{\shortparallel}}{\partial t} = \left[\left(\frac{\ell}{\pi V_A}\right)^2 \frac{\partial^2}{\partial t^2} + 1\right] \nabla_\perp^2 \tilde{\phi} \qquad (9)$$

Assuming a perturbation of the form exp $[i(k_\perp r - \omega t)]$ in (9) yields

$$\left(\frac{2\mu_o \ell}{\pi^2}\right) i\omega \tilde{j}_{\shortparallel} = \left[-\left(\frac{\ell}{\pi V_A}\right)^2 \omega^2 + 1\right] k_\perp^2 \tilde{\phi} \qquad (10)$$

Let us consider two frequency limits of this equation which help us understand the feedback effect of the standing Alfvén mode on the ionospheric perturbation.

1) Low frequency limit: $\omega < \pi V_A/\ell$

In this case, the second term in the R.H.S. of (10) dominates the first term and (10) can be written as

$$\tilde{j}_{\shortparallel}^{\,I.S.} \sim -i \frac{1}{\omega} \frac{\pi^2}{2\mu_o \ell} k^2 \tilde{\phi}^{I.S.} \qquad (11)$$

where the superscript I.S. (ionosphere) means that $\tilde{j}_{\shortparallel}$ and $\tilde{\phi}$ are perturbations at the ionospheric height (110 km). Physically, (11) gives the induction current flowing along the field line and gives $\pi/2$ phase lag of the precipitation of electrons (upward field-aligned current) behind the positive potential perturbation. As we have seen in panel c of Figure 1, this phase lag is caused by the drift of the ionospheric perturbation relative to the magnetic field lines. It is obvious, as discussed in §2, that this phase relationship between $\tilde{\phi}^{I.S.}$ and $\tilde{j}_{\shortparallel}^{\,I.S.}$ is favorable for the growth of the density perturbation, and thus for destabilizing the standing Alfvén mode in the magnetosphere. It is worth noting that (11) describes that a flux tube having unit cross section at the ionospheric level is equivalent to an inductance of $2\mu_o\ell/\pi^2$.

2) High frequency limit: $\omega > \pi V_A/\ell$

In this frequency range the first term in the R.H.S. dominates the second and the response can be written as

$$\tilde{j}_{\shortparallel}^{\,I.S.} \simeq i\,\omega \frac{\ell}{2\mu_o V_A^2} k_\perp^2 \tilde{\phi}^{I.S.} \qquad (12)$$

It is obvious that (12) gives a response when the current is closed only by the polarization current across the magnetic field, because (12) can be derived straightforwardly by integrating the polarization current $j_p = \frac{\partial E}{\partial t}/(\mu_o V_A^2)$ along the field line on the assumption that the field line is equi-potential. By the circuit analogy, (12), says that the flux tube having cross section at the ionospheric level is equivalent to a capacitance of $\ell/(2\mu_o V_A^2)$. Owing to the appearance of (+i) in (12), giving the opposite phase relationship to that of induction current expressed by (11), this capacitive current gives no linearly unstable solution in the coupling system. In this regard it is worth mentioning that the Atkinson's model (1971) of auroral arc includes only the polarization current as the closure current in the magnetosphere.

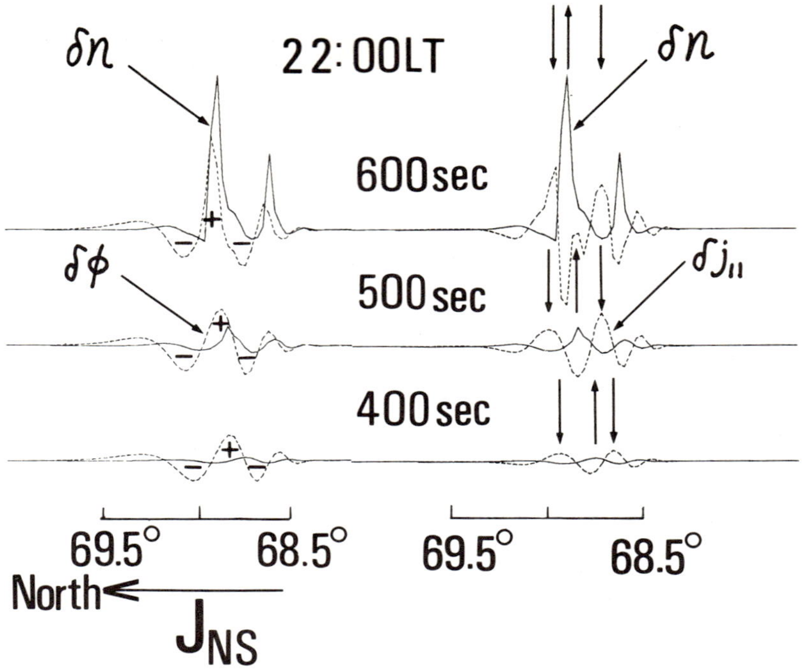

Fig. 5. Latitudinal profiles of the density (solid line) - potential (dotted line) (left panel) and the density (solid line) - field-aligned current (dotted line) (right panel) for three consecutive times at 2200 LT.

Model and Stationary State of the Auroral Oval

We solve the closed set of equations (3), (4), and (7) for a realistic model of the auroral oval, which is confined latitudinally from 67° to 70°. We include three important observational results in this model: First, we take into account a local time dependence of the electron density. Shown in the lower panel of Figure 2 is a perspective view of the density model, where the density is decreasing monotonically from N_{noon} at noon to $N_{midnight}$ at midnight. In the following we express the ionospheric density variation by the parameter $N_o = N_{midnight} - N_{noon}$. Second, two pairs of large-scale stationary Birkeland currents into and out of the ionosphere are included. On the basis of the data analysis by Iijima and Potemra (1976) we give a sinusoidal variation of $j_{\parallel o}$ as shown in the upper panel of Figure 2. Here, current is defined positive for downward current and the peak current density is taken to be 1 $\mu A/m^2$. Third, the potential difference of several tens of kilovolts across the polar cap (Heppner, 1977) is included by imposing a sinusoidal variation of the potential on the poleward boundary of the oval; the potential has the positive peak at 0600 LT and the negative peak at 1800 LT, giving a total potential difference of 50 kV.

The stationary current and $\vec{E} \times \vec{B}$ drift (convection) patterns at T = 0 calculated by solving an elliptic partial differential equation (4) on ϕ are shown in left and right panels of Figure 3, respectively, for two ionospheric density models with $N_o = 4 \times 10^{10} - 4 \times 10^{11}$ m^{-3} (upper panels) and $N_o = 10^{10} - 10^{11}$ m^{-3} (lower panels).

The right panels show $\vec{E} \times \vec{B}$ drift (convection) is sunward in the auroral oval; this indicates that electric field is predominantly northward on the dusk and southward on the dawn primarily for satisfying the current closure of the stationary field-aligned current in Figure 2. The left panels show the ionospheric current is antisunward with a substantial north-south component, which is mainly the ion Pedersen current. Note that this north-south current can feed energy for driving the induction current in the coupling system as we have seen in Figure 1. The north-south electric field takes a peak value of 107 m V/m in the upper right panel and 138 m V/m in the lower right panel; this difference of the electric field strength is due to the factor of four difference in the conductivity of two models. An important difference between the two ionospheric models is a remarkable dawn-dusk asymmetry of the current pattern in the upper left panel in comparison with a rather symmetrical pattern in the lower left panel.

Dynamic Formation of Auroral Arcs

Let us now examine how and where a broad, small amplitude density perturbation, initially given at the mid latitude of the oval, develops with time.

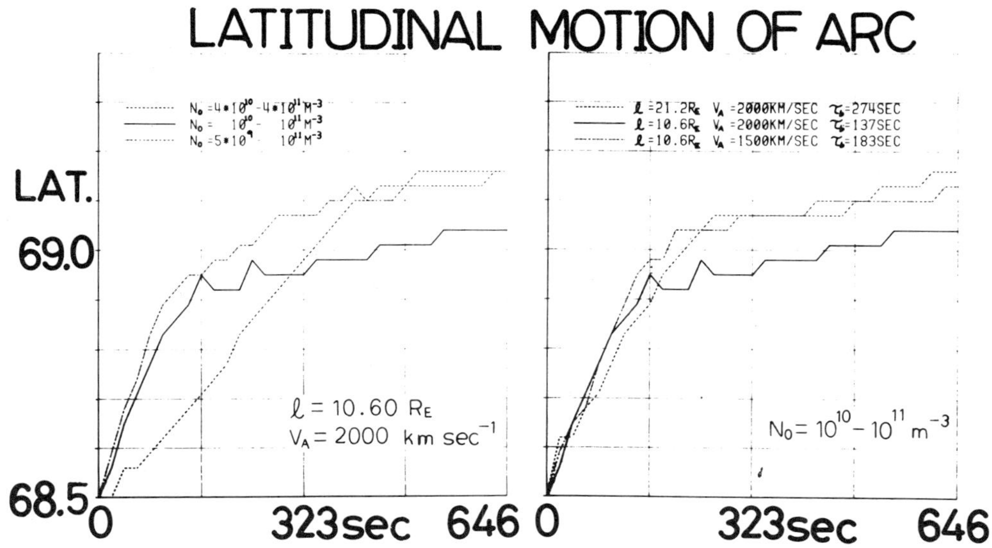

Fig. 6. Latitudinal position of an arc at 2200 LT against time for three different ionospheric density models (left panel) and for three different magnetospheric models (right panel).

Figure 4 shows the temporal evolution of the peak amplitudes of the density (solid line), field-aligned current (dotted line) and potential (dot-dash) disturbances for two ionospheric density models with $N_o = 10^{10} - 10^{11}\ m^{-3}$ (upper panel and $N_o = 4 \times 10^{10} - 4 \times 10^{11}\ m^{-3}$ (lower panel). The same magnetospheric parameters, $\ell = 10.6\ R_E$ (field line length of the dipole magnetic field at INV = 70°) and $V_A = 2000$ km/s, are used in both models. Each panel shows initial decay of the density disturbance by the recombination loss and subsequent linear growth of all disturbances by the feedback instability. As was stated in the scenario, when the field-aligned current reaches the critical current density j_{cr} represented by the dotted line, precipitation of auroral electrons begins and the density starts to grow more rapidly than the previous phase. This occurs in about 7 minutes from the start in the lower panel, however, only in 2 minutes in the upper panel. This marked contrast between the two panels is attributable to the factor of four difference in the recombination loss rate $2\alpha n_o$. Note in both panels that the field-aligned current cannot immediately respond to the rapid growth of the density, as it maintains a rather constant value during the sudden increase of the electron density. This is because the toroidal magnetic field associated with the field-aligned current is not allowed to change suddenly.

In the final stage all disturbances tend to saturate. A good agreement between the amplitude of δn at the final stage and that calculated from $\gamma_{gr}\delta n_{sat} = \alpha\delta n_{sat}^2$ (dashed line) in the lower panel might explain the nonlinear saturation of the instability by the balance between the linear growth and the nonlinear recombination damping. Here, γ_{gr} is a growth rate determined from a slope of the density evolution after the precipitation of hot electrons, which was found to be ≃ 10 (equal to the ion pair production rate) times larger than the growth rate in the initial linear stage.

The left panel of Figure 5 shows the latitudinal profiles of the density (solid line), the potential (dotted line) and the right panel shows the density (solid line), the field-aligned current (dotted line) for three consecutive times at 22:00 LT. In the bottom left a small density undulation can be seen, and it is polarized so that the potential is positive on the poleward side of the density peak and negative on its equatorward side. This potential distribution is created by the polarization of the density enhancement by the stationary preexisting electric field (current), which is northward at this local time (see Figure 3). As the physical picture in Figure 1 says, the whole distrubance is moving northward in the direction of the stationary current J_{NS} relative to the field line, and this relative drift causes the position of the upward field-aligned current (downward electron flow) to coincide with the position of the density enhancement (see right panels).

Finally, in order to demonstrate the close relationship between arc motion and arc growth we show in Figure 6 the latitudinal arc position at 2200 LT against time for three different ionospheric density models (left panel) and for three different magnetospheric models (right panel). A remarkable fact is that regardless of the ionospheric or magnetospheric models, the latitudinal speed of auroral arcs becomes nearly zero in the final saturation stage; this is consistent with the previous physical picture that the drift of the ionospheric disturbance relative to the magnetic field line is driving the feedback instability. In this regard, note also that the velocity of the poleward arc motion during the early stage of evolution increases with the growth rate of the auroral arcs (compare Figure 6 with Figure 4).

Electrodynamic Property Associated with Auroral Arcs

We summarize in Figure 7 latitudinal profiles of the field-aligned current (top), electron density (middle), electric fields, and the height integrated horizontal currents (bottom) in the eveing sector (2200LT) for two models; the left panel shows those profiles at T = 600 sec, corresponding to the lower panel of Figure 4 and the right panel shows those at T = 150 sec, corresponding to the upper panel of Figure 4. Note that the arc is in the saturation stage in the left panel; however, the arc is in a premature stage in the right panel (see Figure 4). The top of each panel shows the presence of several pairs of well developed small-scale upward and downward field-aligned currents embedded in a large-scale upward Birkeland current. The dashed line represents the threshold field-aligned current density for excitation of the current-driven instability in the topside ionosphere. Remember that auroral electrons are assumed to precipitate where the upward current density exceeds the threshold value. Thus, wherever the upward field-aligned current density exceeds this threshold value, the large electron density enhancement appears. The peak electron density in the most developed density spike reaches 1.8×10^{11} m^{-3} in the left panel and 3×10^{10} m^{-3} in the right panel. It is worth noting here that the downward field-aligned current is associated with substantial depletion of the ionospheric plasma density; this is a reasonable consequence of the present simulation, since we have assumed that the downward current is carried by the upward motion of ionospheric electrons.

Two solid lines in the bottom of each panel represent the northward (E_{NS}) and eastward (E_{EW}) (actually westward, since $E_{EW} < 0$) components of the electric field. Although the background electric fields show rather different latitudinal profiles between the two panels owing to the difference of the background electron density, it is commonly seen in both panels that the northward component of the electric field is reduced inside the auroral arc and intensified in between the arcs. The intensification and reduction of the electric field associated with each arc is interpreted as a result of the polarization of the density enhancement by the preexisting electric field. The east-west component of the electric field (E_{EW}), on the other hand, shows no remarkable change, because the arc is aligned nearly in the east-west direction. The northward (j_{NS}) and eastward (j_{EW}) components of the height integrated current are shown by dotted

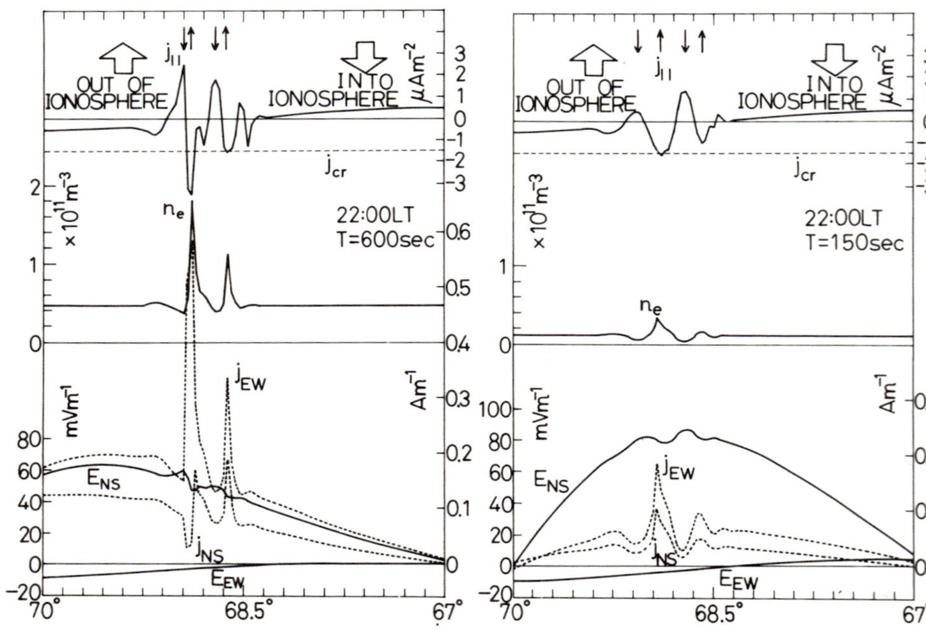

Fig. 7. Electrodynamic properties associated with auroral arcs at 2200 LT against time for three different ionospheric density models (left panel) and for three different magnetospheric models (right panel).

lines in the bottom of each panel. The eastward current is strongly enhanced inside the auroral arc; the peak current intensity in the left panel reaches 0.6 A/m, which is about four times larger than the background value. Note that this is not due to the Cowling effect, since in the present configuration $|E_{EW}| << |E_{NS}|$, which is not in favor of the Cowling effect. The northward component (j_{NS}) is also enhanced inside the auroral arc, but it is not so remarkable as the enhancement of the eastward current. Because of the enhancement of j_{NS} the electrojet direction is not necessarily parallel to the arc alignment in the east-west direction.

Global Development of Auroral Arcs

The time development of contours of the precipitating auroral electrons for three different models are shown in Figure 8. Parameters for simulation runs are (N_o = 4 x 10^{10} - 4 x 10^{11} m^{-3}, ℓ = 10.6 R_E, V_A = 2000 km/sec), (N_o = 10^{10} - 10^{11} m^{-3}, ℓ = 10.6 R_E, V_A = 2000 km/sec) and (N_o = 10^{10} - 10^{11} m^{-3}, ℓ = 21.2 R_E, $V_A \simeq$ 2000 km/sec) from the top row. In the top model, the arc is not seen at T = 400 sec, since the field-aligned current density is still under the threshold value for precipitation of auroral electrons. However, two distinguishable arcs can be seen in the evening sector at T = 500 sec, and they develop into more elongated forms (∿ 1500 km in the east-west direction) at T = 600 sec.

The overall distribution of auroral arcs in Figure 8 is found to depend importantly on the distribution of the electron density, the Birkeland current, and the horizontal current distribution in the auroral oval. The following three effects are of particular importance: First,

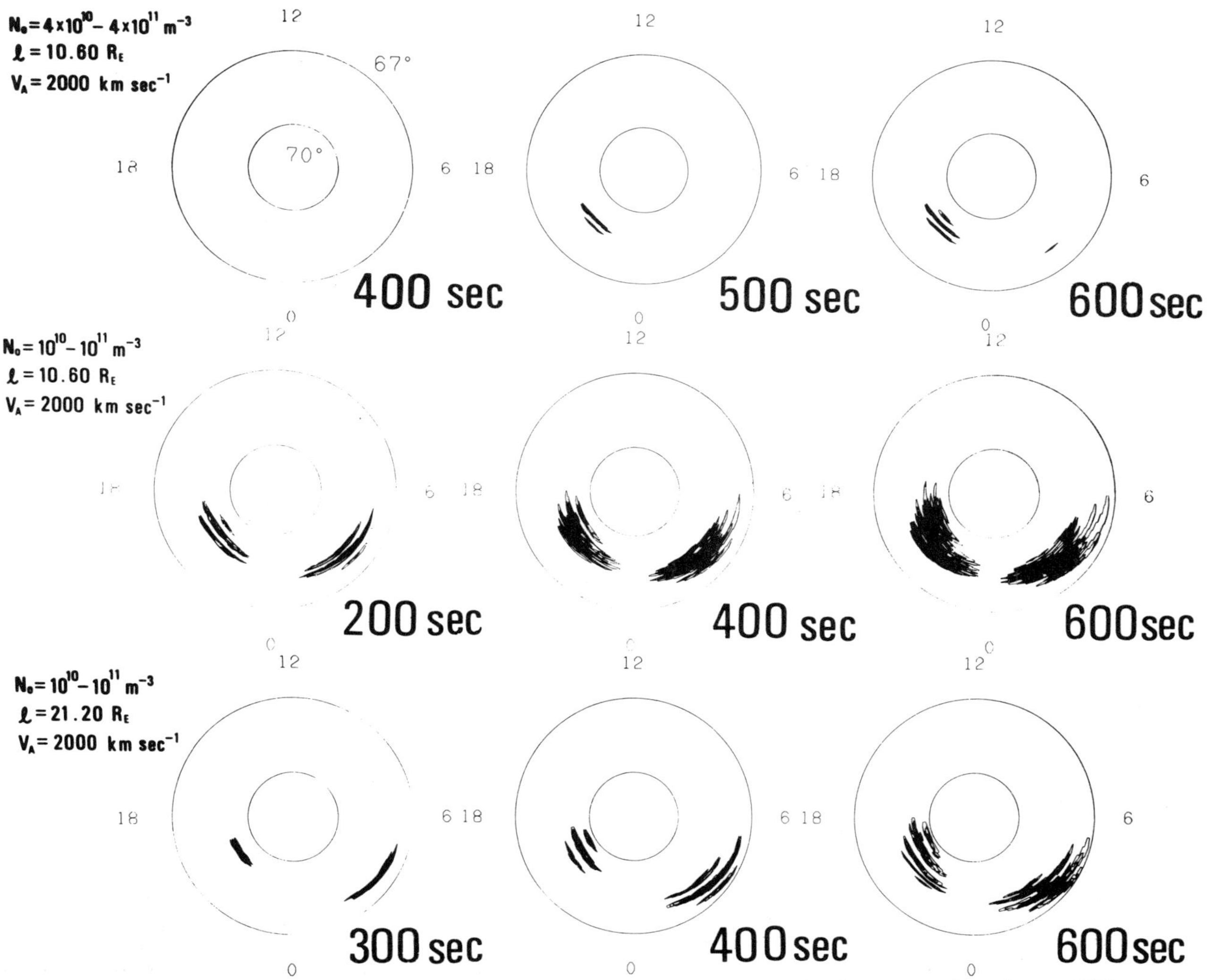

Fig. 8. Time development of contours of the precipitating auroral electrons for three different models.

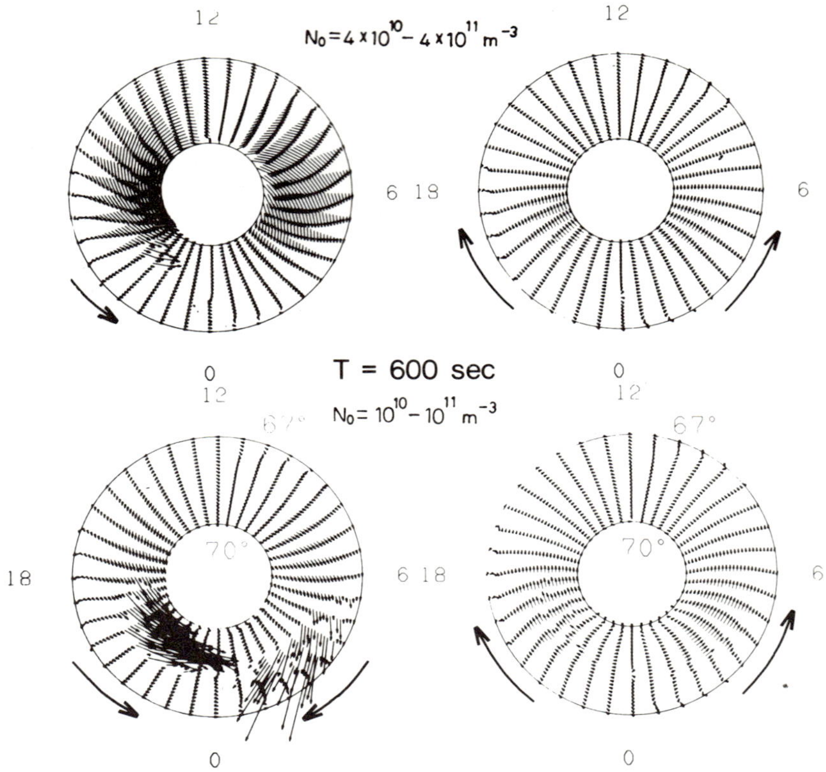

Fig. 9. Current (left panels) and E x B drift (right panels) patterns at T = 600 sec for the same two models as shown in Fig. 3.

auroral arcs appear only in the dark hemisphere and not in the sunlit hemisphere, because the large electron density and the recombination loss in the day hemisphere (see Figure 2) keeps auroral arcs from developing. Secondly, auroral arcs appear preferentially in the background upward Birkeland current region, i.e. the poleward region at dusk and the equatorward region at dawn. This is interpreted as follows: Figure 1 and the physical picture therein say that auroral arcs develop while they drift in the direction of the north-south preexisting ionospheric current. Since the initial perturbation is located at mid-latitude of auroral oval and j_{NS} is northward in the evening and southward in the morning, the observed latitudinal distribution of arcs is a reasonable consequence of this drift. Third, a marked dawn-dusk asymmetry of auroral appearance in the top model is due to a dawn-dusk asymmetry of the north-south, large-scale ionospheric current $|j_{NS}|$, which is the energy source of the arc formation. The asymmetry arises because the Hall current driven by the westward electric field in the dark hemisphere is northward and thus adds to the northward ion Pedersen current in the evening, but is opposite to the southward Pedersen current in the morning. In this regard, it is also worth noting that no arcs develop in the local time around midnight because of the lack of free energy to overcome the damping.

Another important fact to be noted in this figure is that the arc becomes wider when the field line length is doubled (compare the middle and the bottom panels). This is consistent with the quantitative model that the latitudinal scale of the auroral arc (width and/or spacing) is determined by the resonance between the standing Alfvén oscillation in the magnetosphere and the ionospheric perturbation drifting relative to the field line. If we use V_A = 2000 km/s, ℓ = 10.6 R_E, and V_D = 100 m/sec in (2) we obtain $\lambda_\perp \simeq$ 20 km, which agrees well with the simulation result.

Figure 9 shows the current (left panels) and electric drift (right panels) patterns at T = 600 sec for two models corresponding to the stationary current and E x B drift patterns in Figure 4. The bottom left panel shows a strong modification of the initial stationary current patterns by the appearance of multiple auroral arcs; intense auroral electrojets develop in the evening and morning sectors by the conductivity enhancement inside auroral arcs. The right panels show that, owing to the development of polarization electric field associated with auroral arcs, the E x B drift is reduced inside the auroral arcs but is intensified in between them. Note that a remarkable contrast between upper and lower panels is due to difference of the ionospheric density model.

Conclusion

A dynamic process of auroral arc formation by the ionosphere-magnetosphere coupling is demonstrated by numerical simulation. The ionosphere and the magnetosphere are coupled by the shear Alfven wave, which is subject to the feedback instability (Sato, 1978); this instability occurs under the presence of the large-scale stationary convection electric field by polarization and drift of the ionospheric conductivity enhancement and the associated field-aligned induction current.

The simulation result demonstrates many of the important characteristics of quiet auroral arcs to the extent that they can be compared quantitatively with the extensive observational results (see other papers in this issue); for example, the time scale of arc formation, the drift speed of arcs, the arc widths and spacing, the intensity of associated small-scale field-aligned currents and electrojet, electric field decrease inside auroral arcs, etc. (see Figure 7). The simulation result also provides interesting results concerning the global distribution of auroral arcs, which is found to be a direct consequence of the feedback instability.

Two points should be noted here regarding the present simulation: First, the important assumption in the present simulation, i.e. the acceleration of auroral electrons at 1 R_E altitude by the anomalous resistivity and/or double layer, has recently been attacked extensively by the numerical simulation (Sato and Okuda, 1980, Okuda and Ashour-Abdalla, 1981, and Pritchett et al., 1981). Although we did not include this effect in the circuit representation of the magnetosphere, it might be possible to include this as an equivalent resistive element consistent with the parallel electric field (Miura et al., 1980). Second, the background electric field in the present simulation is predominantly north-south direction; however, if the east-west component is dominant, as is observed before the substorm onset in the midnight local time (Mozer, 1971), the feedback instability under the configuration favorable for the Cowling effect might cause a stronger auroral development. In this respect, it is also worth mentioning that the neutral wind in the polar ionosphere might drive the arc formation by dragging ions and thus polarizing the ionospheric conductivity enhancement.

In conclusion, the present numerical simulation is the first demonstration of the formation of auroral arcs, in which the energy is continuously supplied by the large-scale electric field (current) in the auroral ionosphere.

Appendix A

Motion of the ionospheric density perturbation δn. Using typical parameters associated with the auroral arc, $|E_\perp| \sim 10$mV/m and $\nabla_\perp \sim 1/10$km, the ionospheric charge density $|n^+-n^-|$ associated with the arc is calculated as $|n^+-n^-| \sim \epsilon_0|\vec{\nabla}_\perp \cdot \vec{E}_\perp|/e \sim 5\times10^{-3}$, where $|\vec{\nabla}_{\shortparallel} \cdot \vec{E}_{\shortparallel}| << |\vec{\nabla}_\perp\cdot\vec{E}_\perp|$ is assumed in the ionosphere. Since $|n^+-n^-| << n^+ \sim n^- \sim 10^6 m^{-3}$ in the ionosphere, the quasi-neutrality assumption $n^+ \sim n^-$ is well satisfied. Therefore after making use of $n^+ \sim n^- = n$ the continuity equations of electrons and ions can be written as (see Sato, 1978).

$$\frac{\partial n}{\partial t} + \mu_H^- \vec{\nabla}n\cdot \frac{\vec{E}\times\vec{B}_o}{B_o} - \mu_p^- \vec{\nabla}\cdot(n\vec{E}) = -\frac{\gamma+1}{eh} j_h - \frac{1}{eh} j_c - \alpha\,(n^2 - n_o^2) \qquad (A.1)$$

$$\frac{\partial n}{\partial t} + \mu_H^+ \vec{\nabla}n\cdot \frac{\vec{E}\times\vec{B}_o}{B_o} + \mu_p^+ \vec{\nabla}\cdot(n\vec{E}) = -\frac{\gamma}{eh} j_h - \alpha(n^2 - n_o^2) \qquad (A.2)$$

where the charge separation term $n\vec{\nabla}\cdot\vec{E}$ is responsible for keeping the plasma quasi-neutral.

In order to clarify the motion of δn it is more convenient to use two equations derived from (A.1) and (A.2). The first equation is derived by multiplying (A.1) by μ_p^+ and (A.2) by μ_p^- and then adding the resulting two equations:

$$\frac{\partial n}{\partial t} + \mu_H^- \vec{\nabla}n\cdot\frac{\vec{E}\times\vec{B}_o}{B_o} = -\frac{\gamma}{eh} j_h - \frac{j_{\shortparallel}}{eh} - \alpha(n^2-n_o^2) \qquad (A.3)$$

This is the equation (3). By multiplying (A.1) by μ_H^+ and (A.2) by μ_H^- and then subtracting one from the other, we obtain the second equation:

$$\frac{\partial n}{\partial t} + \mu_p^+ \vec{E}\cdot\vec{\nabla}n = -\frac{\gamma}{eh} j_h - \mu_p^+ n\,\vec{\nabla}\cdot\vec{E} - \alpha(n^2-n_o^2) \qquad (A.4)$$

In the above equations we have assumed $\mu_p^+ >> \mu_p^-$ and $\mu_H^- >> \mu_H^+$. Note that in (A.3) the effect of the ionosphere-magnetosphere coupling is included in $j_{\shortparallel}$, although it is included in $\vec{\nabla}\cdot\vec{E}$ in (A.4). Let us consider two limiting cases of the coupling under the assumption $j_h = 0$: (1) No coupling, i.e., $j_{\shortparallel} = 0$, (2) No charge-up of the ionosphere, i.e., $\vec{\nabla}\cdot\vec{E} = 0$. For the first case, (A.3) says that δn moves with the E x B drift (convection) velocity. For the second case, in which $\vec{\nabla}\cdot\vec{E}$ in the ionosphere is completely cancelled out by $j_{\shortparallel}$, (A.4) says that δn moves with the ion Pedersen drift velocity. In the general case both $\delta j_{\shortparallel}$ and $\vec{\nabla}\cdot\vec{E}$ remain finite in the ionosphere and related to each other by the magnetospheric impedance $Z(\omega)$ as $\vec{\nabla}_\perp\,\delta\vec{E} = -Z\delta j_{\shortparallel}$. In this case the phase velocity of δn in the ionospheric frame is given by (see Sato, 1978 and Miura and Sato 1980)

$$\frac{\omega}{k_\perp} = M_p E_{xo} \frac{1}{1+\Sigma_p Z} + M_H E_{yo} \frac{\Sigma_p Z}{1+\Sigma_p Z} + i\,\frac{2\alpha n_o}{k_\perp} \qquad (A.5)$$

Taking $Z = \infty$ or $Z = 0$ in (A.5) yields the phase velocity for the previous two limiting cases. (A.5) means that unless $Z = \infty$, there is always a relative drift between the ionospheric density perturbation and the magnetic field lines convecting with E x B drift velocity.

Appendix B

Numerical procedure solving (7). If we define U and W, such that

$$\frac{dU}{dt} = \nabla_{\perp}^{2}\tilde{\phi} \qquad (B.1)$$

$$W = \frac{d}{dt}\,\nabla_{\perp}^{2}\tilde{\phi} \qquad (B.2)$$

the equation (7) can be written as

$$\frac{d\tilde{j}_{\parallel}}{dt} = C\,\frac{dW}{dt} + \frac{1}{L}\,\frac{dU}{dt} \qquad (B.3)$$

where $C = \ell/(2\mu_o V_A^2)$ and $L = 2\,\mu_o \ell/\pi^2$. Since (B.3) means that $\tilde{j}_{\parallel} - CW - U/L$ is a characteristic of the convection motion, $j_{\parallel}$ can be calculated from

$$\tilde{j} = CW + U/L \qquad (B.4)$$

where $\tilde{j}_{\parallel}, \tilde{\nabla}_{\perp}^{2}\,\tilde{\phi} \rightarrow 0$ at $t \rightarrow \infty$. The time integration of U in (B.1) and the time differentation in (B.2) are made by using upstream-downstream difference scheme to prevent a spurious numerical dispersion.

Acknowledgments. One of the authors (A.M.) would like to thank W. Lotko, R. M. Robinson, and T. Tamao for useful discussions and D. W. Swift for valuable comments and the careful reading of the manuscript. This work was partly supported by National Aeronautics and Space Administration contract #NAGW-78.

References

Atkinson, G., Auroral arcs: Result of the interaction of a dynamic magnetosphere with the ionosphere, J. Geophys. Res., 75, 4746-4755, 1970.

Heppner, J. P., Empirical models of high-latitude electric fields, J. Geophys. Res., 82, 1115-1125, 1977.

Iijima, T. and T. A. Potemra, The amplitude distribution of field-aligned currents at northern high latitudes observed by Triad, J. Geophys. Res., 81, 2165-2174, 1976.

Kindel, J. M. and C. F. Kennel, Topside current instabilities, J. Geophys. Res. 76, 3055-3078, 1971.

Miura, A. and T. Sato, Numerical simulation of global formation of auroral arcs, J. Geophys. Res., 85, 73-91, 1980.

Miura, A., T. Sato, and T. Tamao, Effect of anomalous resistivity on the ionosphere-magnetosphere coupling, EOS, Trans. AGU, 61, 1074, 1980.

Mozer, F. S., Origin and effects of electric fields during isolated magnetospheric substorms, J. Geophys. Res., 76, 7595-7608, 1971.

Okuda, H. and M. Ashour-Abdalla, Formation of a conical distribution and intense ion heating in the presence of hydrogen cyclotron waves, submitted to Geophys. Res. Lett., 1981.

Pritchett, P. L., M. Ashour-Abdalla, and J. M. Dawson, Simulation of the current-driven electrostatic ion cyclotron instability, submitted to Geophys. Res. Lett., 1981.

Sato, T., A theory of quiet auroral arcs, J. Geophys. Res., 83, 1042-1048, 1978.

Sato, T. and H. Okuda, Ion-acoustic double layers, Phys. Rev. Lett., 44, 740-743, 1980.

ION ACOUSTIC DOUBLE LAYERS

Tetsuya Sato

Institute for Fusion Theory, Hiroshima University, Hiroshima 730, Japan

and

H. Okuda

Plasma Physics Laboratory, Princeton University, Princeton, NJ 08544 USA

Abstract. Recent work on ion acoustic double-layers has been reviewed. It is emphasized that the spontaneous generation of negative potential spikes are responsible for ion acoustic double layers.

Introduction

It is quite possible that auroral particles are generated near the magnetospheric equator and ejected toward the earth along the magnetic field. Observations, however, often go against this possibility and suggest acceleration by a field-aligned electric field at a limited low altitude region (see, e.g., Arnoldy, 1974; Mozer et al., 1977).

The possibility of accelerating electrons by a localized potential difference along the magnetic field, double layer, has been recognized for some time (see, Block, 1975). In recent years considerable progress has been made on the understanding of the dynamical processes associated with the formation of double layers by means of numerical simulations (Goertz and Joyce, 1975; Joyce and Hubbard, 1978; Hubbard and Joyce, 1979; DeGroot et al., 1977; Sato and Okuda, 1980, 1981; and Wagner et al., 1980) along with experimental efforts (Torven and Anderson, 1979; Quon and Wong, 1976; Coakley and Hirshkowitz, 1979; and Stenzel et al., 1980).

The primary concern of this paper is to summarize the recent computer simulation results by the present authors (Sato and Okuda, 1980, 1981) with particular emphasis on the formation mechanism of double layers and on the possibility of double layer formation in the ion acoustic regime which was not recognized in the previous studies.

We have carried out one-dimensional particle simulations for two different boundary conditions, periodic and nonperiodic. In the latter case, an external battery and a resistance are connected in series to the simulation system to model the magnetospheric convection electric field and the ionospheric Pedersen resistance. In the magnetosphere an electric field can exist perpendicular to the earth's magnetic field whereby magnetospheric plasmas undergo convective motion. In the flow reversal region, space charges accumulate and drive field-aligned currents into (or away from) the ionosphere connecting to the ionospheric Pedersen current (see, Sato and Iijima, 1979). In this simulation we wish to demonstrate that a potential difference can spontaneously emerge across the system starting from a zero potential difference.

The following parameters are assumed throughout the simulations: initial electron drift speed $v_d = 0.6\ v_{eth}$ (v_{eth} is the electron thermal speed), the temperature ratio $T_e/T_i = 20$, the mass ratio $m_i/m_e = 100$, the plasma parameter $n\lambda_D = 100$, the time step $\Delta t = 0.2\ \omega_{pe}^{-1}$ and the mesh size $\Delta = \lambda_D$.

Periodic System

Initially, the system length L was chosen conventionally to be 128 λ_D and 256 λ_D. In these cases an anomalous resistivity was observed to arise during the growing phase of ion acoustic instability in agreement with the previous results (Biskamp and Chodura, 1971), but formation of double layers were not found for these simulations. As shown later, this is because the system length was so short that the downstream electons were completely mixed with the upstream electrons by the time the double layers developed. Then the system length was increased to L = 512 λ_D with the other parameters remaining the same. A drastic change appeared in the nonlinear evolution of ion acoustic waves. A weak but marked double layer-like structure developed as shown in Fig. 1. In order to confirm this phe-

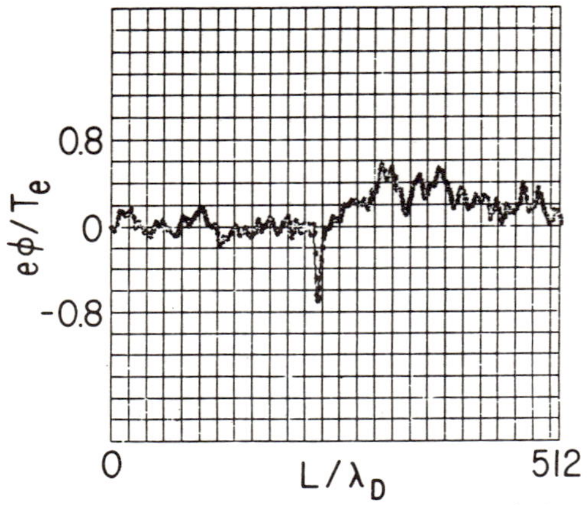

Fig. 1. **Potential structure at $t\omega_{pe}$=1432 for a periodic system with L=512λ_D**

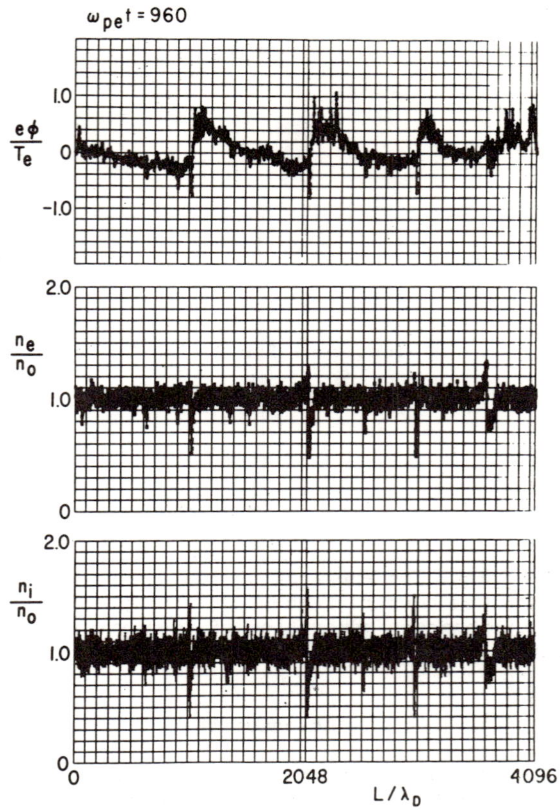

Fig. 3. **Potential structure(top) at $t\omega_{pe}$=960 for a periodic system with L=4096λ_D. The middle and bottom panel show the corresponding electron and ion density distributions.**

nomenon, another simulation was carried out with L = 1024 λ_D. Then, a more prominent potential structure resulted as shown in Fig. 2. It is found that the double layer-like structure is not stable but is destroyed by emitting large amplitude ion acoustic solitons toward downstream. To distinguish this new finding from stable strong double layers we call it "ion acoustic double layers."

The transit time τ for a typical electron ($v \approx 0.6\ v_{eth}$) to traverse the system is given by $\tau = L/0.6\ v_{eth}$. Thus, $\tau \approx 850\ \omega_{pe}^{-1}$ for L = 512 λ_D and 1700 ω_{pe}^{-1} for L = 1024 λ_D, guarantees that periodicity would not have affected the development of the ion acoustic double layer for the case with L = 1024 λ_D.

It is then natural to ask what happens if the system length is increased further. Will stronger double layers be formed, or will double layers be generated in multiples? To answer this question, we ran a simulation in which the system length was increased to L = 4096 λ_D with the other parameters remaining the same. It was found that double layers are generated in multiples with a rather regular spacing of 1000 λ_D as shown in Fig. 3. Upon examination of the time development it was found that double layers develop rather independently if other layers are formed beyond the distance of roughly 1000 λ_D. In other words, an upstream double layer gives an influence upon the development of downstream double layers if they locate within 1000 λ_D and prohibit them from growing. It can be concluded, therefore, that double layers are formed independently with each other at an interval of about 1000 λ_D. It is worth noting that the behavior and structure of each layer are almost identical with each other except that there are some differences in their formation time.

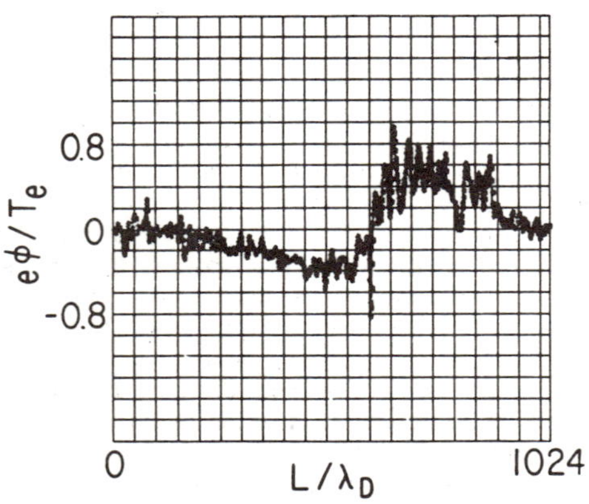

Fig. 2. **Potential structure at $t\omega_{pe}$=960 for a periodic system with L=1024λ_D**

It is interesting to observe that each double layer is formed only after a negative potential spike develops in a plasma. This fact strongly suggests that a negative potential spike must be the direct cause of an ion acoustic double layer. Suppose that a negative potential spike is formed in an electron stream. Then streaming electrons with energy less than the potential

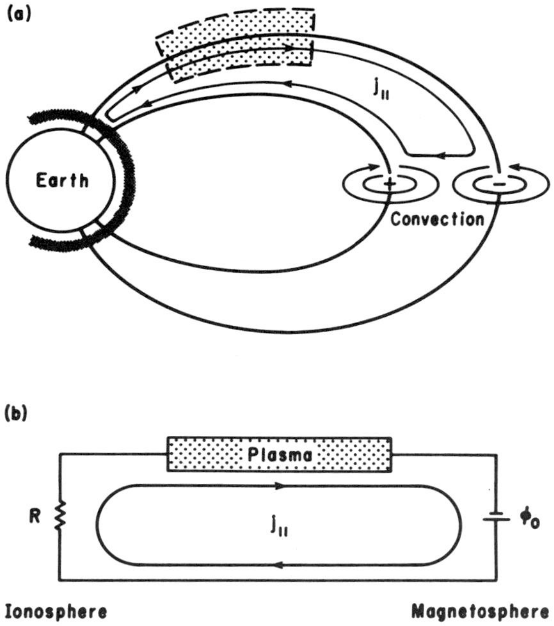

Fig. 4. Model of a coupled magnetosphere-ionosphere system (a) and its equivalent circuit (b) employed in the non-periodic simulations.

energy cannot pass the potential barrier and they are reflected, thereby decreasing the electron population at the downstream side. This implies that the potential must increase after the location of the negative spike. In this respect, we note that the existence of a steady solution of a negative potential spike is shown to exist and that a BGK type double layer solution can be obtained in the presence of such a negative potential spike (Hasegawa and Sato, 1981). One of the discrepancies between theory and simulation is that the width of the negative potential spike is predicted to be of the order of $2\pi\lambda_i$ (λ_i ion Debye length) while the simulations show it is about 5 ~ 10 electron Debye length wide. It is possible that the width can be substantially larger than the ion Debye length for a large amplitude spike because of the nonlinear effects.

Nonperiodic System

We have shown that ion acoustic double layers can indeed be generated in a periodic system. In the periodic system, however, no net potential difference appears across the system. It is interesting to attempt to simulate for a nonperiodic system where a field-aligned current is driven by a magnetospheric battery and is closed by the ionspheric Pedersen current. In order to avoid the excessive accumulation of charges at the boundary, particles leaving one boundary are reinserted from the other boundary as if the system is periodic. The model is shown in Fig. 4. Temporal development of the potential structure is shown in Fig. 5 where $e\phi_o$ = 200 T_e and L = 1024 λ_D with the other parameters remaining the same. The resistance R is chosen to satisfy that j(current) = $-n_o e v_d$ where v_d = 0.6 v_{eth}. As can be seen from Fig. 5, the potential difference across the system rises as a stepwise positive potential jump develops behind a negative potential spike. We performed another simulation with $e\phi_o$ = 10 T_e but no significant difference was observed.

Shown in the top panels of Fig. 6 is the temporal increase of the total kinetic energy for $e\phi_o$ = 10 T_e (left) and $e\phi_o$ = 200 T_e (right). It is seen that the total energy of the system increased by 15% due to the external energy source. Most of the increased energy went to the acceleration of electrons. The middle panels show the evolution of the wave energy and the bottom panels show the current interruption associated with the formation of double layers.

Figure 7 shows the potential structure for the case where L = 4096 λ_D and $e\phi_o$ = 200 T_e. A remarkable point is that compared with the pre-

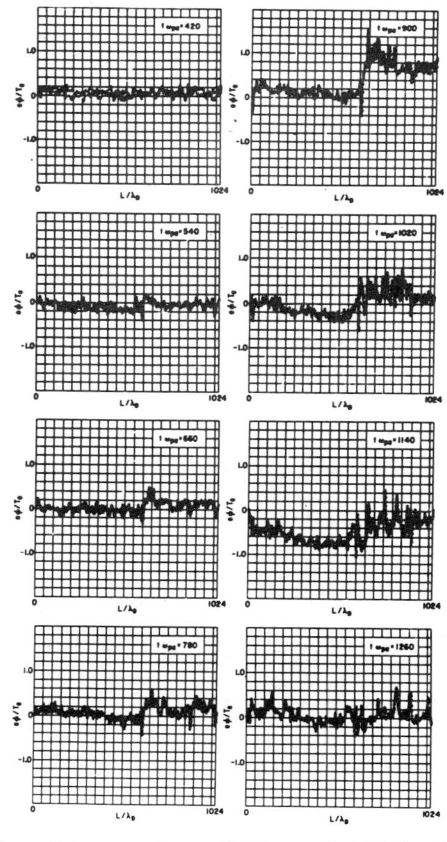

Fig. 5. Time series of the potential structure for the non-periodic system with L=1024λ_D and ϕ_0=200T_e/e.

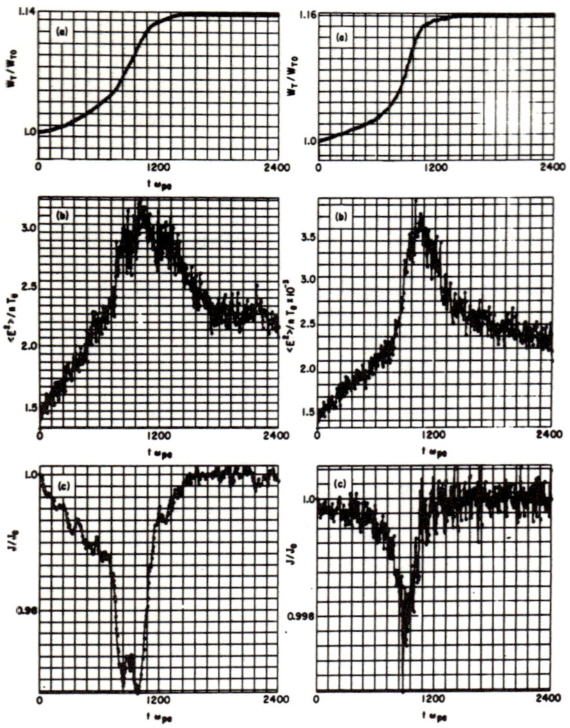

Fig. 6. Temporal changes of the total kinetic energy(top), the wave energy(middle) and the current(bottom) for the non-periodic system with $L=1024\lambda_D$. The left panels correspond to the case of $\phi_0=10T_e/e$ and the right ones to the case of $\phi_0=200T_e/e$.

vious case with $L = 1024\ \lambda_D$, the potential difference across the system is roughly doubled when $L = 4096\ \lambda_D$. This fact suggests that the total potential difference could become larger and larger as the system length increases.

Discussions

The present simulations have shown that the formation of a negative potential spike would play a key role in the formation of an ion acoustic double layer. At this stage, it would be premature to conclude that every double layer is associated with the negative potential. It would also be true, however, that the existence of a negative potential spike reflects low energy streaming electrons leaving a positive potential jump behind it. Therefore, it would be an important task to reveal the role of the negative potential spike in the formation of double layers.

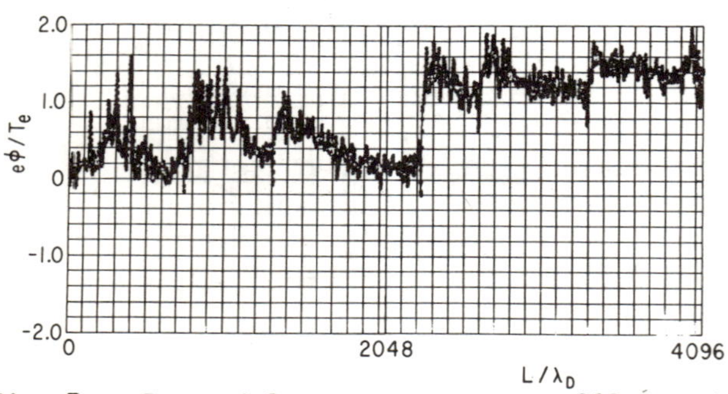

Fig. 7. Potential structure at $t\omega_{pe}=960$ for the non-periodic system with $L=4096\lambda_D$ and $\phi_0=200T_e/e$.

Acknowledgment. This work was supported by United States Department of Energy Contract No. DE-AC02-76-CHO3073.

References

Arnoldy, R. L., Auroral Particle Precipitation and Birkeland Currents, Rev. of Geophys. Space Phys., 12, 217 (1974).

Biskamp, D. and R. Chodura, Computer Simulation of Anomalous DC Resistivity, Phys. Rev. Lett., 27, 1553 (1971).

Block, L. P., Double Layers, in Physics of the Hot Plasma in the Magnetosphere, edited by B. Hultquist and L. Stenflo, Plenum Pub. Co., New York, p. 229, (1975).

DeGroot, J. S., C. Barnes, A. E. Walstead, and O. Buneman, Localized Structures and Anomalous DC Resistivity, Phys. Rev. Lett., 38, 1283 (1977).

Goertz, C. K. and G. Joyce, Numerical Simulation of the Plasma Double Layer, Astrophys. Space Sci., 32, 165 (1975).

Hasegawa, A. and T. Sato, Generation of Field-Aligned Currents During Substorm, in Dynamics of the Magnetosphere, edited by S.-I. Akasofu, D. Reidel Pub. Co., Boston, p. 529 (1979).

Hasegawa, A. and T. Sato, Negative Potential Structures and Formation of Double Layers, Phys. Fluids, (to appear 1981).

Hubbard, R. and G. Joyce, Simulation of Auroral Double Layers, J. Geophys. Res., 84, 4297 (1979).

Joyce, G. and R. Hubbard, Numerical Simulation of Plasma Double Layers, J. Plasma Phys., 20, 391 (1978).

Mozer, F. S., C. N. Carlson, M. K. Hudson, R. B. Torbert, B. Parady, J. Yatteau, and M. C. Kelley, Observations of Paired Electrostatic Shocks in the Polar Magnetosphere, Phys. Rev. Lett., 38, 292 (1977).

Sato, T. and T. Iijima, Primary Sources of Large-Scale Birkeland Currents, Space Sci. Rev., 24, 347 (1979).

Sato, T. and H. Okuda, Ion Acoustic Double Layers, Phys. Rev. Lett., 44, 740 (1980).

Sato, T. and H. Okuda, Numerical Simulations on Ion Acoustic Double Layers, J. Geophys. Res. (in press, 1981).

Stenzel, R. L., M. Goyama, and Y. Nakamura, V-Shaped Double Layers Formed by Ion Beam Reflection, Phys. Rev. Lett. 45, 1498 (1980).

Wagner, J. S., T. Tajima, J. R. Kan, J. N. Leboeuf, S.-I. Akasofu, and J. M. Dawson, V-Potential Double Layers and Formation of Auroral Arcs, Phys. Rev. Lett. 45, 803 (1980).

MAGNETIC X-POINTS, ISLANDS COALESCENCE AND INTENSE PLASMA HEATING

J. N. Leboeuf[1], T. Tajima[2] and J. M. Dawson[1]

[1]Department of Physics, University of California, Los Angeles, CA 90024

[2]Institute for Fusion Studies, University of Texas, Austin, TX 78712

Abstract. Release of a large amount of magnetic energy and intense plasma heating take place and an extremely hot ion tail is created, if and when tearing instability induced magnetic islands merge, according to fully self-consistent particle simulations. Although the tearing instability exponentiates the kinetic energy, it does not give rise to intense heating because of the small amount of magnetic energy available by reconnection. The stability and topology of the x-point depend on the crossing angle of the field lines at the x.

Introduction

It has been postulated that tearing instabilities play an important role to explain such explosive phenomena as disruptions in Tokamaks (Hosea et al., 1971; Mirnov el al., 1971), solar flares (Vasyliunas, 1975), and geomagnetic substorms (Parker, 1979). It is, however, recognized (Parker, 1979) that these instabilities are too mild to account for the large energy required by these phenomena. Here, we report on investigations using particle simulations that show that large amounts of magnetic energy are released and intense plasma heating occur when coalescence of the magnetic islands produced by the tearing instability takes place.

Some aspects of the coalescence process have already been studied both in theory (Finn et al., 1977) and MHD simulations (Pritchett et al., 1979); Biskamp et al., 1980); the procedure has been to study the stability to coalescence of an already formed chain of magnetic islands. For our present particle simulations, saturation of the tearing instability self-consistently creates the chain of islands; this situation turns out to be more unstable to coalescence. The final saturated state is the merged one. Magnetostatic particle simulations (Dickman et al., 1969; Lin, 1978); Katanuma and Kamimura, 1980) have thus far concentrated on the topological changes induced by the attraction of the current filaments. Wu et al., (1980) discussed the violent fluid flow, particles response and energization when coalescence occurs through combined research by MHD and electromagnetic particle simulations.

Model

Our electromagnetic particle simulation follows the evolution of an externally created magnetic x-point and its subsequent development, i.e., formation of a neutral sheet, induction of magnetic islands, and island coalescence. The electromagnetic code (Lin et al., 1974) is 2-1/2 dimensional and periodic. The case we focus on here is one where the rectangular system with $L_x \times L_y = 32\lambda_{De} \times 128\lambda_{De}$ has external current sheets uniformally distributed from $x = 5\lambda_{De}$ to $x = 123\lambda_{De}$ at $y = 0$ (and L_y) where λ_{De} is the Debye length of the electrons at $\lambda_{De} = \Delta$ at $t = 0$. Small gaps exist in the driving current at the ends of the system which create mirror fields. Typical parameters are: 4 particles of each species per cell, ion-to-electron temperature and mass ratios $T_i/T_e = 0.5$ and $M_i/m_e = 10$, the collisionless skin depth $c/\omega_{pe} = 5\lambda_{De}$, where c and ω_{pe} are the speed of light and the electron plasma frequency. The external current rises in time from zero and is crowbarred at $\omega_{pe}t = 31.4$ at a constant value; the ratio of the initial plasma pressure to the magnetic pressure next to the driving current gives $\beta = 0.2$. This setup is similar to the experimental one of Stenzel and Gekelman (1979). Straight-forward calculations of the magnetic Reynolds number from the finite-size particles collision frequency give $S \sim 800$. However, since the current flow direction is the ignorable one here, the actual collisionality in this direction is much smaller than the in-plane one and S is greater than the above value.

Results

Since the crossing angle of the field lines at the x-point is much less than 90° for the present long external sheets setup with narrow separation, the magnetic tension cannot hold the original x-point configuration as the external current rises

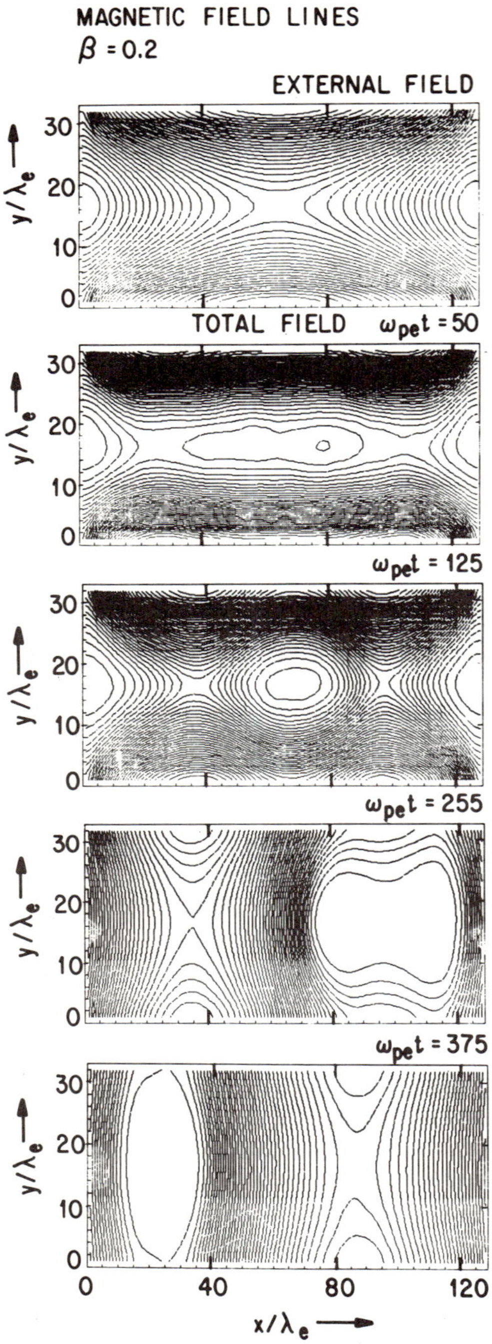

Fig. 1. Field line patterns in time.

(Leboeuf et al., 1980). A neutral sheet is formed within the rise time of the current with an island stably pinned in the gaps between the sheets. The neutral sheet tears and an o-point is induced at the place of the original x-point within an Alfven transit time $\tau_H \simeq 100\omega_{pe}^{-1}$. At $t = 225\omega_{pe}^{-1}$, this o-point starts to move towards the island in the gaps. Then rapid island coalescence ensues, accompanied by quasi-explosive growth of the energy of the coalescence mode ($k_x L_x/2\pi = 1$, $k_y L_y/2\pi = 0$) with a scaling $(t_o - t)^{-1}$ where $t_o = 248\omega_{pe}^{-1}$. The flow patterns and topological changes at merging are similar to the MHD ones (Wu et al., 1980) (see Fig. 1).

The kinetic energy gain at coalescence over the value at the time of island formation is a factor 4 for electrons and 10 for the ions. The temperature for the particles in the x-direction (on-axis temperature) is a factor 9 over thermal for the electrons and a factor of 60 for the ions (see Fig. 2). An extremely hot tail in the ion distribution in the x-direction is formed at coalescence (Fig. 3). The y and z-directions get heating too, but by a lesser amount. After the coalescence, the kinetic energy and the magnetic energy (out of phase with the kinetic energy) exhibit amplitude oscillations similar to the ones in the MHD simulation (Wu et al., 1980). Here, the oscillations are associated with pulsation of the now coalesced island induced by the overshoot of the plasma inflows and by the fact that at the same time the island magnetic field lines stretch (to an oblate shape) and contract (to a cylindrical shape). The period of the island pulsation may be estimated in a simple way. Since the oscillations may be basically regarded as a compressional Alfven wave, their frequency is then kV_A, where $k = 2\pi/\lambda \sim 2\pi/d$ with d the island diameter. In our simulation, the island magnetic field is of the order of the original near-sheet value and the maximum plasma density is about 4 times its original value. Then the estimated pulsation period is $\tau_p \sim 60\omega_{pe}^{-1}$ in agreement with the observed value, where we used $d \sim 32\lambda_{De}$; ω_{ce} (due to the island magnetic field) is $0.75\omega_{pe}$.

An estimate of the gain in kinetic energy during coalescence is as follows. As the current fila-

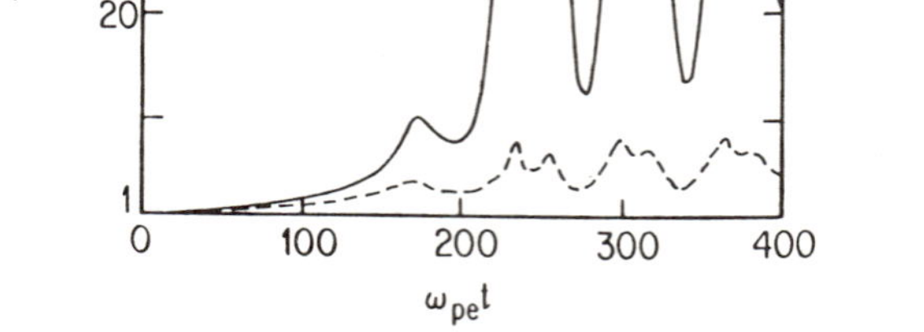

Fig. 2. On-axis (x-direction) ion temperature (upper curve) and electron temperature (lower curve) in the particle simulation.

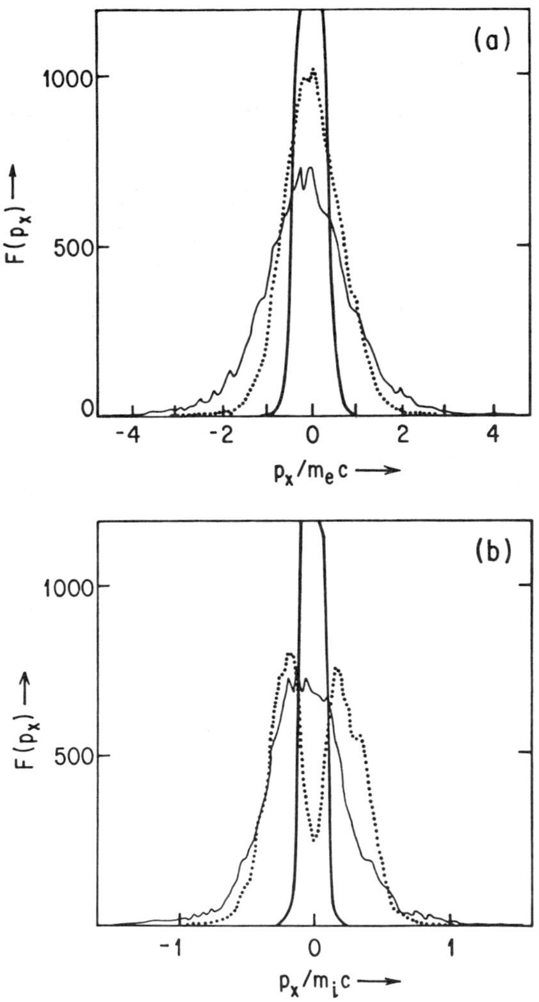

Fig. 3. Snapshots of the on-axis distribution function in the island formation phase as $\omega_{pe}t = 50$ (bold curve), at the coalescence time $\omega_{pe}t = 225$ (dashed curve) and in the merged phase for (thin curve) at $\omega_{pe}t = 375$ a) electrons and b) ions. Hot tails appear in both distributions. At $\omega_{pe}t = 0$, the distributions were Maxwellian with shapes similar to those at $\omega_{pe}t = 50$, but slightly narrower.

ments approach, the maximum potential energy ΔU that could be released to the particles is $\Delta U = (2I^2/c^2)\ \ell n\ (y_s/y_c)$, since the current I is roughly constant, where y_s and y_c are the initial and coalescence separations. The total energy increase is $\sim \Delta U$ ($\simeq 330\ N_o \kappa T$ in the simulation, N_o being the total particle number). The fraction actually converted into kinetic energy from the island phase to the coalescence phase is ΔW $\Delta U/6$ ($\simeq 60\ N_o \kappa T$), the rest going to the fields. Half of this energy goes to the ions; the coalescence velocity on the average is $V_c \simeq 1.7\ v_{te}$, and if all this motion is converted into heat, then the on-axis ion temperature should be $\kappa\Delta T \sim MV_c^2 = 30$ consistent with the above observations.

When the initial crossing angle of the separatrix at the x-point is 90° or close to that, the magnetic flux can go through the x-point without change in the topology of the fields. The run was performed on 32Δ x 32Δ spatial grid with other parameters fixed. The o-point size is only a few times the collisionless skin depth $c/\omega_{pe} = 5\Delta$ and also a few times the ion Larmor radius at the current sheets or of this order in the o-point. On the other hand, when we reduce the crossing angle to smaller than the value for the case 32Δ x 128Δ, the neutral sheet appears as for the 32 x 128 case, tears and now forms more magnetic islands. For the case 16Δ x 256Δ with $c/\omega_{pe} = 3\Delta$ and $\lambda_{De} = 0.5\Delta$, we observed 16 islands in chain. These islands coalesce and eventually wind up into a single island pinned at the gap. These details will be reported elsewhere.

Discussion

We have found through fully self-consistent electromagnetic particle simulations that (i) a large amount of magnetic energy release to kinetic energy takes place when magnetic islands coalesce, while much less energy conversion happens through the tearing instability or island formation phase; (ii) the explosive energy conversion primarily goes to the creation of extremely hot ions; (iii) after merging of the islands, the coalesced island pulsates and eventually the amplitude oscillations dissipate into heat. We note that although the MHD study (Wu et al., 1980) and the particle simulation handle very different regiomes, the emerged physical pictures have a lot in common. This is somewhat surprising since the MHD study has zero Larmor radius and zero collisionless skin depth approximations, while in the particle simulations the ion Larmor radius for the hot ions after merging is practically infinite; they are, however, confined electrostatically. A close correspondence should not be expected between our results and the experiments of Stenzel and Gekelman (1979) due to differences in parameters, even though the set up is similar. For instance, their geometry of the x-point is always a 2-to-1 ratio, their plasma $\beta \sim 1$ and their experiments are primarily in the collisional regime ($S \sim 20$) as well as three-dimensional. Flow patterns as well as distribution functions are relatively similar to our cases of no island merging. They have not observed strong tail heating, only some bulk heating. Again, this is consistent with our results that unless o-points merging takes place, there is neither intense heating nor hot tail formation.

The large kinetic energy increase as a result of coalescence and the resulting energetic tails in the ions and electrons distributions may play an important role in the generation of geomagnetic substorms by topological modification of the deep tail and in the solar flares dynamics.

Acknowledgments. The authors would like to thank Dr. C.C. Wu, Dr. P. Pritchett and Mr. K. Quest for fruitful discussions. This work was supported by the National Science Foundation Contract #NSF-PHY-79-01319 and ATM-79-26492 and the United States Department of Energy Grant #DE-FG05-80ET 53088.

References

Biskamp, D. and H. Welter, Coalescence of Magnetic Islands, Phys. Rev. Lett. 44, 1069, 1980.

Dickman, D.O., R.L. Morse, and C.W. Nielson, Numerical Simulation of Axisymmetric, Collisionless, Finite-β Plasma, Phys. Fluids 12, 1708, 1969.

Finn, J.M. and P.K. Kaw, Coalescence Instability of Magnetic Islands, Phys. Fluids 20, 72, 1977.

Hosea, J.C., C. Bobeldijk, and D.J. Grove, Stability Experiments on the ST Tokamak, in Plasma Physics and Controlled Nuclear Fusion Research (International Atomic Energy Agency, Vienna, 1971), Vol. II. p. 425.

Katanuma, I. and T. Kamimura, Simulation Studies of the Collisionless Tearing Instabilities, Phys. Fluids 23, 2500, 1980.

Leboeuf, J.N., T. Tajima, J.M. Dawson, and A.T. Lin, Particle Simulations of Time-Varying X-Points, Proc. Int. Conf. Plas. Phys. 1, 65, 1980.

Lin, A.T., Nonlinear Evolution of Drift Alfven and Double Tearing Instabilities, Phys. Fluids 21, 1026, 1978.

Lin, A.T., J.M. Dawson, and H. Okuda, Applications of Electromagnetic Particle Simulation to the Generation of Electromagnetic Radiation, Phys. Fluids 17, 1995, 1975.

Mirnov, S.V. and I.B. Semenov, On the Nature of Large-Scale Instabilities in Tokamak, Zh. Eskp. Teor. Fiz. 60, 2105, 1971 [Sov. Phys. - JETP 33, 1134, 1971.

Parker, E.N., Cosmical Magnetic Fields, Clarendon Press, Oxford, 1979.

Pritchett, P.L. and C.C. Wu, Coalescence of Magnetic Islands, Phys. Fluids 22, 2140, 1979.

Rutherford, P.H., Nonlinear Growth of Tearing Modes, Phys. Fluids 16, 1903, 1973.

Stenzel, R. and W. Gekelman, Experiments on Magnetic-Field-Line Reconnection, Phys. Rev. Lett. 42, 1055, 1979.

Vasyliunas, V.M., Theoretical Models of Magnetic Field Line Merging 1, Rev. Geophys. Space Phys. 13, 303, 1975.

Waddell, B.V., M.N. Rosenbluth, D.A. Monticello and R.B. White, Non-Linear Growth of the m=1 Tearing Mode, Nucl. Fusion 16, 528, 1976.

Wu, C.C., J.N. Leboeuf, T. Tajima and J.M. Dawson, Magnetic Islands Coalescence and Intense Plasma Heating, submitted to Phys. Rev. Lett. 1980.

THE KILOMETRIC RADIO EMISSION SPECTRUM: RELATIONSHIP TO AURORAL ACCELERATION PROCESSES

D. A. Gurnett and R. R. Anderson

Department of Physics & Astronomy, The University of Iowa, Iowa City, IA 52242

Abstract. Satellite measurements have now confirmed that intense radio emissions called auroral kilometric radiation are generated at altitudes of 2 to 3 R_E over the auroral regions. These radio emissions are believed to be generated by Doppler-shifted cyclotron radiation from the electrons involved in the auroral acceleration process. Using current theories for the generation of this radiation we discuss the possibilities for obtaining information on the auroral acceleration processes from the spectrum of this radiation. For example, under certain conditions it is shown that the low frequency cutoff provides a direct indication of the upper altitude limit of the acceleration region. Also, certain discrete features in the kilometric radiation spectrum show a very close similarity to whistler-mode emissions, thereby suggesting that the basic instability mechanisms are quite similar. Some of these discrete features have frequency drifts which can be interpreted in terms of the propagation of shock-like disturbances along the auroral field lines at velocities near the ion-acoustic speed.

Introduction

Spacecraft observations over the last decade have revealed that the earth is an intense radio source in the kilometer wavelength range and that the generation of this radiation is closely associated with the auroral acceleration process. The first reports of this radiation were by Benediktov et al. [1965] and Dunkel et al. [1970] who identified intense radio emissions which were closely correlated with high latitude magnetic disturbances. Later studies by Gurnett [1974] showed that the peak intensity occurred at kilometer wavelengths and that the radio emissions were generated at low altitudes over the auroral regions in association with discrete auroral arcs. During bright auroral displays the maximum power radiated was found to be very large, $\sim 10^9$ watts. The earth is, therefore, a very intense radio emitter, comparable in many respects to Jupiter, which has long been recognized as an intense planetary radio source. Because of the association with discrete auroral arcs, Gurnett [1974] suggested that the radiation is associated with inverted-V electron precipitation bands detected by low-altitude satellites [Frank and Ackerson, 1971]. This relationship has now been confirmed by direct in situ measurements in the auroral regions [Benson and Calvert, 1979; Green et al., 1979]. Polarization measurements by Gurnett and Green [1978] and Kaiser et al. [1978] showed that kilometric radiation is generated in the right-hand extraordinary mode. Of the many theories which have been proposed, the observed polarization and the absence of intense electrostatic waves in the source region have greatly restricted the possible generation mechanisms. The most promising mechanisms at the present time appear to be the direct electromagnetic instabilities proposed by Melrose [1976] and Wu and Lee [1979].

In this paper we consider the relationship of the spectrum of the auroral kilometric radiation to the spatial structure and processes occurring in the auroral acceleration region. The basic objective is to explore the possibilities of using the emission spectrum to provide information about the details of the acceleration process. The methods employed may also have much broader applications. It now appears that the radio emission mechanism operative in the terrestrial auroral regions is a universal process occurring in association with the formation of auroral arcs throughout the solar system. For example, at both Jupiter and Saturn the same basic radio emission process is believed to be associated with auroral processes occurring at these planets [Warwick et al., 1979; Kaiser et al., 1980]. Furthermore, radio emissions of the same type have also been tentatively identified from the planet Uranus [Brown, 1976]. Since it will not be possible to conduct direct in situ measurements of aurora at any of these planets in the near future, any advances which can be made in understanding the relationship of the radio emission spectrum to the auroral acceleration process may provide a valuable tool for studying auroral acceleration processes in other regions of the solar system.

Upper and Lower Cutoff Frequencies

A basic characteristic of the kilometric radio emission is the existence of well-defined upper and

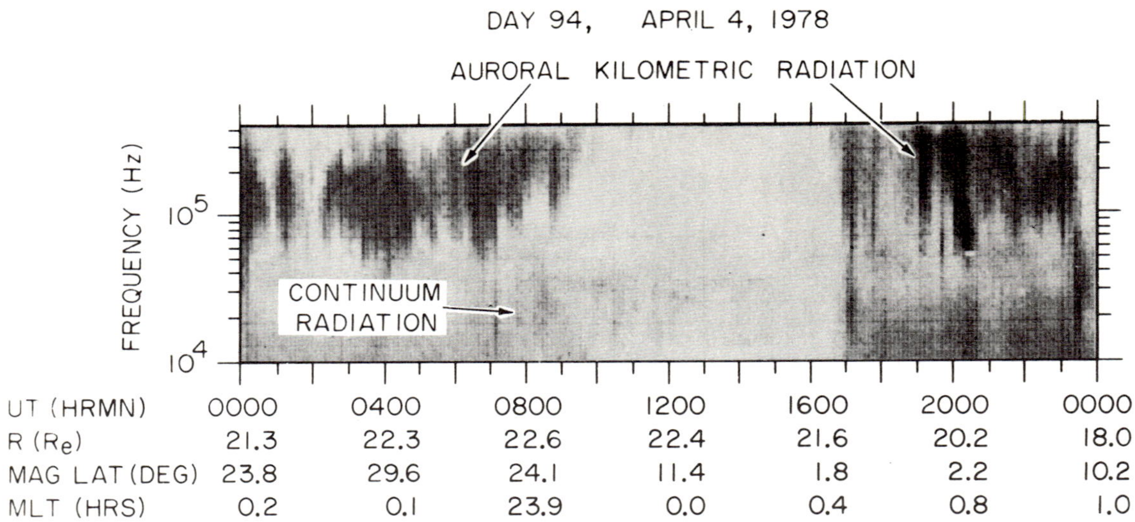

Fig. 1. A representative spectrum of auroral kilometric radiation. Note the upper and lower cutoff frequencies of the spectrum and the occurrence of intervals when the radiation disappears completely.

lower cutoff frequencies. Figure 1, for example, shows a typical auroral kilometric radiation event observed by the ISEE 1 spacecraft in the local evening region of the magnetosphere. As can be seen the radio emission spectrum has clearly identified upper and lower cutoffs which fluctuate over a wide range on a time scale of ten minutes or less. Except for fine structure, which will be discussed later, the radio emission normally extends continuously over the entire frequency band between the upper and lower cutoffs. Occasionally the radiation disappears for short periods, for example, from about 0145 to 0210 UT and from about 1000 to 1630 UT in Figure 1. The onset and termination of the radiation often appears to result from a decrease in the bandwidth, with the upper and lower cutoffs pinching together as the radiation disappears.

If the radio emission is generated by Doppler-shifted cyclotron radiation, as in the models of Melrose [1976] and Wu and Lee [1979], a simple explanation can be advanced for the upper and lower cutoff frequencies. Before discussing the origin of these cutoffs, it is useful to first review the essential features of the cyclotron resonance mechanism. From very general considerations the growth rate for the right-hand extraordinary mode can be written

$$\omega_i = \frac{\pi^2 c^2 \omega_p^2}{4\omega_g} \frac{N_H}{N_c} \int_C dv_\parallel v_\perp \left[\frac{\partial F}{\partial v_\perp} + n \text{Cos}\theta \frac{\omega}{\omega_g} \frac{v_\perp}{c} \frac{\partial F}{\partial v_\parallel}\right] \quad (1)$$

where N_H and N_C are the hot and cold electron densities, ω_p and ω_g are the electron plasma frequency and gyrofrequency, and $v_\parallel$ and $v_\perp$ are the parallel and perpendicular velocities relative to the magnetic field. The integral must be evaluated over a contour C in velocity space defined by the cyclotron resonance condition,

$$\omega - \omega_g \sqrt{1 - \frac{v_\parallel^2 + v_\perp^2}{c^2}} - v_\parallel \frac{\omega n}{c} \text{Cos}\theta = 0 \quad , \quad (2)$$

where n is the index of refraction, θ is the wave normal angle with respect to the magnetic field and ω is the wave frequency. As pointed out by Wu and Lee [1979], it is important to keep the relativistic term in Equation 2 even for velocities much less than the speed of light.

Two free-energy sources can be identified in Equation 1, one corresponding to regions of positive $\partial F/\partial v_\perp$ and the other corresponding to regions of positive $\partial F/\partial v_\parallel$. The condition $\partial F/\partial v_\perp > 0$ is characteristic of a loss-cone distribution function, and $\partial F/\partial v_\parallel > 0$ is characteristic of a beam-type distribution. A loss-cone distribution was the basis of Wu and Lee's [1979] theory of auroral kilometric radiation and a beam-type distribution was the basis of Melrose's [1976] theory. These two types of distribution functions are illustrated in Figure 2. In both cases the presence of a field-aligned electrostatic potential gradient plays a crucial role in establishing the essential features of the distribution function.

As discussed by Chiu and Schulz [1978], in the presence of a parallel electric field the velocity distribution is divided into five regions by a loss-cone boundary, shown by the solid hyperbola-shaped lines in Figure 2, and an electrostatic acceleration boundary, shown by the dashed elliptical curve. The hyperbola-shaped loss cone arises because of the acceleration caused by the parallel electric field between the spacecraft and the atmosphere. The elliptical electrostatic acceleration boundary represents the boundary for magnetospheric electrons accelerated by parallel electric fields above the spacecraft. In the absence of scattering, magnetospheric electrons are excluded from the region inside the dashed ellipse. The

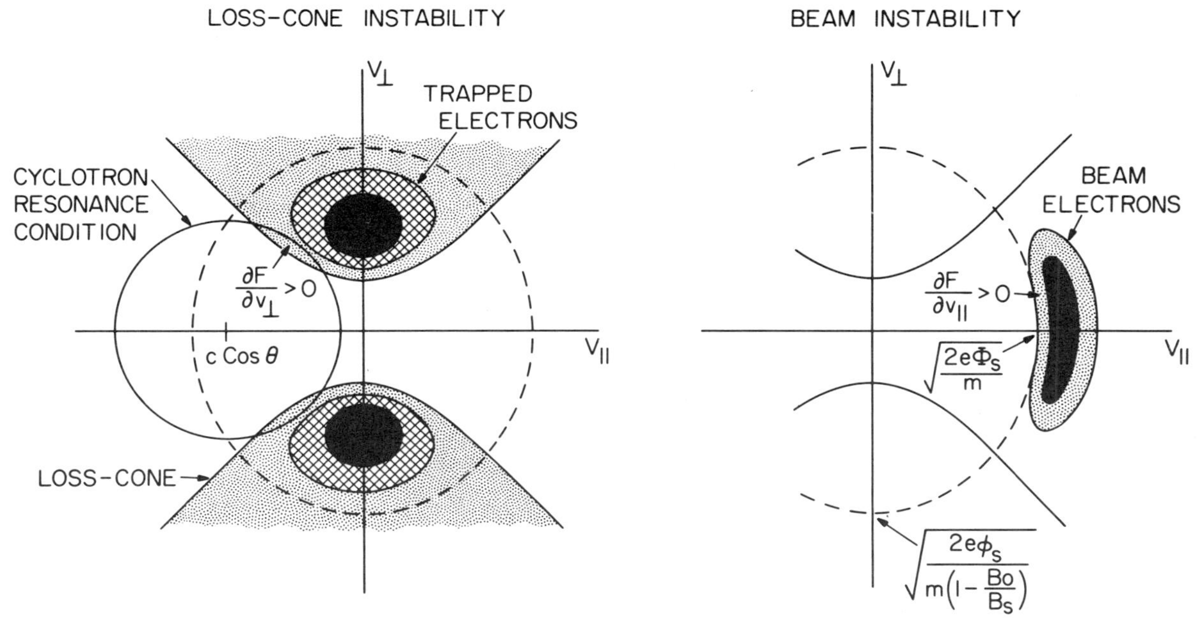

Fig. 2. Idealized velocity distributions illustrating the loss-cone and beam-type free-energy sources. Cyclotron resonance interactions occur along a circle in velocity space with a center at $v_\parallel = c\text{Cos}\theta$ and a radius $v_R = c[\text{Cos}^2\theta - 2\Delta\omega/\omega_g]^{1/2}$.

region bounded by the hyperbola and the ellipse consists of particles trapped in the potential well formed by the mirror force below the satellite and the electrostatic potential barrier above the satellite. For the loss-cone instability it is sufficient for the moment to point out that a large $\partial F/\partial v_\perp$ is expected along the loss-cone boundary. In the case of the beam-type instability a large $\partial F/\partial v_\parallel$ occurs near the electron acceleration boundary because of the acceleration of low energy magnetospheric electrons by parallel electric fields above the spacecraft.

Inspection of distribution functions measured at various points along the auroral field lines shows evidence of both the beam and loss-cone free-energy sources. Figure 3, for example, shows a typical electron velocity distribution function from S3-3 [Mizera and Fennell, 1977] which has regions of large $\partial F/\partial v_\parallel$ and $\partial F/\partial v_\perp$. The region of large $\partial F/\partial v_\parallel$ occurs for downward moving electrons near the elliptical electron acceleration boundary (dashed line) and inside the loss cone. The distribution function in this region is somewhat different than the bimaxwellian used by Melrose for the electrostatically accelerated electron beam, but the basic free-energy source is the same. The region of large $\partial F/\partial v_\parallel$ occurs at a velocity of approximately $v_\parallel = \sqrt{2e\Phi_S/m}$, where Φ_S is the electrostatic potential at the spacecraft measured with respect to the potential above the acceleration region. The large parallel velocity gradient in this region occurs because of the markedly decreased particle population inside the electrostatic acceleration and loss-cone boundaries. This region of velocity space is only accessible to electrons from the ionosphere, which have much lower energies and phase-space densities than the magnetospheric electrons.

As can be seen in Figure 3, the region of large $\partial F/\partial v_\perp$ occurs for upward moving electrons just inside the loss-cone boundary. The perpendicular velocity gradients are particularly large near and

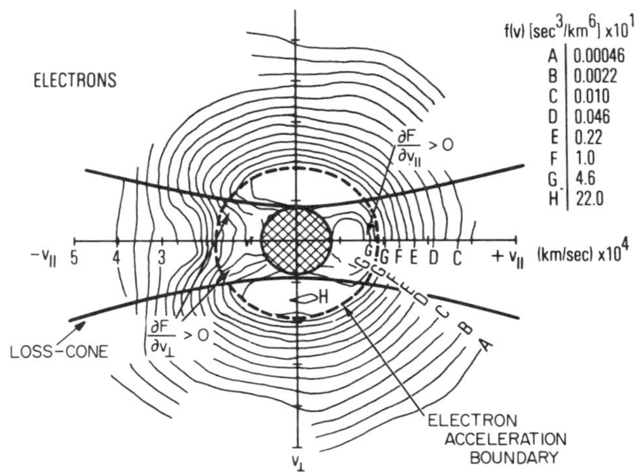

Fig. 3. A velocity distribution function obtained by S3-3 in the auroral acceleration region [Mizera and Fennel, 1977]. Note the regions of large positive $\partial F/\partial v_\perp$ near the loss cone boundary and large positive $\partial F/\partial v_\parallel$ just inside the elliptical electrostatic acceleration boundary.

inside the elliptical electron acceleration boundary, at parallel velocities $v_\parallel \lesssim \sqrt{2e\Phi_s/m}$. Examination of other distribution functions shows that this relationship appears to be a consistent feature of the S3-3 observations. The $v_\parallel \lesssim \sqrt{2e\Phi_s/m}$ constraint on the region of large $\partial F/\partial v_\perp$ apparently arises because of the increased electron intensities in the loss cone at higher energies, probably due to backscattered beam electrons, which tend to have velocities greater than $\sqrt{2e\Phi_s/m}$. At the present time it is not known which of the two free-energy sources is most important. A decision on this question awaits a detailed evaluation of the growth rate from Equation 1 using the measured distribution function. Simple considerations, however, suggest that the loss-cone source is probably more important because the factors $v_\perp/c$ and $\cos\theta$ strongly reduce the contribution due to $\partial F/\partial v_\parallel$ in Equation 1.

Having discussed the essential features of the electron distribution function required for instability we can now discuss the physical reasons for upper and lower cutoffs in the radio emission spectrum. In order to have an instability it is essential that the cyclotron resonance velocity given by Equation 2 be at sufficiently low velocities to be in the region of the distribution function where the velocity gradients are large. At low velocities, $v \ll c$, and for an index of refraction $n \simeq 1$, Wu and Lee [1979] have shown that the cyclotron resonance condition is a circle in the $v_\parallel$, $v_\perp$ plane with a center at $v_\parallel = c\cos\theta$ and a radius of $v_R = c[\cos^2\theta - 2\Delta\omega/\omega_g)]^{1/2}$, where $\Delta\omega = \omega - \omega_g$. A representative resonance circle is shown in Figure 2. Instability occurs when ω and θ are such that the resonance circle passes through a sufficiently large region of positive $\partial F/\partial v_\parallel$ or $\partial F/\partial v_\perp$, thereby giving a positive value for the integral in Equation 1. The largest intensities occur for the particular combination of initial ω and θ which give the largest growth rate integrated along the ray path. It is evident that a low resonance velocity implies propagation angles close to $\pi/2$ and wave frequencies close to the electron gyrofrequency. From Equation 2 one can show that the minimum resonance velocity occurs at $v_{\parallel Res}(\mathrm{Min}) = c(\Delta\omega/\omega)$. The minimum resonance velocity is therefore a very sensitive function of the wave frequency, decreasing rapidly as ω approaches ω_g. The wave frequency, however, is constrained to be above the extraordinary mode cutoff which occurs at

$$\omega_{R=0} = \frac{\omega_g}{2} + \sqrt{\left(\frac{\omega_g}{2}\right)^2 + \omega_p^2} \qquad (3)$$

The minimum resonance energy is then approximately

$$W_{\parallel RES}(\mathrm{Min}) = \frac{1}{2} m c^2 \left(\frac{\Delta\omega_{R=0}}{\omega_g}\right)^2 \qquad (4)$$

where $\Delta\omega_{R=0} = \omega_{R=0} - \omega_g = -\omega_g/2 + \sqrt{(\omega_g/2)^2 + \omega_p^2}$. Equation 4 can then be rewritten in the form

$$W_{\parallel RES}(\mathrm{Min}) = \frac{1}{8} m c^2 \left[-1 + \sqrt{1 + 4(\omega_p/\omega_g)^2}\right]^2 , \qquad (5)$$

which shows that the minimum resonance energy is controlled solely by the ratio of the plasma frequency to the gyrofrequency. Low resonance energies only occur when $\omega_p/\omega_g \ll 1$.

The variation of the minimum resonance energy with radial distance along an auroral field line is illustrated in Figure 4. As can be seen from Equation 3 the extraordinary mode cutoff, $\omega_{R=0}$, tends to be at either ω_p or ω_g, which ever is larger. When $\omega_p > \omega_g$ the resonant energies are very large because a very large velocity is required to Doppler shift the emission frequency above the extraordinary mode cutoff. As ω_p decreases below ω_g, the minimum resonance energy decreases rapidly, to approximately $W_{\parallel Res}(\mathrm{Min}) \simeq (1/2)mc^2(\omega_p/\omega_g)^4$ when $\omega_p \ll \omega_g$. As shown in Figure 4 the condition $\omega_p < \omega_g$ tends to occur at intermediate altitudes between the ionosphere and the outer magnetosphere. If the resonance energy drops below the characteristic energy $e\Phi_s$ of the free-energy source in the electron distribution (where the velocity gradients are large) then instability can occur. For a given electrostatic potential distribution along the field line this condition implies upper and lower boundaries for the unstable region, for example, A-A′ and B-B′ in Figure 4. These boundaries in

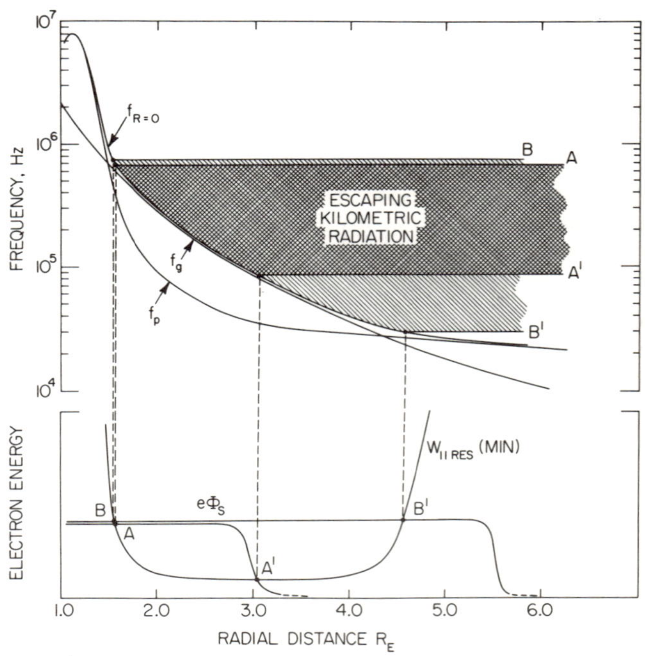

Fig. 4. An ionospheric model illustrating the control of the upper and lower cutoffs of the kilometric radiation by the electron density profile and the electrostatic potential distribution. Instability only occurs if the minimum resonance energy, $W_{\parallel RES}(\mathrm{Min})$, is below the electrostatic potential energy $e\Phi_s$.

turn imply upper and lower cutoffs for the radio emission spectrum. In practice, of course, these boundaries are not as precisely defined as indicated in Figure 4 because the growth rate given by Equation 1 involves an integral over the resonance velocity. This model does, however, provide us with a qualitative guide to analyze how the bandwidth of the emitted radiation depends on the parameters of the system.

We now return to the interpretation of the observed cutoffs of the auroral kilometric radiation. It is evident from Figure 4 that two basic factors control the bandwidth of the emitted radiation: (1) the plasma density distribution via the dependence of $W_{\parallel Res}(Min)$ on ω_p/ω_g, and (2) the detailed profile of the electrostatic potential distribution Φ_s. Two extreme situations can be identified. If the main electron acceleration occurs at very high altitudes, as in case B-B´, then the bandwidth is controlled almost completely by the points where the plasma frequency profile crosses the electron gyrofrequency profile. In this case, the cutoff frequencies are almost completely determined by the plasma density distribution. On the other hand if the main electron acceleration occurs at lower altitudes, as in case A-A , then the lower cutoff frequency is controlled by the electron gyrofrequency at the upper boundary of the acceleration region. The upper cutoff is determined by a combination of factors. If the potential intersects on the steep part of the $W_{\parallel Res}$ curve then the upper cutoff is essentially determined by the plasma density profile at the point where $\omega_p = \omega_g$. If the potential is lower then the upper cutoff frequency depends on the details of the potential profile, tending to decrease as Φ_s decreases.

Not much is known about the variability of the plasma density profile at high altitudes. The main information comes from the S3-3 satellite at about 2.0 R_E and various eccentric orbiting spacecraft at somewhat greater radial distances. Present indications are that $\omega_p << \omega_g$ at 2.0 R_E (R. Torbert, personal communication), more or less in agreement with the model in Figure 4. At larger radial distances, > 5 R_E, the plasma density is quite low, N_e < 10 cm^{-3}, which implies that the plasma frequency profile probably stays below the electron gyrofrequency out to about 5 R_E, as in Figure 4. Since the low frequency cutoff of the kilometric radiation usually corresponds to radial distances less than 5 R_E, it is tempting to assume that variations in the low frequency cutoff are caused by variations in the upper altitude limit of the acceleration region, as in case A-A´ of Figure 4. Unfortunately, so little is known about the density distribution along the auroral field lines one cannot be certain whether this assumption is valid or not. If it is, then the low frequency cutoff gives a direct indication of the upper altitude limit of the acceleration region.

A probability distribution for the upper and lower cutoff frequencies obtained from a portion of the event in Figure 1 is shown in Figure 5. A radial distance scale derived from the electron gyrofrequency, which is always very close to the emission frequency, is shown at the bottom of this illustration. If the plasma frequency remains sufficiently low at high altitudes, as in case A-A´, these low frequency cutoffs imply an upper limit to the electron acceleration region of about 2.0 to 3.5 R_E. Alternatively, these low frequency cutoffs may be representative of the radial distance at which the plasma frequency crosses the gyrofrequency, as in case B-B´. The interpretation of the high frequency cutoff is more certain. At frequencies above 300 kHz the high frequency cutoff is almost certainly controlled by the crossover between the plasma frequency and gyrofrequency profiles which occurs at about 1.5 R_E. At lower frequencies the cutoff becomes more dependent on the electrostatic potential near the crossover point, with a decreasing upper cutoff frequency indicating a smaller accelerating potential.

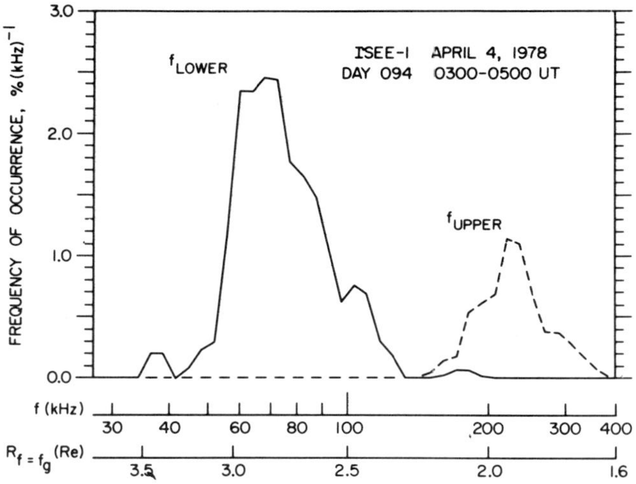

Fig. 5. The distribution of upper and lower cutoff frequencies for Figure 1, and the corresponding radial distance, assuming that the emission occurs very close to the electron gyrofrequency.

As discussed earlier, the radiation sometimes disappears completely during certain periods. At the beginning and end of these periods, the upper and lower cutoffs appear to merge as though the bandwidth of the emission decreased to zero at these points. Again two extreme interpretations are possible. The termination and onset of the emission could be due to changes in the plasma density profile or to changes in the electrostatic potential distribution. If changes in the plasma density are the controlling effect then the emissions would terminate completely as soon as the plasma frequency exceeds the gyrofrequency at all points along the magnetic field line. If such a density control does occur it seems almost certain that it would be imposed by density variations at high altitudes rather than by variations in the ionosphere. Since rather high densities, > $10^2 cm^{-3}$,

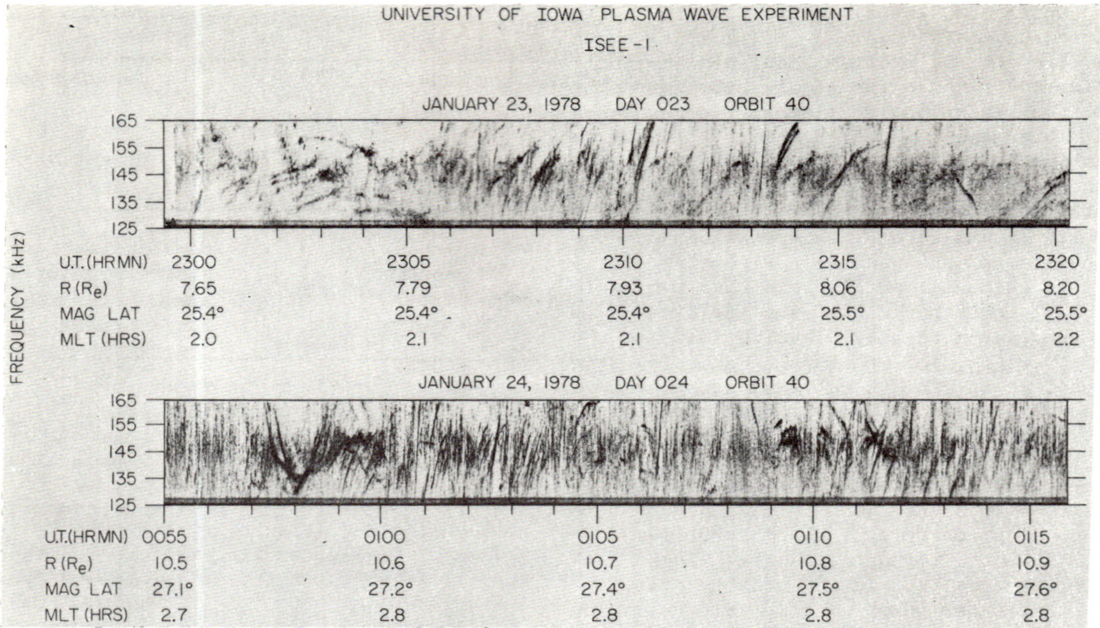

Fig. 6. High resolution spectrograms showing the occurrence of many narrowband emissions drifting predominantly upward in frequency.

would be required in the outer magnetosphere to completely quench the instability, it seems unlikely that changes in the density profile could play the dominant role in switching the radiation on and off. Temporal variations in the accelerating potential appear to be a more likely alternative. This viewpoint is supported by the fact that the kilometric radiation tends to disappear during periods of low magnetic activity [Voots et al., 1977] when field-aligned currents and the accelerating potential are expected to be the smallest. As can be seen from Figure 4, the basic model predicts that the upper and lower cutoffs, A and A′, approach each other as the accelerating potential decreases, thereby shrinking the bandwidth to zero at the termination and onset of the emission, as is observed.

It is interesting to note that the above considerations also account for the bandwidth of the decametric radiation from Jupiter, which is thought to be very similar to the terrestrial auroral kilometric radiation. At Jupiter the bandwidth of the decametric radiation is very large, with the emission frequency extending from about 1 to 40 MHz [Warwick et al., 1979]. The upper frequency limit is in close agreement with the maximum electron gyrofrequency at the surface of the planet. The bandwidth of the radio emission is much larger at Jupiter than at Earth because of the larger magnetic field strength at Jupiter, which keeps the ratio ω_p/ω_g much less than one over a large range of radial distances, even in the ionosphere very close to the planet.

Fine Structure

Recently wideband measurements from the ISEE 1 and 2 spacecraft have provided very high resolution spectrum measurements of auroral kilometric radiation [Gurnett et al., 1979]. The spectrums obtained are remarkable because they show that the radiation consists of many discrete narrowband emissions, rather than a continuous broadband emission as suggested by the lower resolution measurements in Figures 1. Two representative high-resolution frequency-time spectrograms of auroral kilometric radiation are shown in Figure 6. These spectrograms cover a frequency range of 40 kHz, from 125 to 165 kHz, and a time interval of about 20 minutes. As can be seen, the spectrum is extremely complex. The most striking characteristic is the occurrence of many narrowband emissions with a bandwidth of 1 kHz or less. The entire auroral kilometric radiation spectrum appears to be made up of many discrete narrowband tones. Often the center frequency of the individual emissions varies in a systematic manner, sweeping either upward or downward across the spectrum. The spectrum in Figure 6, for example, is dominated by emissions drifting upward in frequency with increasing time. Comparable spectrums can also be found in which the emissions are sweeping downward in frequency. Careful examination shows that the drifting features are not always continuous tones, but sometimes consist of short bursts occurring on time scales of only a few seconds.

A contrasting example in which the drifting fea-

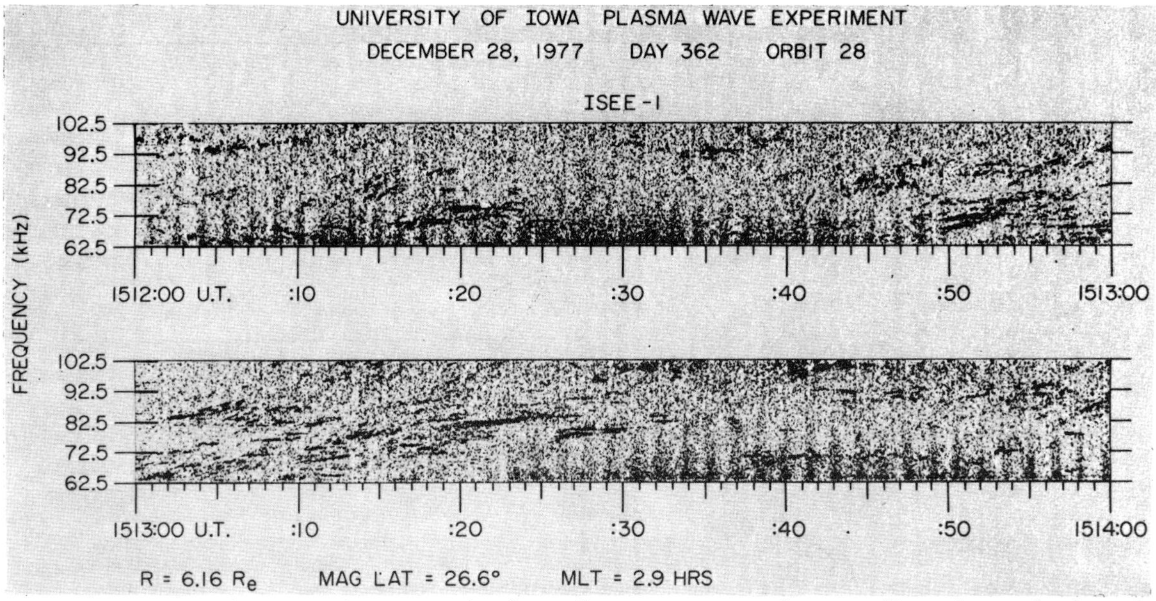

Fig. 7. High resolution spectrograms showing the occurrence of many discrete emissions in the auroral kilometric radiation spectrum. These emissions have many features similar to discrete whistler-mode emissions.

tures are nearly absent is illustrated in Figure 7, which shows a 40-kHz frequency range, from 62.5 to 102.5 kHz, for two successive 1-minute intervals. The spectrum in this case consists of a very large number of brief narrowband bursts. The duration and frequency range of the individual bursts is very small, typically 1 sec duration and 3 kHz frequency range, or less. The spectral shapes of the individual bursts vary over a wide range, sometimes sweeping upwards or downwards in frequency and forming various curved features on the frequency-time diagram.

The extremely complex spectral structure of the auroral kilometric radiation presents some difficult interpretational questions. The occurrence of many discrete narrowband emissions, as in Figure 7, is strongly reminiscent of chorus and other types of discrete whistler-mode emissions observed in the earth's magnetosphere [Helliwell, 1965]. Cases can be found, as in Figure 8, where slowly drifting narrowband features trigger discrete emission (risers) similar to the triggering of whistler-mode emissions by ground transmitter signals [Helliwell and Katsufrakis, 1974]. The close similarity of the dynamic spectral characteristics may be more than purely coincidental, since both the auroral kilometric and whistler-mode instability mechanisms may be of essentially the same type. If the kilometric radiation is produced by a loss cone distribution then both instabilities involve a cyclotron resonance interaction with an anisotropic electron distribution. The index of refraction and mode of propagation are, of course, quite different, but otherwise the physical processes are quite similar.

For whistler-mode emissions it is generally agreed that the dynamic evolution of the spectrum is a highly nonlinear process involving electrons trapped by the rotating wave field [Inan et al., 1978]. It is possible that similar nonlinear processes may be able to account for the discrete structure of the auroral kilometric radiation. It is interesting to note that if the generation of auroral kilometric radiation is fundamentally similar to the whistler-mode instability, then the kilometric radiation may play a significant role in the pitch-angle scattering of auroral electrons, similar to the pitch-angle scattering of radiation belt electrons by whistler-mode emissions in the inner regions of the magnetosphere. Because the auroral kilometric radiation is generated in the right-hand polarized extraordinary mode the radiation carries away right-hand angular momentum, thereby reducing the pitch angle of the interacting electrons, very similar to the situation with whistler-mode interactions. Since the radiation intensities in the source are undoubtedly very large (10 to 100 mVm^{-1}), the pitch-angle diffusion rates could be a very significant factor in controlling the equilibrium electron velocity distribution in the auroral acceleration region. If such scattering occurs it should be characterized by time scales comparable to the time scale of the individual bursts, which is only a few seconds. Rapid fluctuations in inverted-V electron precipitation fluxes have been reported by various investigators, including for example Lin and Hoffman [1979], which could possibly be related to the fine structure of the kilometric radiation.

Probably the most important single feature which

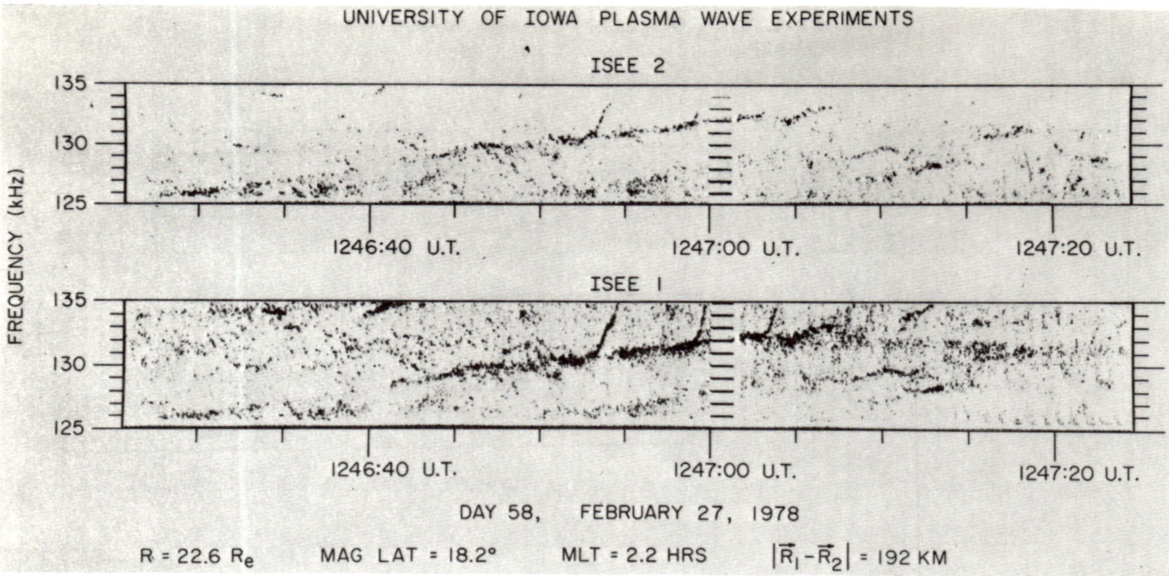

Fig. 8. High resolution spectrograms obtained simultaneously from ISEE 1 and 2 showing the apparent triggering of rising emissions by a narrowband drifting feature. These triggering effects are remarkably similar to the triggering of whistler-mode emissions by ground VLF transmitters.

must be explained in the fine structure of the auroral kilometric radiation is the organization of the discrete bursts into upward and downward drifting bands of the type illustrated in Figure 6. The occurrence of quasi-monochromatic emissions sweeping through a large range of frequencies is a common feature of solar radio bursts [Kundu, 1965] and Jovian decametric radio emissions [Warwick, 1967]. These drifting features are usually interpreted as being due to the motion of the emitting particles, or a propagating disturbance, through a plasma which has a spatial gradient in the characteristic emission frequency. This model is used, for example, to explain the characteristic frequency drifts of type III and type IV solar radio bursts. In the case of the auroral kilometric radiation the characteristic emission frequency is the electron gyrofrequency. Since the resonance energy is very sensitive to the frequency difference, $\Delta\omega$, between the emission frequency and the electron gyrofrequency, it is difficult to see how one can account for the large frequency range of the drifting emissions without involving a motion of the source.

Given that the emission takes place near the gyrofrequency, it is relatively easy to determine the radial component of the source velocity, dR/dt, from the observed frequency drift rate, df/dt. For motion along a dipole field line at high latitudes, the velocity of the source is approximately

$$\frac{dR}{dt} = - \frac{R_E}{3}\left(\frac{f_{go}^{1/3}}{f^{4/3}}\right)\left(\frac{df}{dt}\right) \qquad (6)$$

where $f_{go} \simeq 1.7$ MHz is the electron gyrofrequency at the surface of the earth in the auroral region. Using this equation we have estimated the source velocity using a random sample of the drifting emissions observed in the frequency range from 125 to 500 kHz. The distribution of source velocities is illustrated in Figure 9. As can be seen, the source velocities vary over a large range, from about 3 to 300 km/sec, with a median value of about 30 km/sec. Downward and upward source motions are about equally common.

To aid in the interpretation of these apparent source velocities corresponding electron and proton energy scales are shown at the top of Figure 9. As can be seen, the observed source velocities are much smaller than the typical electron and proton energies, 100 eV to 10 keV, observed along the auroral field lines. The observed drift rates cannot, therefore, be identified with the actual velocities of the auroral particles. In considering other characteristic velocities which could be involved, two possibilities immediately come to mind: the Alfvén speed, $V_A = B/\sqrt{\mu_o \rho_m}$, and the ion-acoustic speed, $V_C = \sqrt{kT_e/m_i}$. Because the plasma density is quite low in the resonance interaction region ($\omega_p < \omega_g$) the Alfvén speed is too large, $V_A \gtrsim 2 \times 10^3$ km/sec, to be considered. The ion-acoustic speed is mainly determined by the electron temperature, T_e, and the ion mass, m_i. Using a temperature of 10^4 °K and assuming that O^+ is the dominant ion species, the ion-acoustic speed in the ionosphere is estimated to be about $V_C \simeq 2.3$ km/sec. For H^+ the ion-acoustic speed is about $V_c = 9.2$ km/sec. At higher altitudes the ion-acoustic speed tends to increase. An upper limit is given by the ion-acoustic speed in the plasma sheet, which is approximately $V_C \simeq 200$ km/sec. This range of ion-acoustic speeds is indicated in Figure 9.

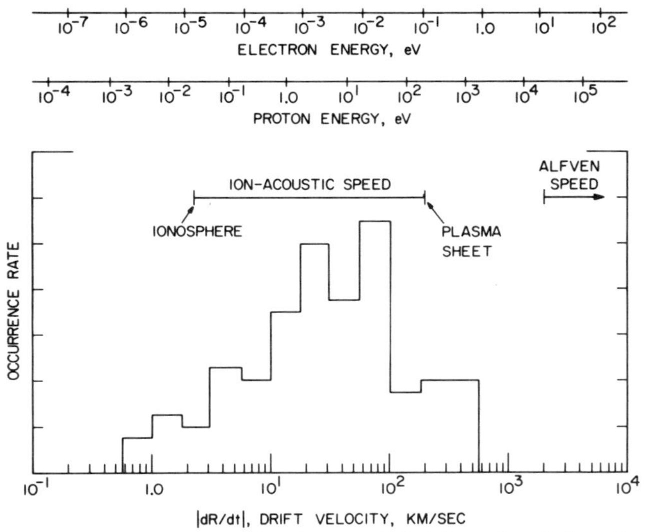

Fig. 9. The distribution of drift velocities for narrowband drifting features similar to those illustrated in Figure 6.

As can be seen, the observed source velocities are in approximately the same range as the estimated ion-acoustic speeds. This agreement suggests that the drifting tones observed in the auroral kilometric radiation may be caused by a discrete structure propagating along the auroral field lines at the ion-acoustic speed. This mechanism could be somewhat similar to the generation of type IV solar radio bursts which are produced by shock waves propagating through the solar corona [Kundu, 1965]. It has long been suspected from computer simulations [Hubbard and Joyce, 1979] and laboratory experiments [Carlqvist and Bostrom, 1970] that electrostatic structures in the auroral acceleration region may be highly turbulent, possibly consisting of many short duration transient disturbances propagating at speeds near the ion-acoustic speed. The exact mechanism by which these disturbances could control the spectrum of the auroral kilometric radiation is not known, although numerous possibilities exist. If the potential perturbations are sufficiently large, then the disturbances may significantly modify the local plasma density or other properties of the auroral plasma distribution via the dependence on Φ_s. These changes could in turn affect the growth rate and spectrum of the radio emission, thereby producing drifting spectral features which appear to move with the disturbance.

Conclusion

In this paper we have discussed the principal features of the spectrum of auroral kilometric radiation. Using current theories we have tried to relate these features to processes occurring in the auroral acceleration region. At the present time, it appears that the upper and lower cutoff frequencies of the auroral kilometric radiation can be readily understood if the radiation is produced by a Doppler-shifted cyclotron emission driven by either a loss-cone or beam distribution. The interpretation of the detailed variations of the upper and lower frequency cutoffs is complicated by uncertainties about the relative importance of variations in the plasma density profile and variations in the electrostatic potential distribution along the auroral field lines. If the electron plasma frequency is sufficiently low, $\lesssim$ 20 kHz, at high altitudes in the magnetosphere, then the lower frequency cutoff of the radio emission spectrum gives the upper limit to the auroral acceleration region. Typically, the upper limit of the acceleration region determined in this way is located at radial distances from about 2.0 to 3.5 R_E. During some intervals the kilometric radiation completely disappears. These intervals are believed to correspond to either periods when the plasma density is unusually high, $\omega_p \gtrsim \omega_g$ at all points along the auroral field lines, or (more likely) to periods of very small accelerating potential.

High resolution frequency-time spectrums of auroral kilometric radiation show that the radiation consists of many discrete narrowband emissions, some of which drift upwards or downwards over a large frequency range. The existence of this very complex fine structure indicates the presence of corresponding complex spatial and temporal structure in the auroral acceleration region. The drifting narrowband emissions indicate the presence of very compact source regions, with radial scale sizes of only 50 km, which move upward or downward along the magnetic field line with velocities ranging from 3 to 300 km/sec. These source velocities are comparable to the ion-acoustic speed in the auroral ionosphere, which suggests the existence of discrete structures moving upward and downward along the auroral field lines. These structures may possibly originate from unstable fluctuations in the electrostatic potential distribution, forming transient double-layers or shock-like disturbances which propagate along the auroral field lines. The fine structure of the auroral kilometric radiation also has features which are remarkably similar to discrete whistler-mode emissions, suggesting that somewhat similar nonlinear processes may be involved in both types of radio emissions.

Acknowledgements. The authors would like to express their thanks to Y. Chiu, J. Fennell and D. Croley of the Aerospace Space Science Laboratory for several useful discussions concerning the S3-3 electron distribution functions and possible relationships to auroral kilometric radiation.

The research presented in this paper was supported by NASA through Contract NAS5-20093 from Goddard Space Flight Center and by Grants NGL-16-001-043 and NGL-16-001-002 from NASA Headquarters, and by the Office of Naval Research.

References

Benediktov, E. A., G. G. Getmantsev, Yu. A. Sazonoy and A. F. Tarasov, Preliminary results of measurement of the intensity of distributed extraterrestrial radio-frequency emission at 725 and 1525-kHz frequencies by the satellite elektron-2, Kosmicheskie Issledovaniya, 3, 614, 1965.

Benson, R. F., and W. Calvert, Isis I observations at the source of auroral kilometric radiation, Geophys. Res. Lett., 6, 479, 1979.

Brown, L. W., Possible radio emission from Uranus at 0.5 MHz, Astrophys. J., 207, L202, 1976.

Carlqvist, P., and R. Bostrom, Space-charge regions above the aurora, J. Geophys. Res., 75, 7140, 1970.

Chiu, Y. T., and M. Schlulz, Self-consistent particle and parallel electrostatic field distributions in the magnetospheric-ionospheric auroral region, J. Geophys. Res., 83, 629, 1978.

Dunkel, N., B. Ficklin, L. Rorden, and R. A. Helliwell, Low-frequency noise observed in the distant magnetosphere with OGO 1, J. Geophys. Res., 75, 1854, 1970.

Frank, L. A., and K. L. Ackerson, Observations of charged-particle precipitation into the auroral zone, J. Geophys. Res., 76, 3612, 1971.

Green, J. L., D. A. Gurnett, and R. A. Hoffman, A correlation between auroral kilometric radiation and inverted-V electron precipitation, J. Geophys. Res., 84, 5216, 1979.

Gurnett, D. A., The earth as a radio source: Terrestrial kilometric radiation, J. Geophys. Res., 79, 4227, 1974.

Gurnett, D. A., and J. L. Green, On the polarization and origin of auroral kilometric radiation, J. Geophys. Res., 83, 697, 1978.

Gurnett, D. A., R. R. Anderson, F. L. Scarf, R. W. Fredricks, and E. J. Smith, Initial results from the ISEE-1 and -2 plasma wave investigation, Space Sci. Rev., 23, 103, 1979.

Helliwell, R. A., Whistlers and related ionospheric phenomena, Stanford Univ. Press, Stanford, CA, 206, 1965.

Helliwell, R. A., and J. P. Katsufrakis, VLF wave injection into the magnetosphere from Siple Station, Antartica, J. Geophys. Res., 79, 2511, 1974.

Hubbard, R. F., and G. Joyce, Simulation of auroral double layers, J. Geophys. Res., 84, 4297, 1979.

Inan, U. S., T. F. Bell, and R. A. Helliwell, Nonlinear pitch-angle scattering of energetic electrons by coherent VLF waves in the magnetosphere, J. Geophys. Res., 83, 3235, 1978.

Kaiser, M. L., J. K. Alexander, A. C. Riddle, J. B. Pearce, and J. W. Warwick, Direct measurements by Voyagers 1 and 2 of the polarization of terrestrial kilometric radiation, Geophys. Res. Lett., 5, 857-860, 1978.

Kaiser, M. L., M. D. Desch, J. W. Warwick, and J. B. Pearce, Voyager detection of nonthermal radio emission from Saturn, Science, 209, 1238, 1980.

Kundu, M. R., Solar Radio Astronomy, Interscience, New York, 1965.

Lin, C. S., and R. A. Hoffman, Fluctuations of Inverted-V electron fluxes, J. Geophys. Res., 84, 6547, 1979.

Melrose, D. B., An interpretation of Jupiter's radiation and the terrestrial kilometric radiation as direct amplified gyroemission, Astrophys. J., 207, 651, 1976.

Mizera, P. F., and J. F. Fennell, Signatures of electric fields from high and low altitude particle distributions, Geophys. Res. Lett., 4, 311, 1977.

Voots, G., D. A. Gurnett, and S.-I. Akasofu, Auroral kilometric radiation as an indicator of auroral magnetic disturbances, J. Geophys. Res., 82, 2259, 1977.

Warwick. J. W., Radiophysics of Jupiter, Space Sci. Rev., 6, 841, 1967.

Warwick, J. W., J. B. Pearce, A. C. Riddle, J. K. Alexander, M. D. Desch, M. L. Kaiser, J. R. Thieman, T. D. Carr, S. Gulkis, A. Boischot, C. C. Harvey, B. M. Pedersen, Voyager 1 planetary radio astronomy observations near Jupiter, Science, 204, 995, 1979.

Wu, C. S., and L. C. Lee, A theory of terrestrial kilometric radiation, Astrophys. J., 230, 621, 1979.

Plasma Waves on Auroral Field Lines

M. A. Temerin

Space Sciences Laboratory,
University of California, Berkeley, CA 94720

Abstract. The dominant waves below the plasma frequency on auroral field lines are whistler mode waves propagating near the resonance cone (VLF hiss) and electrostatic hydrogen cyclotron waves. The type of wave that is observed depends on whether the measurement is made above or below the region of electric fields parallel to the magnetic field. Above the region of the parallel electric field the dominant mode is the electrostatic hydrogen cyclotron wave found in association with upwardly accelerated ion beams. Below the parallel electric field region the dominant waves are in the whistler mode found in association with anisotropic electron distributions. Within the parallel electric field region both modes can be observed. The above statements are true for both the parallel electric field that is associated with the broad in latitude inverted-V features in the electron distribution and the parallel electric field that is associated with the narrow in latitude electrostatic shocks responsible for discrete auroral arcs. However, within the electrostatic shock, that is, the region of large perpendicular electric fields, the waves are often substantially more intense -- both the VLF hiss and the electrostatic hydrogen cyclotron waves can exceed 40 mV/m RMS. These general points are illustrated with data from the S3-3 satellite. In addition, it is shown that electrostatic shocks often occur at either the poleward or equatorward boundary of the region of parallel electric field below the satellite.

The main theme of this paper is the relation of the electric equipotential structure of the aurora to waves. This relation is rather simple: above the parallel acceleration region one sees electrostatic hydrogen cyclotron waves and below the parallel acceleration region one sees VLF hiss. I will illustrate this relation with examples of data from above and below "inverted-V" regions and also show wave data within electrostatic shocks. I will also comment on the relation of electrostatic shocks to "inverted-V's" and discrete auroral arcs.

Now before one can describe waves above and below parallel electric field regions, one has to be able to recognize such regions. Figure 1 shows the particle data from a S3-3 satellite pass through the auroral zone. There are several ways to recognize the presence of a parallel electric field below the satellite. The most obvious way is by means of the ion beams which can be seen at 1120:45 UT, at 1124:40 UT, at 1125:20 UT, at 1127:20 UT and between 1128:00 UT and 1130:40 UT [*Shelley et al.*, 1976; *Ghielmetti et al.*, 1978; *Gorney et al.*, 1981]. Another way is by the depletion of the low energy electrons which can be seen at the same times [*Sharp et al.*, 1979]. Still another way is by a widening of the electron loss cone, though on the scale of Figure 1 this is not easily noticeable [*Mizera and Fennell,* 1981; *Croley et al.*, 1978; *Cladis and Sharp,* 1979].

One can also recognize the presence of a parallel electric field just above the satellite. One way is by the lack of electrons at pitch angles of 90°. Electrons which have just passed through the parallel electric field region from above have acquired parallel momentum and cannot have pitch angles near 90°. Likewise, electrons backscattered from the ionosphere or mirroring below the satellite cannot have pitch angles at 90° by virtue of the conservation of the first adiabatic invariant in a decreasing magnetic field. In Figure 1 one can recognize this effect in the picket fence structure of the higher energy electrons between 1121:50 UT and 1122:30 UT and in the vicinity of 1128:00 UT. This structure is due to absence of electrons at 90°. (I have already assumed that the reader is either already familiar with the topic or has guessed that the S3-3 satellite is spinning and thus samples different pitch angles as given by the bottom trace in Figure 1). Also notable in the region below the parallel field in the interval between 1121:50 UT and 1122:30 UT is the presence of conics (perpendicularly accelerated ions mirroring below the satellite) [*Sharp et al.,* 1977] and counterstreaming electrons (field-aligned electrons both at 0° and 180° pitch angle at low energy) [*Sharp et al.*, 1980].

Where were the shocks (enhanced electrostatic regions) in this pass of the auroral zone? They occurred every time the electric equipotential contours due to the parallel electric field dipped below the satellite. That is, they occurred on both sides of parallel field regions at 1119:40 UT, 1124:55 UT, 1126:00 UT, and 1127:30 UT and throughout the large region of parallel field regions below the satellite between 1128 UT and 1130:40 UT. In each case, the direction of the electric field in the enhanced electrostatic region was such that, if the potential contours closed below the satellite in a U-shaped configuration, ions would have been accelerated upward and electrons downward toward the earth along the magnetic field. That is, the electric field was pointing poleward at the equatorward boundary of the parallel field region and equatorward at the poleward boundary. Figure 2, panel A shows the equatorward component of the electric field for the interval around the first region of parallel field below the satellite at 1119:40 UT.

What were the waves? In the region of the ion beams the main low frequency wave was the electrostatic hydrogen cyclotron mode. In the region immediately below the parallel electric field the main mode was VLF hiss -- whistler mode waves propagating near the resonance cone. This can be seen in the top spectrogram in Figure 1. Because this spectrogram is based on only 13 logarithmically-spaced filter channels, the frequency resolution is poor. Nevertheless, the enhancement at low frequency (below 200 Hz) when the parallel electric field is below the satellite and at high frequency (above 500 Hz) when the parallel electric field is above the satellite is visible.

Likewise, in Figure 2, panel A, one can see the effect of the

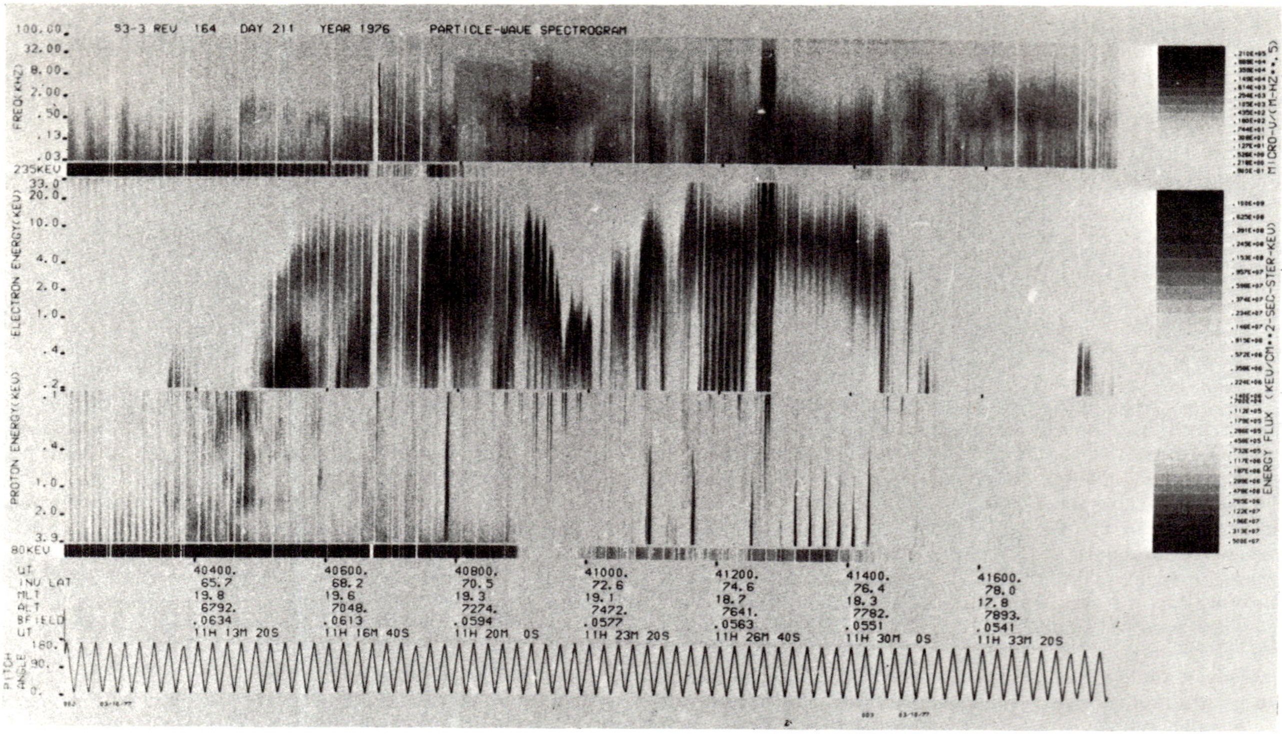

Fig. 1. Aerospace particle and wave spectrograms for orbit 164. (Courtesy of Joe Fennell).

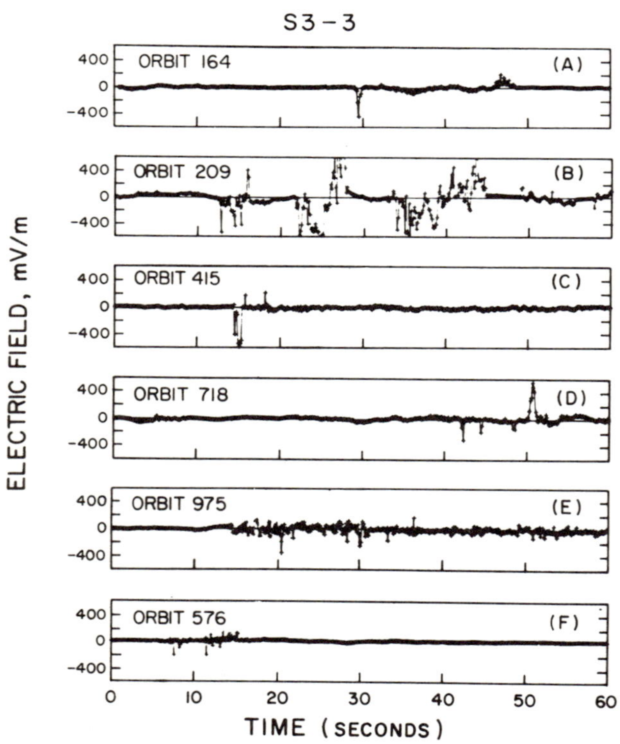

Fig. 2. Examples of enhanced electrostatic regions.

enhanced electrostatic region. Notice that between the oppositely pointing electric fields there is increased electric field turbulence, due, probably, to aliasing of the d.c. detector by the electrostatic ion cyclotron waves which are always found in regions of upward going ion beams.

I will show some more detailed examples of waves in and near enhanced electrostatic regions which we at Berkeley have been calling "electrostatic shocks" [*Mozer et al.,* 1977, 1979]. Figure 3 shows four examples of wave data in the common spectrogram format and the associated enhanced electrostatic regions. Note the three enhanced electrostatic regions between 0121 and 0122 UT for orbit 209. The VLF hiss seen in correlation with the electric field in the spectrogram in each case occurred just equatorward of the enhanced electric field regions. Figure 2, panel B, shows the equatorward component of the electric field. Note the same general pattern: poleward pointing electric field at the equatorward side and equatorward pointing electric field at the poleward side of each enhanced electrostatic region.

Figure 4 shows six seconds of detailed low-frequency wave data corresponding to the first enhanced electrostatic region during orbit 209. The arrows above the wave data show the simultaneous value of the quasi-static electric field. The electrostatic hydrogen cyclotron waves are seen as the nearly sinusoidal but saturated fluctuation at about 120 Hz starting at 0121:12.6 UT in association with the large poleward electric field and ending somewhat after 0121:15 UT in association with the large equatorward electric field. The left margin of Figure 4 shows the angle of the double probe with respect to the magnetic field. The exponential decay-like forms are due to saturation of the receiver electronics.

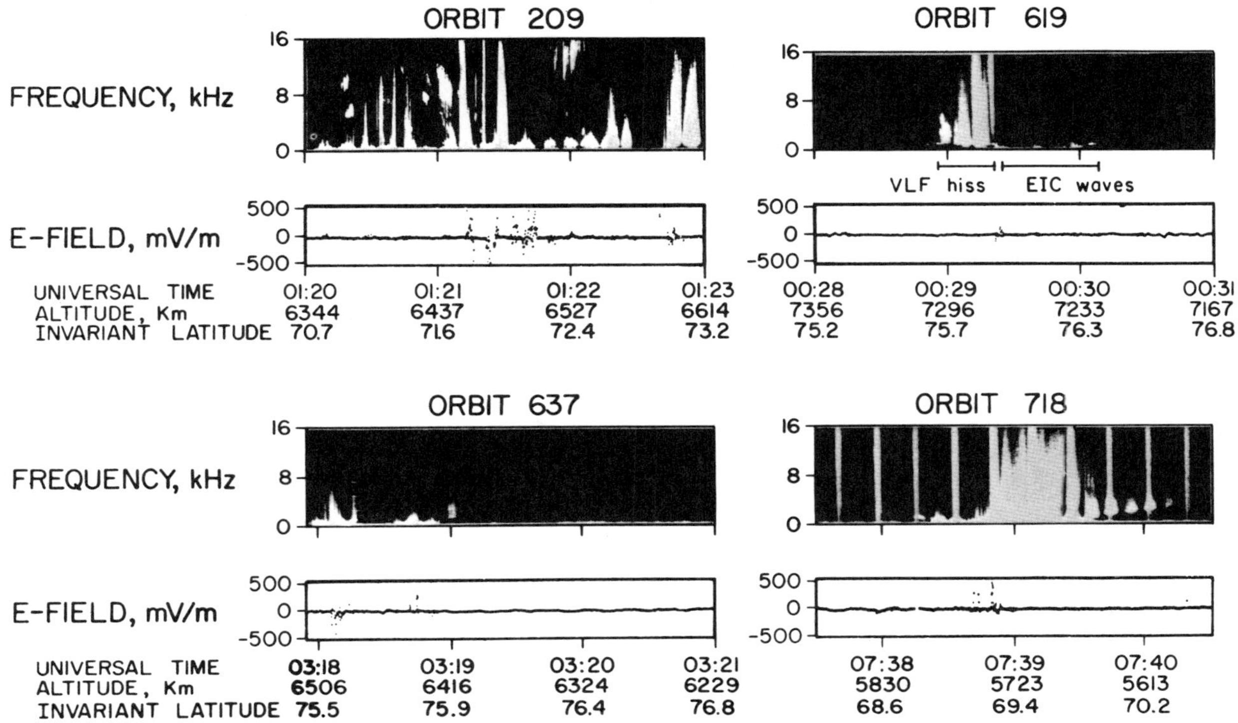

Fig. 3. Four examples of waves correlated with enhanced electrostatic region.

Figure 5 shows the same interval. The data are now from a separate double probe oriented along the spin axis which was approximately in the east-west direction. The underlined portions of the data are shown in an expanded version (every point-resolution is approximately 6 kHz in this channel instead of the averaged data at 1100 points per second) in Figure 6. Note how in interval (B) the wave changes from a frequency near the second harmonic of the hydrogen cyclotron frequency to a frequency just above the first harmonic. This interval corresponds closely to the beginning of an enhanced electrostatic region. Note also the presence of the second and third harmonics in intervals (C) and (D). Note especially the large amplitude of the wave. The maximum peak-to-peak amplitude that can be seen due to instrumental saturation is 55 mV/m. A sine wave that had as much saturation as the wave at the end of interval (D) would have had an amplitude of over 350 mV/m peak-to-peak. The actual wave amplitude was probably somewhat less -- the presence of the harmonics makes a sine wave fit somewhat inappropriate. In all three enhanced electrostatic regions there was little noticeable doppler shift in the electrostatic hydrogen cyclotron waves. even when the simultaneous electric field was in excess of 400 mV/m, corresponding to a $c\,|\mathbf{E} \times \mathbf{B}|\,/\,B^2$ convection velocity in excess of 40 km/s. This is a rather remarkable result, since it implies wavelengths in excess of 2 km.

The particle data show intense upward ion beams up to maximum energy -6 keV- of the Lockheed detector inside the second and third enhanced electrostatic region. Inside the first region the detector was not pointing in the right direction to see the ion beam but, nevertheless, there is good reason to believe it was there. This is because the correlation between upward ion beams and electrostatic ion cyclotron waves is very good, as shown in Figure 7 [*Mozer et al.*, 1977; *Kintner et al.*, 1979]. Figure 7 shows the density fluctuation in the 128 Hz filter channel (the one closest to hydrogen cyclotron frequency) *versus* the maximum energy channel in which the upgoing (within 22.5° of 180° pitch angle) exceeded the perpendicular and downgoing ion fluxes by at least an order of magnitude. Some additional scatter is introduced into the data just from averaging procedures used to determine the density fluctuations from the filter channel. Nevertheless, the trend is clear: not only are there electrostatic ion cyclotron waves in association with ion beams, but also the more intense beams are associated with the more intense waves.

The next example is orbit 619, shown in the top right of Figure 3. Here, there is only a rather weak enhanced electrostatic region near 0029:22 UT. The particle spectrograms show that the shock occurred in the middle of an inverted-V region that lasted from 0028:10 to 0030:30 UT. Note that the enhanced electrostatic region divided the region of VLF hiss from the region of electrostatic hydrogen cyclotron waves. (The cyclotron waves can be seen only as a narrow, unresolved line at the bottom of the spectrogram in Figure 3). Correspondingly, the enhanced electrostatic region divided the region of electrons with the characteristic signature of a parallel field above the satellite from the region of ion beams. The ion beams could be seen in the five lowest Aerospace ion channels (Figure 8) the first time the detector looked down the field line after passing the enhanced electrostatic region. The signature of the parallel field above the satellite can be seen best in the 1.6 - 5 keV electrons in the second electron flux panel (Lockheed electrons). The gentle minima in this channel occur at 90° pitch angles. (The pitch angle is indicated by the magnetometer -- maxima and minima in the magnetometer correspond to 90° pitch angle for Lockheed electrons and 0° or 180° in the Aerospace electrons and ions. The whole event thus shows some nice systematic

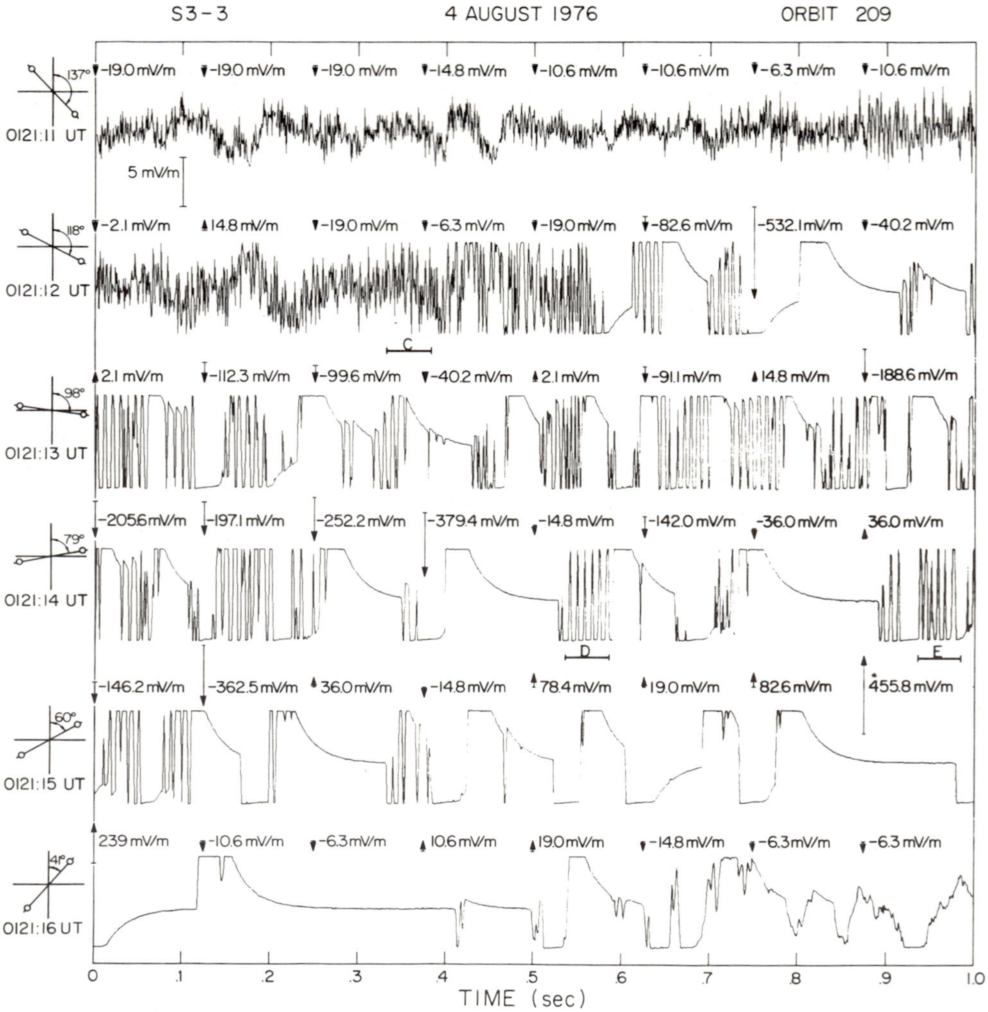

Fig. 4. Six seconds of detailed wave data in the first enhanced electrostatic region on orbit 209 from the radially-oriented double probe on the S3-3 satellite.

behavior. Note also from the top panel of Figure 8 that while the inverted-V started at 0028:10 UT, only the data after 0029:00 has the signature of the parallel field above the satellite and only this portion of the inverted-V is associated with the VLF hiss. The electron density data also shows an interesting correlation which is common in regions where a portion of the parallel electric field is below the satellite. In the region of upward ion beams and electrostatic hydrogen cyclotron waves, the density had a value of about $6 cm^{-3}$, whereas outside it had a value of about $15 cm^{-3}$.

A possible electrostatic equipotential structure that could explain the data is shown in Figure 9. A satellite moving from the left would first see the effect of parallel electric fields above. As the potential contours dipped below the satellite, there would be an enhanced electrostatic region, followed by upwardly accelerated ions. The region need not end, as it did not in this case, with an oppositely directed enhanced electrostatic region.

Such a configuration, with an enhanced electrostatic region at only one side of the region of parallel field below the satellite, is quite common. Figure 2, panel C, shows another example of such an electric field. The equatorward component is shown. The electric field in the enhanced electrostatic region points poleward at the equatorward edge of a region of upward ion beams. Such configurations have been called "S-shocks" as opposed to "V-shocks."

Large S-shocks must be associated with regions broader than just discrete arcs (a discrete arc is a structure having the width in the ionosphere of the order of one kilometer.) For instance, if the potential across the S-shock is 2 kV, then the corresponding potential across the ionosphere must also be 2 kV. Since a large field in the ionosphere is 100 mV/m, this would necessarily correspond to a distance in the ionosphere of at least 20 km, which is not the scale size of discrete auroral arcs. V-shocks, on the other hand, are not limited in this manner.

Another example is shown at the bottom right of Figure 3. (The periodic vertical stripes are telemetry noise.) The intense VLF hiss occurred just after the shock. The corresponding electron data is shown in Figure 10. Note the large minima at 90° in

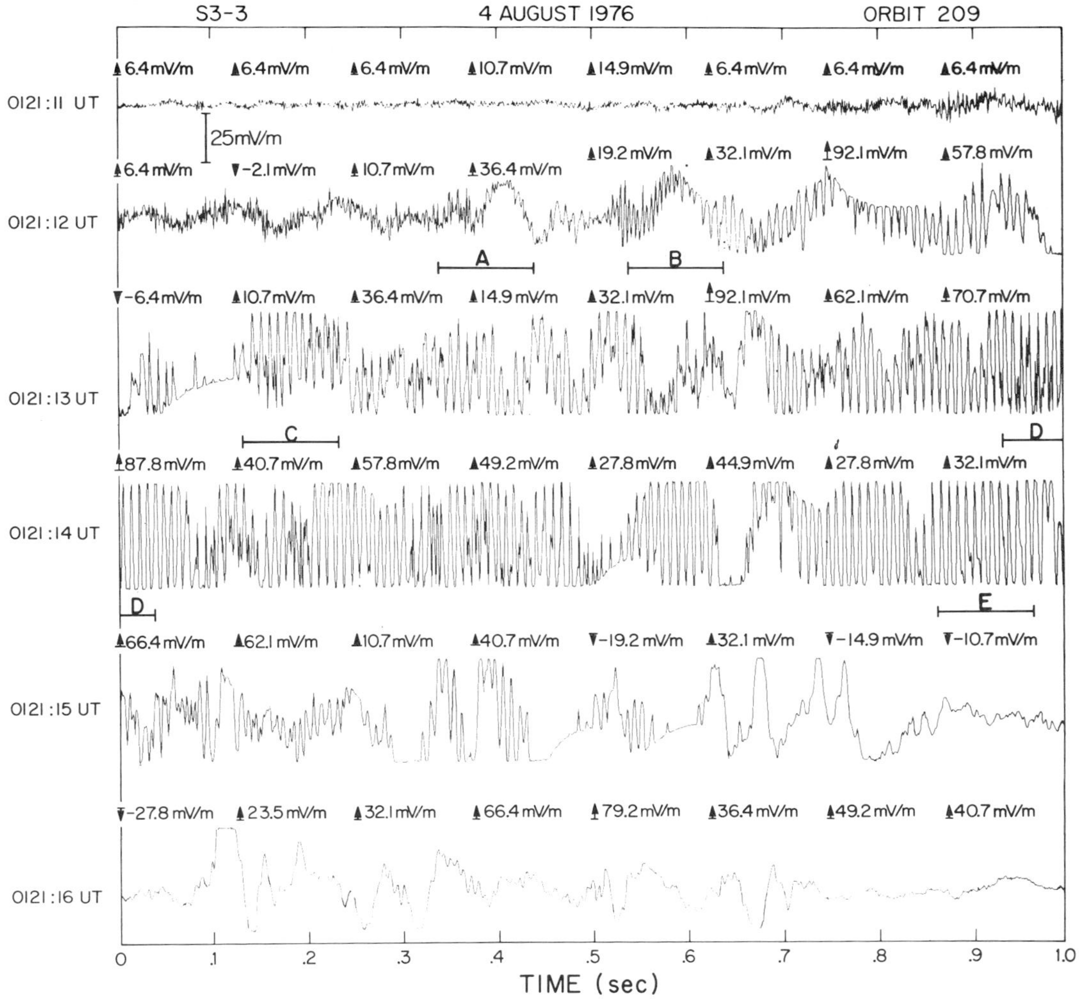

Fig. 5. Same as Figure 4 for the axial double probe.

the 1.6 - 5 keV channel, signifying parallel electric fields above the satellite between 0738:55 and 0739:30 UT. The large enhanced electrostatic region at 0738:51.6 UT separates the region of parallel field predominately below the satellite (note especially the drop-out in the electrons equatorward of the enhanced electrostatic region) from regions of parallel fields predominately above the satellite. The electric field was oriented equatorward, corresponding to the poleward boundary of the parallel field region below the satellite. (See Figure 2, panel D, which shows the equatorward component of the electric field.) The ion beam reappeared again near the end of the inverted-V at 0739:30. Correspondingly, there were both ion cyclotron waves and VLF hiss simultaneously in the data. This indicates that there can be parallel electric fields both above and below the satellite and that parallel fields below the satellite can occur without corresponding enhanced electrostatic regions.

So, notwithstanding various problems, high altitude data in the auroral zone give a fairly consistent picture of the interactions of enhanced electrostatic regions (i.e. electrostatic shocks), inverted-V's, discrete arcs, particle distributions and waves both above and below the acceleration region. Not so for lower altitude enhanced electrostatic regions. Above 6500, nearly all large enhanced electrostatic regions are associated with upwardly moving ion beams. On the other hand, below 4500 km, enhanced electrostatic regions are practically never associated with ion beams. (Not all local times, altitudes and seasons have been surveyed.) Since fields as large as those in enhanced regions are never found in the ionosphere, this suggests that either the parallel fields at low altitudes do not produce distinctive particle signatures at the measured energies or that such parallel fields do not exist. One problem with lower altitude observations is that the satellite moves faster, and the same number of degrees of invariant latitude corresponds to smaller absolute distances, so that many of the signatures of parallel fields below the satellite could have been missed. This should not account, however, for the complete lack of such events. For instance, ion conics are seen near most low altitude enhanced electrostatic regions, as, for instance, with every example in Figure 11.

Some lower altitude events have an extremely turbulent character. In Figure 2, panel E, which shows the east-west com-

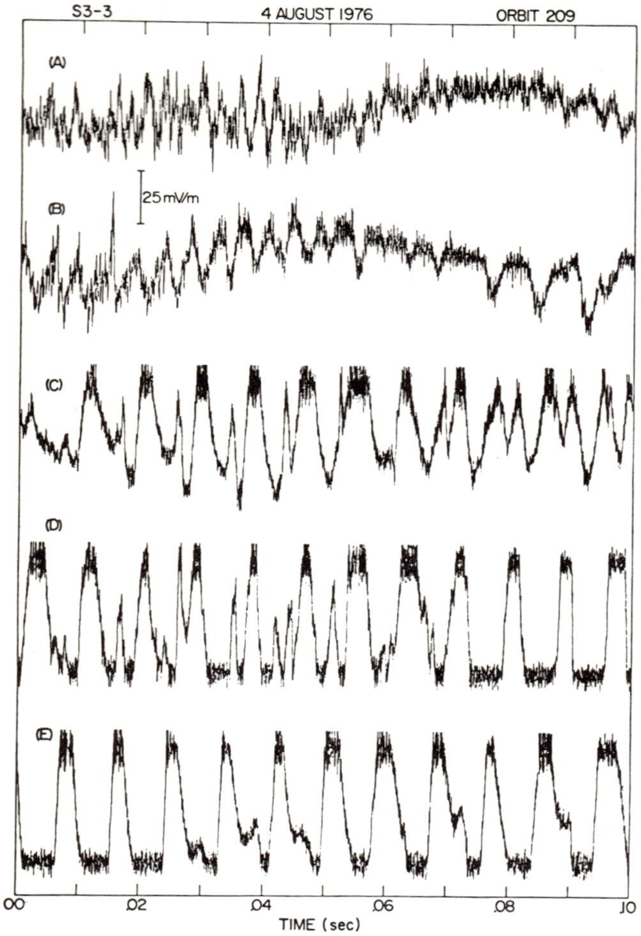

Fig. 6. The underlined portions of Figure 5 shown in greater detail.

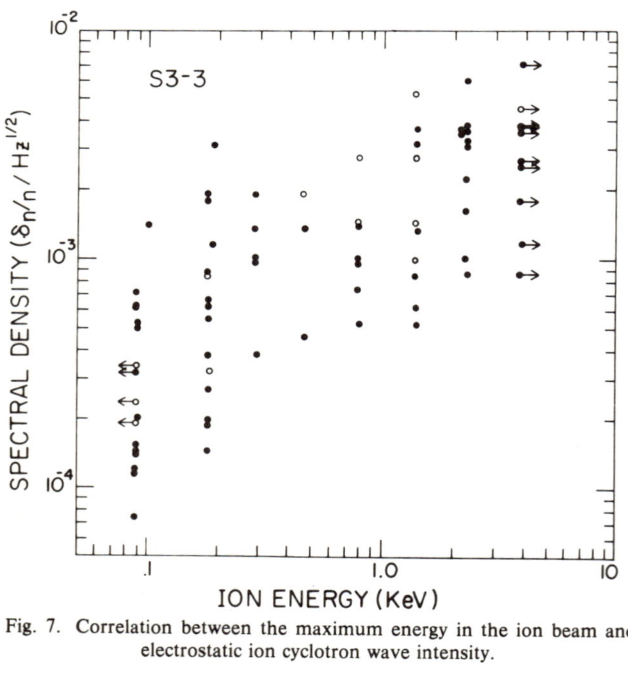

Fig. 7. Correlation between the maximum energy in the ion beam and electrostatic ion cyclotron wave intensity.

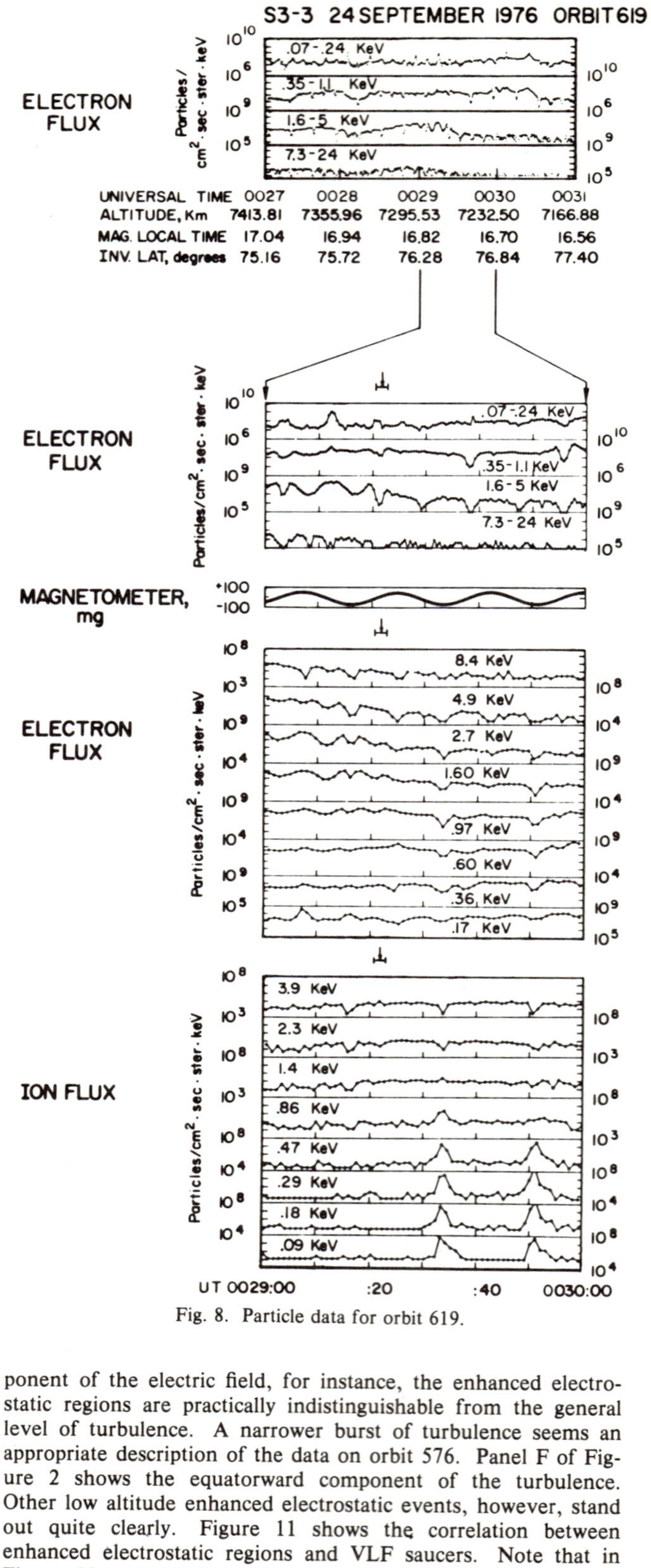

Fig. 8. Particle data for orbit 619.

ponent of the electric field, for instance, the enhanced electrostatic regions are practically indistinguishable from the general level of turbulence. A narrower burst of turbulence seems an appropriate description of the data on orbit 576. Panel F of Figure 2 shows the equatorward component of the turbulence. Other low altitude enhanced electrostatic events, however, stand out quite clearly. Figure 11 shows the correlation between enhanced electrostatic regions and VLF saucers. Note that in Figure 11 each of the enhanced regions is the center of the corresponding VLF saucer emission.

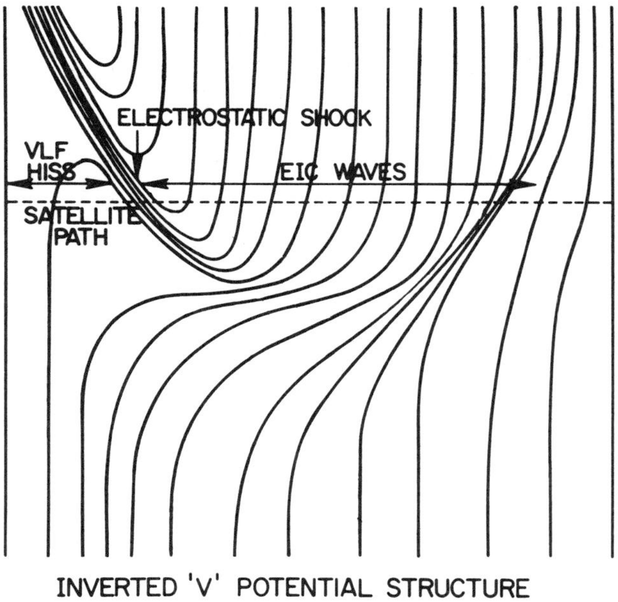

Fig. 9. Possible electric field potential contours for orbit 619.

How is it possible to avoid parallel electric fields and still have large perpendicular electric fields, as seems to be required at lower altitudes? One way is through inductive fields. If the perpendicular electric field is due to an Alfvén wave propagation from some source, as described by *Mallinckrodt and Carlson* [1978] or *Goertz and Boswell* [1979], then the corresponding parallel field need not be large, since $dB/dt \neq 0$ and hence $\nabla \times E \neq 0$. The large electric field in this case would be transient (~1 sec). The parallel current associated with the enhanced electric field need not, however, be a transient phenomenon [*Goertz and Boswell,* 1979]. This would explain the relative permanence of the associated VLF saucers, provided that such large transient electric fields are continuously produced on nearly the same field lines.

Another possibility is that there are large field-aligned currents in the enhanced electrostatic structures. If this is the case, two field lines which are close together in the magnetosphere need not be near each other in the ionosphere and $\nabla \times E = 0$ can be retained without implying large ionospheric fields. For this to work, there must be an electric field parallel to the longitudinal direction of the auroral arc or other structure associated with the large field. Normally, in the auroral zone, this would be the east-west direction.

In this presentation, I have tried to emphasize only the main features of the relation between waves, enhanced electrostatic regions (i.e., electrostatic shocks), and regions of parallel electric fields. The reader is warned that a more quantitative characterization of some of the relationships that have been implied herein has yet to be done.

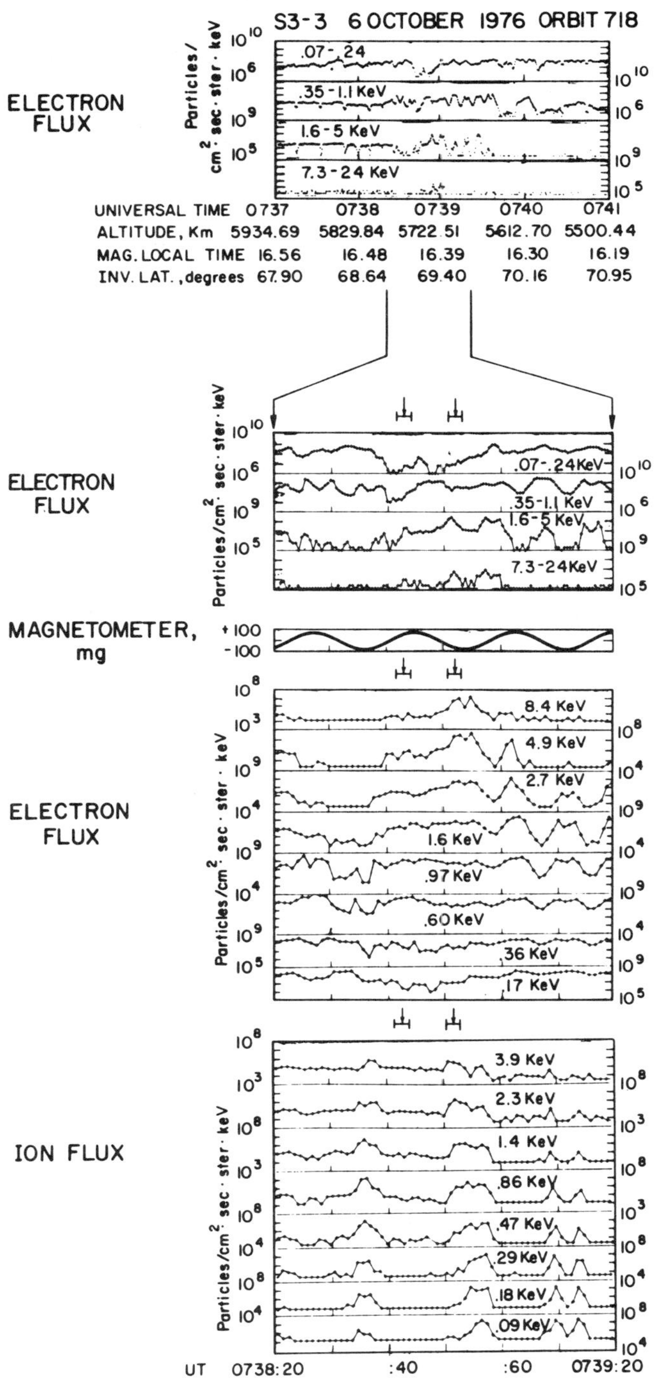

Fig. 10. Particle data for orbit 718.

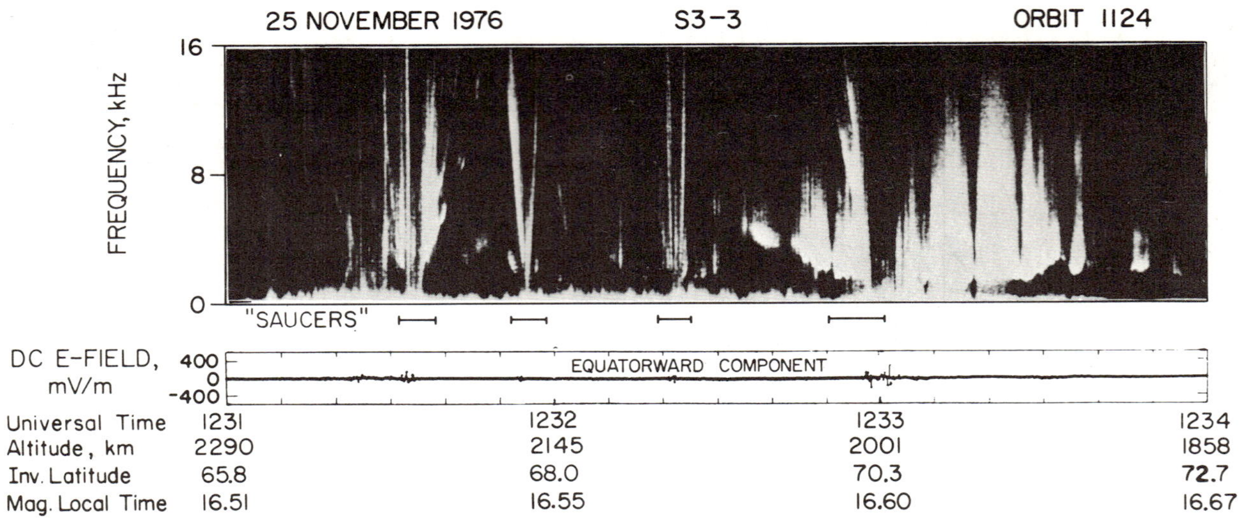

Fig. 11. VLF saucers and enhanced electrostatic regions.

Acknowledgments. This work was performed under Office of Naval Research Contract N00014-75-C-0294.

References

Cladis, J. B., and R. D. Sharp, "Scale of electric field along magnetic field in an inverted-V event," *J. Geophys. Res.*, **84**, 6564, 1979.

Croley, Jr., D. R., P. F. Mizera, and J. F. Fennell, "Signature of a parallel electric field in ion and electron distributions in velocity space," *J. Geophys. Res.*, **83**, 2701, 1978.

Ghielmetti, A. G., R. G. Johnson, R. D. Sharp, and E. G.Shelley, "The latitudinal, diurnal, and altitudinal distributions of upflowing energetic ions of ionospheric origin," *Geophys. Res. Lett.*, **5**, 59, 1978.

Goertz, C. K., and R. W. Boswell, "Magnetosphere-ionosphere coupling," *J. Geophys. Res.*, **84**, 7239, 1979.

Gorney, D. J., A. Clarke, D. Croley, J. Fennell, J. Luhmann, and P. Mizera, "The distribution of ion beams and conics below 8000 km," *J. Geophys. Res.*, **86**, 83, 1981.

Kintner, P. M., M. C. Kelley, R. D. Sharp, A. G. Ghielmetti, M. Temerin, C. Cattell, P. F. Mizera, and J. F. Fennell, "Simultaneous observations of energetic (keV) upstreaming and electrostatic hydrogen cyclotron waves," *J. Geophys. Res.*, **84**, 7201, 1979.

Mallinckrodt, A. J. and C. W. Carlson, "Relations between transverse electric fields and field-aligned currents," *J. Geophys. Res.*, **83**, 1426, 1978.

Mizera, P. F., and J. F. Fennell, "Signatures of electric fields from high and low altitude particle distributions," *J. Geophys. Res.*, **4**, 311, 1977.

Mozer, F. S., C. W. Carlson, M. K. Hudson, R. B. Torbert, B. Parady, J. Yatteau, and M. C. Kelley, "Observations of paired electrostatic shocks in the polar magnetosphere," *Phys. Rev. Lett.*, **38**, 292, 1977.

Mozer, F. S., C. A. Cattell, M. Temerin, R. B. Torbert, S. Von Glinski, M. Woldorff, and J. Wygant, "The DC and AC electric field, plasma density, plasma temperature, and field-aligned current experiments on the S3-3 satellite," *J. Geophys. Res.*, **84**, 5875, 1979.

Sharp, R. D., R. G. Johnson, and E. G. Shelley, "Observations of an ionospheric acceleration mechanism producing energetic (keV) ions primarily normal to the magnetic field direction," *J. Geophys. Res.*, **82**, 3324, 1977.

Sharp, R. D., R. G. Johnson, and E. G. Shelley, "Energetic particle measurements from within ionospheric structures responsible for auroral acceleration," *J. Geophys. Res.*, **84**, 480, 1979.

Sharp, R. D., E. G. Shelley, R. G. Johnson, and A. G. Ghielmetti, "Counterstreaming electron beams at altitudes of ~ 1 R_E over the auroral zone," *J. Geophys. Res.*, **85**, 92, 1980.

Shelley, E. G., R. D. Sharp, and R. G. Johnson, "Satellite observations of an ionospheric acceleration mechanism," *Geophys. Res. Lett.*, **3**, 654, 1976.

See also further references in Mozer, F. S., C.A. Cattell, M. K. Hudson, R. L. Lysak, M. Temerin and R.B. Torbert, "Satellite measurements and theories of low altitude auroral particle acceleration," *Space Sci. Rev.*, **27**, 155, 1980.

OBSERVATIONS PERTAINING TO THE GENERATION OF AURORAL KILOMETRIC RADIATION

James Lauer Green

NASA/MSFC, Huntsville, AL 35812

Abstract. Auroral kilometric radiation (AKR) is an extremely intense radio emission generated at frequencies from 30 to 700 kHz within the earth's magnetosphere. Although the first reported detection of AKR was made in 1965, it wasn't until nearly 10 years later that any significant observational progress was made in understanding the generation mechanism of this radiation. Within the last 5 years, satellite observations have determined that intense AKR is generated from altitudes of about 0.5 to 2.5 earth radii in the region 19 to 01 hours magnetic local time along auroral field lines. AKR observations which have determined the propagation mode or polarization of the radiation and the detailed intensity distribution of the AKR emission cone are discussed. In addition, correlations between AKR with discrete field-aligned currents will also be illustrated. These observations have helped to identify the auroral particle population most likely responsible for the generation of AKR and the possible sources of free energy which drives the instability.

Introduction

Auroral kilometric radiation or AKR is the most intense electromagnetic radiation generated within the earth's magnetosphere. The first reported observations of AKR were made by Benediktov et al. [1965] at 0.725 and 1.5 MHz from observation made by the Elektron satellites. However, it wasn't until nearly 10 years later that AKR became actively studied in the United States. Since 1974 significant progress has been made on determining the possible generation mechanism for this radiation. The purpose of this paper is to review AKR observations which help to identify and characterize the emission mechanism along with reviewing important characteristics of this electromagnetic emission.

Basic Characterisitcs of AKR

An example of kilometric radiation observed near apogee (∿22 R_e, R_e = earth radius) by the Hawkeye spacecraft is illustrated in Figure 1. This figure shows the electric field intensities (height of vertical bars) in units of power flux of AKR at frequencies 178, 100, 56.2 and 42.2 kHz. As can be seen in Figure 1, AKR occurs in distinct bursts or storms lasting typically from tens of minutes to hours and extending above the receiver noise level by 60 dB or more. AKR is observed in the frequency range from 30 kHz to about 700 kHz.

Typical spectra of AKR are shown in Figure 2 from Kaiser and Alexander [1977]. These medium power flux spectra are observed by IMP 6 from radial distances >25 R_e, organized in 3-hour blocks of magnetic local time, and further sorted in three bins of the AE index. Generally, AKR has a spike appearance in its spectrum with a width (bandwidth) of 200 kHz at half of the intensity of the spectral peak. As can easily be seen in Figure 2, the frequency of the spectral peak of AKR varies inversely with AE (geomagnetic activity) from 300 kHz during very quiet times to below 200 kHz during very disturbed times. Gurnett [1974] estimated that at peak intensity

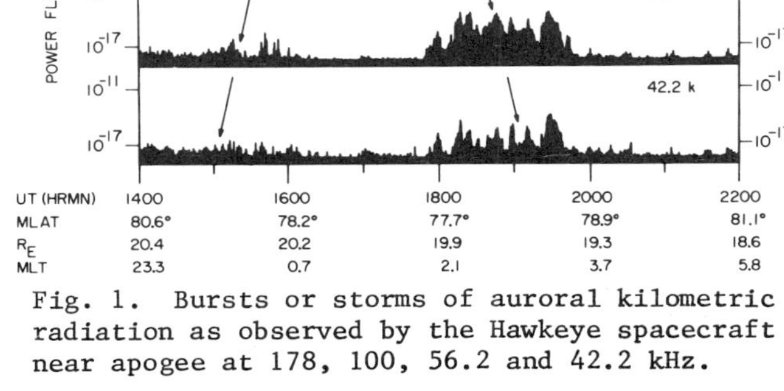

Fig. 1. Bursts or storms of auroral kilometric radiation as observed by the Hawkeye spacecraft near apogee at 178, 100, 56.2 and 42.2 kHz.

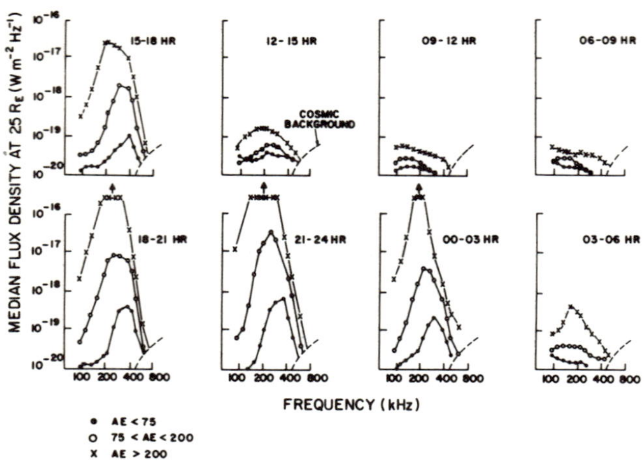

Fig. 2. Median power flux spectra of AKR observed by IMP-6 (from Kaiser and Alexander, 1977). The observations are sorted into 3 hour local time bins and organized by the AE index. IMP-6 had a near equatorial orbit and was not able to observe intense AKR on the dayside (see panels with local times from 6 to 15 hours).

the total power of AKR is sometimes as high as 10^9 Watts. Also, note that in Figure 2 the radiation is the strongest in the nighttime and is most intense around 21-24 hours magnetic local time. This suggests that sources of AKR are confined mostly to the nightside.

Source Region and Polarization Observations

Much of the scientific research on AKR in the past 5 years has centered on determining the source location(s) for this radio emission. The origin of this radiation has been difficult to determine since AKR is extremely sporadic and can vary in intensity by as much as 80 dB on a time scale of minutes and possibly seconds (see Figure 1). One interesting method was the use of the moon literally occulting the source of AKR as observed by the lunar orbiting RAE II satellite (see, for example, Kaiser and Alexander [1976]). Source regions at 250 kHz from successive lunar occultations with the moon at about 17.5 hours local time are shown in Figure 3. These sources are projected on a noon-midnight meridian plane with magnetic field lines from the Mead-Fairfield model. Note that the source regions are seen in the northern and southern hemisphere between 65° and 75° magnetic latitude field lines (auroral field lines) and in this example at distances less than about 6 R_e. The lunar occultation method of Alexander and Kaiser [1976] has determined that source regions of AKR vary with wavelength such that the higher frequency sources (>300 kHz) are usually observed closer to the earth than the lower frequency sources. There are several other independent methods of determining AKR source regions, such as direction-finding techniques using the spin modulation of the observed AKR intensity from Hawkeye and IMP 8 (Kurth et al. [1975]), ray path modeling matching the observed angular distribution of AKR (Green et al. [1977]), studying the AKR global pointing flux and expected $1/R^2$ fall off of the emission (Gallagher and Gurnett [1979]), and, most recently, radio interferometer technique using identical wave receivers on ISEE 1 and 2 (Shawhan and Baumback, private communication [1980]). Recently, a wealth of information has been reported from in-situ observations in the AKR source region by the Hawkeye and ISIS I satellites. Since the ISIS I observations are being discussed by Benson in this volume, only the Hawkeye observations will be discussed in detail.

A typical example of intense AKR observed by Hawkeye in the auroral zone of the southern hemisphere is shown in Figure 4 (from Gurnett and Green [1978]). The dashed line labeled f_g^- is the local electron gyrofrequency calculated from measurements made by the onboard triaxial fluxgate magnetometer which measured the ambient magnetic field in the southern hemisphere to ±125γ on each axis. The dashed line labeled f_p^- is an estimate of the total electron plasma frequency

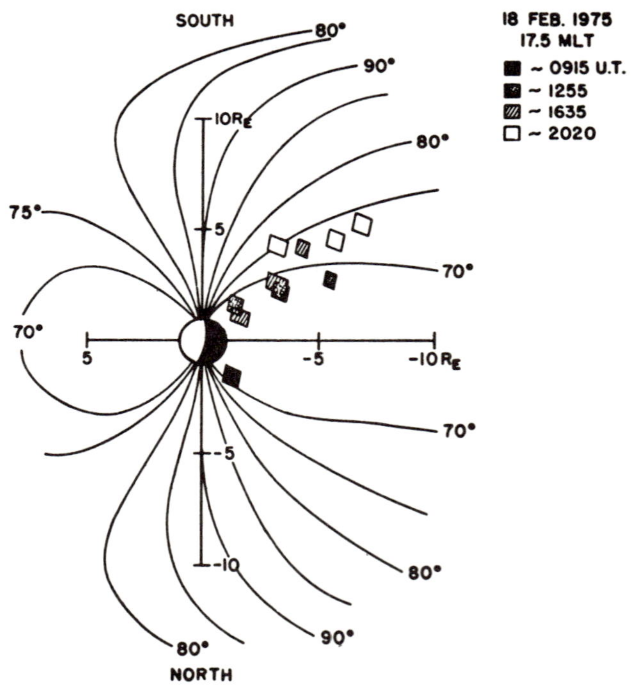

Fig. 3. AKR source regions at 250 kHz as determined by occultations of the source by the moon, as observed by RAE II (Alexander and Kaiser, 1976). The moon is at 17.5 hours MLT. This figure illustrates that sources of AKR are found along 65° to 75° invariant latitude in the Northern and Southern Hemisphere auroral zones.

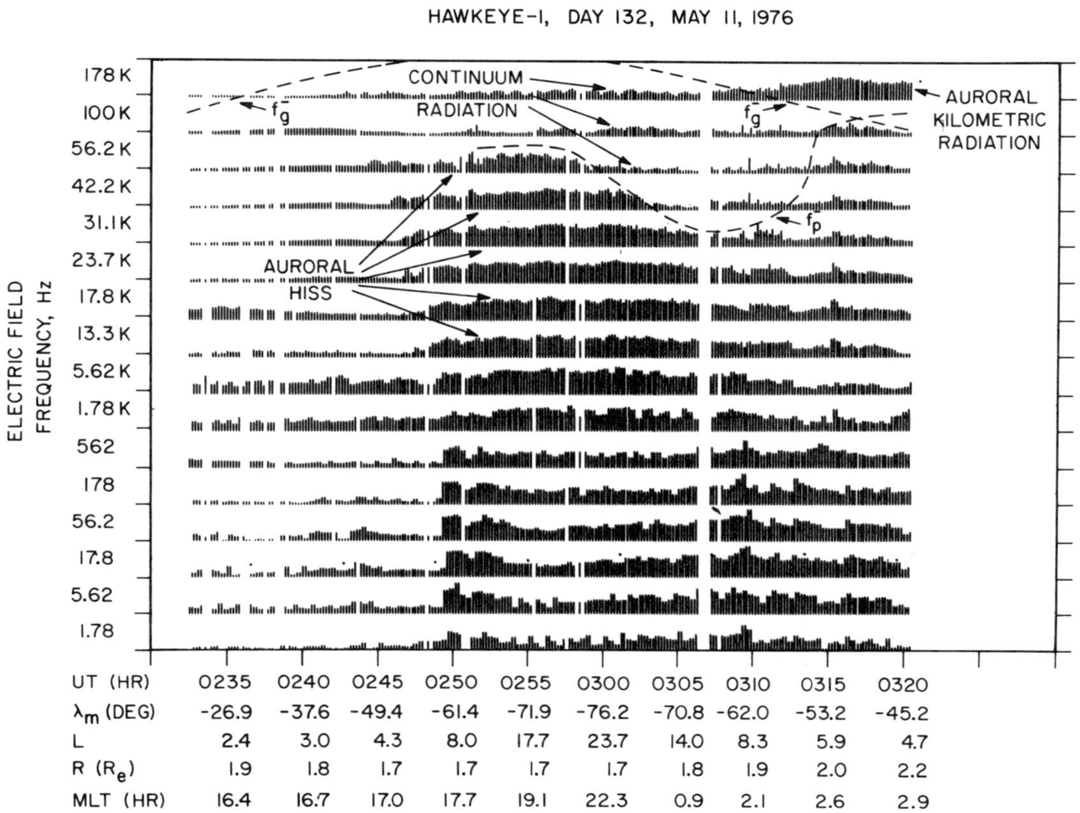

Fig. 4. The main feature of intense auroral kilometric radiation observed by Hawkeye during low altitude auroral zone passes is that it occurs in regions where the wave frequency (178 kHz in this case) is greater than both the local electron gyrofrequency and the local electron plasma frequency.

determined by the low frequency cutoff of continuum radiation and the high frequency cutoff of auroral hiss. The identification of AKR and the nonthermal continuum radiation in the Hawkeye southern hemisphere data is based on the comparison of their observed characteristics as seen at much longer radial distances (see Gurnett [1974] and [1975]). It has been well established in other regions of the magnetosphere by Gurnett and Shaw [1973], Gurnett and Frank [1974], and Shaw and Gurnett [1980] that the low frequency cutoff of the left-hand ordinary polarized (L-O) component of continuum radiation can be used as a reliable (at times within ±1 percent) indicator of the local electron plasma frequency. Both f_g^- and f_p^- are plotted over the electric field intensities on a frequency versus time scale where the frequency scale at the baseline of each channel is that channel's frequency. Thus, at 0312 UT in Figure 4, f_g^- = 178 kHz and f_p^- = 39 kHz. Note that in Figure 4 intense AKR is observed only in regions where the local electron gyrofrequency and local electron plasma frequency are below the wave frequency (178 kHz in this case). A general feature of all the Hawkeye AKR southern hemisphere observations is that no intense AKR is observed below altitudes where the wave frequency is less than the local electron gyrofrequency. In addition, AKR was only observed at low altitudes (2.0 to 2.5 R_e) on the nightside from 18 to 06 hours magnetic local time.

The low frequency cutoff of intense AKR at the local electron gyrofrequency is further illustrated in Figure 5. This figure is a scatter plot of Hawkeye southern hemisphere electric field measurements at 178 kHz (over 225 passes) from 18 to 06 hours magnetic local time plotted as a function of the ratio of the wave frequency to the local electron gyrofrequency (f/f_g^-).

To clearly illustrate the cutoff of AKR at f_g^-, passes for which intense auroral hiss emissions at 178 kHz occurred were eliminated. Virtually all the data below 10^{-16} Watts/m^2Hz) is continuum radiation which is observed not only in regions where the wave frequency is above, but also below, the local electron gyrofrequency. This figure shows a sharp cutoff of nearly five orders of magnitude in the AKR power flux at the local electron gyrofrequency.

The maximum measured electric field strength of

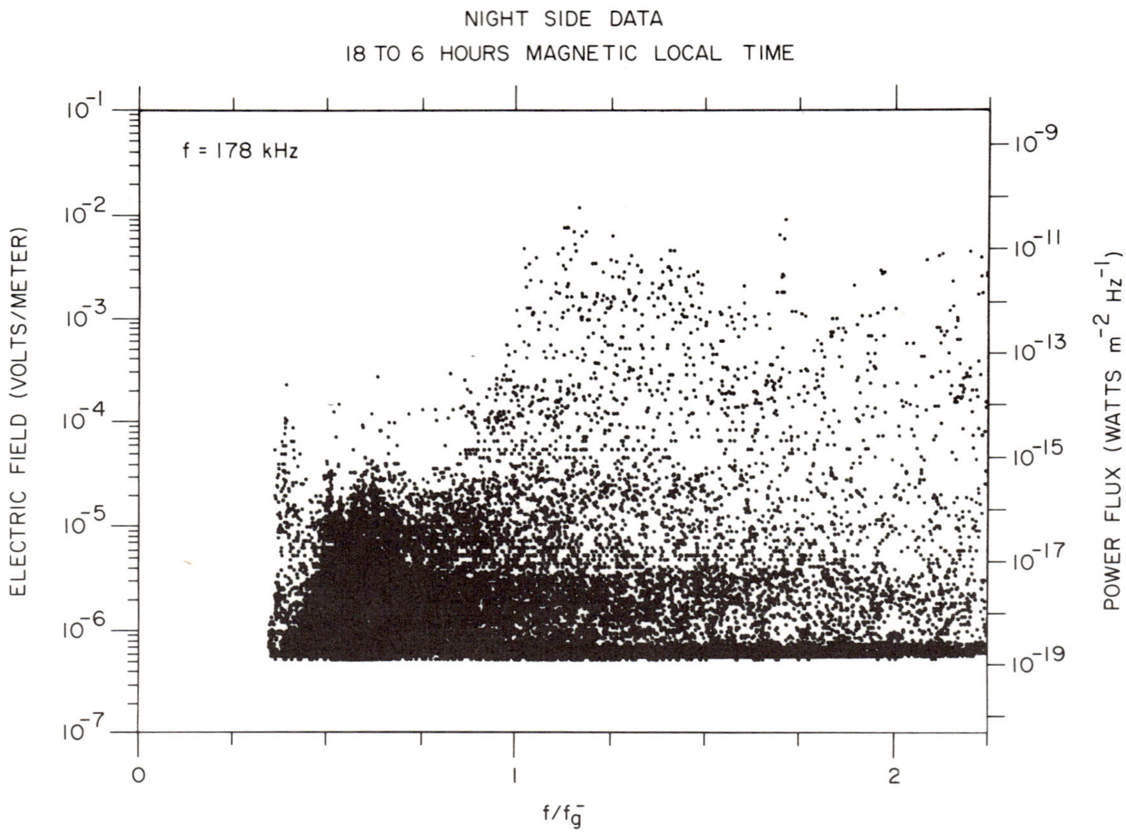

Fig. 5. A scatter plot of electric field measurements at 178 kHz from Hawkeye southern hemisphere passes on the nightside of the earth plotted as a function of the ratio of the wave frequency to the local electron gyro-frequency (f/f_g^-). The maximum electric field of AKR measured is nearly 12 mV/m.

AKR in Figure 5 occurs where $f \simeq f_g^-$ and is nearly 12 mV/m. Since Hawkeye does not have the capability of detecting any magnetic component associated with the electric field observations at 178 kHz, it is not known what fraction (if any) of the 12 mV/m fields observed in or near the AKR source region are electrostatic. Note also that the large electric fields in Figure 5 correspond to power fluxes greater than 10^{-11} Watts/(m^2 Hz). In order to estimate what fraction of the large ($\simeq$ 12 mV/m) electric fields in Figure 5 is electrostatic, it is instructive to compare the observed power fluxes with previous measurements of intense AKR bursts observed far from the earth. Using a $1/R^2$ extrapolation, escaping electromagnetic radiation with an intensity of 10^{-11} Watts/(m^2 Hz) generated at low altitudes would be observed far from the source (7 R_e) with an intensity of $\sim 10^{-13}$ Watts/(m^2 Hz), which is in excellent agreement with the maximum intensity of AKR bursts found in other studies (see Figures 12 and 13). This close agreement provides additional evidence of the low altitude ($\sim$2.5 R_e) origin of AKR but, more importantly, illustrates that large electrostatic fields necessary for AKR generation in some theories, such as proposed by Barbosa [1976] (1 to 3 V/m) and Roux and Pellat [1979] ($\sim$100 mV/m), are not found in association with AKR in the auroral region sampled by Hawkeye.

An interpretation of the observed in-situ f_g^- cutoff of AKR, as discussed by Gurnett and Green [1978], is illustrated in Figure 6 (taken from that paper). This figure shows the expected radial variations of the cutoff frequencies f_p^- and $f_{R=0}$ for the left-hand (L-O) and right-hand (R-X) polarized modes of propagation, respectively, along a 70° invariant latitude auroral field line. The relationship between f_g^- and $f_{R=0}$ in the source region is believed to be a sensitive parameter in determining the spectral shape and intensity of AKR and is discussed extensively by Gurnett and Anderson in this volume. The local electron plasma frequency (f_p^-) in Figure 6 is qualitatively consistent with the plasma density observations made below 10,000 km by ISIS I (see Benson and Calvert [1979] and Benson et al. [1980]), near 14,000 km by S3-3 (see Torbert et al. [1979]), and from the low frequency cutoff of continuum radiation observed by Hawkeye between 14,000 to 16,000 km geocentric radial distance.

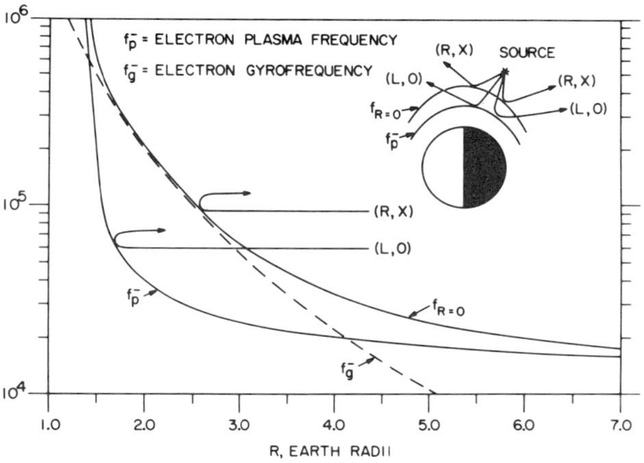

Fig. 6. The expected radial variations of f_p^-, f_g^-, and $f_{R=0}$ along an auroral magnetic field line. The frequencies f_p^- and $f_{R=0}$ present propagation barriers or cutoffs to downward propagating radiation with left-hand (L-0) and right-hand (R-X) polarizations, respectively. Note that in regions where $f_p^- << f_g^-$ then $f_g^- \simeq f_{R=0}$.

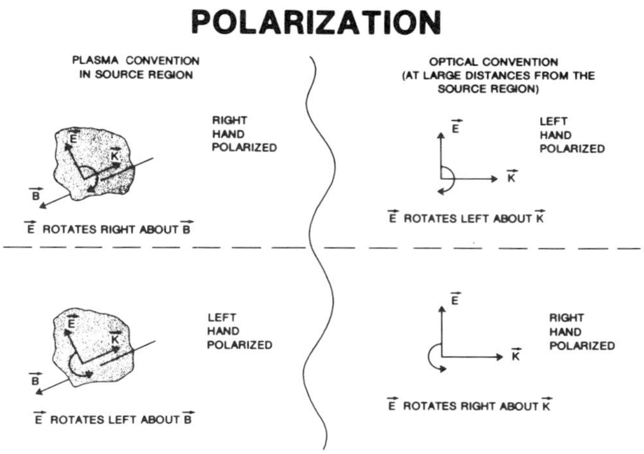

Fig. 8. Polarization definition is illustrated in the plasma convention (left-hand side) as used by Gurnett and Green (1978) and as measured in the optical convention (right-hand side) reported by Kaiser et al., 1978.

The cutoff frequency for the R-X mode ($f_{R=0}$) is calculated from the cold plasma formulation by Stix [1962] and involves knowledge of f_p^- and f_g^-:

$$f_{R=0} = \frac{f_g^-}{2} + \left[\left(\frac{f_g^-}{2}\right)^2 + (f_p^-)^2\right]^{1/2}$$

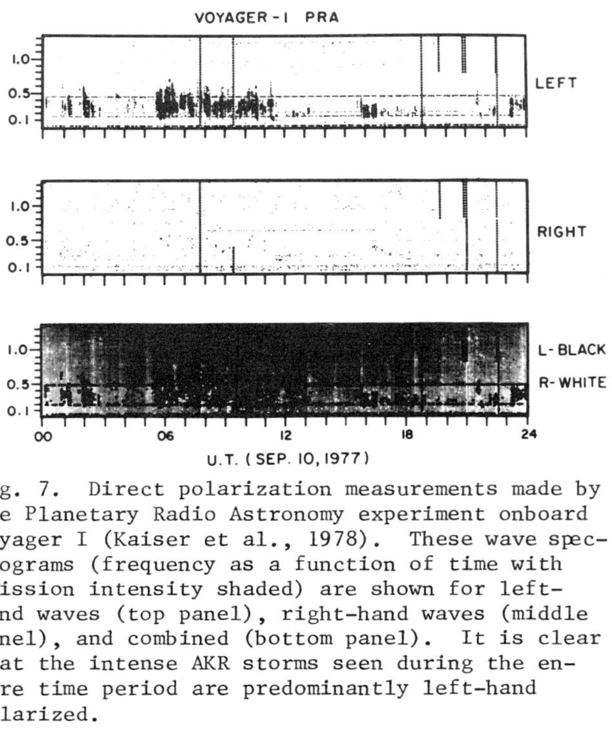

Fig. 7. Direct polarization measurements made by the Planetary Radio Astronomy experiment onboard Voyager I (Kaiser et al., 1978). These wave spectrograms (frequency as a function of time with emission intensity shaded) are shown for left-hand waves (top panel), right-hand waves (middle panel), and combined (bottom panel). It is clear that the intense AKR storms seen during the entire time period are predominantly left-hand polarized.

Under low density conditions the $f_{R=0}$ cutoff for R-X mode radiation is nearly equal to the local electron gyrofrequency in the AKR frequency range from about 50 kHz to 500 kHz in Figure 6. As illustrated in the upper right corner of Figure 6, if both R-X and L-0 mode radiation is generated above the $f_{R=0}$ cutoff, only the R-X mode would not be able to penetrate into regions below the $f_{R=0}$ cutoff, or below the local electron gyrofrequency. Since $f_g^- \simeq f_{R=0}$, the observed cutoff at f_g^- provides strong evidence that the intense AKR is propagating in the right-hand (R-X) polarized mode. It is important to note that the cutoff at f_g^- ($\simeq f_{R=0}$) of AKR has also been observed by the ISIS I satellite (see Benson and Calvert [1979]) at frequencies >450 kHz. These authors also concluded that the intense AKR must be generated in the R-X mode. The R-X mode was originally predicted by Green et al. [1977] to be the most likely generation mode of AKR from matching the observed angular distribution of AKR with modeled ray tracing calculations.

Direct polarization measurements of AKR have been made by Voyagers 1 and 2 (see Kaiser et al. [1978]) and can be compared with the inferred polarization methods discussed above. A three-panel wave spectrogram (frequency as a function of time with emission intensity shaded) is shown in Figure 7. The top panel shows the wave intensity for electromagnetic waves having their electric field vector $\vec{E}$ rotating in a clockwise or left-handed sense about the propagation vector $\vec{k}$. The middle panel shows the wave intensity for waves with $\vec{E}$ rotating counterclockwise about $\vec{k}$. The bottom is the sum of the two polarizations. These are typical of the observations when the Voyagers were in the northern (>18°) magnetic hemisphere of the earth with the probable AKR

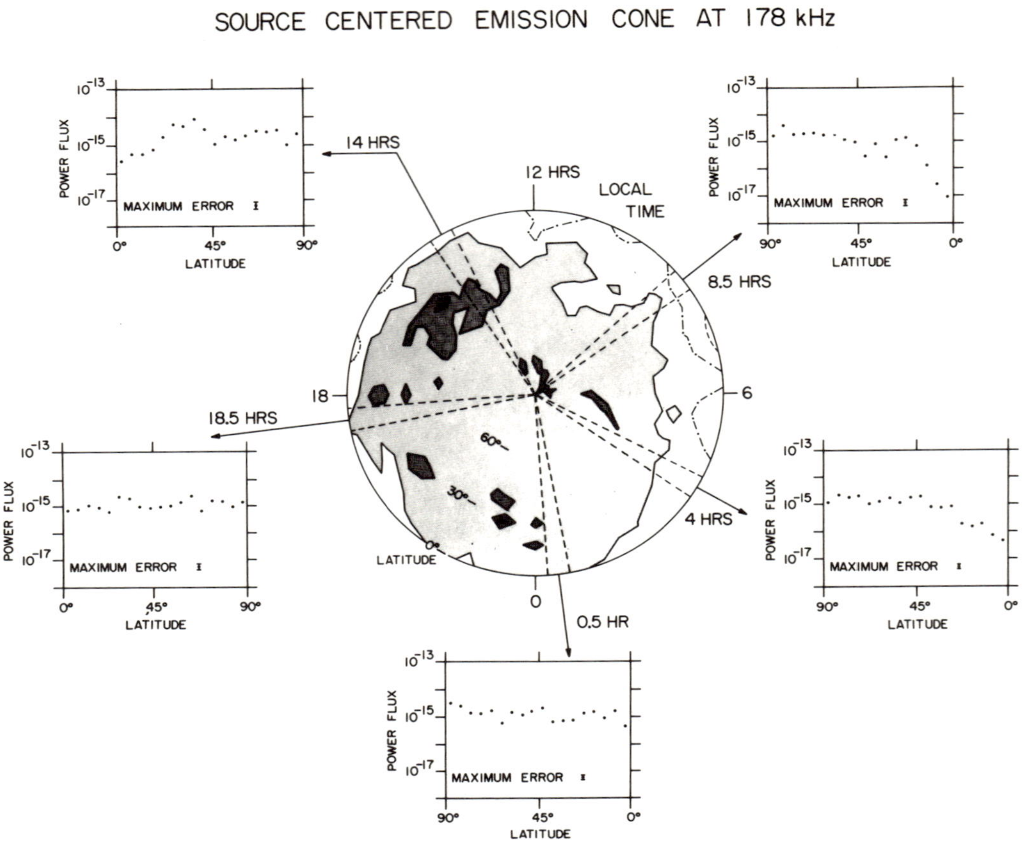

Fig. 9. Over 645,000 spin-averaged power flux observations at 178 kHz from Hawkeye and IMP-6 went into this emission cone survey. Each observation is scaled to take into account the $1/R^2$ drop in intensity relative to the average AKR source region, and placed into one of the 864 latitude and local time bins where it is averaged. Five representative local time slices show the AKR emission cone in greater detail.

source also in the northern hemisphere. Figure 7 illustrates that the degree of left-handed polarization may be as high as 95 percent.

Although there appears to be a contradiction between the inferred polarization of AKR and that measured, closer examination reveals that they are in complete agreement. The discrepancy lies in the definition of polarization as illustrated in Figure 8. The mode of propagation or polarization as reported by Gurnett and Green [1978], Green et al. [1977], and Benson and Calvert [1979] is in the plasma convention (left-hand side of Figure 8). The plasma convention (PC) applies to the direction the wave $\vec{E}$ field rotates about the ambient geomagnetic field in the source region (northern or southern hemisphere). At large radial distances from the source region where the geomagnetic field direction is irrelevant, the sense of the polarization measured is the direction the wave $\vec{E}$ field rotates about its $\vec{k}$ vector, illustrated on the right side of Figure 8. This is called the optical convention (OC). Figure 8 schematically illustrates the geometry of a hypothetical northern hemisphere source emitting right- and left-hand waves (PC) and what should be measured by the Voyagers at large radial distances. The high percentage of left-hand waves as measured by the Voyagers (OC) therefore is in complete agreement with the inferred R-X polarized waves (PC).

Emission Cone Observations

In-situ observations of AKR by the ISIS I satellite indicate that the emission is generated primarily perpendicular to the geomagnetic field ($\vec{B}$) (see Benson and Calvert [1979] and James [1980]) at frequencies >450 kHz. This important observation provides additional constraints on the AKR generation mechanism. The far field radiation pattern or emission cone of AKR at frequencies < 178 kHz has recently been studied in detail by Green and Gallagher [1980]. In the absence of propagation effects it is expected that if AKR is generated perpendicular to $\vec{B}$, a hollow emission cone would be observed at large radial distances

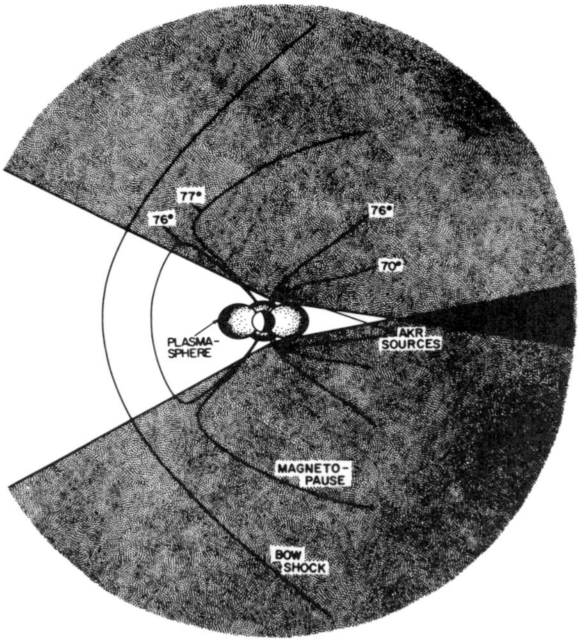

Fig. 10. The average source region of auroral kilometric radiation at 178 kHz as found in the 11- to 23-hour magnetic local time meridian at approximately 2.5 R_e (earth radius) geocentric radial distance along auroral field lines (70° invariant latitude) on the nightside from Gallagher and Gurnett (1979).

from the earth. Preliminary results indicate that the emission cones at 100 kHz and 56.2 kHz are complex islands of emission, but the emission cone at 178 kHz appears to be completely filled with electromagnetic radiation. The AKR emission cone at 178 kHz is shown in Figure 9 intersecting a hemisphere of 7 R_e radius centered on the average source region determined by Gallagher and Gurnett [1979]. In addition, Figure 9 shows the AKR power fluxes at five representative local times. There is less than 10 dB variation in the averaged intensity in almost all of the emission cone, as illustrated in the 18.5 and 0.5 hour local time side panels in Figure 9. It is unclear at this point whether the filled-in emission cone at 178 kHz is a result of propagation effects and/or preferred wave launch angles in the source region.

At 178 kHz a schematic representation or summary of AKR observations so far discussed is shown in Figure 10 (from Gallagher and Gurnett [1979]). In Figure 10, the sun is on the left, and this is a meridian projection into a plane containing the average AKR source region and the earth. Note the low altitude sources in the nightside auroral zone and the large illumination extent of the emission cone. Unlike the intense decametric radio emissions from Jupiter which are believed to be generated into a large hollow cone, its terrestrial counterpart, AKR (at 178 kHz) is beamed into a nearly homogeneous emission cone. Figure 10 also helps us understand the different local time spectra of AKR shown in Figure 2. Since IMP-6 was in the near equatorial plane, the dayside spectra were far less intense than the nightside spectra since the spacecraft was rarely at high enough magnetic latitude to see AKR.

Correlative Studies

Since the AKR emission cone at 178 kHz is uniformly illuminated (within 10 dB), correlative studies using simultaneous measurements from widely separated satellites can be interpreted more easily. Using photographs produced from the optical scanner aboard the DMSP satellite (bottom panels of Figure 11) and IMP-6 electric field measurements (top panel), Gurnett [1974] showed a strong correlation of AKR with discrete auroral arcs. Since IMP-6 was in the AKR emission cone (described in Figure 9), it is in a position most likely to observe AKR if it is being generated. Since AKR is observed when DMSP observes discrete auroral arcs (orbits 1094 and 1096) and not when DMSP observes diffuse aurora (orbits 1095 and 1097), this provides compelling evidence that AKR is generated by the precipitating particles which, at low altitudes, are also involved with generating the discrete aurora. It has been known for some time that precipitating inverted-V electrons are the probable cause of discrete aurora arcs.

Green et al. [1979], Benson and Calvert [1979], and Benson et al. [1980] have recently provided evidence that AKR is correlated with precipitating inverted-V auroral electrons. Inverted-V electron precipitation events are characterized by electron fluxes which increase from a few hundred eV to keV energies and then return to a few hundred eV energies as the spacecraft crosses the band of precipitation. This type of precipitation causes an inverted-V shaped intensity band in energy-time spectrograms. Figure 12 from Green et al. [1979] is a scatter plot of simultaneous AKR power flux measurements at 178 kHz versus the inverted-V peak energy which is used here to characterize the particle energy in the event. The AKR measurements were made by the Hawkeye plasma wave receiver when the spacecraft was in the AKR emission cone. The inverted-V measurements were made by the low energy electron experiment on board AED from over 60 auroral zone passes in the same local time as the average AKR source region. A $1/R^2$ normalization has been applied to the AKR power flux measurements in Figure 12 (and 13) to take into account the radial dependence of this radiation. There appears to be an obvious relationship between the two parameters in Figure 12; that is, as the AKR power flux increases so does the peak energy of the precipitating inverted-V.

For a review of in-situ measurements of AKR and inverted-V electron precipitation see Benson [1981] in this publication. Another interesting correlation is shown in Figure 13 between AKR at

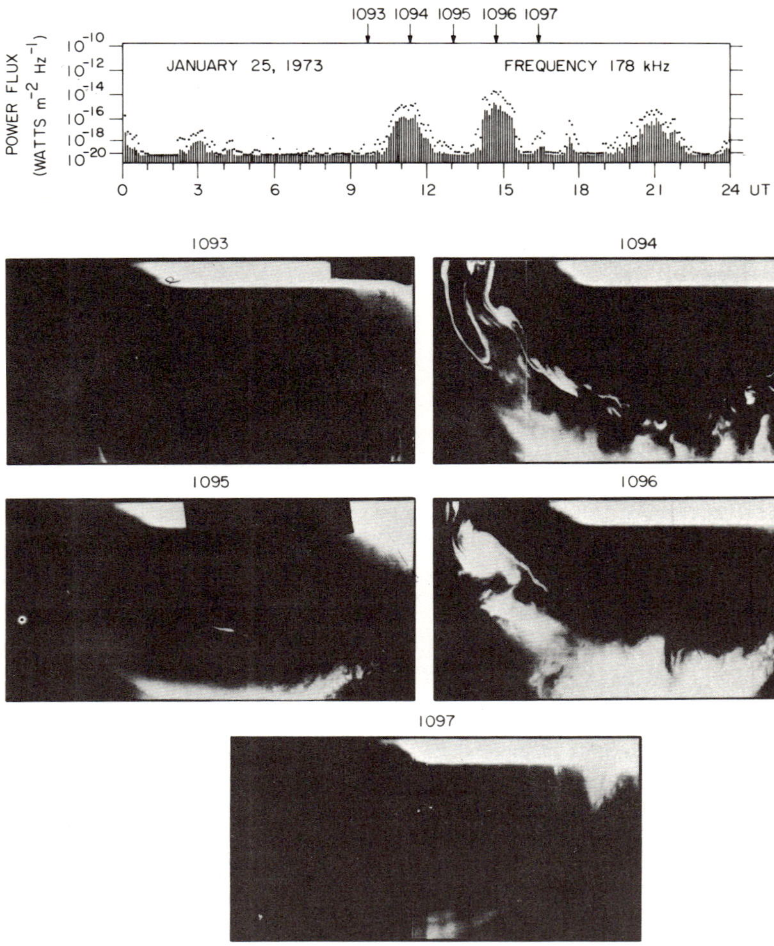

Fig. 11. Auroral photographs from the DMSP satellite compared with auroral kilometric radiation as observed from IMP-6. Note that AKR bursts appear to be closely related to discrete auroral arcs (from Gurnett, 1974).

178 kHz (observed by Hawkeye in the emission cone) and nightside auroral field-aligned currents measured by Triad from Green et al. [1980]. It appears that as the AKR power flux increases so does the nightside field-aligned current. At this point much work needs to be done to better understand the role that field-aligned current may play in the generation of auroral kilometric radiation.

Conclusions

Many recent observations of AKR indicate that stringent requirements or conditions are necessary for the generation of this intense electromagnetic radiation. A reasonable theory on the generation of AKR must take into account these observations. In summary, some of these observations are:

1. Intense AKR is generated in the R-X mode,
2. In regions where $f_g^- \gg f_p^-$,
3. At low altitude (1.5 to 4.0 R_e) in the nightside auroral zone,
4. AKR has not been (so far) observed simultaneously with large electrostatic waves,
5. AKR appears to be generated perpendicular to $\vec{B}$ at least at frequencies >450 kHz,
6. Auroral zone current systems appear to be intimately involved in the generation,
7. In particular, the most likely source of energy for AKR is in the precipitating inverted-V auroral electron distribution.

It is exciting to know that soon, with the launch of the Dynamics Explorer satellites, simultaneous in-situ measurements of AKR and particle distributions in the source region will be made in greater detail than with the previous ISIS I and Hawkeye satellites. It is anticipated that these new measurements will greatly extend our knowledge about AKR and possibly lead to an in-depth understanding of how it is generated.

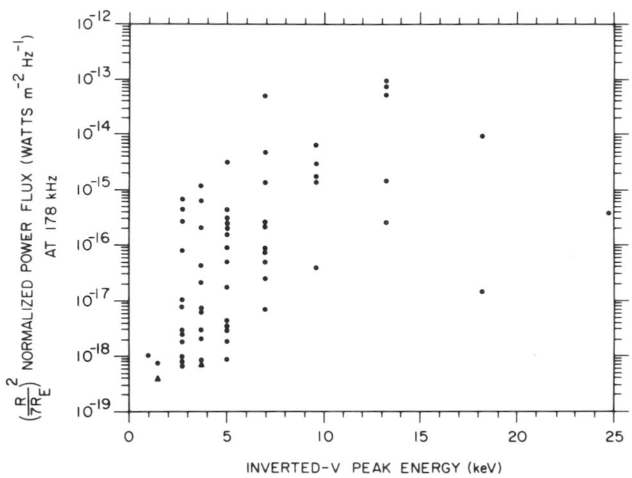

Fig. 12. Scatter plot of simultaneous measurements of AKR intensity made by Hawkeye in the emission cone (normalized to 7 R_e and having a $1/R^2$ scaling) and the peak energy in inverted-V events observed by the low altitude AE-D.

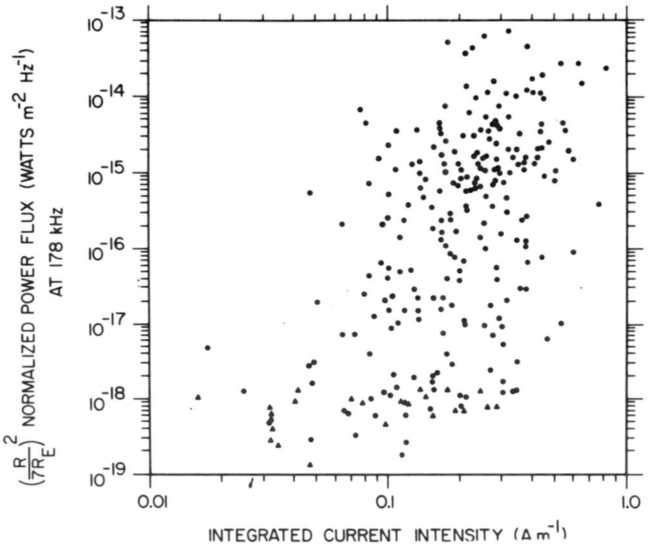

Fig. 13. Simultaneous power flux measurements of AKR at 178 kHz versus the integrated current intensity of auroral zone field-aligned currents. A $(R/7\ R_e)^2$ normalization to the power flux measurements is applied to take into account the radial dependence of this radiation. The R variable is the distance from the satellite to the average souce region of AKR (see text).

References

Alexander, J. K., and M. L. Kaiser, Terrestrial kilometric radiation 1. Spatial structure studies, J. Geophys. Res., 81, 5948, 1976.

Barbosa, D. D., Electrostatic mode coupling at $2w_{uH}$: A generation mechanism for auroral kilometric radiation, Ph.D. Dissertation, Dept. of Physics, Univ. of California, Los Angeles, 1976.

Benediktov, E. A., G. G. Getmantsev, Yu. A. Sazonov, and A. F. Tarasov, Preliminary results of measurement of the intensity of distributed extraterrestrial radio-frequency emission at 725 and 1525-kHz frequencies by the satellite Elektron-2, Cosmic Research, English transl., 3, 492, 1965.

Benson, R. F., Auroral kilometric radiation source region observations, this issue, 1981.

Benson, R. F., W. Calvert, and D. M. Klumpar, Simultaneous wave and particle observations in the auroral kilometric radiation source region, Geophys. Res. Lett., 7, 959, 1980.

Benson, R. F., and W. Calvert, ISIS I observations at the source of auroral kilometric radiation, Geophys. Res. Lett., 6, 479, 1979.

Gallagher, D. L., and D. A. Gurnett, Auroral kilometric radiation: Time-averaged source location, J. Geophys. Res., 84, 6501, 1979.

Green, J. L., and D. L. Gallagher, The detailed intensity distribution of the AKR emission cone. (To be submitted to J. Geophys. Res.), 1980.

Green, J. L., D. A. Gurnett, and S. D. Shawhan,

The angular distribution of auroral kilometric radiation, J. Geophys. Res., 82, 1825, 1977.

Green, J. L., N. A. Saflekos, D. A. Gurnett, T. A. Potemra, A correlation between auroral kilometric radiation and field-aligned currents, J. Geophys. Res. (submitted for publication), 1980.

Gurnett, D. A., The earth as a radio source: Terrestrial kilometric radiation, J. Geophys. Res., 79, 4277, 1974.

Gurnett, D. A., The earth as a radio source: the non-thermal continuum, J. Geophys. Res., 80, 2751, 1975.

Gurnett, D. A., and L. A. Frank, Thermal and suprathermal plasma densities in the outer magnetosphere, J. Geophys. Res., 76, 2355, 1974.

Gurnett, D. A., and R. A. Shaw, Electromagnetic radiation trapped in the magnetosphere above the plasma frequency, J. Geophys. Res., 78, 8136, 1973.

Gurnett, D. A., and J. L. Green, On the polarization and origin of auroral kilometric radiation, J. Geophys. Res., 83, 689, 1978.

Gurnett, D. A., and R. R. Anderson, The kilometric radio emission spectrum: Relationship to auroral acceleration processes, this issue, 1981.

James, H. G., Direction-of-arrival measurements of auroral kilometric radiation and associated ELF data from ISIS I, J. Geophys. Res., 85, 3367, 1980.

Kaiser, M. L., and J. K. Alexander, Source location measurements of terrestrial kilometric radiation obtained from lunar orbit, Geophys. Res. Lett., 3, 37, 1976.

Kaiser, M. L., and J. K. Alexander, Terrestrial kilometric radiation 3. Average spectral properties, J. Geophys. Res., 82, 3273, 1977.

Kaiser, M. L., J. K. Alexander, A. C. Riddle, J. B. Pearce, and J. W. Warwick, Direct measurements by Voyagers 1 and 2 of the polarization of terrestrial kilometric radiation, Geophys. Res. Lett., 5, 857, 1978.

Kurth, W. S., M. M. Baumback, and D. A. Gurnett, Direction finding measurements of auroral kilometric radiation, J. Geophys. Res., 80, 2764, 1975.

Roux, A., and R. Pellat, Coherent generation of the auroral kilometric radiation by nonlinear beatings between electrostatic waves, J. Geophys. Res., 84, 5189, 1979.

Shaw, R. R., and D. A. Gurnett, A test of two theories for the low-frequency cutoffs of nonthermal continuum radiation, J. Geophys. Res., 85, 4571, 1980.

Stix, T. H., The Theory of Plasma Waves, McGraw-Hill, New York, 1962.

Torbert, R. B., F. S. Mozer, S. von Glinski, and M. Temerin, Observations of the local plasma density below 8000 km on the S3-3 satellite (abstract), EOS Trans. AGU, 58, 1214, 1977.

AURORAL KILOMETRIC RADIATION SOURCE REGION OBSERVATIONS FROM ISIS 1

Robert F. Benson

Laboratory for Planetary Atmospheres, Goddard Space Flight Center,
Greenbelt, Maryland 20771

Abstract. The ISIS 1 satellite encounters the lower altitude portion of the auroral kilometric radiation (AKR) source region during high latitude apogee (3500 km) passes. The ISIS 1 ionospheric sounder detects AKR while determining both ambient and remote plasma conditions. This paper will review the ISIS 1 observations of the high frequency portion of the AKR spectrum, i.e., from the minimum frequency encountered for the extraordinary mode cut-off (~450 kHz) to the upper frequency cut-off (~800 kHz). AKR is found to be generated in the extraordinary mode just above the local cutoff frequency and to emanate nearly perpendicular to the magnetic field. It occurs within local depletions of electron density, where the ratio of plasma frequency to cyclotron frequency is less than 0.2. The density depletion is restricted to altitudes above about 2000 km, and the upper AKR frequency limit corresponds to the extraordinary cutoff frequency at this altitude. AKR is observed from ISIS 1 above the nighttime auroral zone over a wider extent in longitude than in latitude with an intense source region observed most often near 2200 magnetic local time and 70° invariant latitude. It is directly related to inverted V electron precipitation events, and particular features within the electron energy spectra are directly correlated with source region AKR encounters. Simultaneous wave and particle measurements within the source region indicate an electron-to-wave energy conversion efficiency of the order of 0.1 to 1%.

Introduction

To a celestial passer-by, our planet earth would appear as one of the dominant radio sources in the solar system. It is an emitter of intense electromagnetic waves at kilometer wavelengths. The radiation originates high above the auroral regions and is directly associated with discrete auroral arcs; hence the name auroral kilometric radiation (AKR). In addition to providing a sensitive indication of magnetic substorms [Voots et al. 1977], it provides an important clue to the type of processes involved in the coupling of energy from the magnetosphere to the auroral ionosphere. Since the radiation intensity peaks at several hundred kilohertz, it is not observed on the ground (due to ionospheric shielding) and can only be observed by spacecraft above the ionosphere that are equipped with receivers. For a summary of AKR observations from distant satellites, see Green [1981]. For a summary of the highly dynamic features of AKR, see Gurnett [1981]. For a summary of the theoretical aspects of AKR see Wu [1981]. All of these reviews appear in this Conference Proceedings.

The purpose of the present paper will be to review the ISIS 1 observations obtained from within the AKR source region. The Hawkeye satellite also encounters the AKR source region. The orbital differences between these two satellites makes their respective coverage of the AKR source region quite complementary in that Hawkeye samples the upper altitude portion of the source region during the perigee portion of its orbit ($\gtrsim 2\ R_E$) while ISIS 1 samples the low altitude portion during its apogee ($\lesssim 1.5\ R_E$). The situation is schematically illustrated in Figure 1.

Data from a number of spacecraft have provided information pertaining to the polarization of AKR. The Hawkeye observations of Gurnett and Green [1978] showed that intense AKR was not observed at frequencies below the local electron gyrofrequency f_H. These observations supported conclusions inferred earlier from ray tracing calculations (performed to explain the angular distribution of AKR detected by Hawkeye, IMP 6 and IMP 8), i.e., that intense AKR propagates in the extraordinary right-hand mode [Green et al., 1977]. Direct polarization measurements from the Voyager spacecraft (Kaiser et al., 1978) confirmed that intense AKR originated in the extraordinary mode. The cutoff frequency f_x, for this mode, is given by the expression

$$f_x = (f_H/2) + [(f_H/2)^2 + f_N^2]^{1/2}$$

where f_N is the plasma frequency. The value of f_x increases with decreasing altitude in the topside ionosphere.

The lowest altitude where AKR is generated should correspond to the altitude where f_x is

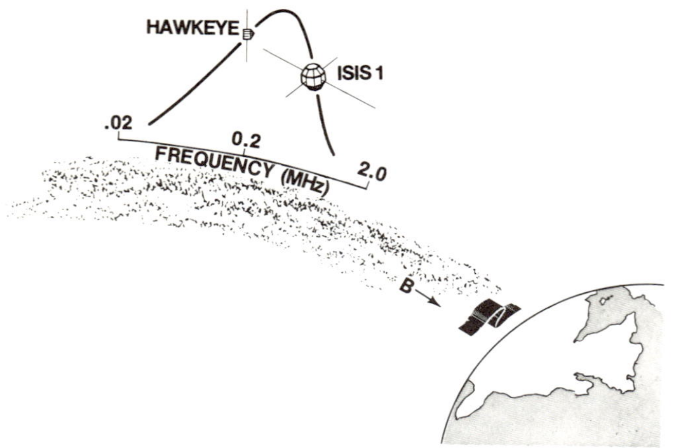

Figure 1. Schematic illustration of the ISIS 1 and Hawkeye satellites encountering the AKR source region. The positioning of each satellite is intended to illustrate the closest approach to the spectral peak.

equal to the highest frequency observed for AKR. This upper frequency cutoff has been determined to be about 600 kHz from an analysis of data from the IMP 6 radio astronomy instrument [Kaiser and Alexander, 1977]. Thus in order to make in-situ observations in the bottom of the AKR source region, it is necessary to achieve an altitude where $f_x \lesssim 600$ kHz.

The above condition is frequently encountered by ISIS 1 in the nighttime auroral ionosphere within several hundred kilometers of apogee. In the next section, the results of several investigations of AKR using ISIS 1 ionogram data obtained under these conditions will be reviewed. This section will be followed by a review of wave/particle and wave/wave ISIS 1 correlation studies, and then by a disucssion and summary.

ISIS 1 Ionogram Observations

One of the primary modes of the ionospheric sounder on-board ISIS 1 provides swept frequency ionograms covering the frequency range from 100 kHz to 10 MHz (see Franklin and Maclean, [1969] and other related articles in the Special Issue of the Proc. IEEE, June, 1969 on topside sounding and the ionosphere). The ionospheric soundings can be used to determine the distribution of the electron density N_e between the satellite location and the peak of the ionospheric F layer. This information is vital to theories of the AKR generation mechanism. In addition to detecting reflected sounder pulses, the sounder receiver detects natural radio noise and radio signals from ground based transmitters.

Natural noise in several characteristic frequency regions has been observed by the swept frequency sounder receivers on the Alouette and ISIS satellites [Hartz, 1969; Muldrew, 1970; James et al., 1974]. An example of noise observed in the frequency region below f_N is presented in Figure 2 where 4 consecutive ISIS 1 ionograms are displayed. Two of these ionograms (a and d) were obtained when the sounder transmitter was operating and two (b and c) when it was not. In the former case ionospheric echoes and plasma resonances

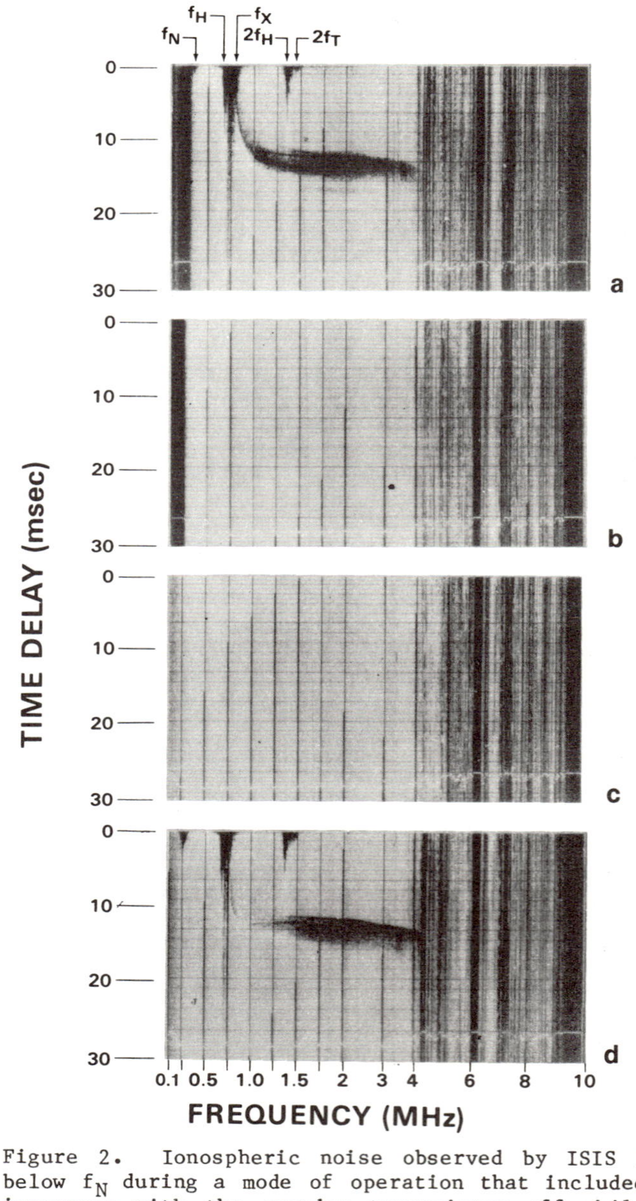

Figure 2. Ionospheric noise observed by ISIS 1 below f_N during a mode of operation that included ionograms with the sounder transmitter off while the sounder swept-frequency reveiver was on (see b and c). The consecutive sequence of ionograms was recorded around 0306 UT on 15 June 1974 (2330 local time) while the satellite altitude varied from 2060 to 1990 km and the invariant latitude ranged from 68.1 to 66°.

stimulated by the sounder pulses are detected, in addition to radio signals unrelated to the sounder operation, whereas in the latter case only non-sounder related signals are detected.

During the periods of active operation, the ambient and remote plasma conditions can be obtained. The ambient conditions are obtained from the plasma resonances (resulting from stimulated electrostatic waves) and the observed cutoffs of the ionospheric reflection traces (resulting from stimulated electromagnetic waves). The remote conditions are determined from inversion techniques used to convert the recorded frequency-time delay (or virtual range) electromagnetic reflection traces into electron density-true height profiles. For additional discussion on the stimulated resonances see the review of Benson [1977]; for a description of the inversion techniques see Jackson [1969, 1972].

In the case of Figure 2, the electron density is so low that f_N ($\sim 9[N_e(\mathrm{cm}^{-3})]^{1/2}$ kHz) is near the lower frequency limit (100 kHz) of the swept frequency portion of the ionogram. Under these conditions, the resonance observed at the upper hybrid frequency f_T, where $f_T^2 = f_N^2 + f_H^2$, is concealed between f_H and f_x. The resonance at the second harmonic of f_T, however, can sometimes be detected slightly above $2f_H$ (see Figure 2a). During the periods of sounder-on operation (a and d in Figure 2), f_N is determined to be 250 kHz and the F region penetration frequency f_xF_2 is determined to be 4.1 MHz. During the periods of sounder-off operation (b and c in Figure 2), the noise below f_N and above f_xF_2 can be confidently attributed to ionospheric origin and radio breakthrough from ground-based transmitters, respectively. This noise of ionospheric origin is not AKR, however, because it is not in the AKR frequency domain, i.e., it is observed below f_x. Since ISIS 1 was in the evening auroral zone during the time of these observations, the observed noise of ionospheric origin could be a hint that AKR was present above the satellite. If this was the case, ISIS 1 would have been too low in altitude to detect such radiation because $f_x \simeq 800$kHz which is above the cutoff frequency normally observed for AKR.

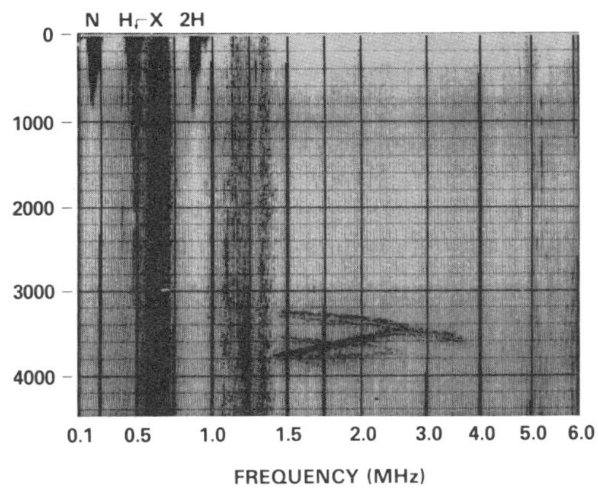

Figure 3. ISIS 1 AKR observations remote from the source region on 7 August 1969 (1326 UT, 2042 magnetic local time, 3465 km altitude, 61.6°Λ). The symbols N, H, X, 2H refer to f_N, f_H, f_x and $2f_H$, respectively.

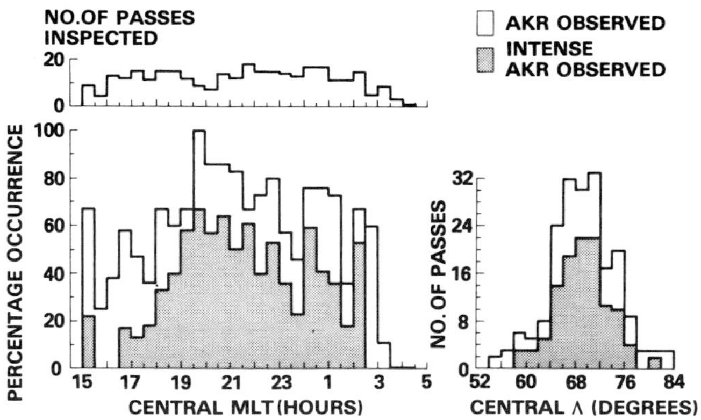

Figure 4. Percentage occurrence of AKR against magnetic local time (left) and the number of ISIS 1 passes revealing AKR against invariant latitude (right) corresponding to 312 apogee nighttime passes over Ottawa centered on the month of November for the four years from 1969-1972. (After Benson and Calvert [1979]).

An example of an ISIS 1 ionogram where AKR was observed is presented in Figure 3. In this case the satellite was at a higher altitude over the auroral zone and f_x was low enough so as to be within the AKR frequency domain. The AKR in this figure is the intense noise band covering the entire virtual range scale, i.e., the entire time delay between sounder pulses, in the frequency range from about 60 kHz above f_x to about 730 kHz. The second harmonic noise band is attributed to a nonlinear instrumental response to the intense wide band bursty AKR fundamental (Benson and Calvert, 1979). This bursty nature of the signal, which is consistent with the ISEE-1 plasma wave receiver [Gurnett et al., 1979], is often more apparent on the harmonic signal where it is seldom intense enough to saturate the receiving system.

In addition to the similarities in the frequency and intensity characteristics of AKR observed by ISIS 1 and the AKR observed by more distant satellites, the ISIS 1 location of the AKR source region (in time and space) also agrees with the previous results obtained from these satellites, e.g., see the review by Gurnett [1978]. The results of a statistical investigation of the ISIS 1 data by Benson and Calvert [1979] to determine the source location is presented in Figure 4. Invariant latitudes Λ between 40 and 84° were sampled. Since AKR is only observed by ISIS 1 on high altitude passes over the nighttime aurora zone when the extraordinary cutoff is in the AKR

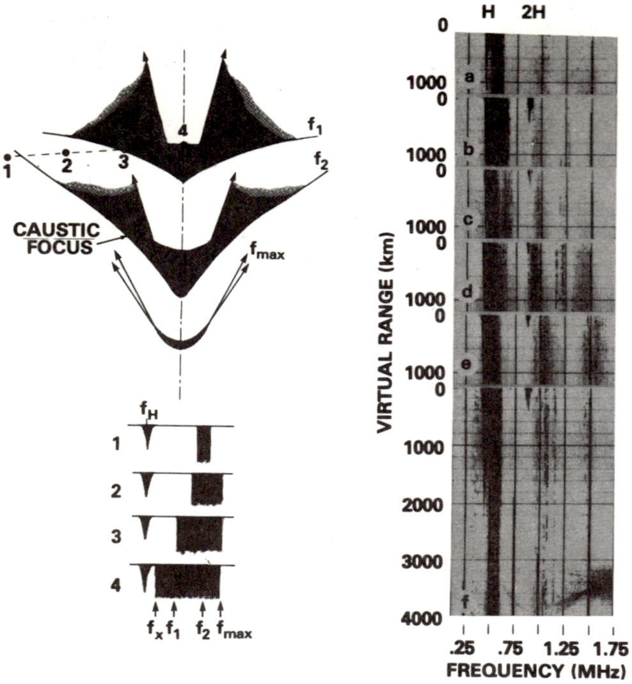

Figure 5. Schematic illustration of the penetration of ISIS 1 through a source region distributed along a vertical magnetic field line (top left), and the expected ionograms (lower left). The appropriate consecutive ionograms (right) were recorded at Ottawa on 26 November 1969 between 0619:15 and 0621:35 UT (centered on 0029 MLT, 3490 km, 71°Λ) with f_H between 0.44 and 0.45 MHz and $f_x - f_H \leq 21$ kHz on b and ≤ 11 kHz on the other 5 ionograms. (After Benson and Calvert [1979].)

frequency domain, a satellite pass with no AKR was included in the statistics only if the altitude h > 2600 km at 70° Λ and if f_x < 700 kHz was encountered on several consecutive ionograms. When AKR was observed, the position recorded was that corresponding to the center of the sequence of ionograms which revealed AKR. Thus AKR was often observed over a wide range of Λ and magnetic local time MLT about the recorded values. Most of the activity was centered between 68 and 72° Λ and was observed over a wide MLT range. Intense AKR activity, which was considered to be a better indication of source encounters than the sequence center time, was somewhat more confined in MLT and was centered near 2200.

These results indicate that the region where AKR is observed by ISIS 1 is extended more in longitude than in latitude (assuming that the likelihood of radiation reaching the satellite is the same from either direction). If 70° is taken as the average source region invariant latitude, then the intense AKR observed over 12° Λ and 8.5 hours MLT would be extended almost 4 times more in longitude than in latitude. Since the method of recording AKR central locations tends to spread the data more in latitude than in longitude (for a polar orbiting satellite), the ratio for individual sources could be even greater.

When ISIS 1 is within the AKR source region, the ionogram signature of the phenomenon is distinctly different from that presented outside the source region [Benson and Calvert, 1979, Calvert, 1981]. The ionogram presented in Figure 3 is an example of intense AKR observed outside the source region. In this case there is a definite frequency gap between f_x and the lower frequency edge of the observed AKR band. Since $f_x \simeq f_H$, this gap is most evident, i.e., twice as large, between the $2f_H$ plasma resonance and the instrumental harmonic of the fundamental AKR noise band. As the satellite approaches the source region, the gap decreases. In addition to the reduction in the gap, the upper frequency edge of the intense AKR noise band also decreases. Within the source region, both the gap and the upper frequency edge are a minimum and intense harmonic noise bands are usually observed. This picture can be complicated by the presence of multiple sources or by a source extended in latitude and/or longitude.

A clear example of an ISIS 1 encounter with a well defined AKR source of limited extent in the horizontal plane is presented in Figure 5. The encounter is presented schematically on the left corresponding to a particular set of observations on the right. The satellite positions 1-4 at the top left correspond to the schematic ionograms 1-4 at the lower left and to the actual ionograms a-d at the right, respectively. The AKR source is

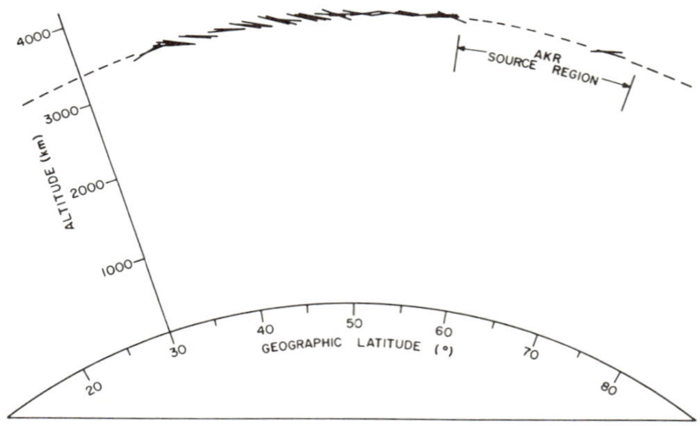

Figure 6. Observed dipole orientations from the ISIS 1 pass at 0815 UT on December 14, 1972, corresponding to spin minima in the fixed-frequency automatic gain control (AGC) level. The broken line is the spacecraft trajectory projected onto the latitude-altitude plane. The instantaneous projections of the long sounder dipole onto the same plane at the times of the spin minima are given by the short line segments. The average angle between the dipole and the horizontal for the cases presented is 12°. (Adapted from James [1980]).

assumed to be aligned along the vertical magnetic field line threading through satellite position 4. The emission is assumed to be distributed in altitude and the shaded areas represent the regions illuminated by intense AKR energy at three representative frequencies. These frequencies decrease with increasing altitude, i.e., $f_1 < f_2 < f_{max}$. In each case, radiation in the extraordinary mode is assumed to originate perpendicular to the magnetic field from a range of altitudes above the extraordinary cutoff level for that frequency, i.e., $f = f_x \simeq f_H (1 + f_N^2/f_H^2)$ when $f_N^2 \ll f_H^2$. The lowest illumination region in figure 5 is drawn narrower to illustrate the expectation that the source is more restricted in altitude just prior to extinction of the radiation mechanism (near the 700 kHz cutoff of the phenomenon). In each illumination diagram, the ray originating

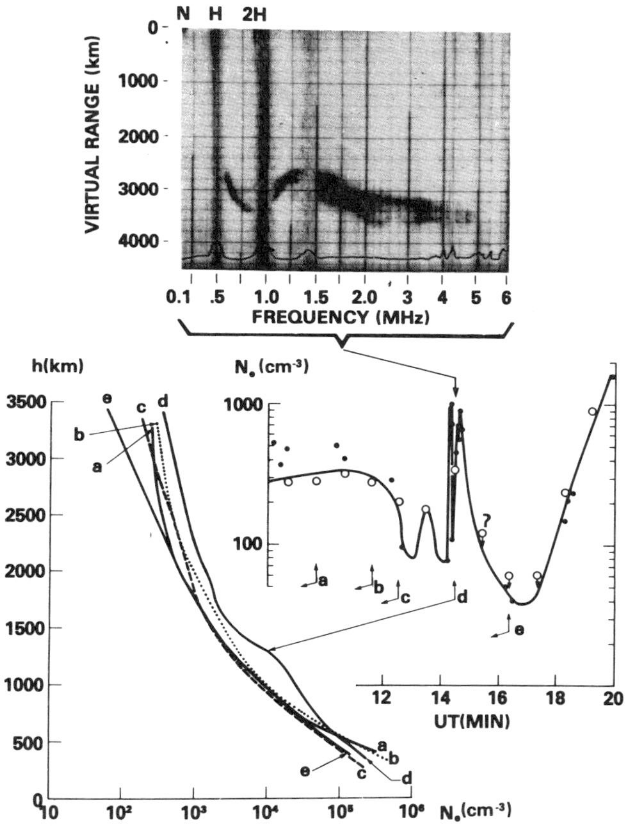

Figure 7. (Top) an ionogram revealing enhanced N_e in an AKR source region of depleted N_e, (middle) N_e along the satellite trajectory as measured by the cylindrical electrostatic probe (solid points) and the sounder (open circles), and (bottom) corresponding N_e profiles calculated from the ionospheric reflection traces. The top ionogram was recorded at Ottawa on 6 November 1970 at 0714:27 (2249 MLT, 3403 km, 66° Λ with f_H = 0.44 MHz. (After Benson and Calvert [1979].)

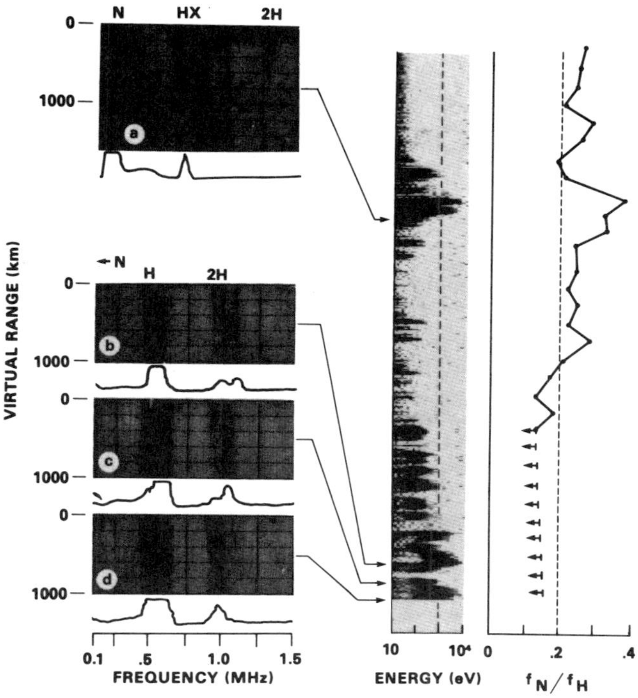

Figure 8. Comparison between ionogram data (left), electron energy data (center), and f_N/f_H as deduced from the ionogram data (right) during AKR source region encounter of 7 August 1969. At the bottom of each ionogram the corresponding AGC trace has been included (1926:50 UT, 1316 MLT, 2340 km, 76°∓ with f_H = 0.64 MHz in a; 1936:39 UT, 1847 MLT, 2940 km, 70°∓, with f_H = 0.46 MHz in d). (After Benson and Calvert [1979].)

from the lowest altitude has a frequency only slightly above the local f_x value, and is subject to the greatest refraction. The ray originating at the highest altitude is at the same frequency but at a position where the local f_x value is less and thus it experiences less refraction. The satellite at position 1 first detects intense AKR at the frequency f_2 corresponding to a caustic focus of all the rays between these two extremes. Emissions at higher frequencies, which are generated at lower altitudes, are not received until a later ionogram (e.g., corresponding to position 2). As the satellite proceeds further and encounters the source region at position 4, however, the intense component of these higher frequency emissions miss the satellite. The gap between f_H ($\simeq f_x$) and the lower frequency edge of the intense component of AKR is widest in ionograms a and f and closes on ionograms d and e. The brief gap dissappearance indicates that ISIS 1 is in and above the source region, rather than receiving noise generated from much greater altitudes which is merely being cut off at f_x.

The above model, which assumes radiation emanating perpendicular to the magnetic field direction, is consistent with the ionogram signature of AKR. A recent study by James [1980] using ISIS 1 fixed-frequency AKR observations indicates that any deviations from a perpendicular origin are in the downward direction. This initial AKR propagation direction was determined from the spin modulated signal resulting from the rotating ISIS 1 dipole antenna, together with ray path calculations. The spin modulation data were used to infer the initial wave vector $\vec{k}$ for ray tracing. It was assumed that the AKR spin minima occurred when the dipole antenna was parallel to the wave electric field $\vec{E}$. The geometrical relationship between $\vec{E}$ and $\vec{k}$ was then considered in order to obtain ray tracing initial conditions. The observed ISIS 1 dipole orientations corresponding to AKR spin minima for one event are presented in Figure 6. After ray tracing, it was concluded that the original wave vectors ranged from being perpendicular to 30° away (and downward) from perpendicular to the magnetic field.

Benson and Calvert [1979] found that the AKR source region coincided with regions of depleted electron density. The electron density was always so low that the condition $f_N^2 << f_H^2$ was well satisfied in the vicinity of AKR sources. Occasionally, however, a sharp N_e enhancement was detected within such source region depletions. This enhancement is indicated by a bulge on the ionospheric reflection trace, as illustrated near 1.25 MHz in the ionogram at the top of Figure 7; and sometimes also by a brief emergence of the f_N resonance above the 100 kHz frequency marker. Since f_N is seldom above 100 kHz during source region encounters, the AKR in this case is attributed to the low density regions (N_e < 100 cm^{-3}) adjacent to the sharp enhancement (Figure 7, center). This fine scale N_e structure was determined from an inspection of the ISIS 1 cylindrical electrostatic probe data. The electron density vs. height profiles calculated from this ionogram, as well as from adjacent ionograms, indicate that the enhancment in N_e extends far down into the ionosphere with a bulge below the AKR source region (Figure 7, lower left). It is possible that such an enhancement is a common source region feature but not large enough to always be detected on the ionograms. In contrast to the deep penetration of this enhancement, the broad denisty depletion is limited to higher altitudes (h>2000 km).

Since the upper frequency limit of the AKR corresponds to waves originating from the bottom of the AKR source region, such density profiles can be used to determine this altitude [Benson and Calvert, 1979]. The upper frequency limit during the series of ionograms used to make up Figure 7 was observed to be between 650 and 700 kHz. After allowing for ½ of the receiver bandwidth, the upper frequency limit at the AKR source would be in the 625 to 675 kHz range which would correspond to f_x below the satellite in the altitude range from 2450 to 2250 km. Thus it appears that the lower altitude limit of the AKR generation was in the vicinity of 2350 km during this event.

Correlated ISIS 1 Wave/Particle Observations

All attempts to explain the origin of the AKR involve the presence of energetic particles along auroral field lines. It is necessary to relate the observed waves to the source particles in order to determine which of the numerous source mechanisms is dominant. Gurnett [1974] took the first steps in this direction by comparing the AKR observed by the IMP 6 spacecraft with the aurora observations obtained from the U.S. Air Force DAPP satellite. The AKR was found to correlate with the occurrence of discrete auroral arcs rather than diffuse aurora. Since discrete auroral arcs had been directly associated with intense inverted V electron precipitation by Frank and Ackerson [1971] and Ackerson and Frank [1972], Gurnett inferred a relationship between inverted V electron precipitation and the generation of AKR. Green et al [1979] conducted a two satellite correlation investigation which involved direct measurements of auroral electrons. They used simultaneous observations of particles from AE-D and the remote detection of waves from Hawkeye to infer that AKR is statistically more closely associated with inverted V electron precipitation than with plasma sheet precipitation. With the ISIS 1 satellite, it is possible to conduct wave/particle correlation studies where both the waves and particles are measured simultaneously on the same vehicle within the AKR source region. Such investigations have been conducted by Benson and Calvert [1979] and Benson et al. [1980] and will be reviewed below.

The ISIS 1 wave/particle correlation studies are based on simultaneous measurements of AKR using the sounder receiver and direct particle measurements using the soft particle spectrometer (SPS) of Heikkila et al. [1970]. Figure 8 presents a case where the AKR is found to correlate best with the peak intensity of the inverted V electron precipitation activity. It also illustrates the importance of low electron density and high altitude in the detection of AKR from ISIS 1. Ionogram a, recorded during an electron precipitation event, shows intense noise when $f \leq f_N$ but no AKR. This ionogram was recorded at a relatively low altitude during daytime when AKR would not be expected to be observed. In addition, the satellite altitude was so low (as in the case of Figure 3) that f_x exceeded the normal AKR upper frequency limit. Intense AKR is apparent during the inverted V detected during the recording of ionogram b where f_x < 700 kHz, f_N < 100 kHz, and f_N/f_H is well below 0.2. The f_H-to-AKR frequency gap is closed during this inverted V (indicating a source region encounter), open on the following ionogram (c) which was recorded farther from the inverted V peak, and closed again on the next ionogram (d) which is coincident with another increase in the number of high energy electrons detected. The frequency band is wider in d than b indicating the reception of AKR from a distant source in addition to the local source. Unfortunately, the data terminated during the recording of ionogram d while electron precipitation and AKR activity were still in progress.

Figure 9 presents a case where the AKR is found to correlate with the poleward edge of the inverted V electron precipitation rather than with the inverted V peak. Portions of 4 topside sounder ionograms are presented together with an insert containing an SPS electron spectrogram. The AKR source region signature is apparent on ionogram b, which was recorded near the poleward edge of the inverted V electron precipitation, but not on ionogram c which was recorded just after the inverted V peak.

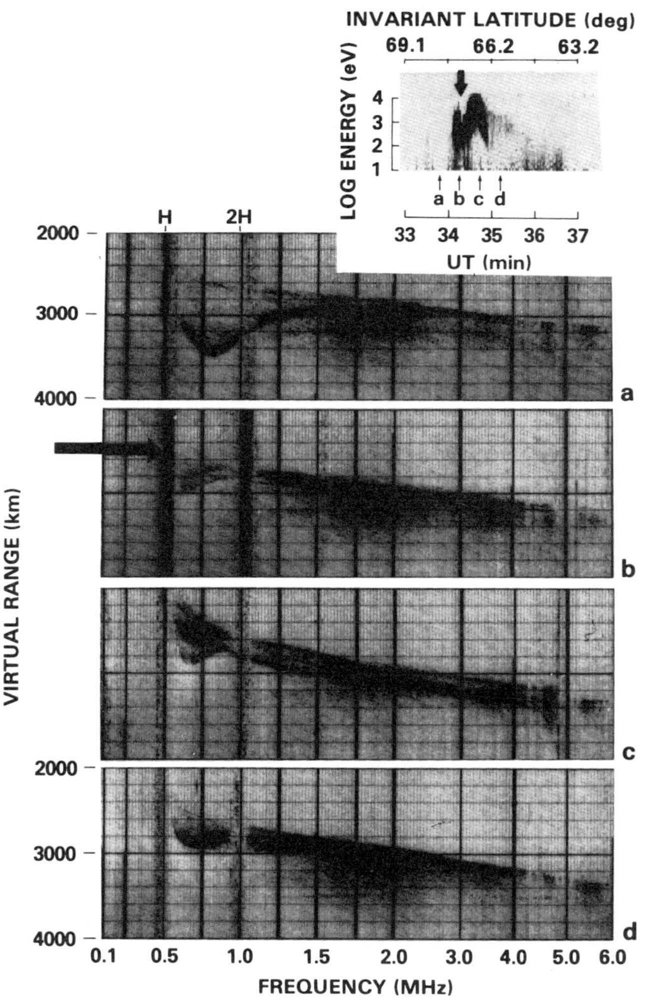

Figure 9. ISIS 1 wave and electron observations in the AKR source region during an AKR event on 8 August 1969. The fundamental AKR noise band, as observed in the source region, is identified by a bold arrow on ionogram b; the corresponding time is similarly identified on the SPS spectrogram insert (0834:16 UT, MLT = 22 hrs, h = 3273 km and f_H = 0.48 MHz). (After Benson et al. [1980].)

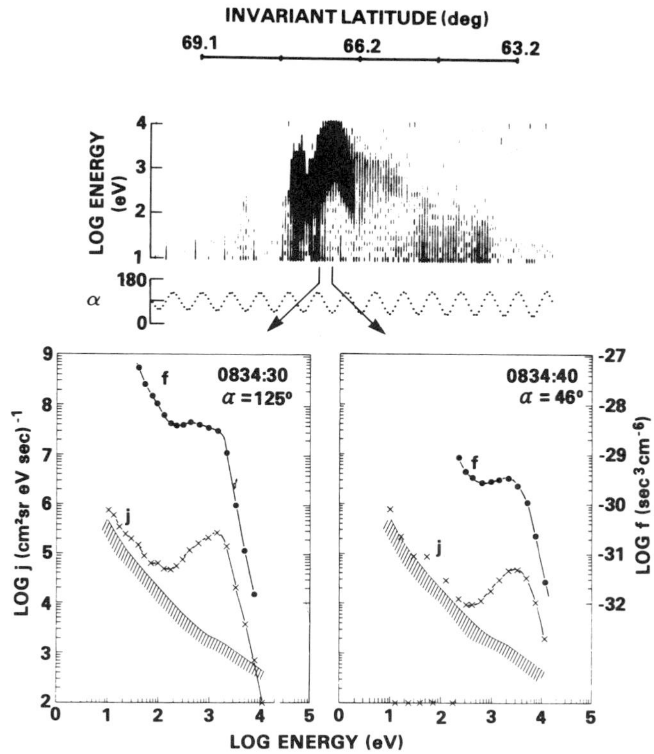

Figure 10. Differential electron energy spectra j and the corresponding electron velocity distribution functions f at two pitch angles near the peak of the inverted V on the SPS spectrogram of Figure 9 which is reproduced here. (After Benson et al. [1980].)

Selected electron energy spectra from the SPS are presented together with the corresponding distribution functions, for this event, in Figure 10. The SPS spectrogram insert of Figure 9 is reproduced to indicate the location of these spectra. The selected spectra, recorded near the peak of the inverted V, reveal a positive slope in the distribution function leading to a secondary enhancement in electron flux at a few keV. Such an electron distribution is a necessary (but not sufficient) condition for the instabilities in several AKR theories. The spectra presented in Figure 10 represent examples recorded near the extreme range of pitch angles α that could be sampled by the SPS during the inverted V encounter. Since the spectrometer was not able to sample electrons traveling within 45° to the magnetic field (due to the orientation of the satellite spin axis), all of the spectra recorded during this inverted V correspond to electrons mirroring at altitudes above 1400 km, thus clearly outside the loss cone.

The electron density measured along the satellite track during this event is presented in Figure 11 together with the SPS spectrogram. A deep minimum in N_e just after 0834 UT is apparent. The presence of intense AKR source region signals

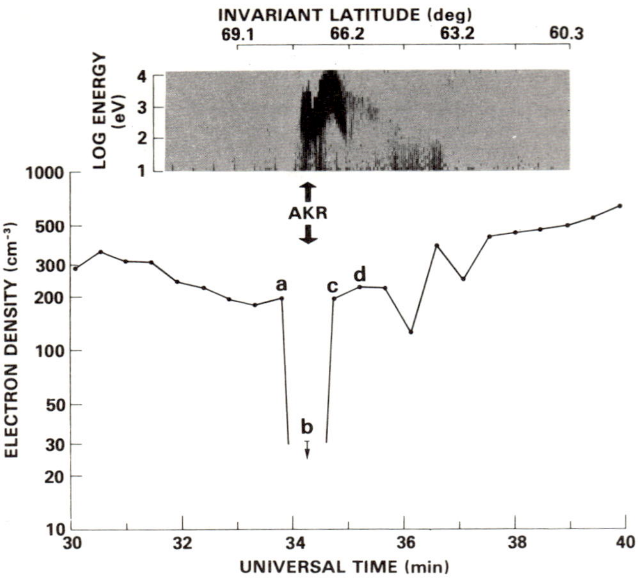

Figure 11. Electron density along the satellite track, as determined from the ionograms, and the SPS spectrogram for the event of Figure 9. The AKR source region encounter is indicated by an arrow. The electron density was assumed to be ≤30 cm^{-3} on iongrams where no frequency component of the plasma resonance occurs above the 0.1 MHz frequency marker; i.e., f_N was assumed to be ≤50 kHz. (After Benson et al. [1980].)

on ionogram b and their disappearance on c is attributed to this variation in local electron density. Thus, even though the distribution function is appropriate for plasma instability, the electron density is too high at c to satisfy the requirement of a low electron density in the source region. Even the intense AKR encountered earlier at b is not detected at the location corresponding to c. Although this null result could be due to the time variability of the source, it is more likely due to wave refraction at the sharp density gradient encountered between b and c along the satellite track. Similar sharp density gradients were detected equatorward of the AKR source regions encountered in two additional events investigated by Benson et al. [1980] where the AKR source location also correlated best with the poleward edge of the inverted V. Such a latitudinal N_e variation is not always associated with the generation of AKR, as can be seen from an inspection of Figure 12 where the N_e variation along the satellite track is presented for all the ISIS 1 AKR source region encounters which have been studied in detail. Figure 12 reveals that of the three ISIS 1 AKR source region encounters studied in detail by Benson and Calvert [1979], two had broad density minima extending throughout the vicinity of the source region and one had sharp density enhancements superimposed on a broad density minimum. It also reveals that two of the three ISIS 1 AKR source region encounters investigated by James [1980] had extended regions of depleted N_e equatorward of the AKR source. Thus the apparent association of the AKR source with the inverted V poleward edge found by Benson et al. [1980] (rather than also with the peak or the equatorward edge, for example) may be the result of the particular latitudinal N_e distributions for the three events considered.

Electron density contours, which were derived from the ionospheric reflection traces, corresponding to the AKR event of Figure 9 are presented in Figure 13. The corresponding SPS spectrogram is reproduced again, together with the total electron number flux for this event. The ionospheric imprint created by the inverted V event is clearly evident down to below 2000 km. The low altitude limit of the AKR generation region can be determined (as it was for the AKR event of Figure 7) by combining the N_e information, the projected value of f_H, and the upper AKR frequency limit observed from the series of ionograms associated with the source region. The results indicate that the bottom of the AKR cavity for the case of Figure 13 was at an altitude of 2750 km where f_N/f_H = 0.16. Thus, even though the density depletion associated with an inverted V event can be detected deep in the ionosphere the AKR generation is limited to the upper altitude region of severe density depletion.

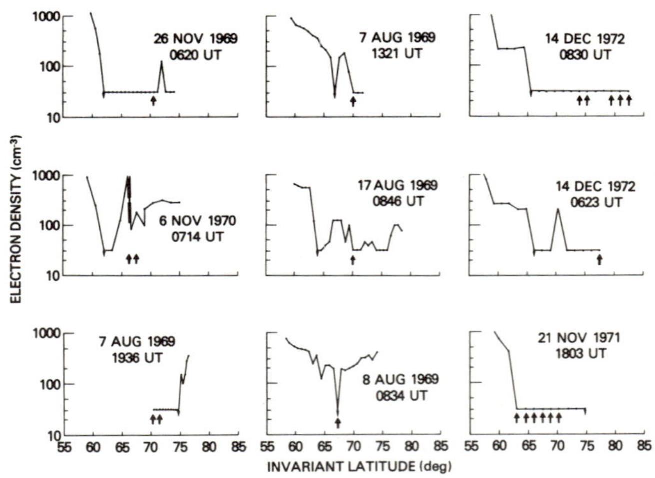

Figure 12. The electron density along the satellite track for 9 ISIS 1 AKR source region encounters. The left, middle and center columns correspond to the events investigated by Benson and Calvert [1979], Benson et al. [1980] and James [1980], respectively. The N_e values were determined from the ionograms; the level values at 30 cm^{-3} represent an observational lower limit as indicated by the small downward pointing arrow head on each frame. The upward pointing vertical arrows indicate regions where ISIS 1 was within the AKR source region.

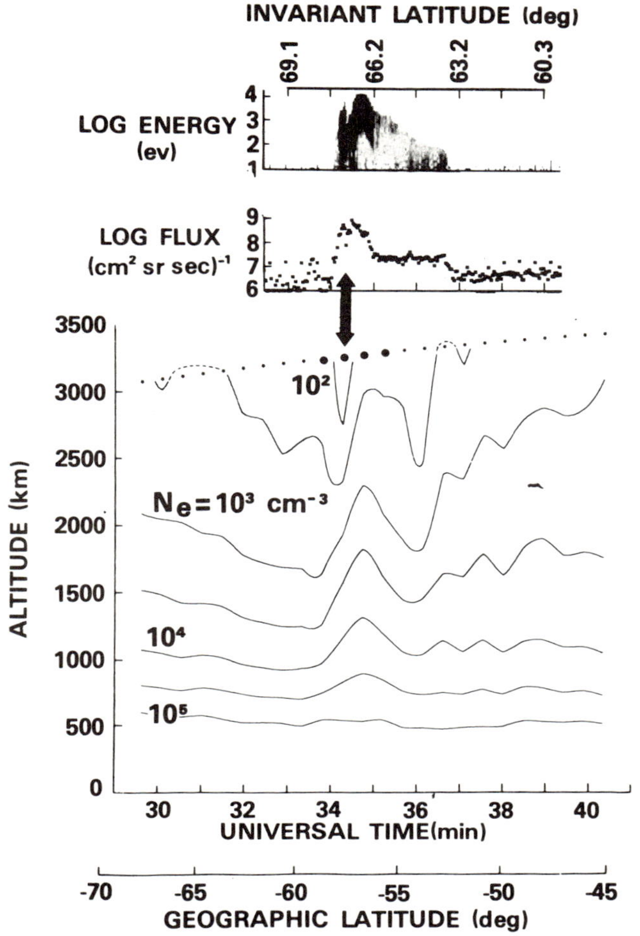

Figure 13. Contours of constant N_e aligned with the SPS spectrogram and the total electron number flux for the event of Figure 9. The small dots at the top of the contours indicate the satellite position at the times when the sounder frequency corresponds to f_x at the satellite. The four large dots correspond to ionograms a-d on Figure 9. (After Benson et al. [1980].)

Correlated ISIS 1 Wave/Wave Observations

A VLF/HF wave/wave correlation investigation was conducted by James [1980] in order to test the AKR theory of Roux and Pellat [1979]. This theory is based on a coherent three-wave interaction involving a quasi-electrostatic wave at f_T, a wave at the lower hybrid resonance frequency f_{LHR} and the resulting AKR wave. An analysis of simultaneous data from the ELF/VLF receiver and the sounder receiver on ISIS 1 indicated that intense emissions at frequencies near and above f_{LHR} are sometimes observed in the AKR source region. Intense ELF/VLF and AKR were generally colocated within 100 km. The ELF/VLF emissions are often observed near the auroral regions, however, and their times of observation do not appear to be confined to AKR events. James concluded that there is no identifiable ELF/VLF phenomenon in the ISIS 1 data that is peculiar to the AKR source regions.

Discussion and Summary

The ISIS 1 observations obtained near the lower altitude boundary of the AKR source region and the Hawkeye observations obtained near the upper altitude boundary both indicate that the radiation is generated in the extraordinary mode [Benson and Calvert, 1979; Gurnett and Green, 1978, respectively]. This consistency suggests that a single generation mechanism is involved over the entire altitude region. The lower altitude boundary can be located from the remote measuring capability of the ISIS 1 sounder. It was determined to be in the 2000 to 3000 km altitude range during two events investigated in detail [Benson and Calvert, 1979; Benson et al. 1980]. From a statistical investigation of several hundred satellite passes, the source region was found to extend farther in latitude than in longitude over the nighttime auroral zone [Benson and Calvert, 1979].

The ISIS 1 observations indicate that the AKR source region coincides with an electron density cavity where $f_N/f_H < 0.2$ [Benson and Calvert, 1979; James, 1980; Benson et al., 1980]. Such a low source region density is required by the AKR theories of Melrose [1976], Wu and Lee [1979] and Grabbe et al. [1980].

ISIS 1 simultaneous wave/particle observations in the AKR source region indicate that AKR is closely correlated with inverted V electron precipitation [Benson and Calvert, 1979; Benson et al. 1980]. This correlation, however, is not always best with the inverted V peak. Often the correlation is best with the edge of the inverted V structure [Benson et al., 1980].

The importance of the edge of the inverted V has been emphasized by several experimenters at this conference. Fennell [1981] conducted an investigation using data from the P72-1, DMSP and S3-3 satellite which found enhancements in field-aligned electrons over a wide range of energies at the poleward edge of inverted V structures. These enhancements correlated with brightening of discrete auroral arcs associated with the inverted V edge. Bryant [1981] recorded field-aligned electrons at the inverted V edge rather than the peak during rocket flights above auroral arcs. On a smaller scale, i.e., the scale size of individual auroral arcs, Arnoldy [1981] and Anderson [1981] discussed rocket observations of field-aligned electrons at the edges of individual discrete auroral arcs.

The electron velocity distribution functions recorded by ISIS 1 near the peak of an inverted V associated with an AKR source region encounter always had a positive slope in the keV energy range [Benson et al., 1980]. This observation held true even when the AKR source location correlated best with the edge of the inverted V. Such electron velocity distributions are required in the AKR theories of Melrose [1976], Wu and Lee [1979], and Grabbe et al. [1980]. In addition, Melrose (1976) requires a hot electron beam with a temperature anisotropy, Wu and Lee require a loss-cone distribution and Grabbe et al. (1980) require both a cold electron beam and coherent ion cyclotron waves. Unfortunately, the ISIS 1 data would not allow a test for these additional conditions.

The simultaneous wave/particle source region measurements from ISIS 1 indicate that the electron/wave energy conversion efficiency can be as large as 1% [Benson and Calvert, 1979]. This result agrees with the conclusions of Gurnett [1974] and Green et al. [1979], which were based on the remote detection of AKR.

References

Ackerson, K. L. and L. A. Frank, Correlated satellite measurements of low-energy electron precipitation and ground-based observations of a visible auroral arc, J. Geophys. Res., 77, 1128-1136, 1972.

Anderson, H. R., Birkeland current and electric field in auroral arcs, this Proceedings, 1981.

Arnoldy, R. L., Review of auroral particle precipitation, this Proceedings, 1981.

Benson, R. F., Stimulated plasma waves in the ionosphere, Radio Sci., 12, 861-878, 1977.

Benson, R. F., W. Calvert, and D. M. Klumpar, Simultaneous wave and particle observations in the auroral kilometric radiation source region, Geophys. Res. Lett., 7, 959-962, 1980.

Benson, R. F. and W. Calvert, ISIS 1 observations at the source of auroral kilometric radiation, Geophys. Res. Lett., 6, 479-482, 1979.

Bryant, D. A., Auroral particles and visual auroral arcs, this Proceedings, 1981.

Calvert, W., The signature of auroral kilometric radiation on ISIS 1 ionograms, J. Geophys. Res., 86, 76-82, 1981.

Fennell, J. F., Auroral particle distribution functions and their relationship to inverted V's and auroral arcs, this Proceedings, 1981.

Frank, L. A., and K. L. Ackerson, Observations of charged-particle precipitation into the auroral zone, J. Geophys. Res., 76, 3612-3643, 1971.

Franklin, C. A., and M. A. Maclean, The design of swept-frequency topside sounders, Proc IEEE, 57, 897-929, 1969.

Grabbe, C. L., K. Papadopoulos, and P. J. Palmadesso, A coherent non-linear theory of auroral kilometric radiation 1. Steady state model, J. Geophys. Res., 85, 3337-3346, 1980.

Green, J. L., D. A. Gurnett, and R. A. Hoffman, A correlation between auroral kilometric radiation and inverted V electron precipitation, J. Geophys. Res., 84, 5216-5222, 1979.

Green, J. L., D. A. Gurnett, and S. D. Shawhan, The angular distribution of auroral kilometric radiation, J. Geophys. Res., 82, 1825-1838, 1977.

Gurnett, D. A., The earth as a radio source: terrestrial kilometric radiation, J. Geophys. Res., 79, 4227-4238, 1974.

Gurnett, D. A., Electromagnetic plasma wave emissions from the auroral field lines, J. Geomag. and Geoelectricity, 30, 257-272, 1978.

Gurnett, D. A., and J. L. Green, On the polarization and origin of auroral kilometric radiation, J. Geophys. Res., 83, 689-696, 1978.

Gurnett, D. A. R. R. Anderson, F. L. Scarf, R. W. Fredricks, and E. J. Smith, Initial results from the ISEE 1 and 2 plasma wave investigation, Space Sci. Rev., 23, 103, 1979.

Hartz, T. R., Radio noise levels within and above the ionosphere, Proc. IEEE, 57, 1042-1050, 1969.

Heikkila, W. J., J. B. Smith, J. Tarstrup, and J. D. Winningham, The soft particle spectrometer in the ISIS 1 satellite, Rev. Sci. Instrum., 41, 1393-1402, 1970.

Jackson, J. E., The reduction of topside ionograms to electron-density profiles, Proc. IEEE, 57, 960-976, 1969.

Jackson, J. E., The P'(f) to N(h) inversion problem in ionospheric soundings, in Mathematics of Profile Inversion, NASA TMX-62, 150, edited by L. Colin, pp. 4-2 to 4-14, National Aeronautics and Space Administration, Washington, D.C., 1972.

James, H. G., Direction-of-arrival measurements of auroral kilometric radiation and associated ELF data from ISIS 1, J. Geophys. Res., 85, 3367-3375, 1980.

James, H. G., E. L. Hagg, and D. L. P. Strange, Narrowband radio noise in the topside ionosphere, in AGARD conference proceedings No. 138 on Nonlinear Effects in Electromagnetic Wave Propagation, Edinburgh, Scotland, 12-15 November, 1973, Technical editing and Reproduction, Ltd., London, pp. 24-1 to 24-17, May 1974.

Kaiser, M. L., and J. K. Alexander, Terrestrial kilometric radiation 3. Average spectral properties, J. Geophys. Res., 82, 3273-3280, 1977.

Kaiser M. L., J. K. Alexander, A. C. Riddle, J. B. Pearce, and J. W. Warwick, Direct measurements by Voyagers 1 and 2 of the polarization of terrestrial kilometric radiation, Geophys. Res. Lett., 5, 857-860, 1978.

Melrose, D. B., An interpretation of Jupiter's decametric radiation and the terrestrial kilometric radiation as direct amplified gyroemission, Astrophys. J., 207, 651-662, 1976.

Muldrew, D. B., Preliminary results of ISIS 1 concerning electron-density variations, ionospheric resonances and Cerenkov radiation, in Space Research X, Edited by T. M. Donahue, P. A. Smith, and L. Thomas, pp. 786-794, North Holland Publishing Company, Amsterdam, 1970.

Roux, A., and R. Pellat, Coherent generation of the auroral kilometric radiation by nonlinear beatings between electrostatic waves, J. Geophys. Res., 84, 5189-5198, 1979.

Voots, G. R., D. A. Gurnett, and S.-I. Akasofu,

Auroral kilometric radiation as an indicator of auroral magnetic disturbances, J. Geophys. Res., 82, 2259-2266, 1977.
Wu, C. S., The source mechanism for Auroral kilometric radiation, this Proceedings, 1981.
Wu, C. S. and L. C. Lee, A theory of the terrestrial kilometric radiation, Astrophys. J., 230, 621-626, 1979.

LABORATORY STUDY ON ION CYCLOTRON WAVES AND THEIR RELATION TO AURORAL ARCS

H. Böhmer and A. Lang

Department of Physics, University of California, Irvine, California 92717

Abstract. In a laboratory study, the response of a collisionless, fully ionized and magnetized plasma to an electron drift ($v_D \lesssim v_{eth}$) is investigated. A barium Q-machine plasma with $T \lesssim T_e$ is used. Electron current is drawn through a localized, field aligned channel with $R \simeq \rho_i$. The ion response; i.e., density and energy, is determined optically, electron energy with an analyzer and potential fluctuation with a capacitive probe. The electron drift destabilizes ion cyclotron waves which heat the ions and cause a density depression [Mozer et al., 1977; Torbert and Mozer, 1978; Wescott et al., 1976]. The plasma potential is drastically changed both in space and time. Electrons are found to be accelerated both on the time scale of the wave period as well as on a steady state basis. Apart from the wave potential fluctuation having $\varphi \gtrsim T_e$ a weak longitudinal and a strong perpendicular electric field was observed. The results are compared with measurements in the polar magnetosphere above auroral arcs.

It has been suggested that electron acceleration in the polar magnetosphere above auroral arcs is due to electric fields supported by electrostatic shocks or double layers. Satellite measurements [Mozer et al., 1977; Torbert and Mozer, 1978; Wescott et al., 1976] as well as theoretical considerations [Block and Falthammer, 1976; Swift, 1975; Kan, 1975] support this assumption. Double layers have also been observed in laboratory plasmas [Quon and Wong, 1976; Coakley et al., 1978], although only for weakly ionized plasmas that were either not magnetized or only weakly magnetized. In contrast, the magnetospheric plasma is fully ionized and the zero order particle orbits are determined by the earth magnetic field. We report here on an experimental investigation of anomalous potential distributions in a fully ionized and strongly magnetized plasma in the presence of electron drift excited electrostatic ion cyclotron wave (EICW) turbulence.

The experimental configuration is shown schematically in Fig. 1. The barium plasma is generated by surface ionization of a neutral beam at the hot plate, which is heated to 2500 K. The fully ionized plasma drifts along the confining magnetic field of up to 5 kG. At the operating density of 10^9 cm^{-3} the electron mean free path is of the order of the machine length; the plasma, therefore, is collisionless. While the electron temperature remains 0.25 eV under all conditions, because the electron thermal conductivity is high , the ions are heated in the presence of EICW from 0.25 to above 1 eV.

Axially and radially movable Langmiur and capacitive probes are used to measure plasma density and potential variations respectively. An optical diagnostic technique using laser induced resonance fluorescence of barium ions is available to measure ion density fluctuations from the total light output and perpendicular or parallel ion velocity distribution functions from the Doppler broadening and shift of the optical lines. The optical lines used in this diagnostic have energy levels of 2.5 and 2.7 eV above ground state. Therefore, with an electron thermal energy of 0.25 eV, the collisional excitation probability is low and under normal plasma conditions collisions do not contribute to the observable light level.

While most of the plasma column is terminated by a cold end plate where ions and electrons recombine, a coaxial plasma channel is terminated by a small diameter hot plate. This plate is an electron emitter and will reionize any ions that come in contact with it. In other words, within this channel, electron and ion flux is possible in either direction. Normally, this channel is used to draw electron drift to the small plate which renders the plasma within that channel unstable to EICW.

The dispersion relation describing the electron drift excited ion cyclotron was first derived by Drummond and Rosenbluth [Drummond and Rosenbluth, 1962]. For $k_\perp >> k_\parallel$ and $k_\perp \rho_i \approx 1$ the solution can be written approximately as

$$\frac{\omega - \Omega_{ci}}{\Omega_{ci}} = \frac{T_e}{T_i} e^{-s} I_1(s) \qquad (1)$$

$$\gamma = \left(\frac{\pi}{2}\right)^{\frac{1}{2}} (\omega - \Omega_{ci}) \left(\frac{-\omega + k_\parallel v_{De}}{k_\parallel v_{eth}} \right) \qquad (2)$$

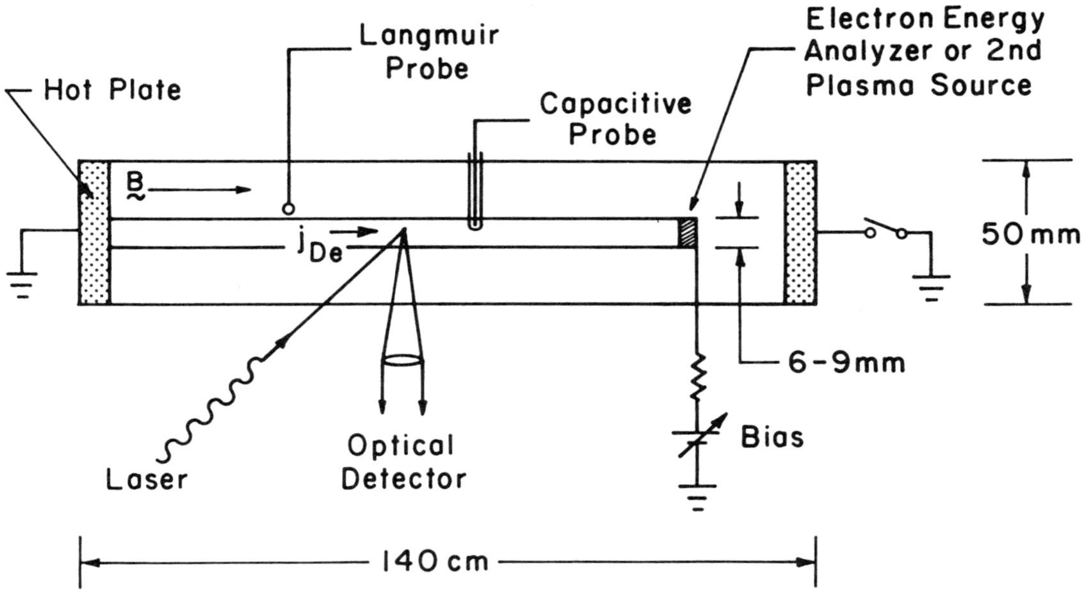

Fig. 1. Schematic of the Experiment.

where

$$s = k_\perp^{\;2} \rho_i^{\;2} \quad .$$

If the electron drift velocity becomes larger than the parallel phase velocity, the EIC waves become unstable ($\gamma > 0$).

It was shown by Kindel [1971] that in a magnetized plasma, under conditions where $T_e \leqslant T_i$, the EICW has the lowest threshold of excitation among the electron drift excited instabilities, namely, the EICW, the ion acoustic and the Buneman instability. Experimentally, we find that the EICW is excited by an electron drift as low as .3 v_{eth}. In contrast, particle simulations frequently assume $v_{De} \gtrsim v_{eth}$, resulting in the excitation of ion acoustic or Buneman instabilities.

The linear properties of the EICW have been investigated in detail by Correll et al. [1977]. They showed that the linear properties of the EICW are well described by Eqs. (1) and (2) over a wide range of parameters. It was also shown that the EICW heats the plasma ions from the initial value of 0.25 eV to more than 1 eV. The Larmor radius of the hot ions becomes of the order of the current channel size. Because of the ion heating and the presence of the strong EIC turbulence with $k_\perp >> k_\parallel$, part of the ion constituent becomes transported across the magnetic field lines, resulting in a density depression inside and a density increase outside the current channel.

The foregoing description of the experiment shows that the situation investigated in the laboratory resembled closely the polar magnetosphere above auroral arcs: in a fully ionized, collisionless plasma, magnetic field aligned electron current excites strong ion cyclotron wave turbulence. The ions are magnetized and are heated by the EICW. Their Larmor radius is of the order of the width of the current channel which means that finite size effects can be important for the phenomena related to the EIC instability.

The question arises, what are the microscopic and macroscopic potential distributions in this current channel containing strong EIC turbulence? Are potential distributions present that allow, because of their size and spatial distribution, the acceleration of electrons to many times their thermal velocity? In order to investigate these questions, the plasma potential was measured both time resolved within a cyclotron wave period and time averaged.

Figure 2 shows for different electron drift currents the current fluctuation in the upper trace and the plasma potential fluctuations in the lower trace of each panel. It is evident that, as the potential fluctuations become a sizable fraction of the electron thermal energy of .25 eV, the waves become highly nonlinear, containing higher harmonics phase-locked with the fundamental. Temerin, Woldorf, and Mozer [1979] have shown that such wave forms, which also have been observed by the S3-3 satellite, can be obtained as a solution of the nonlinear hydrodynamic dispersion relation of the EIC waves. A requirement for the existence of such solutions is that the waves are spatially growing. This is indeed the case in the present experiment as can be seen from Fig. 3 where ion

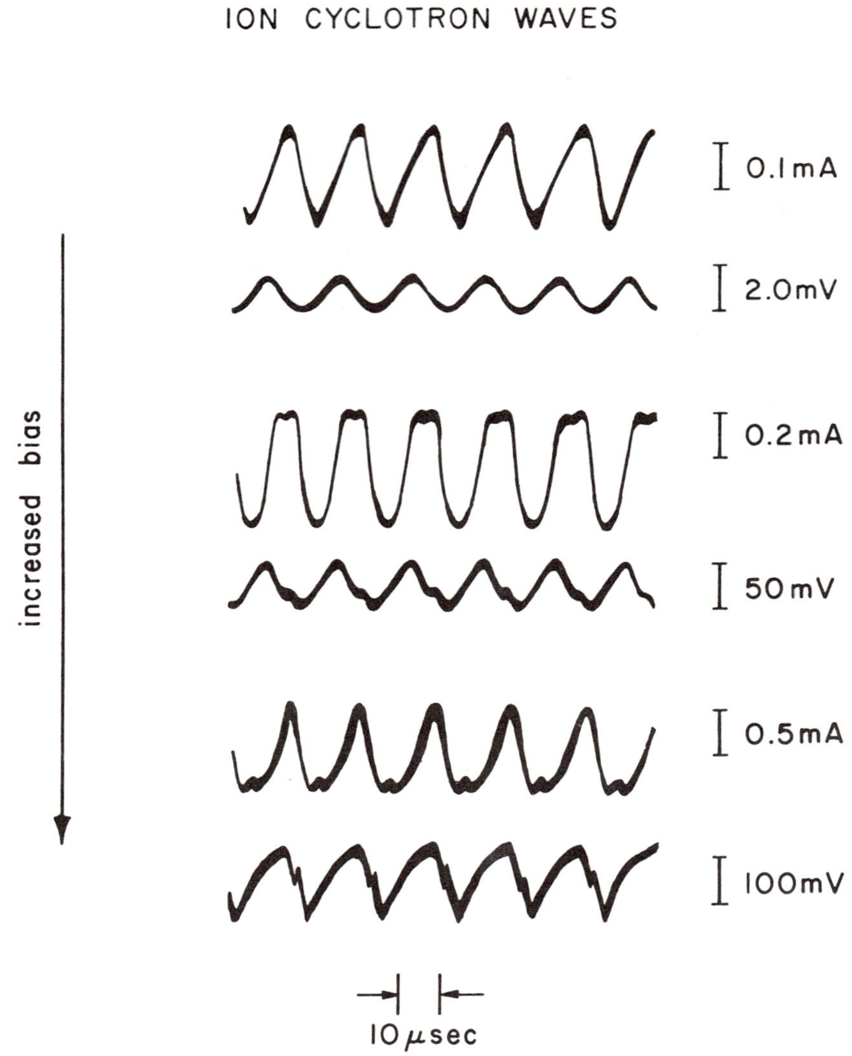

Fig. 2. EIC wave for three different electron drift currents, observed as electron drift fluctuations (upper trace of each panel), and potential fluctuations (lower trace).

density fluctuations, measured optically, are shown for two points along the drift channel containing the waves. While the wave has a small amplitude and is fairly sinusoidal at short distances, it grows to an amplitude of $\delta n/n \simeq .5$ at 104 cm and becomes triangular in shape. This observation agrees phenomenologically with the solution of the nonlinear dispersion relation as described by Temerin, Woldorf, and Mozer [1979]. The question arises, why the EIC waves can obtain an amplitude of the observed magnitude; i.e., an amplitude comparable with the electron thermal energy. Since this instability is driven by the positive slope of the displaced electron distribution function, a quasilinear saturation of the wave at low amplitude would have been expected. As discussed by Davidson [1972], quasilinear saturation does not occur in a two or three dimensional problem. In the case of the EIC waves, the waves are propagating oblique to the magnetic field. On the other hand, $k_{\parallel}$ is interacting with the strongly magnetized electrons, reducing the problem to the one dimensional case and quasilinear saturation should be possible.

In this experiment, as well as in the magnetosphere, several processes can prevent the quasilinear plateau formation from occurring. Since the parallel group velocity of the EIC wave is small compared to the phase velocity and the electron drift velocity, the distribution function, especially in short systems like this experiment, can continuously be replenished from the source.

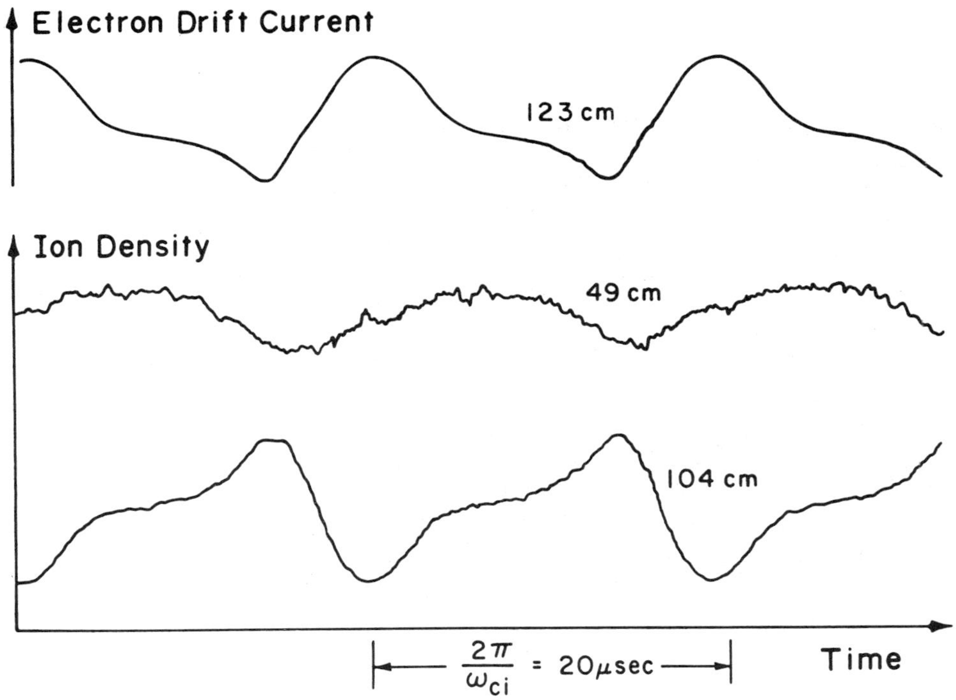

Fig. 3. Ion density fluctuations at two points along the axis of the system and electron drift fluctuations.

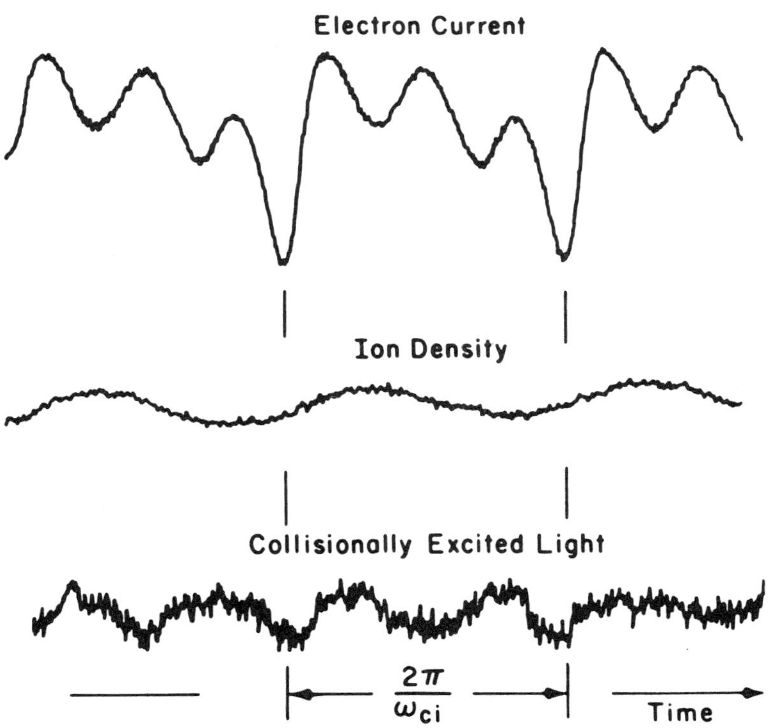

Fig. 4. Electron drift current, ion density and collisionally excited light associated with an EIC wave.

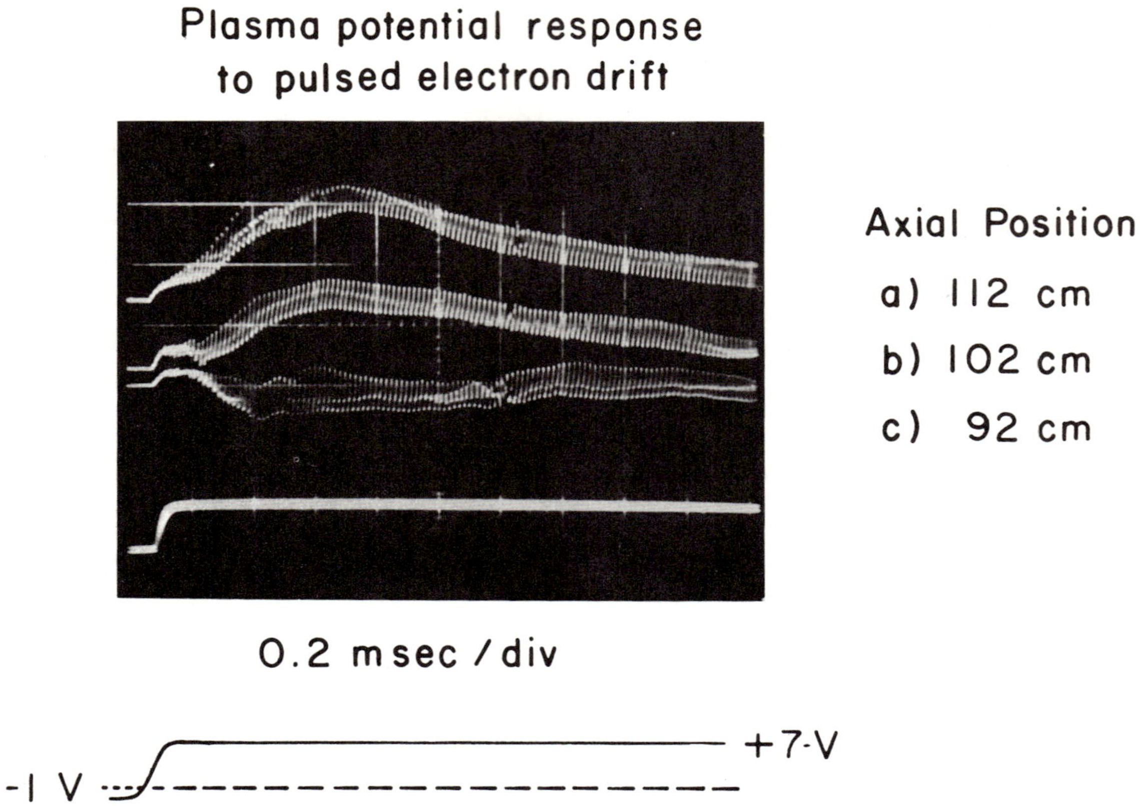

Fig. 5. Plasma potential response to a step function increase in electron drift current measured at three axial positions.

Equally important is the radial transport in radially finite, turbulent systems. The drifting electron distribution in this experiment as well as in the current sheet in the magnetosphere above auroral arcs is bordered by a non-drifting electron component. At the same time, the EIC waves propagate transverse to the boundary into the stationary plasma.

Drummond and Rosenbluth [1962] have already considered the anomalous transport due to the EIC turbulence. They derive for the diffusion coefficient perpendicularly to the magnetic field

$$D_{\perp} = \frac{1}{t}\left(\frac{c}{B}\right)^2 \frac{k_{\perp}^2}{k_{\parallel}^2} \left\langle \left(\int E_{\parallel}(t')dt' \right)^2 \right\rangle .$$

While Drummond and Rosenbluth express the fluctuation spectrum in terms of the given plasma parameters, it is more appropriate here to substitute it with the measurable quantity φ/T_e. φ is the potential fluctuation amplitude of the wave. Assuming that the time scale t is equal to the inverse growth rate γ, one finds

$$D = \gamma \rho_e^2 \frac{k_{\perp}^2}{k_{\parallel}^2} \frac{\varphi}{T_e}$$

Assuming a reasonable density gradient at the edge of the current channel [Böhmer and Fornaca 1979], one finds from this expression that the radial electron flux into the current channel is of the order of the axial flux within that channel. It should be noticed that the diffusion process is selective. The electrons of the non-drifting adjacent plasma which are close to resonance with the parallel phase velocity of the wave, are mostly affected.

There is experimental evidence that the turbulent radial transport does occur. The electron saturation current which is normally determined by flux parallel to the magnetic field, is enhanced by the presence of EIC turbulence.

From the above arguments it seems reasonable that quasilinear saturation does not occur. Indeed, it was shown earlier [Correll, Rynn, and Böhmer, 1977], that the saturation wave amplitude is consistent with resonance broadening. It was also shown [Böhmer and Fornaca, 1979] that electron trapping is likely to occur be-

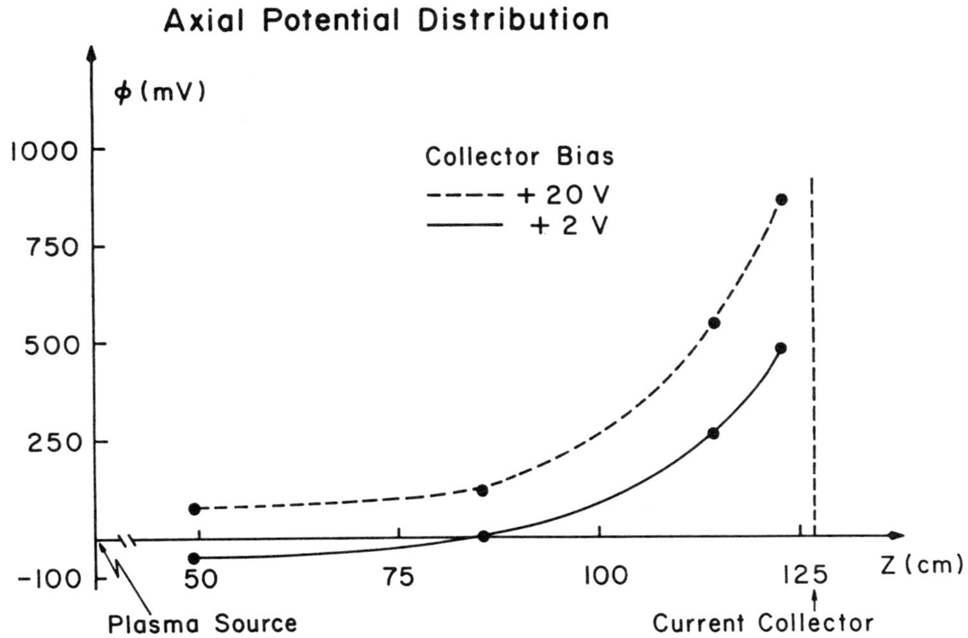

Fig. 6. Time averaged plasma potential variation along the system axis in the presence of strong EIC turbulence.

cause of the large wave amplitude of $\varphi/T_e \simeq 1$ and the large auto-correlation time of $\approx 20/\omega_{ci}$.

It was also shown earlier [Böhmer and Fornaca, 1979], and Fig. 4 gives another example, that the large amplitude EIC waves can, periodically with the wave, accelerate electrons to an energy sufficiently large to collisionally excite the barium ions to an energy level of 2.7 eV above ground state. The subsequently emitted light at 4934Å is plotted in Fig. 4 as a function of time together with the electron drift current and the ion density fluctuations due to the EIC wave.

Because of the large wave amplitude of $\varphi/T_e \approx 1$ and the observed back-action of the wave on the electrons, it is not surprising that the D.C. potential within the current channel is modified by the presence of the wave. Figure 5 shows examples of the plasma potential after the pulsed onset of the EIC wave, measured with a capacitive probe and displayed on an oscilloscope. Potential changes in a laboratory plasma of finite length can either be caused by changes in the sheath potential at the plasma boundaries or by bulk property changes within the plasma like change in plasma resistivity, establishment of double layer, etc. It can be seen from Fig.5 and is more clearly shown in Fig. 6, that the potential changes, time averaged over the EIC wave period, vary between the observation points within the plasma; i.e., an electric field within the plane is established. The on-axis potential changes gradually with distance so one must exclude the possibility of a double layer formation which would manifest itself in a potential change in the form of a step-function. The axial onset of potential changes coincides roughly with the point of amplitude saturation of the axially traveling wave.

This structure of gradually increasing potential is imbedded in a larger diameter plasma column of constant plasma potential. Consequently, a three-dimensional equipotential structure is generated. A potential cross section in the r-z plane, again measured with capacitive probes after a pulsed onset of the EIC waves, is shown in Fig. 7. Note that, for a better graphical presentation, the axial (z) scale is compressed compared to the radial (r) scale. The measured structure has a weak axial and a strong radial electric field. The radial scale length corresponds to the Larmor radius of the hot ions.

The structure investigated here, having a strong oblique electric field at the surface of a current channel containing strong EIC waves, closely resembles the magnetosphere observations by the S3-3 satellite. The picture emerging from these measurements is, that the EIC waves with an amplitude of the order of the electron thermal velocity, disrupts the normal axial electron flux resulting in a weak axial potential gradient. Since this current channel is embedded in a plasma of constant potential, a strong radial electric field is generated at the edge.

The plasma within the potential structure of Fig. 7 must be non-neutral. Therefore it should be subject to azimuthal drift or diochotron in-

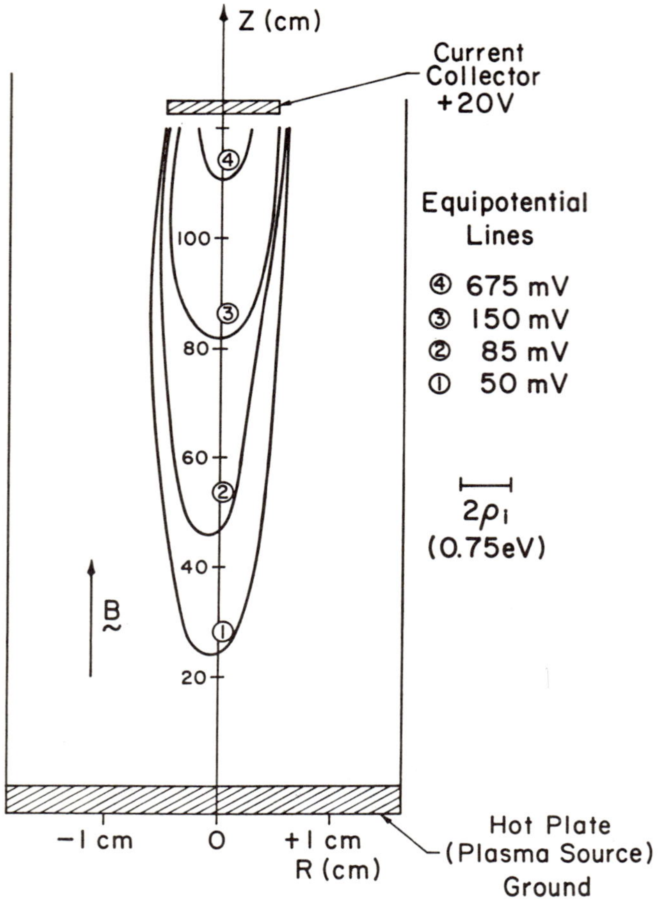

Fig. 7. Two dimensional equipotential contour plot associated with the +20 Volt collector bias case of Fig. 6.

stability. The azimuthal drift, is indeed, observed from the Doppler shift of the fluorescent Ba^+ ion optical lines. Simultaneously, low frequency waves are nearly always present at the surface of the current channel. The wave frequency, consistent with diochotron or drift waves, scales inversely with the external magnetic field.

References

Block, L. P., and C. G. Falthammar, Mechanisms that may support field-aligned electric fields in the magnetosphere, Ann. Geophys. 32, 161 1976.

Böhmer, H., and S. Fornaca, Experiments on nonlinear effects on strong ion cyclotron wave turbulence, J. Geophys. Res. 84, 5243, 1979.

Coakley, P. G., N. Hershkowitz, R. Hubbart, and G. Joyce, Experimental observations of strong double layers, Phys. Rev. Lett. 40, 230, 1978.

Correll, D. L., N. Rynn, H. Böhmer, Onset, growth and saturation of the current-driven ion cyclotron instability: theory and experiment, Phys. Fluids 20, 822, 1977.

Davidson, R. C., Methods in nonlinear plasma theory, Academic Press, New York, 1972.

Drummond, W. E., and M. N. Rosenbluth, Anomalous diffusion arising from microinstabilities in plasmas, Phys. Fluids 5, 1507, 1962.

Kindel, J. M., and C. F. Kennel, Topside current instabilities, J. Geophys. Res. 76, 3055, 1971.

Kan, J. R., Energization of auroral electrons by electrostatic shock waves, J. Geophys. Res. 80, 2089, 1975.

Mozer, F. S., C. W. Carlson, M. K. Hudson, R. B. Torbett, B. Parady, J. Yatteau, and M. C. Kelley, Observations of paired electrostatic shocks in the polar magnetosphere, Phys. Rev. Lett. 38, 292, 1977.

Quon, B. H., and A. Y. Wong, Formation of double layers in plasmas, Phys. Rev. Lett. 37, 1393, 1976.

Swift, D. W., On the formation of auroral arcs and acceleration of auroral electrons, J. Geophys. Res. 80, 2096, 1975.

Temerin, M., M. Woldorf, and F. S. Mozer, Nonlinear steepening of the electrostatic ion cyclotron wave, Phys. Rev. Lett. 43, 1941, 1979.

Torbert. R. B., and F. S. Mozer, Electrostatic shocks as a source of discrete auroral arcs, Geophys. Res. Lett. 5, 135, 1978.

Wescott, E. M., H. C. Stenback-Nielsen, T. J. Hallinan, T. N. Davis, and H. M. Peek, The Skylab barium injection experiment, 2, Evidence for a double layer, J. Geophys. Res. 81, 4495, 1976.

OBSERVATION OF ELECTROMAGNETIC RADIATION AT $2\omega_p$ GENERATED BY BEAM-PLASMA INTERACTIONS

P. Leung, J. Santoru, A. Y. Wong, and P. Y. Cheung

Department of Physics, University of California, Los Angeles, California 90024 USA

Abstract. Electromagnetic radiation at twice the electron plasma frequency ($2\omega_p$) is generated by the interaction of electron plasma waves excited by electron beams in a laboratory experiment. The power of the $2\omega_p$ radiation scales with the fourth power of the electrostatic wave amplitude and is consistent with the prediction of a wave-wave coupling model.

We wish to report experiments in a large unmagnetized, electron beam-plasma system in which the interaction of electron plasma (Langmuir) waves at the plasma frequency (ω_p) generates electromagnetic radiation at the second harmonic of the electron plasma frequency ($2\omega_p$). Previous experiments (Bollinger and Bohmer, 1971; Bollinger and Bohmer, 1972; Apel, 1967; Apel, 1969) have observed the generation of $2\omega_p$ radiation by a single electron beam interacting with a background plasma. This $2\omega_p$ radiation was believed to be generated from $2\omega_p$ electrostatic oscillations. In our experiments two counterstreaming electron beams are injected into a background plasma. The beam generated spatially growing antiparallel Langmuir waves through the beam-plasma instability (Quon et al., 1974; O'Neil and Malmberg, 1968). We observed the generation of enhanced levels of $2\omega_p$ radiation under these conditions. In addition, enhanced levels of $2\omega_p$ radiation are also seen with only one beam when the beam density is high ($n_b/n_o \geq 0.8\%$). The high beam density causes the unstable electron plasma waves to grow to large amplitude. The waves then undergo parametric decay generating backscattered waves and the resulting oppositely propagating wave topology is similar to the one beam case.

Previous theoretical studies of electromagnetic radiation from plasmas at harmonics of the electron plasma frequency have found that one process capable of generating $2\omega_p$ radiation is wave-wave coupling (Aamodt and Drummond, 1964; Sturrock et al., 1965; Sturrock, 1961; Boyd, 1964; Papadopoulous, 1979). In the wave-wave coupling process, generation of $2\omega_p$ radiation can be described as the collision of two Langmuir waves producing a transverse wave. In this process wave energy and momentum are conserved. Consequently

$$\underline{k}_{L1} + \underline{k}_{L2} = \underline{k}_{LT}, \quad (1)$$

$$\omega_{L1} + \omega_{L2} = \omega_T , \quad (2)$$

where $\underline{k}_{L1}$, ω_{L1}, $\underline{k}_{L2}$, ω_{L2} are the wave numbers and frequencies, respectively, of the Langmuir waves and $\underline{k}_T$, ω_T are the wave number and frequency, respectively, of the transverse waves. The waves satisfy the following dispersion relations:

$$\omega_{L1,2}^2 = \omega_{pe}^2 + 3k_{L1,2}^2 v_e^2 \quad (3)$$

$$\omega_T^2 = \omega_{pe}^2 + k_T^2 c^2 \quad (4)$$

where v_e is the electron thermal velocity and c is the speed of light. Since $\underline{k}_{L1}$ and $\underline{k}_{L2}$ are much larger than k_T, equation (1) can be satisfied only if $\underline{k}_{L1}$ and $\underline{k}_{L2}$ are almost antiparallel to each other. Equations (1) and (2) can now be satisfied for $\omega_T \simeq 2\omega_p \simeq 2\omega_{L1} \simeq 2\omega_{L2}$. Substituting these values into equation (4) yields $k_T \simeq \sqrt{3}\,\omega_p/c$. The electromagnetic radiation is generated by a nonlinear source current J_{NL},

$$J_{NL} = \delta n_e \, \delta ve, \quad (5)$$

where δn_e and δv are the first order perturbed density and velocity, respectively. Both δn_e and δv vary linearly with the electric field of the Langmuir wave E_{es}. Since J_{NL} is proportional to E_{es} and the power of electromagnetic radiation at $2\omega_p$, P_{em} $(2\omega_p)$ is proportional to J_{NL} , P_{em} $(2\omega_p)$ scales as E_{es}^4.

Our experiments were performed in a large vacuum chamber (180 cm diameter and 180 cm length). The experimental set up is shown in Fig. 1. The main background plasma is produced independently of the electron beams by pulsed

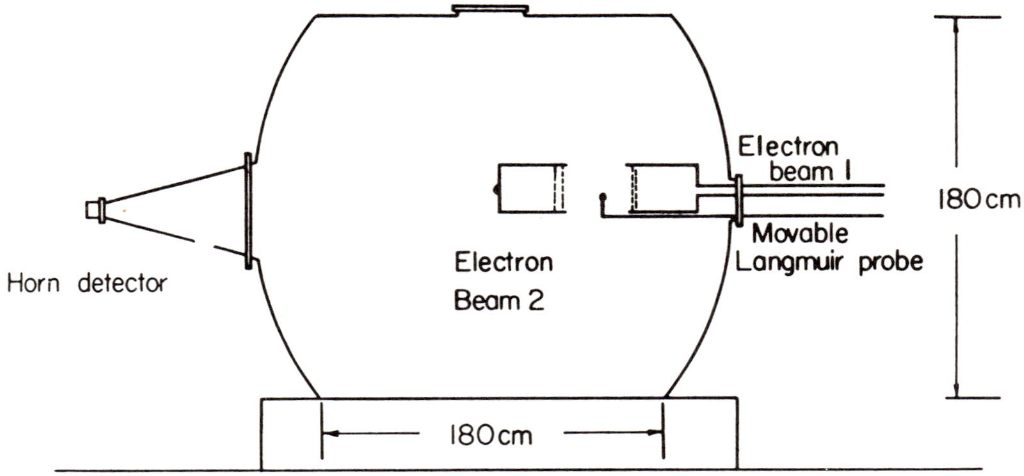

Fig. 1. Schematic of the experimental setup showing the vacuum chamber, two electron beams, Langmuir probe, and microwave horn detector.

filament discharge. The experiment is performed in an afterglow plasma to eliminate noise generated by fast primary electrons during the discharge. Typical plasma parameters are $kT_e = 2$ eV, $T_e/T_i \simeq 10$, and $n_o = 10^{10}$ cm^{-3}. The electron beams are injected simultaneously into the background plasma 900 μsec after the background plasma discharge is shut off. Each electron beam is an independent filament discharge system placed within a stainless steel enclosure and injected into the background plasma through preferential biasing of a two grid system. Each beam is mounted on a movable shaft so that the separation distance between the beams can be varied. Typical beam parameters for each beam are $E_b = 100$ eV, $n_b/n_o = 0.1$ to 1%, $\Delta E_b/E_b = 0.05$ and $\tau_b = 1$ msec, where E_b is the beam energy, n_b is the beam density, n_o is the background density, ΔE_b is the beam spread in energy and τ_b is the duration of the beam pulse. Measurements with single sided disc probes indicate that the electron beams penetrate 15 cm relative to their injection point before they are thermalized. For these experiments the separation between the beam injection points is 20 cm.

Each electron beam excites electron plasma waves through the beam-plasma instability (Quon et al., 1974). The geometry of our system results in essentially antiparallel propagating electron plasma waves. The plasma waves are studied using electrostatic probes with good high frequency response. The wavelengths of the Langmuir waves are measured by launching test waves. They are launched by applying low level signals near the plasma frequency to the inner grid of each beam during beam injection. The wavelength ($\lambda = 0.6$ cm) is consistent with the resonant relation $\omega_L = k_L V_b$. This is the only time external rf signals are applied in our system. All other experiments are performed with only the spontaneous noise generated by the beam plasma interaction.

The spatial variation of the amplitude of the electron plasma waves for two different frequencies is shown in Fig. 2a. For f = 1.1 GHz, the amplitude of the electron plasma waves grows spatially in the direction of the beam velocity with a spatial growth rate of $k_i/k_r = 4 \times 10^{-2}$. The amplitude peaks and then decreases toward the center. However, for f = 1.15 GHz, the amplitude peaks at the center of the interaction region. Test wave studies indicate that the excited plasma waves propagate beyond the center of the interaction region but they do not reach the other beam source. Consequently, the waves excited by one beam are not reflected by the grids of the other beam. Studies of the spectra of the electrostatic waves are made by capacitively coupling the signal from the high frequency probe into a spectrum analyzer. The vertical output of the spectrum analyzer is then sampled with a boxcar integrator. The total energy of the electrostatic waves is determined as follows. The spectra of the electron plasma waves are taken at 40 equally spaced intervals. The data are stored and then analyzed with the aid of a PDP 11/10 computer. The normalized total energy of the electrostatic waves is determined by the following equation

$$W_{es} = \frac{\int_L A \left[\int \frac{1}{4\pi} \hat{E}^2_{es}(\omega)\, d\omega\right] dx}{n_o k T_e} , \tag{6}$$

where $\hat{E}^2_{es}(\omega)$ is the measured amplitude of the electrostatic waves, L is the spatial extent of the interaction region and A is the cross-sectional area.

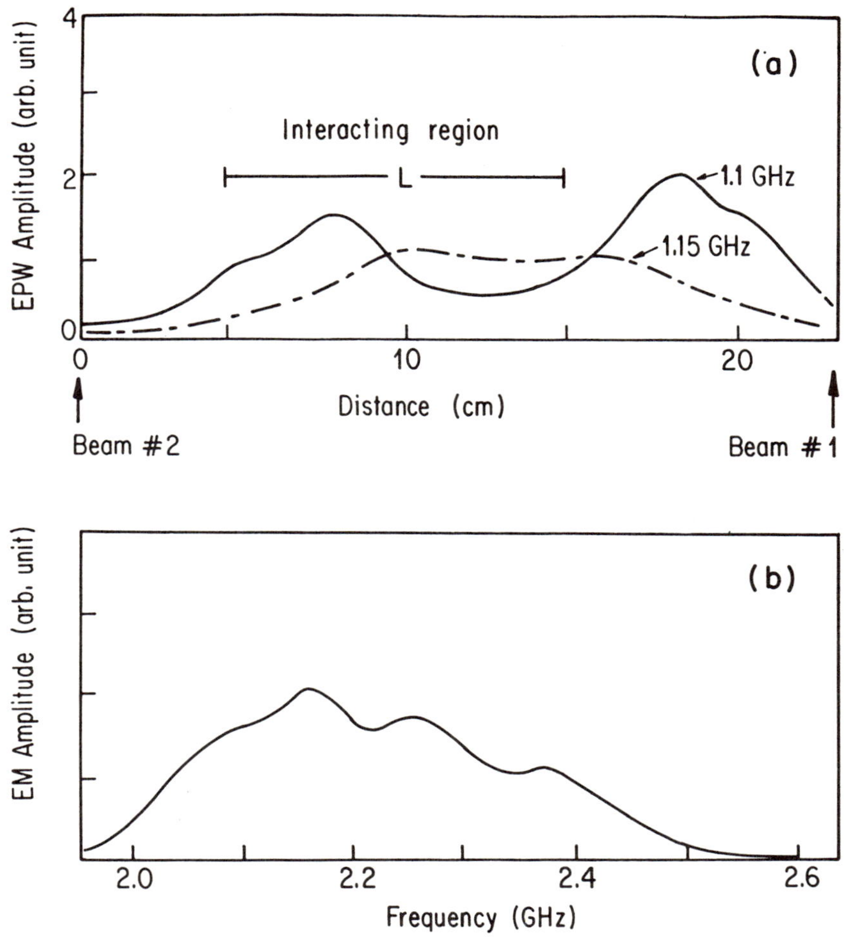

Fig. 2. Electromagnetic spectrum at $2\omega_p$ and corresponding electrostatic wave spectra. a) Spatial amplitude profile of the electron plasma waves excited by counterstreaming electron beams. b) Spectrum of the corresponding electromagnetic radiation at $2\omega_p$.

A sensitive detection system is used to measure the $2\omega_p$ radiation spectrum and power level. The radiation is detected with a L band microwave horn located at the end of the chamber. The radiation source is located in the far field region of the horn. A band pass filter (2-4 GHz) is used to eliminate spurious high and low frequency noise. A heterodyne system is then utilized to process the signal. The IF detector is a spectrum analyzer tuned to 300 MHz with a 3 MHz bandwidth. The vertical signal from the spectrum analyzer is sampled during the beam pulse with a boxcar integrator and stored in the computer. Full spectra are obtained by sweeping the local oscillator. The signal level of the received $2\omega_p$ signal is usually less than -60 dBm. A typical $2\omega_p$ radiation spectrum is shown in Fig. 2b. The data in figures 2a and 2b were obtained under the same experimental conditions. The $2\omega_p$ spectrum is broad with a typical width (full width at half maximum) of 300 MHz. This correlates well with the 100 MHz bandwidth of the electron plasma waves. The total electromagnetic power at $2\omega_p$, P_{em} $(2\omega_p)$, is calculated by performing the following integral:

$$P_{em}\,(2\omega_p) = \int_{\omega_1}^{\omega_2} \hat{P}(\omega)\,d\omega\ , \qquad (7)$$

where $\hat{P}(\omega)$ is the measured spectral intensity of the electromagnetic radiation, and ω_1 and ω_2 are the lower and upper limits of the $2\omega_p$ spectrum.

The power of the electromagnetic radiation at $2\omega_p$ was measured as a function of various beam parameters. The emission process has the following characteristics. (1) In the presence of counterstreaming beams with identical beam parameters, the dependence of P_{em} $(2\omega_p)$ on beam density is strongly nonlinear. (2) For the same total electron beam power (same n_b/n_o and E_b/kT_e) the power of electromagnetic radiation generated by counterstreaming beams is substantially higher (for certain beam densities, up to an order of magnitude) than that generated by a single beam. (3) The total power of $2\omega_p$ radiation scales nonlinearly with the total electrostatic power at ω_p. The appropriate scaling law is $P_{em}(2\omega_p) \propto E_{es}^{\gamma}$ where $\gamma = 4 \pm 1$.

Figure 3a shows the variation of the detected $2\omega_p$ radiation as the density of beam 1 is increased to $n_{b1}/n_o = 0.8\%$. The beam energy and density of beam 2 were maintained at $E_{b2}/kT_e = 50$ and $n_{b2}/n_o = 0.8\%$. Initially when only beam 2 is on, the detected power level is very low (< -70 dbm). As the density of beam 1 is increased, the electromagnetic power begins to increase. When the power of both beams are equal, the electromagnetic power is approximately 20 times above the power level observed with only one beam. The results indicate that the generation of $2\omega_p$ radiation is enhanced in the presence of counterstreaming electron beams. In Fig. 3b the observed $2\omega_p$ radiation is plotted as a function of the beam densities. The parameters of each beam were the same with the beam energies fixed at $E_{b1}/kT_e = E_{b2}/kT_e = 50$. In the same figure the sum of the power of the electromagnetic radiation at $2\omega_p$ obtained by using only one electron beam at a time [keeping the other beam off and then summing $P_{em}(2\omega_p)$ for each beam] is also displayed. This figure indicates that the electromagnetic power generated at $2\omega_p$ for the counterstreaming case is always higher than

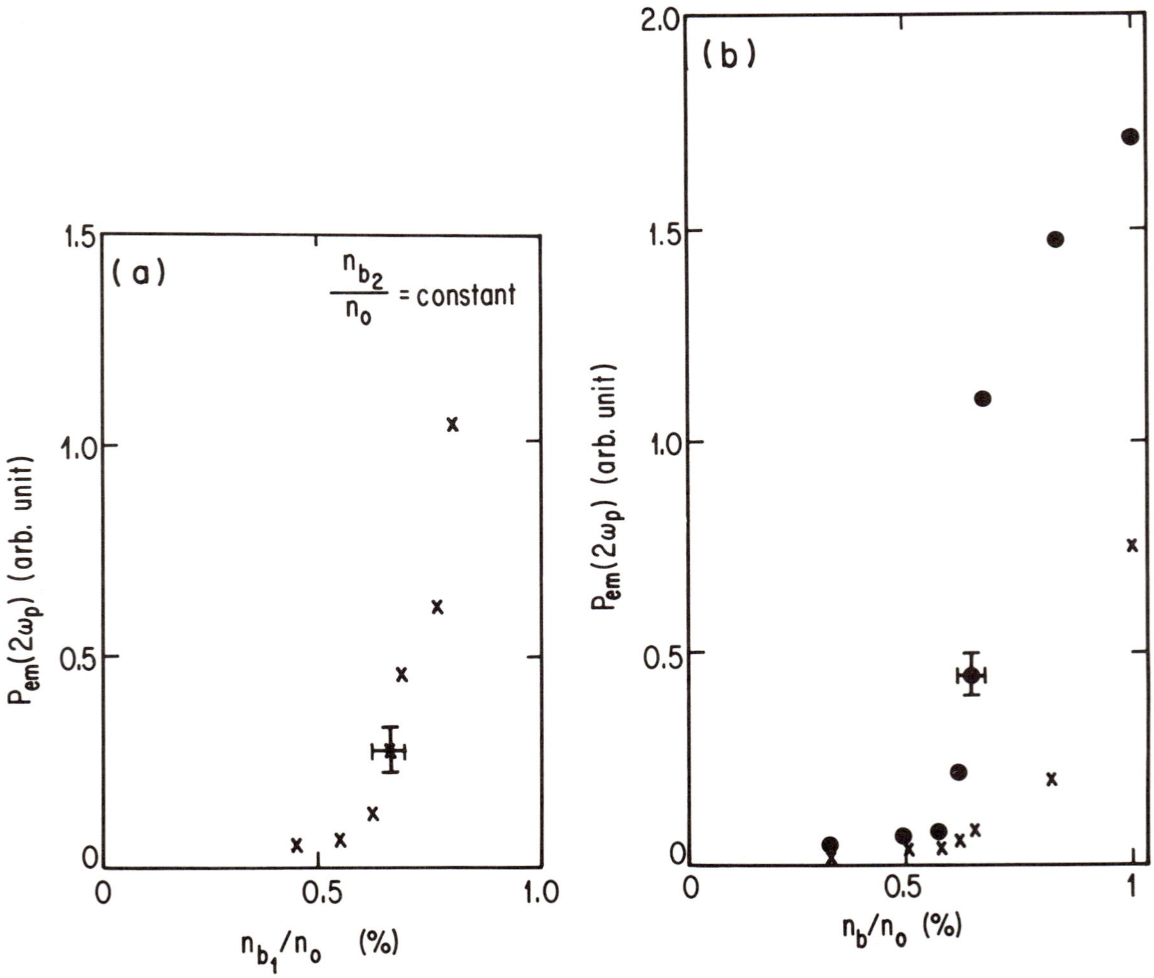

Fig. 3. Variation of $P_{em}(2\omega_p)$ as a function of beam density. a) The parameters of beam 2 are kept constant at $E_{b2}/kT_e = 50$ and $n_{b2}/n_o = 0.8\%$ while for beam 1, E_{b1}/kT_e is held constant at 50 as the beam density is varied (n_{b1}/n_o = 0 to 0.8%). This data is consistent with a nonlinear dependence of $P_{em}(2\omega_p)$ on beam density. b) The • corresponds to the counterstreaming beams. The x points are computed by first using beam 1 (with beam 2 off), noting $P_{em}^{(1)}(2\omega_p)$ and then using beam 2 (with beam 1 off), noting $P_{em}^{(2)}(2\omega_p)$ and then summing $P_{em}^{(1)}(2\omega_p)$ and $P_{em}^{(2)}(2\omega_p)$.

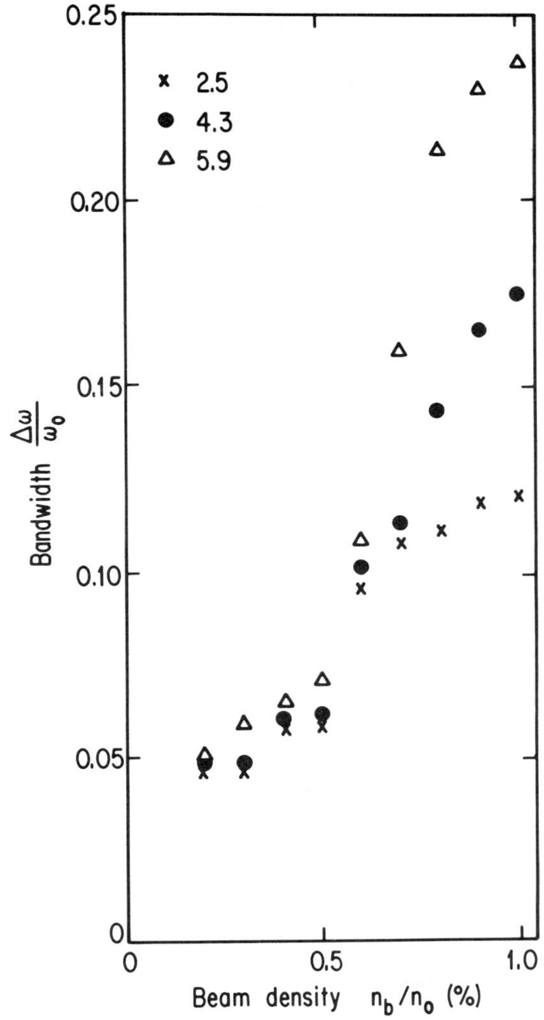

Fig. 4. The bandwidth $\Delta\omega/\omega_p$ of the unstable electron waves at different locations relative to beam 1. The numbers at the upper left corner indicate the distance (in cm) from the beam injection point. The beam energy is $E_{b1}/kT_e = 50$. Beam 2 is turned off.

that for the one beam case. In particular for $n_b/n_o = 0.6\%$, the $2\omega_p$ power generated by counterstreaming beam is 10 times higher than the one beam case. Further, we observe a beam density threshold in the counterstreaming beam case above which the P_{em} $(2\omega_p)$ increases nonlinearly with beam density.

At high beam densities ($n_b/n_o > 0.8\%$) significant electromagnetic power is generated when one beam interacts with the plasma. Figure 4 shows the bandwidth (full width at half maximum) of the electron plasma waves at various distances away from the injection point for the one beam case for different beam densities. At high beam densities ($n_b/n_o = 0.75\%$), significant broadening of the electrostatic wave spectrum occurs. This indicates that the electrostatic waves undergo parametric decay at high beam densities (Quon et al., 1974). One of the possible results of the parametric decay process is a backward propagating electron plasma wave. This wave can then couple with the forward propagating waves producing experimental conditions similar to the counterstreaming beam case thus resulting in enhanced levels of $2\omega_p$ radiation. From the data in Fig. 3a and 3b we conclude that enhanced levels of $2\omega_p$ radiation are generated when antiparallel propagating electron plasma waves are present.

Figure 5 shows the variation of P_{em} $(2\omega_p)$ with the energy of the electrostatic waves W_{es} (Eq. 6) for the counterstreaming beam case. The data was obtained by varying the densities of the counterstreaming beams. The spatial extent of the interaction region is taken to be 10 cm, centered between the two beams as shown in Fig 2. We observe P_{em} $(2\omega_p)$ α $W_{es}{}^{\beta}$ where $\beta = 2 \pm 0.5$. Since W_{es} α E_{es}^2, P_{em} $(2\omega_p)$ α $E_{es}{}^{\xi}$ where $\xi = 4 \pm 1$. This scaling is consistent with the wave-wave coupling process leading to radiation generation.

We define the coupling coefficient ε for the radiation generation process as the ratio of the total power radiated at $2\omega_p$ to the electrostatic power

$$\varepsilon = \frac{P_{em}\ (2\omega_p)}{P_{es}} , \tag{8}$$

where

$$P_{es} = \frac{1}{2} W_{es} n_o kT_e \ V_g A . \tag{9}$$

In this formula, $W_{es}/2$ is the normalized electrostatic energy density (Eq. 6), V_g (= 3 x 10^6 cm/sec) is the group velocity of the excited electron plasma wave for $E_b/kT_e = 50$, and A is the interaction area which is taken to be the cross-sectional area of the beam (100 cm^2). The coupling coefficient is calculated by estimating $W_o \simeq 0.01$ based on similar beam-plasma experiments (Leung et al., 1981) in which absolute measurements of the electrostatic field were made using diagnostic electron beam probes (Harp et al., 1965; Kim et al., 1974). The total power radiated at $2\omega_p$ is found from the power received at the horn. Using a typical value of received power of -70 dbm, we obtain $\varepsilon \simeq 4 \times 10^{-6}$. This is an order of magnitude estimate of the conversion efficiency. Multiple reflections from the chamber walls can scatter radiation into the microwave horn, making it difficult to make a precise determination of what fraction of the emitted $2\omega_p$ radiation is detected.

In summary, we have observed experimentally the generation of electromagnetic radiation at

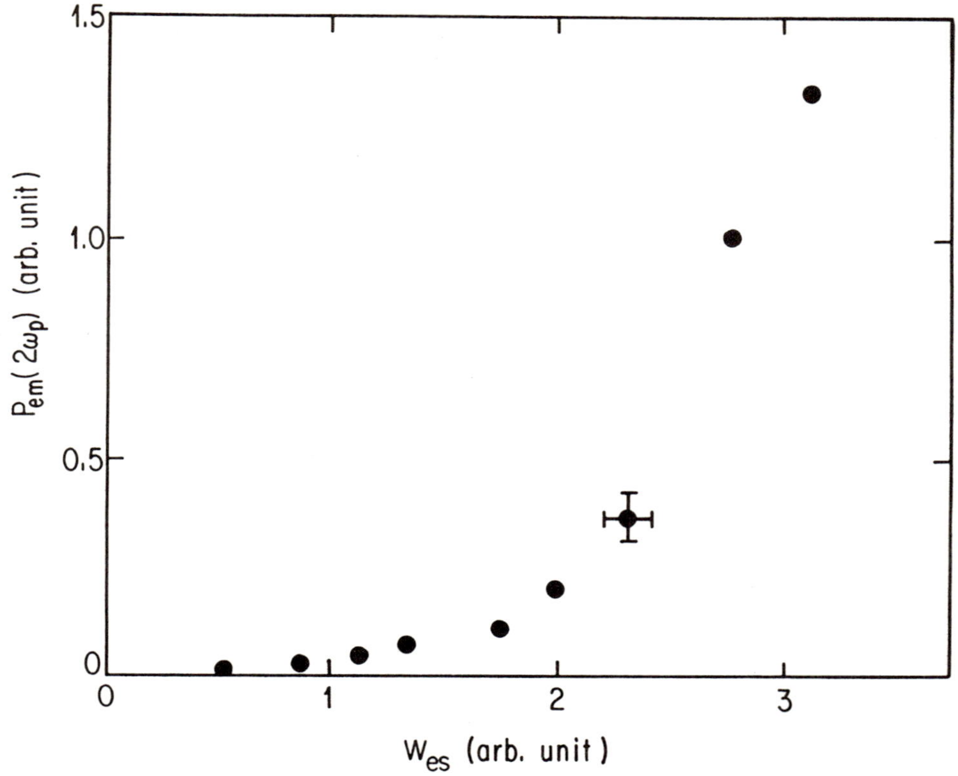

Fig. 5. Variation of P_{em} $(2\omega_p)$ with normalized electrostatic wave energy W_{es}. A least square fit of this data indicates P_{em} $(2\omega_p)$ α $W_{es}{}^{\beta}$ where $\beta = 2 \pm 0.5$.

the second harmonic of the electron plasma frequency. The process is identified to be the coupling of two antiparallel electron plasma waves to form an electromagnetic wave at $2\omega_p$. The observed electromagnetic power scaling of P_{em} $(2\omega_p)$ α E^4 is consistent with wave-wave coupling models. At high beam density ($n_b/n_o \geq 0.5\%$), $2\omega_p$ radiation has also been observed with a single electron beam. In this case, the Langmuir wave spectrum is observed to broaden, indicating that the waves undergo parametric decay. In this way the forward propagating waves are backscattered by ion acoustic or zero frequency density perturbations to generate backward propagating waves (Quon et al., 1974; Wong and Quon, 1975). The result is that the antiparallel plasma waves can interact to generate $2\omega_p$ radiation, as they do in the two beam case.

Acknowledgments. We wish to acknowledge helpful discussions with Professor D. Gurnett. We also aaknowledge the expert technical assistance of Bing Lum. This work was supported by the National Science Foundation Grant PHY79-08480.

References

Aamodt, R. E., and W. E. Drummond, Plasma Phys. 6, 147 (1964).

Apel, J. R., Phys. Rev. Lett. 19, 744 (1967).

Apel, J. R., Phys. Fluids 12, 640 (1969).

Bollinger, L. D. and H. Bohmer, Phys. Rev. Lett. 26, 535 (1971).

Bollinger, L. D. and H. Bohmer, Phys. Fluids 15, 693 (1972).

Boyd, T. J. M., Phys. Fluids 7, 59 (1964).

Harp, R. S., W. B. Cannara, F. W. Crawford and G. S. Kim, Rev. Scient. Inst. 36, 960 (1965).

Kim, H. C., R. L. Stenzel and A. Y. Wong, Phys. Rev. Lett. 33, 886 (1974).

Leung, P., M. Q. Tran and A. Y. Wong, UCLA Plasma Physics Group Report #551, April 1981.

O'Neal, T. M. and J. M. Malmberg, Phys. Fluids 11, 1754 (1968).

Papadopoulous, K., Space Science Rev. 24, 511 (1979).

Quon, B. H., A. Y. Wong and B. H. Ripin, Phys. Rev. Lett 32, 406 (1974).

Sturrock, P. A., Plasma Phys. 2, 158 (1961).

Sturrock, P. A., R. H. Ball and D. E. Baldwin, Phys. Fluids 8, 1509 (1965).

Wong, A. Y. and B. H. Quon, Phys. Rev. Lett. 34, 1499 (1975).

EXPERIMENTAL INVESTIGATIONS OF TURBULENCE IN MAGNETIZED, PARTIALLY IONIZED PLASMAS

T. Mikkelsen*, H.L. Pécseli*+, M. Alport+, and N. D'Angelo+

*Assoc. Euratom-Risø National Laboratory, DK-4000 Roskilde, Denmark
+Dept. of Physics and Astronomy, The University of Iowa, Iowa City, Iowa 52242, USA

Abstract. Experimental investigations of turbulent fluctuations associated with gradient cross-field and Farley-Buneman instabilities are reported. The experiments were carried out in magnetized plasma-columns with an imposed radial electric field. Microwave scattering and probe measurements provided the plasma diagnostics. Particular attention was given to the varying characteristics of the turbulent spectra for both potential and density as a function of neutral gas pressure.

I. Introduction

Radar experiments at Jicamarca, near the geomagnetic equator, demonstrated that the equatorial electrojet has embedded in it irregularities in plasma density, now interpreted as a saturated stage of gradient drift and Farley-Buneman instabilities [Farley, 1963; Buneman, 1963]. The same instabilities were later shown to be operative also in the auroral electrojet. The fluctuations propagate in the direction of the $\underline{E}\times\underline{B}$ electron drift and are excited when $\nu_i \gtrsim \omega_{ci}$, but $\nu_e \ll \omega_{ce}$ ($\nu_{i(e)}$ are the ion (electron)-neutral collision frequencies and $\omega_{ci(e)}$ the corresponding cyclotron frequencies). In this case the electrons experience essentially the full $\underline{E}_o\times\underline{B}$ drift, while the ions are effectively "held back" by collisions with neutrals. A linear stability analysis gives for the case where $\nu_i \sim \omega_{ci}$ (corresponding to the upper part of the electrojet) the real part of the dispersion relation

$$\omega = \underline{k} \cdot \underline{V}_{eo} \qquad (1)$$

and the imaginary part (where we ignore a velocity shear contribution)

$$\gamma = (\underline{k} \cdot \underline{V}_{eo} - \underline{k} \cdot \underline{V}_{io}) \frac{\nu_i}{\omega_{ci}} \frac{k_y}{k_\perp^2} \frac{1}{n_o} \frac{\partial n_o}{\partial \xi}$$
$$+ \frac{\nu_e}{\omega_{ci}\omega_{ce}} [(\underline{k} \cdot \underline{V}_{eo} - \underline{k} \cdot \underline{V}_{io})^2 - k_\perp^2 C_s^2] \qquad (2)$$

where the notation is the same as in Rogister and D'Angelo [1970], assuming that the fluctuations are strongly B-field aligned. This condition is somewhat relaxed by, e.g., Lee and Kennel [1973]. With the dc-electric field parallel to the density gradient, i.e., $\nabla n_o \cdot \underline{E}_o > 0$, the first term in (2) accounts for the gradient instability. When $|\underline{V}_{eo} - \underline{V}_{io}|$ exceeds the ion-sound speed C_s, the second term in (2) describes the Farley-Buneman instability which may be operative also in the absence of gradients. Note that the electron-neutral collison frequency appears explicitly only for this instability, which is thus a resistive electrostatic mode. The nonlinear, turbulent stage of the gradient instability has been investigated in some detail both theoretically and numerically. In this work we report the results of two laboratory experiments dealing with different aspects of the two instabilities in question. Since the instabilities are convective, we found it advantageous to perform the experiments in cylindrical geometry. Therefore, the experiments were carried out in magnetized plasma columns, where a dc-electric field (corresponding to the $\underline{E}$-field of the electrojet) was imposed in the radial direction. In Section II we report results on microwave scattering from the plasma fluctuations, while Section III presents detailed investigations of the spectral characteristics of the plasma turbulence.

II. Microwave Scattering

These measurements were carried out in a discharge-produced argon plasma in the setup described by Alport et al. [1980] and Alport and D'Angelo [1980], modified by the introduction of a variable radial electric field inwardly directed ($\underline{E}_o \cdot \nabla n_o > 0$) in a manner very similar to the one described by John and Saxena [1975] and Saxena and John [1975]. The radial potential variation was roughly parabolic, i.e., the electric field was linearly increasing with radius, thus giving rise to an almost "solid body" $\underline{E}_o\times\underline{B}$ rotation of the column. Plasma densities were $\sim 10^{11}$ cm^{-3}, the electron temperature $T_e \sim 1.5$ eV, ion temperature $T_i \sim 0.2$ eV, while the neutral argon pressure was $p_N \sim 2\times10^{-3}$ mmHg. The plasma column had a length $L \sim 60$ cm, and a diameter $D \sim 12$ cm. The confining magnetic field (homogeneous in the scattering region) was held constant at B ≃ 200 gauss. While the electrons are strongly magnetized, the ions have a gyroradius comparable to D, and are thus confined both electrostatically and by the magnetic field. The ions therefore experience only little $\underline{E}_o \times \underline{B}$ motion, even in the absence of neutral collisions. The effect of the neutral gas pressure is therefore mainly to maintain a non-vanishing electron-neutral collision frequency ν_e to make the second term in (2) significant. A rotatable microwave horn was mounted as shown schematically in Figure 1 (insert). The 3 dB width of the horn was $\sim 24^o$ rather than the few degrees of some ionospheric radars. We operated at a frequency $\sim$ 9.4 GHz (i.e., free space wavelengths $\lambda_\mu \sim 3$ cm), using the same horn both as emitter and receiver. Rotating the horn in the plane perpendicular to $\underline{B}$, we obtain backscattered signals like the ones shown in Figure 1, presented for various angles of θ (see insert). The signal consists of a large peak corresponding to the signal backscattered from the walls and a Doppler-shifted contribution corresponding to the backscattering from fluctuations in the plasma with wavelengths $\sim \lambda_\mu/2$. The Doppler shift is a measure of the phase velocity of these fluctuations. Since the electron velocity is roughly proportional to radius, we get a broad frequency band in the reflected signal due to the finite angular resolution of the microwave horn. We note that the Doppler shift changes sign with the sign of θ, as expected for waves propagating azimuthally. The direction of propagation agrees with the $\underline{E}_o \times \underline{B}$ electron velocity. This observation was further confirmed by reversing the $\underline{B}$-field direction, giving rise to a reversal in Doppler shift (see dotted line in Figure 1). Increasing the radial $\underline{E}_o$-field above a well-defined threshold, we observed an increase in the backscattered amplitude at large Doppler shifts (at frequencies corresponding to velocities above C_s). This observation agrees with the expectations based on Equation (2). At small $\underline{E}_o$, $|\underline{V}_{eo} - \underline{V}_{io}| < C_s$ for all radial positions in the plasma column, the waves being predominantly excited where the density gradient is large. For increasing $\underline{E}_o$, the second term in (2) becomes effective, with an onset of Farley-Buneman type waves, first at that radial distance where the electron velocity is largest. These waves propagate with a phase velocity above C_s. The resulting backscattered signal contains contribution from both wave types due to the finite angular resolution of the microwave horn.

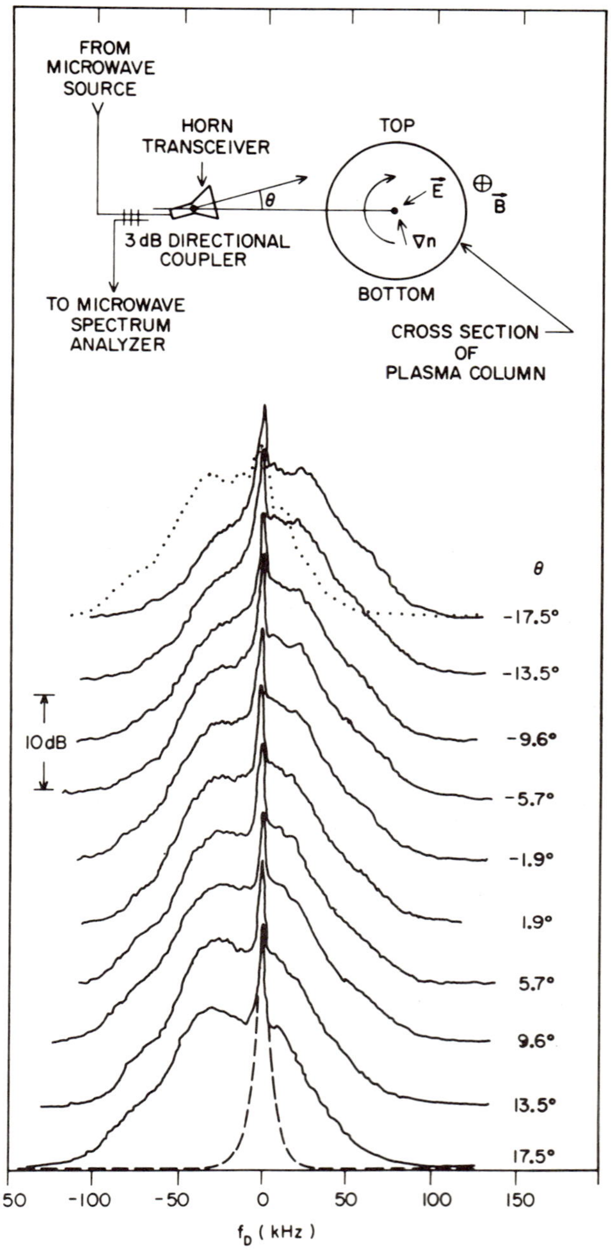

Fig. 1. Backscattered microwave spectra for various angles, θ, of incidence. Insert shows the setup schematically.

Radar scattering from similar wave types demonstrate a high aspect sensitivity, i.e., strongly $\underline{B}$-field aligned fluctuations. In order to give a similar demonstration in our setup, we fixed the horn at a certain value of θ = 17.5^o (see Figure 1) and varied the angle, ψ, between the horn axis and the plane normal to $\underline{B}$. We found only little variation in the shape of the microwave spectrum, but its peak amplitude showed a pronounced variation as shown in Figure 2. For ideally field aligned plasma fluctuations, the reflected signal should simply map the radiation pattern of the horn, given by the dotted line in Figure 2. We find the agreement with expectations convincing.

We have thus demonstrated that this simple setup gives an easy, convenient, and inexpensive way of investigating the gross features of the fluctuations in our plasma. However, because of the frequency and wave number matching conditions on

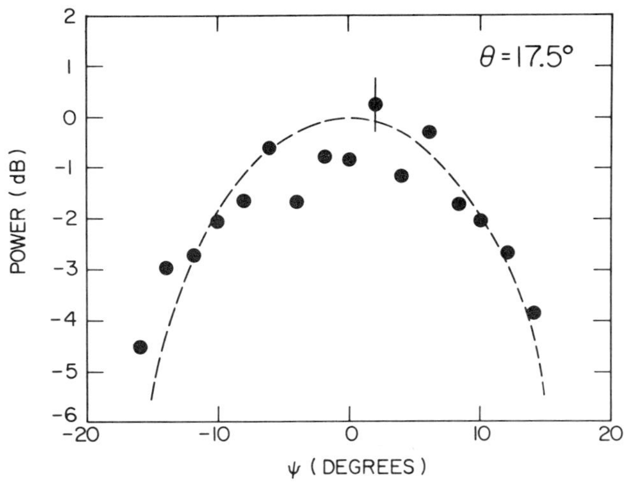

Fig. 2. Measurements of aspect sensitivity. Dashed line shows the radiation pattern of the microwave horn, as measured by another, similar, horn.

the scattering (referred to at times as the Bragg condition), we can investigate only one plasma wavelength with the present setup. Given the "solid body" rotation of the plasma, the Doppler-shifted signal provides an approximate radial amplitude distribution of the scattering (wavelength $\sim \lambda_\mu/2$) plasma fluctuations. For an investigation of the spectral properties of the turbulence, obviously we require information at all wavelengths for any given radial position. Such measurements are presented in the following section.

III. Spectral Measurements

Spectral measurements were performed in a magnetized cesium plasma column produced by surface ionization on a hot ($\sim$ 2000 K) tantalum spiral. For details of the setup, see D'Angelo et al. [1974] and Mikkelsen and Pécseli [1978, 1979, 1980]. Plasma densities were in the range 10^9 - 10^{10} cm^{-3} and temperatures $T_i \approx T_e \approx 0.2$ eV determined by the hot spiral. A magnetic field homogeneous over a length $\sim$ 80 cm and variable in the range 0.5-3 kG confined the plasma radially. The neutral background pressure was varied in the range 10^{-6} - 10^{-1} mmHg by argon or nitrogen inlet. Because of its spiral structure, the hot filament imposed an almost parabolic radial potential variation in the plasma column, giving rise to an imposed electric field, $\underline{E}_o$, that increased linearly with radius. The only significant difference from the experiment described in Section II is thus the increased $\underline{B}$-field. Now also, the ions are magnetized and we were able to observe the effect of ion-neutral collisions on the ion fluid velocity.

We observed the development of strongly turbulent fluctuations in the plasma column. Spectra of potential fluctuations were detected as described by Mikkelsen and Pécseli [1978], while the corresponding spectra for fluctuations in plasma density were detected by measuring the electron saturation current to a small Langmuir probe, located at the edge of the plasma column. Since the low-frequency (i.e., long wavelength) part of the spectrum is strongly affected by the radial geometry and is thus a characteristic of the particular experiment, we pay attention only to the high-frequency, "universal" part of the spectrum, which we characterize by a spectral index, α, by fitting a power law $f^{-\alpha}$. Thus, Figure 3(a) gives examples for potential (solid line) and density (dotted line) spectra for a fixed $\underline{B}$-field and varying neutral background pressure obtained by a spectrum analyzer. Figure 3(b) shows in detail the variation of the spectral index for the po-

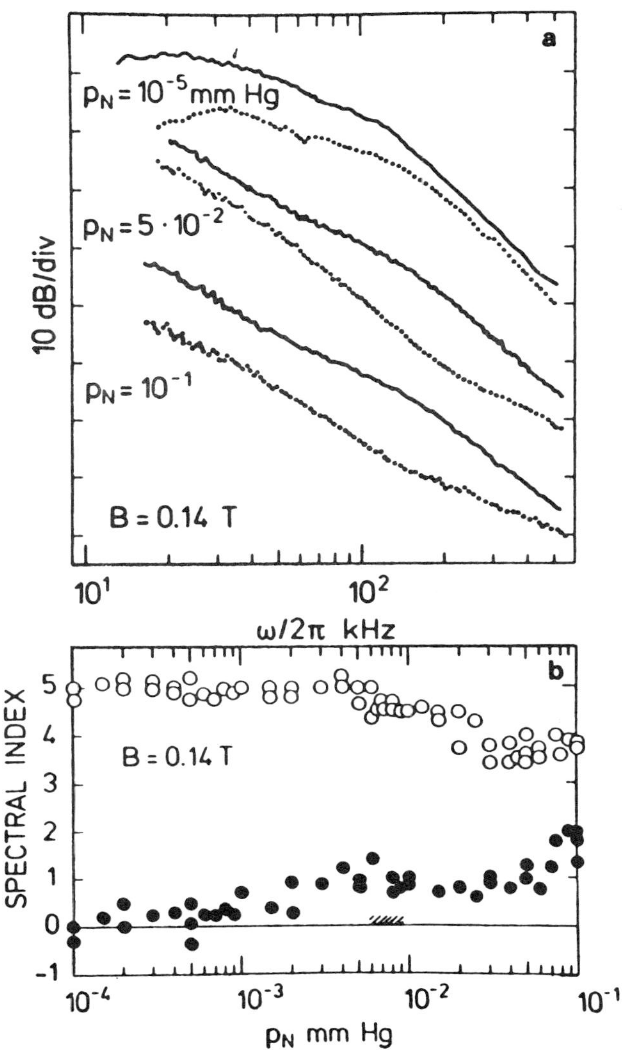

Fig. 3(a). Turbulent spectra for fluctuations in potential (solid line) and density (dotted line) for various neutral pressures. 3(b). Variation of spectral index with neutral pressure.

tential (o) and for the difference in spectral index between potential and density fluctuations (•), the density spectrum being, in general, more or less, flatter than that for the potential. From a theoretical point of view, wave number spectra are more interesting than frequency spectra such as those obtained here. In order to derive the former from the latter in the general case one has to invoke a nonlinear dispersion relation. In particular cases, however, where the plasma flows sufficiently rapidly past the detecting probe, one may apply Taylor's hypothesis (or the "frozen turbulence" approximation) for large ω, giving a simple relationship. We emphasize that the spectra G(k), obtained by this procedure are defined so that, e.g.,

$$\langle\tilde{\phi}^2\rangle = \int_0^\infty G(k)dk.$$

We investigated the applicability of this hypothesis by cross correlating the potential signal from two Langmuir probes. By varying the distance between the two probes, we obtained cross correlation functions $R(\tau)$ like those shown in Fig. 4(a). The probe separation divided by the time lag of the maximum of $R(\tau)$ is a measure for the component of the flow velocity in the direction determined by the probe tips. These measurements showed that the fluctuations were elongated and well correlated along the B-field lines and propagating in the azimuthal direction. Figure 4(a) thus shows $R(\tau)$ for varying azimuthal probe separations, ΔX, but fixed axial and radial positions. The envelope (dotted line on Figure 4(a)) is thus the "corotating-frame autocorrelation function". For a perfectly frozen flow, the cross correlation is a shifted copy of the autocorrelation. Obviously, our case is not ideal, but we find that the "moving frame" correlation time, t_c, is so long that the largest energy-containing eddies move several times their scale size, L, within t_c, i.e., Taylor's hypothesis may be applied with some confidence. We may estimate L from the "laboratory frame correlation time", τ_c, by $L \sim V\tau_c$, and find $L \sim 3$ cm, i.e., comparable to the column radius as expected from simple dimensional arguments.

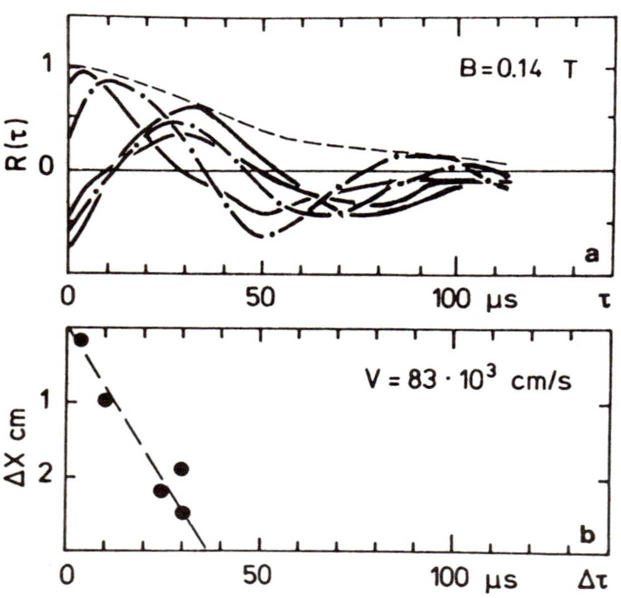

Fig. 4(a). Cross correlation measurements for varying azimuthal probe separation ΔX, but at a fixed radial position. 4(b). Propagation velocity, V, of the fluctuations, deduced from Figure 4(a).

In the present context Figure 3(b) presents our major results. For a discussion we consider ranges corresponding to (i) low, (ii) high, and (iii) intermediate neutral pressures separately.

(i) Low pressure regime. The index $\alpha \sim 5$ for both potential and density spectra agrees well with expectation for gradient driven instabilities [e.g., Tchen et al., 1980]. Measurements at a radial position corresponding to the stronger density gradient gives only slightly flatter spectra [Mikkelsen and Pécseli, 1978]. In only weakly collisional plasmas the electrons can flow along B-field lines even for large B-parallel wavelengths, $\lambda_\parallel$, to maintain (almost) a Boltzmann equilibrium so $\tilde{n}/n_o \sim e\tilde{\phi}/T_e$, thus explaining the similarity between potential and density spectra. Note that due to the finite length of the device, $\lambda_\parallel$ is also finite, although the terminating cold sheet at the endplate allows an effective $\lambda_\parallel$ larger than twice the length of the column.

(ii) At high pressures electron-neutral collisions inhibit the field aligned electron motion, while on the other hand, electrons can easily diffuse across B-field lines. A simplified linear analysis gives for this case, as an approximation, $\tilde{n} \sim k_\perp\tilde{\phi}$, i.e., fluctuations in density and electric field are roughly proportional (John and Saxena, 1975). The difference in the power-spectral index for density and potential is thus expected to be around two, in agreement with our results (see Figure 3(b)).

(iii) As seen from Figure 3(b), the transition of, e.g., potential spectral index ~ 5 to ~ 3.8 occurs at a relatively well defined neutral pressure, around 9×10^{-3} mmHg. Taking the corresponding ion (and electron) neutral collision frequency and calculating ν_{io} and ν_{eo}, we found that the transition coincides with the onset of the Farley-Buneman instability (the shading on the axis of Figure 3(b) gives the uncertainty). By varying B we found a variation of the transition pressure, consistent with this interpretation. The gradient instability excites predominantly long wavelengths, while the Farley-Buneman instability on the other hand directly excites short wavelengths (see equation 2). A flatter spectrum when both instabilities are excited is then to be expected.

IV. Conclusions

We have presented laboratory observations of turbulent fluctuations associated with instabilities playing an important role for the irregularities in the electrojet. Microwave scattering and probe measurements verify that the resulting fluctuations are strongly B-field aligned and propagate essentially with the electron fluid velocity in agreement with theoretical expectations. We found turbulent spectra well characterized by a spectral index ∿ 3.8 and ∿ 2 for potential and density fluctuations respectively. These indexes seem characteristic for the instabilities in question since the character of the turbulent spectra changes significantly when the Farley-Buneman instability is ineffective. The question remains whether the observed spectra are universal. Obviously, our experiments alone cannot clarify this problem but we find it interesting, also, that Langmuir probe measurements performed in connection with the experiment described in Section II gave spectra for density fluctuations in agreement with the results of Figure 3(b). More important, however, we note that in situ rocket measurements in the equatorial electrojet reported by, e.g., Gupta et al. [1977] give spectra very similar to the ones we observe in regions where both the gradient and Farley-Buneman instability are operative.

Acknowledgments. This work was in part supported by NSF grant ATM-7800529 and NASA grant NGL-16-001-043.

References

Alport, M., and N. D'Angelo, Bragg scattering of EM waves from ion-beam mode wave packets, Plasma Phys., 23, 1, 1981.

Alport, M., N. D'Angelo, and M. Khazei, Bragg effects in microwave transmission through stationary plasma structures, IEEE Trans. on Plasma Sci., PS-8, 111, 1980.

Buneman, O., Excitation of field aligned sound waves by electron streams, Phys. Rev. Lett., 10, 285, 1963.

D'Angelo, N., H.L. Pécseli, and P.I. Petersen, The Farley instability: A laboratory test., J. Geophys. Res., 79, 4747, 1974.

Farley, D.T., A plasma instability resulting in field aligned irregularities in the ionosphere, J. Geophys. Res., 68, 6083, 1963.

Gupta, S.P., S. Prakash, and B.H. Subbaraya, Spectral characteristics of cross-field and two stream instabilities as revealed by rocket borne studies, J. Geophys., 43, 681, 1977.

John, P.I., and Y.C. Saxena, Observation of the Farley-Buneman instability in a laboratory plasma, Geophys. Res. Lett., 2, 251, 1975.

Lee, K., and C.F. Kennel, Effects of propagation parallel to the magnetic field on the type I electrojet irregularity instability, Planet. Space Sci., 21, 1339, 1973.

Mikkelsen, T., and H.L. Pécseli, Investigation of strong turbulence in a low-β plasma, Phys. Rev. Lett., 41, 951, 1978.

Mikkelsen, T., and H.L. Pécseli, Strong turbulence in partially ionized plasmas, J. Phys. (Paris) Colloq. 40 (C7), 569, 1979.

Mikkelsen, T., and H.L. Pécseli, Strong turbulence in partially ionized plasmas, Phys. Lett., 77A, 159, 1980.

Rogister, A., and N. D'Angelo, Type 2 irregularities in the equatorial electrojet, J. Geophys. Res., 76, 3879, 1970.

Saxena, Y.C., and P.I. John, Observation of the cross-field instability in laboratory plasmas, Geophys. Res. Lett., 2, 492, 1975.

Tchen, C.M., H.L. Pécseli, and S.E. Larsen, Strong turbulence in low-β plasmas, Plasma Phys., 22, 817, 1980.

LABORATORY EXPERIMENTS ON MAGNETIC FIELD LINE RECONNECTION

R. L. Stenzel, W. Gekelman and N. Wild

Department of Physics, University of California, Los Angeles, California 90024

Abstract. A laboratory experiment on the problem of magnetic field line reconnection in plasmas has been performed. Comprehensive measurements of the magnetic field topologies as well as the plasma dynamics have been performed. The formation of neutral magnetic sheets, tearing, island coalescence and magnetic turbulence will be shown first. Then the force density and ion acceleration have been measured. An anomalous ion collision frequency ν^* has been derived from these quantities and found to be large, $\langle \nu^*/f_{pi} \rangle \lesssim 0.5$, when turbulence is present. Detailed time-dependent measurements of the total electric field and current density vectors have been made from which the plasma resistivity is calculated. The resistivity is found to be spatially inhomogeneous with average values exceeding the classical resistivity on the average by one order of magnitude and locally by two orders. Finally, a careful account of the energy flow is presented showing an efficient transfer from electromagnetic to kinetic energy.

Introduction

The reconnection of magnetic field lines at neutral points in plasmas is a problem of fundamental interest in magnetospheric physics (Dungey, 1958; Vasyliunas, 1975). In both solar flares and magnetic substorms reconnection is considered to explain the energization of particles at the expense of stored magnetic field energy. The relation between magnetic substorms and auroral arcs indicates the importance of reconnection in auroral physics. Unfortunately, direct observations in the neutral sheet region are very difficult to perform and their interpretation is less conclusive than those of auroral potential structures. In order to explore the physics of reconnection processes we have, therefore, built a large basic research laboratory plasma (Stenzel and Gekelman, 1979a). Time varying antiparallel magnetic fields are applied to an initially uniform, nearly collisionless, large magnetoplasma which is highly reproducible and carefully diagnosed.

The crucial aspects which are investigated are the self-consistent magnetic field topologies inside a high beta plasma, particle heating and acceleration, anomalous scattering, resistivity and fluctuation phenomena, and the energy transfer processes from magnetic fields to particles.

Experimental Setup and Measurement Techniques

The experiment is performed in a linear device in which a large (1 m diam., 2 m length) uniform, low pressure (10^{-4} Torr Argon or Helium) discharge plasma is produced with an oxide coated cathode (Stenzel and Gekelman, 1981; Gekelman and Stenzel, 1981). The pulsed plasma ($n_e \approx 10^{12}$ cm^{-3}, $kT_e \approx 10\ kT_i = 5$ to 30 eV) is immersed in a constant uniform axial bias magnetic field ($B_{yo} \simeq 20$ to 100 G) as shown in Fig. 1 (a). As

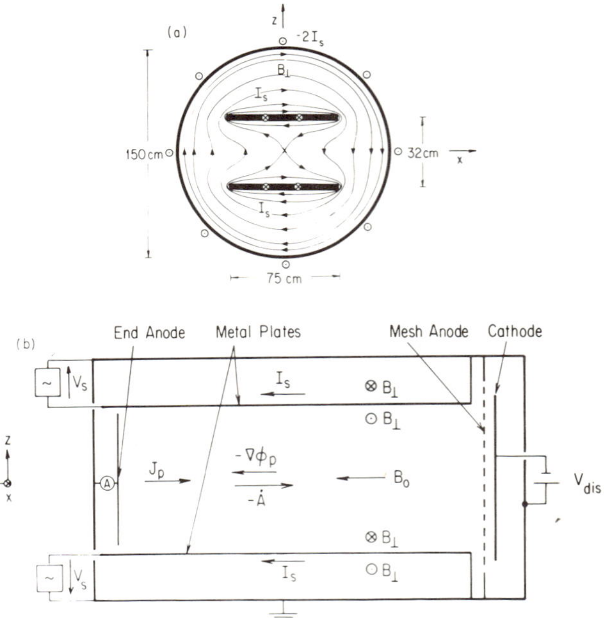

Fig. 1. Schematic view of the experimental setup. (a) Cross sectional view with magnetic field topology in vacuum. (b) Side view with characteristic currents and fields.

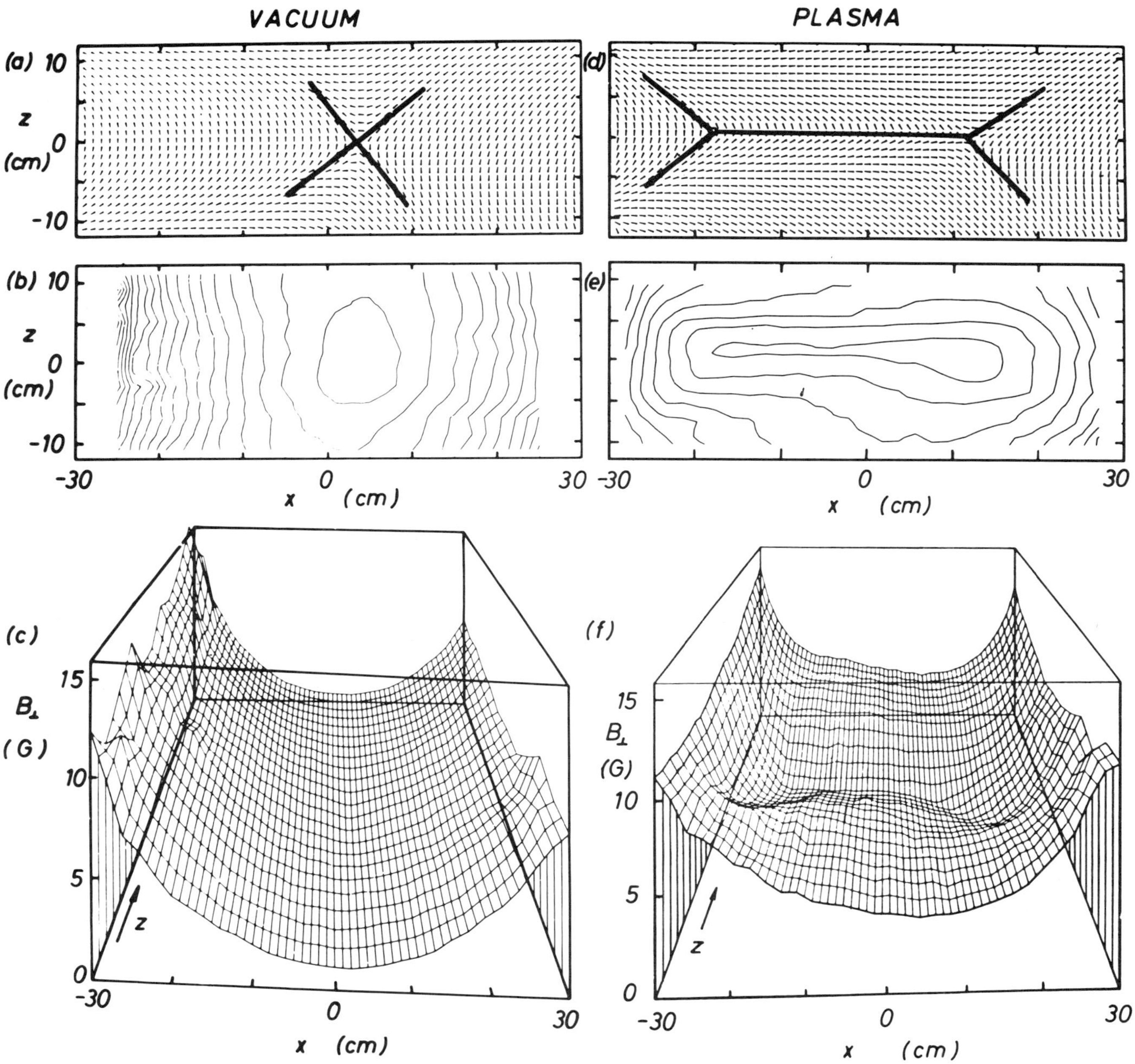

Fig. 2. Comparison of the magnetic field topologies in vacuum (a-c) and in plasma (d-f) at t = 40 μsec. (a), (d) Unit vectors indicating local direction of $\vec{B}_\perp$. Solid lines represent parts of the separatrix. (b), (e) Contours of constant field strength, $|\vec{B}_\perp|$. Incremental change 0.5 G/contour in (b), 2 G/contour in (e). (c), (f) Three-dimensional display of $|\vec{B}_\perp|$. Argon, 2 x 10^{-4} Torr, $B_{yo} \simeq 20$ G.

shown in Fig. 1 (b), a time dependent (damped sinusoidal) transverse magnetic field, $\vec{B}_\perp = (B_x, B_z)$, is established by pulsing axial currents ($I_s \lesssim 20$ kA, $t_r \simeq 80$ μsec rise time) through two parallel aluminum plates (75 cm wide, 32 cm spacing, 200 cm length) located on either side of the plasma column. In vacuum the magnetic field geometry exhibits an X-type neutral point on the axis of the device (x = z = 0). In the plasma, however, axial currents are induced ($I_p > 1000$ A) which modify the magnetic field topology until a self-consistent current and field pattern is established.

The magnetic field is measured point-by-point

with small magnetic loop probes. At each of typically 300 points in the transverse x-z plane the time history of the vector components, $B_x(t)$, $B_z(t)$, is recorded digitally with 20 MHz, 8 bit A-D converters. Subsequently, at a given time, the spatial field pattern is assembled in the form of vector fields or contour maps. Typically, the time can be chosen in 0.2 μsec time intervals from 1024 quantization steps. All three vector components have been recorded in three orthogonal directions. Thus, the complete time space history of the magnetic field is known. Since the reconnection events lead to fluctuations an ensemble average over typically 10 to 25 shots is formed. RMS fluctuations around the mean values of magnitude and angle of $\vec{B}$ are calculated on-line with a fast array processor.

The digital data analysis is also applied to the general plasma diagnostics performed with Langmuir probes and velocity analyzers. Probe measurements compare favorably with microwave interferometer and test wave diagnostics.

Experimental Results

Magnetic Field Topologies

During the rise of the applied magnetic field ($0 < t < 80$ μsec) plasma currents opposite to the applied currents are induced which slow down the flux transfer across the separatrix. Whereas in

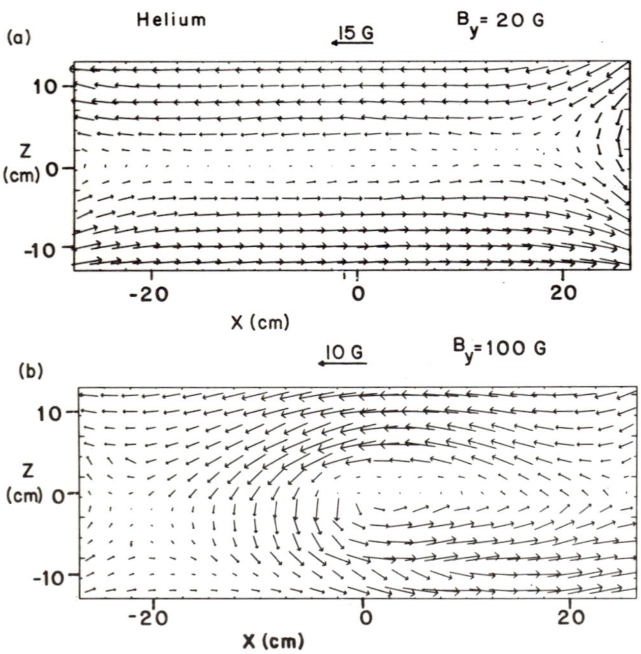

Fig. 3. Magnetic field topologies at different axial fields B_{yo} in Helium at $t \simeq 20$ μsec. (a) A neutral sheet arises at small $B_{yo} \simeq 20$ G. (b) Neutral X and O - points are found at large $B_{yo} = 100$ G.

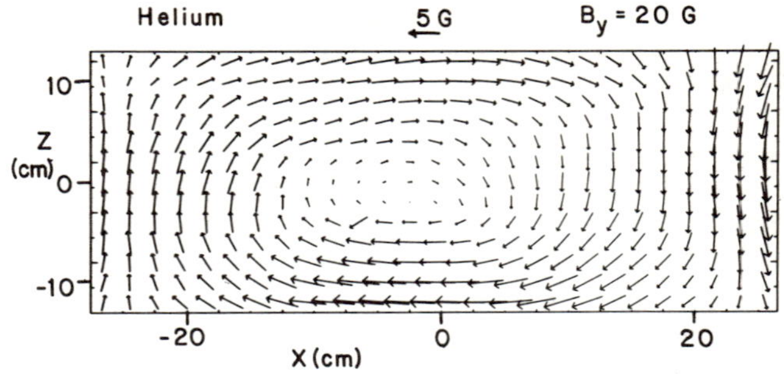

Fig. 4. A large magnetic island is formed at $t \simeq 160$ μsec when the applied plate current vanishes and the induced plasma current determines the field topology. Helium, 3×10^{-4} Torr, $B_{yo} = 20$ G.

vacuum the separatrix intersects at right angles, in the plasma it has two contact points joined by a common line along which $B = 0$. This neutral sheet (in 3 dimensions) predicted to develop self-consistently (Dungey, 1958) is observed in our experiment and shown in Fig. 2. The sheet of length to thickness ratio $\Delta x/\Delta z > 10$ is found to be stable against tearing for over 4 Alfvén transit times from metal plates to neutral sheet. The thickness Δz, defined by the half width of the current sheet $J_y = (\nabla \times \vec{H})_y$ approaches the inertial limit ($\Delta z \approx 3\ c/\omega_{pe}$).

While in an argon plasma the ions are essentially unmagnetized we have also used hydrogen and helium plasmas at larger field strengths where the ion Larmor radius is small compared with the transverse plasma dimensions. Neutral sheets are also observed with light ions. A major influence on the magnetic field topology comes from the axial magnetic field component B_{yo}. As shown in Fig. 3, for $B_{yo} \lessapprox \langle B_\perp \rangle$ neutral sheets evolve, but for $B_{yo} \gg \langle B_\perp \rangle$ magnetic islands form during the period where the external current rises. In the transition region a close inspection of the neutral sheet often shows the presence of multiple small X-and O-points along the X-axis in the region where $B_\perp \approx 0$.

Different field topologies arise after the applied current has reached its maximum and begins to decrease. Then the induced electric field, $-\partial A_y/\partial t$, changes sign and drives plasma currents in the same direction as the applied current. For a brief period a neutral sheet in vertical direction z is formed. It slows down the return of the stored common flux in contrast to the initial horizontal neutral sheet which inhibited the build-up of common flux. However, as the external current (or common flux) decreases the field topology is more and more determined by the plasma current alone. The vertical sheet tears and forms magnetic islands. The islands coalesce and grow into a single large O-point at a time when the applied current crosses zero. Figure 4 shows the field configuration at t = 160 μsec which displays the O-type neutral point topology.

The complete time evolution of the magnetic field topology has been displayed in the form of a computer generated movie. It displays real data and is not a simulation. Further discussions of the field topologies are given in a recent publication (Stenzel and Gekelman, 1981).

There is also considerable interest in the properties of the magnetic fluctuations which are observed but not displayed in the above figures since they represent ensemble averages. At a given position the magnetic field, $\vec{B}_\perp(t)$, from an individual shot is displayed in polar coordinates tracing out the trajectory of the vector tip in time (Fig. 5a). Fluctuations are observed to form multiple loops in the vector display (Fig. 5b) characterizing them as propagating wave phenomena, presumably of MHD nature ($\omega_{ci} < \omega << \omega_{ce}$). Superposition from many shots (Fig. 5c) indicates the randomness of the $\vec{B}$-field oscillations which can be evaluated statistically.

With the help of our on-line digital data analysis we have evaluated at every position and time the fluctuation in magnitude B_{rms} and angle θ_{rms} from ensembles of 25 shots. At a given time during the current rise the fluctuation level vs. position is shown in Fig. 6. In the neutral sheet region the fluctuations maximize reaching $B_{rms}/B_{ave} = O(1)$ and $\theta_{rms} = O(\pi/2)$. Temporally, the turbulence level is high during the formation of neutral sheets and tearing.

Correlation studies between fluctuating magnetic fields observed with two probes at different positions are in progress. From correlation lengths and fluctuation levels we will be able to establish the statistical properties of random magnetic islands formed in dynamic reconnection processes. Such processes are clearly relevant to magnetic substorm phenomena (McPherron, 1979). Satellite data of magnetic fields in the magnetotail show a high level of turbulence (Coroniti, et al., 1977). But a comprehensive picture of $\vec{B}(\vec{r},t)$ like in a laboratory plasma cannot be obtained in space.

Plasma Dynamics

The plasma parameters have been measured with the use of rapidly swept Langmuir probes inserted into the plasma without causing noticeable perturbations. At seven separate times within the applied magnetic field pulse the plasma potential ϕ_p, electron density n, and temperature T_e are determined on-line from computer evaluation of ten shot ensemble averages. This is done at several hundred positions in the transverse x-z plane typically involving the evaluation of about 20,000 Langmuir traces in one data run.

The fluid or ion drift velocities are measured with miniature orthogonal pairs of plane particle detectors collecting ion saturation currents in ±x, ±y, and ±z direction. From the differential

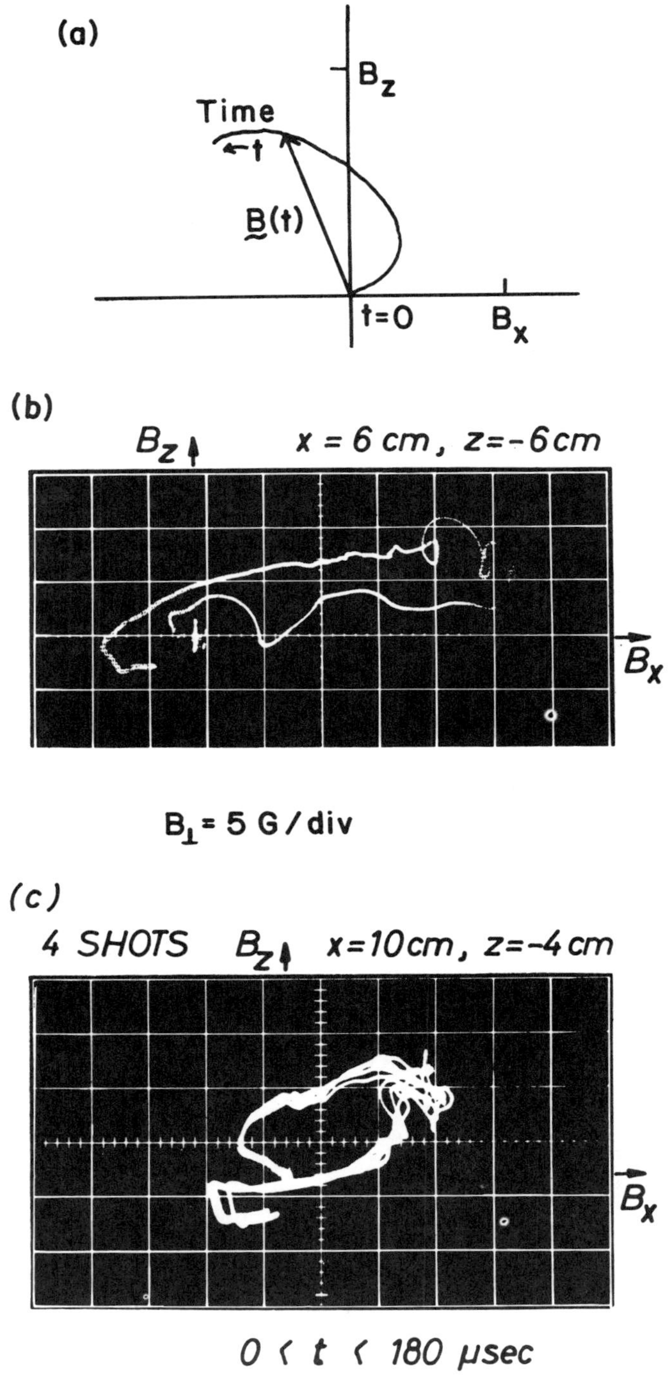

Fig. 5. Trajectories of the magnetic field vector $\vec{B}_\perp$ in time at a given position. (a) Schematic sketch of trajectory in time. (b) Observed trajectory during one experimental shot. Fluctuations appear as loops in $\vec{B}_\perp(t)$. (c) Superposition of trajectories from four individual shots indicating growth and decay of magnetic fluctuations. Helium, B_{yo} = 20 G, $B_\perp \simeq$ 5 G/div.

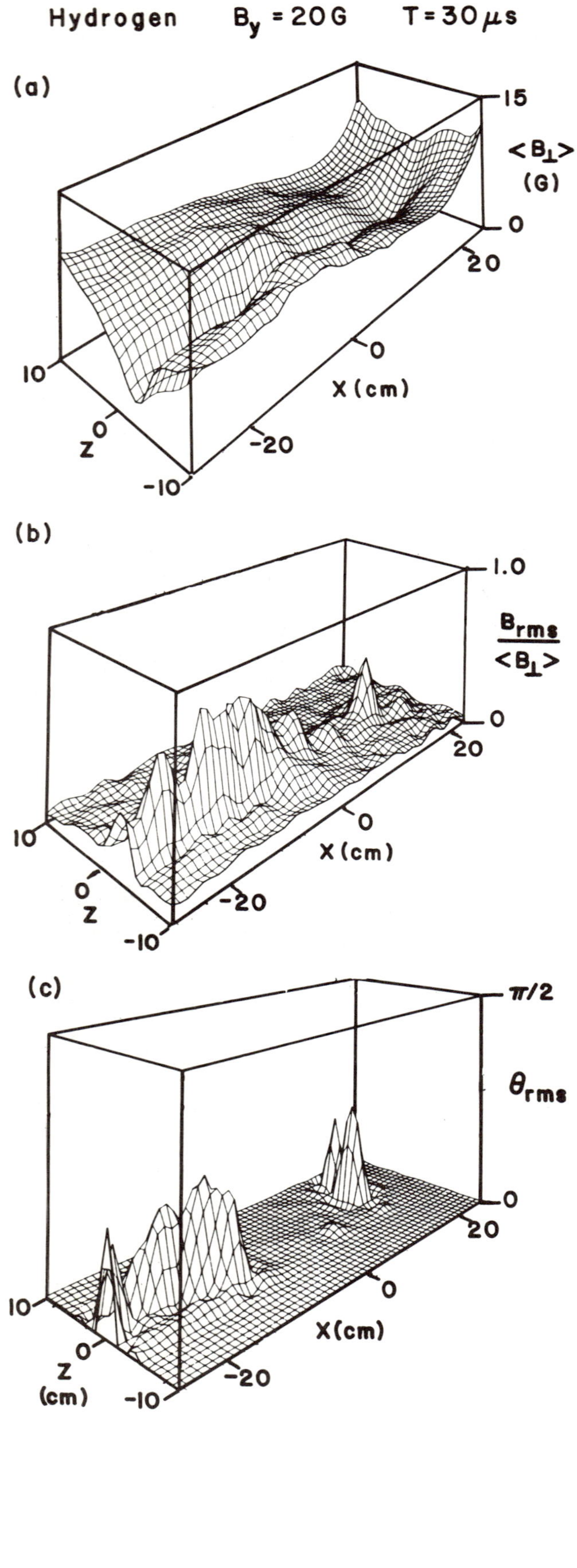

Fig. 6. Statistical properties of $\vec{B}_\perp (x,z)$ at t = 30 μsec evaluated from 25 shots per position. (a) Mean value of the field strength $<|\vec{B}_\perp|>$. (b) Normalized root-mean-square of fluctuations around the mean, $B_{rms}/<B_\perp>$. (c) Fluctuations in the angle of the field vector, θ_{rms}.

and average ion currents one obtains the relative ion drift velocity vector, $\vec{v}_d/C_s$ which, analogous to the magnetic field is recorded digitally vs. time and position. Combined with the temperature measurements we obtain a quantitative measurement of the fluid flow field, $\vec{v}_d(\vec{r},t)$.

Having measured currents, magnetic fields and plasma parameters we can calculate the force density on the fluid, $\vec{J}x\vec{B} - \nabla nkT_e$, from fluid theory. The results will be compared with the fluid acceleration, mn $\partial\vec{v}/\partial t$, calculated from the independent velocity measurements. This will form the basis for understanding the plasma dynamics involved in reconnection events (Gekelman et al., 1980).

Figure 7 shows the essential terms involved in the force density. The magnetic force density $\vec{J}x\vec{B}$, has two contributions due to (i) axial currents and transverse magnetic fields ($\vec{J}_y \times \vec{B}_\perp$, shown in Fig. 7a) and (ii) transverse currents and axial fields ($\vec{J}_\perp \times \vec{B}_y$). Although in the second term the currents are small the magnetic field B_{yo} is large so that the products in (i) and (ii) are comparable. The plasma pressure, $p \simeq nkT_e$, shown in Fig. 7b exhibits, surprisingly, a maximum not in the middle of the neutral sheet, but two peaks at the edges of the sheet near the two contact points of the separatrix. These are largely caused by temperature peaks, ($kT_{e\ max} \simeq$ 15 eV in A, 30 eV in He). The initial temperature ($kT_e \simeq 5$ eV) and density were highly uniform throughout the plane. Taking the spatial derivative of the plasma pressure and combining it with magnetic force density yields the net force density $\vec{f}_\perp$ shown in Fig. 7c.

The measured flow velocity at t = 80 μsec is shown in Fig. 8a. It exhibits the classical fluid motion across the separatrix, i.e. slow inflow along ±z direction toward the neutral sheet and faster outflow in ±x direction ($v_{d\ max} \simeq 0.5\ v_A$, where v_A is the local Alfvén speed). At earlier times the flow field is more complicated exhibiting vortices and flows in unexpected directions. From the time derivative of the flow velocity and the density data the fluid acceleration is obtained and shown in Fig. 8b.

Comparison between force and momentum change show that the fluid does not move as predicted by simple fluid theory but that an effective force term $\Delta\vec{f}$ must be added to account for the observed smaller acceleration, mn $\partial\vec{v}/\partial t = \vec{f} - \Delta\vec{f}$. Postulating that the force term represents a drag exerted by fluctuating fields on the ion motion we can write $\Delta\vec{f}$ = mn $\nu^*\vec{v}$ where ν^* is an effective collision frequency between waves and particles.

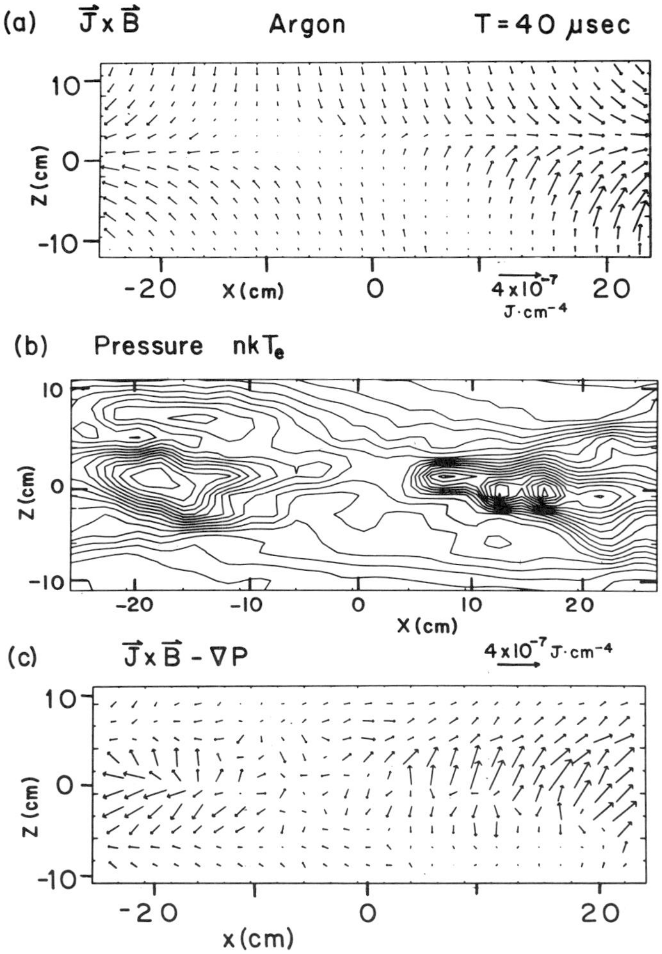

Fig. 7. Physical quantities determining the fluid dynamics. (a) Vector field of the magnetic force density $\vec{J}\times\vec{B}$. (b) Contours of constant plasma pressure, $p = nkT_e$. Maximum at $(x,z) = (-2,16)$ cm is $p_{max} = 6.4 \times 10^{-6}$ J cm^{-3}; average $\langle p\rangle = 2.1 \times 10^{-6}$ J cm^{-3}; contour spacing 3×10^{-7} J cm^{-3}. (c) Vector field of resultant force density $\vec{J}\times\vec{B} - \nabla p$. Argon, 2×10^{-4} Torr, t = 40 μsec.

Since the drag force $\Delta\vec{f}$ and fluid velocity $\vec{v}$ are, in general, not parallel the collision frequency is a tensor quantity. Physically, such scattering properties can be expected from anisotropic wave turbulence. The characteristics of the turbulence spectrum are still under investigation so that the detailed physics of the scattering process is not yet clear. Nevertheless the measurements of $\Delta\vec{f}$ and $\vec{v}$ permit us to determine the diagonal ($\nu^*_{xx} = \nu^*_{zz}$) and off-diagonal ($\nu^*_{xz} = -\nu^*_{zx}$) tensor components which are displayed in Fig. 9. The scattering rate is largest in the neutral sheet region with peak values approaching the ion plasma frequency. In time the collision rate varies as the plasma current does.

The present observations show that fluctuations are a very important part of the reconnections problem. They are not just a perturbation in the plasma dynamics but a first order parameter.

Electric Fields, Currents and Resistivity

One of the critical questions in the problem of magnetic field line reconnection is the plasma resistivity which determines dissipation and size of the diffusion region. Previously published results of resistivity measurements (Baum et al., 1973) are not very trustworthy since they were derived from an incomplete measurement of electric fields. In the framework of fluid theory the relation between currents and electric fields is given by the generalized Ohm's law (Delcroix, 1965):

$$\eta\,\vec{J} + \frac{1}{ne}\,\vec{J}\times\vec{B} = \vec{E} + \vec{v}\times\vec{B} + \frac{1}{ne}\,\nabla\,nkT_e \qquad (1)$$

Ohmic and Hall currents are driven by pressure gradients, Hall fields and electric fields which may be rotational (induced) or irrotational

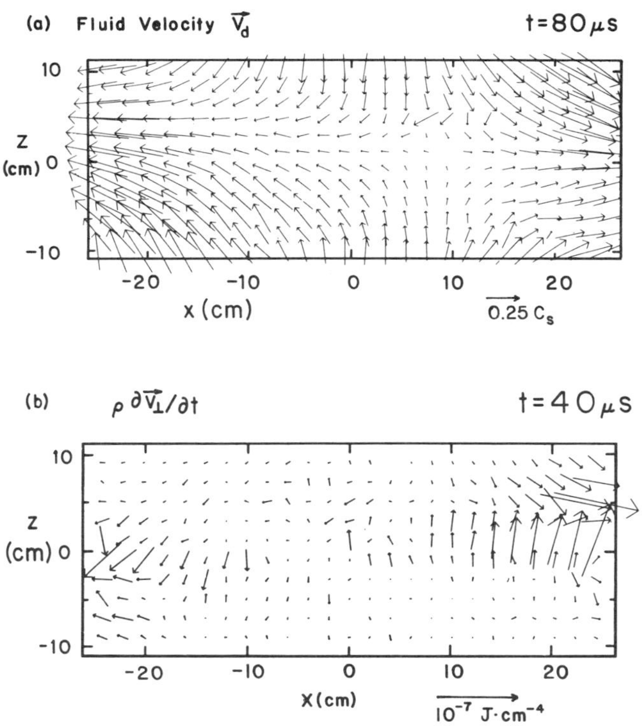

Fig. 8. Properties of the fluid motion. (a) Flow field of the ion drift velocity $\vec{v}_d$ at t = 80 μsec showing the nearly classical fluid flow in a neutral sheet topology. (b) Fluid momentum change, mn $\partial\vec{v}/\partial t$, calculated from time and space resolved density and flow velocity measurements. t = 40 μsec, Argon 2×10^{-4} Torr.

(space charge) fields. By measuring all fields, currents and plasma parameters we have calculated the resistivity from eq. (1).

The electric field measurement is performed as schematically shown in Fig. 10 (Stenzel and Gekelman, 1979b). There is an induced electric field, $\vec{E}_i = -\partial\vec{A}/\partial t$, associated with the applied time-varying magnetic fields, $\nabla \times \vec{E}_i = -\partial\vec{B}/\partial t$. It is directed along the neutral line (separator)

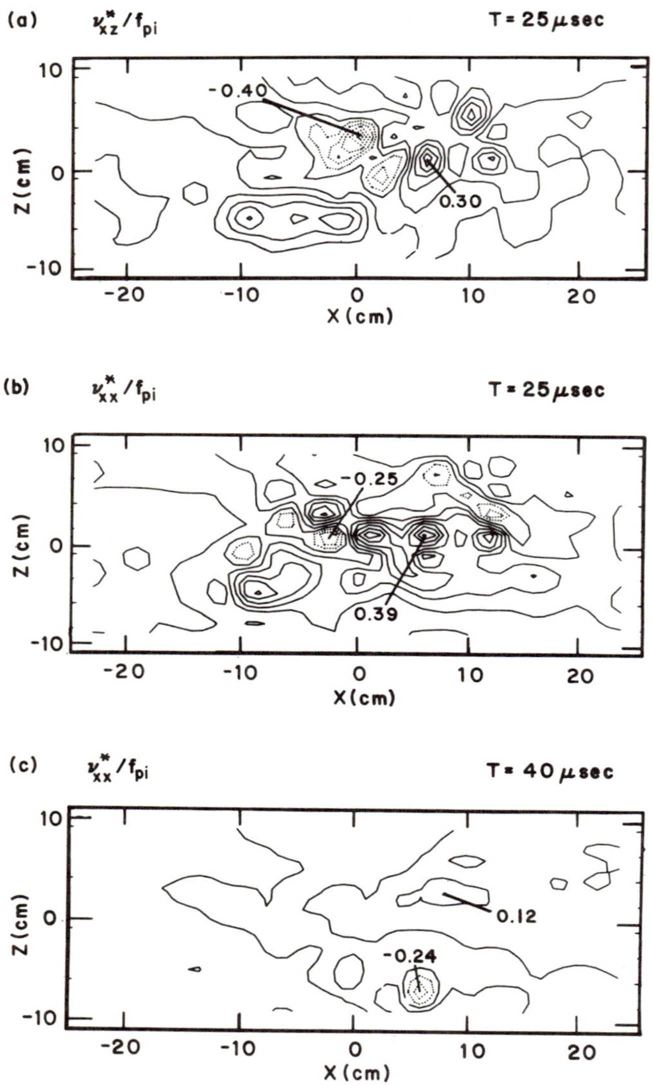

Fig. 9. Contour maps of the anomalous ion scattering coefficients $\underline{\underline{\nu}}^*$ normalized to the ion plasma frequency f_{pi}. (a) Off-diagonal tensor component, ν^*_{xz}/f_{pi}, at t = 25 μsec. Contour spacing 6.25%, solid contours $\nu^*_{xz}/f_{pi} > 0$, dashed contours $\nu^*_{xz}/f_{pi} < 0$. (b) Diagonal tensor component, ν^*_{xx}/f_{pi}, at t = 25 μsec. (c) Diagonal component at t = 40 μsec showing the decay of the scattering coefficient in time.

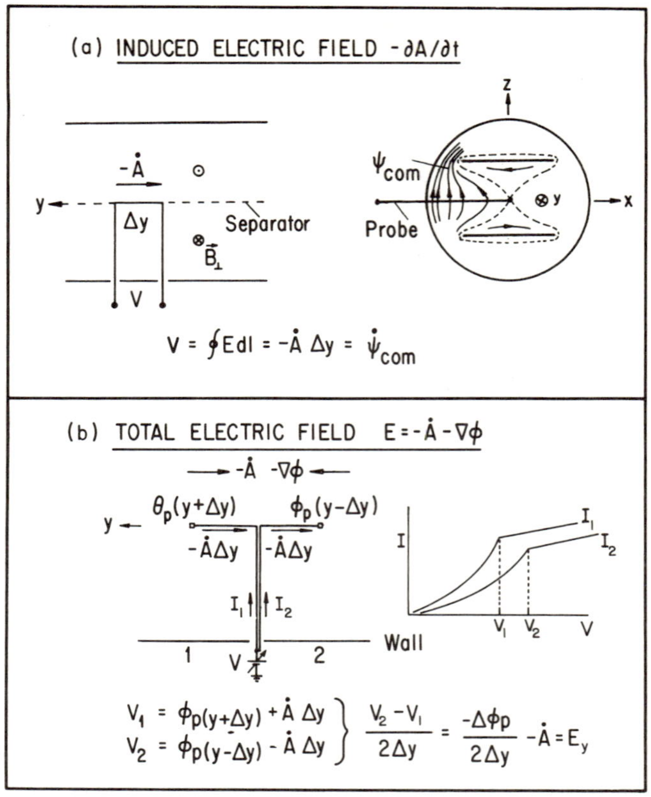

Fig. 10. Measurement methods to determine (a) the induced (rotational) electric field $-\partial A/\partial t$, (b) the total electric field E consisting of induced and space charge $(-\nabla\phi)$ fields.

and is relatively constant between the metal plates since most of the common flux ψ_c is concentrated between the plates and the wall. The induced electric field along the neutral line, is measured with a loop probe shown in Fig. 10a which links the time-varying common flux over a length Δy so as to develop an open-loop voltage $V = -\dot{A}_y\,\Delta y$.

Measurements of the plasma potential with radially inserted Langmuir probes reveal a strong axial gradient toward the cathode. Thus, by space charge separation, an electrostatic field, $\vec{E}_s = -\nabla\phi$, develops which is directed opposite to the applied inductive field and nearly cancels it. The resultant small net electric field is best measured with a differential probe shown in Fig. 10b. It consists of two radially inserted Langmuir probes which are separated axially by short distances $\pm\,\Delta y$. Both identical probes are swept with the same fast ramp voltage V(t) and their currents are recorded simultaneously. The knee in the two probe characteristics is observed at voltages V_1 and V_2, respectively, which differ not only by the local plasma potentials but also by the voltages $\pm\,\dot{A}\,\Delta y$ induced on the insulated axial wire sections. The voltage

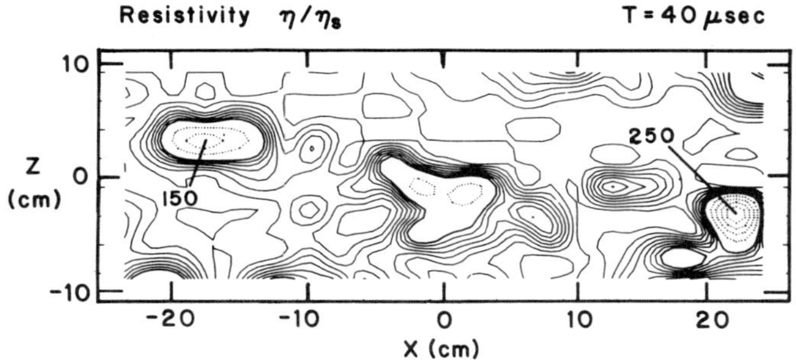

Fig. 11. Contour map of the plasma resistivity normalized to the classical Spitzer value, $\eta(x,z)/\eta^s(x,z)$, at t = 40 μsec. The resistivity is calculated from the generalized Ohm's law with all electric field and current components measured. Contour spacing 20 (solid), 40 (dashed).

difference per axial length yields the net electric field, $\Delta V/2\Delta y \simeq -\nabla\phi - \dot{A} = E_{net}$. This instantaneous difference measurement is far more accurate than subtracting the two field contributions from separate measurements. We also found it important to obtain the space charge field from gradients of the plasma potential and not the floating potential. The latter method gives different erroneous results due to the presence of fast electrons and drifting ions.

The Hall electric field, $\vec{v}\times\vec{B}$, is obtained from the measured fluid velocity and magnetic field. Its contribution is small. Pressure gradients are calculated from electron density and temperature data. All driving terms on the right hand side of eq. (1) are combined to form the total electric field, $\vec{E}_{tot}$.

The current density $\vec{J}$ is either measured directly with differential Langmuir probes or calculated from the magnetic field data, $\vec{J} = \nabla\times\vec{B}/\mu_o$. Both results are in good agreement which permits us to choose the latter one for simplicity.

By dividing the total electric field component $\vec{E}_{tot}||\vec{J}$ and the current density the resistivity is obtained, $\eta = \vec{E}_{tot}\cdot\vec{J}/J^2$. Figure 11 shows the resistivity normalized to the classical Spitzer resistivity in the x-z plane at a time t = 40 μsec when the neutral sheet is present. The resistivity is highly inhomogeneous with average values $\langle\eta/\eta_s\rangle \simeq 25$ and peak values $(\eta/\eta_s)_{max} \simeq 250$. This anomalous resistivity indicates again that a high level of microturbulence is generated by the induced currents in the plasma. Conditions for current-driven ion sound instabilities are satisfied but detailed turbulence studies have not yet been performed.

In earlier reconnection experiments Bratenahl and Baum (1976) have claimed that the onset of anomalous resistivity causes current disruptions and "implusive flux transfer events" (IFTE). Their resistivity calculation is based only on the inductive electric field component and J_y, hence incomplete and erroneously high. Recently, Beeler (1979) repeated the electric field measurements according to our method and found the near cancellation of inductive fields by electrostatic fields during IFTE, hence no anomalous resistivity. Thus, IFTE are caused by other mechanisms than proposed by the authors, the simplest being the loss of plasma of an expanding double inverse pinch. In our dc discharge plasma current disruptions are not observed.

Energy Flow

The conversion of electromagnetic energy into kinetic particle energy is the most important aspect of the magnetic field line reconnection problem. We have analyzed the energy flow in our experiment by considering a control volume inside the plasma and calculated for it both inflow and outflow of the Poynting vector $\vec{E}\times\vec{H}$, the internal dissipation $\vec{E}\cdot\vec{J}$, and the time rate of change of the stored field energy, $\partial/\partial t\ (B^2/2\mu_o)$. These calculations do not only show consistency of the independent measurements with the continuity of power flow given by:

$$\nabla\cdot(\vec{E}\times\vec{H}) = -\vec{E}\cdot\vec{J} - \frac{\partial}{\partial t}(B^2/2\mu_o) \quad (2)$$

but also indicate the fraction of energy dissipated in the plasma. The temporal variation of the terms in eq. (2) are shown in Fig. 12. The applied power vanishes at t = 0 and t ≃ 80 μsec since either currents or voltages have zero crossings. The power densities are integrated over the transverse plane (20 x 50 cm^2) and expressed

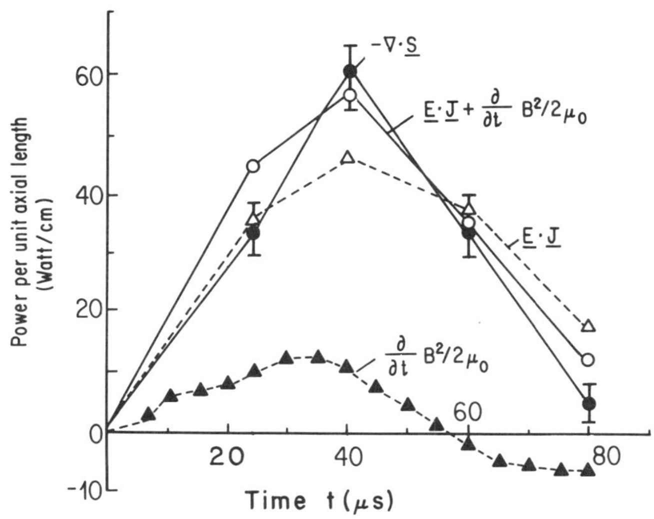

Fig. 12. Temporal variation of the different power densities involved in the energy flow equation (2). Note that all quantities are integrated over the x-z cross section and expressed as power per axial length y. The independently measured terms confirm the conservation of energy flow and show the high conversion rate from electromagnetic to kinetic energy.

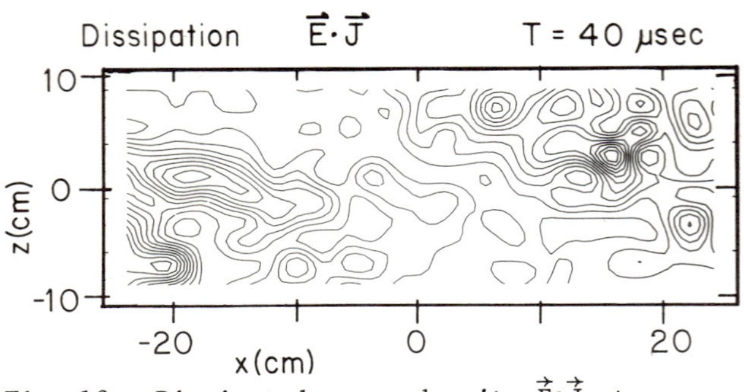

Fig. 13. Dissipated power density $\vec{E}\cdot\vec{J}$ at $t = 40$ μsec. The dissipation does not maximize in the center of the neutral sheet but near its edges where the energy density nkT_e peaks (see Fig. 7b). Contour spacing 20 mW/cm^3, maximum at $x = 16$ cm, $z = 3$ cm is $(\vec{E}\cdot\vec{J})_{max} = 215$ mW/cm^3, average over x-z plane $\langle\vec{E}\cdot\vec{J}\rangle = 46$ mW/cm^3.

as power per axial length y. Most interesting is the fact that for the first 50 μsec about 80% of the electromagnetic energy is dissipated in the plasma and only 20% goes into the build-up of magnetic field energy. After $t \simeq 60$ μsec, even though the external current is still rising the internal magnetic field energy density decreases, thereby sustaining an enhanced level of dissipation in the last 20 μsec.

The dissipation $\vec{E}\cdot\vec{J}$ goes largely into electron heating rather than into directed kinetic particle energy as verified by comparing $1/2\ nm\ v_d^2$ with $nk\Delta T_e$. Even though electron heating is the dominant dissipation mechanism the temperature rise is limited by inelastic processes. In the first 20 μsec there is enhanced ionization up to neutral burn-out, and there is constantly atomic excitation and line radiation in inert gases. Electrons also convect heat to the ends of the device. Knowing the power densities required for producing a discharge plasma of given nkT_e we find that the additional dissipation due to the reconnection process is consistent with the observed increase $\Delta(nkT_e)$. The spatial distribution of the dissipation, shown in Fig. 13 exhibits maxima at approximately the same locations where the plasma energy density, nkT_e, peaks (see Fig. 7b).

Turbulence due to microinstabilities is thought to be responsible for the effective energy conversion. However, the simplified picture from computer simulations (Sato et al., 1978) that the resistivity scales with current hence the dissipation peaks in the current sheet $[\vec{E}\cdot\vec{J} = \eta_{(J)}\ J^2]$ is not confirmed by our observations. Due to the complexity of the reconnection region a detailed, careful instability analysis still has to be performed.

Conclusion

This ongoing laboratory experiment has shown far more detailed data on reconnection events than previous observations. The advantages in terms of diagnostics, flexibility, reproducibility, time and cost of a laboratory experiment over a satellite mission are obvious. Of course, the plasma parameters and boundary conditions are different in both cases which have to be carefully considered in any comparison. In spite of these differences the basic physics of the reconnection processes in the laboratory and the magnetotail should be similar.

The major new observations in the present experiments are the following: (a) Formation of a long thin neutral sheet stable for many Alfven times; (b) A strong magnetic field component along the separator results in magnetic island formation; (d) A high level of magnetic turbulence exists in the null regions; (d) Anomalous ion scattering is inferred from measuring the fluid dynamics; (e) The classical ion flow pattern is established at late times or low levels of turbulence; (f) Strong nonuniform electron heating is observed at the edges of the neutral sheet but not in the center; (g) Pressure gradients give rise to transverse currents J_x, J_z; (h) An electrostatic field develops in the direction opposite to the applied inductive field such that the net electric field in the plasma is greatly reduced; (i) Careful resistivity measurements indicate enhancements over the classical value by one to two orders of magnitude; (j) The anomalous resistivity leads to an efficient dissipation of magnetic field energy into electron heating; (k) The energy input into fast runaway ions and electrons is negligible compared with the increase in thermal energy.

Acknowledgments. The authors wish to thank Mr. T. McFadden for his help with the digital data acquisition system, Mr. M. Urrutia for general assistance in the work, and Dr. F. Coroniti for helpful discussions.

This work was supported by the Division of Atmospheric Sciences, National Science Foundation, under grant ATM79-21331 and the Division of Atomic, Molecular and Plasma Physics, National Science Foundation, under grant PHY79-21387.

References

Baum, P. J., A. Bratenahl, M. Kao, and R. S. White, Plasma Instability at an X-type Magnetic Neutral Point, Phys. Fluids, 16, 1501, 1973.

Beeler, R. G., "The Impulsive Flux Transfer Event in the Double Inverse Pinch Device", Ph.D. dissertation, University of California, Riverside, 1979.

Bratenahl, A., and P. J. Baum, Impulsive Flux Transfer Events and Solar Flares, Geophys. J. R. A. S., 46, 259, 1976.

Coroniti, F. V., F. L. Scarf, L. A. Frank, and R. P. Lepping, Microstructure of a Magnetotail Fireball, Geophys. Res. Lett., 4, 219-222, 1977.

Delcroix, J. L., Plasma Physics, p. 238, Wiley, London, 1965.
Dungey, J. W., Cosmic Electrodynamics, Cambridge U. Press, London, 1958.
Gekelman, W., and R. L. Stenzel, Magnetic Field Line Reconnection Experiments. 2. Plasma Parameters, J. Geophys. Res., 86, 659-666, 1981.
Gekelman, W., R. L. Stenzel and N. Wild, Magnetic Field Line Reconnection Experiments. 3. Ion Acceleration, Flows and Anomalous Drag, UCLA Plasma Physics Group Report PPG 479, 1980.
McPherron, R. L., Magnetospheric Substorms, Rev. Geophys. Space Phys., 17, 657-681, 1979.
Sato, T., T. Hayashi, T. Tamao, and A. Hasegawa, Confinement and Jetting of Plasmas by Magnetic Reconnection, Phys. Rev. Lett., 41, 1548-1551, 1978.
Stenzel, R. L., and W. Gekelman, Experiments on Magnetic Field Line Reconnection, Phys. Rev. Lett., 42, 1055-1057, 1979a.
Stenzel, R. L., and W. Gekelman, Electrostatic and Induced Electric Fields in a Reconnection Experiment, UCLA Plasma Physics Group Report PPG 414, 1979b.
Stenzel, R. L., and W. Gekelman, Magnetic Field Line Reconnection Experiments. 1. Field Topologies, J. Geophys. Res., 86, 649-658, 1981.
Vasyliunas, V. H., Theoretical Models of Magnetic Field Line Merging 1, Rev. Geophys. Space Phys., 13, 303, 1975.

ANOMALOUS RESISTIVITY AND PLASMA DYNAMICS

Christian T. Dum

Max-Planck-Institut für Physik und Astrophysik
8046 Garching, W.-Germany

Abstract. The concept of anomalous resistivity and the calculation of anomalous transport from turbulent wave spectra is reviewed. Detailed investigations always reveal a complex interplay between microscopic and macroscopic dynamics, rather than a quasi-static enhancement of resistivity. This is especially true for turbulence associated with field aligned currents, due to electron runaway and formation of potential structures. The dynamics of these processes is discussed for ion sound, ion cyclotron and Buneman instabilities and related turbulence, with the aim of developing methods for a realistic incorporation of the microscopic physics into global models of plasma evolution.

Introduction

There are so many papers related to anomalous resistivity that they could not even be adequately listed here. With the importance of the subject thus being proven and after much limiting the range of topics, the number of relevant references for a particular point still had to be restricted to the most recent papers which also contain extensive references to important earlier work [see preamble in Papadopoulos, 1977]. This paper primarily examines the coupled microscopic and macroscopic processes connected with anomalous resistivity, with the aim of developing methods for a realistic description of the complete system dynamics. Anomalous resistivity is usually stated in terms of an Ohm's law

$$E = \eta j = -(m/|e|)\nu_{eff} u \qquad (1)$$

where the resistivity $\eta = m\nu_{eff}/ne^2$ increases, often dramatically, for current densities j or drifts $u = -j/n|e|$ above a critical threshold. Ohm's law derives primarily from metallic conductors and most often is even less adequate for anomalous resistivity then it is already for classical resistivity due to Coulomb collisions. The applied electric field E or drift is a negligible perturbation to a metallic conductor, typically $u/v_e \approx 10^{-9}$ (Cu, 20 A/mm^2), where v_e is the effective thermal velocity. In the case of anomalous resistivity much larger drifts, $u/v_e \lesssim O(1)$ are required in the first place for the excitation of instabilities. The fluctuations must then grow far above the thermal level, unlike classical resistivity where the scattering agent preexists and remains nearly unchanged by the current [Papadopoulos, 1977]. A quasi-static and local enhancement of classical resistivity is thus rather unlikely. It is more likely to have some periods of nearly free acceleration and a strong coupling of microscopic processes with the macroscopic dynamics and external boundary conditions. Turbulent heating experiments may show repeated selfquenching of the instability that can be modified by changing external inductance and capacitance [Robertson et al., 1973], cf. Figure 1 a. The turbulent spectrum during a single pulse evolves in a complicated way, involving several instability modes, such as the dominant ion sound, ion cyclotron waves, plasma waves, and initially possibly also the Buneman instability [Wharton et al., 1971, and related papers in that volume], cf. Figure 1 b. Despite the many experimental efforts in approaching the ideal conditions implied by (1) no such universal law in which resistivity depends only on the local plasma parameters and the current or electric field has ever emerged. Moreover, ν_{eff} often refers to the maximum resistivity during a heating pulse in which plasma parameters change by orders of magnitude. If attempts are made at all to express results in terms of dimensionless parameters then this is most often done in terms of ν_{eff}/ν_{cl} and E/E_D, where ν_{cl} and E_D are respectively the collision frequency and runaway field for Coulomb collisions. Demonstrating, as desired, that both parameters can rise several orders of magnitude above unity, also implies that they are totally irrelevant for fully developed turbulence. In the many devices with field aligned current an understanding of temperature increases, often by several orders of magnitude, could never be entirely divorced from the questions as to as to what extent current and turbulence were due to a small group of runaway electrons and to what extent electric field and current density were uniform. In linear devices typically 80 % of the supplied energy is carried towards the

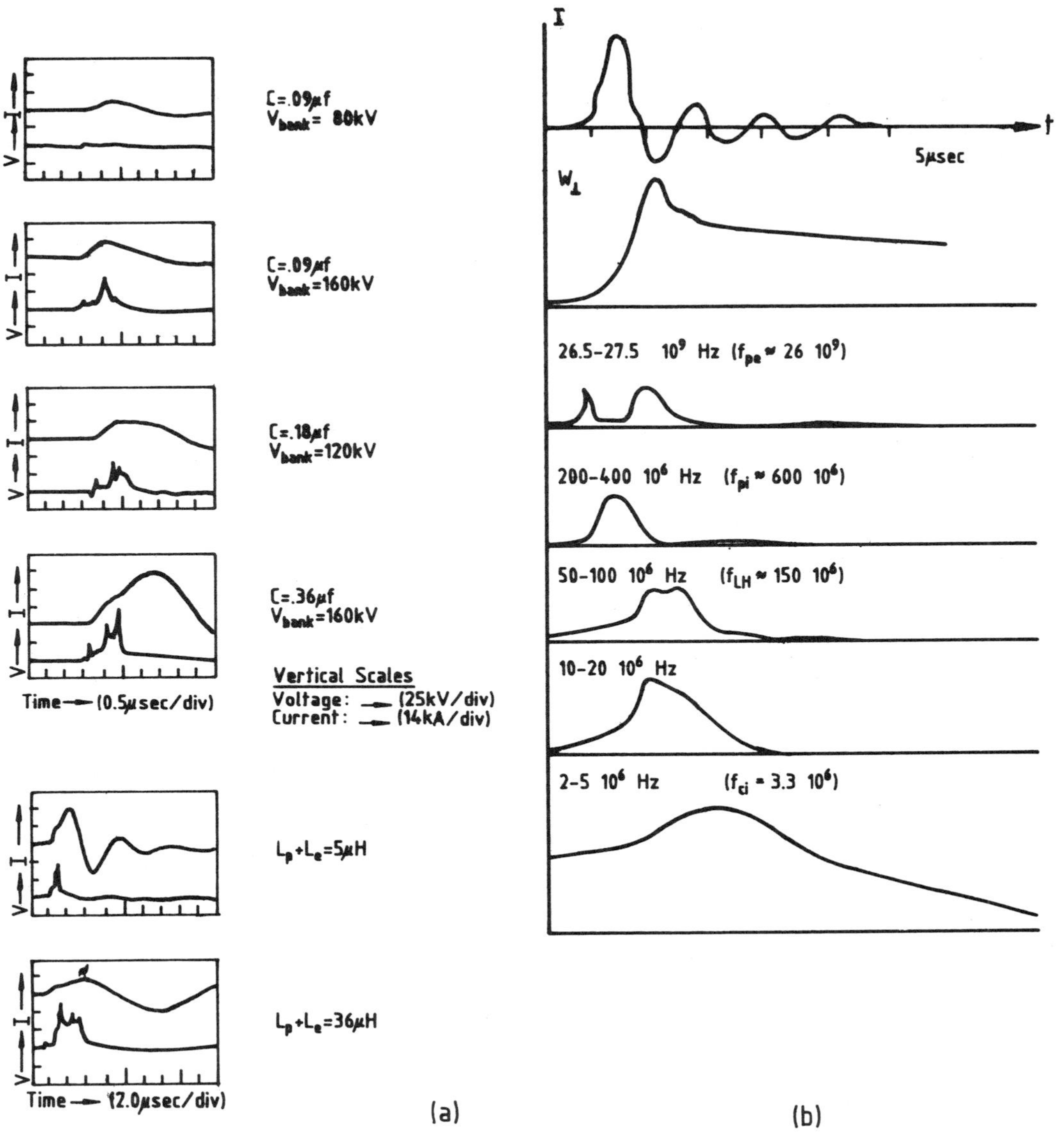

Fig. 1. Bursts of anomalous resistivity in a turbulent heating experiment [after Wharton et al., 1971, and Robertson et al., 1973].

a) Dependence of current and voltage across plasma on the external circuit. $V_{bank} = V + R_e\ I/L_e\ dI/dt$; $V = R_p I + d(L_p I)/dt$. Note the repeated selfquenching.

b) Current, perpendicular thermal energy, and turbulent wave spectrum during a single burst (schematically, relative units).

anode. Very energetic x-rays and high frequency radiation are often present, signaling acceleration of runaway electrons. Sometimes, macroscopic instability connected with the enhanced ν_{eff} abruptly increases these effects, terminates heating, violantly redistributes the current and spontaneously creates double layer potential structures [Kalinin et al., 1971], cf. Figure 2. These processes are obviously very undesirable for turbulent heating but may be of interest to particle acceleration along auroral field lines or in solar flares. Laminar double layers and other mechanisms for creating potential drops and acceleration have been favored by some authors, mainly on the account that anomalous resistivity produces too much local dissipation [Fälthammar, 1977]. However, the electrostatic potential structures that have recently been observed in a number of experiments, simulation studies and along auroral field lines are generally not laminar but turbulent. In some cases it may be shown that turbulence actually plays a crucial role in establishing these structures [cf. contributions to this volume and also Iizuka et al., 1979; Hollenstein et al., 1980; De Groot et al., 1977; Sato and Okuda, 1980; Hudson and Mozer, 1978]. In view of the complicated coupled dynamical processes that can arise for field aligned currents and potential drops it is necessary to be open minded about possible mechanisms, depending much upon the status of microscopic and macroscopic diagnostics. This still applies to resistivity along auroral field lines, even after the most recent observations [Mozer et al., 1980]. Instead of yet another resistivity formula or double layer mechanism a concurrent development of experiments, microscopic theory and numerical transport codes for the global problem is required. Perpendicular shocks, perhaps, provide the best example for such a development. It started with the observation that shock widths and dissipation could not be explained by Coulomb collisions.

Weak Turbulence, Nonlinear Theories And Marginal Stability

Much emphasis has been put on determining the maximum critical drift velocity for instability as a function of the electron (or ion) thermal velocity and various dimensionless parameters such as the ratio between electron and ion temperatures T_e/T_i, plasma (ω_e) and gyro frequency (Ω_e) etc. [e.g. Kindel and Kennel, 1971]. In a range $u_c < u < u_s$, near this threshold it appears plausible that due to nonlinear effects, the fluctuation spectrum saturates, giving rise to an Ohm's law

$$\frac{\nu_{eff}}{\omega_e} = \alpha A \frac{u-u_c}{v_e} = \frac{\alpha A}{2} [(1+4\hat{E}/\hat{E}_A)^{1/2} -1] \frac{u_c}{v_e} \quad (2a,b)$$

where α relates fluctuation level W/nT_e and ν_{eff} with $\alpha = O(1)$ for ion sound. Equation (1) was used in (2b) with $\hat{E} = |e| E/mv_e\omega_e$ and $\hat{E}_A = A(u_c/v_e)^2$ in terms of the critical drift velocity u_c and A, which describes nonlinear saturation. Ion sound turbulence provides the best known theory of this kind [Galeev and Sagdeev, 1979] with nonlinear ion Landau damping as the saturation mechanism. In this case $A = A_o T_e/T_i$ where A_o was originally estimated by Sagdeev in 1967 as 10^{-2}. In experiments, such as perpendicular shocks, however, A_o varies by orders of magnitude, as a function of time [Eselevich et al., 1971] or electric field [Galeev, 1976]. Since all elements of the theory, Landau damping by beat waves included, have been tested independently in a number of experiments [Stenzel, 1978], the many attempts on improving over the original theory that are still made today do not seem to relate to a pressing problem. The problem really is that it is assumed that the shape of the distribution functions, Maxwellian, and the plasma parameters are somehow held fixed as the waves grow to saturation. It is worthwhile to examine this point in detail, since almost without exception or justification the first or both of these assumptions are also made in other nonlinear theories of turbulence and anomalous resistivity. In many experiments involving ion sound the ratio u/v_e approaches a constant value which is much above instability threshold for Maxwellians but is in reasonable agreement with $u_c/v_e \sim (m/M)^{1/4}$ for high energy ion tails [e.g. Eselevich et al., 1971]. Galeev [1976] discusses an Ohm's law for ion sound in which he assumes this result to hold for small E, followed by the nonlinear regime $u \gg u_c$ and a strong current, $u > u_s$, regime which eventually is governed by the Buneman instability. The general relation (2) combines the first two regimes. A mere observation of increasing ν_{eff} with increasing current or applied voltage, as in many turbulent heating experiments, is not a verification of nonlinear saturation or (2), as is sometimes claimed. In addition to a measurement of the local electric field and current density, measurements of the detailed particle distribution functions would be required in order to determine how far the system really is from the linear instability threshold. Not only that, for given plasma parameters, ion tails increase u_c, but distortions of the electron distribution also may lead to an effective drift u^* for instability and resistivity which differs substantially from the mean velocity $u = -j/n|e|$. More generally, quasilinear theory suggests an approach towards marginal stability $u^* \approx u_c$. The plasma parameters also evolve. For a steady state to exist, as is often assumed, heating must be balanced by losses, in the simplest form by

$$\frac{Ej}{nT_e\omega_e} = \frac{\nu_{eff}}{\omega_e} \left(\frac{u}{v_e}\right)^2 = \frac{\delta}{\omega_e} \quad (3)$$

The loss rate δ and the external circuit then determine the operating point on a resistivity curve corresponding to (2), cf. Figure 3. We can often assume that this point is in the quasilinear regime $u^* \approx u_c$. Supronenko et al. [1973] assume that electron heating and a loss rate $\delta = 2c_s/L$, where $c_s = T_e/M)^{1/2}$ is the sound speed and L the system length, drive the system towards marginal stability $u \approx c_s$. The relation $\nu_{eff} \sim E^{1/2}$ in this case has nothing to do with a similar law which may be derived from (2) for $\hat{E}/\hat{E}_A \gg 1$ or $u \gg u_c$, but results from the increase of u_c with E by the heating! For ion cyclotron waves driven by currents along auroral field lines, Palmadesso et al. [1974] assume that marginal stability or selfquenching is the result of ion heating. In addition to increased u_c due to heating, the drift velocity may also be reduced, e.g. by enhanced magnetic diffusion, as in perpendicular shocks and skin current turbulent heating. A logical consequence of such behavior is, on the one hand, that anomalous resistivity

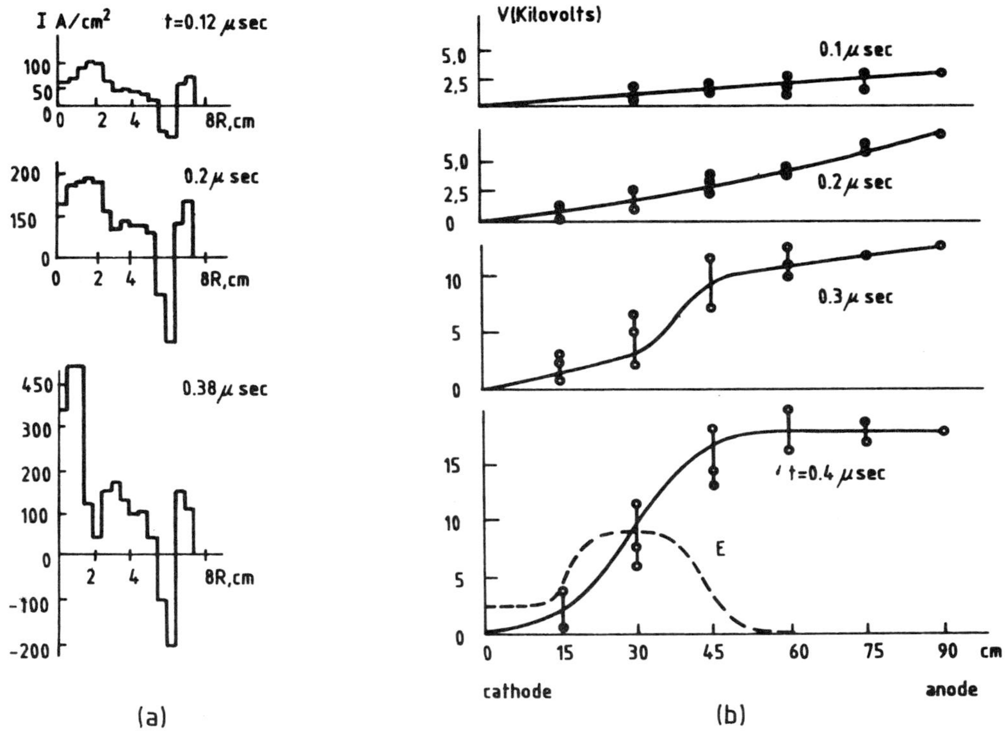

Fig. 2. Abrupt redistribution of the current (a) and formation of a potential double layer (b) by the current convective instability, for a plasma with anomalous resistivity [after Kalinin et al., 1971].

can only be studied in conjuction with macroscopic plasma dynamics and external conditions. On the other hand, one may try to abandon some of the complicated microscopic physics, using only the marginal stability condition $u^* \approx u_c$ as a first step and in the next step, macroscopic fluid equations with an anomalous resistivity of the type (2) and related anomalous transport coefficients which describe heat loss etc. This marginal stability approach to anomalous transport [Manheimer and Boris, 1977] has been used since the early days of plasma physics, with increasing degrees of sophistication and success. The main idea is that due to the selfregulatory effects we discussed, the detailed form of wave growth and saturation does not strongly affect the final macroscopic parameters. Only the fact that above instability threshold, resistivity and other transport processes are drastically modified from their classical values should matter to a first approximation. The validity of this assumption can be tested e.g. by varying arbitrarely but within reasonable limits, the constant A and the cutoff u_s in (2). Kalfsbeek [1978] used various theoretical nonlinear resistivity formulas of the type (2) in establishing the applicability of marginal stability to skin current turbulent heating. By the same token this implies, of course, that the nonlinear theories for ion sound cannot be tested in this case. A calculation of u^* and transport relations for self-consistent non-Maxwellian electrons and u_c for non-Maxwellian ions [Dum, 1978 a,b] proved to be more important in a transport code for perpendicular shocks [Söldner et al., 1977; Chodura et al., 1977] than an actual calculation of spatial [Manheimer and Boris, 1972] or temporal wave growth [Davidson and Krall, 1977] and resistivity with Maxwellian electrons and ions. The effective drift velocity u^* could be much larger than the drift $u = -j/n|e|$ in the shock front. Skewing of the electron distribution by temperature and density gradients is responsible for this effect. The effective drift for instability is related to the transport coefficients for momentum transfer (ρ) and an effective collision frequency (ν^*) by

$$\underline{u}^* = (\underline{\underline{I}} - \underline{\underline{\rho}}_u) \cdot \underline{u} + \frac{v_e^2}{\nu^*} (\underline{\underline{\rho}}_T \cdot \nabla \ell n T_e - \underline{\underline{\rho}}_n \cdot \nabla \ell n\, n) \qquad (4)$$

[see Dum, 1978 a,b] and (14). An exponential switch on -off condition of classical and anomalous transport that is based on u^*/u_c and which simulates wave growth and decay was chosen. Anomalous heat conduction, which is eliminated by assuming Maxwellians, was also very important. Ion dynamics, including ion reflection and ionization, was described by a particle code. Such hybrid codes represent a very powerful tool for studying plasma dynamics [Davidson and Krall, 1977]. A marginal stability condition $u = u_c$ for ion sound and Max-

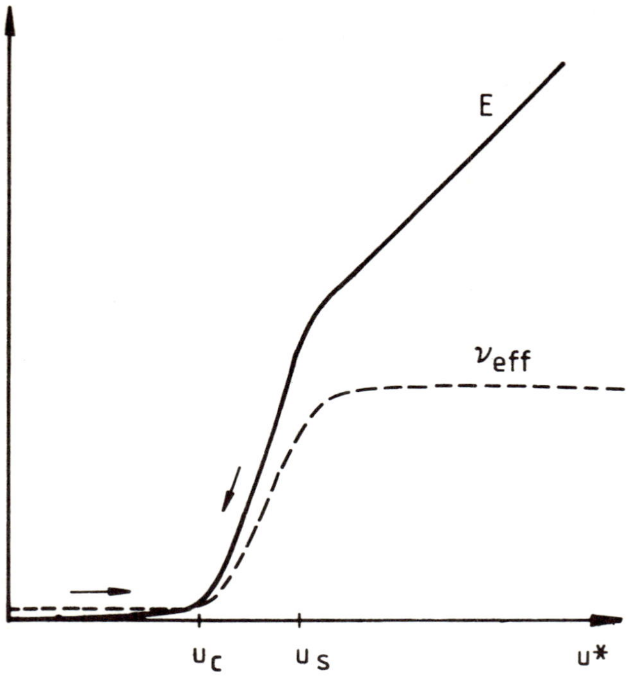

Fig. 3. Possible form of a nonlinear Ohm's law, $E = -(m/|e|)\nu_{eff}\, u^*$. The operating point may be driven towards marginal stability by self-regulatory anomalous transport and quasi-linear effects, often in the form of relaxation oscillations which also depend on the external circuit. Transition to other instability modes and threshold-like wave saturation are also possible. (u^* and u_c are, respectively, the effective and critical drift velocity for instability).

wellian distributions has been applied to the bow shock for a comparison with observed shock widths and ν_{eff} [Manheimer and Boris, 1972; Morse and Greenstadt, 1976; Greenstadt et al., 1978]. For the marginal stability approach it actually is not important that a quasi-steady fluctuation level and resistivity, in the sense of (2), is established. Such a matching of microscopic and macroscopic dynamics is quite unlikely. The instability must only last long enough or be turned on often enough to account on the average for the observed macroscopic effects. A formula (2) may then still be used, even though nonlinear effects may be quite negligible during the wave growth and decay phases. Clearly, individual wave bursts and relaxation oscillations about marginal stability, as observed for turbulent heating (Figure 1 a) may not be described by this method. The short individual bursts of ion sound observed throughout the solar wind [Kurth et al., 1979] most likely also correspond to such relaxation oscillations about marginal stability, but instead of controlling the macroscopic dynamics, as for turbulent heating, correspond to very minute changes in the distribution functions [Dum et al., 1980 a; Dum et al., 1980 b]. Calculation of temporal or spatial growth for sample wave modes is necessary in such cases. One may consider the following model equations for relaxation oscillations in a turbulently heated plasma, cf. Section 3:

$$\frac{\delta \underline{u}}{\delta t} = -\frac{|e|}{m}\underline{E} - \underline{\underline{\alpha}}\cdot(\underline{u}^* - \underline{u}_w)\frac{W}{nT_e} \tag{5}$$

$$\frac{\delta \ell n W}{\delta t} = \sigma_1 \frac{u^* - u_c}{v_e} - \sigma_2 \frac{W}{nT_e} - \sigma_3 \frac{\delta \ell n T_e}{\delta t} \tag{6}$$

$$\frac{\delta \ell n T_e}{\delta t} = \frac{2}{3}\frac{\underline{u}-\underline{u}_w}{v_e}\cdot\underline{\underline{\alpha}}\cdot\frac{\underline{u}^*-\underline{u}_w}{v_e}\frac{W}{nT_e} \tag{7}$$

$$\delta T_i \approx \frac{u_w}{\langle\hat{\underline{k}}\rangle\cdot\underline{u}-u_w}\,\delta T_e \tag{8}$$

where u_w is the mean phase velocity and $\langle \hat{\underline{k}} \rangle$ the mean direction of wave propagation, $\underline{\underline{\alpha}}$ describes anomalous resistivity, σ_1 linear wave growth, σ_2 nonlinear wave damping and σ_3 adiabatic changes by heating. The last relation uses the conservation laws of weak turbulence for an estimate of relative heating rates near marginal stability. The various coefficients represent spectral averages and also depend on the detailed distribution functions. This fact explains selfquenching of ion sound driven by a current perpendicular to the magnetic field, which we observed in a particle simulation with u/v_e held constant by adjusting the applied electric field [Dum et al., 1974]. The fluid parameters became selfsimilar (e.g. $u/v_e=0.75$; $T_e/T_i=10$), as is tacitly assumed in formulas of the type (2) for nonlinear saturation. The electron distribution very rapidly reached its final shape, but the ion tail continued to increase and the spectrum shifted to larger and larger $k\lambda_{De}$ ($\lambda_{De} = v_e/\omega_e$) as a result of heating. The fluctuation level goes through a maximum which is not related to nonlinear saturation, but simply to the point at which quasilinear wave growth is overtaken by heating

$$\frac{\delta \ell n W/nT_e}{\delta t} = \frac{\delta \ell n W}{\delta t} - \frac{\delta \ell n\, nT_e}{\delta t} \tag{9}$$

This happens much before W begins to decrease as a result of increased ion Landau damping. That selfquenching is observed for simulations with u=const is more obvious. For E=const, relaxation oscillations about marginal stability, with increasing $u^* \approx u_c$ can take place if $u_c > u_w$. This is easily seen by drawing a phase space diagram for (5) and (6). Negligible heating may be a reasonable assumption for the first pulse in which, depending on E, a relatively large overshoot $u^* \gg u_c$ may occur before the fluctuation level becomes sufficiently high for braking of the drift [Field and Fried,

1964]. Transfer relations equivalent to (5-8), but actually evaluated for Maxwellian electrons and ions, and a sufficiently large sample of wave modes were coupled by Mondelli and Ott [1974] to equations describing the external circuit in a turbulent heating experiment. This model reproduced the observed relaxation oscillations (Figure 1 a) at least qualitatively. Nonlinear damping (σ_2) leads to no qualitative changes of the results derived from (5-8), with the actual effect depending of course on the relative size of the terms in (6). For electrostatic ion cyclotron waves the situation may be entirely different. In the turbulent heating experiment of Wharton et al., [1971] these waves exist before application of the current and change relatively little during the heating pulse (see Figure 1 b). Although amplitudes n_1/n of density fluctuations are large, the estimated $\nu_{eff} \approx \Omega_i$ is negligible compared with ν_{eff} for ion sound. Marginal stability considerations obviously do not apply in this case, but rather a nonlinear theory with threshold-like saturation. With resonance-broadening, damping by the wave ion interaction indeed sets in abruptly above a critical fluctuation level $n_1{}^o/n$ which is independent of the linear growth rate, or nearly so, if there is also some linear ion Landau damping and damping by ion-ion collisions [Dum and Dupree, 1970].

$$\frac{n_1^o}{n} \approx \frac{e\,\phi^o_{rms}}{T_e} = \frac{\omega-\Omega_i}{\Omega_i}\,\frac{T_i}{T_e}\,\frac{1}{s\langle F_1\rangle^{1/2}} \quad ;$$

$$\frac{n_1^o}{n} \approx 0.3 \text{ for } s = (k_\perp v_{i\perp}/\Omega_i)^2 = 1 \tag{10}$$

where F_1 is expressed by Bessel functions evaluated at $\langle k_\perp v_\perp/\Omega_i\rangle = x \approx (2s)^{1/2}$. Saturation above this threshold is possible by balancing the electron growth term. The scaling of the fluctuation level with the electron drift that results, thus differs substantially from a law based on mode coupling, such as (2). In sharp contrast to ion sound waves, cyclotron waves can largely avoid linear Landau damping by propagating at sufficiently large angles to the magnetic field. It is possible for the nonlinear interaction to remain also dominant in the course of ion heating since according to general energy conservation laws, which also hold for resonance broadening, a fraction $\Omega_i/\omega \lesssim 1$ that is most of the energy will go into directions perpendicular to the magnetic field. The number of linearly resonant particles, $v_\parallel = (\omega-\Omega_i)/k_\parallel$ is thus expected to remain insignificant, in contrast to tail formation for ion sound. We [Dum and Dupree, 1970] felt that, because of the nature of the quasilinear wave - ion interaction, the relatively narrow gyro-resonance structure, and the proven similar, strong stabilizing effect of ion-ion collisions [Wong et al., 1966], a semi-quantitative resonance broadening theory may be adequate for saturation of ion cyclotron waves, but refrained from an apparently more straightforward application to ion sound saturation. In the latter case, resonance broadening could possibly play some role in the initial ion tail formation process [Dum, 1978 a] but requires a more quantitative complete theory, which among other effects, accounts for the strong speed dependence of resonance broadening and the adiabatic interaction. A quantitative theory for resonance broadening by ion sound, which correctly accounts for the speed dependence, shows that corrections of the quasi-linear result are negligible for the electrons [Dum, 1975]. The rate at which ions are heated is governed by conservation laws, such as (8), and for a quasi-steady state is independent of the particular wave saturation mechanism. The shape of the heated ion distribution gives much more detailed information about the nature of the wave - ion interaction. The speed dependence of resonance broadening by ion cyclotron waves is essentially given by $F_1\ (k_\perp v_\perp/\Omega_i)$, cf. (10), and is such that interaction with the bulk of the distribution should dominate [Dum and Dupree, 1970]. A two stage process in which resonance broadening is followed by quasi-linear wave - ion interaction and isotropization of $T_\perp/T_\parallel$ by ion-ion collisions was proposed by Ionson et al. [1976]. Whether or not this actually happens depends strongly upon macroscopic conditions. Although we made a considerable effort in trying to justify the assumptions of a semi quantitative resonance broadening theory for ion cyclotron waves, experimental confirmation under quite different conditions, in particular the detailed agreement with wave saturation and also ion heating reported by the U.C. Irvine Group [Correll et al., 1975; Correll et al., 1977; Böhmer et al., 1979], still comes as somewhat of a surprise. It gives us hope, however, that a more complete, quantitative theory, of the ion distribution and the wave-electron interaction in particular, can withstand an even closer scrutiny. Numerical investigations, such as reported by Lysak [1980] and Okuda et al. [1981], which allow very detailed studies of the particle distribution functions are especially important. We have discussed in this section how these microscopic studies should at some point be coupled with the global dynamics in order to definitively establish the role of ion cyclotron waves in potential drops and particle acceleration. The recent observations of these waves on auroral field lines [Mozer et al., 1980] are especially stimulating for this work.

Anomalous Resistivity And Heating Rates For Various Instabilities

In weak turbulence theory [Galeev and Sagdeev, 1979] one derives the transfer rates

$$\underline{R}_j = -\int \frac{d\underline{k}}{(2\pi)^3}\; 2\,\gamma_{\underline{k}j}\, N_{\underline{k}}\, \underline{k} \tag{11 a,b}$$

$$K_j = - \int \frac{d\underline{k}}{(2\pi)^3} \, 2 \, \gamma_{\underline{k}j} \, N_{\underline{k}} \, \omega_{\underline{k}}$$

for momentum and energy, where $\gamma_{\underline{k}j}$ is the total, linear and nonlinear contribution of species j to the growth rate γ and $N_{\underline{k}} = (\partial\epsilon/\partial\omega_{\underline{k}})W_{\underline{k}}$ in terms of the electric field energy density $W_{\underline{k}}$. After defining the number of waves by $N_{\underline{k}} = \hbar s_{\underline{k}} \hat{N}_{\underline{k}}$; $s_{\underline{k}} = \mathrm{sgn}\,\partial\epsilon/\partial\omega_{\underline{k}}$, equations (11 a,b) have an obvious quantum mechanical interpretation. Using $-\gamma_{\underline{k}j} N_{\underline{k}} = \mathrm{Im}\epsilon_j W_{\underline{k}}$, these relations are seen to be equivalent to (actually more general) relations in terms of the quasi-linear and nonlinear dielectric constants, $\epsilon = 1 + \Sigma_j \epsilon_j$. Moments of the turbulent collision term in the kinetic equation for the particle distributions are, of course, also expected to give these results. Relations of the form (11 a,b) loose much of their actual usefulness, other than as a constraint, if the difficult problem of determining distribution functions is not simply circumvented by assuming Maxwellian distributions in evaluating the growth rates and dielectric constants. This is not even justified in classical transport theory where Coulomb collisions try to maintain Maxwellians against the perturbations by drifts and gradients. For a rough estimate it may be permissible to assume Maxwellians, which after all, in the sense of information theory, represent the state of maximum uncertainty consistent with the fluid parameters n_j, u_j, T_j. This assumption, however, eliminates interesting and often dominant effects such as anomalous heat conduction, thermoelectric effect and heat flux driven instability, e.g. for ion sound turbulence [Dum, 1978 b]. Evaluation of the transfer rates for a combination of Maxwellians cannot describe the formation of electron runaway tails or high energy ion tails, as is sometimes thought, when, as usual, the increase in number density rather than tail temperature is the dominant feature [Dum, 1978 a]. If distribution functions and wave spectra are known, evaluation of the transfer rates from (11 a,b) is indeed straightforward. For ion sound and related short wavelength $(kv_e/\Omega_e) \gg 1$, low phase velocity $(\omega/kv_e) \ll 1$ fluctuations, the imaginary part of the dielectric constant may be written as [Dum, 1978 a],

$$- \mathrm{Im}\epsilon_e = \frac{1}{(k\lambda_{De})^2} \left(\frac{\pi}{2}\right)^{1/2} a_{-3,\hat{\underline{k}}} \frac{\hat{\underline{k}}\cdot\underline{u}^*_{\underline{k}} - \omega/k}{v_e} \tag{12}$$

where $\hat{\underline{k}} = \underline{k}/k$. For a drifting Maxwellian $a_{-3} = 1$, $\underline{u}^*_{\underline{k}} = \underline{u}$, and more generally, (12) may be obtained by expanding the reduced distribution function $F_{\hat{\underline{k}}} = \int d\underline{v}\,\delta(\hat{\underline{k}}\cdot\underline{v} - v_{\hat{\underline{k}}}) f(\underline{v})$ about its maximum at $\hat{\underline{k}}\cdot\underline{u}^*_{\underline{k}}$. It was already shown by Field and Fried [1964] that quasilinear effects distort the electron distribution in such a way that growth of modes that are oblique to the drift direction actually becomes dominant. Particle simulation for ion sound driven by field aligned currents shows that this effect increases with increasing magnetic field, until for $\Omega_e/\omega_e \gtrsim 2$ electrons are essentially magnetized [Biskamp and Chodura, 1971]. In the case $(kv_e/\Omega_e) \ll 1$, $\mathrm{Im}\epsilon_e$ is determined by the reduced distribution function in the magnetic field direction,

$$- \mathrm{Im}\epsilon_e = \frac{1}{(k\lambda_{De})^2} \left(\frac{\pi}{2}\right)^{1/2} a_{-3,\parallel} \frac{k}{|k_\parallel|} \frac{\hat{\underline{k}}\cdot\underline{u}^* - \omega/k}{v_e} \tag{13}$$

Equation (13) applies to ion cyclotron waves, but also to drift waves, etc., with magnetized electrons. Quasi-linear plateau formation by diffusion along the magnetic field, given the time, makes $a_{-3,\parallel}$ and thus $\mathrm{Im}\epsilon_e$ arbitrarely small, unless this is prevented by pitch angle scattering from Coulomb collision or other short wavelength fluctuations such as ion sound, influx of new particles etc. These processes are thus crucial for anomalous resistivity, but unfortunately also mask its effect. From (11) and (12) we obtain immediately

$$\underline{R}_e = \int \frac{d\underline{k}}{(2\pi)^3} \, 2 \, \mathrm{Im}\epsilon_e \, W_{\underline{k}} \, \underline{k} =$$

$$= -(2\pi)^{1/2} \left\langle a_{-3,\hat{\underline{k}}} \frac{1}{k\lambda_{De}} \hat{\underline{k}}\,\hat{\underline{k}}\cdot\left(\underline{u}^*_{\underline{k}} - \frac{\omega_{\underline{k}}}{k}\hat{\underline{k}}\right) \right\rangle \times$$

$$\times \frac{W}{nT_e} \, nm\omega_e = - nm\underline{\underline{\nu}}^*\cdot(\underline{u}^* - \underline{u}_w); \tag{14}$$

$$\underline{\underline{\nu}}^* = \int \frac{d\underline{k}}{(2\pi)^3} \underline{\underline{\nu}}^*_{\underline{k}} = \underline{\underline{\alpha}} \, \omega_e \frac{W}{nT_e}$$

where the quantities in the second line are defined by the spectral averages. Similarly, by transforming (11 b) to the electron rest frame $\underline{u}$ we obtain the electron heating rate

$$Q_e = K_e - \underline{R}_e\cdot\underline{u}; \quad K_e = \underline{R}_e\cdot\underline{u}_w + Q_{\Delta w} \tag{15}$$

$$Q_{\Delta w} = \int \frac{d\underline{k}}{(2\pi)^3} \, \Delta\underline{u}_w \cdot \underline{\underline{\nu}}^*_{\underline{k}} \cdot (\Delta\underline{u}_w - \Delta\underline{u}^*)$$

where $\Delta\underline{u}_w = (\omega_{\underline{k}}/k)\hat{\underline{k}} - \underline{u}_w$; $\Delta\underline{u}^* = \underline{u}^*_{\hat{\underline{k}}} - \underline{u}^*$.
are the 'fluctuations' of the phase velocity and effective drift relative to the spectral averages defined by (14). By conservation of energy, $-K_e$ is the ion heating rate for steady state turbulence, or more generally the rate at which energy is transferred from the electrons to waves and ions. Its relation to Q_e is approximately expressed by (8). Note that we have not put the transfer relations into the straight jacket of an Ohm's law, (1), as is usually done. We can have potential drops without a drift and vice

versa, e.g. in the presence of temperature gradients. Also $\underline{\nu}^*$ as defined by (14) and not ν_{eff} is the exact analog of the electron-ion collision frequency ν_{ei} for anisotropic short wavelength turbulence. This analogy includes the same v^{-3} speed dependence of scattering. There is no corresponding enhancement of electron-electron collisions which try to maintain a Maxwellian. Distortion of the distribution function from a drifting Maxwellian and electron runaway are thus relatively even more important than in the classical case, where they lead to a reduction of longitudinal resistivity by 0.51 for Z = 1 and arbitrarely small electric fields. For instability, this corresponds to a reduction of the effective drift velocity u^* from u, see (14) and the first term of (4). For magnetized electrons, (13) leads to obvious replacements in the transfer rates (14-15) which, however, as discussed, correspond to very significant differences in the stochastic scattering process. The formulation (11 a,b) applies to linearly or nonlinearly resonant wave particle interactions but does not include adiabatic terms connected with changing wave amplitudes and plasma parameters. For the electrons this is permissible if, as usual, the sloshing energy in the wave motion is very small compared to the thermal energy. For the ions, this may not be the case, at least not in the initial phase of wave growth. It is certainly not the case for non-resonant or reactive instabilities, such as the Buneman instability for $u \gg v_e$. Although an effective collision frequency can formally be defined by (1) for the usually very efficient momentum and energy transfer in this case, it does not correspond to a stochastic scattering process in the sense of collisions. Buneman [1959] was very careful in making this distinction. It is certainly most inappropriate to speak of an effective collision frequency for saturated turbulence since a reactive instability mode exists only as long as it is growing. The increase in the sloshing motion gives rise to the transfer rates

$$\underline{R}_e \approx - nm\underline{u} \frac{\delta W/\delta t}{nmu^2/2} \; ; \quad Q_e \approx \frac{\delta W}{\delta t} \; ; \quad Q_i \approx \left(\frac{4m}{M}\right)^{1/3} \frac{\delta W}{\delta t} \qquad (16\ a,b,c)$$

for $u \gg v_e$. The complicated dynamical processes connected with this instability and its nonlinear transition to ion sound turbulence are best discussed using particle simulation [Dum and Chodura, 1979]. Instead of the transition to ion sound turbulence at $u < v_e$, seen in most experiments and in our simulations with u=const., simulations with E=const. and (too) small system size have lead to a single mode, while maintaining $u \approx v_e$. Propagation of the applied voltage pulse along the plasma column and sheath effects were discussed for one of the most recent experiments related to Buneman instability and anomalous resistivity [Clark and Hamberger, 1979]. Significant distortions of the electron distribution from a drifting Maxwellian, similar to the ones in our simulations, were also inferred for this experiment. The Buneman mode is also related to formation of electrostatic potential structures. It may thus be interesting to note that we saw tendencies in this direction and trapping only for one dimensional periodic systems, but not for the more stochastic and physically more relevant two dimensional system. Undoubtely, driving conditions, system size and preexisting inhomogeneities play a role in the formation of potential structures [De Groot et al., 1977; Iizuka et al., 1979; Sato and Okuda, 1980].

Conclusion

In view of new observations of anomalous resistivity, heating and acceleration, along auroral field lines in particular, the dynamics of various wave-particle interactions was critically examined. The actual importance of such mechanisms in a particular case, unfortunately, cannot be established without due regard of macroscopic (fluid) dynamics and external boundary conditions. In addition to theoretical investigations of potentially relevant mechanisms, refined experimental studies and, most important, theoretical and numerical studies of the global problem, which include anomalous transport, are required. Important developments in this direction have already been initiated for the auroral plasma [see Akasofu, 1977; Haerendel, 1980; Lysak and Carlson, 1980; Wagner et al., 1980; Fedder, 1976; Fontheim et al., 1978, and contributions to this volume]. Possible methods for reducing the complexity of a realistic model were pointed out, drawing mainly on the experience with turbulent heating and shock waves, where the development has already reached a high state of maturity.

References

Akasofu, S.-I., Physics of Magnetospheric Substorms, Reidel, Hingham, Mass., 1977, especially Chapters 3 and 7.

Biskamp, D., and R. Chodura, Computer simulation of anomalous resistance, Plasma Physics and Controlled Nuclear Fusion Research, Vol.II, 265, IAEA, Vienna, 1971.

Böhmer, H. and S. Fornaca, Experiments on nonlinear effects of strong ion cyclotron wave turbulence, J. Geophys. Res., 84, 5234, 1979.

Buneman, O., Dissipation of currents in ionized media, Phys. Rev., 115, 503, 1959.

Chodura, R., C.T. Dum, F. Söldner, and K.-H. Steuer, Anomalous transport in shock produced plasma, 8th Europ. Conf. on Controlled Fusion and Plasma Physics, Vol. I, 93, Prague, 1977.

Clark, W.H.M., and S.M. Hamberger, Anomalous conductivity and electron heating in a plasma unstable to the two-stream instability, Plasma Phys., 21, 943, 1979.

Correll, D.L., N. Rynn, and H. Böhmer, Onset, growth and saturation of current-driven ion cyclotron instability, Phys. Fluids, 18, 1800, 1975.

Correll, D.L., H. Böhmer, N. Rynn, and R.A. Stern, Temporal evolution of ion temperatures in the presence of ion cyclotron instabilities, Phys. Fluids, 20, 822, 1977.

Davidson, R.C., and N.A. Krall, Anomalous transport in high temperature plasmas with application to solenoidal fusion systems, Nuclear Fusion, 17, 1313, 1977.

De Groot, J.S. C. Barnes, A.E. Walstead, and O. Buneman, Localized structures and anomalous dc resistivity, Phys. Rev. Lett., 38, 1283, 1977.

Dum, C.T., and T.H. Dupree, Nonlinear saturation of high frequency instabilities in a magnetic field, Phys. Fluids, 13, 2064, 1970.

Dum, C.T. and R. Chodura, and D. Biskamp, Turbulent heating and quenching of the ion sound instability, Phys. Rev. Lett., 32, 1231, 1974.

Dum, C.T., Strong turbulence theory and the transsition from Landau to collisional damping, Phys. Rev. Lett., 35, 947, 1975.

Dum, C.T., Anomalous heating by ion sound turbulence, Phys. Fluids, 21, 945, 1978 a.

Dum, C.T., Anomalous electron transport equations for ion sound and related turbulent spectra, Phys. Fluids, 21, 956, 1978 b.

Dum, C.T., and R. Chodura, Anomalous transition from Buneman to ion sound instability, Wave Instabilities in Space Plasmas, Palmadesso and Papadopoulos, eds., Reidel, 1979.

Dum, C.T., E. Marsch, and W. Pilipp, Determination of wave growth from measured distribution functions and transport theory, J. Plasma Phys. 23, 91, 1980 a.

Dum, C.T., E. Marsch, K.-H. Mühlhäuser, W. Pilipp, D.A. Gurnett, H. Rosenbauer, and R. Schwenn, Free energy source for ion sound turbulence in the solar wind, Eos Trans. AGU, 61, 1097, 1980 b.

Eselevich, V.G., A.G. Eskov, R. Kh. Kurtmullaev, and A.I. Malyutin, Fine structure of shock waves in a plasma and the mechanism of saturation of ion-acoustic turbulence, Zh. Eksp. Theor. Fiz. 60, 1658, 1971 (Sov. Phys. JETP, 33, 898, 1971).

Fälthammar, C.-G., Problems related to macroscopic electric fields in the magnetosphere, Rev. Geophys. Space Phys., 15, 457, 1977.

Fontheim, E.G., R.S.B. Ong, R.G. Roble, H.G. Mayr, M.J. Baron, W.H. Hoegy, V.B. Wickwar, R.R. Vondrak, and J.A. Ionson, Effect of anomalous transport coefficients on the thermal structure of the storm time auroral ionosphere, J. Geophys. Res., 83, 4831, 1978.

Fedder, J.A., Effects of anomalous resistivity on auroral Birkelund current systems, Ann. Geophys., 32, 1975, 1976.

Field, E.C., and B.D. Fried, Solution of the kinetic equation for an unstable plasma in an electric field, Phys. Fluids, 7, 1937, 1964.

Galeev, A.A., Collisionless shocks, Physics of solar planetary environments, D.J. Williams, ed., AGU, Washington, 1976.

Galeev, A.A., and R.Z. Sagdeev, Nonlinear Plasma Theory, Rev. of Plasma Physics, 7, 1, Consultants, 1979.

Greenstadt, E.W., V. Formisano, C.T. Russell, M. Neugebauer, and F.L. Scarf, Geophys. Res. Letter, 5, 399, 1978.

Haerendel, G., Auroral particle acceleration - An example of a universal plasma process, ESA J., 4, 197, 1980.

Hollenstein, Ch., M. Guyot, and E.S. Weibel, Stationary potential jumps in a plasma, Phys. Rev. Lett., 45, 2110, 1980.

Hudson, M.K., and F.S. Mozer, Electrostatic shocks, double layers, and anomalous resistivity in the magnetosphere, Geophys. Res., Lett., 5, 131, 1978.

Iizuka, S., K. Saeki, N. Sato, and Y. Hatta, Buneman instability, Pierce instability, and double layer formation in a collisionless plasma, Phys. Rev. Lett., 43, 1404, 1979.

Ionson, J.A., R.S.B. Ong, and E.G. Fontheim, Anomalous resistivity of the auroral plasma in the top side ionosphere, Geophys. Res. Lett., 3, 549, 1976.

Kalfsbeek, H.W., Numerical calculations on the anomalous current penetration in a toroidal Hydrogen plasma, Rijnhuizen Report, 78-113, 1978.

Kalinin, Yu.G., D.N. Lin, L.I. Rudakov, V.D. Ryutov and V.A. Skoryupin, Observation of current convective instability during turbulent heating of plasma by a current, Zh. Eksp. Teor. Fiz., 59, 1056, 1970 (Sov. Phys. JETP, 32, 573, 1971).

Kindel, J.M., and C.F. Kennel, Topside current instabilities, J. Geophys. Res., 76, 3055, 1971.

Kurth, W.S., D.A. Gurnett, and F.L. Scarf, High resolution spectrograms of ion acoustic waves in the solar wind, J. Geophys. Res., 84, 3413, 1979.

Lysak, R.L., Electrostatic ion cyclotron turbulence and auroral particle acceleration, Ph.D. Dissertation, U.C. Berkeley, 1980.

Lysak, R.L., and C.W. Carlson, The effect of microscopic turbulence on magnetosphere-ionosphere coupling, U.C. Berkeley Report, 1980.

Manheimer, W.M., and J. Boris, Selfconsistent theory of a collisionless resistive shock, Phys. Rev. Lett., 28, 659, 1972.

Manheimer, W.M., and J. Boris, Marginal stability analysis - A simpler approach to anomalous transport in plasmas, Comments Plasma Phys. Cont. Fusion, 3, 15, 1977.

Mondelli, A., and E. Ott, The application of quasilinear theory to the Cornell turbulent heating experiments, Plasma Phys., 16, 413, 1974.

Morse, D.L., and E.W. Greenstadt, Thickness of magnetic structures associated with the earth's bow shock, J. Geophys. Res., 81, 1791, 1976.

Mozer, F.S., C.A.Cattell, M.K. Hudson, R.L. Lysak, M. Temerin, and R.B. Torbert, Satellite measurements and theories of low altitude auroral particle acceleration, Space Science Rev., 27, 155, 1980.

Okuda, H., C.Z. Cheng, and W.W. Lee, Anomalous diffusion and ion heating in the presence of electrostatic hydrogen cyclotron instabilities, Phys. Rev. Lett., 46, 427, 1981.

Palmadesso, P.J., T.P. Coffey, S.L. Ossakov, and K.Papadopoulos, Topside ionosphere ion heating due to electrostatic ion cyclotron turbulence, Geophys. Res. Lett., 1, 105, 1974.

Papadopoulos, K., A Review of anomalous resistivity in the ionosphere, Rev. Geophys. Space Phys., 15, 113, 1977.

Robertson, S., P. Korn, and C.B. Wharton, Parametric study of turbulent heating in a linear discharge, Plasma Science, 2, 17, 1973.

Sato, T., and H. Okuda, Numerical Simulations on ion acoustic double layers, Princeton Report PPPL-1681, 1980.

Soeldner, F., C.T. Dum, and K.-H. Steuer, Electron and ion heating in a high-voltage belt pinch, Phys. Rev. Lett., 39, 194, 1977.

Stenzel, R.L., Experiments on current-driven three-dimensional ion sound turbulence, Part II, Wave dynamics, Phys. Fluids, 21, 99, 1978.

Supronenko, V.A., E.A. Sukhomlin, and V.T. Tolok, Collective interactions and electrical conducditivty of a plasma in a strong electric field, Plasma Phys., 15, 353, 1973.

Wagner, J.S., T. Tajima, J.R. Kan, J.N. Leboeuf, S.I. Akasofu, and J.M. Dawson, V Potential double layers and the formation of auroral arcs. U. Alaska Report, 1980.

Wharton, C.B., P. Korn, D. Prono, S. Robertson, P. Auer, C.T. Dum, New results in turbulent heating, Plasma Physics and Controlled Nuclear Fusion Research, Vol. II, 25, IAEA Vienna, 1971.

Wong, A.Y., O. Judd, and F. Hai, Effects of collisions on electrostatic ion cyclotron waves, Phys. Lett., 21, 157, 1966.

THE SOURCE MECHANISM OF AURORAL KILOMETRIC RADIATION

C. S. Wu

University of Maryland, College Park, Maryland 20742

Abstract. The cyclotron maser mechanism for the auroral kilometric radiation (AKR) proposed by Wu and Lee (1979 and Lee et al. (1980) is reviewed. Moreover, some recent developments concerning the quasilinear saturation of AKR and its energy conversion efficiency are also discussed. These discussions are concerned with the case in which the population of the secondary electrons prevails over that of the energetic electrons. In this paper the limit of zero secondary electrons is discussed briefly. It is found that the dispersion relation of the radiation is significantly modified.

I. Introduction

One of the most fascinating problems in the magnetospheric physics is the phenomenon of auroral kilometric radiation (AKR) (Gurnett, 1974). There are several reasons why this subject has attracted much attention in recent years. First, since the intense nonthermal emissions corelate very well with inverted-V events (Green, et al. 1979), researchers believe that the phenomenon of AKR is inseparable from the physics of the auroral acceleration region. Second, AKR itself reflects evidently a very intriguing plasma process which may have profound and practical implications. Third, the study of the physical mechanism of AKR may enable us to understand the emission of decametric radiation from Jupiter, since the two phenomena have many similar features. In short, the consensus is that AKR occupies a special place in the plasma physics of the polar region.

Theoretically, the source mechanism responsible for the generation of AKR has been the center of discussions. Although, since 1975, a number of models have been suggested in the literature (Benson, 1975; Barbosa, 1976; Palmadesso et al. 1976; Melrose, 1976; Roux and Pellat, 1979; Wu and Lee, 1979; Lee et al.1980; Grabble et al. 1980), the subject still needs further investigation and many questions remain to be answered. Most of these theories rely on low-efficiency and indirect amplification processes and, therefore, have difficulty in explaining the high efficiency of AKR.

Among the existing theories, the models suggested by Melrose (1976) and Wu and Lee (1978) predict direct amplification of fast extraordinary mode radiation which agrees with the observed polarization (Gurnett and Green, 1978; Kaiser et al. 1978; Benson and Calvert, 1979). Melrose's theory assumes that the beaming electrons can possess an exceedingly large thermal anisotropy which has not been confirmed by avilable observational results. The theory proposed by Wu and Lee (1979) stresses the importance of the reflected electrons due to a magnetic mirror effect. These electrons are expected to have a loss-cone distribution which can result in the excitation of electromagnetic waves. Other than these two theories, a very different notion is suggested in a recent theory by Grabble et al., (1980). These authors argue that a parametric process involving coherent ion cyclotron waves may produce AKR. There are a few questions about their theory. First, observations by the S3-3 satellite have found that hydrogen ion cyclotron waves are consistently absent at low altitudes, say from 3×10^3 to 5×10^3 km, (Kintner et al 1979), whereas AKR has been frequently seen by ISIS-1 around 3×10^3 km. Thus it implies that ion cyclotron waves are not indispensible in the generation of AKR. Second, the theory by Grabble et al. assumes that the plasma-sheet electrons are present in the AKR source region in the form of a beam with a small velocity spread. The assumption is by no means realistic. It is known from S3-3 observations that the precipitating electrons usually have a velocity spread comparable to, or even larger than, the beam velocity. The effect of this large velocity spread needs to be included in the analysis. It is expected that this effect tends to reduce the growth rate very significantly. Moreover, in many cases the electron distribution function does not indicate the existence of a beam at high altitude.

The main purpose of this paper is to review the AKR theory within the context of two earlier publications (Wu and Lee, 1979; Lee et al., 1980) and some recent developments which we have accomplished lately. The organization of the paper may be described as follows. In section II we describe the basic physical model and related assumptions. Section III is concerned with some further theoretical considerations and developments. Concluding remarks are presented in Section IV.

II. Physical Model

In the discussion by Wu and Lee (1979) we emphasize two points. First, the reflected electron due to the mirror effect of the convergent magnetic field are expected to possess inherently a loss-cone distribution function. Second, since AKR and inverted-V events corelate (Green et al. 1979) it is reasonaable to hypothesize that a parallel electric field exists in the AKR source region. This electric field can, in turn, reduce the density of the background low energy electronc. These two points are of fundamental importance in our theory. The former is the cause of amplification and the latter influences the dispersion properties of the radiation. For the fast extraordinary mode, which is the polarization of AKR, we find that one of the critical parameters which can influence the amplification process is the ratio of the local electron plasma frequency, ω_e , to the local electron gyrofrequency, Ω_e . Therefore, the density depletion due to the parallel electric field can play a very important role in the generation of AKR.

Both the loss-cone distribution of the reflected keV electrons and the density depletion of ionospheric electrons are conjectured in the initial discussion (Wu and Lee, 1979). These conjectures have been varified by the S3-3 results (D. R. Croley, J.F. Fennell and D.J. Gorney, private comminication) and the ISIS-1 observations (Benson and Calvert, 1979), respectively. However, the $v_\parallel$-dependence of the energetic electron distribution function postulated by Wu and Lee (1979) is not realistic according to available S3-3 data. This motivated us to look into the problem further. In the analysis by Lee et al. (1980) a more complete plasma model is adopted. What we have done is to incorporate the theory of the extended double layer by Kan et al. (1979) into the AKR theory. The implication is that the source of AKR is in the auroral acceleration region which may be attributable to the extended double layer. This is admittedly a hypothesis but the model is consistent with observations.

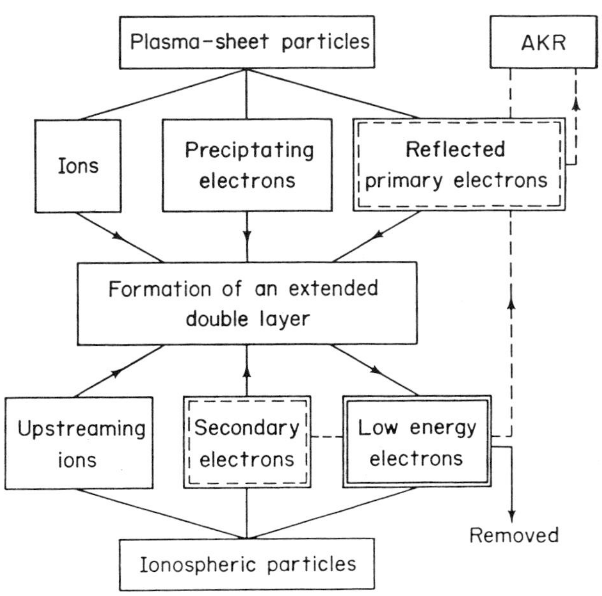

Fig 1. A schematic presentation of the relation between an extended double layer and AKR.

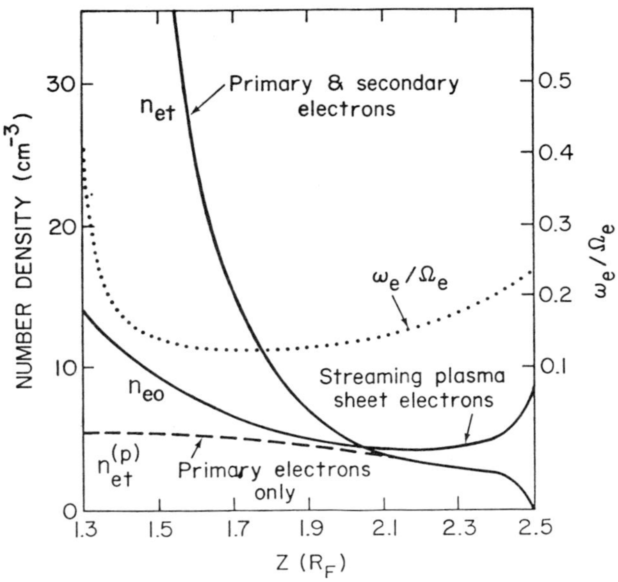

Fig 2. An example of the double layer model. The electron densities n_{eo} , n_{et} and the local frequency ratio ω_e/Ω_e are plotted as functions of z.

1. Extended Double Layer

Based on the above-mentioned consideration, the acceleration region is discussed for our AKR theory (Lee et al. 1980). The physical model is described schematically in Fig. 1. The block diagram shows that the formation of an extended double layer is due to the mutual interaction of the plasmas from two different sources; the plasma sheet and the ionosphere. The plasma sheet electrons with initial energies about a couple of keV again consist of two basic components; one is the precipitating electrons and the other is the reflected electrons. The reflection is due to the mirror effect of the convergent magnetic field. The reflected electrons may be further divided into two groups; namely, the trapped and untrapped. The trapping of reflected electrons by the self-consistent electric field is due to non-adiabatic processes which may occur at lower altitudes.

On the other hand, the electrons from the ionosphere also consist of two components; the low energy ($\sim$ 1 to 2 eV) ionospheric electrons and the so-called secondary electrons. The latter is produced by the precipitating keV electrons which interact with the upper atmosphere and the ionosphere. The secondary electrons may have energies from several tens of eV to several hundred eV. During the formation of the double layer the low energy electrons are removed self-consistently in the acceleration region. Thus only the secondary electrons can survive in the double layer region. Yet their population decreases with altitude because only the energetic secondary electrons can overcome the potential difference associated with the parallel electric field and reach the high altitude region.

In our model only the reflected plasma sheet electrons and the secondary electrons play a significant role in the AKR process. The reflected electrons which have a loss-cone distribution can excite electromagnetic waves, whereas the secondary electrons can influence the dispersion characteristics of the radiation. Both groups can affect the amplification rate.

In Fig. 2 we illustrate one example which is calculated based on the extended double layer model. The electron gyrofrequency is calculated based on a simple dipole-like model, i.e.,

$$B/B_0 = (z_0/z)^3$$

in which $B = 0.6$ Gauss at $z = 1\ R_E$. Other boundary conditions are discussed by Lee et al. (1980). Here we must point out that what we have shown is only a permissible solution which satisfies the desired boundary conditions. Moreover, in this particular case we have assumed that all reflected electrons are trapped. In general this condition is not necessarily required to be true. In Fig. 2, the electron plasma frequency ω_e is calculated based on the total electron density. The ratio ω_e/Ω_e serves as an important parameter in the present theory, as will be discussed later.

2. Cyclotron Maser Instability

One of the pertinent question concerning the theory of AKR is what mechanism is responsible for the amplification of the radiation. The present model advocates that a cyclotron maser process plays the important role. A crude analysis is presented in the early publication (Wu and Lee, 1979) in which the effect of the secondary electrons and that of electrostatic perturbations on the instability are ignored. Later these effects are included in a more elaborate analysis (Lee and Wu 1980; Lee et al. 1980).

In this section we review briefly this instability. However, the emphasis is on the physical nature of the instability not on the detailed analysis. The discussion is based on the assumption that the electrons consist of two essential species; the low energy background electrons, say the secondary electrons, and the reflected energetic electrons such that the population of the former is much larger than that of the latter. Thus the wave dispersion relation may be determined by ignoring the reflected electrons. Furthermore, the cold plasma approximation may be used in the calculation of the real frequency, ω_r .

On the other hand, the temporal growth rate, ω_i , of the waves can be calculated by considering that waves are only emitted by the reflected energetic electrons and assuming that absorption by the low energy background electrons is negligible. The expression for the growth rate is expected to be proportional to the following quantity

$$\frac{\omega_i}{\Omega_e} \sim \frac{\omega_e^2}{\Omega_e^2}\frac{n_r}{n_e}\Omega_e \int d^3v \frac{\partial F}{\partial v_\perp} v_\perp^2$$

$$\delta\left[\omega_r - \Omega_e\left(1 - \frac{v^2}{2c^2}\right) - k_\parallel v_\parallel\right] \qquad (1)$$

where n_r and n_e denote the density of energy and background, respectively, and we have imposed the weakly relativistic approximation. Notice that in Eq. (1) the real frequency ω_r is supposed to be known, as determined by the background electrons. Evidently the argument of the delta function implies that the wave-particle interaction is dictated by the resonance condition

$$\omega_r - \Omega_e\left(1 - \frac{v^2}{2c^2}\right) - k_\parallel v_\parallel = 0 \quad , \qquad (2)$$

from which we obtain

$$\frac{v_\perp}{c} = \sqrt{2}\left(\frac{k_\parallel v_\parallel}{\Omega_e} - \frac{v_\parallel^2}{2c^2} + 1 - \frac{\omega_r}{\Omega_e}\right)^{1/2} \qquad (3)$$

and

$$v_1 \leq v_\parallel \leq v_2$$

$$v_{1,2} \equiv \left(\frac{k_\parallel c \mp \sqrt{k_\parallel^2 c^2 - 2\Delta\omega\Omega_e^2}}{\Omega_e}\right)c \qquad (4)$$

$$\Delta\omega \equiv \frac{\omega_r}{\Omega_e} - 1 \quad . \qquad (5)$$

Of course, Eq. (4) is meaningful only if

$$k_\parallel^2 c^2/\Omega_e^2 > 2\Delta\omega \quad .$$

Figure 3 illustrates the resonance region in velocity space. For a given frequency ω_r and $k_\parallel$, there exists a maximum $v_\perp$ at $v_\parallel = c(k_\parallel c/\Omega_e)$.

$$\left(\frac{v_\perp}{c}\right)_{max} = \left(\frac{k_\parallel^2 c^2}{\Omega_e^2} - 2\Delta\omega\right)^{1/2} \quad . \qquad (6)$$

From Eqs. (1) and (6) we see readily that $\omega_i > 0$, if $(v_\perp)_{max} \simeq \alpha$ which indicates the peak of $F_{e\perp}$ as shown in Fig. 3.

At this point several remarks on the growth rate ω_i may be appropriate. First, as can be seen readily from Eq. (1), for given ω_e/Ω_e and n_r/n_e, the growth rate ω_i depends upon the integral

$$\int_{v_1}^{v_2} dv_\parallel \frac{\partial f}{\partial v_\perp} v_\perp \Bigg|_{v_\perp=\sqrt{2}c\left(\frac{k_\parallel v_\parallel}{\Omega_e} - \frac{v_\parallel^2}{2c^2} - \Delta\omega\right)^{1/2}} \quad . \qquad (7)$$

Thus not only the size of the tesonance region as shown in Fig. 3 is significant, but the magnitude of $v_\perp \partial F/\partial v_\perp$ over the resonance region is decisive. Second, for the fast extraordinary mode, we require

$$\omega_r - \Omega_e > \omega_x - \Omega_e \simeq \omega_e^2/\Omega_e \quad , \qquad (8)$$

where ω_x denotes the cutoff frequency. In Eq. (8) we have implicitly assumed $\omega_e^2 << \Omega_e^2$ which is true in the AKR source region. Thus $\Delta\omega$ has a lower bound. It is apparent now that the larger the value of ω_e^2/Ω_e^2, the higher resonance energy is required. Consequently, fewer electrons can interact with the wave. In other words, if we increase the value of ω_e^2/Ω_e^2, we expect that the value of the integral described by Eq. (7) should decrease. On the other hand if ω_e^2/Ω_e^2 is too small the growth rate ω_i again becomes small because it is explicitly proportional to ω_e^2/Ω_e^2 as indicated in Eq. (1). Hence we expect ω_i has a maximum, if ω_e^2/Ω_e^2 is sutiably choosen. Numerical calculations demonstrate that this expectation is indeed true. A complete expression for ω_i is given by Lee et al. (1980), from which one can easily see that the growth rate ω_i is further affected by the dispersion properties of the radiation through a quantity.

$$\omega_r \, \partial/\partial\omega_r \, \mathrm{ReD}(\underset{\sim}{k},\omega_r) \quad ,$$

where $\mathrm{ReD}(k,\omega_r) = 0$ is the dispersion equation which determines the real frequency ω_r. We have found that when ω_r approaches ω_x, this quantity which appears in the denominator of ω_i becomes large. As a result, it tends to reduce the growth rate significantly from the value which is obtained by completely ignoring the effect of low energy background electrons.

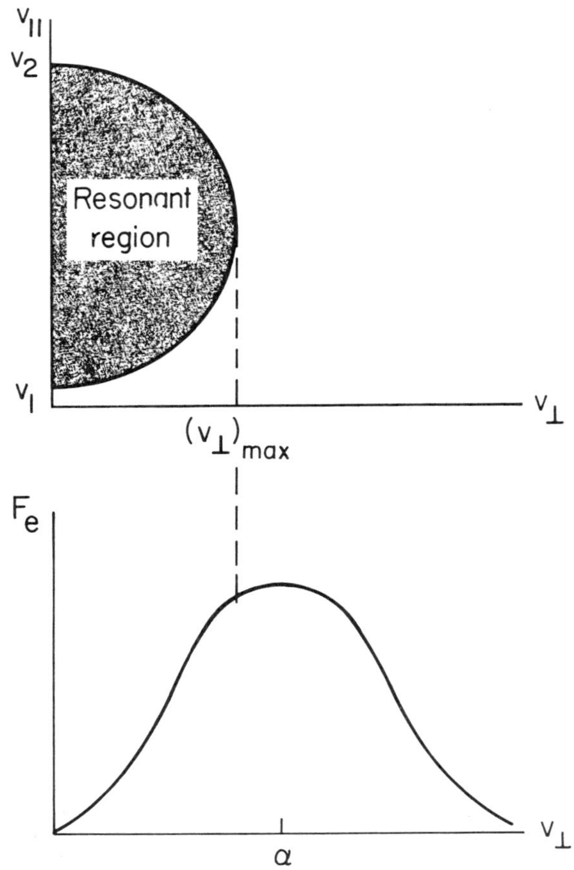

Fig 3. Resonant region in velocity space and the cause of instability.

3. Distribution Function of Reflected Electrons

As seen from the preceding discussion, instability can occur, if $\partial F/\partial v_\perp > 0$ at the resonance velocity. Still we need to consider a model distribution function which may approximately represent the real situation.

Let us consider a satellite passing through the auroral flux tube at a given altitude. If we assume that the incoming electrons from the plasma sheet region have an isotropic distribution and there is no parallel electric field below, the distribution function measured may be schematically described by that shown in Fig. 4(a). On the other hand, if there is a parallel electric field below the satellite, the loss-cone is broadened (Chiu and Schultz 1978) and the distribution function may take the form shown in Fig. 4(b). If the parallel electric field extends to altitude higher than the satellite, then electrons with energies lower than the threshold value will be reflected somewhere above the satellite. These electrons become trapped (by the convergent magnetic field below and by

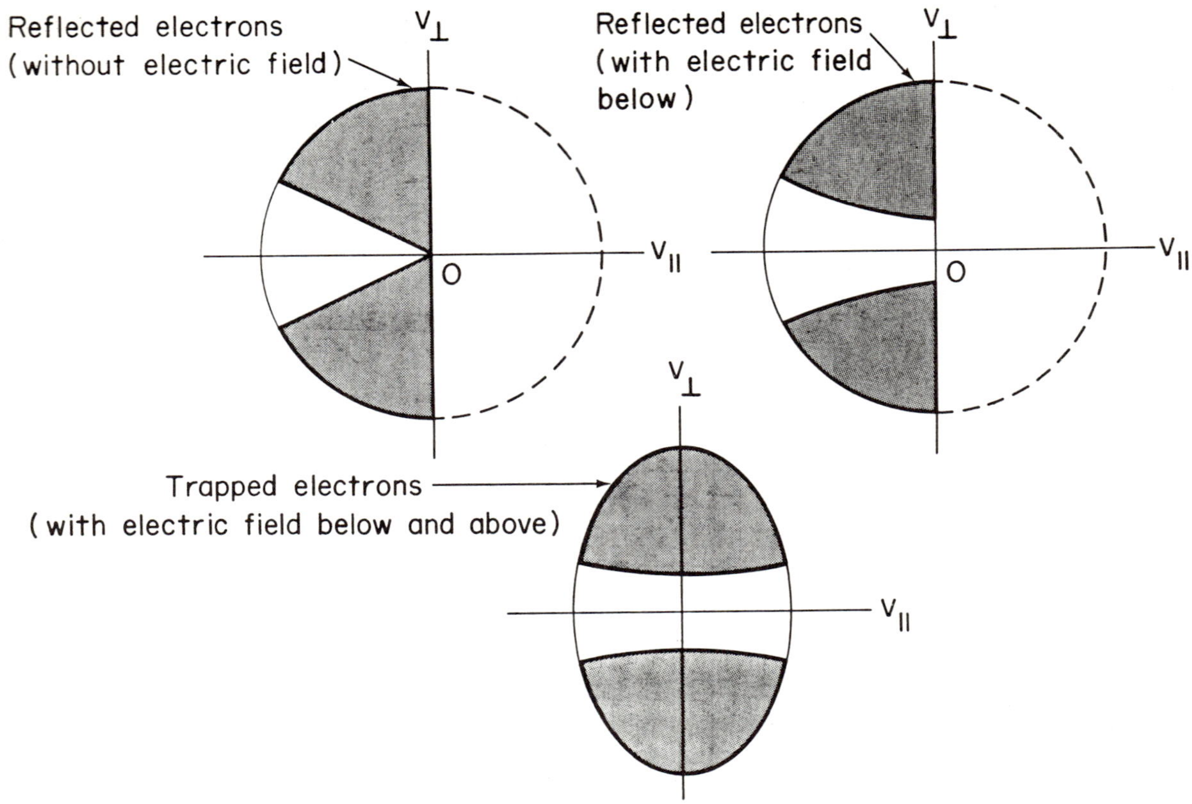

Fig 4. Distribution function of reflected electrons and trapped electrons under different conditions.

the electrostatic potential above) in an extended double layer region. This case is shown in Fig. 4(c).

For mathematical simplicity, we have considered a distribution function which is frequently studied in plasma stability analyses (Lee, et al. 1980). Admittedly, this distribution function does not accurately describe the real situation. However, it does have the essential characteristics of an inverted population in $v_\perp$ space. The shortcoming of using such an approximation is that the analysis tends to significantly underestimate the growth rate. For this reason we are currently investigating the problem with a distribution function which can describe the broadened loss-cone effect more accurately. For example, we consider

$$F_e(v) = A \exp\left(-\frac{v^2}{\alpha^2}\right) \quad \text{for } v_\perp^2 \geq \frac{v_\parallel^2+\Phi}{\theta}$$
$$= 0 \quad \text{for } v_\perp^2 < \frac{v_\parallel^2+\Phi}{\theta}, \tag{9}$$

where A is a normalization constant, Φ and θ are two positive numbers which are determined by the potential difference and the mirror effect.

$$\Phi \equiv \frac{2e\delta\phi}{m_e} = \frac{2e(\phi_1 - \phi)}{m_e}$$

$$\theta \equiv \frac{B_{max}}{B} - 1$$

(ϕ_1 and B_{max} are the electrostatic potential and magnetic field at the mirror point respectively). The distribution given by Eq. (9) is essentially that shown in Fig. 4(b). The result of our analysis will be reported in a forthcoming paper.

III. Further Theoretical Considerations

So far as the amplification of the radiation is concerned, the preceding discussion is restricted to a linear theory. It is also implied that the instability mechanism cannot be easily suppressed by nonlinear effects. In other words, the saturation level of the radiation is assumed to be high. Whether this assumption is justifiable deserves further study.

Another point which needs investigation is the following. In the stability analysis by Lee et al. (1980) we have assumed that the population of the secondary electrons prevails over that of the reflected plasma sheet electrons. This condition may not be true at high altitude. How does this situation affect the amplification process?

In the following two subsections we discuss these points separately. The discussion is based on some recent studies made by the author and his colleagues.

1. Quasilinear Saturation of AKR

In the theory of the cyclotron maser instability we show that a loss-cone type distribution can result in the amplification of the fast-extraordinary mode radiation. It is also anticipated intuitively that the amplified radiation can in turn affect the initial distribution. This process may gradually suppress the instability and eventually the amplification stops. This maximum level of wave energy is called the saturation level which should be comparable to or higher than the observed levels.

Concerning the saturation mechanism we assume that a quasilinear process prevails over other nonlinear processes. This hypothesis is justified, if the calculated saturation level based on the quasilinear theory is such that other nonlinear interactions cannot be significant. The governing equations of the quasilinear theory are

$$\frac{\partial F_e}{\partial t} = \frac{1}{v_\perp}\frac{\partial}{\partial v_\perp}\left(v_\perp D \frac{\partial F}{\partial v_\perp}\right) \qquad (10)$$

$$\frac{\partial \mathcal{E}_{\underset{\sim}{k}}}{\partial t} = 2\omega_i(\underset{\sim}{k},t)\mathcal{E}_{\underset{\sim}{k}} \qquad (11)$$

where D is a diffusion coefficient

$$D = \frac{4\pi^2 e^2}{m^2}\int d^3k \mathcal{E}_{\underset{\sim}{k}}(t)\,\delta\left[\omega_r - \Omega_e\left(1 - \frac{v^2}{2c^2}\right) - k_\parallel v_\parallel\right] \qquad (12)$$

ω_i is the growth rate and $\mathcal{E}_k$ is the wave energy per mode. Equations (10) and (11) should be investigated with a set of initial conditions for $F_e(v)$ and $\mathcal{E}_k$. The initial $F_e(v)$ may be that considered in the linear theory and $\mathcal{E}_k$ may be the energy associated with thermal fluctuations.

For purpose of illustration we chose for the sake of convenience $F_e(v)$ to be a loss-cone distribution at $t = 0$ which may be expressed as

$$F_e = \frac{\exp - \left(\frac{v_\parallel^2}{\alpha^2}\right)}{\pi^{3/2}\alpha(\alpha_1^2 - \alpha_2^2)}\left[\exp\left(-\frac{v_\perp^2}{\alpha_1^2}\right) - \exp\left(-\frac{v_\perp^2}{\alpha_2^2}\right)\right] \qquad (13)$$

with $\alpha_1^2 > \alpha_2^2$. Then from Eq. (10) we can obtain

$$F_e(\underset{\sim}{v},t) = \frac{\exp\left(-\frac{v_\parallel^2}{\alpha^2}\right)}{\pi^{3/2}\alpha(\alpha_1^2-\alpha_2^2)}\left[\frac{1}{1+\frac{\tau}{\alpha_1^2}}\exp\left(-\frac{v_\perp^2}{\alpha_1^2+\tau}\right) - \frac{1}{1+\frac{\tau}{\alpha_2^2}}\exp\left(-\frac{v_\perp^2}{\alpha_2^2+\tau}\right)\right] \qquad (14)$$

where $\tau = \int_0^t dt' D(t',\underset{\sim}{v})$.

At the same time, from Eq. (11)

$$\mathcal{E}_{\underset{\sim}{k}}(t) = \mathcal{E}_{\underset{\sim}{k}}(0)\exp 2\int_0^t dt'\omega_i(\underset{\sim}{k},t') \quad . \qquad (15)$$

Hereafter, we are only interested in a qualitative analysis. Remember that $\partial F_e/\partial v_\perp^2|_{v_\perp=0} \simeq 0$ is a sufficient condition for stability. If t_s denotes a saturation time, an approximation criterion for saturation is

$$\frac{\omega_e^2\pi}{n_e m\alpha_1\alpha_2}\int_0^{t_s} dt'\int d^3k\mathcal{E}_k(t')\int d^3v F_e(v) \times \delta\left[\omega_r - \Omega_e\left(1-\frac{v^2}{2c^2}\right) - k_\parallel v_\parallel\right] = 1 \quad . \qquad (16)$$

Consequently, this condition leads to an estimate of the saturation level of the wave energy. We find

$$\frac{W_s}{n_r m\alpha_1^2} \simeq \frac{1}{\pi}\frac{\alpha_2}{\alpha_1}\frac{\Omega_e^2}{\omega_e^2}\left(\frac{k_{o\parallel}\alpha}{\Omega_e}\right)\left(\frac{\omega_{i,max}}{\Omega_e}\right)\left(\frac{n_e}{n_r}\right) \qquad (17)$$

where n_e is the total electron density and n_r is the density of the reflected electrons; $\omega_{i,max}$ is the maximum growth rate; $k_{o\parallel}$ is the wave number of $k_\parallel$ at which the spectrum $\mathcal{E}_k$ peaks. The details of this analysis is presented in a separate paper (Wu et al. 1980).

To illustrate the case further, we consider $\alpha_2/\alpha_1 = 0.5$, $\Omega_e^2/\omega_e^2 = 50$, $n_r \simeq 1\ cm^{-3}$ and $m\alpha_1^2 \simeq 3$ keV. Numerical computations give

$$\frac{n_e}{n_r}\frac{\omega_{i,max}}{\Omega_e} \simeq 10^{-2}$$

$$\frac{k_{o\parallel}}{\Omega_e} \simeq 10^{-1}$$

for $m\alpha^2 \simeq 2$ keV. Thus

$$W_s \simeq 10^{-10} \quad ergs/cm^3 \quad .$$

This crude estimate agrees with the result obtained by accurate numerical computation reasonably well.

We now discuss the efficiency of AKR which is defined as

$$\varepsilon = \frac{\text{estimated radiation power}}{\text{power available from auroral electrons}} \quad .$$

If we assume that the radiation has a group velocity very close to the speed of light when it leaves the source region, the power flux per unit area at the boundary of the source region is

$$S = W_s v_g = 10^{-10} \times 3 \times 10^{10} \times 10^{-7} \times 10^{4}$$
$$= 3 \times 10^{-3} \text{ watts/m}^2 \ .$$

The AKR source region along the auroral flux tube can have a surface area of the order 10^5 km^2, say thousand kilometers along the magnetic field lines and hundred kilometers transverse to the field limes. Thus the total power is rougly 10^8 watts which is in agreement with the value estimated by other authors (Gurnett, 1974 and others). On the other hand, the available power of the auroral electrons during disturbed times is estimated to be of the order of 10^{11} watts, as suggested by Axford (1967). Thus we have

$$\varepsilon \simeq 0.1 \ \%$$

which is comparable to that discussed by Benson and Calvert (1979).

A very important requirement of the present model is that the amplification process must be highly efficient such tahat waves can grow to large amplitude before the ambient magnetic field has changed significantly. This requirement seems to be satisfied, as discussed qualitatively by Wu et al. (1980). An accurate and quantitative study is not easy, because it depends upon the location of the ray paths and other geometrical effects.

2. Limit of Zero Secondary Electrons

At sufficiently high altitudes the density of secondary electrons is expected to be very small. In this case only those electrons originating in the plasma sheet are important. When this situation is true, the approach used by Lee et al. (1980) is no longer meaningful, because in this case the dispersion relation should also be determined by the plasma-sheet electrons. The thermal effect of the energetic electrons can significantly modify the cold plasma dispersion relation which was considered previously [Lee and Wu, 1980; Lee et al., (1980)]. In other words, the dispersion equation must be rederived for the present case. This investigation has been carried out recently. The results will be reported in two forthcoming articles (Tsai et al. 1980; Wu et al., 1980). In the following we summarize briefly what have been found.

First of all, if we consider a distribution function of the form

$$F_e = \frac{1}{\pi^{3/2}\alpha^3 m_{eo}^3}\left(\frac{p_\perp^2}{m_{eo}^2\alpha^2}\right)\exp\left(-\frac{p^2}{m_{eo}^2\alpha^2}\right)$$

where p is the momentum and m_{eo} is the rest mass of electrons, and if we restrict our discussion to a weakly relativistic plasma and the radiation frequency is close to the electron gyrofrequency, the dispersion equation can be expressed in the following form.

$$\left|\frac{c^2k^2}{\omega}\left(\delta_{ij} - \frac{k_ik_j}{k^2}\right) + 1 + \chi_{ij}\right| = 0 \qquad (18)$$

where

$$\chi_{ij} = -i\,\frac{\omega_e^2 c^2}{\omega^2\alpha^2}\int_0^\infty d\tau\,\frac{M_{ij}}{(1-i\tau)^{5/2}}$$
$$\times \exp\left(-\frac{h^2\tau^2}{1-i\tau} + i\,\frac{2c^2}{\alpha^2}\,\frac{\omega-\Omega}{\omega}\,\tau\right) + \eta_{ij}$$

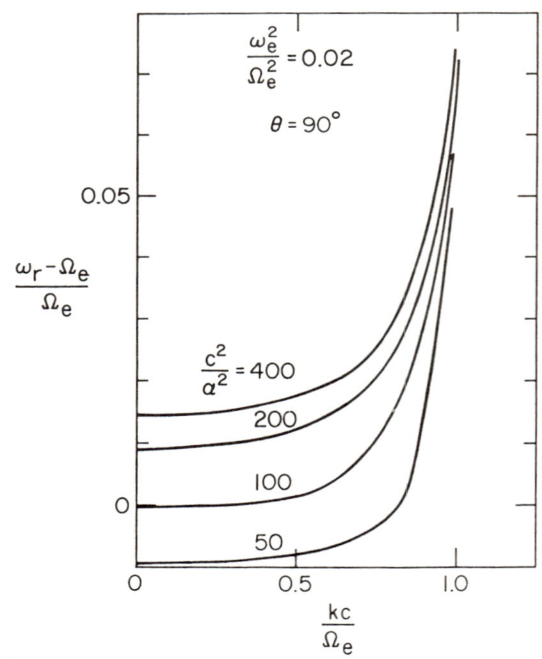

Fig 5. Dispersion relation of the fast extraordinary mode in the absence of the cold background electrons. The ratio c^2/α^2 is a measure of the energy of the reflected electrons. For example if $c^2/\alpha^2 = 50$, the energy is of 5 keV.

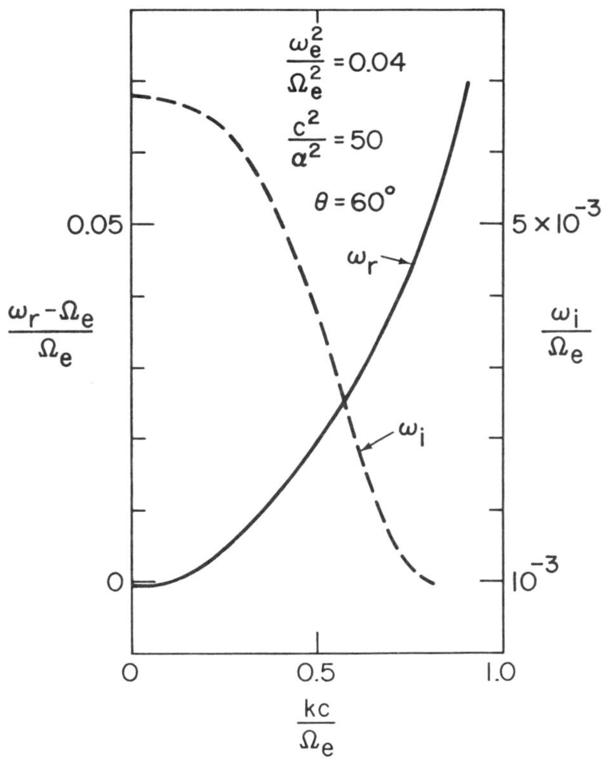

Fig 6. A plot of the real frequency ω_r and growth rate ω_i as functions of k for $\omega_e^2/\Omega_e^2 = 0.04$, $c^2/\alpha^2 = 50$, and $\theta = 60°$.

where

$$\eta_{ij} = \begin{pmatrix} -\frac{\omega_e^2}{2\omega(\omega+\Omega)} & i\,\frac{\omega_e^2}{2\omega(\omega+\Omega)} & 0 \\ -i\,\frac{\omega_e^2}{2\omega(\omega+\Omega)} & -\frac{\omega_e^2}{2\omega(\omega+\Omega)} & 0 \\ 0 & 0 & -\frac{\omega_e^2}{\omega^2} \end{pmatrix}$$

$$M_{xx} = 1 - \frac{2}{1-i\tau} - h\,\frac{k_{\parallel}\alpha}{\omega}\left(1 - \frac{1}{1-i\tau}\right) = M_{yy}$$

$$M_{xy} = -\,i\,M_{xx} = -\,M_{yx}$$

$$M_{xz} = +\,\frac{k_{\perp}\alpha}{\Omega}h\left(\frac{1}{1-i\tau} - 1\right)\left(\frac{2}{1-i\tau} - \frac{\Omega}{\omega}\right) = M_{zx} = iM_{yz} = -iM_{zy}$$

$$M_{zz} = -\,\frac{k_{\perp}^2\alpha^2}{\Omega^2}\left[\frac{h^2}{(1-i\tau)} + \frac{1}{2}\,\frac{(1-4h^2)}{1-i\tau} + h^2\left(\frac{2}{1-i\tau} - \frac{\Omega}{\omega}\right)\right.\ .$$

$\Omega = |\Omega_e|$ and $h \equiv k_{\parallel}c^2/\omega\alpha$. Furthermore, χ_{ij} can be expressed in terms of a basic integral

$$F_q(z) = -\,i\int_0^{\infty} d\tau\,\frac{\exp(iz\tau)}{(1-i\tau)^q}$$

where $z = 2c^2/\alpha^2(\omega - \Omega_e/\omega) - k^2c^4/\alpha^2\omega^2$, which can be related to the usual plasma dispersion function Z , as pointed out by Shkarofsky (1966). The relation is

$$F_q(z) = \sum_{m=0}^{q-3/2} i(-z)^m\,\frac{\Gamma(q-1-m)}{\Gamma(q)} + \frac{\pi^{1/2}}{\Gamma(q)}\,(-z)^{q-3/2}\left[iz^{1/2}Z(iz^{1/2})\right]$$

which facilitates the numerical study of the dispersion equation (18).

From the preliminary results which we have obtained lately a few highlight points may be mentioned as follows: First, it is found that when $m\alpha^2/2 > 2$ keV, the dispersion relation is significantly modified in the vicinity of $\omega \simeq \Omega_e$. The cutoff frequency predicted by the usual cold plasma approximation is no longer true. In fact, as $k \to 0$, the real frequency ω_r of the fast extraordinary mode passes through the point $\omega_r = \Omega_e$ and reaches to the region $\omega_r < \Omega_e$. Second, the maximum growth rate is found to be approximately one order of magnitude higher than that computed for the case in which the secondary electrons are not negligible. Third,

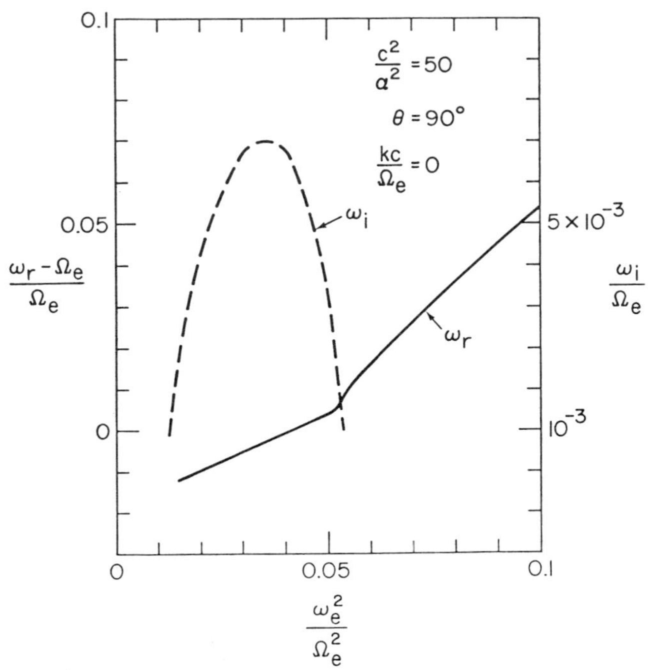

Fig 7. A plot of the real frequency ω_r and growth rate ω_i as functions of ω_e^2/Ω_e^2 for $c^2/\alpha^2 = 50$, $\theta = 90°$ nad $kc/\Omega_e = 0$.

for a fixed energy, i. e., α^2 is given, if we compute ω_i/Ω_e for various ω_e^2/Ω_e^2, we find that ω_i/Ω_e peaks at certain ω_e^2/Ω_e^2, a result similar to that found in the case with secondary electrons. These results can be appreciated from Figs. 5, 6, and 7.

Summary and Concluding Remarks

In this paper we have reviewed the AKR theory based on the cyclotron maser process suggested by Wu and Lee (1979) and Lee et al. (1980). Furthermore, the review also includes some recent theoretical developments along the same line; for example, the discussion of the quasilinear saturation process, the energy conversion efficiency and the limit of zero secondary electrons.

Despite existing theories of AKR in the literature, a complete understanding of the phenomenon is still far away. Since AKR is obviously related to the inverted V events and the parallel electric fields, the study of the basic physics of AKR depends very much upon the understanding of these processes in the polar region. In view of the limited knowledge of the polar region plasma during the disturbed period, one can easily understand that theoretical study of the AKR is rather restricted and hindered. We can only make plausible arguments based on available models concerning the other processes, for example, the inverted V events, etc.

In conclusion we feel that because of the lack of sufficient observational data acquired directly from the source region of AKR, theoretical studies of the subject have encountered numerous basic difficulties. We only have a rather limited amount of information which is needed for theoretical modeling. The ISIS-1 results are valuable but restricted to low altitudes ($\sim$ 3500 km). On the other hand, S3-3 satellite has reached the right altitude but the satellite has no AKR measurement. We hope that the forthcoming OPEN program and Dynamic Explorer may help us to resolve many pertinent questions concerning the physics of the polar region which are crucial and urgently needed for our research of AKR.

Acknowledgments. The present research was supported by the National Aeronautics and Space Administration grant NGL 21-002-005. The author wants to thank many of his colleagues and collaborators for fruitful and stimulating discussions.

References

Axford, W.I., Magnetic storm effects associated with the tail of the magnetosphere, Space Sci. Rev., 149, 1967.

Barbosa, D.D. Ph.D. thesis, UCLA, 1976.

Benson, R.F., Sources mechanism for terrestrial kilometric radiation, Geophys. Res. Lett. 2, 52, 1975.

Benson, R.F. and W. Calvert, Isis I observations at the source of auroral kilometric radiation, Geophys. Res. Lett., 6, 479, 1979.

Chiu, Y.T. and M. Schulz, Self-consistent particle and parallel electric field distributions in the magnetospheric-ionospheric auroral region, J. Geophys. Res. 83, 629, 1978.

Grabble, C., Palmadesso, and K. Papadopoulos, A coherent nonlinear theory of auroral kilometric radiation: I. Steady state model, J. Geophys. Res., 85, 659, 1980.

Green, J.L., D.A. Gurnett, and R.A. Hoffman, A correlation between auroral kilometric radiation and inverted V electron precipitation, J. Geophys. Res. 84, 5216, 1979.

Gurnett, D.A., The earth as a radio source: terrestrial kilometric radiation, J. Geophys. Res. 79, 4227, 1974.

Gurnett, D.A. and J.L. Green, On the polarization and origin of the auroral kilometric radiation, J. Geophys. Res., 83, 697, 1978.

Kaiser, M. and J. Alexander, Direct measurements by Voyager 1 and 2 of the polarization of the terrestrial kilometric radiation, Geophys. Res. Lett., 5, 857, 1978.

Kan, J.R., L.C. Lee, and S.I. Akasofu, Two dimensional potential double layer and discrete auroras, J. Geophys. Res. 84, 4305, 1979.

Kintner, P.M., M.C. Kelley, R.D. Sharp, A.G. Ghielmetti, M. Temerin, C. Catell, P.F. Mizera and J.F. Fennell, Simultaneous observations of energetic upstreaming ions and electrostatic hydrogen cyclotron waves, J. Geophys. Res. 84, 7201, 1979.

Lee, L.C. J.R. Kan, and C.S. Wu, Generation of auroral kilometric radiation and the structure of auroral acceleration region, Planet. Space Sci., 28, 703, 1980.

Lee, L.C. and C.S. Wu, Amplification of radiation near cyclotron frequency due to electron population inversion, Phys. Fluids, 23, 1348, 1980.

Melrose, D.B., An interpretation of Jupiter's radiation as direct amplified gyroemission, Astrophys. J., 207, 651, 1976.

Palmadesso, P., T.P. Coffey, S. Ossakov, and K. Papadopoulos, Generation of terrestrial radiation by a beam-driven electromagnetic instability, J. Geophys. Res., 81, 1762, 1976.

Roux, A. and R. Pellat, Coherent generation of the terrestrial kilometric radiation by nonlinear beating of electrostatic waves, J. Geophys. Res., 84, 5189, 1979.

Shkarofsky, I.P., Dielectric tensor in Vlasov plasmas near cyclotron harmonics, Phys. Fluids, 9, 561, 1966.

Tsai, S.T., C.S. Wu, Y.D. Wang, and S.W. Kang, Dielectric tensor of a weakly relativistic, nonequilibrium and magnetized plasma. Submitted to Phys. Fluids, 1980.

Wu, C.S., C.S. Lin, H.K. Wong, S.T. Tsai, and

R.L. Zhou, Absorption and Amplification of radiation near cyclotron frequency in a non-equilibrium plasma, submitted to Phys. Fluids, 1980.

Wu, C.S., S.T. Tsai, M.J. Xu, and J.W. Shen, Saturation and energy-conversion efficiency or auroral kilometric radiation, submitted to Astrophys. J., 1980.

ALTITUDE DEPENDENCE OF AURORAL BEAM GENERATED ELECTROSTATIC NOISE

J.E. Maggs

Institute of Geophysics and Planetary Physics, University of California
Los Angeles, California 90024

Abstract. A specific altitude dependent model of the auroral beam is used to investigate the effects of changing ionospheric parameters on the power flux spectrum of beam generated electrostatic noise. In the model the beam pitch angle distribution evolves conserving the first adiabatic (flux) invariant. The beam generates electrostatic whistler noise through Landau resonance over most of the field line. Cyclotron resonances can drive narrow band upper hybrid noise at low altitudes. Only a relatively narrow range of beam source densities generates noise linearly along the entire field line. Generally at altitudes of a few thousand kilometers the beam generated whistler noise is intense enough that nonlinear plasma processes can be expected to be important. The evolution of the beam in the presence of this intense electrostatic noise is not investigated using this model.

The loss cone in the upgoing mirrored beam flux will support weakly growing electromagnetic noise in the R-mode.

Introduction

This paper presents a brief review of a theoretical investigation into the plasma physical processes involved in the generation of radio noise in the auroral ionosphere. The bulk of the paper is devoted to a description of a theory for the convective amplification of electrostatic noise by the precipitating electron fluxes responsible for the visual aurora. Only brief mention will be made of some of the implications of this theory for the production of electromagnetic modes.

Historically the interest in radio noise produced by auroral processes was stimulated by the disruption of communications in the polar regions caused by this natural interference. With the advent of polar orbiting satellites it was observed that a broadband, relatively low frequency, radio noise, dubbed VLF hiss, was very closely associated with auroral precipitation (Barrington et al., 1971; Hoffman and Laaspere, 1972). Initially VLF hiss was attributed to the emission of Cerenkov radiation in the whistler mode by the individual electrons in the precipitating auroral flux (Jorgensen, 1968). VLF hiss was pictured as arising from an incoherent quasi-thermal process. The inability of this simple theory to explain the observed power fluxes in the VLF hiss (Taylor and Shawhan, 1974) prompted investigations into the production of this noise via a collective plasma process - the "bump in tail" instability (Swift and Kan, 1975; Maggs, 1976). Subsequently other plasma modes have been observed generated by the auroral beam. Most notable, in terms of total radiated power, are the electromagnetic modes of auroral kilometric radiation (Gurnett, 1974).

The investigation of auroral beam produced noise may even be an important part of the study of the mechanism that produces the auroral beam. It is now generally acknowledged that the auroral precipitation is caused by a field aligned electric field. The exact structure of this field is not known. The electric potential can be regarded as a specific example of a BGK mode, however, so that trapped particle populations could play an important role in determining the potential structure. Since waves can scatter trapped particles into untrapped orbits, or vice versa, the presence of intense waves on auroral field lines may alter the global structure of the electric potential by modifying the trapped particle population.

Theoretical Method

The general theoretical approach used to estimate the power flux spectra of waves produced by the precipitating electron flux is based on WKB theory (Maggs, 1976). To calculate the wave noise at a point in space at a given frequency, the noise from all ray paths of the wave mode of interest is summed. Along a given ray path the noise is found by integrating the wave kinetic equation

$$\frac{dP}{d\omega}(\omega,\underline{k},x) = \int_{s_1}^{\infty} \varepsilon\, F \exp 2\left(\int_{s_1}^{s} \frac{\gamma}{v_g}\, ds'\right) ds$$

where s_1 is the point of observation and s denotes distance along the ray path. Each unit volume of plasma along the path emits a flux of waves, ε, assumed to be given by the incoherent radiation levels. F is a factor representing the spreading of ray paths. The emitted noise then convectively amplifies or decays along the ray path at a local rate of γ/v_g. The temporal growth rate, γ, is determined by the state of the plasma. If the wave amplitude is small, γ is the linear growth rate. If the amplitude is large, γ must be found by including nonlinear effects. However, in a strongly nonlinear plasma the WKB concept is nugatory. The group velocity of the waves, v_g, depends upon the dispersion relation of the wave mode under investigation.

The precipitating auroral electrons are pictured as a spatially limited nonthermal population that generates a nonthermal wave flux. The wave flux can be calculated using the concepts of wave propagation. The wave flux calculation then requires knowledge of the properties of the ionospheric plasma as well as details of the nonthermal precipitating electron flux because wave propagation plays a major role.

Properties of Auroral Precipitation

A detailed calculation of power flux spectra using the wave kinetic theory requires a complete model of the precipitating auroral flux and auroral ionosphere. Representative properties of the auroral flux have been established using satellite and rocket observations. The auroral precipitation typically shows a peak in the differential intensity between one and ten keV (Meng, 1976; Liu and Hoffman, 1979). The fluxes below this peak increase with decreasing energy usually in a power law fashion. The population of electrons following a power law distribution will be referred to as warm electrons. The peak in the differential intensity often represents a region of positive slope in velocity space and thus is a potential source of wave generation. The pitch angle distribution of the electrons near the peak intensity varies with altitude. At altitudes of a few hundred kilometers the intensities in the downward flux are nearly uniformly distributed about the magnetic field (Kaufman et al., 1978). At higher altitudes the intensity peak is confined to small pitch angles (Frank and Gurnett, 1971; Eliasson et al., 1979). The distribution of energetic electrons is important in determining the temporal growth rate, γ. The complete electron distribution includes the relatively cold dense ionospheric population. It is this cold population that determines the wave dispersion properties and enters into the convective growth rate primarily through v_g. The scale lengths of the ionospheric electron population and the geomagnetic field, along with the global geometry of the auroral precipitation, also play important roles in determining the detailed forms of wave power flux spectra.

Altitude Dependent Model

The auroral precipitation is modeled as an electron beam with a drifting Maxwellian distribution produced at some source altitude on auroral field lines (Maggs and Lotko, 1981). The beam pitch angle distribution is then assumed to evolve according to conservation of total energy and first adiabatic invariant $\mu = v_\perp^2/2B$. Below the source altitude the parallel electric field is assumed zero. The population of warm electrons is assumed to be isotropic in pitch angle and the number density constant with altitude. The entire warm electron population is modeled by a sum of six Maxwellian distributions of various number densities and temperatures. In addition, a single hot Maxwellian plasma is included to better model the electron population in the vicinity of the beam. The velocity space distribution as a function of α is shown in Figure 1. The parameter α is the ratio of the source field strength B_s to the local field strength B, and thus decreases with decreasing altitude away from the source. The beam electron pitch angle distribution tends toward isotropy in the downward hemisphere as α tends to zero. In the examples shown in this paper the source altitude is taken as 7500 km and the smallest value of α in the model is .12. However, the source altitude need only be increased to over 10,000 km in order to obtain a value of α less than .05. We will see that small values of α are important for generating some types of noise. In Figure 2 the model of the warm electrons is shown to agree very well with observations from ISIS and is used for this reason.

Three different models of the topside ionospheric number density profile have been chosen based on Maeda's model (Maeda, 1975) and S3-3 data. The models were chosen to represent a range of ionospheric conditions from dense to depleted. The ionospheric number density is assumed to vary in the magnetic meridian plane (roughly north-south) with a scale length equal to the vertical scale length at that altitude. This is an ad hoc assumption made in the absence of direct observation. The magnetic field strength is assumed to vary as r^{-3}, as in a dipole field. The vertical profiles and scale lengths used in the model are shown in Figures 3 and 4.

The precipitation region itself (the auroral arc) is taken to be uniform in the magnetic meridian plane, (north-south direction) and infinitely extended in the direction perpendicular to this plane (east-west direction). Effects of arc curvature in the east-west direction can be approximately modeled by introducing a finite extent to the arc in this direction. The arc width is assumed to vary with altitude as the cross section of a flux tube (viz, $B^{\frac{1}{2}}$). In the model the arc width is taken as 20 km at the source altitude.

Wave Mode Characteristics

The various wave modes produced by auroral electrons can be roughly divided into two types: electrostatic and electromagnetic. Electrostatic waves have large indicies of refraction ($\mu = ck/\omega$), and the wave electric field is closely aligned

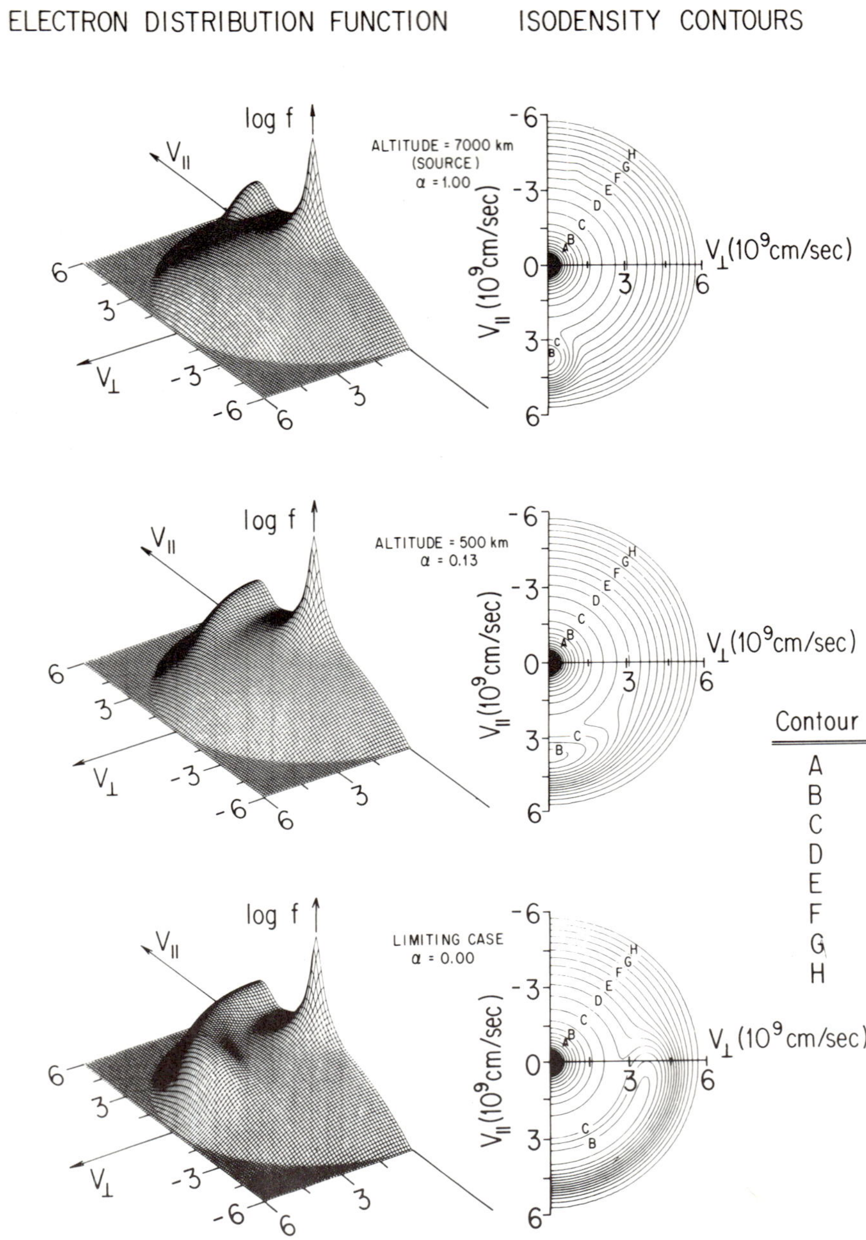

Contour	f (10^{-31} cm^{-6} sec^{3})
A	$10^{3.0}$
B	$10^{2.6}$
C	$10^{2.2}$
D	$10^{1.8}$
E	$10^{1.4}$
F	$10^{1.0}$
G	$10^{0.6}$
H	$10^{0.2}$

Fig. 1. Contours of constant density for the model of the electron velocity distribution are shown for various values of the parameter $\alpha = B_s/B$. The electron beam is a drifting Maxwellian at the source ($\alpha = 1$) and changes towards a hemispherical distribution at lower altitudes. The smallest value of α in the model is .12 at 300 km. The value of α at low altitudes would decrease if the source were at higher altitudes so that the distribution at low altitudes would resemble the limiting case of $\alpha = 0$

along the wave number vector. Electromagnetic waves on the other hand generally have indices of refraction less than unity (phase velocities faster than the speed of light) and the electric field vector is nearly perpendicular to the wave number vector. For waves interacting resonantly with the electron distribution the waves either interact directly with the particles through the Landau resonance (the wave phase velocity along the magnetic field, $\omega/k_{\parallel}$, is equal to the particle velocity along the magnetic field, $v_{\parallel}$) or indirectly through the doppler shifted cyclotron resonances

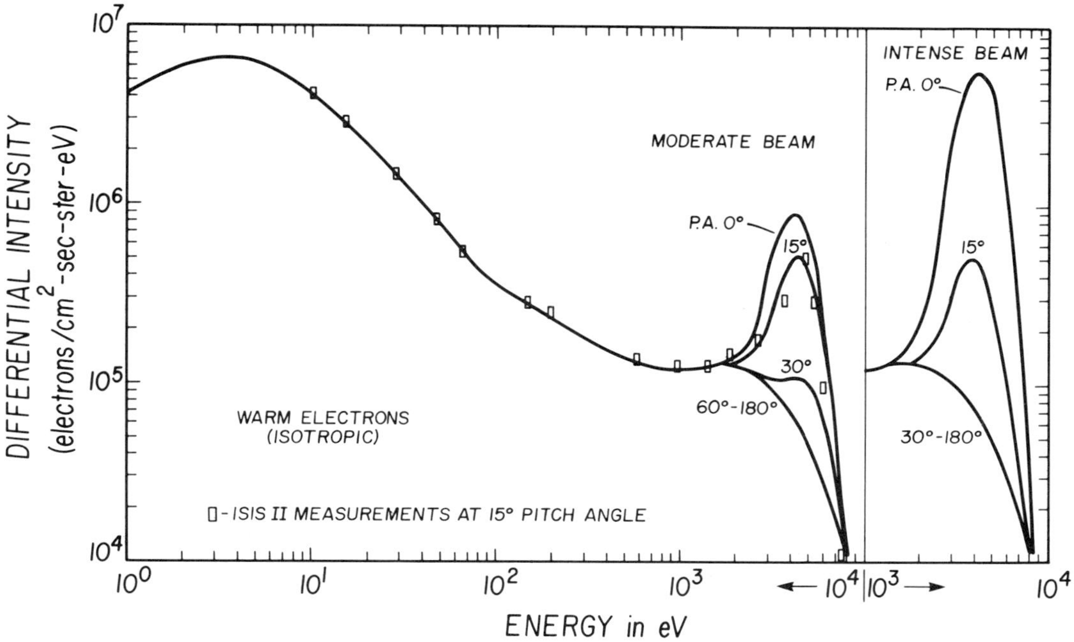

Fig. 2. The squares show electron measurements from ISIS-II. The solid line is the model distribution which can be seen closely models the measured distribution. Fluxes of the beam at the source altitude are shown for various pitch angles.

$\omega - n\omega_{ce} = k_{\parallel}v_{\parallel}$. Waves interacting with electrons through the Landau resonance alter the distribution of parallel velocities while the cyclotron resonant waves alter the distribution of perpendicular velocities. Only electrostatic waves can interact directly with the auroral electrons because the velocity of the auroral beam, v_b, is such that $v_b/c << 1$ so that any wave interacting with the beam through the Landau resonance must have a very large index of refraction ($\mu >> 1$). Electromagnetic waves can only interact with the auroral electrons through the cyclotron resonances. Because of their direct interaction with the beam particles and their slow group velocities electrostatic waves tend to be readily amplified by the auroral beam and can reach large amplitudes.

The dispersion characteristics of electrostatic waves arises largely from the thermal properties of the plasma. Thermal terms in the plasma dispersion relation introduce an infinite number of thermal waves called Bernstein modes - one for each cyclotron harmonic band. Two of these modes are noteworthy, however, because they couple at short wavelengths to two cold plasma modes. These two cold plasma modes are called: 1) the whistler mode propagating at the resonance cone angle; and 2) the upper hybrid resonance mode. The electrostatic whistler propagates at frequencies between the lower hybrid frequency, f_{LH}, and the smallest of the plasma frequency, f_{pe}, or the electron gyrofrequency, f_{ce}. The upper hybrid mode propagates at frequencies between the larger of f_{pe} or f_{ce} and the upper hybrid frequency f_{UH}.

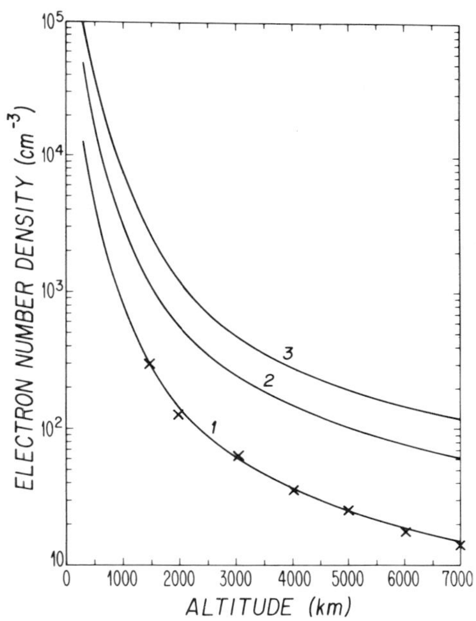

Fig. 3. The ionospheric electron density as a function of altitude for the three models of the topside ionosphere. Model 1 represents a low density ionosphere and is one fourth the value of Model 2. The x's correspond to measurements from the satellite S3-3 (Mozer et al., 1980). Model 2 is based on Maeda's ionospheric model, while Model 3 represents a dense ionosphere and is simply twice as dense as Model 2.

The electromagnetic waves of interest are light waves in regions of low plasma density. They are conveniently classified by their polarization, either right or left circularly polarized. The modes become distinguishable near the major plasma resonances. The right hand circularly polarized wave has a cutoff above the upper hybrid frequency while the left mode cuts off near the plasma frequency for nonparallel propagation.

Predictions of Power Spectra

The basic goal of this theoretical research into beam generated noise is to develop the ability to predict wave power flux spectra given the initial electron distribution at the source. This not only entails a means for calculating the wave spectra but also requires evaluating the effects of the wave noise on the electron distribution and ultimately the accelerating electric potential. The evaluation of these effects depends a great deal on the mechanism limiting the amplitude of the beam generated noise. The noise amplitude is limited by two very disparate means. It can be limited by wave propagation out of the spatially limited auroral beam; or it can be limited by nonlinear mechanisms: transfer of wave energy out of the beam region by wave-wave interactions, alteration of the beam distribution, alteration of the wave dispersion characteristics, or all of these. The wave spectra and subsequent beam dynamics are considerably different depending upon which limiting mechanism is operating.

The propagational limitation of amplitudes is much more closely aligned with the weakly turbulent WKB approach espoused above and most of the results reported below pertain to this limiting mechanism. When amplitudes are limited by ray refraction out of the beam, the beam is said to be below the "beam persistence limit" (Maggs, 1978). In this regime nonlinear effects are assumed negligible, and the beam distribution is then only affected by the wave noise through the weakly turbulent process of quasilinear diffusion. The wave spectra produced are broad, ranging over most of the propagating bandwidth of the modes of interest. The arc geometry and ionospheric properties play key roles in determining the details of the wave power flux spectra.

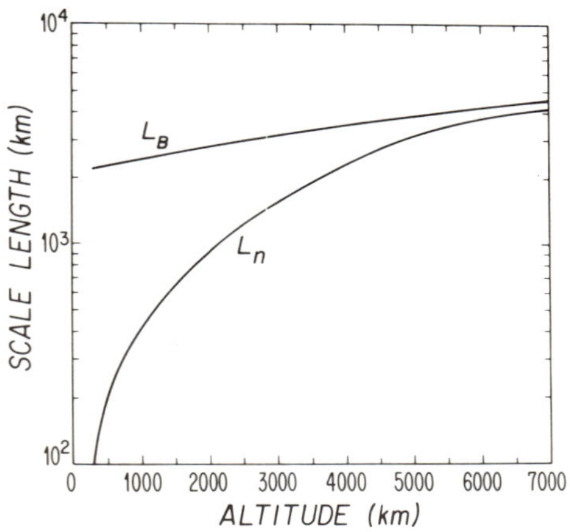

Fig. 4. The scale sizes of the magnetic field and ionospheric vertical density profile increase steadily with altitude. The magnetic field is a dipole model while the number density scale sizes are taken from Model 2.

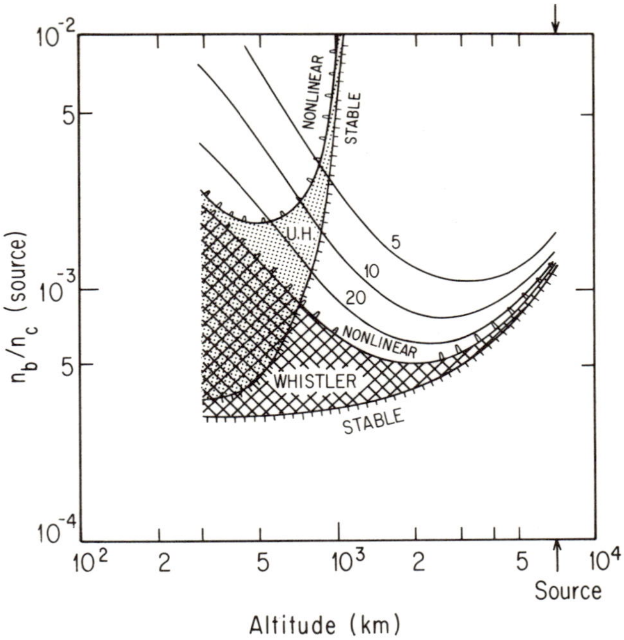

Fig. 5. The regions of linear growth as a function of source density and altitude for the upper hybrid and whistler modes are shown cross-hatched. The line marked "nonlinear" indicates the source density at a given altitude that would produce a wave energy density the same order of magnitude as the beam energy density. The curves marked 5, 10, 20 denote this same limit provided the rays are assumed to be limited to a propagation distance in the east-west direction of 5, 10, or 20 times the arc width.

The method used to study beams below the persistence limit consists of the following steps (only some of which have been followed to completion):

1) The power flux spectra are calculated using linear growth rates to evaluate the wave kinetic equation along ray paths. The plasma emissivity is assumed given by incoherent Cerenkov radiation levels. The wave-particle effects on the electrons are suppressed so that the beam evolves adiabatically with altitude.

2) The electron distribution is evaluated as a function of altitude assuming the distribution is altered by quasilinear diffusion in the presence of the wave noise calculated from the adiabatically evolved beam in step 1.

3) The power flux is recalculated using the altered beam of step 2.

4) Steps 2 and 3 are repeated until the desired accuracy is achieved.

The calculation of the power flux spectra, assuming that nonlinear mechanisms limit the wave amplitude, necessarily requires very different techniques than those outlined above because the WKB concept is no longer applicable. Ion dynamics probably play an important role because the pondermotive force drives ion waves and density fluctuations which, in turn, alter the high frequency dispersion and wave number spectra. These processes lead to rapid transfer of wave energy to short wave lengths and subsequent absorption by electrons of energies below the beam energy (see e.g. Papadopoulos and Coffey, 1974). These complex, interrelated, nonlinear processes are probably best studied by computer simulation of the auroral plasma.

The results of using the theory of convective amplification of incoherent Cerenkov radiation at the linear growth rate with the amplitude limited by ray refraction out of the beam is illustrated in Figure 5. The power levels are calculated using a local approximation algorithm that agrees well with full ray tracing calculations (Maggs, 1978). The regions of linear wave growth for the whistler and upper hybrid modes, growing through the Landau resonant interaction, are shown shaded. The regions of linear growth are functions of altitude and beam number density at the source altitude. The thresholds for wave growth are indicated by the line marked "stable." The thresholds for wave growth of the whistler decreases with altitude roughly as B^{-1} because the beam number density increases almost as rapidly as the cross section of a flux tube. The upper hybrid mode grows only at lower altitudes where $f_{pe} > f_{ce}$; because, at higher altitudes, the entire wave band is quite close to f_{ce}, and cyclotron damping by electrons of hundreds of eV energy is strong.

The line marked "nonlinear" in Figure 5 roughly indicates the upper limit to the linear growth regime. On this line the beam generated power flux is approximately equal to the beam energy density. The line specifically does not indicate the threshold for any particular nonlinear process but rather only indicates that, at the very least, local quasilinear diffusion is strong enough to scatter the beam. The lines marked "5," "10," and "20" illustrate the effect of limiting propagation in the east-west direction to 5, 10, or 20 times, respectively, the local width of the arc. These lines are intended to indicate the effect of arc curvature on the calculation of the power flux.

Individual whistler and upper hybrid spectra typical of those found in the arc are illustrated in Figure 6. The whistler spectra tends to peak near the lower hybrid frequency, especially at higher altitudes. The spectra is broadband and may be relatively flat across the entire bandwidth depending on the ionospheric parameters chosen. The effects of propagation outside the arc are not shown. These propagation effects tend to enhance the peak at the lower hybrid resonance (Yamamoto, 1979) because as the whistler noise propagates downward into the auroral ionosphere it is reflected at the lower hybrid resonance leading to a sharp cutoff in the spectrum at the local value of f_{LH}. These features of the whistler spectrum agree well with the observed properties of VLF hiss. The upward propagating upper hybrid noise, on the other hand, tends to peak at frequencies in the middle of the propagation band. Noise near the plasma frequency in either mode is limited by rapid wave refraction away from parallel propagation. Thus the magnetic field is important in limiting the growth of the, typically, very rapidly growing parallel propagating plasmons. Regions of growth for Bernstein modes are not shown because the growth of these modes is strongly limited by refraction effects and they do not grow at the beam densities shown in Figure 5.

The evolving pitch angle distribution of the beam increases the energy available for wave instabilities interacting with the beam through the cyclotron resonance. This source of wave instability that relys on a positive perpendicular slope ($\partial f/\partial v_{\perp} > 0$) proves useful for driving upper hybrid noise, but it is not very effective in generating Bernstein modes or whistler waves (Lotko and Maggs, 1981). In fact, effective growth of upper hybrid waves is achieved only in the regions of small α (the ratio of the magnetic field strength at the source altitude to the local field strength, B_s/B) or large beam density because upper hybrid waves are stronlgy refracted. The magnitude of upper hybrid wave refraction is illustrated by Figure 7 which shows the ratio of the linear growth time to the wave resonance time as a function of beam density. The threshold beam density for wave growth on the perpendicular energy source is denoted by n_{cr}. The detrimental

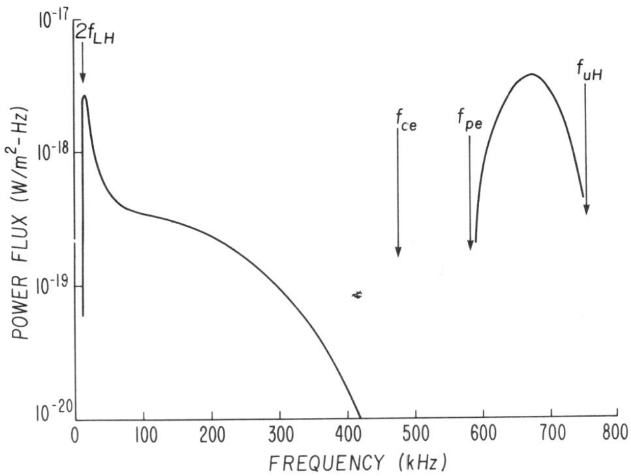

Fig. 6. Examples of the power flux spectra for the whistler and upper hybrid modes obtained using the theory of convective amplification at the linear growth rate. Wave amplitudes are limited by wave refraction out of the beam.

effects (for growth) of wave refraction (changing $\underline{k}$) can be alleviated by choosing large beam densities. Wave growth is then sufficiently rapid to ensure that the plasma parameters do not change significantly over distances on the order of the wave growth length. Figure 8 illustrates that wave growth is also substantially enhanced by decreasing α. Decreasing α corresponds to increasing the source altitude. Thus upper hybird waves can grow from the perpendicular beam energy with dense source beams generated at high altitudes.

In summary, the electrostatic whistler wave grows along the entire field line from source to atmosphere for the model parameters used here. The upper hybrid wave, on the other hand, only grows at low altitudes, where $f_{pe} > f_{ce}$, or α is small. The region of linear growth of the whistler mode is very narrow (in terms of beam source densities) unless the arc geometry is distorted through curvature so that propagation in the east-west direction is limited. Thus it appears that for most auroral beam densities a region of the magnetic field line will exist along which nonlinear interactions become important in the wave and beam dynamics. This region is most likely in the altitude range from 2000 - 4000 kilometers.

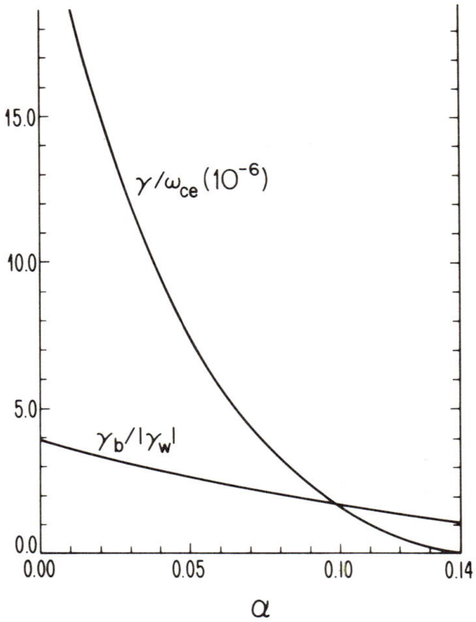

Fig. 8. Dependence of the temporal growth rate on $\alpha \equiv B_s/B$ for density profile 2 at 500 km. Decreasing α corresponds to moving the beam origin to higher altitudes.

Electromagnetic Waves

The auroral beam model developed for the study of electrostatic noise has interesting implications for some of the mechanisms suggested for the generation of electromagnetic radiation by the auroral precipitation. Barbosa (1976) and Roux and Pellat (1979) have argued that Auroral Kilometric Radiation (AKR) is generated from the nonlinear interactions between two electrostatic upper hybrid modes. The electromagnetic wave is generated at twice the upper hybrid frequency. Since the peak AKR flux occurs at frequencies between 150 - 250 kHz in the source region (i.e., greater than about one earth radius in altitude (1 R_E)), strong upper hybrid noise can be generated in the region from only the perpendicular energy source of the beam and then only if $\alpha (B_s/B)$ is small, say $\alpha < .05$. This requires the source altitude be beyond 5 R_E. Alternatively upper hybrid noise may be generated at these altitudes from the parallel beam energy (so the source altitude could be much lower) if the warm electron population is assumed not to be present.

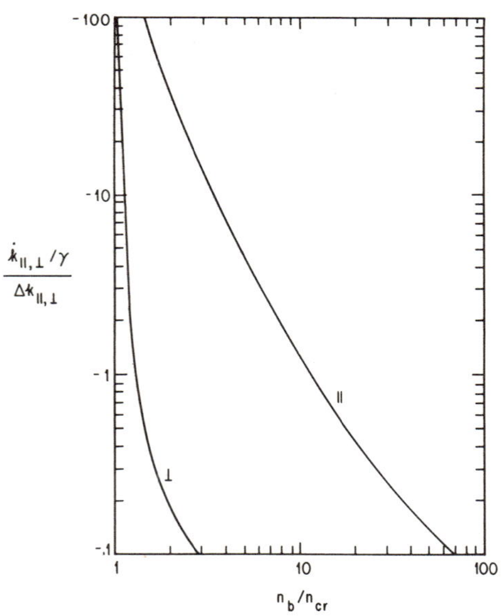

Fig. 7. Ratio of the linear growth time (γ^{-1}) to the resonance time ($\Delta k_{\parallel,\perp}/\dot{k}_{\parallel,\perp}$) as a function of beam number density for density profile 2 at 500 km. "$\parallel$" ($\perp$) refers to the resonance time associated with refraction of parallel (perpendicular) wavenumbers. The model distribution function has $n_b/n_{cr} \simeq 1.14$ and $k/\gamma\Delta k \simeq 10$.

Melrose (1976) has suggested that AKR is generated directly in the downward propagating right hand polarized electromagnetic mode through a cyclotron interaction with the beam electrons. Melrose assumed the beam had a temperature anisotropy. The model beam has a source of perpendicular energy at small α and thus might generate AKR through the cyclotron resonance interaction as suggested by Melrose. As it turns out amplification of the R-mode is possible but only for an extremely small region of ($\omega,\underline{k}$) space. This is better illustrated by a related mechanism suggested by Wu and Lee (1979) in which an upgoing R-mode cyclotron resonates with the beam popula-

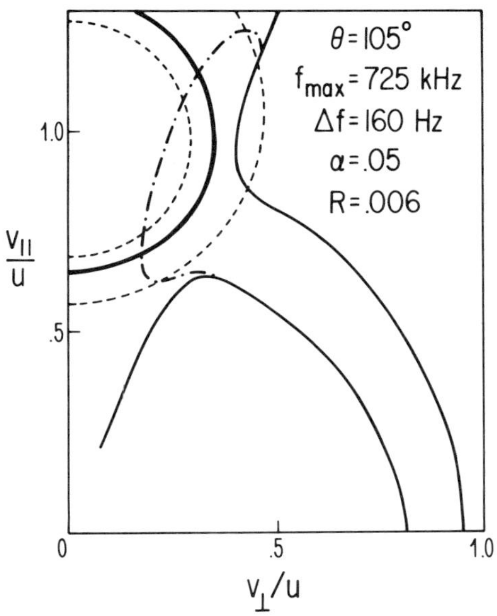

Fig. 9. The region of positive perpendicular slope lies between the solid funnel shaped lines. The contours of particle resonance are circles. The dash-dot curve illustrates the limits of the region of positive slope if only the beam electrons have a loss cone. The dashed circles delimit the region of growth lengths smaller than 1500 km for the upward propagating R-mode. The minimum growth length of 300 km occurs along the heavy circular contour.

tion as well as the warm population. The upgoing warm electrons have a loss cone feature and can thus assist in driving the instability. This mechanism has a much broader region of growth in (ω,$\underline{k}$) space than that suggested by Melrose, but it is still extremely limited in range.

Figure 9 illustrates the limited range of growing waves in parameter space for the Wu and Lee mechanism. The region of positive perpendicular slope lies between the curved, funnel-shaped solid lines. The dash-dot curve illustrates the positive slope region if only the beam particles are assumed to have a loss cone feature. The circular contours are the resonant interaction curves. They are circular because relativistic corrections have been taken into account. For this particular example the minimum growth length is less than 1500 kilometers. These growth lengths are about doubled if only the beam population is assumed to have the loss cone feature. The total bandwidth of this region is only 160 Hz. Thus the ionosphere would have to be very homogenerous in order that the waves not move out of the growth region. Two orders of magnitude longer scale lengths than used in the model ionosphere would be needed to obtain one e-folding of the instability. Not all of the possible "reasonable" density configurations have been investigated, but it is clear that refractive effects will be important for this process. The process can grow R-mode waves off the model beam and thus merits further study as a candidate for AKR.

Concluding Remarks

Work to date indicates that, under certain conditions, the power flux spectra of electrostatic noise generated by the precipitating auroral electrons can be understood in terms of convective amplification at the linear growth rate. Wave noise amplitudes are limited by refraction out of the spatially limited growth region. It is also quite clear that a great deal of the time the auroral beam should be dense enough to produce nonlinear interactions in the amplitude region of a few thousand kilometers along auroral field lines. The effects of this wave noise on the beam electron distribution and any subsequent effects on the parallel electric field remain to be evaluated. Post acceleration wave particle dynamics may be a major part of the steady state (if one exists) auroral acceleration mechanism.

References

Barbosa, D.D., Electrostatic mode coupling at $2\omega_{UH}$: A generation mechanism for auroral kilometric radiation, Ph.D. thesis, Univ. of Calif., Los Angeles, 1976.

Barrington, R.E., T.R. Hartz and R.W. Harvey, Diurnal distribution of ELF, VLF and LF noise at high latitudes as observed by Alouette 2, J. Geophys. Res., 76, 5278, 1971.

Eliasson, L., Holmgren, L.-A. and K. Rönnmark, Pitch angle and energy distributions of auroral electrons measured by the ESRO 4 satellite, Planetary and Space Science, 27, 87, 1979.

Frank, L.A. and D.A. Gurnett, Distributions of plasmas and electric fields over the auroral zones and polar caps, J. Geophys. Res., 76, 6829, 1971.

Gurnett, D.A., The earth as a radio source: Terrestrial kilometric radiation, J. Geophys. Res., 79, 4227, 1974.

Hoffman, R.A. and T. Laaspere, Comparison of very low frequency auroral hiss with precipitating low-energy electrons by the use of simultaneous data from two OGO 4 experiments. J. Geophys. Res., 77, 640, 1972.

Jorgensen, T.S., Interpretation of auroral hiss measured on OGO 2 and at Byrd Station in terms of incoherent Cerenkov radiation, J. Geophys. Res., 73, 1055, 1968.

Kaufmann, R.L., Dusenbury, P.B., Thomas, B.J. and R.L. Arnoldy, Auroral electron distribution function, J. Geophys. Res., 83, 586, 1978.

Lin, C.S. and R.A. Hoffman, Characteristics of the inverted-V event, J. Geophys. Res., 84, 1514, 1979.

Lotko, W. and J.E. Maggs, Amplification of electrostatic noise in cyclotron resonance with an adiabatic auroral beam, J. Geophys. Res., 86, May 1981.

Maeda, K., A calculation of auroral hiss with improved models for geoplasma and magnetic field, Planet. Space Sci., 23, 843, 1975.

Maggs, J.E., Coherent generation of VLF hiss, J. Geophys. Res., 81, 1707, 1976.

Maggs, J.E., Electrostatic noise generated by the auroral electron beam, J. Geophys. Res., 83, 3173, 1978.

Maggs, J.E. and W. Lotko, Altitude dependent model of the auroral beam and beam generated electrostatic noise, J. Geophys. Res., 86, May 1981.

Melrose, D.B., An interpretation of Jupiter's decametric radiation and the terrestrial kilometric radiation as direct amplified gyroemission, Astrophys. J., 207, 651, 1976.

Meng, C.-I., Simultaneous observations of low energy electron precipitation and optical auroral arcs in the evening sector by the DMSP 32 satellite, J. Geophys. Res., 81, 2771, 1976.

Mozer, F.S., C.A. Cattel, M.K. Hudson, R.L. Lysak, M. Termerin and R.B. Torbert, Satellite measurements and theories of low altitude auroral particle acceleration, Space Sci. Reviews, 27, 155, 1980.

Papadopoulos, K. and T. Coffey, Anomalous resistivity in the auroral plasma, J. Geophys. Res., 79, 1558, 1974.

Roux, A. and R. Pellat, Coherent generation of auroral kilometric radiation by nonlinear beatings between electrostatic waves, J. Geophys. Res., 84, 5189, 1979.

Swift, D.W. and J.R. Kan, A theory of auroral hiss and implications on the origin of auroral electrons, J. Geophys. Res., 80, 985, 1975.

Taylor, W.W.L. and S.D. Shawhan, A test of incoherent Cerenkov radiation for VLF hiss and other magnetospheric emissions, J. Geophys. Res., 79, 105, 1974.

Wu, C.S. and L.C. Lee, A theory of the terrestrial kilometric radiation, Astrophysical J., 230, 621, 1979.

Yamamoto, T., On the amplification of VLF hiss, Planet. Space Sci., 27, 273, 1979.

STATIONARY ELECTROSTATIC SOLITARY WAVES IN THE AURORAL PLASMA

W. Lotko and C. F. Kennel

Department of Physics and Institute of Geophysics and Planetary Physics, University of California, Los Angeles, California 90024

Abstract. We use time-stationary fluid equations to describe electrostatic solitons in an auroral plasma of cold ionospheric and hot plasma sheet particles. Plasma sheet electrons drift along the magnetic field to simulate a field-aligned current. Two weakly damped, small amplitude solitons propagate in the direction of the electron drift. The slower mode is a generalization of an ion-acoustic soliton. Since it can be a negative pulse when cold electrons are depleted relative to cold ions and it can stand in a polar wind, this mode may be associated with auroral double layer phenomena. The faster mode resembles a clump of shielded negative space charge convected by the drifting plasma sheet electrons.

Introduction

Observations of intense, quasi-static, localized electric fields on auroral field lines indicate that the auroral plasma can evolve to a highly coherent and nonlinear state. The spatial profiles of the observed fields (Mozer et al., 1979) sometimes resemble those of shock waves, thus encouraging their identification as electrostatic shocks (Hudson and Mozer 1978) and double layers (Shawhan et al., 1978). The variety of observed spatial profiles suggests however that more general nonlinear structures may be present. For example, the so-called V-shocks (Mozer et al., 1977) detected by the S3-3 electric field antenna have the appearance of negatively polarized solitary waves propagating nearly perpendicular to the geomagnetic field. Large amplitude localized oscillatory waves (Torbert and Mozer 1978) have also been found. Theory has concentrated largely on time-stationary, laminar, electrostatic potential transitions (Block 1972; Swift 1975, 1979; Levine and Crawford 1980; Kan and Lee 1980). The various theoretical profiles reflect different choices of trapped particle distributions and transitional boundary conditions. While laminar transitions certainly occur at times, there has been little theoretical investigation of other nonlinear structures that occur in the auroral plasma. Of those which might arise, we will consider small amplitude solitary waves in the electrostatic potential with polarization along the geomagnetic field. These modes have scale lengths along the magnetic field which are much less than the perpendicular scale length.

Our interest in small amplitude solitons is twofold. First, the theory of these nonlinear waves is analytically tractable. Second, four independent, one-dimensional computer simulations (Sato and Okuda 1980; Hudson and Potter 1980; Chanteur et al. 1980; Barnes 1980) indicate that a double layer evolves from an initially small amplitude, negatively polarized soliton which apparently is amplified by coupling to the electron current. We will not consider the nonlinear evolution of the small amplitude waves but do describe the time-stationary properties of electrostatic solitons allowed in the auroral plasma. Of course, by restricting the analysis to small-amplitude, parallel polarized modes, we will not be able to describe the obliquely polarized modes which are the easiest to observe at high altitudes.

We model the auroral plasma as a four fluid medium. Two originate in the plasma sheet where $T_{ih} \sim 5T_{eh} \sim 5$ keV and n_{ih}, $n_{eh} \sim 1\text{-}10$ cm^{-3} (Garrett 1979). The other two originate in the ionosphere. At lower altitudes the cold ionospheric protons and electrons have $T_{ic} \sim T_{ec} \sim 0.1 - 1$ eV and n_{ic}, $n_{ec} \sim 10^3 - 10^4$ cm^{-3} (Banks and Kockarts 1973). At higher altitudes a small, upward-directed, quasi-neutral parallel electric field (Chiu and Schulz 1978) leads to $T_{ec} \sim \alpha T_{ic}$ and $n_{ic} \sim \alpha n_{ec}$ where α is of order 10. The fact that $n_{ic} \neq n_{ec}$ (implying also $n_{ih} \neq n_{eh}$) has significant consequences for the soliton properties. We give the plasma sheet electrons a small zeroeth-order drift to model a field-aligned current. The ionospheric plasma may also have an upward bulk drift resulting from pressure imbalance between the plasma sheet and ionosphere (Banks and Holzer 1969; Lemaire and Scherer 1973).

Small-Amplitude Fluid Analysis

For parallel propagating waves, the analysis is identical to the unmagnetized case. Since the scale lengths of the geomagnetic, gravitational and quasi-neutral electric fields are large compared to the soliton scale length (∿ Debye lengths), zeroeth-order inhomogeneities and fields can be neglected. The system is described by the equations of continuity, momentum, and energy, together with Poisson's equation. Except for the hot electron fluid, which is always taken to be isothermal, the ratio of specific heats for a given species is set to $\gamma = 1$ ($\gamma = 3$) when the wave phase velocity is less (greater) than the species thermal speed $a_s = \sqrt{\gamma_s T_s / ms}$. Unperturbed fluid drifts are taken parallel to the magnetic field. We assume that all quantities vary as $\xi = x - st$ where s is the wave speed. Upon choosing dimensionless variables n, u, T and ϕ representing species number density, velocity, temperature and electrostatic potential, the stationary fluid and Poisson equations become

$$(u - s)n' + nu' = 0, \qquad (1)$$

$$n(u - s)u' + \mu(nT)' = -\varepsilon\mu n\phi', \qquad (2)$$

$$(u - s)T' + (\gamma - 1)Tu' = 0, \qquad (3)$$

$$1/2(\phi')^2 = -\sum\varepsilon\int_0^\phi n(\phi)\, d\phi \equiv -V(\phi), \qquad (4)$$

where $\mu = 1$ or m_e/m_i and $\varepsilon = -1$ or $+1$ for electrons or ions respectively and ' refers to $d/d\xi$. Densities have been normalized to the total unperturbed density $n_T = n_{eo} = n_{io}$, temperatures to the hot electron temperature T_{eh}, velocities to $(T_{eh}/m_e)^{1/2}$, electrostatic potential to T_{eh}/e, distance to $(T_{eh}/4\pi n_T e^2)^{1/2}$ and time to $(m_e/4\pi n_T e^2)^{1/2}$. The reference (unperturbed) potential has been chosen as $\phi_o = 0$.

The Sagdeev potential, $V(\phi)$, characterizes the system. For small amplitudes ($\phi << 1$), $V(\phi)$ may be expanded in a Taylor series. To third order, the lowest nonlinear order we have

$$V(\phi) = \frac{\phi^2}{2} V_o'' + \frac{\phi^3}{6} V_o''' \qquad (5)$$

V_o and V_o' are zero because of the boundary conditions for a localized disturbance, quasi-neutrality and zero electric field in the unperturbed region. Nonlinear periodic waves can be retrieved when $V_o' \neq 0$, and laminar shock solutions, corresponding to the boundary conditions $V(0) = V'(0) = V(\phi_s) = V'(\phi_s) = 0$, $V''(0)$, $V''(\phi_s) < 0$, arise in fourth order.

V_o'' and V_o''' computed from equations (1) - (3) are

$$V_o'' = \sum \varepsilon \left.\frac{\partial n}{\partial \phi}\right|_{\phi=0} = \sum \frac{\mu n_o}{(s - u_o)^2 - a^2}, \qquad (6)$$

$$V_o''' = \sum \varepsilon \left.\frac{\partial^2 n}{\partial \phi^2}\right|_{\phi=0} = \sum \varepsilon\, \mu^2 n_o \times \frac{3(u_o - s)^2 + (\gamma - 2)\, a^2}{[(u_o - s)^2 - a^2]^3}. \qquad (7)$$

When $\phi'' = -k^2\phi$ and the third order term is neglected, the linear dispersion relation is recovered. In linear approximation, real k requires $V_o'' > 0$. Keeping the third order term, the solution to equation (4) consistent with the boundary conditions $\phi, \phi' \to 0$ as $|\xi| \to \infty$ is:

$$\phi = \frac{-3V_o''}{V_o'''} \operatorname{sech}^2 \frac{\sqrt{-V_o''}}{2}\, \xi, \quad V_o'' < 0. \qquad (8)$$

In contrast to linear waves, stationary solitary waves are characterized by $V_o'' < 0$ and are positive (negative) potential wells depending on whether $V_o''' > 0$ (< 0).

The critical soliton speed (s_o) is determined by $V_o'' = 0$ or

$$\sum_\ell \frac{\mu n_o}{(s_o - u_o)^2 - a^2} = 0, \qquad (9)$$

which is the k = 0 mode or quasi-neutral dispersion relation. For ℓ species, (9) is a $2(\ell - 1)$ order polynomial and therefore $2(\ell - 1)$ wave modes are allowed. When all unperturbed drifts are zero, half of the modes propagate in the positive direction and half in the negative direction with $s_{o-} = -s_{o+}$. When zeroeth order drifts are small ($u_o << a$), the distinction between forward and backward propagating modes still exists; however, the relation $s_{o-} = -s_{o+}$ is only approximate. We will consider only the three forward travelling modes allowed in our four fluid model, since they may be weakly amplified by a resonant process involving the drifting electrons (cf. Taniuti 1974; Karpman 1979).

The deviation of V_o'' from zero can be determined by considering wave speeds which differ slightly from s_o. Let $s = s_o + w$ where $|w| << s_o$ and s_o is a root of (9). To lowest order in w we have

$$V_o'' \simeq \frac{-(M - 1)}{L_s^2}, \qquad (10)$$

where

$$\frac{1}{L_s^2} = 2\, s_o \sum \frac{\mu n_o (s_o - u_o)}{[(s_o - u_o)^2 - a^2]^2}. \qquad (11)$$

$M \equiv s/s_o$ is the soliton Mach number, and according to (8), the width of the localized disturbance scales as L_s. Equation (10) implies that small amplitude disturbances have $M \simeq 1$.

For the stationary fluid theory to be approximately correct, the waves must propagate relatively unattenuated. One measure of the attenuation of a small amplitude soliton is the linear growth rate (deduced from kinetic theory) of the $k = 0$ mode which propagates at about the same phase velocity as the soliton. When the growth rate is small ($|\gamma|/\omega << 1$), attenuation of the solitary wave is negligible. In general, when the soliton propagates at speeds greater than about 2-3 times the thermal speed of a given species, resonant particle effects due to that species are small since the wave interacts primarily with particles in the tail of that distribution function. At the other extreme, for phase speeds less than about 0.1 the thermal speed, the wave speed is in the flat part of the distribution function and again resonant effects are small.

Time-Stationary Wave Properties

We call the three forward propagating solitons slow, intermediate, and fast, according to the magnitude of their propagation speeds.

Slow Mode

Phase velocity. The slow soliton, which propagates at speeds satisfying $a_{ic} << s << a_{ec}, a_{ih}, a_{eh}$, resembles the ion-acoustic soliton (Chen 1974) in a two-component nonisothermal plasma with $T_e >> T_i$. Its critical speed (equation 9) is approximately

$$s_o \simeq a_{ic} \sqrt{1 + x/3} \quad , \tag{12}$$

$$x = \frac{n_{ic}}{T_{ic}} \left(\frac{n_{ih}}{T_{ih}} + \frac{n_{ec}}{T_{ec}} + \frac{n_{eh}}{1 - u^2} \right)^{-1} \quad . \tag{13}$$

We have assumed that u, the unperturbed hot electron drift speed, is not too close to one. In this approximation, apart from direction of propagation, the forward and backward slow modes are identical. When $x >> 3$, the phase velocity is $s \simeq c_s \sqrt{xT_{ic}}$ with $c_s^2 \equiv T_{eh}/m_i$. Since $T_{ec} < 1$ by definition, $xT_{ic} < 1$. Therefore the slow mode critical speed is always less than that of the usual ion-acoustic soliton. In this limit, with $n_{ih} = 0$ and $u = 0$, we recover Buti's (1980) result for ion-acoustic solitons in a two-electron temperature plasma. With $n_{ih} = n_{ec} = 0$ but $u \neq 0$, $s \simeq c_s\sqrt{1 - u^2}$. In fluid theory, an electron drift reduces the critical speed below c_s.

Soliton polarity. Equations (7) and (12) lead to an approximate criterion for a negatively polarized soliton:

$$n_{ic} > 1.5\, n_{ec} r^2 \left[1 + \sqrt{1 + \frac{48}{9r}\frac{T_{ic}}{T_{ec}}} \right] \quad , \tag{14}$$

where

$$r \equiv 1 + \frac{T_{ec}}{T_{ih}}\frac{n_{ih}}{n_{ec}} + \frac{T_{ec}}{1 - u^2}\frac{n_{eh}}{n_{ec}} \quad .$$

(14) implies that negative potential solitons require the cold ion density to significantly exceed the cold electron density. The criterion is less restrictive when T_{ec} is small compared to T_{eh} or T_{ih} or large compared to T_{ic}. Figure 1 plots inequality (15) for $T_{ec}/T_{ic} = 1$ and 10 with $(1 - u^2)T_{eh}/T_{ec}$ and T_{ih}/T_{eh} held constant. Figure 2 shows the dependence of inequality (14) on $(1 - u^2)T_{eh}/T_{ec}$ and T_{ih}/T_{ec}. Increasing the hot electron drift velocity lowers $(1 - u^2)T_{eh}/T_{ec}$ and reduces the range of densities which allow negative potential solitons.

Density and velocity perturbations. Linearization of (1) - (4) leads to the following expressions for the soliton density and velocity perturbations:

$$\delta n = \frac{\mu e n_o}{(s - u_o)^2 - a^2}\,\phi \quad , \tag{15}$$

$$\delta u = (s - u_o)\,\frac{\delta n}{n_o} \quad . \tag{16}$$

For all but the hot ion species, the density perturbation is rarefactive (compressional) for a negative (positive) potential slow soliton. According to (16), a rarefactive slow soliton decelerates the cold ions and electrons while it accelerates the hot ions. Hot electrons are accelerated by the rarefactive slow mode only when the unperturbed electron drift exceeds the soliton speed.

Scale length. With $x >> 3$, (11) implies that $L_s \simeq s_o/\sqrt{2\mu n_{ic}}$, or

$$L_s \simeq \frac{1}{\sqrt{2}} \left[\frac{n_{ih}}{T_{ih}} + \frac{n_{ec}}{T_{ec}} + \frac{n_{eh}}{1 - u^2} \right]^{-1/2} . \tag{17}$$

L_s is the effective Debye length normalized to $(T_{eh}/n_T)^{1/2}$. Colder plasma sheet ions and ionospheric electrons relative to the plasma sheet electrons and a faster hot electron drift reduce the soliton width. Unless n_{ec} is extremely small ($\lesssim 1\%$), or the cold electrons are significantly warmer than the low altitude ionospheric plasma, L_s is approximately the cold electron Debye length based on $(T_{ec}/n_{ec})^{1/2}$.

Polar wind. Generalizing (12) to include a polar wind with ionospheric electrons and ions drifting upward at a speed u_{pw}, we find, with x given by (13)

$$s_o \simeq a_{ic} \sqrt{1 + x/3} - u_{pw} \quad . \tag{18}$$

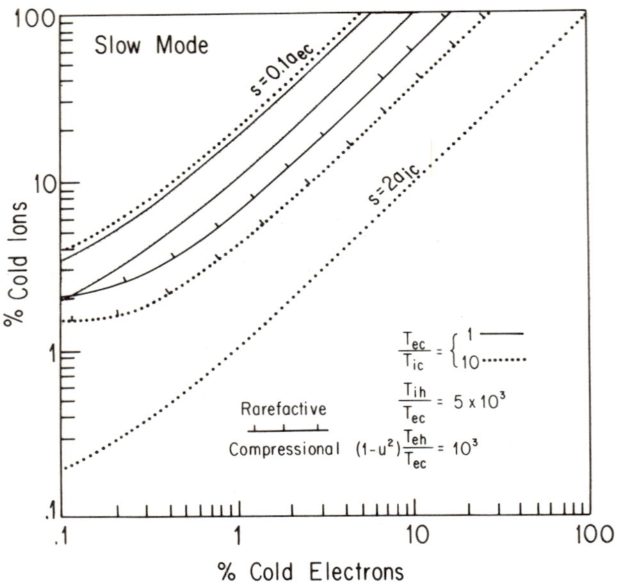

Fig. 1. Dependence of the slow mode characteristics on the ratio of cold electron to cold ion temperature. The slow mode is rarefactive (compressional) above (below) the barbed lines. Resonant particle effects are minimal when $2a_{ic} \lesssim s \lesssim 0.1\, a_{ec}$. In this regime $|\gamma|/\omega \lesssim 0.1$ for the linear $k \cong 0$ kinetic mode propagating at about the soliton phase speed. For $T_{ec}/T_{ic} = 10$, this regime lies between the dotted lines marked $s = 0.1\, a_{ec}$ and $s = 2a_{ic}$. For $T_{ec}/T_{ic} = 1$, the shaded region indicates this parameter regime.

The polarity criterion (14) is unchanged since $u_{pw} << a_{ec}$. (18) implies that the slow mode can stand ($s = 0$) or even propagate upward depending upon u_{pw}.

Numerical example. The slow mode amplitude is approximately

$$\frac{e\phi_{max}}{T_{eh}} \simeq \frac{2x\,(3 + x)\,(T_{ic}/T_{eh})\,(M - 1)}{x + 3 - \frac{x^3}{3}\left(\frac{n_{ec}}{n_{ic}}\right)\left(\frac{T_{ic}}{T_{ec}}\right)^2} \,. \quad (19)$$

For $T_{ec}/T_{ic} = 10$, $n_{ec} = .1$, $n_{ic} = .4$ and the plasma sheet values of Figure 1, we find for a rarefactive soliton that $x \simeq 40$ and $e\phi/T_{eh} \simeq -.03(M - 1)$. The maximum Mach number consistent with the small amplitude expansion is approximately 1.01. This follows from requiring that the first neglected term in (5) be small compared to the first and second order terms. A plasma sheet electron temperature of 1 keV thus yields a maximum wave amplitude of about .3 V. Equation (17) implies that $L_s \simeq .07\, (T_{eh}/4\pi n_T e^2)^{1/2} \simeq 17$ m, corresponding to a total plasma number density of 1 cm^{-3}. The soliton width at half-maximum is $L \simeq L_s/(M - 1)^{1/2} \simeq 170$ m. The maximum electric field is $E \approx \phi_{max}/L \simeq 2$ mV/m. From equation (12), $s_o \simeq 20$ km/sec and the pulse duration is of order $L/s_o \simeq 9$ msec. Longer pulse durations arise when a polar wind is present.

Intermediate Mode

This mode propagates at speeds satisfying $\min(a_{ec}, a_{ih}) < s < \max(a_{ec}, a_{ih})$. The inequalities must be replaced by much greater or less than for the fluid result to be valid. These more stringent conditions cannot be satisfied, however, for typical auroral parameters, and we find from equation (9) that $s_o \simeq a_{ih}$. Since the fluid theory may be inaccurate for waves propagating near one of the fluid thermal speeds, we do not discuss this mode further except to mention that resonant hot ions are likely to dominate its dynamics.

Fast Mode

Phase velocity. The fast soliton speed must satisfy the condition $a_{eh} >> |s - u| >> a_{ic}$, a_{ec}, a_{ih}. In this limit with $\tau \equiv T_{ec}/T_{eh}$, equation (9) yields

$$s_o \simeq n_{ec}u \pm \sqrt{n_{ec}(1 - u^2 n_{eh}) + 3n_{ec}\tau} \,. \quad (20)$$

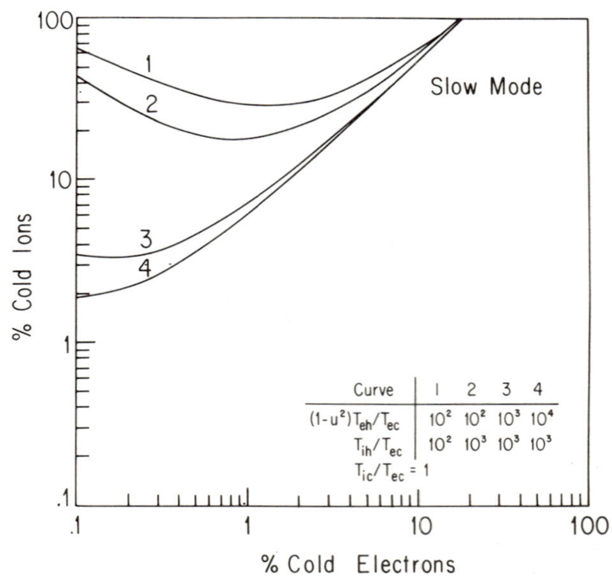

Fig. 2. Dependence of the slow mode rarefactive-compressional line on the ratios $(1 - u^2)T_{eh}/T_{ec}$ and T_{ih}/T_{ec}. The mode is rarefactive above the lines. Although not shown as in Figure 2, only certain values of the parameters give rise to weak resonant particle effects.

This is an electron mode for which ion inertia is unimportant ($m_e/m_i \to 0$). The positive root in (20) yields the fast mode critical speed for propagation parallel to the hot electron drift. The expression is valid for arbitrary u provided that s_o satisfies the above constraints. The negative root yields the critical speed of the backward mode when $u \ll 1$. For $u \gtrsim 1$, this branch can have $s_o > 0$ although for typical auroral parameters, the dynamic evolution in this regime is strongly influenced by resonant interactions with the hot and cold electrons. Subsequently, we will only discuss the properties of the fast mode when the positive root is taken in (20).

In the limits $\tau \to 0$ (dead cold ionospheric electrons) and $s \to u$, which is equivalent to neglecting the hot electron drift and inertia, equation (20) becomes $s_o \simeq a_{eh} \sqrt{n_{ec}/n_{eh}}$. This special case has been examined by Goswami and Sinha (1978) for a plasma with one (cold) ion population (see also Watanabe and Taniuti 1977). When $u^2 > n_{eh}^{-1} + 3\tau n_{ec}^{-1}$, corresponding to the threshold for the linear $k = 0$ mode fluid instability (Buneman 1959; Mantei et al. 1976), the critical speed is imaginary and fluid soliton solutions cannot be constructed.

Soliton polarity. Negatively polarized fast solitons occur when

$$n_{eh}s^2 > \frac{2}{3} n_{ec}^2 + n_{eh}(4n_{ec} - 1)\tau \quad , \qquad (21)$$

with s given approximately by (20). For 0, the criterion is $u_- < u < u_+$ where

$$u_\pm = n_{ec}\left(\frac{2}{3n_{eh}}\right)^{1/2} \pm \left(1 - \frac{2}{3} n_{ec}\right)^{1/2} . \qquad (22)$$

The upper limit is approximately the threshold for linear instability. Both u_+ and u_- are only weakly dependent on the cold electron temperature for $\tau \lesssim 0.01$.

Density and velocity perturbation. According to (15), the negative potential fast mode rarefies hot electrons and compresses cold electrons. We will call a fast, negative potential soliton a rarefactive mode although this name is somewhat ambiguous. Equation (16) indicates that cold electrons are always accelerated by a fast rarefactive disturbance while hot electrons are accelerated only when their unperturbed drift exceeds the soliton speed. Figure 3 shows the fast soliton characteristics and indicates the parameter regime where fluid theory is accurate. Within the confines of fluid theory, a fast soliton is a shielded clump of negative charge density convected by the hot drifting electrons.

Scale length. In the fast mode limit, equation (11) yields

$$\frac{1}{L_a^2} \simeq \frac{2n_{ec}}{u^2}\left[1 + \frac{\delta}{u}\left(\frac{n_{eh}}{n_{ec}} u^4 - 2\right)\right] . \qquad (23)$$

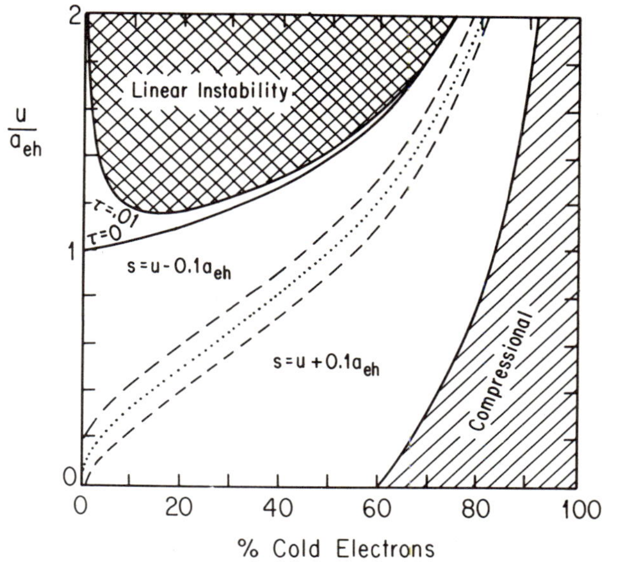

Fig. 3. Fast mode characteristics. The vertical axis is in units of the hot electron fluid streaming speed normalized to its thermal speed and $\tau \equiv T_{ec}/T_{eh}$. Resonant particle effects are minimal between the dashed lines and for $n_{ec} \gtrsim 5\%$ (same criterion as in Figure 2). While the fluid theory predicts that the mode can be compressional for $n_{ec} \gtrsim 60\%$, the soliton phase speed is $\sim u + a_{eh}$ in this parameter regime and therefore hot electron damping may significantly modify the wave properties. The plasma is linearly (fluid) unstable in the upper portion of the figure and the dependence of the threshold for instability on τ is indicated. The rarefactive-compressional line varies little with τ for $\tau \lesssim .01$. The dotted line corresponds to $s = u$ and the dashed lines correspond to $s = u \pm 0.1a_{eh}$.

where $\delta \equiv s_o - u$ measures the deviation of the critical speed from the electron drift speed. A sufficiently small δ implies $L_s \sim u/\omega_{pec}$ which is a hybrid Debye length based on cold electron density and hot electron drift energy.

Numerical example. With $s \simeq u$, the maximum soliton amplitude is approximately

$$\frac{e\phi_{max}}{T_{eh}} \simeq \frac{-2(u^2 - 3\tau)^3}{u^2(u^2 + \tau)} (M - 1) \simeq -2u^2(M - 1) \qquad (24)$$

A current density of $3\mu A/m^2$ at 1000 km altitude maps to $.6\ \mu A/m^2$ at 6400 km assuming a dipole magnetic field. This implies $u/a_{eh} \simeq 0.28$ when $T_{eh} \simeq 1$ keV and $n_{eh} \simeq 1\ cm^{-3}$. For this hot electron drift speed, Figure 3 indicates that the fluid theory can be used when $n_{ec} \simeq 10\%$. With $\tau = 0.001$ the second expression in (24) yields $e\phi_{max} \simeq -160(M - 1)$ eV. The maximum small-amplitude Mach number is $M \simeq 1.04$, yielding a soliton potential of $e\phi_{max} \simeq 6$ eV. With L_s u/ $\sqrt{2n_{ec}}$, the wave scale length is $L \simeq (T_{eh}/4\pi n_T e^2)^{1/2} L_s/(M - 1)^{1/2} \simeq 0.7$ km. The maximum electric field is $E \sim \phi_{max}/L \sim 9$ mV/m, and the pulse duration time is $L/s_o \simeq L/u \sim .2$ msec.

Conclusions

A one-dimensional fluid analysis of a four-component model auroral plasma indicates that at least two different, weakly damped, small amplitude electrostatic solitons can propagate along the geomagnetic field. The slower of the two is a generalization of an ion-acoustic solitary wave in a multi-component plasma. Unlike the usual ion-acoustic soliton, its polarity can be either positive or negative depending on the species concentrations and temperatures. Negative solitons are most likely to occur at altitudes above about 1 R_E. The slow soliton can stand in the polar wind, which may have consequences relevant to steady auroral double layer phenomena. Because we assumed small wave amplitudes and required dissipative effects to be weak, we are restricted to describing slow solitons at altitudes of about 1 R_E or greater with maximum amplitudes of a few tenths of a volt, scale lengths on the order of 100 m and electric fields of a few mV/m. These waves may be observable with a d.c. or low frequency wave antenna, although the associated density and velocity perturbations probably exceed the time resolution of particle detectors.

Ion inertia is negligible for the faster mode, which is supported by the two electron components. Within the validity of our analysis, this nonlinear localized wave is a shielded negative space charge convected by the drifting plasma sheet electrons. For parameters characterizing the plasma above about 1 R_E altitude, the small amplitude fast solitons have amplitudes of a few volts, scale lengths less than a kilometer, electric fields of a few mV/m, and pulse durations somewhat less than a msec. They may appear as oppositely directed electric field pulses in kHz wave records. Secondary electrons, with a temperature between the hot and cold electron temperature, would probably damp the fast mode.

The addition of a heavy ion fluid such as O^+ would allow another mode to propagate at speeds slower than the two which we have described. The oxygen thermal speed would have to be much less than the cold ion thermal speed, however, to avoid strong ion Landau damping.

Although we have not considered the large amplitude properties of these waves, we can indicate qualitatively some of the expected features. The large amplitude versions of both the slow and fast solitons would have smaller scale lengths than those given by small amplitude theory. Particle reflection, trapping, and anomalous dissipation could significantly affect the large amplitude structure. Such effects remove the symmetry of the soliton profile, and in particular, shock-like or double layer structures may arise. If these effects are moderately weak the leading edge of the shock is similar to one-half of the soliton, and the asymptotic (downstream) potential ϕ_s is given by the quasi-neutrality condition $\sum_s qn(\phi_s) = 0$ (Sagdeev 1966), which is just the bottom of the Sagdeev potential well (cf. equation 4). Thus, there may be an analogy between the theory of ion-acoustic shocks (Moiseev and Sagdeev 1963), which are essentially positively polarized solitons that reflect a significant number of ions, and a theory of double layers which requires current-coupled, negatively polarized (slow) solitons to reflect electrons, thereby producing a shock-like profile in the electrostatic potential.

Acknowledgements. We appreciate enlightening discussions with J.E. Maggs, F.V. Coroniti and G.J. Morales. The research has been supported by the Office of Naval Research, Contract Number N00014-76-C-0307 and NASA NGL 05-007-190.

References

Banks, P.M. and T.E. Holzer, High-latitude plasma transport: the polar wind, J. Geophys. Res. 74, 6317, 1969.

Banks, P.M. and G. Kockarts, Aeronomy, Part B, Chap. 10, Academic Press, NY, 1973.

Barnes, C., The formation of double layers, paper presented at Workshop on Space Plasma Physics, Los Alamos Scientific Laboratory, September 1980.

Block, L.P., Potential double layers in the ionosphere, Cosmic Electrodynamics 3, 349, 1972.

Buneman, O., Dissipation of currents in ionized media, Phys. Rev. 113, 503, 1959.

Buti, B., Ion-acoustic holes in a two-electron temperature plasma, Phys. Lett. 76A, 251, 1980.

Chanteur, G.M., J.C. Adam and R. Pellat, Current Interruption by localized ion acoustic waves, poster presented at this meeting.

Chen, F., Introduction to Plasma Physics, pg. 249, Plenum Press, 1974.

Chiu, Y.T. and M. Schultz, Self-consistent particle and parallel electrostatic field distributions in the magnetospheric-ionospheric auroral region, J. Geophys. Res. 83, 629, 1978.

Garrett, H.B., Review of quantitative models of the 0- to 100- keV near earth plasma, Rev. Geophys. Space Phys. 17, 397, 1979.

Goswami, B.N. and M. Sinha, Effect of cold electrons on linear and nonlinear ion acoustic waves, Plasma Phys. 20, 91, 1978.

Hudson, M.K. and F.S. Mozer, Electrostatic shocks, double layers and anomalous resistivity in the magnetosphere, Geophys. Res. Lett. 5, 131, 1978.

Hudson, M.K. and D.W. Potter, Electrostatic shocks in the auroral magnetosphere, paper presented at this meeting.

Kan, J.R. and L.C. Lee, Double layer criterion on

the altitude of the auroral acceleration region, Geophys. Res. Lett. 6, 429, 1980.

Karpman, V.I., The effects of the interaction between ion-sound solitons and resonance particles in a plasma, Sov. Phys. JETP 50, 695, 1979.

Lemaire, J. and M. Scherer, Plasma sheet particle precipitation: a kinetic model, Planet. Space Sci. 21, 281, 1973.

Levine, J.S. and F.W. Crawford, A fluid description of plasma double layers, J. Plasma Phys. 23, 223, 1980.

Mantei, T.D., F. Doveil and D. Gresillon, The large density electron beam-plasma Buneman instability, Plasma Phys. 18, 705, 1976.

Moiseev, S.S. and R.Z. Sagdeev, Collisionless shock waves in plasma in a weak magnetic field, Plasma Phys.(J. Nucl.Energy, part C) 5, 43, 1963.

Mozer, F.S., C.W. Carlson, M.K. Hudson, R.B. Torbert, B. Parady, J. Yatteau and M.C. Kelley, Observations of paired electrostatic shocks in the polar magnetisphere, Phys. Rev. Lett. 38, 292, 1977.

Mozer, F.S., C.A. Cattell, M.K. Hudson, R.L. Lysak, M. Temerin and R.B. Torbert, Satellite measurements and theories of low altitude auroral particle acceleration, Space Sci. Rev. 27, 155, 1980.

Sagdeev, R.Z., Cooperative phenomena and shock waves in collisionless plasmas, in Reviews of Plasma Physics, Vol. 4, edited by M.A. Leontovich, NY, 1966.

Sato, T. and H. Okuda, Numerical simulation of ion-acoustic double layers, paper presented at this meeting.

Shawhan, S.D., C.G. Falthammar and L.P. Block, On the nature of large auroral zone electric fields at 1 R_E altitude, J. Geophys. Res. 83, 1049, 1978.

Swift, D.W., On the formation of auroral arcs and acceleration of auroral electrons, J. Geophys. Res. 80, 2096, 1975.

____________ An equipotential model for auroral arcs: the theory of two-dimensional laminar electrostatic shocks, J. Geophys. Res. 84, 6427, 1979.

Taniuti, T., Ion acoustic solitary waves with effects of resonant particles, Prog. Theor. Phys., Supplement No. 55, 191, 1974.

Torbert, R.B. and F.S. Mozer, Electrostatic shocks as the source of discrete auroral arcs, Geophys. Res. Lett. 5, 135, 1978.

Watanabe, S. and T. Taniuti, Electron acoustic mode in a plasma of two temperature electrons, J. Phys. Soc. Japan 43, 1819, 1977.

ELECTRON AND ION ACCELERATION BY STRONG ELECTROSTATIC TURBULENCE

Robert L. Lysak

Space Sciences Laboratory, University of California, Berkeley 94720

Abstract. The presence of electrostatic turbulence on auroral field lines will cause particle orbits to deviate from adiabatic trajectories. In weak turbulence theory, the turbulence causes a diffusion in velocity space which is quadratic in the amplitude of the waves. When turbulence is strong, particle trapping effects which cause a first order change in the particle orbits become important. For waves observed on auroral field lines, ion trapping may be important and affect models of transverse ion acceleration. The strength of the electrostatic ion cyclotron turbulence observed in the auroral zone may be attributed to the large convective growth rate associated with this mode. These waves mey be destabilized by the enhancement in the field aligned current which is a result of the onset of a magnetospheric substorm. The increase in the field aligned current is associated with the passage of shear mode Alfvén waves betweeen the magnetosphere and ionosphere. When these Alfvén waves have a perpendicular wavelength comparable to the ion gyroradius, they carry a parallel electric field which can accelerate the plasma in the topside auroral ionosphere, destabilizing the electrostatic ion cyclotron mode. The anomalous transport effects associated with the ion cyclotron turbulence can cuase dissipation of the Alfvén wave, leading to an energization of the plasma and the establishment of a double layer on auroral field lines.

Introduction

The acceleration of auroral zone particles to high energies is one of the most central problems of auroral physics. In recent years, the suspicion that electric fields parallel to the magnetic field are responsible for the acceleration of auroral electrons has been strengthened by results from the S3-3 satellite [*Mozer et al.*, 1977, 1979, 1980; *Torbert and Mozer,* 1978; *Mizera et al.*, 1980]. Such an electric field may provide the field aligned acceleration required to account for auroral electrons and injection of ionospheric ions into the magnetosphere. However, there are some features of the observed particles in the auroral zone which cannot be simply accounted for by laminar acceleration in quasistatic electric fields, both in the electrons [*Kaufmann et al.*, 1976; *Lundin,* 1976; *Bryant et al.*, 1978; *Whalen and Daly,* 1979; *Sharp et al.*, 1980] and in the ions, which show signatures of acceleration perpendicular to the magnetic field [*Shelley et al.*, 1976; *Mizera and Fennell,* 1977; *Sharp et al.*, 1977, 1979; *Ghielmetti et al.*, 1978; *Gorney et al.*, 1980]. In addition, the turbulence is commonly observed on auroral flux tubes [*Gurnett and Frank,* 1977; *Temerin,* 1978, 1979; *Kintner et al.*, 1978, 1979; *Temerin et al.*, 1979; *Mozer et al.*, 1979, 1980]. In these cases, it is necessary to examine how the interaction of particles with fluctuating electric fields may change the particle distributions. In addition, this interaction may affect the way in which a parallel potential drop is distributed along the field line, since effective collisions modify the electron equation of motion, and thus the response of the electrons to parallel electric fields. Thus, an understanding of turbulence in the auroral zone is necessary to account for the growing body of particle and field observations in this region of space.

Transverse ion acceleration

In this section, we shall discuss the mechanisms which can accelerate ions across the magnetic field in the auroral zone. The major mechanisms which will be considered are the polarization drift due to varying perpendicular electric field and the interaction of ions with electrostatic waves at frequencies above the ion gyrofrequency.

Although a static electric field perpendicular to the magnetic field causes only a uniform drift of plasma and no particle acceleration, an electric field with time variation or space variation may accelerate particles across the magnetic field. This effect is known as the polarization drift, or acceleration drift. If a particle sees a perpendicular electric field which is varying in its frame of reference, it drifts in the direction of this field by the amount

$$\begin{aligned} v_\perp &= \frac{m_i c^2}{eB^2}\frac{d}{dt}E_\perp \\ &= \frac{m_i c^2}{eB^2}\left(\frac{\partial}{\partial t}+v_\parallel\frac{\partial}{\partial z}+v_\perp\frac{\partial}{\partial x}\right)E_\perp \end{aligned} \tag{1}$$

The first term in the second form of this equation is due to explicit time variations of the perpendicular electric field. This term may be present if short scale time variations such as those envisioned by *Goertz and Boswell* [1979] are present in the auroral zone. Such fields may be associated with time-dependent coupling between the auroral ionosphere and the plasma sheet. The second term is connected with the parallel motion of ions into a region of larger perpendicular electric field. This mechanism has been discussed by *Swift* [1979].

The final term, involving the perpendicular convective derivative, will average out over a gyroperiod if the ion orbit is periodic. This is the case except when the electric field gradient is sufficiently large to destroy the ion gyromotion. This occurs when the condition

$$\frac{m_i c^2}{eB^2}\frac{\partial E_\perp}{\partial x}>1 \quad (2)$$

is met. This condition is quite stringent, implying, for auroral zone parameters, that

$$\frac{\partial E_\perp}{\partial x}>10\frac{\text{mV}}{\text{m}^2}\frac{(B/0.1\text{G})^2}{m_i/m_p} \quad (3)$$

Observations on S3-3 indicate that at some times, electric field gradients of up to 7 mV/m^2 are present [*Mozer et al.*, 1980], indicating that oxygen ions and possibly hydrogen ions may be accelerated by this mechanism.

The maximum energization in the polarization mechanisms is given by the change in the ExB drift energy seen by the particle, $\Delta W_i = \frac{1}{2} m_i c^2 \Delta E_\perp^2/B^2$. For $E_\perp$ given in V/m and B in units of 0.1 G, this energy is 50 eV for H^+ and 800 eV for O^+. These energies should by viewed as upper limits since an ion may not spend enough time in the region of changing electric field to achieve the entire energy gain. Note, however, that the energy gained is proportional to mass, which indicates that oxygen acceleration is favored by this mechanism. In addition, the condition given by (2) above is also 16 times less stringent for oxygen than hydrogen. Thus, ion acceleration by polarization drift is more effective for heavier ions such as oxygen than for hydrogen.

Next we shall consider the acceleration of ions by electrostatic turbulence. The simplest theory for the wave-particle interaction is the unmagnetized quasi-linear theory. In the near earth auroral zone, this theory is applicable only for wavelengths short compared to the ion gyroradius and frequencies much larger than the ion gyrofrequency; i.e., $k_\perp \rho_i >> 1$ and $\omega/\Omega_i >> 1$. These conditions may be met for heavier ions such as O^+ in a hydrogen dominated plasma since, in such a plasma, ion cyclotron waves have wavelengths the order of the proton gyroradius and frequencies at and above the proton gyrofrequency. This theory may also apply for protons in turbulence on the whistler resonance cone at frequencies between the lower hybrid frequency and the electron plasma frequency.

An unmagnetized ion can only interact with waves for which the resonance condition $\omega - \mathbf{k}\cdot\mathbf{v} \approx 0$ is satisfied; that is, waves with a phase velocity roughly equal to the velocity of the ion. Approximately in this sense means within the trapping width, which, in an unmagnetized model, is given by $\Delta v = 2(e\Phi/m_i)^{1/2}$. Since $e\Phi/T_e < 1$ for all auroral zone waves which have been measured, the trapping width is at most equal to the ion acoustic speed. Since hydrogen cyclotron waves have phase velocities the order of the hydrogen acoustic speed which is much greater than the oxygen acoustic speed, some preacceleration of the oxygen ions is required for them to interact with the hydrogen cyclotron waves. A similar condition holds for waves on the whistler resonance cone, since the observed phase velocities of these whistler mode waves are greater than the hydrogen thermal speed [*Mozer et al.*, 1979].

Once an ion has been accelerated to the point where it may interact with these waves, it will be further accelerated to the point where its velocity is roughly equal to the largest phase velocity of a wave in the excited spectrum. Thus, provided the preacceleration mechanism exists, protons and oxygen ions may be accelerated to the same velocity by, for example, whistler mode waves.

For waves with frequencies comparable to low harmonics of the ion gyrofrequency and wavelengths comparable to the ion gyroradius, the magnetized model of quasi-linear heating must be used. This model has been applied to ion cyclotron waves by a number of authors [*Palmadesso et al.*, 1974; *Ionson et al.*, 1978; *Ashour-Abdalla and Thorne*, 1978; *Lysak et al.*, 1980]. Ion heating in the quasi-linear approximation is given by the moment of the quasi-linear diffusion equation with respect to the perpendicular energy. Writing the diffusion equation with respect to the invariants of the ion motion: guiding center position **X** and the momenta p_ϕ and $p_\parallel$, we obtain

$$\frac{\partial f_0}{\partial t} = \sum_{k_\perp, k_\parallel, l}\left(\frac{k_\perp}{m\Omega}\frac{\partial}{\partial X}+k_\parallel\frac{\partial}{\partial p_\parallel}+l\frac{\partial}{\partial p_\phi}\right)e^2|\tilde{\Phi}_{\mathbf{k}}(X)|^2 J_l^2(k_\perp r) \quad (4)$$

$$\times R(\omega_{\mathbf{k}}-k_\parallel v_\parallel - k_\perp v_E - l\Omega)\left(\frac{k_\perp}{m\Omega}\frac{\partial}{\partial X}+k_\parallel\frac{\partial}{\partial p_\parallel}+l\frac{\partial}{\partial p_\phi}\right)f_0(\mathbf{p})$$

Taking the moment of (4) with respect to $W_\perp = p_\phi \Omega_i$ yields

$$\frac{\partial n W_\perp}{\partial t}+\frac{\partial}{\partial \mathbf{x}}\cdot\mathbf{Q}_\perp = -\sum_{\mathbf{k}} m\Omega \int dp_\parallel\, dp_\phi\, d\phi\, l\Omega \quad (5)$$

$$\times|\tilde{H}_{\mathbf{k}}(\mathbf{p})|^2 R(\omega_{\mathbf{k}}-\mathbf{k}\cdot\dot{\mathbf{q}})\mathbf{k}\cdot\frac{\partial}{\partial \mathbf{p}}f_0$$

The perpendicular heating rate for a given $v_\parallel$ may be calculated by defining

$$T_\perp = \frac{m\Omega}{f_\parallel}\int dp_\phi d\phi p_\phi \Omega f_0 \quad (6)$$

where the parallel distribution $f_\parallel$ is defined as

$$f_\parallel \equiv m\Omega \int dp_\phi d\phi f_0 \quad (7)$$

In the limit where the gradient in x and the parallel wave number are neglected except in the resonance function the heating rate becomes:

$$\frac{\partial T_\perp}{\partial t} = \sum_{\mathbf{k}} \frac{e^2|\Phi_{\mathbf{k}}|^2}{T_\perp} l^2\Omega^2 \Gamma_l \left(\frac{k_\perp^2 v_i^2}{\Omega^2}\right) R_{\mathbf{k}}\,(\omega - k_\parallel v_\parallel - l\Omega) \quad (8)$$

Note that this result is independent of the shape of the parallel distribution function. Since ions near gyroresonance $(\omega - k_\parallel v_\parallel - l\Omega \approx 0)$ are heated preferentially, the ion distribution will not remain Maxwellian, and ions in the tail will form a population with a large perpendicular temperature.

(8) describes a diffusion of ions in perpendicular energy. If all parameters other than $T_\perp$ are held constant, and the weak dependence of Γ_l on temperature is ignored, (8) may be written:

$$\frac{d}{dt}(\tfrac{1}{2}T_\perp^2) = D_T \quad (9)$$

where D_T, the diffusivity in temperature, is

$$D_T = \sum_{\mathbf{k}} e^2|\Phi_{\mathbf{k}}|^2 l^2\Omega^2\Gamma_l\left(\frac{k_\perp^2 v_i^2}{\Omega_i^2}\right)R_{\mathbf{k}}(\omega - k_\parallel v_\parallel - l\Omega) \quad (10)$$

Assuming that $\Gamma_1 = .2$ and $R \approx 1/\Omega$, the diffusivity is

$$D_T \approx 190\left(\frac{e\Phi}{1eV}\right)^2\left(\frac{B}{.1G}\right)\left(\frac{m_p}{m_i}\right) \quad eV^2/s \quad (11)$$

The ion heating, when integrated over a Maxwellian distribution, results in roughly three times the parallel electron heating [*Lysak*, 1980]. The ratio is roughly equal to the quantity $(k_\parallel/k_\perp)\sqrt{m_i/m_e}$, which is, in an order of magnitude sense, the ratio $(k_\parallel \partial \ln f_e/\partial p_\parallel)/(l \partial \ln f_i/\partial p_\phi)$, using the fact that $k_\perp \rho_i \approx l$ for waves at the lth harmonic.

In a steady-state situation when the growth rate due to the electrons is balanced by ion damping, the ion heating rate may be related to the parallel resistivity of the electrons. In this situation, the growth rate is zero and therefore Im $\epsilon = 0$, where ϵ is the dielectric function. When the dielectric function is expressed in canonical coordinates as in *Lysak* [1980], the zero growth rate condition implies:

$$\sum_s m_s \Omega_s \int dp_\parallel \, dp_\phi \, d\phi \sum_l J_l^2(k_\perp r_s) R\,(\omega_{\mathbf{k}} - \mathbf{k}\cdot\dot{\mathbf{q}})\mathbf{k}\cdot\frac{\partial}{\partial \mathbf{p}} f_s = 0 \quad (12)$$

Since $m_i \Omega_i = -m_e \Omega_e$, it follows that the integral in (12) is the same for electrons and ions. Comparing (12) with the ion heating equation (5) and the equation found by taking the moment of (4) with respect to the parallel momentum for the electrons yields the relation:

$$\left.\frac{\partial(nW_\perp)}{\partial t}\right|_{turb} = \frac{l\Omega_i}{k_\parallel} \left.\frac{\partial(nm_e v_\parallel)}{\partial t}\right|_{turb} \quad (13)$$

If the ion perpendicular heating rate is assumed to be roughly 100 eV/s (see (11)) and it is further assumed that $l\Omega_i/k_\parallel \approx v_\parallel \approx$ 400 km/s, (13) yields $\nu_\parallel \approx 100\text{sec}^{-1}$, where the effective collision frequency $\nu_\parallel$ is defined by $\nu_\parallel n m_e v_\parallel = (\partial n m_e v_\parallel / \partial t)|_{turb}$. This is consistent with the collision frequency which may be deduced from the work of *Lysak and Hudson* [1979] using the relation $j_z = ne^2 E_z/(m_e \nu_\parallel)$.

The quasi-linear theory described above fails when the turbulence is narrow-banded; that is, when the bandwidth of the turbulent spectrum is less than the transverse trapping frequency $\omega_T = A_l k_\perp c\tilde{E}/B$, where A_l is a constant dependent on the harmonic number of the waves which is equal to 0.265, 0.242, 0.216, 0.200 for the first four harmonics. These numbers come from a consideration of the dynamics of ion motion in a single electrostatic wave [*Lysak et al.*, 1980; *Lysak*, 1980]. As discussed in these works, a ion resonant with a single wave propagating perpendicular to the magnetic field will see an effective potential given by

$$U(p_\phi) = -(e\tilde{\Phi}_{\mathbf{k}})^2 J_l^2 (k_\perp r) \quad (14)$$

The bounce frequency for an ion trapped in the first well of this potential is that given above, with

$$A_l = \frac{lJ_l(x_l)}{x_l}\left(1 - \frac{l^2}{x_l^2}\right)^{1/2} \quad (15)$$

where x_l is the first maxima of the Bessel function J_l. For $k_\perp \rho_i \approx l$, which is typical of ion cyclotron waves, and a bandwidth $\Delta\omega/\omega \approx 0.1$, such as is observed in the auroral zone [*Kintner et al.*, 1978, 1979] ion trapping becomes important for $e\Phi/T_e \approx 0.3$ for the first harmonic.

Numerical calculations showing the acceleration of ions by narrow band waves have been performed by *Lysak* [1980]. These results show that in a single in cyclotron wave, ions are trapped in the potential well and achieve a maximum velocity of $v_\perp \approx 3.8\Omega_i/k_\perp$. Additional heating beyond this point is possible for ion cyclotron waves only if either the amplitude is increased beyond that observed or if additional waves are added to the calculation. A calculation with four ion cyclotron waves with a bandwidth $\Delta\omega/\omega \approx 0.1$ around the maximally growing wave showed that a small number (about 4%) of the ion population was accelerated into a superthermal tail, with the maximum energy in the tail given by

$$\frac{W_\perp}{T_0} = 1 + 0.55(\Omega_i t)^{0.65} \quad (16)$$

For a lower hybrid wave with $\omega = 6\Omega_i$ and $k_\perp \rho_i = 5$, the ion motion was already stochastic at observed wave levels $e\Phi/T_e = 0.5$. Thus, a single lower hybrid wave accelerates nearly all of the resonant ions at a rate which was found to be

$$\frac{W_\perp}{T_0} = 1 + 2.87(\Omega_i t)^{0.33} \quad (17)$$

In order to examine ion heating in the auroral zone, the role of parallel electric fields and the force due to the convergent magnetic field in accelerating ions up the field line must be considered. Consider a field-aligned region of turbulence of length L at an altitude h above the earth's surface. Assuming the turbulence to be quasi-stationary in time and to possess components at various phase velocities to insure continuous heating, the energy an ion can achieve may be determined by how long it stays in the spatial region of turbulence.

As time progresses the perpendicular energy $W_\perp$ and the parallel velocity $v_\parallel$ evolve according to the coupled equations

$$\frac{dW_\perp}{dt} = \left.\frac{dW_\perp}{dt}\right|_{turb} - \frac{3W_\perp v_\parallel}{R} + \frac{e^2 E_\perp}{m\Omega^2} v_\parallel \frac{\partial E_\perp}{\partial z} \quad (18)$$

$$\frac{dv_\parallel}{dt} = \frac{3W_\perp}{mR} + \frac{e}{m} E_\parallel \quad (19)$$

To tackle these equations analytically, the electric field will be set to zero and the second term in (18) will be neglected, since this term is small if the parallel energy gained through the mirror force is small. Neglect of this term limits the analysis to small heating regions. A numerical model without this restriction has been presented by *Lysak et al.* [1980].

As was seen above, heating rates due to turbulence have the form $W_\perp(t) = W_{\perp_0}(1+\alpha t^\gamma)$ where α and γ depend on the turbulence model. Using this form the energy increases may be calculated in cases where spatial scale determines the energization. If the heating region has a length L along the field, the time the ion will spend in the region is given by

$$T = \left[\frac{mR}{3\alpha}(\gamma+1)(\gamma+2)L\right]^{\frac{1}{\gamma+2}} \quad (20)$$

and the energy attained is

$$\frac{W_\perp}{W_{\perp_0}} = \left[\frac{mR}{3}\alpha^{\frac{2}{\gamma}}(\gamma+1)(\gamma+2)L\right]^{\frac{\gamma}{\gamma+2}} \quad (21)$$

This model has been applied to a variety of different examples of heating in the auroral zone, and the results are summarized in Table 1. These results assume a source region of length 10 km along the magnetic field and heating rates given by the quasi-linear theory given by (8) for ion cyclotron waves at both the hydrogen and oxygen cyclotron frequencies and the unmagnetized heating rate given in *Lysak et al.* [1980] for turbulence near the lower hybrid frequency. In addition the typical heating rates for strong waves deduced from the numerical calculations presented above were considered, as well as the upper limit on the heating rates given in *Lysak et al.* [1980]. The table considers heating both at altitudes of 6000 km, where conical ion distributions are observed by the S3-3 satellite, and at 2000 km, where transversely accelerated ions are observed by ISIS. Note that the final column showing final energy is normalized to the initial energy of the ions; thus, for a 1 eV plasma, this column may be read in eV. It can be seen from the table that quasi-linear heating can energize both hydrogen and oxygen by 10-30 eV in a 10 km region. The numerical calculations show heating which is more efficient than in the quasi-linear theory, with energies in the 40-100 eV range being attainable. The upper bound on ion acceleration in a 10 km region is on the order of 200 eV. Note that the numerically calculated energies are much greater than the quasi-linear values and about a factor of 2 below the theoretical upper bound. If the initial temperature of the plasma is greater than 1 eV, proportionately higher energies may be achieved. It should be remembered in interpreting these

Table I. Heating of ions in 10 km region by various turbulence models

Model	Ion	Altitude (km)	α $(s^{-\gamma})$	γ	$W_\perp / W_{\perp 0}$
Hydrogen IC, Quasi-linear	H^+	6000	6.0	0.5	18.8
		2000	11.0	0.5	27.6
Oxygen IC, Quasi-linear	O^+	6000	1.51	0.5	10.6
		2000	2.76	0.5	15.9
HIC, Numerical	H^+	6000	35.1	0.65	93.2
		2000	76.7	0.65	153
OIC, Numerical	O^+	6000	5.8	0.65	47.3
		2000	12.7	0.65	77.6
Lower hybrid, Numerical	H^+	6000	23.7	0.33	41.9
		2000	35.3	0.33	55.9
Upper bound $(k_\perp \rho_i \approx 1)$	H^+	6000	69.7	0.8	186
		2000	183	0.8	332
	O^+	6000	7.54	0.8	83.9
		2000	19.9	0.8	150

numbers that these are the energies achieved by the particles in the superthermal tail of the distribution, and do not represent bulk heating of the distribution.

Properties of the kinetic Alfvén wave

Large-scale electric fields generated near the equatorial plane will propagate toward the ionosphere at the Alfvén speed $V_A = B/(4\pi n m_i)^{1/2}$. When the perpendicular scale length becomes comparable to the ion gyroradius or electron inertial length, however, the Alfvén dispersion relation is modified. When $\beta > m_e/m_i$, this relation becomes [*Hasegawa,* 1976; *Goertz and Boswell,* 1979]

$$\omega = k_\parallel V_A \sqrt{1+k_\perp^2 \rho^2} \tag{22}$$

where $\rho = \rho_i \sqrt{T_e/T_i + 3/4}$. When $\beta < m_e/m_i$:

$$\omega = \frac{k_\parallel V_A}{\sqrt{1+k_\perp^2 c^2/\omega_{pe}^2}} \tag{23}$$

Thus, the Alfvén wave propagates at greater than the Alfvén speed in the higher β regime and slower than the Alfvén speed at very low β. *Lysak and Carlson* [1980] have modeled the density on the auroral flux tube and found that expected plasma sheet temperatures of 100 eV to 1 keV, the transition takes place in the region from 3 to 5 R_E.

Although the shear Alfvén wave is incompressible in the limit of long perpendicular wavelength, there is a density perturbation associated with the wave when the perpendicular wavelength is of the order of the ion gyroradius or electron inertial length. The magnitude of the density perturbation is

$$\frac{\delta n}{n} = \frac{m_i c^2}{eB^2}\frac{\partial E_x}{\partial x} \tag{24}$$

In addition, the magnitude of the field aligned current associated with the kinetic Alfvén wave is given by

$$j_z = \frac{c^2}{4\pi V_A}\frac{\partial E_x}{\partial x} \tag{25}$$

In the regime where $\beta > m_e/m_i$, this current is carried by precipitating electrons, resulting in the broadening of the anti-loss cone of the plasma sheet electron distribution. However, when $\beta < m_e/m_i$, the wave travels at a velocity greater than the electron thermal speed. In this regime, the current is carried by the acceleration of ambient electrons, leading to a relative drift of these electrons with respect to the plasma ions. We shall examine this source of free energy and its effects in the following section.

Effects of microscopic turbulence on kinetic Alfvén waves

In this section, we shall consider the conditions under which a kinetic Alfvén wave may become unstable to microscopic electrostatic turbulence, which may in turn affect the propagation of the kinetic Alfvén wave. The field-aligned current of the Alfvén wave is associated with two sources of free energy for driving this electrostatic turbulence: electron drift and density gradients.

Electron drift is the most effective source of free energy associated with the kinetic Alfvén wave in the region where $\beta < m_e/m_i$, since in this regime a parallel electric field associated with the wave accelerates the cold plasma, leading to a situation in which the electron distribution has a bulk drift with respect to the ions. Thus, the established theory of current driven instabilities [*Kindel and Kennel,* 1971] applies. For a plasma in which ion and electron temperatures are similar in magnitude, which seems to be a reasonable approximation in the low altitude auroral zone [*Mozer et al.,* 1979], the current driven ion cyclotron instability is excited when the drift of the electrons satisfies the condition $V_D/v_i > 13 T_i/T_e$. The Buneman instability, with a frequency near the lower hybrid resonance, is excited for stronger drifts, $V_D > v_e = 43 v_i (T_e/T_i)^{1/2}$. A simple relation between the drift speed and the characteristics of the kinetic Alfvén wave may be found taking parameters typical of 1 R_E altitude:

$$\frac{V_D}{v_i} \approx \frac{1}{B/0.06\mathrm{G}\sqrt{(n/10\mathrm{cm}^{-3})(T_i/1\mathrm{eV})}}\frac{\partial E_x(\mathrm{mV/m})}{\partial x(\mathrm{km})} \tag{26}$$

It can be seen from this equation that both ion cyclotron and Buneman instabilities are possible if, for example, $E_x = 500$ mV/m and $\Delta x = 10$ km in an equal temperature plasma.

When turbulence is important, its effect on the propagation of the Alfvén wave must be considered. The effect of turbulence may be modeled by assigning an effective collision frequency for parallel momentum transfer to the electrons. Following the derivation of *Hasegawa and Mima* [1978], the parallel component of Ohm's Law may be written

$$\delta j_z = \frac{ne^2(\nu_\parallel - i\omega)}{k_\parallel^2 T_e} W\left(\frac{\omega + i\nu_\parallel}{k_\parallel v_e}\right) E_z \tag{27}$$

where W(x) is the plasma dispersion function as defined by *Ichimaru* [1973]. Combining this equation with the ion response function and Maxwell's equations from *Hasegawa and Mima* [1978] the dispersion relation becomes

$$\omega^2 = k_\parallel^2 V_A^2 \left[1 + \frac{3}{4}k_\perp^2\rho_i^2 + k_\perp^2\rho_i^2 \frac{T_e}{T_i}\frac{\omega}{\omega + i\nu_\parallel}\left(W\left(\frac{\omega + i\nu_\parallel}{k_\parallel v_e}\right)\right)^{-1}\right] \tag{28}$$

assuming $k_\perp \rho_i << 1$.

When the effective collision frequency vanishes, (28) reduces to (22) or (23), in the limit $\omega/k_{\parallel}v_e << 1$ or $>> 1$, respectively. When collisions dominate, $|\omega + i\nu_{\parallel}|/k_{\parallel}v_e >> 1$, and (28) becomes

$$\left(\frac{k_{\parallel}V_A}{\omega}\right)^2 = 1 + \frac{k_{\perp}^2c^2}{\omega_{pe}^2}\left(1 + i\frac{\nu_{\parallel}}{\omega}\right) \tag{29}$$

This dispersion relation gives two damped waves propagating toward positive and negative z corresponding to transmitted and reflected waves. In the limit of strong collisions, $\nu_{\parallel} >> \omega$, (29) becomes

$$\frac{k_{\parallel}V_A}{\omega} = \pm\frac{k_{\perp}c}{\omega_{pe}}(1+i)\left(\frac{\nu_{\parallel}}{2\omega}\right)^{1/2} \tag{30}$$

For waves in which $k_{\perp}^2c^2/\omega_{pe}^2 \approx 1$, which corresponds to perpendicular wavelengths of about 1 km for $n = 10$ cm^{-3}, the field aligned current is likely to surpass the threshhold for EIC turbulence. The effective collision frequency due to EIC waves is approximately $\nu_{\parallel} \approx 0.1 - 0.3\Omega_i$ [*Lysak and Hudson,* 1979; *Gary,* 1980], which corresponds to an effective parallel collision frequency of about 200 sec^{-1} at 2 R_E. Since the kinetic Alfvén wave has a frequency $\omega = 1$ sec^{-1}, the strong collision limit is satisfied, and (30) gives a damping length $L_D = \mathrm{Im} k_{\parallel}^{-1} = 200$ km.

This model indicates that a kinetic Alfvén wave may propagate distances of 100-1000 km into regions of turbulence. In the limit where $\nu_{\parallel} >> \omega$, the potential difference along the field line becomes comparable to the potential across the field. The ratio $\Phi_{\parallel}/\Phi_{\perp}$ may be calculated from (27) along with Equation 12 of *Hasegawa and Mima* [1978], and it is found that $\Phi_{\parallel} \rightarrow \Phi_{\perp}$ in the limit $\omega << \nu_{\parallel}k_{\perp}^2c^2/\omega_{pe}^2$. Thus, the parallel potential drop becomes equal to the potential across the field when effective collisions are strong.

Critical damping scale size for kinetic Alfvén waves

If we combine the linear current-voltage relation given in Equation 9 of *Fridman and Lemaire* [1980] with the wave conductivity relation given in (25), and consider the limit $\omega/\nu_{\parallel} \rightarrow 0$, the potential drops out and we find:

$$\frac{k_x^2c^2}{\omega_{pe}^2} = 1.2\frac{B/0.06\mathrm{G}}{(T_{\parallel}/1\mathrm{keV})^{1/2}}\frac{n_S/n}{(n/10\mathrm{cm}^{-3})^{1/2}} \tag{31}$$

where n_S and $T_{\parallel}$ refer to the density and temperature in the source region and n and B refer to the density and magnetic field strength in the region of parallel electric field normalized to typical parameters observed at 6000 km altitude by S3-3 [*Mozer et al.,* 1979]. Defining the scale size $L_x = k_x^{-1}$, we find

$$L_x = 1.5\mathrm{km}\frac{(T_{\parallel}/1\mathrm{keV})^{1/4}(n/n_S)^{1/2}}{(B/0.06\mathrm{G})^{1/2}(n/10\mathrm{cm}^{-3})^{1/4}} \tag{32}$$

This scale size will scale to about 500 m at ionospheric heights if the parallel field region is at an altitude of 6000 km. This corresponds to the scale size of visual discrete arcs [*Maggs and Davis,* 1968; *Davis,* 1978]. This may be compared to the scale size calculated by *Chiu and Cornwall* [1980] of 40 km. This calculation used a similar kinetic model; however, they used an ionospheric conductivity model instead of the wave model considered here.

The scale size given by (32) may be interpreted as the critical damping width for the auroral zone LRC circuit. Alfvén wave pulses of a larger width than this will not be associated with a parallel electric field component, and thus will penetrate to the ionosphere and be reflected as discussed by *Mallinckrodt and Carlson* [1978] and *Goertz and Boswell* [1979]. In this case, many bounces of the Alfvén wave will be required before a steady state current system is set up. Waves with widths shorter than this will be strongly damped and will not penetrate into the near earth region. Thus, the scale size given by (32) represents the maximum possible transfer of energy from the Alfvén pulse to the precipitating electrons.

Discussion

The model presented above is based on that presented by *Goertz and Boswell* [1979] but differs from that model in a number of respects. In the previous model, the parallel electric field was transient, and vanished in the stationary state, whereas the parallel electric field becomes a permanent part of the auroral circuit when turbulence is included. In addition, the steady state current system of *Goertz and Boswell* [1979] is only established after very many bounces of the Alfvén wave, a process which will take tens of minutes to an hour, whereas the turbulence may establish a steady state system in one pass, taking less than a minute. The perpendicular scale size calculated in this paper is also a new feature of this model, resulting from the consideration of the plasma sheet particle distribution as well as the wave nature of current flow in the auroral zone.

The considerations of this paper lead to the following scenario for the response of the auroral zone to a magnetospheric substorm. When the near earth plasma sheet has excess mechanical energy, in the form of the pressure gradient or a high speed flow, a spectrum of Alfvén waves are generated. The response of the flux tube to a particular wave depends on the perpendicular scale size of the wave. Very broad components of the wave propagate to the ionosphere and undergo multiple reflections, as described by the model of *Mallinckrodt and Carlson* [1978], Narrower structures, with scale sizes comparable to the ion gyroradius, carry with them a parallel electric field, and accelerate the cold ionospheric plasma when they reach the regime where the plasma $\beta < m_e/m_i$, which takes place at a geocentric distance of 3 to 5 R_E. In this region, two scenarios may occur. If the structure is relatively broad and weak, the ion cyclotron instability may be excited, leading to a large scale parallel electric field. This type of structure may be associated with the inverted-V structure, which is typically 1-2° wide in invariant latitude [*Lin and Hoffman,* 1979]. In this case the ionospheric conductivity may set the scale size, as described by *Chiu and Cornwall* [1980].

Narrower structures, which carry stronger currents and parallel electric fields, may lead to the Buneman instability [*Buneman,* 1959]. As noted by *Iizuka et al.* [1979] this instability may lead to the formation of a purely growing electric field, or a double layer, since locally the current exceeds the Bohm criterion [*Block,* 1972]. This electric field creates a runaway electron beam, which will interact with the ionospheric electrons via binary collisions. This interaction gives rise to a backscattered electron population, as has been modeled by *Evans* [1974]. The presence of this energetic backscattered population relaxes the criteria for double layer existence [*Kan and Lee,* 1980]; therefore, once the double layer is set up by the Alfvén wave pulse, it becomes self sustaining. Ions which are accelerated by the parallel electric field may generate additional EIC waves, enhancing the effective collision frequency. In this situation, the wave conductivity may set the scale size as calculated above.

Although this model is suggestive, a more complete model should be considered to determine the response of the auroral flux tube to magnetospheric perturbations. The propagation of the kinetic Alfvén wave should be considered including the nonlinear effect of the density perturbation associated with the wave and the electron convective derivative, both of which may be important in double layer formation. In addition, the simple effective collision frequency model considered above, which allowed a calculation of turbulent effects within a linear model, should be replaced with a more realistic picture of the wave-particle interaction. Finally, the generation of the kinetic Alfvén wave in the plasma sheet must be studied to determine the spectrum of waves generated, as well as to include ionospheric feedback effects such as have been considered by *Miura and Sato* [1980].

In conclusion, we have considered the effect of turbulence on the coupling of the magnetosphere and ionosphere by including an effective collision frequency in the equation of motion describing a kinetic Alfvén wave. A static parallel electric field is generated, which, in the limit that the effective collision frequency is much greater than the Alfvén wave frequency, leads to a potential drop along the magnetic field equal to the potential across the field lines. This leads to a scale size for the critical damping of the Alfvén wave which corresponds to the scale size of visual auroral arcs of a few hundreds of meters in the ionosphere.

Acknowledgments. The author would like to thank Chuck Carlson, John Mallinckrodt and Mary Hudson for useful discussions and reading the manuscript. This work was supported in part by NASA Grants NSG-7401 and NAGW-75.

References

Ashour-Abdalla, M., and R. M. Thorne, Toward a unified view of diffuse auroral precipitation, *J. Geophys. Res.*, **83**, 4775, 1978.

Block, L. P., Potential double layers in the ionosphere, *Cosmic Electrodynamics*, **3**, 349, 1972.

Bryant, D. A., D. S. Hall, and D. R. Lepine, Electron acceleration in an array of auroral arcs, *Planet. Space Sci.*, **26**, 81, 1978.

Buneman, O., Dissipation of currents in ionized media, *Phys. Rev.*, **115**, 503, 1959.

Chiu, Y. T. and J. M. Cornwall, Electrostatic model of a quiet auroral arc, *J. Geophys. Res.*, **85**, 543, 1980.

Davis, T. N., Observed characteristics of auroral forms, *Space Sci. Rev.*, **22**, 77, 1978.

Evans, D. S., Precipitating electrons fluxes formed by a magnetic field-aligned potential difference, *J. Geophys. Res.*, **79**, 2853, 1974.

Fridman, M. and J. Lemaire, Relationship between auroral electron fluxes and field-aligned electric potential difference, *J. Geophys. Res.*, **85**, 664, 1980.

Gary, S. P., Wave-particle transport from electrostatic instabilities, *Phys. Fluids*, **23**, 1193, 1980.

Ghielmetti, A. G., R. G. Johnson, R. D. Sharp, and E. G. Shelley, The latitudinal, diurnal, and altitudinal distributions of upward flowing energetic ions of ionospheric origin, *Geophys. Res. Lett.*, **5**, 59, 1978.

Goertz, C. K. and R. W. Boswell, Magnetosphere-ionosphere coupling, *J. Geophys. Res.*, **84**, 7239, 1979.

Gorney, D., A. Clarke, D. Croley, J. Fennell, J. Luhmann, and P. Mizera, Auroral ion beam distributions below 8000 km, in *High Latitude Electric Fields in the Magnetosphere and Ionosphere*, J. C. Foster, ed., Utah State University, Logan, Utah, 1980, p. 73.

Gurnett, D. A. and L. A. Frank, A region of intense plasma turbulence on auroral field lines, *J. Geophys. Res.*, **82**, 1031, 1977.

Hasegawa, A., Particle acceleration by MHD surface wave and formation of aurora, *J. Geophys. Res.*, **81**, 5083, 1976.

Hasegawa, A. and K. Mima, Anomalous transport produced by kinetic Alfvén wave turbulence, *J. Geophys. Res.*, **83**, 1117, 1978.

Ichimaru, S., *Basic Principles of Plasma Physics*, W. A. Benjamin, Reading, Mass., p. 186, 1973.

Iizuka, S., K. Saeki, N. Sato and Y. Hatta, Buneman instability, Pierce instability, and double layer formation in a collisionless plasma, *Phys. Rev. Lett.*, **43**, 404, 1979.

Ionson, J. A., R. S. B. Ong, and E. G. Fontheim, Anomalous resistivity of the auroral plasma in the top-side ionosphere, *Geophys. Res. Lett.*, **3**, 549, 1978.

Kan, J. R. and L. C. Lee, On the auroral double layer criterion, *J. Geophys. Res.*, **85**, 788, 1980.

Kaufmann, R. L., D. N. Walker, and R. L. Arnoldy, Acceleration of auroral electrons in parallel electric fields, *J. Geophys. Res.*, **81**, 1673, 1976.

Kindel, J. M., and C. F. Kennel, Topside current instabilities, *J. Geophys. Res.*, **76**, 3055, 1971.

Kintner, P. M., M. C. Kelley, and F. S. Mozer, Electrostatic hydrogen cyclotron waves near one earth radius altitude in the polar magnetosphere, *Geophys. Res. Lett.*, **5**, 139, 1978.

Kintner, P. M., M. C. Kelley, R. D. Sharp, A. G. Ghielmetti, M. Temerin, C. A. Cattell, P. Mizera, Simultaneous observations of energetic (keV) upstreaming ions and EHC waves, *J. Geophys. Res.*, **84**, 7201, 1979.

Lin, C. S. and R. A. Hoffman, Characteristics of the inverted-V event, *J. Geophys. Res.*, **84**, 1514, 1979.

Lundin, R., Rocket observations of electron spectra and angular characteristics in an inverted-V event, *Planet. Space Sci.*, **24**, 499, 1976.

Lysak, R. L., and M. K. Hudson, Coherent anomalous resistivity in the region of electrostatic shocks, *Geophys. Res. Lett.*, **6**, 661, 1979.

Lysak, R. L., M. K. Hudson, and M. Temerin, Ion heating by strong electrostatic ion cyclotron turbulence, *J. Geophys. Res.*, **85**, 678, 1980.

Lysak, R. L., Electrostatic ion cyclotron turbulence and auroral particle acceleration, Ph. D. Thesis, University of California, Berkeley, 1980.

Lysak, R. L. and C. W. Carlson, The effect of turbulence on magnetosphere-ionosphere coupling, *Geophys. Res. Lett.*, **7**, in press, 1980.

Maggs, J. E. and T. N. Davis, Measurements of the thickness of auroral structures, *Planet. Space Sci.*, **16**, 205, 1968.

Mallinckrodt, A. J. and C. W. Carlson, Relations between transverse electric fields and field-aligned currents, *J. Geophys. Res.*, **83**, 1426, 1978.

Miura, A. and T. Sato, Numerical simulation of global formation or auroral arcs, *J. Geophys. Res.*, **85**, 73, 1980.

Mizera, P. F. and J. F. Fennell, Signatures of electric fields from high and low altitude particle distributions, *Geophys. Res. Lett.*, **4**, 311, 1977.

Mizera, P. F., J. F. Fennell, D. R. Croley, Jr., A. L. Vampola, F. S. Mozer, R. B. Torbert, M. Temerin, R. L. Lysak, M. K. Hudson, C. A. Cattell, R. G. Johnson, R. D. Sharp, P. M. Kintner, and M. C. Kelley, The aurora inferred from S3-3 particles and fields, *J. Geophys. Res.*, **85**, in press, 1980.

Mozer, F. S., C. W. Carlson, M. K. Hudson, R. B. Torbert, B. Parady, J. Yatteau, and M. C. Kelley, Observations of paired electrostatic shocks in the polar magnetosphere, *Phys. Rev. Lett.*, **38**, 292, 1977.

Mozer, F. S., C. A. Cattell, M. Temerin, R. B. Torbert, S. Von Glinski, M. Woldorff, and J. Wygant, The DC and AC electric field, plasma density, plasma temperature, and field-aligned current experiments on the S3-3 satellite, *J. Geophys. Res.*, **84**, 5875, 1979.

Mozer, F. S., C. A. Cattell, M. K. Hudson, R. L. Lysak, M. Temerin, and R. B. Torbert, Satellite measurements and theories of auroral acceleration mechanisms, *Space Sci. Rev.*, **24**, in press, 1980.

Palmadesso, P. J., T. P. Coffey, S. L. Ossakow, and K. Papadopoulos, Topside ionosphere ion heating due to electrostatic ion cyclotron turbulence, *Geophys. Res. Lett.*, **1**, 105, 1974.

Sharp, R. D., R. G. Johnson, and E. G. Shelley, Observations of an ionospheric acceleration mechanism producing energetic (keV) ions primarily normal to the electric field direction, *J. Geophys. Res.*, **82**, 3324, 1977.

Sharp, R. D., R. G. Johnson, and E. G. Shelley, Energetic particle measurements from within ionospheric structures responsible for auroral acceleration processes, *J. Geophys. Res.*, **84**, 480, 1979.

Sharp, R. D., E. G. Shelley, R. G. Johnson, and A. G. Ghielmetti, Counterstreaming electron beams at altitudes of $\sim 1\ R_E$ over the auroral zone, *J. Geophys. Res.*, **85**, 92, 1980.

Shelley, E. G., R. D. Sharp, and R. G. Johnson, Satellite observations of an ionospheric acceleration mechanism, *Geophys. Res. Lett.*, **3**, 654, 1976.

Swift, D. W., An equipotential model for auroral arcs: the theory of two-dimensional laminar electrostatic shocks, *J. Geophys. Res.*, **84**, 6427, 1979.

Temerin, M., The polarization, frequency, and wavelengths of high latitude turbulence, *J. Geophys. Res.*, **83**, 2609, 1978.

Temerin, M. A., A comment on the source region of VLF saucers, *J. Geophys. Res.*, **84**, 6691, 1979.

Temerin, M. A., M. Woldorff, and F. S. Mozer, Nonlinear steepening of the electrostatic ion cyclotron wave, *Phys. Rev. Lett.*, **43**, 1941, 1979.

Torbert, R. B., and F. S. Mozer, Electrostatic shocks as the source of discrete auroral arcs, *Geophys. Res. Lett.*, **5**, 135, 1978.

Whalen, B. A. and P. W. Daly, Do field-aligned auroral particle distributions imply acceleration by quasi-static parallel electric fields?, *J. Geophys. Res.*, **84**, 4175, 1979.

DISCRETE BREAKUP ARCS AND KINETIC ALFVEN WAVES

C.K. Goertz

Department of Physics and Astronomy, University of Iowa, Iowa City, Iowa, USA
and
Max-Planck-Institut für Aeronomie, D-3411 Katlenburg-Lindau 3, FRG

Introduction

Auroral phenomena display an excitingly rich phenomenology. There are narrow, rapidly moving breakup arcs; quiet pre-breakup arcs are somewhat thicker and move relatively slowly and finally there are diffuse post-breakup arcs. Although it may be tempting to reduce all these phenomena to one basic mechanism this has so far not been possible. In a general sense auroral phenomena are all manifestations of the solar wind-magnetosphere interaction. A statement of that generality, however, is quite empty. We believe that real progress can only be made when the differences are properly taken into account.

In this paper we will deal only with the extremely narrow auroral structures whose characteristic dimensions perpendicular to the magnetic fields are of the order of several hundred meters or even less. These structures are not very stable but move rapidly and quite irregularly. According to Akasofu (1977) they are associated with rapid energy release in the magnetotail. They are basically non-stationary structures and a steady state theory involving, for example, electrostatic double layers is not applicable.

The starting point of our model is the assumption that discrete breakup arcs are caused by a change of the magnetosphere convection pattern which is localized in space. Such a change of convection could be due to such popular but little understood processes as magnetic field reconnection, current-interruption, and fire-balls. The perturbation of convection is a disturbance which must be thought of as a packet of mhd waves (provided the time scale for the change is longer than an ion gyroperiod) containing a certain mode structure and spectrum of perpendicular wave-numbers $k_\perp$.

The wave packet will spread out from the origin. Its shear Alfvén wave component will be guided along the background magnetic field $\vec{B}_o$ and far from the origin this will be the only mode which carries a significant power flux. The other modes will not be guided and spread over a large area. We thus concentrate on the shear Alfvén mode. It is a well documented but little appreciated fact that the shear Alfvén wave is dispersive if $k_\perp \neq 0$, i.e., if it propagates obliquely to the magnetic field $\vec{B}_o$. Another important fact relating to this mode is the existence of an electric field $\vec{E}$ parallel to $\vec{B}$ and hence this mode is affected by kinetic effects. This wave is thus rightfully called kinetic Alfvén wave by Hasegawa (e.g., Hasegawa, 1976; Hasegawa and Chen, 1975; Hasegawa and Mima, 1978; but see also Fejer and Lee, 1967; Fejer and Kan, 1968; Stepanov, 1958). This paralle electric field can cause charged particle acceleration and particle precipitation into the ionosphere. In other words, a spatially localized "mhd" disturbance is dissipated via particle acceleration. We propose that discrete breakup arcs are a direct manifestation of this kinetic interaction between charged particles and a "mhd" perturbation. We would like to stress here that our model applies only to narrow structures (thickness of the order of the plasma skin depth) and to short lived phenomena (time scales shorter than a dispersion time). Quiet discrete arcs or diffuse aurora cannot be accounted for in terms of our model.

The Propagation of a Shear Alfvén Wave Packet

A disturbance of the convection in the tail is associated (through the generalized Ohm's law) to a current perpendicular to the magnetic field B_o and vice versa. Let us consider the waves generated by this source-current. We shall assume that the current is given as a function of time and distance $\vec{x}$ perpendicular to $\vec{B}_o$ and write

$$\vec{j} = \vec{j}_\perp (\vec{x}_\perp, t)\, \delta(z) \qquad (1)$$

The Fourier transform $\tilde{j}_i$ in time and space of the current is related to the electric field component $\tilde{E}_i$ by the relation

$$\Lambda_{ij}(\omega, \bar{k})\, \tilde{E}_j(\omega, \bar{k}) = -\frac{4\pi i}{\omega} \tilde{j}_i(\omega, \bar{k}) \qquad (2)$$

where the conductivity tensor Λ_{ij} is related to the dielectric tensor ϵ_{ij} by the well known relation

$$\Lambda_{ij} = n^2(K_iK_j - \delta_{ij}) + \epsilon_{ij}(\omega, \vec{k}) \qquad n^2 = k^2c^2/\omega^2 \qquad \vec{K} = \vec{k}/|\vec{k}| \tag{3}$$

The solution of equation (2) is in a formal sense given by

$$\tilde{E}_i(\omega, \vec{k}) = -\frac{i}{\omega} D_{ij}(\omega, \vec{k}) \tilde{j}_i(\omega, \vec{k}) \tag{4}$$

where the dispersion tensor D_{ij} is the inverse of the conductivity tensor divided by 4 . The time and space variation of E is given by the inverse Fourier transform

$$\tilde{E}_i(\vec{x}, t) = \frac{1}{(2\pi)^3} \int \tilde{E}_i(\omega, \vec{k})\, e^{-i\omega t + i\vec{k}\cdot\vec{x}}\, d\omega d\vec{k} \tag{5}$$

Thus when the perturbation at z = 0 is given and the dielectric tensor known it is in principle possible to calculate the response of the plasma for all later times. The integral of equation (5) involves integration around the singularities of D_{ij} by a procedure well known in plasma physics. There are some subtleties to be considered but these have been discussed in detail by many authors. (See any textbook in Plasma Physics.) Since we are interested in low frequency perturbations only mhd modes are of interest. We use a perturbation with a Gaussian spatial profile

$$j_i(x_\perp, t) = \frac{1}{\pi^{\frac{1}{2}}} j_{oi}\, e^{-(x/d)^2} g(t) \tag{6}$$

$$\tilde{j} = \frac{j_{oi}}{(2\pi)^2} \delta(k_y) e^{-(\frac{1}{4}k_x^2 d^2)} \tilde{g}(\omega) \tag{7}$$

$$\tilde{g}(\omega) = \int_{-\infty}^{\infty} g(t)\, e^{-i\omega t}\, dt \tag{8}$$

In order to solve equations (4) and (5) we utilize the cold plasma dispersion relation for shear Alfvén waves, derived by, e.g., Fejer and Kan (1969) who also derive the dispersion relation for the case of a warm plasma. For our purposes the cold plasma approximation is valid in regions where the electron thermal velocity is much less than the Alfvén velocity. In this case we have for shear Alfvén waves:

$$k_z^2 = \frac{\omega^2}{V_A^2}(1 + k_x^2 c^2/\omega_{pe}^2) \tag{9}$$

where V_A is the Alfvén velocity and ω_{pe} the electron plasma frequency. The integral (6) can be solved (see, e.g., Fejer and Kan, 1967) by standard methods. We find that

$$E_i(x, t) = -\frac{j_{oi}\mu_o \Omega_e c}{2\omega_{pi}} \int \tilde{g}(\omega) e^{i(\omega t + \Phi(z,\omega))} \frac{e^{-x^2/w^2}}{w^{\frac{1}{2}} d^{\frac{1}{2}}} d\omega \tag{10}$$

where the phase $\Phi(z,\omega)$ is given by

$$\Phi(z, \omega) = (\omega z/V_A) + \tfrac{1}{2}\tan^{-1}(2z\omega c^2/V_A\omega_{pe}^2 d) - 2\frac{z\omega}{V_A}\frac{x^2c^2}{\omega_{pe}^2}\frac{1}{w^2} \tag{11}$$

and the width w of the wave packet

$$w = [d^2 + 4t^2\omega^2 c^4/d^2 V_A^2 \omega_{pe}^2]^{\frac{1}{2}} \tag{12}$$

The first term in the expression for the phase dominates for large distances away from the source and indicates propagation at the Alfvén speed. The width w increases with increasing distance z. This is due to the fact that obliquely propagating shear Alfvén waves are dispersive (see equation 9) contrary to some statements made in the literature which usually deal with strictly parallel propagating waves. It can be shown (e.g., Fejer and Lee, 1967) that in addition to being dispersive obliquely propagating Alfvén waves also carry a field aligned electric field E_z which for a cold plasma is

$$\tilde{E}_z = \frac{k_x k_z}{(k_x^2 + \omega_{pe}^2/c^2)} \tilde{E}_x \tag{13}$$

or in real space

$$E_z = \frac{m_e}{m_i}\frac{c}{\omega_{pi}\Omega_i}\frac{d^2}{dtdx} E_x(\vec{x}, t) \tag{14}$$

It should be noted that we have implicitly assumed that the Fourier transform of g(t) exists. For the case of a perturbation being switched on at t = 0 and held constant afterwards the Fourier transform in time must be replaced by the LaPlace transform as e.g., done by Goertz and Boswell (1979). The basic points to be made below are, however, very similar in both cases. In the following we will interpret ω^{-1} as a characteristic time scale for the perturbation.

In addition to the spreading of the wave packet due to dispersion the width of the wave packet is affected by an inhomogeneous magnetic field. In the derivation above we have assumed $dB_o/dz = 0$. In reality magnetic field lines converge. Fejer and Lee (1967) have shown that, unless there are strong perpendicular density gradients, the wavenumber $|k_x|$ is directly proportional to B_o. The width of the Fourier spectrum is thus inversely proportional to the width of the flux tube to which the perturbation is confined. This indicates guidance, $d = d_o B_o/B$. Taking this into account we can write

$$w(z) = [d_o^2 B_o/B + 4z^2 \omega^2 c^4/V_{Ao}^2 \omega_{peo}^2 d_o^2]^{\frac{1}{2}} \quad (15)$$

where the subscript o indicates that the subscripted quantity is to be evaluated at z = 0. We see that guiding decreases w if the wavepacket moves into an increasing magnetic field whereas dispersion always leads to an increase of the width.

We are now in the position to calculate the change of momentum of a particle overtaken by the wave apcket by integrating qE_z over time. We are interested in the maximum value which is proportional to the maximum value of dE_x/dx. For the Gaussian profile we find for the change of electron energy parallel to B_o (taking $\Delta p_{ze} = \int qEdt$ and $\Delta W_{ze} = (\Delta p_{ze})^2/2\,m_e$)

$$\Delta W_{ze} = \frac{c^2}{\omega_{pe}^2 w^2} \frac{m_i}{2} E_x^2/B^2 = \frac{c^2}{\omega_{pe}^2 w^2} \frac{m_i}{2} v_\perp^2 \quad (16)$$

where $v_\perp$ is the velocity amplitude of the wave packet. For $w = c/\omega_{pe}$, the plasma skin depth, there is an equipartition between the perpendicular kinetic energy of the ions in the wave and the field aligned energy of the electrons.

Is the energy calculated above of interest for auroral phenomena? Recently Mozer et al. (1977) have observed large perpendicular electric fields $E \approx 0.5$ V/m at altitudes of roughly 1 R_E. The magnetic field strength in those regions is of the order of $B_o \sim 5 \times 10^{-2}$ G. Thus $m_i/2\ v_\perp^2 \approx 50$ A eV where A is an average ion mass of the plasma. For a density of 1 cm^{-3} the plasma skin depth is 5.3 km. We see that for small structures $w \approx 0.1 \ldots 5$ km the kinetic energy of the electrons which, incidentally, carry a parallel current could easily be of the order of a few hundred eV if even a few keV. Are such small structures likely to be observable or will they spread out rapidly due to dispersion before they reach the ionosphere? Let us take as an example a perturbation in the plasma sheet where the density is roughly 1 cm^{-3} and $V_A \sim 200$ km/s. If the second term in equation (15) is much smaller than the first, even when z equals the length of the field line, dispersive broadening can be neglected. This is the case when

$$d_o^2 > (2\, z\, \omega c^2/\omega_{peo} V_{Ao})(B_o/B)^{\frac{1}{2}}$$

For $B/B_o \sim 2000$ and $z \sim 20\ R_E$ we find that for a perturbation of a characteristic duration of 100 seconds dispersive broadening is negligible if $d_o > 60$ km. For such a perturbation $w \gtrsim 1.3$ km at the point where $B/B_o = 2000$. Thus structures somewhat smaller than the local plasma skin depth can exist and the energy of the accelerated (current carrying) electrons can even exceed the kinetic drift energy of the ions. Electron energies of the order of a few keV seem quite likely to occur within the wave packet.

We note that any structure which is much larger than the local plasma skin depth will lead to an insignificant energy gain and can thus not cause auroral phenomena in the transient stage. In other words, transient auroral phenomena are highly structured and thin because only highly structured and thin perturbations cause sufficient particle acceleration. Structures very much smaller than the local plasma skin depth will broaden rapidly and should not be observed very often. It is, therefore, not surprising that the thickness distributions of auroral arcs has a peak at small thickness.

Summary and Discussion

We have shown above that a change of convection pattern in a spatially limited region leads to the emission of a shear-Alfvén wave packet which propagates along the magnetic field line with a characteristic velocity V_A. Because of guiding the wave packet will contract if it moves into a region of increasing magnetic field strength. Dispersion will lead to dispersive broadening. For perturbations lasting 100 seconds or more dispersive broadening is negligible if the perpendicular (to $\vec{B}$) scale length in the distant magnetosphere is larger than a few tens to hundred km. In the front and back of the wave packet, i.e., where $d^2E_x/dtdx \neq 0$, parallel electric fields exist. The electric field accelerates electrons which carry a parallel current. The electron energy may be quite high for small structures but will be insignificantly small for thick structures. Some electrons may overtake the wave packet and be accelerated to even larger energies as discussed by Goertz and Boswell (1979).

We have not discussed the processes that do occur when the wave packet reaches the ionosphere. This was done in some detail by Goertz and Boswell (1979) who show that partial reflection of the wave will occur if the ionospheric Pederson conductivity Σ_p is not matched to the Alfvén conductance $\Sigma_A = 1/\mu_o V_A$. The return wave will partially cancel $E_\perp$ but carry a parallel electric field in the same direction as the incoming wave. The electron energy and parallel current will be increased again as the return wave moves away from the ionosphere. (See Goertz and Boswell for more details.) If the perturbation lasts for a sufficiently long time so that many reflections can occur eventually a steady state will be established. However, such a steady state will not contain very narrow structures as these are damped out by the particle acceleration described above. Thus we arrive at the following schematic view of the magnetosphere-ionosphere coupling (Figure 1).

Any arbitrary change of convection contains large and small $k_\perp$ components. The large $k_\perp$ components dissipate their energy by accelerating electrons quite rapidly, i.e., $j \cdot E$ is large and positive. It should be noted that $j \cdot E$ can also be negative, namely at the back of the wave

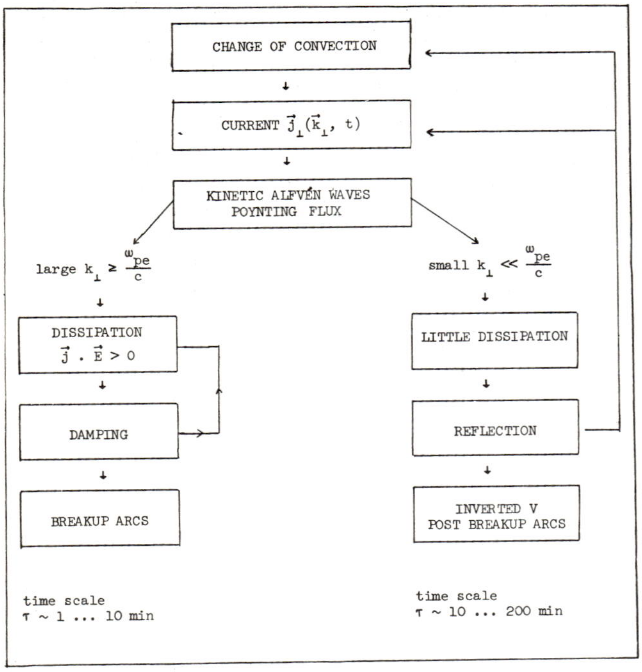

Figure 1 : A schematic view of the response of magnetosphere-ionosphere system to a change in the convection pattern. The left-hand path involves small scale structures which damps rapidly. The right-hand path concerns large scale structures which do not damp.

packet. However, it is quite likely that the accelerated electrons will cause some instability and lose some of their energy to growing waves, in which case even a finite wave packet (finite in duration and distance along field lines) will damp. This damping is, of course, outside the scope of cold plasma theory which we have applied here. Thus the strong acceleration which occurs in narrow structures is limited in duration by dispersive broadening and damping.

The small $k_\perp$ components do not suffer significant dispersive broadening or damping. They will survive many wave bounce periods between $z = 0$ and the ionosphere eventually establishing a standing wave pattern if $g(t)$ is periodic or a steady state current system if $g(t)$ approaches a constant but non-zero value. This steady state may or may not involve a steady state parallel electric field as described by Lyons (this issue). The response of the magnetospheric plasma to a perturbation which is switched on at $t = 0$ and approaches a constant value has been described by Goertz and Boswell (1979).

For completeness we would like to add some observational tests of the ideas expressed in the paper.

If the large electric fields of Mozer et al. (1977) are, indeed, manifestations of a kinetic Alfvén wave they should be accompanied by a magnetic field perturbation b_y given by

$$b_y = E_x c/V_A \, (1 + k_x^2 c^2/\omega_{pe}^2)^{\frac{1}{2}}$$

using $A \sim 2 k_x^{-1} = 2.5$ km, $E_x = 0.5$ V/m, $V_A = 2 \times 10^9$ cm/s and $c/\omega_{pe} = 5$ km we find a magnetic field perturbation $b_y \sim 8\ \gamma$. This field perturbation should correlate with E_x and can be thought of as resulting from a field aligned current flowing in the region where E_x reverses sign. Goertz and Boswell estimate this current to be a few $\mu A/m^2$. This current is, of course, carried by the accelerated electrons discussed above. In other words the "electrostatic shocks" of Mozer et al. should not be exactly electrostatic. The distinguishing feature of this model as opposed to an electrostatic shock model with a field aligned current is the definite relation between this current (or equivalently b_y) and E_x. However, Lysak (this issue) shows that if collisions due to current generated wave turbulence in the Alfvén wave front are included the dispersion relation becomes severely modified and this relation will not be obeyed. The structures should be observed at the onset of auroral break-up and, possibly, at the beginning of substorms. The acceleration of ions in these structures is quite different from that of the electrons because of their large mass. Using the same reasoning that led to equation (16) we find that the paralle acceleration of ions is entirely negligible. On the other hand the perpendicular gyro-energy of the ions will be increased by the polarization drift effect as briefly described by Lysak (this issue) by an amount equal to $m_i/2\ v_\perp^2$. This is in addition to the increase of their guiding center drift energy by the same amount. The structures should thus be accompanied by ions who have been heated perferentially perpendicular to B, i.e. by the famous conic distributions.

Finally we add a note of caution. The entire theory is based on a linear dispersion relation. The derivations break down when (i) the time scale for the variation of the fields becomes comparable to an ion gyroperiod; (ii) the parallel electric field becomes comparable to the perpendicular field and when (iii) the width w becomes so small that the ion gyromotion is destroyed. The first two conditions are not very critical. However, the last condition may be violated in some of the so-called "electrostatic shocks". It is relatively easy to show that this conditions implies that the parallel velocity of the accelerated electrons must be less than the local Alfvén velocity or that the current cannot be super Alfvénic. However, at 1 R_E V_A could be as large as a few tenths of the velocity of light which would correspond to an electron energy of a few keV.

Acknowledgements. Helpful discussions with D. Evans are gratefully acknowledged. This work was supported in part by the National Science Foundation under Grant ATM 76-82739 and by the National Aeronautics and Space Administration under Contract NAS2-6553.

References

Akasofu, S.I., Physics of Magnetospheric Substorms, J. Reidel Publishing Company, Dordrecht, Holland, 1977.

Fejer, J.A., and K.F. Lee, Guided propagation of Alfvén waves in the magnetosphere, J. Plasma Phys., 1, 25, 1967.

Fejer, J.A., and J.R. Kan, A guiding center Vlasov equation and its application to Alfvén waves, J. Plasma Phys., 3, 331, 1969.

Goertz, C.K., and R.W. Boswell, Magnetosphere-ionosphere coupling, J. Geophys. Res., 84, 7239, 1979.

Hasegawa, A., Particle Acceleration by MHD Surface Wave and Formation of Aurora, J. Geophys. Res., 81, 5083, 1976.

Hasegawa, A., and L. Chen, Kinetic Process of Plasma Heating due to Alfvén wave Excitation, Phys. Rev. Lett., 35, 370, 1975.

Hasegawa, A., and K. Mima, Anomalous Transport Produced by Kinetic Alfvén Wave Turbulence, J. Geophys. Res., 83, 1117, 1978.

Mozer, F.S., C.W. Carlson, M.K. Hudson, R.B. Terbert, B. Pasady, J. Yatteau and M.C. Kelley, Observations of paired electrostatic shocks in the polar magnetosphere, Phys. Rev. Lett., 38, 292, 1977.

Stepanov, K.N., Soviet Physics-JETP, Vol. XXXIV, 892, 1958.

THE GENERATION OF ION-CONICS VIA QUASI-LINEAR DIFFUSION

P. B. Dusenbery and L. R. Lyons

Space Environment Laboratory, NOAA, ERL, Boulder, Colorado 80303

Abstract. Downward currents in auroral regions are commonly measured with amplitudes of 1-5 $\mu A/m^2$. Such currents are likely the result of upgoing thermal ionospheric electrons falling through a field-aligned potential drop on the order of their thermal energy. Similar distributions of upgoing ionospheric electrons may also occur in regions of diffuse auroral electron precipitation to preserve current continuity in the presence of the loss of the precipitating electrons to the ionosphere. The drift velocity of the upgoing electrons is sufficient to excite electrostatic ion cyclotron waves. In addition to being in Landau resonance with the upgoing electrons, these waves cyclotron resonate with the upgoing thermal ions with a parallel energy of several eV. Calculated quasilinear diffusion rates using measured wave spectra indicate that resonant ions can be heated to perpendicular energies on the order of 100 times the initial ion thermal energy. The results imply that conic distributions should not occur simultaneously with the keV electron precipitation associated with discrete arcs in regions of upward current but should occur in regions of upgoing ionospheric electrons which may include regions of downward current near the edges of auroral arcs and regions of diffuse auroral electron precipitation.

Introduction

There is strong evidence that the high latitude ionosphere is an important source of magnetospheric ions [McIlwain, 1975; Frank et al., 1977; Cornwall and Schulz, 1978]. Part of the evidence is from the observation of upward flowing accelerated ionospheric ions in the auroral zone [Shelley et al., 1976; Sharp et al., 1977; Mizera and Fennell, 1977]. These up flowing ions can be characterized as either a beam-like distribution of particles with pitch angles centered about 180 degrees or a conic-like distribution with enhanced particle fluxes at pitch angles between 90-130 degrees. Understanding the ion acceleration responsible for the conic-like distributions is the goal of this paper.

Observations from the S3-3 satellite at altitudes of $\sim$ 1 earth radius include both types of upward flowing ions for energies between 90 eV and 16 keV. Gorney et al. [1980] have presented a statistical survey of upward flowing ions as a function of local time, latitude, altitude and degree of magnetic activity. They concluded that ion beams exist only at altitudes above 4 - 5,000 km, above which their occurrence frequency increases with increasing altitude. On the other hand, ion-conics were observed uniformly in altitude above 2000 km, suggesting a low altitude generation region for conics. Furthermore, the vast majority of conic distributions, observed in quiet times, had energies well below 400 eV [Gorney et al., 1980]. This suggests that such distributions may have a significant ionospheric component.

Ion-conics have frequently been observed at lower altitudes ($\sim$ 1500 km) by the Isis 1 and 2 satellites [Klumpar, 1979; Ungstrup et al., 1979]. Observations in the winter nighttime imply an association between ion-conics and a low thermal number density. In fact, ion-conics have been observed on the dayside but at higher altitudes than on the nightside, consistent with a low density [Klumpar, 1979].

The energy spectra of low altitude conics is softer than measured at higher altitude on S3-3. Typically, low altitude conics have energies up to $\sim$100eV. The ion detector on board the Isis 1 and 2 spacecraft measures energies from 5eV to 15 keV, but the fluxes generally decrease with increasing energy and fall below the level of statistically meaningful count rates [D. Klumpar, private communication, 1980]. On the other hand, the S3-3 detector only samples ion energies greater than 90eV and is therefore unable to observe conics at energies $\lesssim$ 90eV. At rocket altitudes, Whalen et al. [1978] observed ions with pitch angles near 100 degrees in the altitude range 400-500 km.

The optical emissions of discrete auroral arcs result from precipitating electrons of 1-10 keV energies which carry an upward field-aligned current. These electrons have presumably been accelerated by a field-aligned potential difference on the order of 1-10 kV [Evans, 1976]. Beams, of both downward electrons and upward ions with fluxes peaked in energy and enhanced in the direction parallel to the ambient magnetic field, are a natural consequence of this acceleration process. Conics, on the other hand, monotonically decrease with increasing energy which strongly suggests energy diffusion by plasma waves. We therefore look to wave observations to see which mode(s) may be significant in the generation of ion-conics.

Wave observations relevant to the generation of ion-conics have come primarily from the S3-3 satellite [Kintner et al., 1979]. They frequently measured electrostatic ion-cyclotron waves and concluded that the waves were often associated with regions of upward flowing ions. These waves were usually excited in bands between harmonics of the proton gyrofrequency. At lower altitudes ($\sim$1500 km), ion-cyclotron wave observations are difficult to make because the satellite velocity is several times larger than the wave phase velocity. Thus, the waves experience a large Doppler shift which greatly distorts the wave spectrum. At rocket altitudes, the Doppler shift of the waves would be less since rocket velocities are much lower than the low altitude satellite velocities. In fact, Fredericks and Russell [1973] claim to have seen ion-cyclotron waves at the edge of an arc on a rocket flight.

Even though there are no definitive measurements of ion-cyclotron waves by the Isis 1 and 2 satellite, observations of V-shaped whistler mode waves called saucers have frequently been made [James, 1976]. These waves often show absorption lines at multiples of the ion-cyclotron frequency. Saucers do not appear to be a likely candidate for the generation of conics for two reasons. One, the wave frequency is, for the most part, too large to resonant with the thermal ions. This is not so for the ion-cyclotron waves which can interact strongly with the ions. Two, the wave energy density is too small (corresponding to 3mV/m electric fields) to be a significant factor in heating the ions. On the other hand, Kintner et al. [1979] frequently measured ion-cyclotron wave amplitudes between 10-50mV/m. Thus, ion-cyclotron waves have a low enough frequency to effectively resonate with the thermal ions and sufficient electric field amplitudes to heat them significantly.

James [1976] concluded that there were no significant electron fluxes above 5eV in the source region of the saucers and that they were collocated with localized regions of upflowing ions with energies < 50eV. The ions had pitch angles $\sim$ 100° and were therefore conics. These observations suggest a strong correlation between low altitude conics and saucer emissions. However, we shall not attempt to give an explanation for this strong correlation in this paper. Instead, we shall investigate the generation of ion-conics in regions of upgoing ionospheric electrons which are unstable to the growth of ion-cyclotron waves. These waves, in turn, resonate with the ambient population of ionospheric ions forming conic-like distributions.

We will explore the quasi-linear relaxation of auroral ion distributions in resonance with electrostatic ion-cyclotron waves to determine which parameters describing either the waves or the particles are significant in the heating and acceleration process. Resonant interactions between charged particles and plasma waves generally cause particle diffusion in pitch angle and energy. For the case of ion-conic diffusion, it is more appropriate to consider the Cartesian components in velocity space ($v_{\parallel}$, $v_{\perp}$) where $\parallel$ and $\perp$ refer to directions parallel and perpendicular to the ambient magnetic field respectively.

Typically, ion-cyclotron wave amplitudes range from 10-30mV/m. The energy density in the wave fluctuations is thus very small ($\sim 10^{-14}$ erg cm^{-3}) compared to the thermal energy density of the particles (10^{-10} erg cm^{-3}), assuming an ion number density equal to 300 cm^{-3} and a thermal energy of 0.3 eV. These are typical thermal parameters at altitudes of $\approx$ 1500 km. This implies [Davidson, 1976] that weak turbulence theory may be applied to the problem. The evidence does not appear to require the use of a nonlinear theory to explain the generation of ion-conics, though nonlinear theories have been proposed [Palmadesso et al., 1974; Lysak et al., 1980].

Quasi-linear theory may be invoked if the time scale for changes in the space averaged distribution function is much longer than the average wave period of the turbulence [Kennel and Englemann, 1966]. We will show that this inequality is satisfied for the interaction of the ion-cyclotron waves with ionospheric ions. This theory allows us to treat wave generation and particle diffusion separately. Thus, direct comparison of theory with measurement is facilitated.

If the linear wave growth, ω_I, is much smaller than the wave frequency, ω_R (i.e., $\omega_I << \omega_R$), then the resonant interactions of the ions with the wave field will dominate the non-resonant interactions and the latter can be ignored [Kennel and Englemann, 1966; Hollweg and Volk, 1971]. In this case the particle diffusion can be treated in the resonant diffusion limit [Kennel and Englemann, 1966]; this is the approach employed in this paper.

Section 2 begins with a discussion of the wave generation process. We find that keV electron precipitation in upward current regions cannot be responsible for the excitation of ion-cyclotron waves. We are therefore led to look at regions of upgoing thermal ionospheric electrons, and we find that such particles can indeed excite ion-cyclotron waves. With a reasonable model for the wave power spectrum, we then evaluate the quasi-linear diffusion coefficients.

In Section 3, we investigate the behavior of a distribution of particles which are constrained to diffuse on surfaces called single-wave characteristics. We also use the calculated diffusion coefficients to determine limits of strong parallel and perpendicular diffusion. Section 4 summarizes our conclusions, the major conclusion being that the quasi-linear interaction of electrostatic ion-cyclotron waves with a thermal distribution of ions offers a quantitative explanation for the generation of ion-conics.

Formulation of the Problem

Both upward and downward field aligned currents may be important in the generation of ion-conics. Discrete arcs are associated with regions of upward current carried by downward precipitating electrons of 1-10 keV which have been accelerated through a field-aligned potential difference ∿1-10 kV.

Adjacent to upward current regions are often found regions of downward current. The current carriers in these two regions are not the same. In the downward current region at low altitudes (∿1500 km) the charge carriers must be low energy, thermal ionopheric electrons (thermal energy ∿0.3eV). If we assume a density depletion in the topside ionosphere with a density of 300 cm^{-3} and a potential, which accelerates the ionospheric electrons upward, on the order of the electron's thermal energy then currents of 1-5 $\mu A/m^2$ will result. Currents of this magnitude have frequently been measured in downward current regions.

In both current regions, electron populations have been accelerated through a potential difference. In addition, upgoing ionospheric electrons may be present in regions of diffuse auroral electron precipitation to preserve current continuity, irrespective of the magnitude and direction of the net current. These upgoing electrons would presumably be accelerated through a potential difference on the order of their thermal energy as is likely the case in the downward current region. The net current depends upon the precipitating, back scattered and secondary electrons as well as the upward accelerated ionospheric electrons. The physics of the wave-particle interaction, however, should be basically the same as in the downward current region. The positive gradients in velocity space, which are a natural result of the acceleration, may drive plasma waves unstable. The wave mode which we feel is responsible for the generation of ion-conic distributions, is the electrostatic ion-cyclotron mode.

Figure 1 is a schematic representation of the accelerated electron and thermal ion distributions which would be expected to be present in each current region. Typical number densities and thermal energies are included in Figure 1 for an altitude ∿6000 km in the upward current region and ∿1500 km in the downward current region. v_a is the drift velocity of the accelerated electron distribution. The superscript r, refers to those particles in resonance with the excited waves.

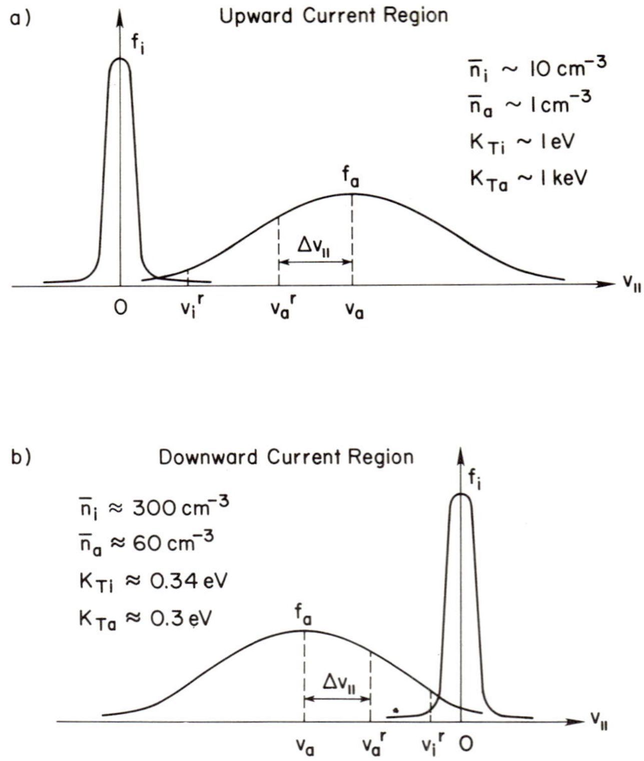

Fig. 1 A schematic of the auroral particle populations that may be important in the generation of ion-conics. The thermal characteristics of upward (∿6000 km altitude) and downward (∿1500 km altitude) current regions are displayed. The subscript "a" refers to the accelerated electrons which have passed through a parallel potential difference on the order of their thermal velocity and the subscript "i" refers to the ambient ion distribution. v_a^r and v_i^r are the Landau and cyclotron resonant velocities of the accelerated electron and ion distributions, respectively. For simplicity, we have neglected the thermal electron distribution function which must be present to preserve charge neutrality.

Free energy in the drifting electrons is fed into the thermal ions via a Landau resonance with the electrons. The parallel electron velocity for Landau resonance is:

$$v_a^r = \frac{\omega_R}{k_{\parallel}} \qquad (1)$$

where ω_R is the real frequency and $k_{\parallel,\perp}$ is the parallel or perpendicular wave number of the excited wave mode respectively. The parallel resonant velocity of the accelerated electrons is too large for thermal ions to be in Landau resonance with the excited waves. However, the thermal ions may cyclotron resonate with ion-cyclotron waves at a parallel velocity that is much smaller than v_a^r, namely v_i^r. Usually v_i^r is a velocity in the tail of the ion distribution. From linear theory [Ichimaru, 1973], the real frequency can be expressed as

$$\omega_R = m\omega_{ci}(1 + \Delta_{mi}) \qquad (2)$$

Because ion-cyclotron waves may be excited at various harmonics we shall label these emissions by the integer m. Δ_{mi} is a positive factor, $0 < \Delta_{mi} < 1$, which shifts the wave frequency away from the mth cyclotron frequency. A typical Δ_{1i} for the m=1 ion-cyclotron emission (ω_R=1.2 ω_{ci}) is Δ_{1i}= 0.2.

In the ion rest frame the condition for a cyclotron resonance between an ion and a plasma wave is

$$v_i^r = \frac{\omega_R - n\omega_{ci}}{k_\parallel}$$

That is

$$v_i^r = \frac{m\omega_{ci}}{k_\parallel}\left(1 + \Delta_{mi} - \frac{n}{m}\right)$$

The above equation may be rewritten in terms of the accelerated electron resonant velocity as

$$\frac{v_i^r}{v_a^r} = \frac{1+\Delta_{mi} - \frac{n}{m}}{1+\Delta_{mi}}$$

An alternative approach is to express the ion resonant velocity in terms of the wave normal angle, θ. This angle is measured with respect to the ambient magnetic field direction.

$$\frac{k_\perp}{k_\parallel} = \tan\theta$$

$$k_\perp^2 \frac{V_{Ti}^2}{\omega_{ci}^2} = \eta$$

where V_{Ti} is the ion thermal velocity and from linear theory η is a number $\sim m^2$ [Kindel and Kennel, 1971; Ashour-Abdalla and Thorne, 1978]. Therefore

$$\frac{v_a^r}{V_{Ti}} = \tan\theta\ (1+\Delta_{mi})$$

or

$$\frac{v_i^r}{V_{Ti}} \equiv z_o = \tan\theta\ \left(1+\Delta_{mi} - \frac{n}{m}\right) \qquad (3)$$

To understand why (3) is important in understanding the generation of ion-conics, consider the following example where we investigate the principal (n=m) harmonic interaction of the m=1 emission (ω_R= 1.2 ω_{ci}). Therefore $\frac{v_i^r}{V_{Ti}}$ = 0.2 $\tan\theta$. For $|\tan\theta| >> 10$, the above equation implies that the ion resonant velocity is too far out on the tail of the thermal ion distribution where there are too few particles to produce significant number fluxes. A large value for $\tan\theta$ is precisely what is expected in the keV electron precipitation region because $\frac{k_\perp}{k_\parallel} >> 10$. For example, let us assume that the accelerated electrons carry a current between 2.5μA/m^2 and 3.5μA/m^2, have a temperature of 1keV, a number density of 1 cm^{-3} and resonate with a 3eV thermal proton distribution. Then $k_\perp/k_\parallel$ will vary between 80 and 300 where we have assumed from linear theory that $v_a - v_a^r = V_{Ta}/\sqrt{2}$ [Dusenbery and Kaufmann, 1980]. These values describing the plasma characteristics in the high altitude ($\sim$6000 km) precipitation region are fairly typical. Because the thermal energy decreases with decreasing altitude, $\tan\theta$ at low altitude will be greater than at higher altitude. We can therefore conclude that in general $\tan\theta$ >> 10 in regions of upward current carried by precipitating keV electrons. On the other hand, a downward current carried by upgoing thermal ionospheric electons at low altitude which varies between -2.5μA/m^2 and -3.5μA/m^2 would excite ion-cyclotron waves such that $\tan\theta$ varies between -5 and -15. For this case, the accelerated electrons were assumed to have a temperature of 0.3eV, a number density of 60 cm^{-3} and resonate with a 0.34eV proton distribution. Thus $\tan\theta \approx -10$ or $v_p^r \approx -2\ V_{Tp}$. Far more particles are now in resonance than in the previous case. We are thus led to investigate the diffusion of thermal ions via a spectrum of ion-cyclotron waves in regions where a thermal distribution of ionospheric electrons has been accelerated upwards by approximately its thermal energy.

To quantitatively evaluate the diffusion of thermal ions, we turn to quasi-linear diffusion theory. Diffusion coefficients are first specified, applicable to any number of ion-cyclotron emissions and are then evaluated for the m=1 emission.

The quasi-linear diffusion equation for an infinite, homogeneous, collisionless plasma immersed in a uniform magnetic field $\underline{B}_o$ for the case when electrostatic waves are present [Kennel and Englemann, 1966] is, in the resonant diffusion limit,

$$\frac{\partial f_i}{\partial t} = \underset{V\to\infty}{Lt} \int \frac{d\underline{k}}{8\pi^2} \left(\frac{q_i}{m_i}\right)^2 \frac{|\delta\underline{E}|^2}{Vk^2} .$$

$$\sum_n \left[\frac{n\omega_{ci}}{v_\perp}\frac{\partial}{\partial v_\perp} + k_\parallel \frac{\partial}{\partial v_\parallel}\right] \delta(k_\parallel v_\parallel + n\omega_{ci} - \omega_R) \cdot$$

$$J_n^2 (k_\perp v_\perp/\omega_{ci}) \left[\frac{n\omega_{ci}}{v_\perp}\frac{\partial}{\partial v_\perp} + k_\parallel \frac{\partial}{\partial v_\parallel}\right] f_i \qquad (4)$$

Here f_i is the space averaged zeroth-order distribution function in velocity space for an ion of mass m_i and charge q_i. V is the plasma volume, $\underline{k}$ is the wave number, $\delta\underline{E}$ is the wave electric field as a function of $\underline{k}$, $\omega_{ci} = eB_o/m_i c$ is the ion-cyclotron frequency and J_n is the Bessel function of the first kind of order n.

Equation 4 may be rewritten in terms of a diffusion tensor, $\underline{\underline{D}}'$ in $(v_\parallel, v_\perp)$ space:

$$\frac{\partial f_i}{\partial t} = \nabla \cdot (\underline{\underline{D}}' \cdot \nabla f_i)$$

$$= \frac{1}{v_\perp}\frac{\partial}{\partial v_\perp}\left[v_\perp D'_{v_\perp v_\perp}\frac{\partial f_i}{\partial v_\perp} + v_\perp D'_{v_\perp v_\parallel}\frac{\partial f_i}{\partial v_\perp}\right] \qquad (5)$$

$$+ \frac{\partial}{\partial v_\parallel}\left[D'_{v_\parallel v_\perp}\frac{\partial f_i}{\partial v_\perp} + D_{v_\parallel v_\parallel}\frac{\partial f_i}{\partial v_\parallel}\right]$$

We will follow the analysis of Lyons [1974] in evaluating the diffusion coefficients in $(v_\parallel, v_\perp)$ Defining $\underline{\underline{D}} = \underline{\underline{D}}'/v^2$ gives diffusion coefficients with dimensions of sec^{-1}. Therefore $D_{v_\perp v_\perp}$ gives the characteristic time scale for diffusion in perpendicular velocity whereas $D_{v_\parallel v_\parallel}$ gives the time scale for diffusion in parallel velocity.

The power spectrum of the electric field fluctuations is taken to be a shifted Gaussian in parallel wave number and to vary as a delta function in perpendicular wave number. The delta function was taken for the variation of $|\delta\underline{E}|^2$ with $k_\perp$ to avoid integrating over $k_\perp$ in the evaluation of $\underline{\underline{D}}$. However, as a result of the use of the delta function, the argument of the Bessel function in (4) causes the diffusion coefficients to oscillate unrealistically with $v_\perp$. Because wave energy will be spread over a finite range of $k_\perp$, the diffusion coefficients will be smooth functions of $v_\perp$. To represent the spread of energy over a spectrum in $k_\perp$ we replace J_n^2 by its average, $\langle J_n^2 \rangle$, which we will define later in this section.

After some algebraic manipulation, the parallel and perpendicular diffusion coefficients may be rewritten as

$$D_{v_\perp v_\perp} = \sum_{n,m} 2\sqrt{\pi}\left(\frac{q_i}{m_i}\right)^2 \frac{1}{\mathrm{erf}(1.0)} \frac{|\delta E_w^m|^2}{k^m_{\parallel 0}} \cdot$$

$$\frac{n^2\omega_{ci}^2}{v^2 v_\perp^2 k^2} \frac{\langle J_n^2(k_\perp v_\perp/\omega_{ci})\rangle}{|v_i^r|} \exp\{-(k_\parallel^r - k^m_{\parallel 0})^2/(\Delta k_\parallel^m)^2\}$$

(6)

$$D_{v_\parallel v_\parallel} = \sum_{n,m} 2\sqrt{\pi}\left(\frac{q_i}{m_i}\right)^2 \frac{1}{\mathrm{erf}(1.0)} \frac{|\delta E_w^m|^2}{k^m_\parallel} \cdot$$

$$\frac{(\omega_R - n\omega_{ci})^2}{v^2 (v_i^r)^2 k^2} \frac{\langle J_n^2(k_\perp v_\perp/\omega_{ci})\rangle}{|v_i^r|} \exp\{-(k_\parallel^r - k^m_{\parallel 0})^2/(\Delta k_\parallel^m)^2\}$$

From linear theory, it is found that the fastest growing mode has a zero group velocity, i.e., it is nonconvective [Ashour-Abdalla and Thorne, 1978]. We have thus neglected the parallel group velocity relative to v_i^r in evaluating the above diffusion coefficients. Further simplifications may be made in the diffusion coefficients that are consistent with measurements. For instance, Kintner et al. [1978] on the S3-3 satellite frequently measured ion-cyclotron waves with frequencies $\approx 1.2\omega_{ci}$ (i.e. m=1, $\Delta_{1i} \approx 0.2$); at times higher order harmonic emissions were measured as well. Thus it is reasonable to look solely at the m=1 contribution in the above sums to investigate the generation of ion-conics in low and high altitude regions of upgoing ionospheric electrons even though at low altitude no definitive wave measurements have been made. It turns out that only the principal harmonic (n=m) needs to be considered because the parallel velocity of the resonant ions is >> ion thermal velocity for all terms satisfying the condition n≠m (see(3)). That is, the resonance is too far out on the tail of the ion distribution to be important except for the n=m term. We may approximate a band of resonant waves by letting $\Delta k_\parallel^1 = \frac{1}{2}|k^1_{\parallel 0}|$ where $k^1_{\parallel 0}$ is the center wave number of the m=1 band. Thus (6) reduces to

$$D_{v_\perp v_\perp} = \frac{10\sqrt{\pi}}{\mathrm{erf}(1.0)} \frac{|\delta E_w^1|^2}{B_0^2} \omega_{ci} \frac{c^2}{v^2} \frac{(v_i^r)^2}{v_\perp^2} \frac{\langle J_1^2(v_\perp/V_{Ti})\rangle}{[z_o^2 + 0.04]}$$

(7)

$$D_{v_\parallel v_\parallel} = 4\mathrm{x}10^{-1} \frac{\sqrt{\pi}}{\mathrm{erf}(1.0)} \frac{|\delta E_w^1|^2}{B_0^2} \omega_{ci} \frac{c^2}{v^2} \frac{\langle J_1^2(v_\perp/V_{Ti})\rangle}{[z_o^2 + 0.04]}$$

It is convenient to define normalized diffusion coefficients given by

$$\mathcal{D} = \underline{\underline{D}} \frac{K_{Ti}}{|\delta E_w^1|^2}$$

and normalized velocities by

$$x = \frac{v_\perp}{V_{Ti}}$$

$$z = \frac{v_\parallel}{V_{Ti}}$$

where K_{Ti} is the ion thermal energy and δE_w^1 is the measured wave electric field for the m=1 emission. After evaluating the numerical constants in (7) the normalized diffusion coefficients reduce to

$$\mathcal{D}_{v_\perp v_\perp} = 3.5\text{x}10^{-1} \frac{z_o^2}{x^2(z_o^2+x^2)} \cdot \frac{\langle J_1^2(x)\rangle}{(z_o^2+0.04)} (\text{sec}^{-1}\cdot\text{eV}\cdot(\text{mV/m})^{-2}) \tag{8}$$

$$\mathcal{D}_{v_\parallel v_\parallel} = 1.4\text{x}10^{-2} \frac{1}{(z_o^2+x^2)} \cdot \frac{\langle J_1^2(x)\rangle}{(z_o^2+0.04)} (\text{sec}^{-1}\cdot\text{eV}\cdot(\text{mV/m})^{-1})$$

where the thermal energy is in eV and δE_w^1 is in mV/m. To evaluate (8) we need an explicit expression for the average of J_1^2 over a spectrum of $k_\perp$. Beyond x=1.24 the asymptotic expansion for $J_1(x)$ is employed and $\langle J_1^2\rangle$ can simply be taken to be $(\pi x)^{-1}$. Between $0 < x < 1.24$ we replace $\langle J_1^2\rangle$ by J_1^2. While the choice of methods for including the spread of wave energy over $k_\perp$ will effect the detailed variation of $\underline{\underline{\mathcal{D}}}$ with $v_\perp$, in an average sense it will have little effect. In Section 3, we will use the parallel and perpendicular diffusion coefficients defined above in investigating the limits of strong parallel and perpendicular diffusion of thermal ions in the presence of a spectrum of ion-cyclotron waves.

Simple Interpretations

Wave-Particle Diffusion

In the ion rest frame, the condition for a cyclotron resonance between an ion and a transverse propagating ion-cyclotron wave is

$$\omega_R - k_\parallel v_i^r = n\omega_{ci};\ n=0, \pm 1, \pm 2, \ldots$$

Considerable qualitative information can be obtained by using the above condition to determine which particles resonate with a given spectrum of waves. The regions of resonance in the $(v_\parallel, v_\perp)$ plane for various excited frequencies, labeled by the integer $m(\omega_R \approx 1.2\ m\omega_{ci})$, are illustrated in Figure 2 for the case when protons are the primary ion species. In this figure, the solid curves represent distribution function contours. For the interaction of ions with ion-cyclotron waves, we have found that the principal harmonic resonance (n=m) determines the smallest parallel velocity an ion will have. Resonant ions too far out on the tail of the ion distribution will have little effect on the generation of conics because there are too few particles to start with. The width of each resonant band is proportional to the bandwidth of the waves which we took to be, $\Delta k_\parallel^m = \frac{1}{2} k_{\parallel 0}^m$, where $k_{\parallel 0}^m$ is the center wave number of the mth band.

Particles in resonance with a single wave or with an ensemble of nondispersive waves are con-

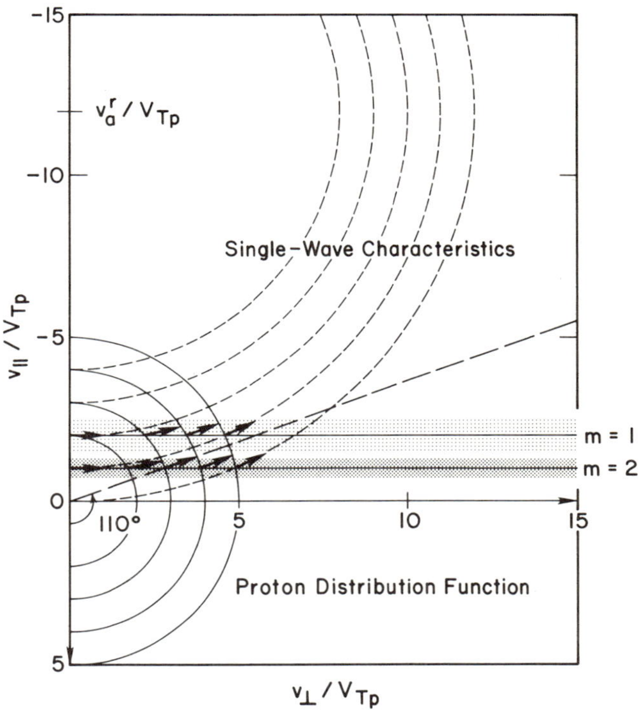

Fig. 2 Regions of resonance in the $(v_\parallel, v_\perp)$ plane for the interaction of a proton distribution with proton-cyclotron waves labeled by the integer m $(\omega_R \approx (m+0.2)\omega_{ci})$. The parallel and perpendicular velocities are normalized to the proton thermal velocity. The dashed contours represent the single-wave characteristics which interact with the particle distribution (solid contours) causing diffusion of particles in the m=1 and 2 bands in the direction of the solid arrows. The wave normal angle was chosen such that $\tan\theta = -10$ for the m=1 and 2 emissions.

strained to diffuse along specific surfaces in velocity space given by [Kennel and Engelmann, 1966]

$$(v_{\parallel} - \frac{\omega_R}{k_{\parallel}})^2 + v_{\perp}^2 = \text{const} \tag{9}$$

These surfaces are called single-wave characteristics (dashed curves in Fig. 2). They merely represent energy conservation in the wave frame.

Many conclusions about the interactions of particles with a spectrum of waves may be made with the use of (9), though only in a qualitative sense. Protons in resonance with an ensemble of ion-cyclotron waves exist over a band that depends upon which harmonic is excited. In the previous section, we found that a principal harmonic resonance (n=m) implies that $v_i^r/V_{Ti} = \Delta_{ni} \tan\theta$. For resonant protons we may assume that $\Delta_{1p} = 0.2$ ($\omega_R = 1.2\ \omega_{cp}$) and $\Delta_{2p} = 0.1$($\omega_R = 2.2\ \omega_{cp}$) for the m=1 an 2 emissions respectively. If $\tan\theta = -10$, then $v_a^r = \frac{\omega_R}{k_{\parallel}} \approx -12\ V_{Tp}$. The single-wave characteristics in Figure 2 are centered on this point. The normalized resonant ion velocities for the m=1 and 2 emissions are therefore -2 and -1 respectively. The arrows, in this figure, represent diffusive velocity changes of the resonant protons. The diffusion, and the arrows, are constrained to lie along the single-wave characteristics, and the direction of the arrows is always chosen to point toward lower values of the distribution function (recall that velocity space diffusion is driven by gradients of f_i in that space).

Figure 2 indicates that the diffusion of low energy protons is primarily in the perpendicular velocity whereas the diffusion of high energy protons has a strong parallel component. The behavior of the calculated diffusion coefficients will be compared with the behavior of the resonant proton diffusion deduced above. Because there are far more low energy protons than high, perpendicular diffusion of resonant protons will dominate parallel diffusion. Due to the band of resonant waves, with different but overlapping wave characteristics, particles can diffuse to large energies via diffusion in their perpendicular velocity. On the other hand, diffusion in their parallel velocity will be limited to the relatively narrow band of resonant parallel velocities [Kennel and Englemann, 1966]. Parallel velocity diffusion will smooth the distribution function as a function of $v_{\parallel}$ over the band of resonant $v_{\parallel}$'s but should not significantly affect the diffusion in the perpendicular direction.

The Limit of Strong Parallel and Perpendicular Diffusion

We may use (12) to insure that a quasi-linear theory is, in fact, valid for the interaction of ions with the observed ion-cyclotron waves. That is, quasi-linear theory may be used if the changes in the one-particle distribution function (f) via a spectrum of waves are slow compared to the period of the waves (T_ω). Let the characteristic time scale for changes in f be given by T_f. Thus

$$\frac{1}{f}\frac{\partial f}{\partial t} \sim \frac{1}{T_f} \sim D_{v_\perp v_\perp}$$

$$T_f >> T_\omega$$

Now $T_\omega \approx 1/\omega_{ci}$, therefore

$$D_{v_\perp v_\perp} << \omega_{ci}$$

or

$$\frac{|E_w^1|^2}{K_{Ti}} \mathcal{D}_{v_\perp v_\perp} << \omega_{ci} \tag{10}$$

From (8), the maximum $\mathcal{D}_{v_\perp v_\perp}$ occurs at x=0 where

$$\mathcal{D}_{v_\perp v_\perp} = 2.2\text{x}10^{-2}\text{sec}^{-1} \cdot \text{eV} \cdot (\text{mV/m})^{-2} \text{ for } \frac{v_i^r}{V_{Ti}} = -2.$$

At an altitude of 1500 km, the magnetic field is 0.3 Gauss. For a distribution of protons, $\omega_{cp} = 2.87\text{x}10^3\ \text{sec}^{-1}$. If we assume that the protons have a thermal energy of 0.34 eV and a wave electric field amplitude of 20mV/m, then (10) reduces to

$$2.5\text{x}10^1 << 2.9\text{x}10^3.$$

For this particular case, quasi-linear theory is clearly applicable. On the other hand, a quasi-linear theory for the generation of oxygen-conics may be inappropriate for large amplitude wave electric fields because oxygen is sixteen times heavier than protons. For instance, if $K_{TO} = 0.34$ eV, $|\delta E_w^1| = 50$mV/m and $B_o = 0.3$ Gauss, then (10) gives

$$1.6\text{x}10^{-2} < 1.8\text{x}10^{-2}$$

Thus the requirement of a quasi-linear theory that the changes of the oxygen distribution function be slow compared to the wave period becomes marginal for this case. However, observations show that wave electric fields of 50mV/m are only an upper limit. More than likely, $|\delta E_w^1| \lesssim 30$mV/m at altitudes of $\sim$ 1500 km.

Figure 3 is a plot of the perpendicular and parallel normalized diffusion coefficients versus normalized perpendicular energy for two normalized resonant velocities (-1, and -2). $v_i^r/V_{Ti} = -1$ and -2 correspond to the m=1 cyclotron emission ($\omega_R = 1.2\ \omega_{ci}$) for two values of $\tan\theta$, -5 and -10 respectively. Notice that the perpendicular co-

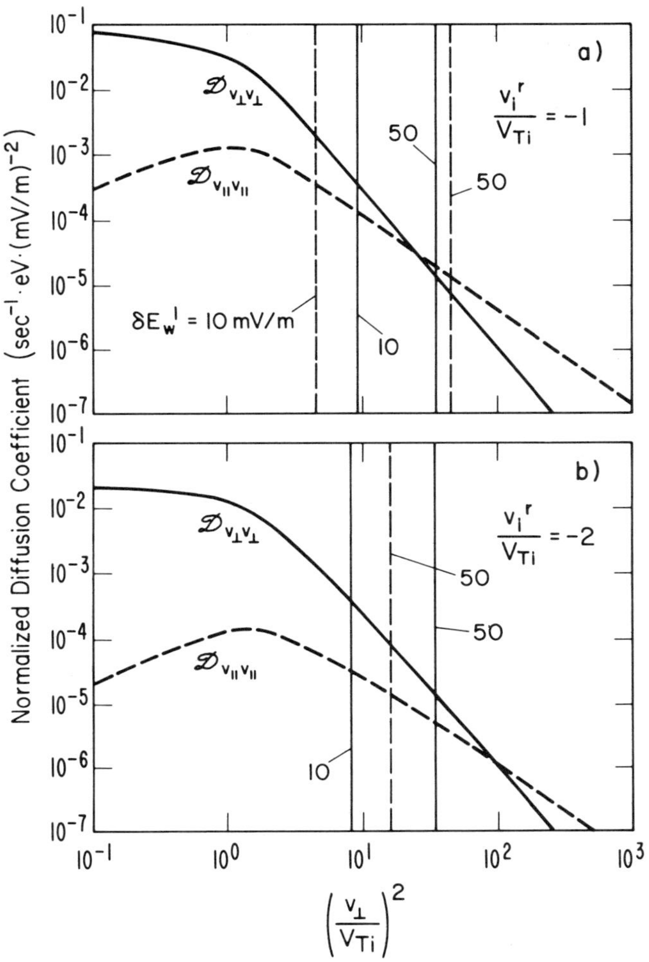

Fig. 3 The perpendicular (solid curves) and parallel (dashed curves) normalized diffusion coefficients versus normalized perpendicular energy for two normalized resonant velocities (-1 and -2) for the m=1 wave. These resonant velocities correspond to tanθ = -5 and -10 respectively. The solid and dashed vertical lines refer to the limit of strong perpendicular and parallel diffusion respectively for two values of the wave electric field (10 and 50mV/m).

efficient is much greater than the parallel coefficient at low perpendicular energies whereas at high energy the parallel coefficient is greater. This behavior was suggested from simply looking at diffusion along single-wave characteristics (Fig. 2).

Constructing the phenomenological diffusion coefficients

$$D_{v_\perp v_\perp} \sim \frac{(\Delta v_\perp)^2}{v^2} \frac{1}{2t_i} \text{ and } D_{v_\parallel v_\parallel} \sim \frac{(\Delta v_\parallel)^2}{v^2} \frac{1}{2t_i}$$

where t_i is the duration of the interaction, we may deduce the perpendicular energy when strong parallel ($\Delta v_\parallel/v \sim 1$) and perpendicular ($\Delta v_\perp/v \sim 1$) diffusion is present. The results are shown in Figure 3 for two values of the wave electric field (10 and 50mV/m). We have assumed that resonant protons (K_{Tp} = 0.3eV) with v_p^r/V_{Tp} = -2 are in an interaction region of 80 km, corresponding to t_p = 5 sec. Solid verticle lines refer to perpendicular diffusion whereas dashed vertical lines refer to parallel diffusion. Notice, in Figure 3b, the absence of the dashed 10mV/m line. This is due to the fact that strong parallel diffusion does not occur at any $v_\perp$ for a wave electric field of 10mV/m when the normalized resonant velocity is -2.

Changing the normalized resonant velocity from -1 to -2 changes the character of strong parallel diffusion far more than strong perpendicular diffusion. The result depends upon where the crossover point ($\mathscr{D}_{v_\perp v_\perp} = \mathscr{D}_{v_\parallel v_\parallel}$) is in relation to the vertical lines. For example, if the lines lie outside the crossover point (Fig. 3a) than the limit of strong parallel diffusion is greater than the limit of strong perpendicular diffusion. This behavior is reversed if the lines lie inside the crossover point (Fig. 3b). Earlier in this section, we found that the center of the single-wave characteristics for the m=1 proton-cyclotron emission when tanθ = -10 was $v_a^r = -12V_{Tp}$. If tanθ had been -5, then v_a^r would have been $-6V_{Tp}$. From Figure 2, the intersection of the single-wave characteristics with the contours of the proton distribution function in the m=1 band has a much larger parallel component when tanθ = -5 than when tanθ = -10. The qualitative results of Figure 2 are thus consistent with the behavior of the diffusion coefficients plotted in Figure 3.

From Figure 3b, we may calculate the value of $(v_\perp/V_{Ti})^2$ for strong perpendicular diffusion. The normalized perpendicular energy ranges between 10-40. Ion thermal energies range from 0.3eV at low altitude (1500 km) to ∿3eV at higher altitude (6000 km). Strong perpendicular diffusion thus varies between 3-10eV at low altitudes and 30-100eV at higher altitude.

Quasi-linear diffusion of resonant ions leads to the formation of a plateau in velocity space in the ion distribution function [Kennel and Engelmann, 1966] and such plateaus should extend to the energies given by the limits of strong perpendicular diffusion. For values of $(\frac{v_\perp}{V_{Ti}})^2$ greater than this limit, the flux of particles should decrease with increasing energy, but may still be well above the fluxes prior to the interaction. This suggests that diffusion of 0.3eV ions up to 100eV at low altitudes and 3eV ions up to ∿ 1 keV at higher altitudes is reasonable. This result is consistent with low altitude Isis 2 observations of ion-conics [Ungstrup et al.,

1979] as well as the higher altitude observations of S3-3 [Mizera and Fennell, 1977].

Summary and Discussion

The principal goal of this paper has been an investigation of the generation of ion-conics by resonant interactions with electrostatic ion-cyclotron waves in regions of accelerated upgoing ionospheric electrons. We investigated the generation of ion-cyclotron waves in regions of upward current carried by precipitating keV electrons and found that resonance with a thermal population of ions was unlikely due to the large drift velocity of the keV electrons. Thus, we were led to look at regions of accelerated upgoing ionospheric electrons, to explain the generation of the ion-cyclotron waves and we found that such particles can indeed generate ion-cyclotron waves that will resonate with the thermal ion population.

While we have found that accelerated upgoing electrons can generate the appropriate ion-cyclotron waves, we have made no attempt to self-consistently account for the amplitude and frequency distribution of the waves. Instead we have calculated the quasi-linear diffusion rates of ions in resonance with ion-cyclotron waves using the spectrum and amplitude of the ion-cyclotron waves that have been observed on auroral field lines. Quasi-linear theory was shown to be appropriate because the energy density in the wave electric field was much smaller than the thermal ion energy density and the time changes in the ion distribution function were slow compared to the wave period.

It was found that the waves should cause strong perpendicular diffusion, leading to plateau formation in the perpendicular distribution function for parallel ion resonant velocities with magnitudes between 1 and 2 times the ion thermal velocity. Such strong diffusion extends to perpendicular energies 10-40 times the initial ion thermal energy. Strong diffusion extends to higher perpendicular energies as the magnitude of the wave electric field is increased, as the length of time the resonant ions are in the interaction region is increased and as the initial thermal energy is decreased.

Ion-cyclotron waves have frequently been measured at altitudes $\sim$ 6000 km by the S3-3 satellite. We assume that such waves will exist at lower altitudes $\sim$ 1500 km even though direct observations have not been made. Furthermore, we assume that the wave frequency will be proportional to the proton cyclotron frequency in a H^+ plasma and the oxygen cyclotron frequency in an O^+ plasma.

No attempt was made in this paper to treat the evolution of the electron distributions self-consistently with the ions and the excited waves. We presume, however, that the resonant electrons will form a plateau in velocity space in the $v_{\parallel}$ direction, simultaneously with the one formed by the resonant ions in the perpendicular direction.

We have not undertaken a detailed analysis of ion-cyclotron heating in a multi-ion plasma. It is unlikely, however, that such an analysis will change our over-all conclusions concerning the generation of ion-conics.

Thus, we feel quasi-linear theory offers an adequate explanation for the generation of conics via the observed spectrum of ion-cyclotron waves. Our results imply that conic distributions should not occur simultaneously with the keV electron precipitation associated with discrete arcs in regions of upward current but should occur in regions of upgoing ionospheric electrons which may include regions of downward current near the edges of auroral arcs and regions of diffuse auroral electron precipitation.

Acknowledgments. This study has greatly benefited from discussions with D. S. Evans, T. Tamao and D. M. Klumpar. Support for P. B. Dusenbery has come from an NRC Resident Research Associateship.

References

Ashour-Abdalla, M., and R. M. Thorne, Toward a unified view of diffuse auroral precipitation, J. Geophys. Res., 83, 4755, 1978.

Cornwall, J. M., and M. Schulz, Physics of Heavy Ions in the Magnetosphere, published in Solar System Plasma Physics, edited by C. F. Kennel and L. J. Lanzerotti, 1978.

Davidson, R. C. Methods in Nonlinear Plasma Theory, p. 133, Academic Press, New York, 1972.

Dusenbery, P. B., and R. L. Kaufmann, Properties of the longitudinal dielectric function: An application to the auroral plasma, J. Geophys. Res. 85, 5969, 1980.

Evans, D. S., The acceleration of charged particles at low altitudes, Physics of Solar Planetary Environments, edited by D. J. Williams, P. 730, AGU, Washington, D. C., 1976.

Frank, L. A., K. L. Ackerson and D. M. Yeager, Observations of atomic oxygen (O^+) in the earth's magnetotail, J. Geophys. Res., 82, 129, 1977.

Fredericks, R. W., and C. T. Russell, Ion-cyclotron waves observed in the polar cusp, J. Geophys. Res., 78, 2917, 1973.

Gorney, D. J., A. Clarke, D. Croley, J. Fennell, J. Luhmann, and P. Mizera, The distribution of ion beams and conics below 8000 km, Submitted for publication in J. Geophys. Res., 1980.

Hollweg, J. V., and H. J. Volk, Energy and momentum exchange in transverse plasma waves, J. Geophys. Res., 76, 7527, 1971.

Ichimaru, S., Basic Principles of Plasma Physics, p. 78, W. A. Benjamin, Inc.,Reading,Mass.,1973.

James, H. G., VLF saucers, J. Geophys. Res., 81, 501, 1976.

Kennel, C. F., and F. Engelmann, Velocity space diffusion from weak plasma turbulence in a magnetic field, Phys. of Fluids, 9, 2377, 1966.

Kindel, J. M., and C. F. Kennel, Topside current instabilities, J. Geophys. Res., 76, 3055, 1971.

Kintner, P. M., M. C. Kelley, R. D. Sharp, A. G. Ghielmetti, M. Temerin, C. Cattell, P. F. Mizera and J. F. Fennell, Simultaneous observations of energetic (keV) upstreaming and electrostatic hydrogen cyclotron waves, J. Geophys. Res., 84, 7201, 1979.

Klumpar, D. M., Tranversely accelerated ions: an ionospheric source of hot magnetospheric ions, J. Geophys. Res., 84, 4229, 1979.

Lyons, L. R., Electron diffusion driven by magnetospheric electrostatic waves, J. Geophys. Res., 79, 575, 1974.

Lysak, R. L., M. K. Hudson, and M. Temerin, Ion heating by strong electrostatic ion cyclotron turbulence, J. Geophys. Res., 85, 678, 1980.

McIlwain, C. E., Auroral electron beams near the magnetic equator, Physics of the Hot Plasma in the Magnetosphere, edited by B. Hultqvist and L. Stenflo, Plenum, New York, 1975.

Mizera, P. F., and J. F. Fennell, Signatures of electric fields from high and low altitude particle distributions, Geophys. Res. Lett., 4, 311, 1977.

Palmadesso, P. J., T. P. Coffey, S. L. Ossakow and K. Papadopoulos, Topside ionosphere ion heating due to electrostatic ion cyclotron turbulence, Geophys. Res. Lett., 1, 105, 1974.

Sharp, R. D., R. G. Johnson, and E. G. Shelly, Observations of an ionospheric acceleration mechanism producing (keV) ions primarily normal to the geomagnetic field direction, J. Geophys. Res., 82, 3324, 1977.

Shelley, E. G., R. D. Sharp, and R. G. Johnson, Satellite observations of an ionospheric acceleration mechanism, Geophys. Res. Lett., 3, 654, 1976.

Ungstrup, E., D. M. Klumpar and W. J. Heikkila, Heating of ions to superthermal energies in the topside ionosphere by electrostatic ion cyclotron waves, J. Geophys. Res., 84, 4289, 1979.

Whalen, B. A., W. Bernstein and P. W. Daly, Low altitude acceleration of ionospheric ions, Geophys. Res. Lett., 5, 55, 1978.